FUEL CELLS

Principles Design and Analysis

MECHANICAL and AEROSPACE ENGINEERING

Frank Kreith & Darrell W. Pepper
Series Editors

RECENTLY PUBLISHED TITLES

Air Distribution in Buildings, *Essam E. Khalil*

Alternative Fuels for Transportation, *Edited by Arumugam S. Ramadhas*

Computer Techniques in Vibration, *Edited by Clarence W. de Silva*

Distributed Generation: The Power Paradigm for the New Millennium, *Edited by Anne-Marie Borbely and Jan F. Kreider*

Elastic Waves in Composite Media and Structures: With Applications to Ultrasonic Nondestructive Evaluation, *Subhendu K. Datta and Arvind H. Shah*

Elastoplasticity Theory, *Vlado A. Lubarda*

Energy Audit of Building Systems: An Engineering Approach, *Moncef Krarti*

Energy Conversion, *Edited by D. Yogi Goswami and Frank Kreith*

Energy Management and Conservation Handbook, *Edited by Frank Kreith and D. Yogi Goswami*

The Finite Element Method Using MATLAB®, Second Edition, *Young W. Kwon and Hyochoong Bang*

Fluid Power Circuits and Controls: Fundamentals and Applications, *John S. Cundiff*

Fuel Cells: Principles, Design, and Analysis, *Shripad Revankar and Pradip Majumdar*

Fundamentals of Environmental Discharge Modeling, *Lorin R. Davis*

Handbook of Energy Efficiency and Renewable Energy, *Edited by Frank Kreith and D. Yogi Goswami*

Heat Transfer in Single and Multiphase Systems, *Greg F. Naterer*

Heating and Cooling of Buildings: Design for Efficiency, Revised Second Edition, *Jan F. Kreider, Peter S. Curtiss, and Ari Rabl*

Intelligent Transportation Systems: Smart and Green Infrastructure Design, Second Edition, *Sumit Ghosh and Tony S. Lee*

Introduction to Biofuels, *David M. Mousdale*

Introduction to Precision Machine Design and Error Assessment, *Edited by Samir Mekid*

Introductory Finite Element Method, *Chandrakant S. Desai and Tribikram Kundu*

Large Energy Storage Systems Handbook, *Edited by Frank S. Barnes and Jonah G. Levine*

Machine Elements: Life and Design, *Boris M. Klebanov, David M. Barlam, and Frederic E. Nystrom*

Mathematical and Physical Modeling of Materials Processing Operations, *Olusegun Johnson Ilegbusi, Manabu Iguchi, and Walter E. Wahnsiedler*

Mechanics of Composite Materials, *Autar K. Kaw*

Mechanics of Fatigue, *Vladimir V. Bolotin*

Mechanism Design: Enumeration of Kinematic Structures According to Function, *Lung-Wen Tsai*

Mechatronic Systems: Devices, Design, Control, Operation and Monitoring, *Edited by Clarence W. de Silva*

The MEMS Handbook, Second Edition (3 volumes), *Edited by Mohamed Gad-el-Hak*

MEMS: Introduction and Fundamentals

MEMS: Applications

MEMS: Design and Fabrication

Multiphase Flow Handbook, *Edited by Clayton T. Crowe*

Nanotechnology: Understanding Small Systems, Second Edition, *Ben Rogers, Sumita Pennathur, and Jesse Adams*

Nuclear Engineering Handbook, *Edited by Kenneth D. Kok*

Optomechatronics: Fusion of Optical and Mechatronic Engineering, *Hyungsuck Cho*

Practical Inverse Analysis in Engineering, *David M. Trujillo and Henry R. Busby*

Pressure Vessels: Design and Practice, *Somnath Chattopadhyay*

Principles of Solid Mechanics, *Rowland Richards, Jr.*

Principles of Sustainable Energy Systems, Second Edition, *Edited by Frank Kreith with Susan Krumdieck, Co-Editor*

Thermodynamics for Engineers, *Kau-Fui Vincent Wong*

Vibration and Shock Handbook, *Edited by Clarence W. de Silva*

Vibration Damping, Control, and Design, *Edited by Clarence W. de Silva*

Viscoelastic Solids, *Roderic S. Lakes*

Weatherization and Energy Efficiency Improvement for Existing Homes: An Engineering Approach, *Moncef Krarti*

FUEL CELLS

Principles Design and Analysis

Shripad Revankar
Pradip Majumdar

CRC Press
Taylor & Francis Group
Boca Raton London New York

CRC Press is an imprint of the
Taylor & Francis Group, an **informa** business

CRC Press
Taylor & Francis Group
6000 Broken Sound Parkway NW, Suite 300
Boca Raton, FL 33487-2742

Printed on acid-free paper
Version Date: 20140411

International Standard Book Number-13: 978-1-4200-8968-4 (Hardback)

Library of Congress Cataloging-in-Publication Data

Revankar, Shripad T.
Fuel cells : principles, design, and analysis / authors, Shripad T. Revankar, Pradip Majumdar.
pages cm. -- (Mechanical engineering series)
Includes bibliographical references and index.
ISBN 978-1-4200-8968-4 (hardback)
1. Fuel cells. I. Majumdar, Pradip, 1954- II. Title.

TK2931.R448 2014
621.31'2429--dc23 2014001352

Visit the Taylor & Francis Web site at
http://www.taylorandfrancis.com

and the CRC Press Web site at
http://www.crcpress.com

In loving memory of my (Shripad T. Revankar) late parents:

Kamakshi and Timmappa Gundu Revankar

To my (Pradip Majumdar) late parents Snehalata and Rati Ranjan, wife Srabani, and children Diya and Ishan

Contents

Preface

Overview

In the 21st century, the demand for clean and sustainable energy sources has become a strong driving force in continuing economic development and hence in the improvement of human living conditions. In that respect, fuel cells have been recognized to form the cornerstone of clean energy technologies due to their high efficiency, high energy density, and low or zero emissions. Recently, fuel cells have seen explosive growth and application in various energy sectors including transportation, stationary and portable power, and micro-power. The rapid advances in fuel cell system development and deployment require basic knowledge of science and technology as well as advanced techniques on fuel cell design and analysis. This book brings together for the first time in a single volume the fundamentals, principles, design, and analysis aspects of the fuel cell and thus would benefit beginners such as undergraduate students as well as practicing engineers and scientists alike. Since design and analysis are interlinked, the book presents the fuel cell design at component and at system levels and then elaborates on the analysis methods for various phenomena associated with component and systems. The material in the book guides the reader from the foundations and fundamental principles through the analysis methods and design of the fuel cell with latest technology and cutting-edge applications, ensuring a logical, consistent approach to the subject.

Themes

The content of the book has three main themes: basic principles, design, and analysis. The theme of basic principles provides the necessary background information on the fuel cells, including the fundamental principles such as the electrochemistry, thermodynamics, and kinetics of fuel cell reactions as well as mass and heat transfer in fuel cells. It also provides an overview of the key principles of the most important types of fuel cells and their related systems and applications. This includes polymer electrolyte membrane fuel cell systems, alkaline fuel cell modules and systems, phosphoric acid fuel cells, direct methanol fuel cells, molten carbonate fuel cells, and solid oxide fuel cells.

The theme of design deals with important characteristics associated with various fuel cell components, electrodes, electrocatalysts, and electrolytes. It also includes fuel cell flow channel characterization and stack design with advances in state of the technology. Both component-level and stack-level characteristics are identified, enabling one to identify the phenomena at component and system levels.

The analysis theme deals with phenomena characterization and modeling both at component and system levels. The analysis component of the book includes modeling of charge transport phenomena, mass and heat transport, computational methods, and system control and dynamic simulation. The book provides a practical account of how to create models, how to manipulate them, and how to interpret results. The dynamic simulation using the MATLAB®/Simulink platform provides design engineers and researchers with a valuable tool to understand and lead the design and construction of the next generation of fuel cells.

Learning Features or Pedagogy

The objective in developing this book was to prepare engineering and science students with an understanding of the fundamentals of fuel cell operation and design and further provide techniques and methods employed to analyze different fuel cell systems based on applications and operating conditions. The authors have been teaching fuel cell courses for both undergraduate and graduate students at their institution for more than 10 years. Based on the authors' teaching experience, the book is structured to include both fundamental principles and advanced analysis methods. The book chapters are designed to teach junior- to senior-level undergraduate technical elective classes and for graduate students pursuing advanced study and research in the fuel cell. The book gives the most fundamental information on the fuel cell, the arts and science of the fuel cell including its components, how and why it operates the way it does, and how one can mathematically model fuel cell behavior so that one can predict its response. The book helps beginners to understand and gain enough knowledge to work in this field and contains material on the fuel cell for scientists and engineers to further advance their knowledge.

Material from Chapters 1–5 and 9 is suitable for beginners and can be covered in a 45-lecture undergraduate course. Chapters 6–8, 11, and 12, which cover advanced topics on transport, modeling, and simulation methods, are suitable for graduate students. In addition, Chapters 13 and 14 can also be covered in both undergraduate-and graduate-level courses. Each chapter in general contains opening paragraphs defining objectives and ends with a chapter summary. As appropriate examples, problems are presented to demonstrate the application of theory or principle, and case studies are given to elaborate on fuel cell analysis. Each chapter has a bibliography that mentions references and material for further reading. Mathematical methods including numerical methods and MATLAB/Simulink techniques are presented in Chapters 11 and 12, where extensive modeling and simulation of the fuel cell are covered. Some chapters have chapter end problems as assignments to test learning skills.

Chapter 1 highlights the current reliance on fossil fuel as the primary energy source and possible renewable alternative energy systems including hydrogen-based energy carriers to address the pollution and greenhouse gas emission problems. The fuel cell seems to be the best energy conversion tool in hydrogen-based energy carrier systems. This chapter gives a summary of fuel cell types and their applications. Chapter 2 presents the basics of electrochemistry to help understand the operation and processes in fuel cell anodes and cathodes. Similarly, Chapter 3 introduces the basics of thermodynamics, the first law and second law of thermodynamics, Gibbs energy of formation, and chemical reaction energy. In Chapter 4, definitions of fuel cell voltage and power, efficiency, reactant consumption, and product generation rates are presented. Detailed kinetics of electrochemical processes at the anode and cathode of fuel cells, Tafel laws, fuel cell irreversibility, and fuel cell polarization curve are discussed in Chapter 5. Chapters 6 and 7 cover the mass transport, charge transport, and water transport in electrolyte and electrodes. The fuel cell system and component characterization and associated techniques and instrumentation are discussed in Chapter 8. Detailed components and design of each type of fuel cell are presented in Chapter 9. Chapters 10–12 deal with numerical and simulation methodology employed for fuel cell flow field analysis, single and multi-dimensional heat and mass transfer analysis, and fuel cell systems analysis. Chapter 13 describes fuel cell power generations systems including fuel processor, hydrogen generation, and storage methods. Chapter 14 describes various fuel cell applications currently in practice, relevant codes and standards for commercial application, and environmental impact owing to the fuel cell technology. Three appendices at the end of the book list various physical constants, conversion factors, useful equations for fuel cell reaction and power, and chemical and thermodynamic data useful in the calculation of reaction rate, power, and efficiency of the fuel cell.

This book is expected to be an invaluable source of reference for all those working directly in this important and dynamic field, for beginners and for scientists, engineers, and educators involved in the quest for clean and sustainable energy sources. Because of their interdisciplinary nature, fuel cells involve and require knowledge of engineering thermodynamics, chemical thermodynamics, electrochemistry, mass transfer, heat transfer, fluid mechanics, manufacturing and design, engineering optimization, materials science and engineering, chemistry and chemical engineering, and electrical engineering. The book supplies basic principles, examples, and models required in the design and optimization of fuel cell systems.

Shripad T. Revankar
Pradip Majumdar

MATLAB® is a registered trademark of The MathWorks, Inc. For product information, please contact:

The MathWorks, Inc.
3 Apple Hill Drive
Natick, MA 01760-2098 USA
Tel: 508-647-7000
Fax: 508-647-7001
E-mail: info@mathworks.com
Web: www.mathworks.com

Acknowledgments

For the past several years, numerous students, professional colleagues, and family members have contributed directly or indirectly to the development and preparation of the material in this book. The authors would like to thank Prof. Frank Kreith, editor-in-chief, for his extremely valuable comments and advice. The authors also wish to express their gratitude to Taylor & Francis/CRC Press for giving them the opportunity to publish this book and particularly to Jonathan Plant, executive editor, for his continued support and patience during the completion of this book.

Pradip Majumdar would like to express his deep appreciation to his wife, Srabani, and children, Diya and Ishan, for their unlimited support, understanding, and patience during the preparation of the manuscript. He is thankful to a number of his graduate students for their work on PEMFC and SOFC over the last 10 years at NIU.

Shripad T. Revankar is thankful to the following:

His former students, particularly Drs. Joshua Walters and Brian Wolf, for their work on PEMFC, DMFC, MCFC, and hydrogen storage.

His family members, wife Jayashree, and children, Vedang, Sachit, and Pavan, for their continued support and encouragement to complete the manuscript.

Authors

Shripad T. Revankar is currently a professor of nuclear engineering at Purdue University, West Lafayette, Indiana, where he has been teaching since 1987. He also holds the position of visiting professorship at Pohang University of Science and Technology (POSTECH), South Korea, in the Division of Advanced Nuclear Engineering. He received his BSc (1975), MSc (1977), and PhD (1983) degrees all in physics from Karnataka University, India, a postgraduate diploma (1978) in vacuum science and technology from University of Bombay, and MEng (1982) degree in nuclear engineering from McMaster University, Canada. He has worked as a postdoctoral researcher (1984–1987) at Lawrence Berkeley Laboratory and at the Nuclear Engineering Department of the University of California, Berkeley. He has completed more than 30 research projects and has published more than 300 refereed research papers in journal and conferences on nuclear thermal hydraulics and reactor safety, multiphase flow and heat transfer, instrumentation, fuel cells, and hydrogen systems. He is a Life Member of the following professional societies: American Nuclear Society (ANS), American Society of Mechanical Engineers (ASME), American Institute of Chemical Engineers (AIChE), Indian Society of Heat and Mass Transfer (ISHMT), and Korean Nuclear Society (KNS). He is also an active member of the American Association for Advancement of Science (AAAS), American Society of Engineering Education (ASEE), and Electro Chemical Society (ECS). He is a Fellow of ASME and has received several awards including the best teacher award, best conference paper awards, appreciation awards, seed for success award for research, and service awards from various sectors including universities, United States–Department of Labor, ASME, and ANS from the countries India, Korea, and the United States. He has served as a research and educational consultant to academia, national laboratories, and industries in Canada, China, Hong Kong, India, South Korea, and the United States. He serves on the editorial boards of seven international journals including *Heat Transfer Engineering*, *Journal of Thermodynamics*, and *ASME Journal of Fuel Cell Science and Technology*.

Pradip Majumdar is currently a professor and chair of mechanical engineering and the director of the Heat and Mass Transfer Laboratory in the Department of Mechanical Engineering, Northern Illinois University, DeKalb, Illinois. He received his BS degree (1975) in mechanical engineering from B.E. College, University of Calcutta, and MS (1980) and PhD (1986) degrees in mechanical engineering from Illinois Institute of Technology, Chicago. He has worked as a design engineer on a nuclear power plant project for DCL in Bombay, India, from 1975 to 1977. His specialties includes heat and mass transfer, fluid mechanics, thermal energy systems, computational fluid dynamics, and heat transfer and experimental techniques. His research interests and experiences are in thermofluid sciences, computational fluid dynamics (CFD), and heat transfer; fuel cell power system, Li-ion battery storage system, and solar thermal energy systems; heat and mass transfer in porous media; micro–nano scale fluid flow and heat transfer; and high heat flux electronics cooling, high energy laser material processing, CFD analysis of scour formation, blood flow in human arteries and stent design, and laser–human tissue interactions. He has worked on a number of federal and industrial research projects and published over 100 refereed research papers in archival journals and conference proceedings. He has received the 2008 Faculty of the Year Award for teaching from Northern Illinois University. He is the author of the book *Computational Methods for Heat and Mass Transfer*. He has served as a guest editor for special issues on advances in electronics cooling in the *Journal of Electronics Manufacturing*. Since 2011, he has been serving as the editor-in-chief of the *Transactions of Fluid Mechanics, International Journal*. As a member of ASME, he serves on the technical committee for Computational Heat Transfer, Heat Transfer on Electronics, and Energy Systems. He has organized and chaired many ASME and InterPack conference sessions on fuel cells, micro–nano scale heat transfer, electronics cooling, and computational heat transfer. He has organized a number of international workshops on fuel cells.

1

Introduction

Reduced resources of fossil fuel and increased demand, cost, and uncertainty in the steady supply of imported oil have sparked the search for alternative electric power generation systems. Additionally, there is an increased global concern for the higher greenhouse gas emission and its effect on global warming and the environment. A fuel cell power generation system is an attractive alternative to conventional power generation systems such as steam-turbine thermal power plant and internal combustion engine because of its higher efficiency, improved fuel economy, reduced dependence on conventional fuel, and lower emission of environmentally unsafe pollutants.

In a conventional power generation system, the chemical energy content of a fuel is first transformed into heat energy in a direct combustion process. The heat energy is then transformed into mechanical energy in a heat engine and finally into electrical energy using an electrical generator. In a fuel cell, these multiple energy conversion steps are avoided and the chemical energy content of the fuel is directly converted into electrical energy using an electrochemical reaction process.

1.1 Primary Energy Sources—Fossil Fuel

A major portion of electric power production in the United States comes from the use of fossil fuels such as coal, natural gas, and nuclear fuel, and a smaller portion comes from renewable sources as shown in the Figure 1.1 (2007 data).

World primary energy sources for stationary power generation as well as for transportations are fossil fuels, which include coal, liquid hydrocarbon (oil), and gaseous hydrocarbon (natural gas). Other potential fossil energy sources are *shale oil* and *shale gas,* which are being explored. One of the most commonly available forms of fuel is *hydrocarbon fuel,* which has carbon and hydrogen as the primary constituents. The hydrocarbon fuel exits in different phases such as *liquid* like gasoline, solid like *coal,* and gas like *natural gas.* Some of the common hydrocarbon fuels are gasoline or *octane, diesel, methyl alcohol* or *methanol* and ethyl *alcohol* or *ethanol.*

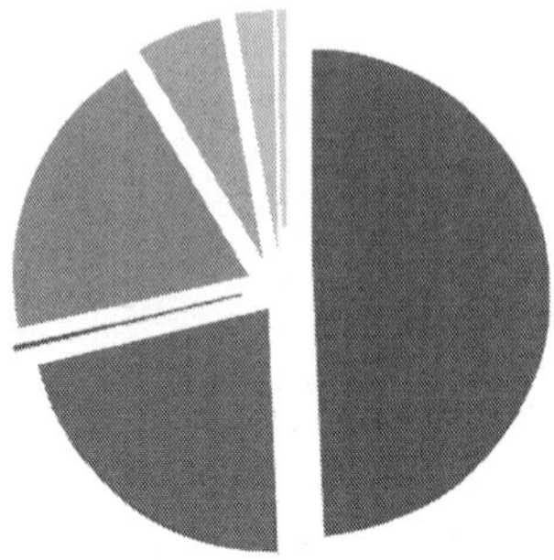

FIGURE 1.1
Energy source for power production in the United States (2007 data).

1.1.1 Coal

Recent data show the following current US reserves of coal, oil, and natural gas:

Total coal reserve: 246,643 million tons; *total crude oil reserve:* 21.3 billion barrels; and *natural gas:* 9.1 billion barrels (liquid) or 237.7 billion cubic feet. Figure 1.2 shows world coal reserves by countries and expected life based on current rate of usage.

Coal is mainly composed of carbon, sulfur, oxygen, and hydrogen with varying compositions. Composition changes from location to location. For application purposes, solid coal is also transformed into *Syngas* (a mixture of mostly carbon monoxide and hydrogen) or into pure hydrogen gas. The gasification of coal in a gasifier involves multiple processes such as (1) reaction with oxygen and water under pressure and addition of heat to form syngas and (2) water gas-shift reaction *to* convert carbon monoxide into hydrogen and carbon dioxide. There is also ongoing effort to produce *clean coal gas* by removing sulfur dioxide, carbon dioxide, ammonia, and other gas species from the syngas for potential use in a fuel cell. Coal is also used in liquid form by transforming it into gasoline or diesel by the Fisher–Tropsch process.

1.1.2 Liquid or Gaseous Hydrocarbons

Normally, liquid and gaseous hydrocarbons are a mixture of many different hydrocarbons. For example, gasoline consists of a mixture of 40 different hydrocarbons. Most liquid hydrocarbons like gasoline, kerosene, diesel, and so on are derived from crude oil by distillation or cracking processes: each type is characterized by its distillation curve. The distillation curve is obtained by slowly heating the crude so that each hydrocarbon element vaporizes and condenses. The more volatile component is vaporized first.

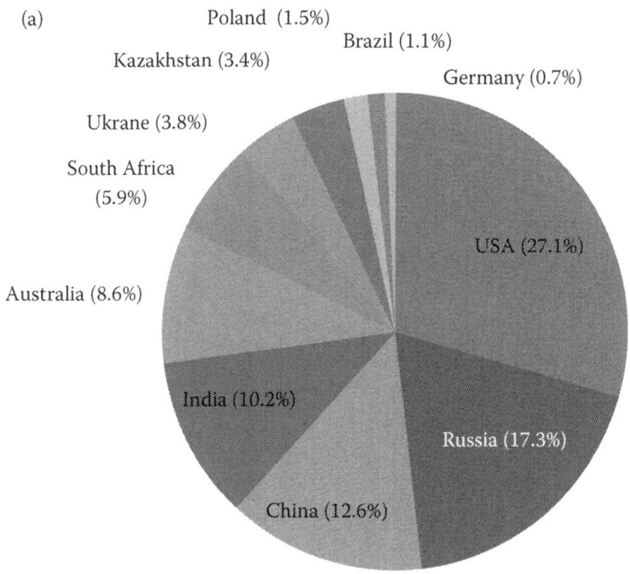

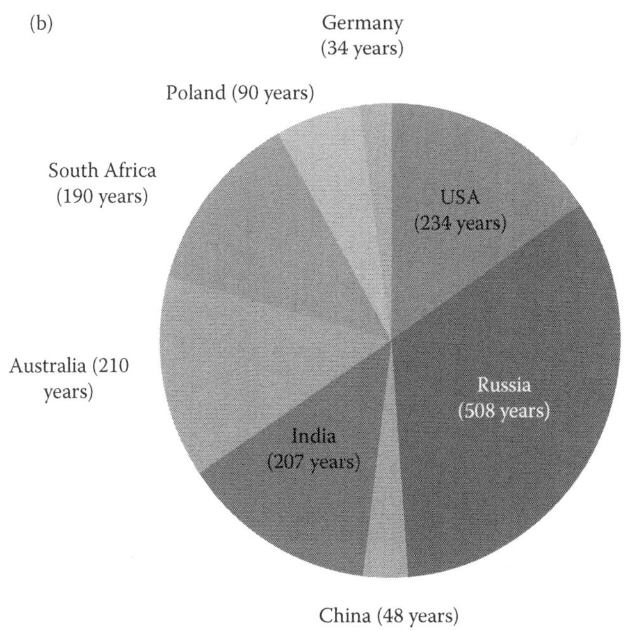

FIGURE 1.2
(a) Current world coal reserve by countries, (b) current world coal reserve and expected life based on current usage.

1.1.3 World Oil Reserve

Figure 1.3 shows current world oil reserves for different countries in the world.

1.1.4 Shale Oil

Oil shale is an organic-rich sedimentary rock from which liquid hydrocarbons are extracted. It is a solid mixture of organic chemical compounds, primarily composed of *kerogen*, quartz, clay carbonate, and pyrate and smaller amounts of uranium, iron, vanadium, nickel and molybdenum. An estimate of global deposits is approximately 3.0 trillion barrels. However, it requires more processing than crude oil and involves higher cost using existing technology.

Shale oil processing involves conversion of *kerogen* into synthetic crude oil by the chemical process of pyrolysis. Oil shales can be heated to a sufficiently high temperature in order to drive off vapor, which can be then be distilled to produce petroleum-like unconventional oil and combustible oil-shale gas. Oil shale can also be directly burned as low-grade fuel for power generation and heating purposes. Major concern for oil-shale mining and processing includes use of land, water consumption, wastewater management, waste disposal, greenhouse gas emission, and air pollution.

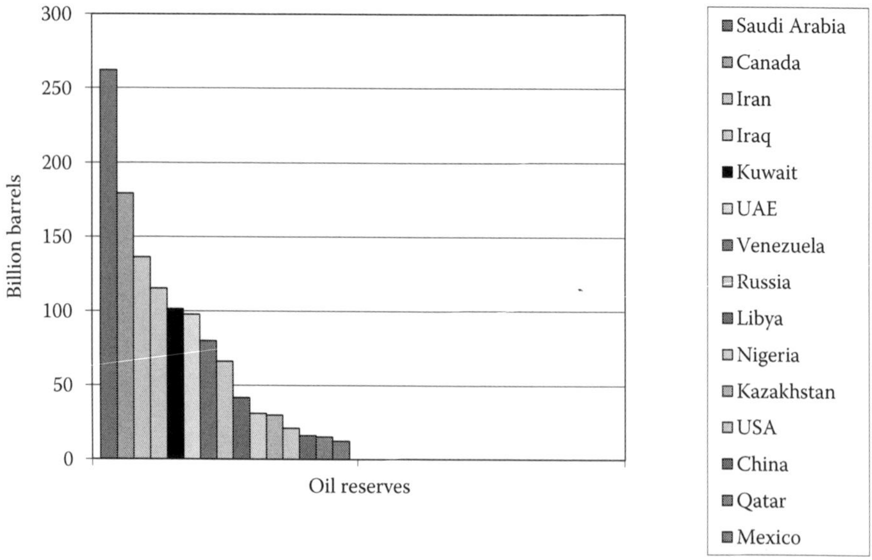

FIGURE 1.3
World oil reserve by countries. (From US Energy Information Administration (EIA) http://tonto.eia.doe.gov/country/index.cfm?view=reserves.)

1.1.5 Gaseous Hydrocarbons

Sources of gaseous hydrocarbons are the natural gas wells and chemical manufacturing processes. Major constituents of natural gas is primarily composed of methane (over 90%) and smaller compositions of other gases like ethane, propane, butane, carbon dioxide, hydrogen, nitrogen, and oxygen.

1.1.6 Shale Gas

Shale gas is natural gas stored in petroleum-rich shale rocks where petroleum is converted into natural gas subjected to high heat and pressure. Shale gas is becoming an important source of natural gas in the United States and in other countries. Major concerns are the cost, effect on environment, complexity, and safety issues of drilling technology, which require both vertical and horizontal drilling of lateral length of thousands of feet.

1.1.7 Biofuel

Biofuels are renewable fuels derived from waste biological materials in contrast to the fossil fuels, which are formed from long dead biological materials over a long period. Biofuels can be of solid, liquid, or gaseous form and can be produced theoretically from any biological or organic carbon-source materials. Most common are the plant-derived materials such as corns, sugar cane, woodchips, and trash. Since plants essentially remove carbon dioxide from the atmosphere, plant-based biofuels essentially recycle existing carbon in the atmosphere and do not release any new carbon in contrast to fossil fuels and are considered endlessly renewable.

There are a number of ways for biofuel production: (1) Grow crops high in *starch* (corn and maize) or grow crops high in *sugar* (sugar cane, sugar beet, and sweet sorghum) and then use a *yeast fermentation process* to produce *ethyl alcohol* or *ethanol*, which is one of the most common form of biofuels. (2) Grow plants that are rich in vegetable oil such as oil palm, soy bean, and algae, which are chemically processed to produce fuels such as *biodiesel*. (3) Wood and its by-products can also be converted into biofuels such as wood gas, methanol, and ethanol fuel.

1.1.7.1 Challenges of Ethanol—Biofuel

Currently, ethanol is produced primarily from food-grade materials such as corn and soybean and requires a considerable amount of energy consuming large amounts of fossil fuel. A recent study also shows that it is not possible to produce enough ethanol from corn to meet the demands in the United States. The current rush to produce ethanol from food-grade material also leads to global food shortages and increased food prices.

New research is now focused on using alternative technologies to develop more efficient processes for ethanol production from a wider range of non-food-grade biomass materials.

1.2 Renewable Energy Resources and Alternative Energy Systems

There is increased demand to reconsider our energy supply system because of the following reasons:

1. Global resources for fossil fuel as well as nuclear fuel are reduced and limited as demonstrated in Figure 1.4.
2. High cost and risk of uninterrupted supply of imported oil.
3. Increased demands for energy and fossil fuels in countries with higher economic growth.
4. Increased concern over climate change and global warming caused by increased level of environmental pollution including greenhouse gas emission from the consumption of fossil fuels used in power generations and transportations. Cleaner forms of energy are essential to reduce carbon and greenhouse gas emissions.
5. There is increased effort to use low-carbon energy to reduce greenhouse gas emission.

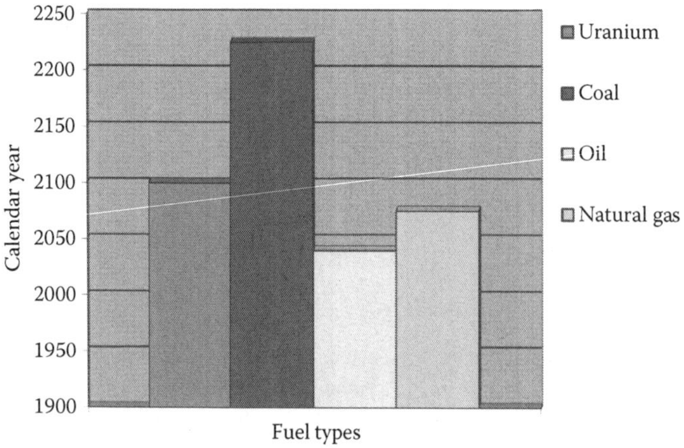

FIGURE 1.4
Global resources of fossil and nuclear fuels.

Renewable energy resources are those that are continuously being replenished by natural processes on a human timescale. In contrast, fossil fuels like coal, oil, and natural gas require millions of years of geological processes to form. Power generation systems that emit little or no carbon and greenhouse gases are (i) solar power, (ii) tidal power, (iii) wind power, (iv) geothermal hydrogen, (v) hydroelectric, (vi) hydrogen-powered engines, and (vii) fuel cell.

1.2.1 Solar Energy

Solar energy is a free source of energy and can be converted directly either into electrical energy using photovoltaic cells or into thermal heat energy using solar collectors. The energy conversion efficiency of solar energy conversion devices is generally low. Solar energy can also be used to produce hydrogen for fuel cells.

1.2.2 Tidal Energy

In a tidal energy system, water is stored during high tide and released during low tide. The power is recovered using water turbine. This energy can also be used to produce hydrogen for fuel cells.

1.2.3 Geothermal Energy

Geothermal energy is the heat generated in the Earth's interior. The heat can be used to run geothermal power plants that include steam generator, turbine, condenser, and electrical generator.

1.2.4 Wind Energy

Wind turbines are used to convert wind energy into mechanical energy and then into electrical energy. Commercial wind turbines have aerodynamic loss at the rotor, which transfers approximately 45%–50% of the wind energy into mechanical energy. Additional losses include transmission and generator losses. In recent times, use of wind turbine has been increased significantly for power generation.

1.2.5 Renewable Energy for Hydrogen Production

One of the major issues with the use of alternative energy is that the energy supply and demand do not coincide. Solar panels do not work at night when we still need power and so may need a storage system. Wind-generated power may be needed in faraway places. Thus, we need convenient and cost-effective storage and transportation. In situations like these, hydrogen may also act as future storage and transport medium. Also, during times

TABLE 1.1

Comparison of Hydrogen Fuel with Other Fuels in Terms of Heating Values

Fuel Type	Chemical Formula	Higher Heating Values (MJ/kg)
Coal		15–45
Hydrogen	H_2	142
Diesel (*N*-cetane–*N*-hexadecane)	$C_{16}H_{24}$	44
Gasoline (octane)	C_8H_{18}	48
Ethanol	C_2H_5OH	30
Methanol	CH_3OH	23
Natural gas	Mostly CH_4	54
Wood	$C_6H_{10}O_5$	21

when solar panels and wind turbines produce more energy than needed, the excess energy can be used to produce hydrogen and run a fuel cell.

1.2.6 Hydrogen Production and Hydrogen Fuel Cell

Hydrogen can be produced in a number of different ways such as electrolysis, reforming fossil and biofuels, hydrocarbon cracking, and iron-water vapor reaction. One of the common methods to produce hydrogen uses *electrolyzers,* which use electricity to split water into hydrogen and oxygen using an electrochemical process known as *electrolysis.* The hydrogen as well as oxygen can be stored and transported as necessary. These gases can be fed into a fuel cell that converts the chemical energy of hydrogen and oxygen into electricity using an electrochemical process and produce water and heat as the only by-products.

A comparison of hydrogen in terms of heating values and densities with other major fuels is given in Table 1.1.

1.3 Electrochemical Device—Basic Components and Operation

An electrochemical device is one in which chemical reactions occur directly with the presence of electrical energy as input energy source or as the output energy (Hamann et al., 2007; Newman and Thomas-Alyea 2004). This device is also referred to as *electrolytic cell, electrochemical cell,* or *galvanic cell* and includes devices such as *electrolyzer, batteries,* and *fuel cells. Batteries* and *fuel cells* are electrochemical cells in reverse, known as *galvanic cells,* producing electricity directly from the chemical energy content of fuel.

A typical electrochemical device is shown in Figure 1.5.

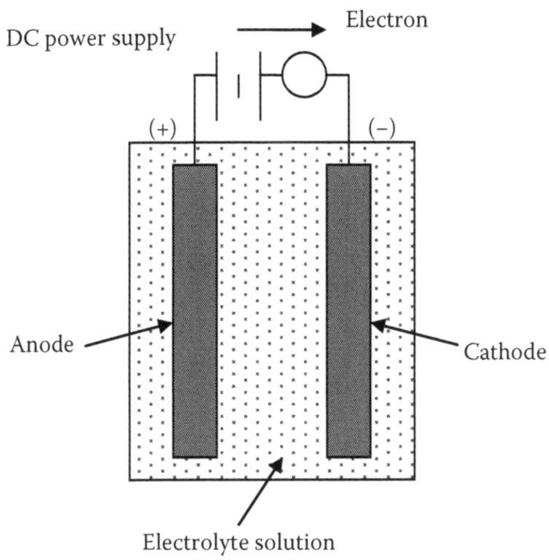

FIGURE 1.5
Basic arrangement of an electrochemical cell.

It primarily consists of two electronic conductors, referred to as *electrodes,* immersed in a bath of electrically conducting liquid called *electrolyte solution* or *electrolyte.* Electrodes are generally solids made of metals, carbon-based, or semiconductor materials. The electrodes are connected outside the bath in an external electrical circuit or *dc power supply.* When an electromotive force or electrical voltage is applied, electrochemical chemical reactions take place at each electrode–electrolyte interface, producing or consuming electrical charge species like *ions* and *electrons.* The charge transfer process that occurs between the electrode and the electrolyte results in a flow of electrons through the external circuit and motion of ions through the electrolyte from one electrode to the other.

Electrical conduction also takes place in the electrolyte owing to the motion of *ion charged species.* These ions are made available from a chemical compound (an electrolyte) either by melting it or by dissolving it in an ionizing solvent. The potential difference between two electrodes acts as one of the driving forces for the transport of the *negatively charged ions, called anions* toward the *positive electrode or anode* and transport of *positively charged ions, called cations* toward the *negative electrode* or *cathode.* The ionic transport is also aided by the presence of the concentration gradient of ions caused by the production and consumption of ions at the two electrodes.

At the interfaces of the electrode–electrolyte or the electronic–ionic conductors, ions are transformed by capture or release of electrons. These reactions at the electrodes are characterized as *oxidation reaction* and *reduction reaction.* In an *oxidation reaction,* electrons are removed or released from reactant species, and in a *reduction reaction,* electrons are added to or consumed by reactant species.

1.3.1 Electrolyzer

An electrolyzer is a device that uses an electrochemical process known as *electrolysis.* In an *electrolysis process,* the chemical bond in a liquid chemical compound is changed and decomposed into its constituents with the application of external current. For example, water can be decomposed into its constituent elements hydrogen and oxygen by the electrolysis process. Thus, electrolysis is a process in which electrical energy is converted into chemical energy.

The liquid chemical compound in an electrolyzer is referred to as an electrolyte, which is decomposed with the application of an external DC voltage between the two electrodes that are in contact with the electrolyte. As a requirement for the electrolysis to happen, the applied voltage or the cell voltage has to be at least greater than a threshold decomposition voltage associated with electrolyte material. For example, in the electrolysis of pure water, hydrogen and oxygen are produced when an electric current is passed through the water with the application of a DC voltage greater than the decomposition voltage of water (a theoretical voltage of 1.23 V).

Major components in an electrolyzer are negatively charged cathode electrode, positively charged anode, and electrolyte. The two half electrochemical reactions and the overall reaction are as follows:

$$\begin{aligned}
&\text{Anode reaction:} && H_2O \rightarrow 2H^+ + 2e^- + \frac{1}{2}O_2 && \text{(a)}\\
&\text{Cathode reaction:} && 2H^+ + 2e^- \rightarrow H_2 && \text{(b)}\\
&\text{Overall reaction:} && H_2O \rightarrow H_2 + \frac{1}{2}O_2 && \text{(c)}
\end{aligned} \tag{1.1}$$

At the anode (right electrode), water is oxidized into oxygen, proton (H^+), and free electrons (e^-). While *oxygen gas* (O_2) is collected directly at the anode, the proton migrates through the *proton conducting electrolyte* to the cathode side. *At the cathode,* proton is reduced to hydrogen (H_2) by combining with incoming free electrons supplied by the DC power source.

1.3.2 Battery

A battery is more like a batch electrochemical device in which electricity is generated from a fixed amount of initially stored fuel by direct conversion of its chemical energy content. The process or the electricity generation ceases when one of the reactants is depleted. A battery electrochemical cell consists of two electrodes, separated by an electrolyte/separator. The electrodes have a different electromotive force based on spontaneous half reactions and the difference of the cell electromotive force defines the batteries' terminal voltage. These two half reactions occur simultaneously and result in a conversion of chemical energy to electrical energy. During this process, the electrons transfer through external circuits performing electrical work as needed by the load. The electrolyte allows ions to transport from one electrode to other. The separator prevents the electrolytes from mixing and prevents contact between the anode and cathode but still allows the ions to flow through it.

One way to classify batteries is by defining them as a primary or secondary battery. A primary battery is referred to as the disposable battery as it cannot be charged once discharged. This is due to the fact that the materials used may not return to its original form because of the irreversible chemical reaction used during discharge. A secondary battery, referred to as the *rechargeable battery,* can be recharged by performing a reversed chemical reaction with the supply of electrical energy from an external source.

Another classification is given on the basis of the cell being wet or dry. A wet cell contains a liquid electrolyte that covers all the internal parts. Wet cells can be either primary or secondary cells. A dry cell contains an electrolyte in a paste or solid layer, which prevents it from moving. This enables the cell to be operated in any orientation without running the risk of the electrolyte spilling.

One of the oldest forms (and is a very common form) of rechargeable and wet type is the lead-acid battery. Alternative chemical reactions have given way to new rechargeable battery cells like lithium ion and metal hydride. Typical lithium ion anodes are based on carbon while the cathode is made from lithium cobalt dioxide, lithium manganese dioxide, or many other chemical combinations.

During the discharge process, the negative electrode that has low positive standard reduction potential undergoes an oxidation reaction producing *cations* (i.e., positively charged ions) and electrons. While cations migrate through the electrolyte toward the positive electrode, electrons travel through an external circuit toward the positive electrode with higher standard reduction potential, which in turn undergoes a reduction reaction with incoming cations and electrons. During the charging of a rechargeable battery, the reverse phenomenon happens; the cations move from the positive electrode toward the negative electrode and the electrons are driven through an external path toward the negative electrode.

The electrode that undergoes oxidation is termed as *anode* or *the negative electrode* while the electrode that accepts electrons and undergoes reduction decreases in oxidation number and is termed as *cathode* or *the positive electrode.* Hence, during charging, the positive electrode becomes the anode and the negative electrode becomes the cathode.

Some of the commonly used secondary rechargeable batteries are lead acid, Li-ion, Li metal, Ni-Cd, NiMH, and Zn-air.

Lithium ion batteries are very attractive because of their high specific energy and power density (Wakihara and Yamamoto 1998). Since lithium is one of the lightest metals, it has been considered as one of the leading candidates in the search in the lower-weight and smaller-size batteries that give higher energy density. Owing to their high performance and minimal volume, they are attractive for use in electric and hybrid vehicles as well as for portable devices.

Lithium metal anodes provide the highest theoretical voltage. However, they are also the most unstable and have issues ranging from highly resistive passivation films to dendrite formation and therefore are rarely used as a metal. Alternatively, lithium is being used as an intercalation material in porous carbon structure as an anode electrode. During discharge, lithium ions just get extracted from the structure of the carbon, leaving the electron, and during charging, they are inserted back. This type of insertion/extraction reaction depends largely upon the crystal structure, morphology, and orientation of the crystallites of a porous base material like carbon. Lithium is used as an insertion material in base materials. This mechanism relies on an open crystal structure to allow the insertion or extraction of lithium ions, as well as the ability to accept compensating electrons at the same time.

Hence, each electrode is made of active materials bound together with an electronically insulating binder and conductive additives. During discharge, Li is removed from the anode (negative electrode), transmitted through the electrolyte, and inserted back into the cathode (the positive electrode). During charging, the reverse process occurs. The difference in voltages between the two electrodes is the cell voltage. The amount of Li stored in the electrodes directly contributes to the energy capacity (product of voltage and capacity), and the rate of transfer of Li from one electrode to the other determines the power.

A large number of carbonaceous materials have been tested as a negative electrode (anode when discharging) and graphite is now the most commonly used for carbon electrodes.

During discharge, lithium ions just get extracted from the structure of the carbon, leaving the electron, and during charging, they are inserted back.

A vast majority of the positive electrodes (cathode when discharging) are also insertion compounds made of lithiated oxides of metals like manganese ($Li_yMn_2O_4$), cobalt (Li_yCoO_2), nickel (Li_yNiO_2), vanadium ($Li_xV_2O_5$ and $Li_yV_6O_{13}$), iron phosphate ($LiFePO_4$), and many other combinations capable of accepting and giving lithium ions.

Electrolytes are classified into three basic categories: liquid, solid, and composite. Organic electrolytes are used in lithium ion batteries as lithium reacts when it comes in contact with water. Most common organic electrolytes are the carbonates or esters of simple alcohol and glycol, for example, ethylene carbonate (EC-$C_3H_4O_3$), dimethyl carbonate (DMC-$C_3H_6O_3$), and diethyl carbonate (DEC). These are generally mixed with salts such as $LiPF_6$, $LiBF_4$, or $LiClO_4$.

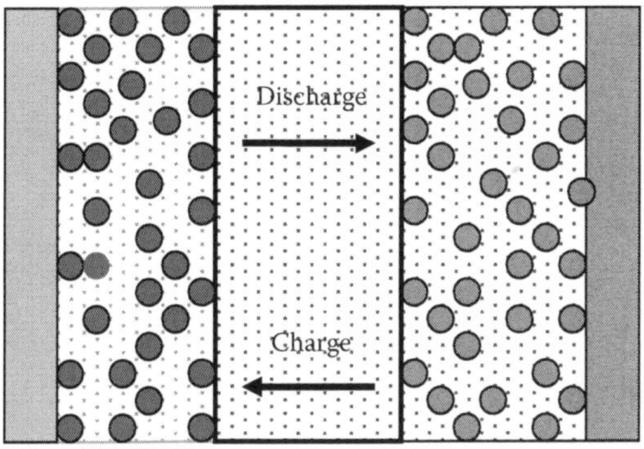

FIGURE 1.6
A typical lithium ion cell.

Figure 1.6 shows a typical lithium ion cell consisting of a negative electrode current collector of copper (Cu), lithiated porous carbon as a *negative electrode* (Li_xC_6), lithiated porous manganese dioxide ($Li_yMn_2O_4$) as a *positive electrode*, a positive electrode current collector of aluminum (Al), and a separator.

The separator is a porous matrix that gives mechanical stability to the cell and is made of a copolymer of vinylidene fluoride and hexafluoropropylene, p (VdF-HFP). All the pores or the voids of the negative electrode, separator, and positive electrode are filled with electrolyte and form the solution or the liquid phase of the cell. The *electrolyte* of the lithium ion battery is a mixture of ethylene carbonate (EC-$C_3H_4O_3$) and dimethyl carbonate (DMC-$C_3H_6O_3$) in a ratio of 1:2 volume/volume with *lithium hexaflorophosphate* ($LiPF_6$) as a salt. In recent times, lithium-polymer cells consisting of lithium metal as negative electrode and solid polymer electrolyte have also been considered (Wang and Sastry, 2007). The polymer electrolyte considered is composed of $LiClO_4$ salt with polyethylene oxide.

The deintercalation/intercalation reactions that occur during the charge and discharge processes are shown below, assuming lithiated porous carbon as a *negative electrode* (Li_xC_6) and lithiated porous manganese dioxide ($Li_yMn_2O_4$) as a *positive electrode* during discharge:

Reactions during discharge:

At the negative electrode (i.e., at the anode)

$$Li_xC_6 \xrightarrow{\text{discharge}} xLi^+ + xe^- + 6C \tag{1.2a}$$

At the positive electrode (i.e., at the cathode)

$$xLi^{+} + xe^{-} + Li_yMn_2O_4 \xrightarrow{\text{discharge}} Li_{x+y}Mn_2O_4 \qquad (1.2b)$$

Reactions during charging:

At the negative electrode (i.e., at the anode)

$$xLi^{+} + xe^{-} + 6C \xrightarrow{\text{charge}} Li_xC_6 \qquad (1.3a)$$

At the positive electrode (i.e., at the cathode)

$$Li_{x+y}Mn_2O_4 \xrightarrow{\text{charge}} xLi^{+} + xe^{-} + Li_yMn_2O_4 \qquad (1.3b)$$

where x and y are the insertion factors for the negative and positive electrodes, respectively. These factors, viz., x and y, are defined as the ratio of the initial lithium concentration to the maximum concentration in the solid phase of the negative and positive electrodes, respectively.

The function of the separator in the whole system is to provide mechanical stability to the liquid electrolytes and to act as a barrier between the anode and cathode to prevent them from short circuiting. Like electrolytes, it should have high conductivity to allow ions to pass through them and should act as an insulator to the electron.

When a cell is at rest, and is neither charging nor discharging, the terminal voltage is known as the open-circuit voltage. When the cell is discharged, the terminal voltage decreases owing to a number of internal resistances like irreversibilities associated with the electrochemical reactions and ohmic loss. The composition of the cell determines this terminal voltage. Since different combinations of chemicals can be used for cells, there are several different open-circuit voltages that can be created.

1.3.2.1 Battery Technology

Battery technology has been under extensive research and development over the last several decades. Technology has moved from the traditional lead-acid battery to the more sophisticated high-energy and high-power batteries like nickel metal hydride (NiMH), nickel cadmium (NiCd), and lithium ion (Li-ion) batteries. Nickel metal hydride batteries have been the choice for many hybrid electric vehicles (HEVs) in the market because of the significant reduction in weight and improved energy density compared to lead-acid batteries. Lithium ion batteries are the most common choice for consumer electronics applications because of their high energy density. Lithium ion batteries are also preferred in many of the newer HEVs because of their energy density advantage over NiMH.

1.3.3 Fuel Cell

A *fuel cell* is an *electrochemical device* that uses reverse electrochemical reactions and continuously converts the chemical energy content of the fuel into electrical energy, water, and some heat as long as fuel and oxidant are supplied. It differs from a battery in the sense that it does not contain all the necessary reactants as initially stored. When electricity is required, fuel and oxidant are fed to the cell continuously and the reaction products are also continuously removed simultaneously. In principle, electricity will be produced indefinitely if the supply and removal of materials are maintained. It operates quietly and efficiently, and when hydrogen is used as a fuel, it generates only electric power and pure water, and so referred to as a *zero emission engine*. It is similar to the operation of a battery in terms of reverse electrochemical reaction without the need of recharging and similar to a conventional heat engine in terms of its ability to produce power continuously as fuel is supplied.

The fuel cell was first invented by *Sir William Robert Grove* and *Christian Friedrich* in 1839 through their discovery of reversed electrolysis process. Grove developed the first cell, which consisted of two glass-enclosed platinum electrodes immersed in a diluted sulfuric acid electrolyte. The glass tube anode and cathode electrodes were supplied with hydrogen and oxygen reactants. The electrodes were connected to an external electrical circuit for electricity generation.

1.4 Basic Components and Operation of a Fuel Cell

A fuel cell is a power generation engine that takes fuels like hydrogen and oxygen as inputs and converts their chemical energy content directly into electrical energy through electrochemical reactions, producing water and heat as the by-products as shown in Figure 1.7.

A fuel cell consists of two electrodes separated by an electrolyte medium, which avoids direct contact of hydrogen and oxygen and direct chemical combustion as shown in a general schematic diagram in Figure 1.8.

Electrodes are usually composed of a gas diffusion layer with thin catalyst coatings at the electrode–electrolyte interfaces. Hydrogen and oxygen reactants are supplied to the anode and cathode electrodes' surfaces. The original rod-type electrodes are generally replaced with flat or circular annular surfaces to increase the contact surface area for reactions. The structure of the electrode is

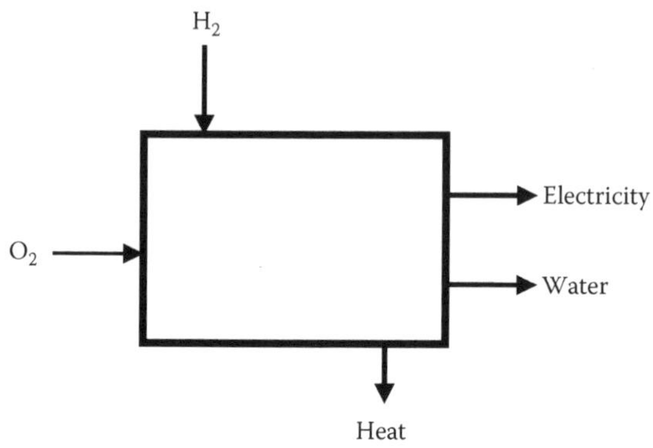

FIGURE 1.7
Fuel cell power generation.

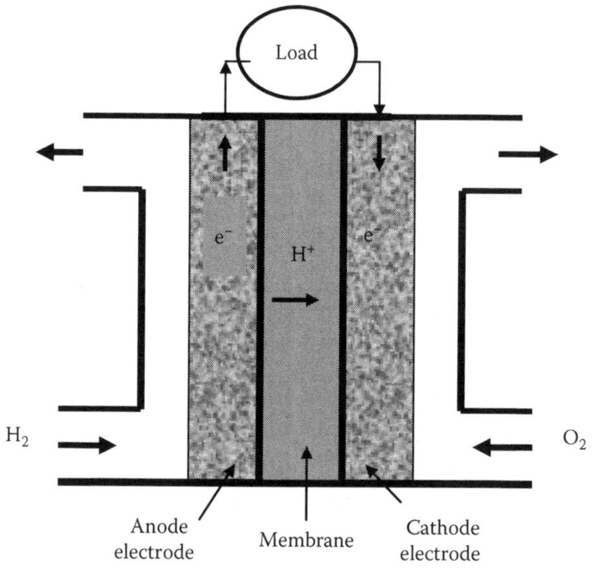

FIGURE 1.8
A three-layer hydrogen–oxygen fuel cell.

made porous for easy transport of reactant gases toward electrolyte. This also leads to higher contact area between the gas, electrode, and the electrolyte.

In a fuel cell, a direct hydrogen–oxygen combustion reaction is replaced by two electrochemical half reactions at two *electrode–electrolyte interfaces*. The hydrogen and oxygen reactants are spatially separated by the electrolyte. The electrons released from breaking the hydrogen bond transfer through an external circuit to recombine with oxygen and form water as the new hydrogen–oxygen bond. The electrochemical reactions at the anode and cathode sides take place simultaneously, producing electricity, water, and heat as the only by-products when hydrogen is used as the fuel. Heat is produced as a result of the irreversibilities associated with the electrochemical reactions at electrodes and by ohmic heating caused by the charge transport through cell components.

In one of the typical fuel cells, for example, a fuel (usually hydrogen in a hydrogen fuel cell) undergoes an electrochemical oxidation reaction and is transformed into hydrogen ion or proton by releasing electrons at the anode. The charged ions or protons transport through the ion-conducting but electronically insulating electrolyte material from the anode side to the cathode side. At the cathode, oxygen undergoes an electrochemical reduction reaction by combining with the incoming protons and electrons, producing water. The electrons flow through the electrically conducting electrodes and the external load circuit, resulting in electricity and performing electrical work. The two electrochemical half reactions and the overall reaction that represents the indirect combustion of hydrogen in the fuel cell are as follows:

$$\begin{aligned}
&\text{Anode reaction:} && H_2 \rightarrow 2H^+ + 2e^- \\
&\text{Cathode reaction:} && \frac{1}{2}O_2 + 2H^+ + 2e^- \rightarrow H_2O \\
&\text{Overall reaction:} && H_2 + \frac{1}{2}O_2 \rightarrow H_2O
\end{aligned} \tag{1.4}$$

1.5 Classification and Types of Fuel Cell

A number of different fuel cells have been developed and are in use. They are generally classified or characterized primarily by the type of electrolyte used, the type of ion transferred, and the range of applicable temperature level. Table 1.2 shows a list of the most commonly used fuel cells along with the type of electrolyte used, migrating ions, operating temperature ranges, and type of fuel used (Breiter, 1969; Larminie and Dicks, 2003; O'Hayre et al., 2006; Xianguo, 2006).

Since the type of electrolyte material dictates operating principles and characteristics of a fuel cell, a fuel cell is generally named after the type of electrolyte used. For example, an alkaline fuel cell (AFC) uses an alkaline solution such as potassium hydroxide (KOH) in water, an acid fuel cell such as phosphoric acid fuel cell (PAFC) uses phosphoric acid as electrolyte, a solid polymer electrolyte membrane fuel cell (PEMFC) or proton exchange membrane fuel cell uses proton-conducting solid polymer electrolyte membrane, a molten carbonate fuel cell (MCFC) uses molten lithium or potassium carbonate as electrolyte, and a solid oxide ion-conducting fuel cell (SOFC) uses ceramic electrolyte membrane.

TABLE 1.2

List of Fuel Cells Classified Based on Electrolytes, Ions, and Temperature Range

Fuel Cell Type	Charge Carrier	Electrolyte Type	Typical Operating Temperature (°C)	Fuel and Oxidant
AFC	OH^+	Potassium hydroxide or sodium solution	≈60–120	H_2, O_2
PAFC	H^+	Phosphoric acid	≈220	Pure H_2
MCFC	CO_3^{2-}	Lithium or potassium carbonate	≈600–700	H_2, CO, CH_4, and other hydrocarbons
PEMFC	H^+	Solid polymer (Nafion)	≈80	Pure H_2
SOFC	O^{2-}	Solid oxide electrolyte (yttria-stabilized zirconia)	≈700–1000	H_2, CO, CH_4, and other hydrocarbons
DMFC	H^+	Solid polymer	≈80	Methanol

Fuel cells are also classified based on type of ion exchange involved in the two half electrochemical reactions and ion transport through the electrolyte. Fuel cells are termed as *cation transfer fuel cells* that involve positively charged ion transporting through the electrolyte. For example, PAFC and PEMFC are referred to as *cation fuel cells* as the transporting ion is the positively charged hydrogen ion or proton (H^+). There are two major characteristics as well as challenges in cation transfer fuel cells. *First,* the oxygen reduction reaction in cathode is relatively slow, causing higher cell voltage loss referred to as activation *overpotential* or *activation voltage loss,* and hence requires expensive catalyst materials to enhance electrochemical kinetics. *Second,* the electrochemical by-product water is formed at the cathode side and requires an effective water removal mechanism in order to avoid the so-called water flooding of the cathode electrode, which prevents oxygen molecules from reaching reaction sites because of its low diffusivity in water. Water flooding in the cathode side results in a reduced oxygen mass transfer and reduced oxygen concentration at the cathode–electrolyte interface to sustain the reaction, and this causes cell voltage loss referred to as *mass transfer voltage loss* or *concentration overpotential.* These are two of the major challenges and foci for cation transfer fuel cell development.

Fuel cells are also termed *anion fuel cells* that involve negatively charged ions transporting through the electrolytes, such as AFC with hydroxyl ion (OH^-), MCFC with carbonate ion (CO_3^{2-}), and SOFC with oxide ion (O^{2-}). The following are two major characteristics of anion fuel cells: (i) the oxygen reduction reaction in the cathode side is relatively fast, resulting in lower activation voltage loss, and hence does not require any noble metals as catalyst, and (ii) the by-product water is formed at the anode side where hydrogen is supplied as the reactant. The mass transfer loss as a result of water flooding at the anode is relatively low and less critical in anion fuel cells because of the higher diffusivity of hydrogen in water.

The fuel cells are also classified into *low-temperature fuel cell, intermediate-temperature fuel cell,* and *high-temperature fuel cell* based on their operating temperature range. Low-temperature fuel cells include AFC and PEMFC with an operating temperature range of 60°C–120°C, intermediate-temperature fuel cells include PAFC with operating temperatures below 220°C, and high-temperature fuel cells include MCFC and SOFC with operating an temperature range of 600°C–1000°C. High-temperature fuel cells are attractive because they do not need expensive precious metal catalysts and are suitable for a wider range of fuel types. It also produces high-temperature exhaust gases, which makes it suitable for integration with other thermal systems for cogeneration or gasification or for better thermal management of the overall system.

All of these fuel cells function in the same manner. At the anode, a fuel, usually hydrogen, produces free electrons, and at the cathode, oxygen is reduced to oxide species. Depending on the electrolyte, either protons or oxide ions are transported through the ion-conducting but electronically insulating electrolyte to combine with oxide or protons to generate water and electric power.

In order for both anode and cathode reactions to proceed continuously, electrons produced at the anode must pass through an electric circuit to the cathode, and ions must migrate through the electrolyte. It is also important that the electrolyte material only allows ion transport and not electrons. Any motion of electrons through the electrolyte will cause reduced electron flow though the external circuit and hence cause performance loss.

The anode and cathode reaction characteristics, however, vary for different types of fuel cells. While a detailed discussion of these fuel cells is given in Chapter 8, brief descriptions of these fuel cells along with their associated reactions are given here.

1.5.1 Alkaline Fuel Cell

An AFC uses either potassium hydroxide solution (KOH) or sodium hydroxide (NaOH) as the electrolyte and operates over a temperature range of 60°C–120°C. The transporting ions through the electrolyte is a hydroxyl (OH^-) ion, moving from cathode to anode. At the anode, hydroxyl (OH^-) reacts with hydrogen and releases electrons and produces water. At the cathode, oxygen reacts with returning electrons taken from the electrode and water from the electrolyte to form new hydroxyl (OH^-) ions:

$$\begin{aligned}
&\text{Anode reaction:} && H_2 + 2OH^- \rightarrow 2H_2O + 2e^- && \text{(a)} \\
&\text{Cathode reaction:} && \frac{1}{2}O_2 + 2e^- + H_2O \rightarrow 2OH^- && \text{(b)} \\
&\text{Overall reaction:} && H_2 + \frac{1}{2}O_2 \rightarrow H_2O && \text{(c)}
\end{aligned} \tag{1.5}$$

Note that while water is consumed at the cathode, two times more water is produced at the anode. The operating efficiency of AFC is quite high, in the range of 60%–70%. However, this fuel cell is currently restricted to the use of pure hydrogen and oxygen only because of the incompatibility of the alkaline electrolyte with other fuels and air that contains carbon dioxide. A strongly alkaline electrolyte like NaOH and KOH absorb CO_2, which reduces electrolyte conductivity considerably. This restricts the use of impure H_2 containing CO_2 (using reformate fuel), and air has to be scrubbed free of CO_2 prior to use as an oxidant.

AFC has the longest history of all fuel cell types. While it was first developed around the 1930s, the technology was further developed by NASA for the Apollo space program.

1.5.2 Proton Exchange Membrane Fuel Cell

The name originated from the use of *polymer electrolyte membrane* with *proton as the migrating ions*. It was originally developed by DuPont for the chlor–alkali industry. Technology combined all three major parts—anode, cathode, and electrolyte—into a tri-layer membrane known as *membrane electrode assembly (MEA)* with thickness on the order of a few microns. This allows for a number of similar MEA cells to be stacked in a compact design and for scale-up operations with higher output voltage and power, and makes them suitable for a wide range of applications. A basic design may include a thin polymer electrolyte membrane coated with two thin layers of catalyst as electrodes. However, electrode design may also include a gas diffusion layer coated with catalyst.

Anode is supplied with hydrogen and cathode is supplied with oxygen or air. The hydrogen undergoes an electrochemical reaction and ionized releasing *electrons* (e^-) and hydrogen ions (H^+) or *proton* at the electrode–membrane interface. The proton transports through the membrane toward the cathode interface. The electrons move through the anode electrode toward the external electrical circuit. The oxygen reacts electrochemically with the returning electrons (e^-) from the anode electrode and proton (H^+) from the electrolyte to form water at the cathode electrode–membrane interface. The electrochemical half and overall PEM fuel cell reactions are as follows:

$$\begin{aligned}
&\text{Anode reaction:} && H_2 \rightarrow 2H^+ + 2e^- && \text{(a)} \\
&\text{Cathode reaction:} && \frac{1}{2}O_2 + 2H^+ + 2e^- \rightarrow H_2O && \text{(b)} \\
&\text{Overall reaction:} && H_2 + \frac{1}{2}O_2 \rightarrow H_2O && \text{(c)}
\end{aligned} \tag{1.6}$$

Attractive features like high efficiency, compactness, and quick and cold start, make PEMFC very appealing for a wide range of applications including vehicle transportation as well as stationary power generation. Major disadvantages of PEM fuel cells are their restricted use with only pure hydrogen with very little trace of carbon monoxide (CO), which is poisonous to the catalyst used in a PEM fuel cell, and the very high cost owing to their use of expensive catalyst material such as platinum. Water management also poses a considerable challenge for PEMFC design because of the requirement for a proper hydration level in the *Nafion* polymer membrane and the issue of drying of the membrane at the anode side and flooding of the membrane near the cathode side.

1.5.3 Phosphoric Acid Fuel Cell

A PAFC uses phosphoric acid (H_3PO_4) as the liquid electrolyte solution and a hydrogen ion (H^+) or *proton* as the migrating ion through it. Hydrogen gas ionizes at the anode, releasing electrons (e^-) and protons (H^+). At the cathode, oxygen reacts with the returning electrons (e^-) from the electrode and protons (H^+) from the electrolyte to form water. The reactions are summarized as follows:

$$
\begin{aligned}
&\text{Anode reaction:} && H_2 \rightarrow 2H^+ + 2e^- && \text{(a)} \\
&\text{Cathode reaction:} && \frac{1}{2}O_2 + 2e^- + 2H^+ \rightarrow H_2O && \text{(b)} \\
&\text{Overall reaction:} && H_2 + \frac{1}{2}O_2 \rightarrow H_2O && \text{(c)}
\end{aligned}
\tag{1.7}
$$

PAFCs are referred to as intermediate-temperature fuel cells with an operating temperature of around 200°C. Typical efficiency is 55%, which is relatively low compared to other types of fuel cell, except the direct methanol fuel cell (DMFC). PAFCs are developed mainly for medium-scale power generation with a unit operating power-up to 200 kW. Applications include stationary power generation as well as combined heat and power (CHP).

1.5.4 Molten Carbonate Fuel Cell

MCFC is a higher-temperature fuel cell that operates at a temperature range of 600°C–700°C with a high operating efficiency of 65%. The electrolyte in an MCFC is a molten alkali carbonate such as lithium or potassium carbonate retained in a ceramic matrix of lithium aluminum oxide. Hydrogen undergoes oxidation reduction by combining with incoming carbonate ions $\left(CO_3^{2-}\right)$, releasing electrons to the external circuit and producing water and carbon dioxide. At the cathode, oxygen undergoes reduction reaction by combining with carbon dioxide and incoming electrons from the external circuit and releasing carbonate ions.

The electrochemical half and overall MCFC reactions are as follows:

$$
\begin{aligned}
&\text{Anode reaction:} && H_2 + CO_3^{2-} \rightarrow H_2O + CO_2 + 2e^- && \text{(a)}\\
&\text{Cathode reaction:} && \frac{1}{2}O_2 + CO_2 + 2e^- \rightarrow CO_3^{2-} && \text{(b)}\\
&\text{Overall reaction:} && H_2 + \frac{1}{2}O_2 \rightarrow H_2O && \text{(c)}
\end{aligned}
\tag{1.8}
$$

It can be noticed that while carbon dioxide is produced at the anode side, it is also consumed at the cathode side. At the cathode side, the inlet oxygen or air gas stream includes a mixture with carbon dioxide, which can be supplied by either recovering carbon dioxide from the anode exhaust gas stream and recirculating it back to the cathode side or just drying and mixing the anode exhaust with the cathode inlet gas stream.

MCFCs can operate not only with hydrogen but also with other fuel types including natural gas, biogas, and clean coal gas by producing hydrogen through a reforming process. Applications include large stationary power generation and CHP.

1.5.5 Solid Oxide Fuel Cell

SOFC is a higher-temperature fuel cell that operates at a temperature range of 800°C–1000°C with a high operating efficiency of 65%. Electrolyte in an SOFC is a solid ceramic-based material like yttrium-stabilized zirconium (YSZ). It can operate with hydrogen fuel as well as with other fuel types such as natural gas, biogas, and coal gas. The basic components and the overall reaction are similar in an SOFC with the exception of the electrochemical reactions at the anode and cathode electrodes.

At the cathode electrode, oxygen picks up electrons and forms a negatively charged *oxygen ion*. The oxygen ion transports through the solid oxide ion-conducting membrane electrolyte toward the anode where it combines with the hydrogen gas producing water and electrons that travel to the cathode side through the external electrical circuit.

The reactions in an SOFC are summarized as follows:

$$
\begin{aligned}
&\text{Anode reaction:} && H_2 + O^{2-} \rightarrow H_2O + 2e^- && \text{(a)}\\
&\text{Cathode reaction:} && \frac{1}{2}O_2 + 2e^- \rightarrow O^{2-} && \text{(b)}\\
&\text{Overall reaction:} && H_2 + \frac{1}{2}O_2 \rightarrow H_2O && \text{(c)}
\end{aligned}
\tag{1.9}
$$

SOFCs can also operate with a number of other hydrocarbon fuels including natural gas and clean coal gas by producing hydrogen through an internal or external reforming process. Applications include large stationary power generation and CHP.

1.5.6 Direct Methanol Fuel Cell

DMFC gets its name from the use of methanol as the fuel instead of hydrogen. It is similar to the design and structures of the PEMFC that includes a proton-conducting solid polymer electrolyte membrane with two catalyst-coated electrodes. Basic operation involves supply of a mixture of methanol and water at the anode side. At the anode side, hydrogen is separated from the mixture and transformed into protons (H^+) and electrons (e^-) with the presence of a catalyst. Oxygen and carbon also react to form carbon dioxide (CO_2) at the anode. Electrons travel through the electrodes and the external power circuit toward the cathode. The proton ions transport through the electrolyte membrane and combine with oxygen and the returning electrons to form water at the cathode. The reactions are as follows:

$$\begin{aligned}
&\text{Anode reaction:} && CH_3OH + H_2O \rightarrow CO_2 + 6H^+ + 6e^- \\
&\text{Cathode reaction:} && \frac{3}{2}O_2 + 6H^+ + 6e^- \rightarrow 3H_2O \\
&\text{Overall reaction:} && CH_3OH + \frac{3}{2}O_2 \rightarrow 2H_2O + CO_2
\end{aligned} \tag{1.10}$$

Since methanol exists as a liquid in the temperature range of –97°C to 64°C at atmospheric pressure, it can be stored, transported, and may be used in liquid form similar to other liquid fuels like gasoline and diesel, and this makes DMFC compact and suitable for portable applications such as a battery substitute in laptop computers. A major disadvantage of DMFC is its low efficiency compared to other types of fuel cells. Additional challenges for the design of DMFC include the corrosiveness and poisonous nature of methanol fuel.

1.5.7 Micro Fuel Cells

A micro fuel cell is a compact miniaturized fuel cell with sizes ranging from a few square millimeters to 1000 mm^2. They are used as a portable power source for cell phones, laptop computers, personal digital assistants, and other portable low-power electronic devices. The power densities of miniature FCs range from a few tens of microwatts per square centimeter up to several hundreds of milliwatts per square centimeter. Miniature fuel cells are not just scaled-down large fuel cells. There are much greater engineering challenges that require a difficult balance of providing sufficient power and convenience while minimizing the size and the cost. One of the technological ways to miniaturize fuel cells is to make use of standard microfabrication techniques mainly used in microelectronics and, more especially, the fabrication of micro- and nano-electromechanical systems (MEMS/NEMS) either directly with silicon substrates or adapting the methods to other substrates such as metals or polymers (Pichonat, 2009). These techniques enable mass fabrication at low cost.

Typically, micro fuel cells use methanol as fuel although hydrogen-fed micro fuel cells have also been developed. The choice of the type of fuel cell to use in portable devices may be limited to low-temperature fuel cells such as PEMFC (proton exchange membrane fuel cell/polymer electrolyte membrane fuel cell) and DMFC. However, micro reformed methanol fuel cells and miniature SOFCs have also been developed.

The basic structure of micro fuel cells has a thin film planar stack generally made of silicon, foils, polymer, or glass with commercial ionomer, most often Nafion, and layers being micromachined (microchannels or porous media) for gas/liquid management and coated with gold for current collecting. Two basic design approaches are employed in micro fuel cells: the classic bipolar design where all the components of the micro fuel cell are stacked together and where fuel and oxidant are separated by the MEA, and the planar design where the fuel and oxidant channels are interdigitated and both electrodes are on the same single side. The bipolar design ensures the separation of fuel and oxidant but requires all components to be fabricated separately and then assembled together. The planar design is more suitable for a monolithic integration but requires a larger surface area to deliver a similar performance. For fabrication, different materials have been considered: silicon, stainless steel or titanium metal foils, and polymers.

Recently, membraneless laminar flow–based fuel cells are being explored and beginning to emerge. They are primarily aimed at avoiding ionomer membranes, which have disadvantages, most notably their change in size with humidification and their incompatibility with microtechnology.

1.5.8 Biological Fuel Cells

Biological fuel cells are different from conventional electrochemical fuel cells in various aspects. Biological fuel cells use biocatalysts to drive oxidation and reduction reactions. A biocatalyst can be used to generate fuel substrates through metabolic processes or biocatalytic transformations, or it could partake in the electron transfer that occurs between the fuel substrate and the electrode's surface. The electrolyte layer typical in conventional fuel cells is replaced by a membrane in the biofuel cell, which still allows ion exchange. Biofuel cells usually operate at ambient temperature, atmospheric pressures, and neutral pH. There are two types of biofuel cells: microbial and enzymatic fuel cells (Aelterman, 2009; Allen and Bennetto, 1993; Atanassov et al., 2007; Logan et al., 2006; Pant et al., 2010).

1.5.8.1 Microbial Biofuel Cells

In microbial fuel cells, bacteria are used as a catalyst to convert bio-convertible substrate into electrons. Bacteria are very small (≈1 μm) organisms that can convert a variety of organic compounds into CO_2, water, and energy. The microorganisms use the produced energy to grow and to maintain their metabolism; however, one can harvest part of this microbial energy in the form of electricity with a microbial fuel cell. A typical microbial fuel cell consists of anode and cathode compartments separated by a cation (positively charged ion)-specific membrane. In the anode compartment, fuel is oxidized by microorganisms, generating electrons and protons. Electrons are transferred to the cathode compartment through an external electric circuit, while protons are transferred to the cathode compartment through the membrane. Electrons and protons are consumed in the cathode compartment, combining with oxygen to form water.

There are two types of microbial fuel cell: *mediator* and *mediator-less microbial* fuel cells. Most of the microbial cells are electrochemically inactive. The electron transfer from microbial cells to the electrode is facilitated by mediators such as thionine, methyl viologen, methyl blue, and humic acid. Mediator-free microbial fuel cells do not require a mediator but uses electrochemically active bacteria to transfer electrons to the electrode (electrons are carried directly from the bacterial respiratory enzyme to the electrode). Among the electrochemically active bacteria are *Shewanella putrefaciens*, *Aeromonas hydrophila*, and others. Some bacteria, which have pili on their external membrane, are able to transfer their electron production via these pili. Mediator-less microbial fuel cells can, besides running on wastewater, also derive energy directly from certain aquatic plants. These include reed sweet grass, cordgrass, rice, tomatoes, lupines, and algae. These microbial fuel cells are called plant microbial fuel cells. Given that the power is thus derived from a living plant (in situ energy production), this variant can provide extra ecological advantages.

A microbial fuel cell consists of an anode, a cathode, a proton or cation exchange membrane, and an electrical circuit. The bacteria live in the anode and convert a substrate such as glucose, acetate, as well as wastewater into CO_2, protons, and electrons as shown in Figure 1.9. Under aerobic conditions, bacteria use oxygen or nitrate as a final electron acceptor to produce water. However, in the anode of an MFC, no oxygen is present and bacteria need to switch from their natural electron acceptor to an insoluble acceptor, such as the MFC anode. Because of the ability of bacteria to transfer electrons to an insoluble electron acceptor, we can use an MFC to collect the electrons originating from the microbial metabolism. The electron transfer can occur

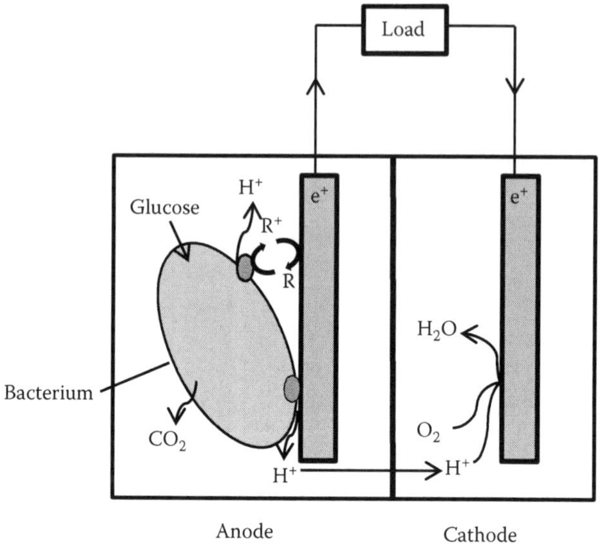

FIGURE 1.9
A microbial fuel cell schematic where bacteria in an anodic compartment can bring about oxidative conversions, while in the cathodic compartment, chemical and microbial reductive processes can occur.

either via membrane-associated components, soluble electron shuttles, or nano-wires. The protons flow through the proton or cation exchange membrane to the cathode.

1.5.8.2 Enzymatic Biofuel Cell

In contrast to microbial biofuel cells, enzymatic biofuel cells utilize the redox enzymes rather than the whole microorganism as a biocatalyst. The redox enzyme, which is separated and purified from an organism, participates in the electron transfer chain that occurs between the substrate and the anode by oxidizing the fuel or between the substrate and the cathode as shown in Figure 1.10.

The basic reaction for a functioning between the enzymatic biofuel cell is a complete circuit composed of the cathodic and anodic enzyme reactions that release and trap electrons, respectively, as shown in Figure 1.10. Glucose oxidase has been very commonly used as an anodic enzyme because of its high stability at a physiological pH of 7.2 and high turnover rates. Bilirubin oxidase has been used as a cathodic enzyme primarily because of its high stability at physiological pH.

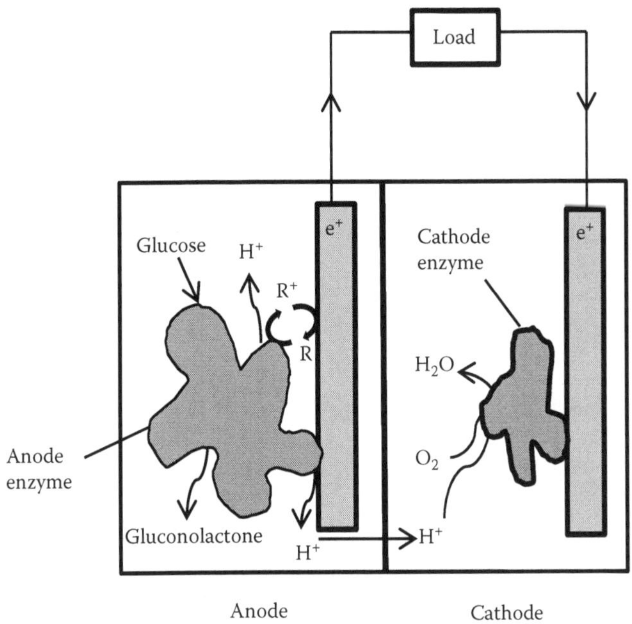

FIGURE 1.10
A schematic for a simple glucose oxidase/bilirubin-based enzymatic biofuel cell.

The redox reactions of these enzymes are shown below:

Glucose oxidase reaction (anode):

$$\text{Glucose} \rightarrow \text{Gluconolactone} + 2\text{H}^+ + 2\text{e}^- \; (\text{E}^{\circ}_{\text{anode}} = -0.36\text{V vs. Ag/AgCl at pH 7.2}) \quad (1.11a)$$

Bilirubin oxidase reaction (cathode):

$$\text{O}_2 + 2\text{H}^+ + 2\text{e}^- \rightarrow \text{H}_2\text{O} \; (\text{E}^{\circ}_{\text{anode}} = 0.58\text{ V vs. Ag/AgCl at pH 7.2}) \quad (1.11b)$$

The electrons exchanged at each electrode need to be transported to the electrode to make the cell active. Redox enzymes are incapable of direct contact with the electrode since their redox centers are insulated from the conductive support by the protein matrices. In order to contact these enzymes with the electrode, mediators are utilized, which are dependent on the class of oxidative enzymes.

1.6 Applications of Fuel Cell

The fuel cell power generation systems are considered for a wide range of applications including transportation, stationary power generation, portable power generation, and space and military applications (Amphiett et al., 1995, Arendas et al., 2012; Bidwai et al., 2012; Hilmansen, 2003; Jones et. al., 1985; Miller et al., 2006, 2007; Roan, 1992, Rose and Geyer, 2000; Scott et al., 1993; Srinivas, 1984).

1.6.1 Transportation

The fuel cell is one of the strongest contenders to replace internal combustion engines for personnel vehicles as well as diesel engines for trucks, buses, and locomotives because of its higher efficiency, improved fuel economy and reduced dependence on conventional fuel, and lower pollution emission.

PEMFCs in particular have been considered to replace internal combustion engines in vehicles because they are compact, are lightweight, and operate relatively at low temperatures. This low-temperature operation makes them suitable for rapid start-up and shutdown and makes them more responsive to load variation during vehicle operation. Fuel cell–powered vehicles have been considered by many major automakers in the world. Major obstacles to the development of a fuel cell–powered electric vehicle are the infrastructure developments for hydrogen transportation, storage and fueling stations, and onboard storage of hydrogen with enough capacity for an average driving distance.

Different types of fuel cells are considered for larger vehicle and locomotive power generation systems. This includes the high-temperature fuel cells such as SOFCs and MCFCs and low-temperature fuel cells such as AFCs and PEMFCs.

PAFC and solid PEMFC have considerable potential for long duty cycle vehicles like buses, trucks, and locomotives using methanol from natural gas and coal. A hydrogen PEM fuel cell hybrid power system is considered for a switcher locomotive or yard locomotive (Miller et al., 2006, 2007). The system includes a 250 kW PEM fuel cell, a lead-acid battery as power storage for peak power and auxiliary power needs, and a roof-mounted compressed hydrogen storage. An onboard PEMFC-powered locomotive with metal hydride–stored hydrogen generated by electrolysis of water by off-board surplus power is also considered (Hasegawa and Ohki, 1995).

High-temperature fuel cells are attractive in locomotives because of their higher power ratings and the potential use of high-temperature exhaust gases for cogeneration or onboard fuel reforming. In addition, because road locomotives tend to be used for extended periods, the start-up time for the fuel cell is not as serious of a concern as it is in other applications. SOFC with onboard gasification of diesel and biodiesel and power generations is considered for long duty cycle locomotives and trucks (Kumar et al., 1993; Schroeder and Majumdar, 2010).

Some of the major contenders for fuel cells in transportations are battery-powered electric vehicles and improved internal combustion engines.

1.6.2 Stationary Power Generation

Stationary applications include large-scale central power generation (1 MW and higher), mid-range commercial and industrial power (10–1000 kW), and small-range residential power (5–10 kW).

Fuel cells are considered for residential application using the existing natural gas supply line and as a CHP system to meet the space heating and water heating needs. The uses of fuel cells for mid-size commercial and industrial applications are also considered along with cogeneration of heat and power.

High-temperature fuel cells are primary contenders for large-scale central power generation systems because they are suitable for a wider range of fuel types. It also produces high-temperature exhaust gases, which makes it suitable for integration with other thermal systems for cogeneration or gasification and for better thermal management of the overall system. A combined cycle power generation system results in a high-energy conversion efficiency using either pure hydrogen as fuel or clean coal gas. Figure 1.11 shows an

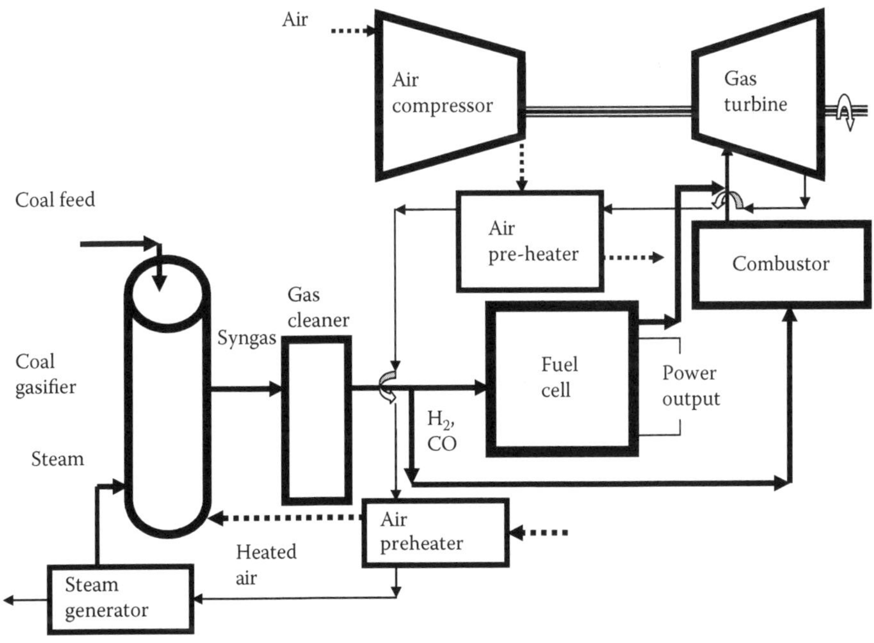

FIGURE 1.11
Advanced combined cycle fuel cell–gas turbine power generation system and using clean coal gas.

advanced combined cycle fuel cell–gas turbine power generation system based on using clean coal gas.

The system includes four major subsystems: (1) coal gasifier and cleaner system, (2) SOFC power generation system, (3) gas turbine power generation system, and (4) thermal heat management system. The diagram shows a coal gasifier supplied with oxygen or air and steam. Generated *syngas* is then cleaned in a gas cleaner to remove any unwanted species such as sulfur and nitrogen. The remaining gas mixture containing primarily hydrogen and carbon monoxide is then delivered to the prime mover. The prime mover shown here is the combined cycle system consisting of an SOFC and gas turbine/generator to generate electric power. While the input fuel gas to the SOFC primarily consists of H_2 and CO, the input to the gas turbine system may be a combination of inputs. The exhaust from the SOFC will be fed to the turbine and a direct feed from the coal gasifier/gas cleaner system may also be added to the fuel input to the gas turbine combustor. As a part of the thermal management system, the exhaust heat from the gas turbine is used to generate steam and preheat the air supplied to both the gas turbine and the gasifier.

1.6.3 Portable Power

Portable applications of fuel cells include auxiliary power unit and emergency power systems, power tools, laptop computers, and other mobile devices including cell phones. Demands are also growing with the increased energy and power requirements for broadband mobile computing (Dyer, 2002). Power requirements may vary from a few watts (<1 W) to a few hundred watts or kilowatts (1–5 kW). Portable fuel cells are often categorized on the basis of power requirements (DOE 2010—Record # 11009) such as applications less than 2 W, applications for 10–50 W, and applications for 100–250 W.

Some of the user-specific requirements for portable applications are as follows: quick turn-on and -off capability, responsive to dynamic variation in power needs of the device, compact, lightweight, and suitable for operation over a wide range of ambient temperature and humidity conditions. Additionally, portable fuel cells are expected to operate safely, providing power without exposing users to hazardous or unpleasant emissions, high temperature, and low noise.

Both DMFC and PEMFC are considered for portable applications.

References

Aelterman, P. Microbial fuel cells for the treatment of waste streams with energy recovery. PhD Thesis, Gent University, Belgium, 2009.

Allen, R. M. and H. P. Bennetto. Microbial fuel cells: Electricity production from carbohydrates. *Applied Biochemistry and Biotechnology* 39–40: 27–40, 1993.

Amphiett, J. C., R. F. Mann, B. A. Peppley, P. R. Roberge and A. Rodrigues. A practical PEM fuel cell model for simulating vehicle power sources. Proceedings of the 10th Annual Battery Conferences on Applications and Advances, pp. 221–226, 1995.

Arendas, A., P. Majumdar, K. Rao and D. Schroeder. Experimental analysis thermal characteristics of li-ion battery for hybrid locomotives. Proceedings of the International Conference on Renewable Energy: Generation and Applications—ICREGA2012, 2012.

Atanassov, P., C. Apblett, S. Banta, S. Brozik, S. C. Barton, M. Cooney, B. Y. Liaw, S. Mukerjee and S. D. Minteer. Enzymatic biofuel cells. *The Electrochemical Society Interface* 16(2): 28–31, 2007.

Bidwai, J., P. Majumdar, D. Schroeder and K. Rao. Electrochemical and thermal run away analysis of lithium-ion battery for hybrid locomotive. Proceedings of the 2012 ASME Summer Heat Transfer Conference, July 2012.

Breiter, M. W. *Electrochemical Processes in Fuel Cell*. Springer-Verlag, Heidelberg, 1969.

Dyer, C. K. Fuel cells for portable applications. *Journal of Power Sources* 106: 31–34, 2002.

Hamann, C. H., A. Hamnett and W. Vielstich. *Electrochemistry*, 2nd Edition. Wiley, 2007.

Hasegawa, H. and Y. Ohki. Development of a model of on-board PEMFC powered locomotive with a metal hydride Cylinder. Proceedings of the Symposium on Material for Electrochemical Energy Storage and Conversion—Batteries, Capacitors and Fuel Cells, pp. 145–150, 1995.

Hilmansen, S. The applications of fuel cell technology to rail transport operations. *Journal of Rail and Rapid Transport* 217: 291–298, 2003.

Jones, L. E., G. W. Hayward, K. M. Kaylyanam, Y. Rotenbero, D. S. Scott and B. A. Steinberg. Fuel cell alternative for locomotive propulsion. *International Journal of Hydrogen Energy* 10(7/8): 505–516, 1985.

Kumar, R., M. Krumpelt and K. M. Myles. Solid oxide fuel cell for transportation: A clean, efficient, alternative for propulsion. Proceedings of the Third International Symposium on Solid Oxide Fuel Cell, pp. 948–956, 1993.

Larminie, J. and A. Dicks. *Fuel Cell System Explained*, 2nd Edition. Wiley & Sons, 2003.

Logan, B. E., B. Hamelers, R. Rozendal, U. Schroder, J. Keller, S. Freguia, P. Aelterman, W. Verstraete and K. Rabaey. Microbial fuel cells: Methodology and technology. *Environmental Science & Technology* 40: 5181–5192, 2006.

Miller, A. R., J. Peters, B. E. Smith and O. Velev. Analysis of fuel cell hybrid locomotives. *Journal of Power Sources* 157: 855–861, 2006.

Miller, A. R., K. S. Hess, D. L. Barnes and T. L. Erickson. System design of a large fuel cell hybrid locomotive. *Journal of Power Sources* 173: 935–942, 2007.

Newman, J. and K. E. Thomas-Alyea. *Electrochemical Systems*, 3rd Edition. Wiley Intersciences, 2004.

O'Hayre, R. O., S.-W. Cha, W. Colella and F. B. Prinz. *Fuel Cell Fundamentals*. John Wiley & Sons, 2006.

Pant, D., G. V. Bogaert, L. Diels and K. Vanbroekhoven. A review of the substrates used in microbial fuel cells (MFCs) for sustainable energy production. *Bioresource Technology* 101(6): 1533–1543, 2010.

Pichonat, T. MEMS-based micro fuel cells as promising power sources for portable electronics. In: *Micro Fuel Cells: Principles and Applications*, Editor T. S. Zhao. Elsevier Inc., 2009.

Roan, V. P. A study of potential attributes of various fuel cell-powered surface transportation vehicle. Proceedings of the 27th Intersociety Energy Conference, 6/159—66, vol. 6, 1992.

Rose, R. and B. Geyer. Opportunities for fuel cells in rail applications and specialty vehicles. Proceedings of the International Conference with Exhibition—FUEL CELL 2000, pp. 283–289, 2000.

Schroeder, D. and P. Majumdar. Feasibility for solid oxide fuel cells as power source for railroad locomotives. *International Journal of Hydrogen Energy* 35: 11309–11314, 2010.

Scott, D. S., H.-H. Roger and M. B. Scott. Fuel cell locomotives in Canada. *International Journal of Hydrogen Energy* 18: 253–263, 1993.

Srinivas, S. Power sources for transportation applications: Types, status, and needed advances in technologies. *Advances in Hydrogen Energy* 4: 1717–1128, 1984.

Wakihara, M. and O. Yamamoto, Editors. Lithium Ion Batteries—*Fundamentals and Performance*. Wiley-VCH, 1998.

Wang, C.-W. and A. M. Sastry. Mesoscale modeling of a li-ion polymer cell. *Journal of the Electrochemical Society* 154(11): A1035–A1047, 2007.

Xianguo, L. *Principles of Fuel Cells*. Taylor & Francis, 2006.

2

Review of Electrochemistry

Electrochemistry is the study of mutual transformation of chemical and electrical energy. Specifically, it deals with chemical reactions driven by an electric current and with the electricity produced by chemical reactions. Examples of electrochemistry are electroplating, iron oxidation (rusting), solar-energy conversion, electrochemical conversions (fuel cells, batteries), photosynthesis, and respiration. In this chapter, the principles of electrochemistry are reviewed. First, let us briefly look into the history of electrochemistry.

The field of electrochemistry was discovered near the beginning of the 19th century. In 1791, Italian physician and anatomist Luigi Galvani observed that while dissecting a frog, a coworker touched the internal crural nerves of the frog with the tip of a scalpel and all the muscles of the frog's limb contracted. This led him to establish a relation between chemical reactions and electricity. In 1880, Alessandro Volta reported on chemical-to-electrical energy conversion to the Royal Society in London. He showed that by placing a brine-soaked membrane in contact with silver and zinc plates, on either side, an electric current would flow in the external circuit connecting the silver and zinc plates. Volta is credited with building the first electrochemical cell, which consisted of two electrodes: one made of zinc, the other of copper and the electrolyte is sulfuric acid or a brine mixture of salt and water. Following this, the same year William Nicholson and Johann Wilhelm Ritter succeeded in decomposing water into hydrogen and oxygen by electrolysis (electricity to chemicals). Further work on electrolysis by Sir Humphry Davy led to the conclusion that the production of electricity in simple electrolytic cells resulted from chemical action and that chemical combination occurred between substances of opposite charge. In 1832, Michael Faraday performed pioneering experiments that led him to state his two laws of electrochemistry.

William Grove produced the first fuel cell in 1839, where he demonstrated that when hydrogen and oxygen were fed to two platinum electrodes separated by electrolyte in individual cells and connected externally in series, an electric current was generated. Figure 2.1 shows a schematic of the Grove fuel cell. In 1884, Svante Arrhenius, through his investigations on the galvanic conductivity of electrolytes, concluded that electrolytes, when dissolved in water, split or dissociated into electrically opposite positive and negative ions at varying degrees. Walther Hermann Nernst developed the theory of the electromotive force of the voltaic cell in 1888 and showed how the characteristics of the current produced could be used to calculate the free

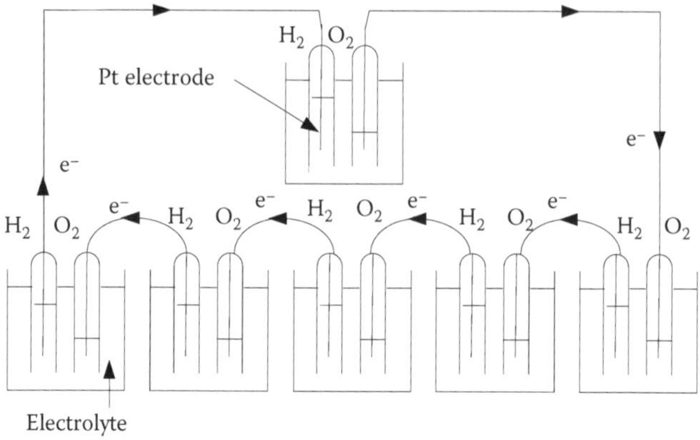

FIGURE 2.1
Demonstration of a fuel cell by Grove in 1839; the electricity produced by five cells was used to decompose water.

energy change in the chemical reaction producing the current. His equation, known as Nernst equation, related the voltage of a cell to its properties.

EARLIER PIONEERS IN ELECTROCHEMISTRY

Luigi Galvani (1737–1798)

Alessandro Giuseppe Antonio Anastasio Volta (1745–1827)

Sir Humphry Davy (1778–1829)

Michael Faraday (1791–1867)

Sir William Robert Grove (1811–1896)

Svante August Arrhenius (1859–1927)

Walther Hermann Nernst (1864–1941)

2.1 Electrochemical and Electrolysis Cell

An electrochemical cell involves a chemical reaction driven by electricity or the cell generates electricity because of spontaneous chemical reaction (Bockris and Reddy, 1970; Bogotsky, 2006). The cell that produces electricity because of spontaneous chemical reaction is called a *voltaic cell* or a *galvanic cell*. The cell in which chemical reaction is nonspontaneous and is forced by external electricity is called an *electrolysis cell*. In the galvanic cell, the current is caused by the reactions releasing and accepting electrons at the different ends of a conductor. A common example of a galvanic cell is a battery. In contrast to this electrolysis, cells decompose chemical compounds by means of electrical energy through electrolysis with a net increase in chemical energy. An example of electrolysis is the decomposition of water into hydrogen and oxygen gas with external voltage supply.

The galvanic cell consists of two separate compartments called half-cells. The two half-cells may use the same electrolyte, or they may use different electrolytes. Each half-cell consists of electrolyte solutions and electrodes that can be connected in a circuit to some voltmeter placed between the two electrodes within the circuit. In one half-cell, an electrode called the anode accepts electrons given by ion species (anions) that migrate to the electrode that are then passed through conducting wires in a circuit. In the other half-cell, the electrode called the cathode attracts ions (cations) where electrons can be gained by the species that migrates to that electrode. In order for this to occur, the circuit is completed by a conducting medium that allows ions or electrons to pass from one half-cell to the other. A salt bridge is often employed to provide electrical contact between two half-cells. The salt bridge can be as simple as an electrical conducting wire such as platinum or it can be a saturated electrolyte gel. When the salt bridge is in place and the electrodes have been connected, then electrical current is produced as the chemical reactions take place within the half-cells. The electrons will flow from the anode to the cathode in the external circuit. Electrons will actually travel from cathode to anode within the half-cell compartments through the salt bridge. The voltaic cell will generate a certain voltage called the cell potential. The cell potential is determined by adding the potential of the two half-cells together.

In Figure 2.2 a schematic of galvanic cell is shown with zinc (Zn) and copper (Cu) electrodes in the aqueous solutions of $ZnSO_4$ and $CuSO_4$ as electrolytes, respectively. A porous partition, typically a permeable membrane, separates the two half-cells. If two half-cells are in two separate containers, they can be connected by a salt bridge with salt solution such as saturated potassium nitrate solution. The metal of a metallic electrode tends to go into solution, thereby releasing positively charged metal ions into the electrolyte and retaining negatively charged electrons on the electrode. Thus, each half-cell has its own half-reaction. When the electrodes are connected externally

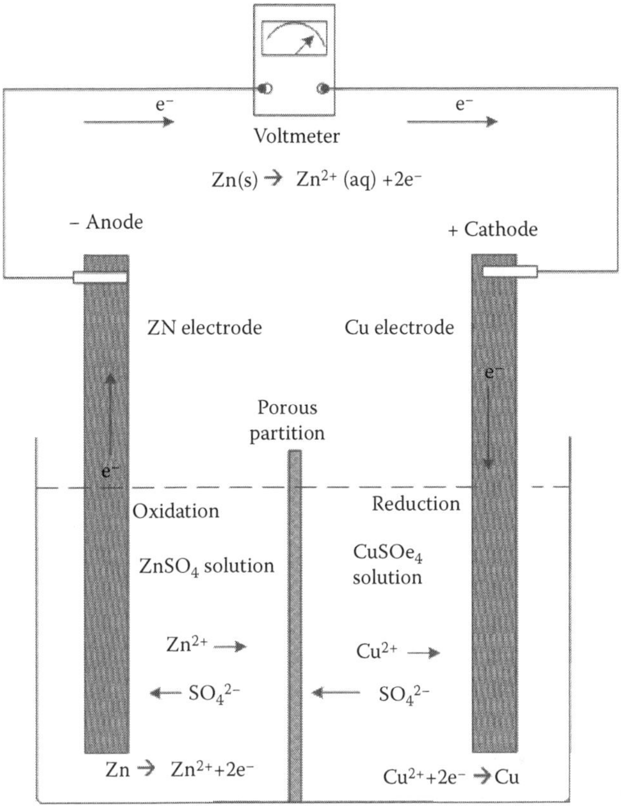

FIGURE 2.2
Schematic of a Zn–Cu galvanic cell.

(as in the figure, with wire and a light bulb/voltmeter), the electrons tend to flow from the more negative electrode (Zn) to the more positive electrode (Cu). Because the electrons have negative charge, this produces an electric current that is opposite to the electron flow. At the same time, an equal ionic current flows through the electrolyte. For every two electrons that flow from the Zn electrode through the external connection to the Cu electrode, on the electrolyte side, a Zn atom must go into solution as a Zn^{2+} ion, at the same time replacing the two electrons that have left the Zn electrode by the external connection. By definition, the *anode* is the electrode where removal of electrons takes place, so in this galvanic cell, the Zn electrode is the anode. Because the Cu has gained two electrons from the external connection, it must release two electrons at the electrolyte side, where a Cu^{2+} ion plates onto the Cu electrode. By definition, the *cathode* is the electrode where gain of electrons takes place, so the Cu electrode is the cathode.

The reactions at each half-cell are given as

$$Zn(s) \rightarrow Zn^{2+}(aq) + 2e^- \quad (2.1)$$

$$Cu^{2+}(aq) + 2e^- \rightarrow Cu(s) \quad (2.2)$$

Here, the symbols in the parentheses refer to the material state: s, solid; aq, aqueous solution. In one half-cell with zinc sulfate, the Zn loses two electrons (oxidizes), and in the other cell, Cu^{2+} accepts two electrons (reduces). The Zn electrode loses Zn as Zn^{2+} ion into the solution and appears as corroded. The Cu electrode is being plated with Cu from the electrolyte. As the cell operates, ions move through the partition or salt bridge to keep the individual cells electrically neutral. Because of the charge accumulation at one electrode, the driving force is created and the electrons travel through the wire generating electrical energy. As the reaction proceeds, the system approaches equilibrium and the generation of electrical energy would eventually stop.

Now let us look at the electrolysis cell, which is also referred to as the electrolytic cell. An electrolysis cell contains three elements: an electrolyte, a cathode, and an anode. The electrolyte is typically a solution of water or other solvent in which ions are dissolved or molten salts such as potassium chloride. Charge-transferring reactions take place at electrodes when an external voltage is applied to the electrodes; the ions in the electrolyte flow to and from the electrodes. The decomposition of a normally stable or inert chemical compound in the solution takes place only if the applied external electrical potential is of correct polarity and large enough magnitude.

Consider the decomposition of water by electrolysis as shown in Figure 2.3b. An electrical power source is connected to two electrodes, or two plates, typically made of inert metal such as platinum, which are placed in the water. Electrolysis of pure water is very slow and can only occur because of the self-ionization of water. Pure water has an electrical conductivity approximately one-millionth that of seawater. It is sped up dramatically by adding an electrolyte (such as a salt, an acid, or a base). When the power is applied, hydrogen will appear at the cathode where electrons are pumped into the water, and oxygen will appear at the anode. With aqueous electrolytes such as sulfuric acid in water, the reactions that occur at each electrode are as follows:

$$\text{Cathode: } 2H^+ + 2e^- \rightarrow H_2 \quad (2.3)$$

$$\text{Anode: } H_2O \rightarrow 1/2\ O_2 + 2H^+ + 2e^- \quad (2.4)$$

$$\text{Overall reaction: } 2H_2O(aq) \rightarrow 2H_2(g) + O_2(g) \quad (2.5)$$

It should be noted that the electrodes are inert; that is, they are not consumed, nor is there any deposition on them. In general, reactions 1 and 2

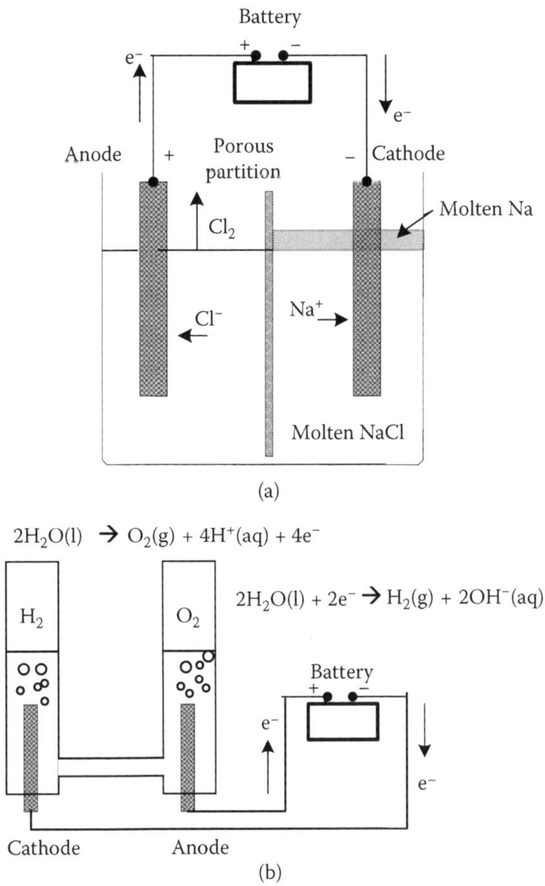

FIGURE 2.3
Electrolysis cell. (a) Molten salt electrolyte. (b) Aqueous electrolytes solution.

are reversible, meaning that the electrochemical reactions can occur in both directions. The generated amount of hydrogen is twice the amount of oxygen, and the total amount of gases produced is proportional to the total electrical charge that was sent through the water.

The following are key differences between a voltaic and an electrolysis cell:

1. In a voltaic cell, electrons flow from the negative electrode to the positive electrode, whereas in an electrolysis cell, the electrons flow from the positive to the negative electrode.
2. The chemical reaction in a voltaic cell is a spontaneous reaction producing electricity and is referred to as battery, while in an electrolysis cell, the reaction is nonspontaneous, driven by electrical power, and needs a battery.

2.2 Oxidation and Reduction Processes

Oxidation and reduction or *redox reactions* involve the exchange of electrons from one chemical species to another. Normally, this is done when the two chemicals contact each other in the activated complex. Generally, the oxidation reaction involves loss of electrons and can occur in a homogeneous medium or at an electrode–electrolyte interface whereas reduction reaction involves a gain of electrons.

The initial models developed for the redox reactions are based on transfer (loss and gain) of electrons. The process of oxidation cannot occur without a corresponding reduction reaction. Oxidation is always coupled with reduction, and the electrons that are lost by one substance must always be gained by another as matter cannot be destroyed or created. The terms lost or gained simply imply that the electrons are being transferred from one particle to another.

In general, an electrode reaction can be written as

$$\text{Oxidation } \nu\text{x} + n\text{e}^- \rightarrow \text{Reduction } \nu\text{x}, \tag{2.6}$$

where x is the reacting species and ν is its stoichiometric coefficient. The parameter ν/n mol, is called *chemical equivalent*. For an electrode reaction, the balance between products, reactants, and electron charges must be conserved.

Now, consider the electrolysis process shown in Figure 2.3b. In the water at the negatively charged cathode, a reduction reaction takes place, with electrons (e^-) from the cathode being given to hydrogen cations to form hydrogen gas (the half-reaction balanced with acid):

$$\text{Cathode (reduction): } 2\text{H}^+\text{ (aq)} + 2\text{e}^- \rightarrow \text{H}_2\text{ (g)} \tag{2.7}$$

At the positively charged anode, an oxidation reaction occurs, generating oxygen gas and giving electrons to the cathode to complete the circuit:

$$\text{Anode (oxidation): } 2\text{H}_2\text{O(l)} \rightarrow \text{O}_2\text{(g)} + 4\text{H}^+\text{(aq)} + 4\text{e}^- \tag{2.8}$$

The same half-reactions can also be balanced with base as listed below. Not all half-reactions must be balanced with acid or base. To add half-reactions, they must both be balanced with either acid or base.

$$\text{Cathode (reduction): } 2\text{H}_2\text{O(l)} + 2\text{e}^- \rightarrow \text{H}_2\text{(g)} + 2\text{OH}^-\text{(aq)} \tag{2.9}$$

$$\text{Anode (oxidation): } 4\text{OH}^-\text{(aq)} \rightarrow \text{O}_2\text{(g)} + 2\text{H}_2\text{O(l)} + 4\text{e}^- \tag{2.10}$$

When a magnesium metal is reacted with oxygen, it is thought to lose electrons (oxidize) to form Mg^{2+} ions and the element or compound that gained these electrons is said to undergo reduction. In this case, O_2 molecules are reduced to form O^{2-} ions.

$$Mg(s) + 1/2\, O_2(g) \rightarrow Mg^{2+} + O^{2-}(s) \tag{2.11}$$

Another example of redox involves placing a piece of copper wire into an aqueous solution of the Ag^+ ion. The reaction involves the net transfer of electrons from copper metal to Ag^+ ions to produce a coating of silver metal on the copper wire and Cu^{2+} ions.

$$Cu(s) + 2Ag^+(aq) \rightarrow Cu^{2+}(aq) + 2Ag(s) \tag{2.12}$$

It was recognized that redox need not always involve the transfer of electrons, as in the case when CO_2 reacts with H_2, there is no change in the number of valence electrons on any of the atoms.

$$CO_2(g) + H_2(g) \rightarrow CO(g) + H_2O(g) \tag{2.13}$$

Therefore, a model-based oxidation number was developed to extend the idea of oxidation and reduction to reactions in which electrons are not really gained or lost. This model of oxidation–reduction reactions is based on the following definitions. Oxidation involves an increase in the oxidation number of an atom. Reduction occurs when the oxidation number of an atom decreases.

The oxidation number of an atom is defined as the number equal to the charge that would be present on the atom if the compound was composed of ions. Now, according to this definition, if we assume that CH_4 contains C^{4-} and H^+ ions, for example, the oxidation numbers of the carbon and hydrogen atoms would be −4 and +1. Consider the reaction shown in Equation 2.12. CO_2 is reduced when it reacts with hydrogen because the oxidation number of the carbon decreases from +4 to +2. Hydrogen is oxidized in this reaction because its oxidation number increases from 0 to +1.

Certain rules govern how the oxidation state is determined. During the oxidation reaction, the oxidation state of the metal always increases from zero to a positive number, such as "+1, +2, +3....," depending on the number of electrons lost. The number of electrons lost and the charge of the cation formed are always equal to the group number of the metal or the valence electrons. For example, for Group I, the charge of the cation is +1, and for Group IV, it is +4.

The electrons lost by the metal are gained by the nonmetal. During this reduction reaction, the oxidation state of the nonmetal always decreases from zero to a negative value (−1, −2, −3...) depending on the number of electrons gained. The oxidation number of hydrogen is +1 when it is combined with a nonmetal. Hydrogen is therefore in the +1 oxidation state in CH_4, NH_3, H_2O, and HCl. The oxidation number of hydrogen is –1 when it is combined with a metal. Hydrogen is therefore in the –1 oxidation state in LiH, NaH, CaH_2, and $LiAlH_4$. Oxygen usually has an oxidation number of –2. Exceptions include molecules and polyatomic ions that contain O–O bonds, such as O_2, O_3, H_2O_2, and the O_2^{2-} ion. The nonmetals in Group VIIA often form compounds (such as AlF_3, HCl, and $ZnBr_2$) in which the nonmetal is in the –1 oxidation state. The sum of the oxidation numbers of the atoms in a molecule is equal to the charge on the molecule. The most electronegative element in a compound has a negative oxidation number.

Example 2.1

Show the oxidation state for calcium.

Answer

By convention, oxidation reactions are written in the following form using the element, calcium, as an example:

Symbol of atom	Symbol of cation formed	Number of electrons lost
Ca $\longrightarrow$	Ca^{+2}	$+2e^-$

Thus, the oxidation state of Ca increases from zero to a "+2."

Example 2.2

Calculate the oxidation number of sulfur in sulfuric acid H_2SO_4.

Answer

Hydrogen = +1 oxidation number, oxygen = –2 oxidation number.

Therefore:
$$(2 \times H) + S + (4 \times O) = 0$$
$$2 + S - 8 = 0$$
$$\underline{S = 6}$$

2.3 Faraday's Laws

Faraday's laws provide relationships between the quantities of charge (current) passed through an electrode–electrolyte interface, and the amount of chemical change that occurs owing to the passage of the current.

2.3.1 Faraday's First Law of Electrolysis

The amount of a substance consumed or produced due to chemical reaction at one of the electrodes during electrolysis is directly proportional to the amount of electricity transferred at that electrode. Amount of electricity refers to electrical charge, typically measured in *coulombs*.

2.3.2 Faraday's Second Law of Electrolysis

For a given amount of electric charge, the substance consumed or produced at one of the electrodes is directly proportional to the equivalent mass. Faraday's laws can be written together as

$$m_x = \frac{\xi QM}{n_e F}, \tag{2.14}$$

where m_x is the mass of a species x consumed or produced at the electrode (g), Q is the total electric charge transfer to or from the electrode (coulomb, C), and F is a Faraday constant, which is a basic unit of electrical charge used by chemists and is defined as the charge on 1 mol of electrons. It is calculated as

$$F = N_a e = \frac{6.022045 \times 10^{23}\,e^-}{1\,mol} \times \frac{1.6021892 \times 10^{-19}\,C}{1\,e^-} = 96,484.56\,C/mol, \tag{2.15}$$

where $N_a = 6.022 \times 10^{23}$= Avogrado number, defined as the number of molecules per mole and e = $1.602 \times 10^{=19}$ C is the charge per electron.

In terms of number Δn_x (moles) produced or consumed at the electrode, Faraday's laws can be written as

$$\Delta n_x = \frac{m_x}{M} = \frac{\xi It}{nF}, \tag{2.16}$$

where constant current $I = Q/t$ (amperes). In terms of rate of mass (g/s) consumed or produced at the electrode, Faraday's laws can be written as

$$\dot{m}_x = \frac{\xi IM}{nF}. \tag{2.17}$$

For Faraday's first law, M, F, and n are constants, so Q changes directly as m changes. For Faraday's second law, Q, F, and n are constants, so that the value of equivalent mass $\xi M/n$ changes as m_x changes.

Consider electrolysis of sodium iodide solution:

$$\left.\begin{array}{r} Na^{+}(l) + e^{-} \rightarrow Na(l) \quad \text{Reduction half-reaction} \\ 2I(l) \rightarrow I_2(s) + 2e^{-} \quad \text{Oxidation half-reaction} \\ \textit{Net reaction}: 2Na^{+}(l) + 2I(l) \rightarrow 2Na(l) + I_2(s)(\text{Redox}) \end{array}\right\} \tag{2.18}$$

In order to produce 1 mol of sodium metal, 1 mol of electrons is required, so one Faraday of charge must pass through the cell. To produce 1 mol of chlorine gas, 2 Faradays of electric charge must pass through the cell. Thus, a passage of 2 Faradays of charge yields 2 mol of sodium metal and 1 mol of chlorine gas.

Now, let us apply Faraday's first law of electrolysis to the example of electrolysis of NaI solution. In this reaction, 1 mol of I_2 was produced by 2 Faradays, which means that to produce 10 mol of I_2 requires the passage of 20 Faradays through the cell.

Consider the reduction of sodium and calcium ions:

$$\left.\begin{array}{r} Na^{+}e^{-} \rightarrow Na \\ Ca^{2+} + 2^{e-} \rightarrow Ca \end{array}\right\} \tag{2.19}$$

According to Faraday's second law, the mass produced is proportional to the equivalent mass $(\nu/n)\ M$. Here, the first reaction requires 1 mol of electrons ($n = 1$) whereas the second reaction requires 2 mol of electrons ($n = 2$). If we fix the total charge to 1 mol of electrons for the reaction, the reduction reaction of Na^+ produces 1 mol of Na. But in case of reduction reaction of Ca^+, only half mole of Ca is produced for 1 mol of electrons.

Example 2.3

Calculate the number of grams of sodium metal that will form at the cathode of electrochemical cell when a 20 A current is passed through molten sodium chloride for a period of 6 h.

Answer

The reduction reaction at the cathode is given as

$$Na^+ + e^- \rightarrow Na.$$

Thus, 1 mol of sodium is produced for every 1 mol of electron.

First, we calculate the amount of electric charge that passes through the cell.

$$20.0\,\text{amp} \times \frac{1\,\text{C}}{1\,\text{amp} \cdot \text{s}} \times 6.0\,\text{h} \times \frac{3600\,\text{s}}{1\,\text{h}} = 432{,}000\,\text{C}$$

Using Faraday's constant *F*, the number of moles of electrons transferred when 432,000 C of electric charge flows through the cell is calculated as follows.

$$432{,}000\,\mathrm{C} \times \frac{1\,\mathrm{mol\,e^-}}{96{,}484.56\,\mathrm{C}} = 4.48\,\mathrm{mol\,e^-}$$

From the reduction reaction, we have 4.48 mol of sodium produced at the cathode. In terms of mass,

$$4.48\,\mathrm{mol\,Na} \times \frac{22.99\,\mathrm{g\,Na}}{1\,\mathrm{mol\,Na}} = 109\,\mathrm{g\,Na},$$

109 grams of sodium is produced when 20 A current is passed in a cell for 6 h.

2.4 Ideal Polarized Electrode

The oxidation and reduction owing to electron transfer at the electrode–electrolyte interface are governed by Faraday's laws and are called *faradaic processes*. However, as noted above, change in the electrode–electrolyte interface can occur without charge transfer taking place. Processes such as adsorption and desorption can result in change in the electrode–electrolyte interface structure such as changes in electrode area, electrolyte composition, and potential. In these cases, external currents can flow at least intermittently even though there is no charge transfer across the interface. These processes are termed *nonfaradaic processes*. Typically, in an electrochemical reaction, both faradaic and nonfaradaic processes take place.

An ideal polarized electrode (IPE) is defined as an electrode at which no charge transfer occurs across the electrode–electrolyte interface regardless of potential applied by an external voltage source. Even though no real electrode behaves as an IPE, there are some electrode solution systems that behave as an IPE over certain limited potential range. One example is a mercury electrode in the solute of deaerated potassium chloride KCl with a potential range of ~2V that approached the behavior of an IPE.

Since no charge transfer occurs across the electrode–electrolyte interface for an IPE, the behavior of such an interface is similar to a capacitor where two metallic sheets are separated by a dielectric material. The behavior of such an electrode, shown in Figure 2.4, is governed by the standard capacitor equation:

$$C = Q/V, \tag{2.20}$$

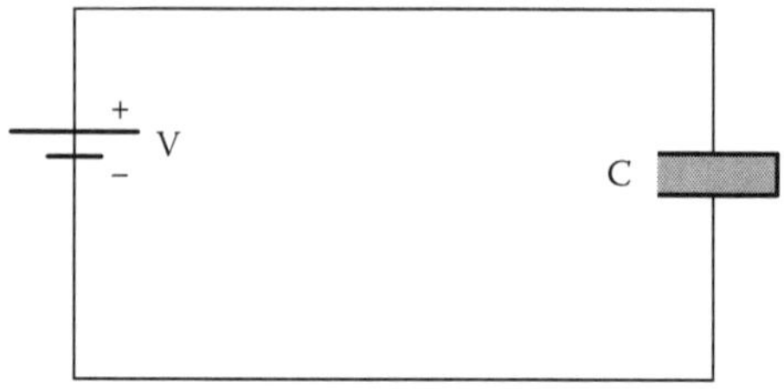

FIGURE 2.4
Capacitor charging with battery.

where C is the capacitance in farads, Q is the charge on the capacitor stored in coulombs, and V is the potential across the capacitor (IPE) in volts. Therefore, when a potential is applied across the capacitor (IPE), charge will build up on its metal plates until Q in Equation 2.20 is fulfilled. As the charge builds up on the capacitor plates, a current known as charging current flows across the circuit. In certain cases, the charging currents may exceed the faradaic current.

2.5 Polarization and Overpotential

The variation of current and voltage characteristics in the form of current–voltage (I–V) curves provides the information on the reaction occurring at the electrode. The motion of current in electrodes depends on the potential applied to it. When a current is passed through an electrode, any deviation of the electrode potential from its equilibrium value (reversible potential) is called polarization. As discussed above, an IPE has negligible current flow for a large change in potential; thus, IPE response can be shown as a horizontal line in an I–V curve (Figure 2.5). The amount of departure of potential from the equilibrium potential is called *overpotential*. If V is electrode potential and V_{eq} is equilibrium potential, then the overpotential η is given as

$$\eta = V - V_{eq}. \tag{2.21}$$

Overpotential η always reduces theoretical cell potential when current is flowing. Thus, current–voltage curves obtained under steady state are also referred to as *polarization curves*.

Basically, there are two types of polarization: (1) activation polarization and (2) concentration polarization. Activation polarization is caused by resistance to the passage of potential-determining ions through phase boundary at the electrode–electrolyte interface. For many electrodes, a large polarization is observed at low current density, mainly because of activation polarization.

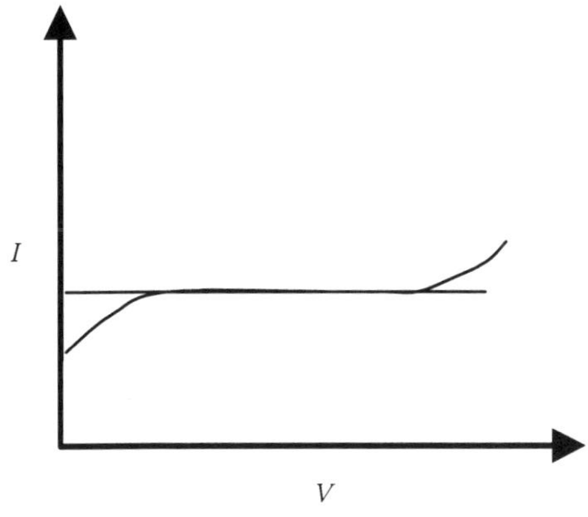

FIGURE 2.5
Ideal polarized electrode.

The inhibition of the transport of an ion through a layer contiguous to the electrode causes resistance polarization. The presence of foreign substances on the cathode surface may consist of electrolyte anions and cations, oxides or hydroxides, or other organic or inorganic components of the electrolyte. These substances are adsorbed at the electrolyte surface, and when the electrolyte is completely covered by foreign substance, it is passive. This gives rise to *resistance polarization*. Resistance polarization is also a type of activation polarization.

Concentration polarization is due to the concentration difference, which develops at the anode and cathode during electrolysis and is caused by the resistance to the transport process owing to diffusion. An increase in the concentration of the dissolved metal ions in the anodic diffusion layer causes diffusion polarization. Diffusion polarization is a type of concentration polarization. Polarization can under certain circumstances be greatly increased by the addition of specific substances. For example, nickel bath has low concentration polarization and high activation polarization, while cyanide baths often have high concentration polarization. Polarization increases with increasing current but decreases with increasing temperature and increasing agitation of the bath.

2.6 Conductivity and Ohm's Law

Electrical conductivity is an index of a material's ability to conduct an electric current. The electric current occurs in a conductor when an electrical potential difference is placed across it. The strength of electric current **I** measured (unit: amperes, A) depends on the conductor material, on the electrostatic field strength **E**, and on conductor cross section A. In electrochemistry, for current, a conductor dimension-independent parameter current density **i** (unit: A/cm^2) is used. The conductivity σ is expressed as the ratio of the current **I** to the electric field strength **E**

$$\sigma = \mathbf{I}/\mathbf{E}. \tag{2.22}$$

The reciprocal of conductivity (1/conductivity) is resistivity (ρ). Thus, resistivity is an index of difficulty of flow of electric current.

$$\sigma = 1/\rho \tag{2.23}$$

In metals, the electrons are the carrier of the electrical charge, and in an electrolyte solution, the ions carry the electrical charge. In a NaCl electrolyte solution, sodium ions (Na^+) and chlorine ions (Cl^-) pass electricity from one to the next. This means that the more Na^+ and Cl^- contained in solution, the more electricity is carried, and the higher the conductivity.

The electrical resistance of a wire would be expected to be greater for a longer wire and less for a wire of larger cross-sectional area and would be expected to depend upon the material out of which the wire is made. Experimentally, the resistance of a wire can be expressed as

$$R = \frac{\rho L}{A}, \tag{2.24}$$

where L is the wire length. The unit of conductivity in SI units is siemens per meter (S/m). Thus, 10 S/m is one millisiemens per centimeter (mS/cm). Typical values of NaCl solution conductivity as a function of salinity (density of salt in salt water) at a liquid temperature of 25°C are given in Table 2.1.

TABLE 2.1

Salinity (Density of Salt in Salt Water) and Conductivity at 25°C

NaCl Density (W/V)%	Conductivity (mS/cm)	NaCl Density (W/V)%	Conductivity (mS/cm)
0.1	2.0	1.1	19.2
0.2	3.9	1.2	20.8
0.3	5.7	1.3	22.4
0.4	7.5	1.4	24.0
0.5	9.2	1.5	25.6
0.6	10.9	1.6	27.1
0.7	12.6	1.7	28.6
0.8	14.3	1.8	30.1
0.9	16.0	1.9	31.6
1.0	17.6	2.0	33.0

Ohm's law states that the current through a conductor between two points, I, is directly proportional to the potential difference or voltage across the two points, V, and inversely proportional to the resistance R between them

$$I = \frac{V}{R}, \tag{2.25}$$

wherein V, I, and R are in SI units of volts (V), amperes (A), and ohms (Ω), respectively. The resistance is constant, independent of the current.

2.7 Mass Transport and Nernst–Planck Equation

There are three modes of mass transport in an electrochemical system: diffusion, migration, and convection. Diffusion is driven by the concentration gradient where the material transfer occurs from a high concentration to a low concentration. Diffusion is particularly significant near the electrode surface where conversion reaction occurs. Consequently, electrode has a lower reactant concentration than in bulk solution. Similarly, product concentration is higher near the electrode than further out into solution. The diffusion flux ($J_{d,j}$ mol/cm^{2}s) for species j in steady state is expressed for a constant viscosity solution using Fick's first law

$$J_{d,j}(x) = -D_j \frac{\partial c_j}{\partial x}, \tag{2.26}$$

where D_j is the diffusion coefficient (cm^2/s) and c_j is the species concentration (mol/cm^3). Here, diffusion is considered normal to an electrode surface (x direction). Thus, the steeper the change in concentration, the greater the rate of diffusion flux. The diffusion coefficient is constant for low concentration of ions. At high concentrations, the proportionality relation between the flux and concentration gradient is not valid since the diffusion coefficient is a function of concentration.

For time-dependent processes, Fick's second law is expressed as

$$\frac{\partial c_j}{\partial t} = \frac{\partial}{\partial x}\left(-D_j \frac{\partial c_j}{\partial x}\right). \tag{2.27}$$

For constant D_j, this reduces to

$$\frac{\partial c_j}{\partial t} = -D_j \frac{\partial^2 c_j}{\partial x^2}. \tag{2.28}$$

Fick's second law allows one to predict how the concentrations of species vary as a function of time and their cell position within the electrochemical cell. In order to solve these expressions, analytical or computational models are usually employed.

Migration is essentially an electrostatic effect that arises because of the application of a voltage on the electrodes. For field of strength **E**, the ions carrying charge (z_jF) are under the effect of an applied electrical driving force ($z_jF\mathbf{E}$) causing the ions to move (migrate) in a direction given by the field. The velocity of migration depends on the electrical driving force ($z_jF\mathbf{E}$) and the retarding force owing to solution viscosity. The velocity of migration corresponding to the unit field strength (1 V/cm) is called the electrical mobility of the ions (u_j) and is given as

$$u_j = \frac{z_jF}{RT} D_j. \tag{2.29}$$

The migratory flux $J_{m,j}$ is proportional to concentration and the potential gradient ($\partial\phi/\partial x$) with proportionality parameter being the electrical mobility, as given by

$$J_{m,j}(x) = -u_j c_j \frac{\partial \phi}{\partial x}. \tag{2.30}$$

Convection is the movement of species owing to mechanical forces such as by stirring or hydrodynamic transport. The fluid flow can occur because of natural convection (because of density gradient) and forced convection and is characterized by stagnation regions, laminar flow, and turbulent flow. Convection usually can be eliminated on a short time scale. The convection mass flux ($J_{c,j}$) is directly related to mass concentration and the solution velocity in the direction of electrode, $v(x)$, and is given as

$$J_{c,j}(x) = v(x)c_j. \tag{2.31}$$

The solution velocity can be calculated in many situations solving the appropriate form of the Navier–Stokes equations.

Now, combining the mass flux caused by diffusion, migration, and convection, the total flux is given by the relation known as the *Nernst–Planck equation*.

$$J_j(x) = -D_j \frac{\partial c_j}{\partial x} - u_j c_j \frac{\partial \phi}{\partial x} + c_j v(x), \tag{2.32}$$

where the first, second, and third terms in the right-hand side represent contribution from diffusion, migration, and convection, respectively. Near the electrode, the mass flux is supported by migration and diffusion. However, the relative contribution of diffusion and migration to the total mass flux of the species differs at a given time for different locations in solution. If the total mass flux is known, then one can calculate the instantaneous current density j_j as

$$j_j = -nFJ_j, \tag{2.33}$$

where n is the number of electrons per 1 mol.

2.8 Standard Hydrogen and Other Reference Electrodes

2.8.1 Standard Hydrogen Electrode and Potentials

The standard hydrogen electrode (SHE), also referred to as normal hydrogen electrode, is the universal reference for reporting relative half-cell potentials. The SHE could be used as either an anode or a cathode depending upon the nature of the half-cell it is used with. SHE is based on the redox half-cell:

$$2H^+(aq) + 2e^- \rightarrow H_2(g). \tag{2.34}$$

The SHE consists of a platinum electrode dipped in an acidic solution (Figure 2.6). The concentration of hydrogen in the solution is 1 molar. The platinum electrode is made of a small square of platinum foil that is platinized (known as platinum black). Pure hydrogen gas, at a pressure of 1 atm, is bubbled around the platinum electrode. The platinum black serves as a large surface area for the reaction to take place, and the stream of hydrogen keeps the solution saturated at the electrode site with respect to the gas. The SHE is an arbitrary reference electrode with a half-cell potential of 0.00 V. Hence, the condition for SHE are

$$\mathrm{H^+_{(aq)}\,1\,mol\,l^{-1}\,(pH = 0), H_{2(g)}\,1\,atm}, T = 25°\mathrm{C} \quad E_0 = 0\ \mathrm{V}.$$

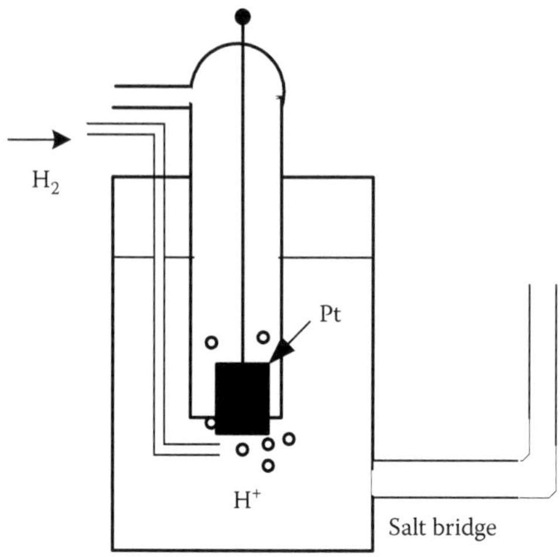

FIGURE 2.6
Standard hydrogen electrode.

Its absolute electrode potential is estimated to be 4.44 ± 0.02 V at 25°C. Potentials of any other electrodes are compared with that of the SHE at the same temperature.

To find the reduction potentials for other electrodes, the cells are formed by connecting the SHE as the anode and the other half-cell as the cathode. Since its potential is assigned the value of 0.000 V, all of the potential difference measured experimentally is attributed to the other test electrode. In Table 2.2, the standard electrode potentials at 25°C for various electrodes are listed.

As can be seen from Table 2.2, F_2 is a stronger oxidant than Ag^+ while Cu^{2+} is a weaker oxidant than Ce^{4+}. Similarly, Ag is a stronger reductant than Au and Co^{2+} is a weaker reductant than Sn^{2+}.

Consider a copper electrode in solution with Cu^{2+} ions. The test cell is

$$\text{Pt} \mid \text{H}_2\,(1.00\text{ atm}) \mid \text{H}^+\,(1.00\text{ M}) \parallel \text{Cu}^{2+}\,(1.00\text{ M}) \mid \text{Cu}.$$

The measured potential for this cell is +0.337 V. Since the activity of all components in the Cu cell is standard, +0.337 V is the standard reduction potential of the Cu^{2+}/Cu electrode system.

The convention for connecting the cell is that the positive input terminal to the voltmeter is connected to the test cell and the negative terminal is connected to the SHE. This is how to make connections when the test cell is a cathode. If the cell potential is measured to be positive, then the process under test is reduction, and if the measured potential is negative, then the test cell was operating as an oxidation.

TABLE 2.2

Standard Potentials of Electrode Reactions at 25°C

$Li^+ + e^- \rightarrow Li$	−3.045	$Ni^{2+} + 2e^- \rightarrow Ni$	−0.25
$Rb^+ + e^- \rightarrow Rb$	−2.98	$Sn^{2+} + 2e^- \rightarrow Sn$	−0.136
$Cs^+ + e^- \rightarrow Cs$	−2.92	$Pb^{2+} + 2e^- \rightarrow Pb$	−0.126
$K^+ + e^- \rightarrow K$	−2.935	$Sn^{2+} + 2e^- \rightarrow Sn$	−0.14
$Ba^{2+} + 2e^- \rightarrow Ba$	−2.92	$D^+ + e^- \rightarrow 1/2\,D_2$	−0.003
$Sr^{2+} + 2e^- \rightarrow Sr$	−2.89	$H^+ + e^- \rightarrow 1/2\,H_2$	0.000
$Ca^{2+} + 2e^- \rightarrow Ca$	−2.866	$Sn^{4+} + 2e^- \rightarrow Sn^{2+}$	0.15
$Na^+ + e^- \rightarrow Na$	−2.714	$Cu^{2+} + 2e^- \rightarrow Cu$	0.337
$Mg^{2+} + 2e^- \rightarrow Mg$	−2.375	$1/2\,O_2 + H_2O + 2e^- \rightarrow 2OH^-$	0.401
$Be^{2+} + 2e^- \rightarrow Be$	−1.70	$Cu^+ + e^- \rightarrow Cu$	0.52
$Al^{3+} + 3e^- \rightarrow Al$	−1.662	$Hg^{2+} + 2e^- \rightarrow 2Hg$	0.789
$Ti^{2+} + 2e^- \rightarrow Ti$	−1.628	$Ag^+ + e^- \rightarrow Ag$	0.799
$Mn^{2+} + 2e^- \rightarrow Mn$	−1.180	$Pd^{2+} + 2e^- \rightarrow Pd$	0.987
$Zn^{2+} + 2e^- \rightarrow Zn$	−0.764	$AuCl_4^- + 3e^- \rightarrow Au + 4Cl$	1.00
$Ga^{3+} + 3e^- \rightarrow Ga$	−0.52	$2Br + 2e^- \rightarrow 2Br^-$	1.065
$Fe^{2+} + 2e^- \rightarrow Fe$	−0.441	$O_2 + 4H^+ + 4e^- \rightarrow 2H_2O$	1.229
$Cd^{2+} + 2e^- \rightarrow Cd$	−0.403	$Cl_2 + 2e^- \rightarrow 2Cl^-$	1.36
$In^{3+} + 3e^- \rightarrow In$	−0.34	$Ce^{4+} + e^- \rightarrow Ce^{3+}$	1.61
$Tl^+ + e^- \rightarrow Tl$	−0.34	$Au^+ + e^- \rightarrow Au$	1.69
$Co^{2+} + 2e^- \rightarrow Co$	−0.277	$F_2 + 2e^- \rightarrow 2F^-$	2.65

Every cell reaction consists of a reduction reaction and an oxidation reaction. Hence, every cell must be constructed to employ one reaction and another reaction written opposite. The cell potential is calculated as

$$E_{\text{cell}} = E_{\text{cathode}} - E_{\text{anode}}. \tag{2.35}$$

For example, the reduction reaction $Au^+ + e^- \rightarrow Au$ and oxidation reaction $Cu \rightarrow Cu^{2+} + 2e^-$ gives the overall reaction after taking a balance on electrons: $2Au^+ + Cu \rightarrow Cu^{2+} + 2Au$. The potential for this cell is $E = +1.69 - 0.34 = +1.35$ V. Thus, a cell constructed using an Au and a Cu electrode and solutions of Au(I) and Cu(II) ions, when at unit activity (1.00 M), has a potential of +1.35 V. This indicates that indeed the Au electrode is the cathode for the reduction process occurring there. Conversely, the copper electrode is the anode and oxidation is occurring there.

Example 2.4

Calculate the cell potential for reaction: $Fe^{2+} + V \rightarrow Fe + V^{2+}$

Answer

The half-cell reactions are $Fe^{2+} + 2e^- \rightarrow Fe$ and $V^{2+} + 2e^- \rightarrow V$.

From Equation 2.34 and Table 2.2, we have the cell potential

$$E_{cell} = -0.44 - (-1.19) = +0.75 \text{ V}.$$

Example 2.5

Calculate the cell potential for reaction, $2Ag^+ + Sn \rightarrow 2Ag + Sn^{2+}$

Answer

The half-cell reactions are $Sn^{2+} + 2e^- \rightarrow Sn$ and $Ag^+ + e^- \rightarrow Ag$.

A more negative potential reaction is seen in the anode.

Multiply the Ag reaction by 2, but do not modify the cell potential. The cell potential is given by

$$E_{cell} = +0.80 - (-0.14) = +0.94 \text{ V}.$$

2.8.2 Reference Electrodes

In electrochemical experiments, often the interest is on one of the electrode reactions. Since for measurement a complete cell with two electrodes must be used, it is common practice to employ a reference electrode as the other half of the cell. It is desired that the reference electrode should be easy to prepare and maintain and has stable potential. This is accomplished by using an electrode reaction involving a saturated solution of an insoluble salt of the ion. The common reference electrodes and potential with respect to the SHE are saturated calomel electrode (E = +0.242 V saturated), copper–copper(II) sulfate electrode (E = −0.314 V), and silver chloride electrode (E = 0.225 V saturated). The silver–silver chloride electrode is

$$Ag \mid AgCl(s) \mid Cl^-(aq) \mid\mid \ldots \text{ with reaction } Ag(s) + Cl^-(aq) \rightarrow AgCl(s) + e^-.$$

Typically, this electrode is silver wire coated with AgCl. The coating is done by making the silver the anode in an electrolytic cell containing HCl; the Ag^+ ions combine with Cl^- ions as fast as they are formed at the silver surface.

The *calomel electrode* has the following half-cell and reaction:

$$Hg \mid Hg^{2+}(aq) \mid KCl \mid\mid \ldots Hg(l) + Cl^- \rightarrow 1/2\, HgCl_2(s) + e^-.$$

2.9 Cyclic Voltammetry

Cyclic voltammetry is one of the most versatile electroanalytical techniques for the study of electroactive species (Bard and Faulkner, 1980; Prentice, 1991).

Cyclic voltammetry has the capability for rapidly observing redox behavior over a wide potential range. Cyclic voltammetry is generally used to study the electrochemical properties of an analyte in solution.

In typical cyclic voltammetry, a solution component is electrolyzed (oxidized or reduced) by placing the solution in contact with an electrode surface and then making that surface sufficiently positive or negative in voltage to force electron transfer. The controlling potential applied across these two electrodes is an excitation signal. The excitation signal is a linear potential scan with a triangular waveform. In simple cases, the surface is started at a particular voltage with respect to a reference half-cell such as calomel or Ag/AgCl, the electrode voltage is changed to a higher or lower voltage at a linear rate, and, finally, the voltage is changed back to the original value at the same linear rate. The electrode potential is ramped linearly versus time as shown in Figure 2.7. This ramping is known as the scan rate (V/s). When the surface becomes sufficiently negative or positive, a solution species may gain electrons from the surface or transfer electrons to the surface. This results in a measurable current in the electrode circuitry. However, if the solution is not mixed, the concentration of transferring species near the surface drops, and the electrolysis current then falls off. When the voltage cycle is reversed, it is often the case that electron transfer between electrode and chemical species will also be reversed, leading to an "inverse" current peak. In the experiment, the potential is measured between the reference electrode and the working electrode and the current is measured between the working electrode and the counter electrode. The typical cyclic voltammogram is a plot of current (i) versus potential (E) as shown in Figure 2.8.

The forward scan produces a current peak for any analytes that can be reduced through the range of the potential scanned. The current will increase rapidly, reaching peak value until the concentration of the analyte at the electrode surface approaches zero. The current then decreases as the solution surrounding the electrode is depleted of analytes. When the applied potential is reversed, anodic current is generated as the electrode becomes

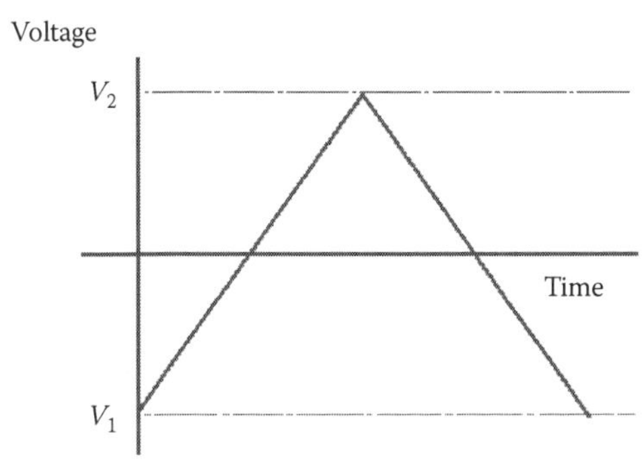

FIGURE 2.7
Cyclic voltammetry waveform.

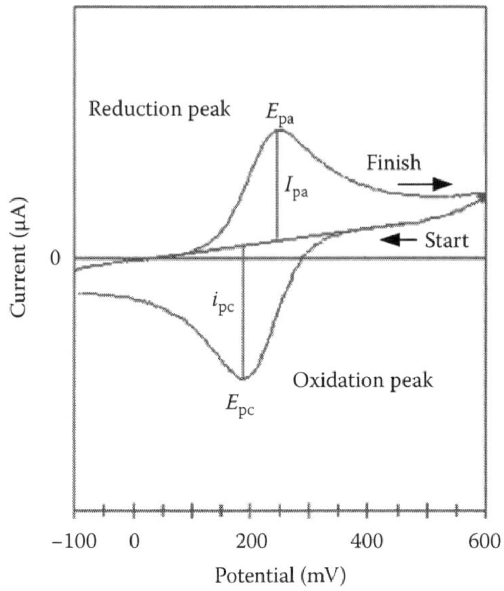

FIGURE 2.8
A typical cyclic voltammogram showing reduction and oxidation current peaks.

a sufficiently strong oxidant and the product formed in the first reduction reaction re-oxidizes. The anodic current increases until the surface concentration of the product approaches zero and the current peaks. The current then decreases as the solution surrounding the electrode is depleted of product formed in reduction reaction. This oxidation peak will usually have a similar shape to the reduction peak.

The cyclic voltammogram provides magnitudes of anodic peak current (i_{pa}), cathodic peak current (i_{pc}), anodic peak potential (E_{pa}), and cathodic peak potential (E_{pc}). These parameters can be used to obtain information on the redox potential and detection of chemical reactions that precede or follow the electrochemical reaction and evaluation of electron transfer kinetics.

The formal reduction potential E_{red} for an electrochemically reversible couple is given as

$$E^0 = (E_{pa} + E_{pc})/2. \tag{2.36}$$

For a reversible couple, $i_{pa} = i_{pc}$.

Peak ratios are often strongly affected by chemical reactions coupled to the redox process.

For a reversible redox couple, the number of electrons transferred in the electrode reaction can be determined by the Randles–Sevcik equation for the forward sweep of the first cycle

$$i_p = 0.4463\, n\, F\, A\, C\, (n\, F\, v\, D/R\, T)^{1/2}, \tag{2.37}$$

where n is the number of electrons (stoichiometry), v is the scan rate (V/s), F is Faraday's constant (96,485 C/mol), A is the electrode area (cm^2), R is the universal gas constant (8.314 J/mol K), C is the concentration (mol/cm^3), T is the absolute temperature (K), and D is the analyte's diffusion coefficient (cm^2/s). Note that if the temperature is assumed to be 25°C (298.15 K), the Randles–Sevcik equation can be written in a more concise form,

$$i_p = (2.687 \times 10^5)\, n^{3/2}\, v^{1/2}\, D^{1/2}\, A\, C, \tag{2.38}$$

where the constant is understood to have units (i.e., 2.687×10^5 C mol^{-1} V$^{-1/2}$).

Furthermore, i_p increases with $v^{1/2}$ and is directly proportional to concentration. This relationship becomes particularly important in the study of electrode mechanisms.

PROBLEMS

1. Determine the oxidation number of each element in the following compounds:

   (a) BaO_2, (b) $(NH_4)_2MoO_4$, (c) $Na_3Co(NO_2)_6$, (d) CS_2

2. Classify each of the following as either a metathesis or an oxidation–reduction reaction. Note that mercury usually exists in one of three oxidation states: mercury metal, Hg_2^{2+} ions, or Hg^{2+} ions.

   (a) $Hg_2^{2+}(aq) + 2\ OH(aq)\ \ Hg_2O(s) + H_2O(l)$

   (b) $Hg_2^{2+}(aq) + Sn^{2+}(aq)\ \ 2\ Hg(l) + Sn^{4+}(aq)$

   (c) $Hg_2^{2+}(aq) + H_2S(aq)\ Hg(l) + HgS(s) + 2\ H^+(aq)$

   (d) $Hg_2CrO_4(s) + 2\ OH^-(aq)\ \ Hg_2O(s) + CrO_4^{2-}(aq) + H_2O(l)$

3. Calculate the weight of copper produced by the reduction of copper(II) ions during the passage of 2.50 A of current through a solution of copper(II) sulfate for 45.0 min.

4. Consider electrolysis of water. $2H_2O_{(l)} \rightarrow 2H_{2(g)} + O_{2(g)}$. Determine the net voltage required for this reaction.

5. What is the spontaneous electrochemical reaction that occurs when a standard copper half-cell is combined with a standard silver half-cell and what is E^0 for the resulting cell?

6. Will tin(IV) ions, Sn^{4+}, oxidize gaseous nitrogen oxide NO, to nitrate ions, NO_3^-, in acidic solution or will NO_3^- oxidize Sn^{2+}?

7. Calculate the volume of H_2 gas at 25°C and 1.00 atm that will collect at the cathode when an aqueous solution of Na_2SO_4 is electrolyzed for 2.00 h with a 10.0 A current.

8. Determine the oxidation number of the chromium in an unknown salt if electrolysis of a molten sample of this salt for 1.50 h with a 10.0 A current deposits 9.71 g of chromium metal at the cathode.

References

Bard, A. J. and L. R. Faulkner. *Electrochemical Methods*. John Wiley & Sons, New York, 1980.

Bockris, J. O'M and A. K. N. Reddy. *Modern Electrochemistry*. Plenum Press, New York, 1970.

Bogotsky, V. S. *Fundamentals of Electrochemistry*, 2nd Edition. John Wiley & Sons, New Jersey, 2006.

Prentice, G. *Electrochemical Engineering Principles*. Prentice Hall, New Jersey, 1991.

3

Reviews of Thermodynamics

In this chapter, we have presented some of the basic principles, laws, and relations of engineering thermodynamics. This includes thermodynamic properties; energy forms such as heat, work, internal energy, enthalpy, and Gibbs function; the first and the second law of thermodynamics; ideal gas equation of states; and relations of change in enthalpy, entropy, and Gibbs function. A prior knowledge and a review of these materials are essential for clear understanding of the principles, analysis, and design of fuel cells that are presented in the next chapters. For more comprehensive reviews of engineering thermodynamics, the books by Moran and Shapiro (2008), Sonntag et al. (2003), and Cengel and Boles (2006) can be referred.

3.1 State, Phase, and Properties

The *state* of a system represents the condition of the system as defined by the properties. *Properties* are macroscopic quantities that are perceived by our senses and can be measured by instruments. A quantity is defined as the *property* if it depends only on the state of the system and independent of the process by which it has reached at the state. Some of the common thermodynamic properties are pressure, temperature, mass, volume, and energy. Properties are also classified as intensive and extensive. Intensive properties are independent of the mass of the system and a few examples of this include pressure, temperature, specific volume, specific enthalpy, and specific entropy. Extensive properties depend on the mass of the system. All properties of a system at a given state are fixed. For a system that involves only one mode of work, two independent properties are essential to define the thermodynamic state of the system and the rest of the thermodynamic properties can be determined on the basis of the two known independent properties and using thermodynamic relations. For example, if pressure and temperature of a system are known, the state of the system is then defined. All other properties such as specific volume, enthalpy, internal energy, and entropy can be determined through the equation of state and thermodynamic relations.

3.2 Thermodynamic Process and Cycle

When properties of a system change and the system moves from one thermodynamic equilibrium state to another, the path of succession of states that the system passes through is defined as the *process*. For example, the gas in a cylinder–piston arrangement shown in Figure 3.1 undergoes an expansion process from state 1 with pressure, P_1, and volume, V_1, to state 2 with pressure, P_2, and volume, V_2.

When a system starts from an initial state and undergoes a number of different change of states or processes and returns back to the same initial state, then the system is said to undergo a thermodynamic cycle.

A number of the following thermodynamic processes are defined depending on how a process is controlled:

Isothermal process: A constant temperature process

Isobaric process: A constant pressure process

Isochoric process: A constant volume process

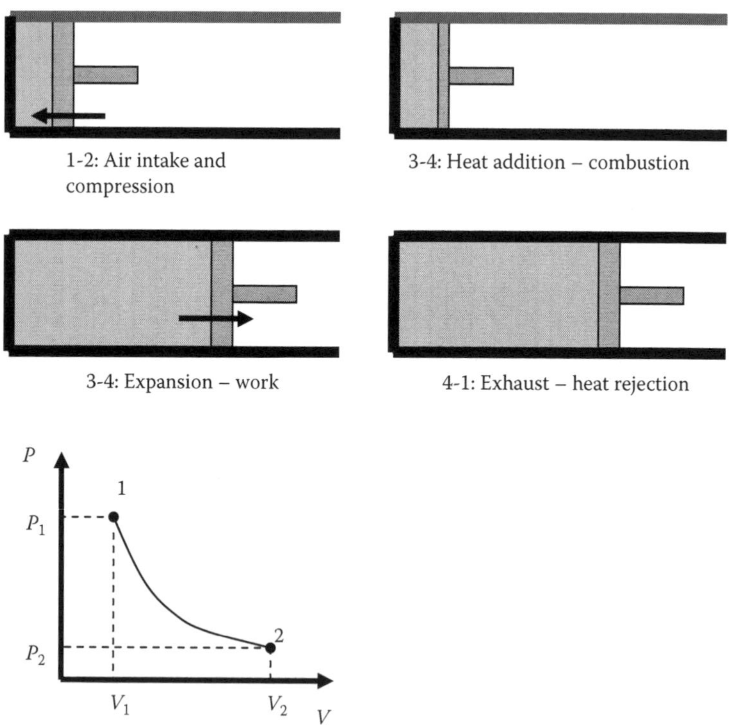

FIGURE 3.1
Thermodynamic process.

Adiabatic process: A process with no heat transfer

Polytropic process: A process in which the relationship between pressure and volume is given as PV^n = constant. The index n may be any value from $-\infty$ to $+\infty$ depending on the process.

Reversible process: The reversible process for a system is defined as a process that once has taken place can be reversed without leaving any changes in the system or the surroundings. It basically represents an ideal process that leads to the maximum possible performance of a system or a control volume. A real process involves a number of factors that cause irreversibilities and lead to a performance less than a reversible process operating between the same two thermodynamics states. Some of these common factors are (1) friction, (2) unrestrained expansion, (3) mixing process, (4) ohmic heating, (5) heat transfer through finite temperature difference, (6) hysteresis effect, and (7) chemical or electrochemical reactions.

Isentropic process: A reversisble and adiabatic or a constant entropy process.

3.3 Ideal Gas Equation of State

An equation of state is a relationship among the three basic measurable intensive properties of gases: temperature, pressure, and volume. This relationship is essential for thermodynamic analysis and determining all other thermodynamic properties including internal energy, enthalpy, and entropy, as well as performing thermodynamic analysis using gases. An *ideal gas equation of state* is one such relationship that is applicable for gases at low densities and defined as

$$P\bar{v} = \bar{R}T, \tag{3.1a}$$

where

$$\bar{R} = \text{Universal gas constant} = 8.3145\frac{\text{kN.m}}{\text{kmol.K}} = 8.3145\frac{\text{kJ}}{\text{kmol.K}}$$

and on a mass basis as

$$Pv = RT, \tag{3.1b}$$

where

$$R = \text{gas constant} = \frac{\bar{R}}{M}.$$

In terms of total volume,

$$PV = n\bar{R}T \tag{3.2a}$$

and

$$PV = mRT, \tag{3.2b}$$

where n and m are the number of mole and mass of the gas, respectively.

Any gas that satisfies this relationship is defined as the ideal gas. At a higher pressure, all gases deviate from ideal gas behavior and a number of other equations of state are derived to represent the relationship of such real gas behavior. One such relationship that represents the degree of deviation from ideal gas behavior is given as

$$Pv = ZRT, \tag{3.2c}$$

where Z is defined as the *compressibility factor* and this factor approaches a value of unity as a real gas approaches ideal gas behavior. Compressibility factors are also presented in the form of a *generalized chart* as a function of reduced pressure $\left(P_r = \dfrac{P}{P_C}\right)$ and reduced temperature $\left(T_r = \dfrac{T}{T_C}\right)$. Detailed discussions of generalized chart and real gas equation of state are given in books by Moran and Shapiro (2008), Sonntag et al. (2003), and Cengel (2006).

3.4 Energy and Energy Transfer

Total energy content of a system, E, is classified into three basic categories: (1) the kinetic energy, $\mathrm{KE} = \frac{1}{2}mV^2$, associated with the translation velocity of the system, (2) the potential energy, $\mathrm{PE} = mgZ$, associated with the elevation of the system from some reference level, and (3) the internal energy, U, that includes all energy forms associated with the atomic and molecular structures and orientations that include translation, rotation, and vibrational motion of atoms and energy associated with the atoms, molecules, and subatomic particles. Internal energy is also classified in different forms such as *latent energy* associated with the phase of the substance, *chemical energy* associated with the atomic bonds in a molecular structure, and *nuclear energy* associated with the binding force within the nucleus of the atom.

The energy content of a system can change because of the transfer of energy in the form of heat and work to or from the surroundings across the system boundary as shown in Figure 3.2.

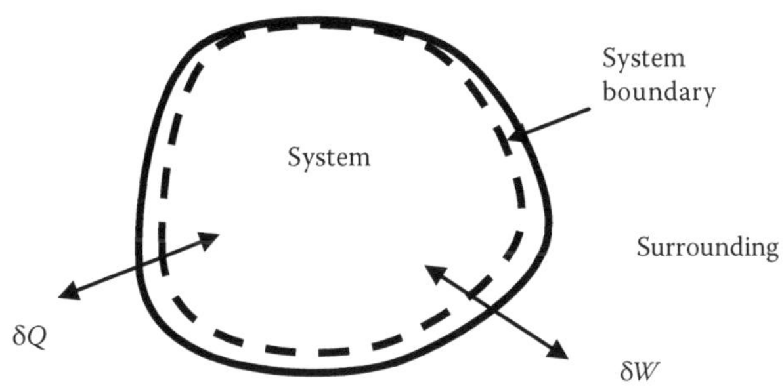

FIGURE 3.2
Interaction between system and surroundings through energy transfer.

3.4.1 Heat and Work

3.4.1.1 Heat Energy

Heat is an energy form that is transferred between two parts of a system or between a system and the surroundings owing to temperature difference. It is an energy form that is in transit and it can be identified only at the boundary of a system. If there is no difference in temperature between the system and the surroundings, then there is no heat transfer.

3.4.1.2 Work

Work in mechanics is defined as the product of force acting through a displacement as

$$\delta W = F\,\mathrm{d}x \tag{3.3}$$

and for a process with change of state from 1 to 2 as

$${}_1W_2 = \int_1^2 F\,\mathrm{d}x. \tag{3.4}$$

In thermodynamics, work is expressed in terms of pressure and change in volume for a simple compressible substance as

$$\delta W = P\,\mathrm{d}V \tag{3.5}$$

and for a process with change of state from 1 to 2 as

$${}_1W_2 = \int_1^2 P\,\mathrm{d}V. \tag{3.6}$$

In an electrical system such as in an electrochemical fuel cell, the electrical work is done as a result of the flow of electron across the system boundary through an electrical load circuit under the effect of an electromagnetic potential difference. When N electrical charges flow through an external load circuit owing to the electromagnetic potential difference E, electrical work is given as

$$\delta W_e = NE. \tag{3.7}$$

The number of electrical charge moving through an electrical circuit is the current flowing over time, dt, and given as

$$N = i\,\mathrm{d}t. \tag{3.8}$$

Combining Equations 3.7 and 3.8, we have

$$\delta W_e = iE\,\mathrm{d}t \tag{3.9}$$

and for a process with change of state from 1 to 2 as

$$W_e = \int_1^2 iE\,\mathrm{d}t. \tag{3.10}$$

3.5 The Conservation of Mass

The law of conservation of mass is a statement of the mass balance for flow in and out and changes of mass storage of a system. Change in mass caused by any energy transfer such as in a chemical reaction or a combustion process is negligibly small and therefore is not taken into account. For analysis purposes, the conservation of mass law is presented for both the system and the control volume.

3.5.1 System

Since a system is defined as a fixed and identifiable quantity of mass, the conservation mass for a system is defined as

$$\left.\frac{\mathrm{d}m}{\mathrm{d}t}\right)_{\text{System}} = 0. \tag{3.11}$$

3.5.2 Control Volume

A control volume is an open system that involves mass flow in and out. The conservation of mass statement for a control volume takes account of all mass flow in and mass flow out as well as change in any mass inside the control volume. The statement is derived as

$$\left.\frac{dm}{dt}\right)_{CV} = \sum m_e - \sum m_i. \tag{3.12}$$

3.6 The First Law of Thermodynamics

The first law of thermodynamics is a statement of conservation of energy taking into account all forms of energy transfer, storage, consumption, and generation.

3.6.1 The First Law of Thermodynamics for a System

The first law of thermodynamics for a system or fixed quantity of mass is defined as the balance of change of total energy content with energy transfer across the system boundary in terms of heat and work. A system interacts with the surroundings through transfer of energy in the form of work energy and heat energy across the system boundary as shown in Figure 3.1.

For a system undergoing a thermodynamic cycle, the first law of thermodynamics is given as

$$\oint \delta Q = \oint \delta W. \tag{3.13}$$

It physically states that the cyclic integral or the sum of heat transfer in all processes is equal to the cyclic integral or the sum of work done in all processes.

The first law of thermodynamics for a specific process states that the change in energy content of a system is caused by net transfer of energy in the form of heat and work across the system boundary and given as

$$\delta Q = dE + \delta W \tag{3.14}$$

and for a process with a change of state from 1 to 2

$${}_1Q_2 = E_2 - E_1 + {}_1W_2, \tag{3.15}$$

where

E = energy content of the system = U + KE + PE

U = internal energy associated with rotational, vibrational, and translational motions and structures of the atoms and molecules

KE = kinetic energy of the system = $\frac{1}{2}mV^2$

PE = potential energy of the system = mgZ

With the substitution of the expressions for different energy forms, Equation 3.15 can be written as

$$ {}_1Q_2 = (U_2 - U_1) + \frac{1}{2}m\left(V_2^2 - V_1^2\right) + mg(Z_2 - Z_1) + {}_1W_2. \tag{3.16}$$

In terms of per unit mass of the system, the equation is expressed as

$$ {}_1q_2 = (u_2 - u_1) + \frac{1}{2}\left(V_2^2 - V_1^2\right) + g(Z_2 - Z_1) + {}_1w_2, \tag{3.17}$$

where

u = specific internal energy = $\frac{U}{m}$

q = heat transfer per unit mass of the system = $\frac{Q}{m}$

w = work done per unit mass = $\frac{W}{m}$

3.6.1.1 Additional Thermodynamic Properties

3.6.1.1.1 Internal Energy (u)

Internal energy, u, is an intensive property of the system and it represents the energy associated with translation, vibration, and rotational motions and structures of the atoms and molecules.

3.6.1.1.2 Enthalpy (h)

Enthalpy is also an intensive property and expressed as a sum of internal energy and product of pressure, p, and specific volume, v

$$h = u + Pv \tag{3.18a}$$

and in terms of total enthalpy

$$H = U + PV. \tag{3.18b}$$

3.6.1.1.3 Gibbs Function or Gibbs Free Energy (g)

The specific Gibbs function or Gibbs free energy is defined as

$$g = h - Ts \tag{3.19a}$$

and in terms of total Gibbs energy

$$G = H - TS. \tag{3.19b}$$

3.6.2 The First Law of Thermodynamics for a Control Volume

For a control volume shown in Figure 3.3, *the first law of thermodynamics for a control volume* is derived on the basis of the conservation of energy across a control volume as

$$\dot{Q}_{\text{CV}} + \sum \dot{m}_{\text{i}}\left(h_{\text{i}} + \frac{V_{\text{i}}^2}{2} + gZ_{\text{i}}\right) = \frac{\text{d}E_{\text{CV}}}{\text{d}t} + \sum \dot{m}_{\text{e}}\left(h_{\text{e}} + \frac{V_{\text{e}}^2}{2} + gZ_{\text{e}}\right) + \dot{W}_{\text{CV}}. \tag{3.20}$$

For a stationary control volume with negligible changes in kinetic energy and potential energy, Equation 3.20 can rewritten as

$$\dot{Q}_{\text{CV}} + \sum \dot{m}_{\text{i}}\left(h_{\text{i}} + \frac{V_{\text{i}}^2}{2} + gZ_{\text{i}}\right) = \frac{\text{d}U}{\text{d}t} + \sum \dot{m}_{\text{e}}\left(h_{\text{e}} + \frac{V_{\text{e}}^2}{2} + gZ_{\text{e}}\right) + \dot{W}_{\text{CV}}. \tag{3.21}$$

3.6.2.1 *Special Cases*

For analysis purposes, the equations for conservation mass and the first law of thermodynamics are simplified with some assumptions. Two such common cases are (i) *the steady-state steady-flow process* and (ii) *the uniform-flow uniform-state process.*

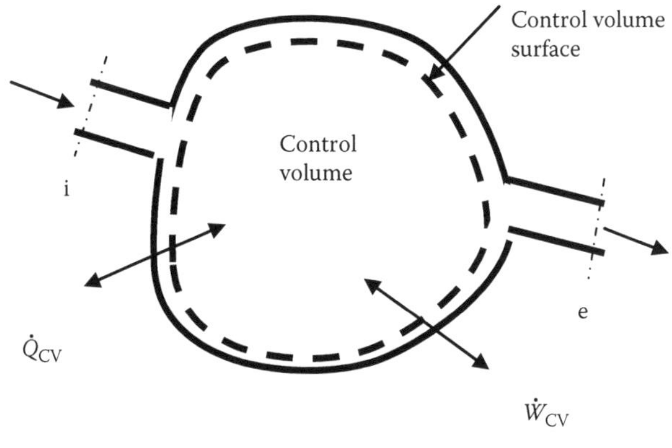

FIGURE 3.3
Flow and energy transfer across a control volume.

3.6.2.2 Steady-State Steady-Flow Process

A steady-state steady-flow (SSSF) process is a simplified model that represents a process in which all properties at each point of the system, all properties and flow rates of flow in and out of the control volume and energy transfer rates across the control volume surface are assumed to be constant and invariable with time. With these assumptions, Equations 3.12 and 3.20 reduce to

$$\sum m_e = \sum m_i \tag{3.22}$$

and

$$\dot{Q}_{CV} + \sum \dot{m}_i\left(h_i + \frac{V_i^2}{2} + gZ_i\right) = \sum \dot{m}_e\left(h_e + \frac{V_e^2}{2} + gZ_e\right) + \dot{W}_{CV}. \tag{3.23}$$

For a single flow in and out, Equations 3.22 and 3.23 reduce to

$$\dot{m}_i = \dot{m}_e = \dot{m} \tag{3.24}$$

and

$$\dot{Q}_{CV} + \dot{m}\left(h_i + \frac{V_i^2}{2} + gZ_i\right) = \dot{m}\left(h_e + \frac{V_e^2}{2} + gZ_e\right) + \dot{W}_{CV}. \tag{3.25a}$$

For unit mass flow rate, the first law equation reduces to

$$q_{CV} + \left(h_i + \frac{V_i^2}{2} + gZ_i\right) = \left(h_e + \frac{V_e^2}{2} + gZ_e\right) + w_{CV}. \tag{3.25b}$$

3.6.2.3 Uniform-Flow Uniform-State Process

A uniform-flow uniform-state (UFUS) process is a simplified model of a transient process. It is derived on the basis of the initial and final states of the control volume as

$$(m_2 - m_1)_{CV} + \sum m_e - \sum m_1 = 0 \tag{3.26}$$

$$\dot{Q}_{CV} + \sum \dot{m}_i\left(h_i + \frac{V_i^2}{2} + gZ_i\right) = \left[m_2\left(u_2 + \frac{V_2^2}{2} + gZ_2\right) - m_1\left(u_1 + \frac{V_1^2}{2} + gZ_1\right)\right]_{CV}$$

$$+ \sum \dot{m}_e\left(h_e + \frac{V_e^2}{2} + gZ_e\right) + \dot{W}_{CV}, \tag{3.27}$$

where states 1 and 2 represent the initial and the final states of the control volume, respectively.

3.7 The Second Law of Thermodynamics

The second law of thermodynamics is stated through the *Kelvin–Planck statement* and the *Clausius statement*. The *Inequality of Clausius* is a consequence of the second law of thermodynamics, and it is stated for a *system* undergoing a thermodynamic *cycle* as

$$\oint \frac{\delta Q}{T} \leq 0 \qquad \text{for a } \textit{cycle} \tag{3.28a}$$

with

$$\oint \frac{\delta Q}{T} = 0 \qquad \text{for a } \textit{reversible cycle} \tag{3.28b}$$

and

$$\oint \frac{\delta Q}{T} < 0 \qquad \text{for an } \textit{irreversible} \text{ cycle.} \tag{3.28c}$$

Application of this statement for a *process* leads to

$$dS \geq \left(\frac{\delta Q}{T}\right), \tag{3.29}$$

where S is the thermodynamic property *entropy*, defined as a consequence of the fact that the quantity $\frac{\delta Q}{T}$ is constant for all thermodynamic processes between the same two thermodynamic states.

For a *reversible process*, this leads to

$$dS = \left(\frac{\delta Q}{T}\right)_{rev} \tag{3.30a}$$

or

$$S_2 - S_1 = \int_1^2 \frac{\delta Q}{T}\Bigg)_{rev}. \tag{3.30b}$$

It states that entropy increases in a process with heat addition to the system and decreases for a process with heat rejection from the system. For a *reversible and adiabatic process*, also known as *isentropic process*, there is no heat transfer and entropy remains constant; that is,

$$S_2 = S_1. \tag{3.30c}$$

Also, for a reversible process, the heat transfer across a system is given by Equation 3.30a as

$$\delta Q = T\mathrm{d}S. \tag{3.31}$$

For an *irreversible process*,

$$\mathrm{d}S > \left(\frac{\delta Q}{T}\right)_{\text{irrev}} \tag{3.32a}$$

or

$$S_2 - S_1 > \int_1^2 \frac{\delta Q}{T}\bigg)_{\text{irrev}}. \tag{3.32b}$$

Considering the entropy generation or production associated with the irreversible process, the entropy change for a process can be expressed as

$$\mathrm{d}S = \left(\frac{\delta Q}{T}\right)_{\text{irrev}} + \mathrm{d}S_{\text{gen}} \tag{3.33a}$$

or

$$S_2 - S_1 = \int_1^2 \frac{\mathrm{d}Q}{T}\bigg) + {}_1S_{2\text{gen}}, \tag{3.33b}$$

where $\mathrm{d}S_{\text{gen}}$ and ${}_1S_2$ are the entropy generation in the process due to irreversibilities caused by various system factors.

The second law of thermodynamics for a control volume is stated as

$$\frac{\mathrm{d}S_{\text{CV}}}{\mathrm{d}t} + \sum \dot{m}_{\text{e}}s_{\text{e}} - \sum \dot{m}_{\text{i}}s_{\text{i}} \geq \sum \frac{\dot{Q}_{\text{CV}}}{T} \tag{3.34a}$$

or

$$\frac{\mathrm{d}S_{\text{CV}}}{\mathrm{d}t} + \sum \dot{m}_{\text{e}}s_{\text{e}} - \sum \dot{m}_{\text{i}}s_{\text{i}} = \sum \frac{\dot{Q}_{\text{CV}}}{T} + \dot{S}_{\text{gen}}. \tag{3.34b}$$

Equation 3.34 is simplified to the SSSF and UFUS processes as follows:

SSSF Process

$$\sum \dot{m}_e s_e - \sum \dot{m}_i s_i \geq \sum \frac{\dot{Q}_{CV}}{T} + \dot{S}_{gen} \quad (3.35a)$$

and for a single flow in and out

$$\dot{m}(s_e - s_i) = \sum \frac{\dot{Q}_{CV}}{T} + \dot{S}_{gen}. \quad (3.35b)$$

For an adiabatic process, $\dot{Q}_{CV} = 0$ and Equation 3.35b can be written as

$$\dot{m}(s_e - s_i) = \dot{S}_{gen} \quad (3.36a)$$

or

$$s_e \geq s_i, \quad (3.36b)$$

where

$$s_e = s_i \text{ for a reversible adiabatic process} \quad (3.37a)$$

and

$$s_e > s_i \text{ for an irreversible process.} \quad (3.37b)$$

UFUS Process

$$[m_2 s_2 - m_1 s_1]_{CV} + \sum m_e s_e - \sum m_i s_i = \int_0^t \frac{\dot{Q}_{CV}}{T} dt + {}_1S_{2gen}. \quad (3.38)$$

3.7.1 Carnot Cycle

The Carnot cycle is an ideal thermodynamic cycle that represents the most efficient cycle for a heat engine and refrigeration machine operating between two temperature limits. It consists of four reversible processes: (1) reversible isothermal heat addition, Q_H, from high temperature, T_H; (2) reversible adiabatic expansion; (3) reversible isothermal heat rejection, Q_L, to a low temperature, T_L; and (4) reversible adiabatic compression.

Thermal efficiency of this Carnot cycle is given as

$$\eta = \frac{\text{Work Out}}{\text{Heat Added}} = \frac{W}{Q_H} = \frac{Q_H - Q_L}{Q_H}. \tag{3.39}$$

With the application of the second law of thermodynamics, it can be shown that for reversible heat addition at high temperature and heat rejection at low temperature,

$$\frac{Q_H}{Q_L} = \frac{T_H}{T_L}. \tag{3.40}$$

Substituting Equation 3.40 into Equation 3.39, the Carnot cycle efficiency is given as

$$\eta_{carnot} = 1 - \frac{T_L}{T_H}. \tag{3.41}$$

Example 3.1

An automobile engine burns a fuel to the combustion product at a temperature of 1100°C and rejects heat in a radiator with exhaust temperature of 200°C. What is the maximum possible efficiency of this engine?

The maximum possible efficiency of the engine is given by the Carnot cycle efficiency and given as

$$\eta_{carnot} = 1 - \frac{T_L}{T_H}$$

or

$$\eta_{carnot} = 1 - \frac{200 + 273}{1100 + 273} = 1 - \frac{473}{1373} = 0.3445$$

or

$$\eta_{carnot} = 34.45\%.$$

3.8 Thermodynamic Relations

Thermodynamic relations are derived to determine changes in properties such as enthalpy, entropy, and Gibbs function based on the known basic properties. The following thermodynamic relations are derived from the first law and by using relations among work, enthalpy, entropy, and Gibbs function:

$$Tds = du + Pdv \tag{3.42a}$$

$$Tds = dh - vdp \tag{3.42b}$$

$$dg = dh - Tds \tag{3.42c}$$

$$dg = vdP - sdT \tag{3.42d}$$

In the molar form:

$$Td\bar{s} = d\bar{u} + Pd\bar{v} \tag{3.43a}$$

$$Td\bar{s} = d\bar{h} - \bar{v}dp \tag{3.43b}$$

$$d\bar{g} = d\bar{h} - Td\bar{s} \tag{3.43c}$$

$$d\bar{g} = \bar{v}dP - \bar{s}dT \tag{3.43d}$$

In terms of total properties:

$$TdS = dU + PdV \tag{3.44a}$$

$$TdS = dH - vdP \tag{3.44b}$$

$$dG = dH - TdS \tag{3.44c}$$

$$dG = VdP - SdT \tag{3.44d}$$

3.9 Specific Heat

Specific heat is defined as the energy needed to raise the temperature of a unit mass by a unit degree temperature. For constant volume process $\delta w = pdv = 0$ and the first law reduces to $\delta Q = dU$. The *constant volume specific heat* is expressed as

$$C_v = \frac{1}{m}\frac{\delta Q}{\delta T}\bigg)_v = \frac{\partial u}{\partial T}\bigg)_v. \tag{3.45}$$

For a constant pressure process, $\delta w = pdv$ and the first law reduces to $\delta Q = dU + PdV = dH$. The constant pressure specific heat is expressed as

$$C_p = \frac{1}{m}\frac{\delta Q}{\delta T}\bigg)_p = \frac{\partial h}{\partial T}\bigg)_p. \tag{3.46}$$

3.10 Estimation of Change in Enthalpy, Entropy, and Gibbs Function for Ideal Gases

For ideal gases, enthalpy and internal energy are a function of temperature only. Hence, change in enthalpy and internal energy for a change of state or process is derived from the definition of specific heat and expressed as

$$h_2 - h_1 = \int_1^2 C_{po}\, dT \tag{3.47a}$$

and

$$u_2 - u_1 = \int_1^2 C_{vo}\, dT, \tag{3.47b}$$

where C_{po} and C_{vo} represent ideal gas specific heat values. These equations can be evaluated for constant specific heat values and for temperature-dependent specific heat values.

3.10.1 Case I: Constant Specific Heat

For constant specific heat values, Equation 3.47 can be evaluated as

$$h_2 - h_1 = C_{po}(T_2 - T_1) \tag{3.48a}$$

and

$$u_2 - u_1 = C_{vo}(T_2 - T_1). \tag{3.48b}$$

3.10.2 Case II: Temperature-Dependent Specific Heat Values

For temperature-dependent specific heat functions, equations can be evaluated by simply substituting the functional relations and carrying out the integrations term by term.

$$h_2 - h_1 = \int_{T_1}^{T_2} C_{\mathrm{P}0}(T)\,\mathrm{d}T, \tag{3.49}$$

where $C_{\mathrm{P}0}(T)$ is a functional relationship of the specific heat as a function of temperature. Table C.4 presents such functional relationship for some of the common ideal gases.

3.10.3 Case III

In order to simplify the computations, the integral equation (Equation 3.49) is written by computing enthalpy change from a reference temperature as

$$h_2 - h_1 = \int_{T_0}^{T_2} C_{\mathrm{P}0}(T)\,\mathrm{d}T - \int_{T_0}^{T_1} C_{\mathrm{P}0}(T)\,\mathrm{d}T. \tag{3.50}$$

By defining

$$h_T = \int_{T_0}^{T} C_{\mathrm{P}0}(T)\,\mathrm{d}T, \tag{3.51}$$

the change in enthalpy equation is written as

$$h_2 - h_1 = h_{T_2} - h_{T_1}. \tag{3.52}$$

The integral given by Equation 3.51 is evaluated for different gases over a range of temperature and assuming a reference temperature of T_0 = 20°C or 298 K. Table C.7 presents such integral values for some of the common ideal gases.

3.10.4 Entropy Change in Process

Entropy change of an ideal gas is derived from the thermodynamic relations (Equation 3.42a or 3.42b) and expressed as

$$s_2 - s_1 = \int_1^2 C_{vo} \frac{dT}{T} + R\ln \frac{v_2}{v_1} \tag{3.53a}$$

and

$$s_2 - s_1 = \int_1^2 C_{po} \frac{dT}{T} - R\ln \frac{P_2}{P_1}. \tag{3.53b}$$

3.10.5 Special Cases

3.10.5.1 Case I: Constant Specific Heat Values

$$s_2 - s_1 = C_{vo} \ln \frac{T_2}{T_1} + R\ln \frac{v_2}{v_1} \tag{3.54a}$$

and

$$s_2 - s_1 = C_{po} \ln \frac{T_2}{T_1} - R\ln \frac{P_2}{P_1}. \tag{3.54b}$$

3.10.5.2 Case II: Temperature-Dependent Specific Heat Values

$$s_2 - s_1 = \int_1^2 C_{vo}(T) \frac{dT}{T} + R\ln \frac{v_2}{v_1} \tag{3.55a}$$

and

$$s_2 - s_1 = \int_1^2 C_{po}(T) \frac{dT}{T} - R\ln \frac{P_2}{P_1}. \tag{3.55b}$$

With the substitution of the functional relations for specific heats, these equations can be evaluated by carrying out the integration term by term.

3.10.5.3 Case III

Using the procedure outlined for enthalpy, the integral equation (Equation 3.55b) is written by computing entropy change from a reference temperature as follows:

$$s_2 - s_1 = \left(s^0_{T_2} - s^0_{T_1}\right) - R\ln\frac{P_2}{P_1}, \tag{3.56a}$$

where

$$s^0_T = \int_{T_0}^{T} \frac{C_{po}}{T}\,dT. \tag{3.56b}$$

The integral given by Equation 3.48 is evaluated for different gases over a range of temperatures and assuming a reference temperature of $T_0 = 20°C$ or 298 K. Table C.7 presents such integral values for some of the common ideal gases.

3.10.6 Change of Gibbs Function

Change in Gibbs function can be derived from the thermodynamic relations (Equation 3.42c) and expressed as

$$g_2 - g_1 = (h_2 - h_1) + \int_1^2 T\,ds. \tag{3.57a}$$

For an isothermal process, we have

$$g_2 - g_1 = (h_2 - h_1) + T(s_2 - s_1), \tag{3.57b}$$

where change in enthalpy and change in entropy can be estimated on the basis of the procedure outlined in the previous section.

Example 3.2

Oxygen gas stream is heated from 300 K to 900 K with pressure dropping from 300 kPa to 150 kPa. Calculate the change in enthalpy, change in entropy, and change in Gibbs function of the gas stream.

Solution

Enthalpy Change:
Enthalpy change based on constant specific heat at an average temperature is given by

$$\Delta h = h_2 - h_1 = C_{po}(T_2 - T_1).$$

The constant specific heat can be evaluated at an average temperature of $T_{av} = \dfrac{T_1 + T_2}{2} = \dfrac{300 + 900}{2} = 600\,\text{K}$ from the functional relation for specific heat.

From Table C.4, the function relation for oxygen is given as

$$C_{po} = 0.88 - 0.0001\left(\frac{T_{av}}{1000}\right) + 0.54\left(\frac{T_{av}}{1000}\right)^2 - 0.33\left(\frac{T_{av}}{1000}\right)^3.$$

Evaluating at the average temperature

$$C_{po} = 0.88 - 0.0001\left(\frac{600}{1000}\right) + 0.54\left(\frac{600}{1000}\right)^2 - 0.33\left(\frac{600}{1000}\right)^3$$

$$= 0.88 - 0.00006 + 0.1944 - 0.07128$$

$$C_{po} = 1.00306\ \text{kj/kg}\cdot\text{K}$$

$$\Delta h = h_2 - h_1 = 1.00306 \times (900 - 300) = 601.836\ \text{kj/kg}\cdot\text{K}$$

Enthalpy change based on variable specific heat value is given as

$$\Delta h = h_2 - h_1 = h_{T_2} - h_{T_1}.$$

Using enthalpy data for oxygen given in Table C.7, the change in enthalpy is given by

$$\Delta h = h_{T_2} - h_{T_1} = \frac{1}{M}\left(\bar{h}_{T=900\,\text{K}} - \bar{h}_{T=300\,\text{K}}\right) = \frac{19{,}241 - 54}{31.999}$$

$$\Delta h = 599.61\ \text{kJ/kg}.$$

Entropy Change:
Entropy change based on Equation 3.54b:

$$\Delta s = s_2 - s_1 = C_{po}\ln\frac{T_2}{T_1} - R\ln\frac{P_2}{P_1}$$

$$\Delta s = s_2 - s_1 = C_{po} \ln \frac{900}{300} - 0.2598 \ln \frac{150}{300}$$

$$= 1.00306 \times 1.0986 + 0.1800$$

$$\Delta s = 1.2819 \text{ kJ/kg} \cdot \text{K}.$$

Entropy change based on Equation 3.56a:

$$\Delta s = s_2 - s_1 = \frac{1}{M_{O_2}} \left[\left(\bar{s}_{T_2}^0 - \bar{s}_{T_1}^0 \right) - \bar{R} \ln \frac{P_2}{P_1} \right].$$

From Table C.7: $\bar{s}_{T=900\,\text{K}}^0 = 239.931$ kJ/kmol and $\bar{s}_{T=300\,\text{K}}^0 = 239.931$ kJ/kmol

$$\Delta s = \frac{1}{M_{O_2}} \left[\left(239.931 - 205.329 - 8.3145 \ln \frac{150}{300} \right) \right]$$

$$\Delta s = \frac{1}{31.999} (34.602 + 5.7631) = 40.3651 \text{ kJ/kg} \cdot \text{K}$$

$$\Delta s = 1.2614 \text{ kJ/kg} \cdot \text{K}.$$

Change of Gibbs Free Energy:
Change in Gibbs energy based on Equation 4.42c:

$$\Delta g = \Delta h - T \Delta s$$

$$\Delta g = 599.61 \text{ kj/kg} - \frac{900 + 300}{2} \times 1.2614 \text{ kJ/kg} \cdot \text{K}$$

$$\Delta g = 157.23 \text{ kJ/kg K}.$$

3.11 Mixture of Gases

Operations of most fuel cell power systems involve a mixture of gases. Therefore, we need to perform thermodynamic analysis and transport phenomena analysis with a mixture of gases. The gas mixture may be a mixture of ideal gases or a mixture of real gases. In this book, the presentation of fuel cell analysis is restricted to the mixture of ideal gases only.

3.11.1 Basic Mixture Parameters

Let us define some of the common mixture parameters assuming the mixture to be composed of N number of components.

3.11.1.1 Mass Fraction and Concentration

The mass fraction of a component in a mixture is defined as the ratio of mass of the component gas to the total mass as

$$\text{Mass fraction,} \quad x_i = \frac{m_i}{\sum_{i=1}^{N} m_i} = \frac{\text{Mass of component } i}{\text{Total mass of the mixture}}. \tag{3.58a}$$

Mass concentration or *mass density* is defined as the mass of the species i per unit volume of the mixture and expressed as

$$\rho_i = \frac{m_i}{\forall} (\text{kg/m}^3). \tag{3.58b}$$

3.11.1.2 Mole Fraction and Concentration

The mole fraction of a component in a mixture is defined as the ratio of number of moles of the component to the total number of moles in the mixture as

$$\text{Mole fraction,} \quad y_i = \frac{n_i}{\sum_{i=1}^{N} n_i} = \frac{\text{Number of moles of component } i}{\text{Total number of moles in the mixture}}. \tag{3.59a}$$

Mole concentration is defined as the ratio of number of moles of species i per unit volume of the mixture and expressed as

$$C_i = \frac{n_i}{\forall} (\text{kmol/m}^3). \tag{3.59b}$$

The relation between mass fraction and mole fraction is given as

$$x_i = \frac{y_i M_i}{\sum_{i=1}^{N} y_i M_i}. \tag{3.60}$$

The molecular weight of the mixture is given as

$$M_{\text{mix}} = \frac{m_{\text{mix}}}{N} = \sum_{i=1}^{N} y_i M_i. \tag{3.61}$$

Application of the ideal gas law model leads to the following two important relations for ideal gas mixtures:

Dalton's law assumes each component of the mixture exists at the same temperature and total volume of the mixture and it leads to

$$P = \sum_{i=1}^{N} P_i, \tag{3.62}$$

where

$$P_i = \frac{n_i \bar{R} T}{V} = \text{partial pressure of the component gas.} \tag{3.63}$$

Amagat's law assumes that the each component of the mixture exists at the same temperature and total pressure of the mixture and it leads to

$$V = \sum_{i=1}^{N} V_i, \tag{3.64}$$

where

$$V_i = \frac{n_i \bar{R} T}{P}. \tag{3.65}$$

Additionally, it can be shown that the volume fraction, the mole fraction, and the ratio of partial pressure to the total pressure are all equal; that is,

$$y_i = \frac{n_i}{n} = \frac{P_i}{P} = \frac{V_i}{V}. \tag{3.66}$$

3.11.2 Ideal Gas Mixture Properties

Ideal gas mixture properties are represented on the basis of Dalton's law as the sum of contributions from all components of the mixture. The following is a list of some of the basic mixture properties:

Mixture Gas Constant

$$R_{\text{mix}} = \frac{\bar{R}}{M_{\text{mix}}} = \sum_{i=1}^{N} x_i R_i \tag{3.67}$$

Mixture Specific Heat

$$C_p = \sum_{i=1}^{N} x_i C_{p_i}, \bar{C}_p = \sum_{i=1}^{N} y_i \bar{C}_{p_i} \tag{3.68a}$$

$$C_v = \sum_{i=1}^{N} x_i C_{v_i} \bar{C}_v = \sum_{i=1}^{N} y_i \bar{C}_{v_i} \tag{3.68b}$$

Total Properties

$$U = \sum_{i=1}^{N} U_i, H = \sum_{i=1}^{N} H_i \tag{3.69a}$$

$$S = \sum_{i=1}^{N} S_i, G = \sum_{i=1}^{N} G_i \tag{3.69b}$$

Specific Properties

$$u = \sum x_i u_i, \bar{u} = \sum y_i \bar{u}_i \tag{3.70a}$$

$$h = \sum x_i h_i, \bar{h} = \sum y_i \bar{h}_i \tag{3.70b}$$

$$s = \sum x_i s_i, \bar{s} = \sum y_i \bar{s}_i \tag{3.70c}$$

$$g = \sum x_i g_i, \bar{g} = \sum y_i \bar{g}_i \tag{3.70d}$$

It is important to note that all component gas properties are evaluated at the mixture temperature, T, and component partial pressure, P_i. For the ideal gas mixture, however, enthalpy and internal energy are a function of temperature, and hence component gas enthalpy and internal energy are estimated as a function of mixture temperature only as given by Equations 3.47a and 3.47b, respectively. However, entropy of an ideal gas is a function of temperature and pressure, and so the component gas entropy is estimated as a function of partial pressure of the component in the mixture and the gas mixture temperature as given by Equation 3.53.

Example 3.3

The volume composition of a gas mixture is given as H_2: 78%, CO_2: 20%, and H_2O: 2%. Determine (a) the mass fraction of the component gasses in the mixture, (b) the gas constant of the mixture, (c) the constant pressure specific heat of the mixture, and (d) the heat transfer to cool the mixture from 500°C to 100°C.

Solution

Molar composition is given as

$$y_{H_2} = 0.78$$

$$y_{CO_2} = 0.2$$

$$y_{H_2O} = 0.02.$$

The molecular weight of the mixture or the mass of the mixture per kilomole of mixture is

$$\begin{aligned} M_{mix} &= \sum_{i=1}^{N} m_i = m_{H_2} + m_{CO_2} + m_{H_2O} \\ &= M_{H_2} \times y_{H_2} + M_{CO_2} \times y_{CO_2} + M_{H_2O} \times y_{H_2O} \\ &= 2.016 \times 0.78 + 44.0 \times 0.2 + 18.016 \times 0.02 \\ &= 1.5724 + 8.8 + 0.3603 \\ M_{mix} &= 10.7327 \text{ kg/kmol of mixture.} \end{aligned}$$

Mass fraction is given as

$$x_{H_2} = \frac{m_{H_2}}{\sum_{i=1}^{N} m_i} = \frac{1.5724}{10.7327} = 0.1465$$

$$x_{CO_2} = \frac{m_{CO_2}}{\sum_{i=1}^{N} m_i} = \frac{8.8}{10.7327} = 0.8199$$

$$x_{H_2O} = \frac{m_{H_2O}}{\sum_{i=1}^{N} m_i} = \frac{0.3603}{10.7327} = 0.0335.$$

The gas constant of the mixture is

$$R_{mix} = \frac{\bar{R}}{M_{mix}} = \frac{8.3144}{10.7327} = 0.7746 \text{ kj/kg}\cdot\text{K}.$$

The constant pressure specific heat of the mixture is

$$\begin{aligned} C_{P_{mix}} &= \sum x_i \, C_{P_i} \\ &= x_{H_2} \times C_{P_{H_2}} + x_{CO_2} \times C_{P_{CO_2}} + x_{H_2O} \times C_{P_{H_2O}} \\ &= 0.1465 \times 14.209 + 0.8199 \times 0.842 + 0.0335 \times 1.872 \\ &= 2.0816 + 0.69035 + 0.0627 \\ &= 2.835 \text{ kJ/kg}\cdot\text{K}. \end{aligned}$$

Heat transfer to cool the mixture

$$\begin{aligned} {}_1q_2 &= h_2 - h_1 = C_{P_{mix}}(T_2 - T_1) = 2.835 \times (500 - 100) \\ {}_1q_2 &= 1134 \text{ kJ/kg}. \end{aligned}$$

3.11.3 Transport Properties of Gas Mixture

Transport properties such as viscosity and thermal conductivity of a gas mixture are estimated based on the mixture rules. A simplified mixture model developed based on the kinetic theory model is widely used (Bird et. al., 1960; Mills, 2001; Wilke, 1950). These formulae are given as follows:

3.11.3.1 Viscosity of Gas Mixture

Viscosity is a measure of fluid resistance to motion, and it relates the strain rate to applied shear stress. A functional dependence of gas viscosity on temperature at low density is given by Chapman-Enskog based on kinetic theory (Bird et al., 1960) using Lennard-Jones potentials. The theory has been also extended to multicomponent gas mixtures. For most common applications, however, a simplified semiempirical formula of Wilke (1950) is used:

$$\mu_{mix} = \sum_{i=1}^{N} \frac{y_i \mu_i}{\sum_{j=1}^{N} y_j \phi_{ij}}, \tag{3.71}$$

where μ_i is the viscosity of individual gas element in the mixture.

A temperature-dependent expression for gas viscosity of a pure monoatomic gas is given by Chapman-Enskog's kinetic theory as

$$\mu = 2.6693 \times 10^{-5} \frac{\sqrt{MT}}{\sigma^2 \Omega_\mu}, \tag{3.72}$$

where M is the molecular weight, T is the absolute temperature (K), σ is the characteristic diameter of gas molecule (Å), and Ω_μ is the slow varying function of dimensionless temperature KT/ε, given in Table C.9.

The values of Lennard-Jones parameters σ and ε/K are given in Table 6.3.

Equation 3.72 is also found to be applicable to polyatomic gases. Viscosity of gases at low density increases with temperature in a power law with power index in the range of 0.6–10. The simple power law expression is given as

$$\frac{\mu}{\mu_0} \cong \left(\frac{T}{T_o}\right)^n. \tag{3.73}$$

A more comprehensive relation is given by Sutherland's law as

$$\frac{\mu}{\mu_0} \cong \left(\frac{T}{T_o}\right)^n \frac{T_o + S}{T + S}. \tag{3.74}$$

The constants values of n, μ_o, T_o, and S are obtained by experiments or from kinetic theory of gases and are given in Table C.10.

3.11.3.2 Thermal Conductivity of Gas Mixture

Thermal conductivity of a gas mixture is given as

$$k_{\text{mix}} = \sum_{i=1}^{N} \frac{y_i k_i}{\sum_{j=1}^{N} y_j \phi_{ij}}, \tag{3.75}$$

where ϕ_{ij} in Equations 3.71 and 3.75 is a dimensionless number obtained from

$$\phi_{ij} = \frac{1}{\sqrt{8}} \left(1 + \frac{M_i}{M_j}\right)^{-\frac{1}{2}} \left[1 + \left(\frac{\mu_i}{\mu_j}\right)^{\frac{1}{2}} \left(\frac{M_j}{M_i}\right)^{\frac{1}{4}}\right]^2, \tag{3.76}$$

where
N = total number of species in the mixture
y_i, y_j = mole fractions of species i and j
M_i, M_j = molecular weights (kg/mol) of species i and j.

3.12 Combustion Process

In a chemical reaction, the bond structure of the reactants is modified to form a new bond structure, and in the process, electronic configuration within the atoms is changed and chemical energy is released. The thermal heat energy is released owing to changes from the initial bonds and associate electronic configurations of the fuel and oxygen to the regrouped bond and electronic configuration of the products. The amount of chemical energy converted to thermal energy is the difference between the internal energy content of the original bond structure of the reactants and the internal energy content of the regrouped bond structures of the products. Combustion process is chemical reaction in which a fuel is oxidized and a large quantity of chemical energy is released.

One of the most commonly available forms of fuel is *hydrocarbon fuel*, which has carbon and hydrogen as the primary constituents. Some of the common hydrocarbon fuels are *gasoline* or *octane*, *diesel*, *methyl alcohol* or *methanol*, and *ethyl alcohol* or *ethanol*. The hydrocarbon fuels exist in different phases such as liquid like gasoline, solid like coal, and gas like natural gas. Coal is mainly composed of carbon, sulfur, oxygen, and hydrogen with varying composition. Natural gas consists of methane, carbon dioxide, hydrogen, nitrogen, and oxygen with varying composition.

In the combustion of hydrocarbon fuel, carbon, hydrogen, and any other constituents in the fuel that are capable of being oxidized react with oxygen. For example, a typical combustion reaction of carbon as a fuel with oxygen is shown in Figure 3.4.

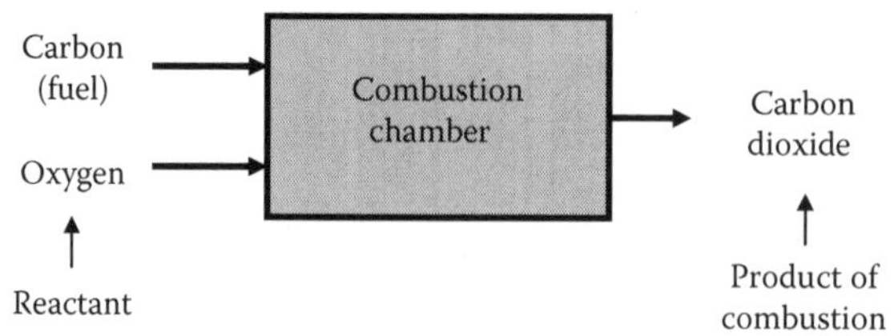

FIGURE 3.4
Typical schematic representation of a combustion reaction.

It is represented in the following manner:

$$C + O_2 \rightarrow CO_2, \tag{3.77}$$

where carbon and oxygen on the left-hand side of the reaction equation are referred to as reactants and carbon dioxide on the right hand side is referred to as the product.

In this reaction, 1 kmol (32 kg) of oxygen reacts with 1 kmol (12 kg) of carbon and forms 1 kmol (44 kg) of carbon dioxide. In a typical combustion reaction of hydrocarbon fuel such as methane gas (CH_4), carbon and hydrogen combine with oxygen to form carbon dioxide and water.

$$CH_4 + 2O_2 \rightarrow CO_2 + 2H_2O \tag{3.78}$$

The chemical reaction is written by simply applying a mole balance of each component of the reactants and of the products. For example, total number of moles of oxygen in the left-hand side of the reaction in Equation 3.78 is two and it balances with the total number of oxygen moles in the product, which is the sum of one oxygen moles in carbon dioxide and one oxygen mole in water. In a similar manner, hydrogen is also balanced.

Oxygen is often supplied as air rather than in a pure form as it is free and available in abundance. Even though air is composed of a number of different gases such as oxygen, nitrogen, argon, and so on, it is assumed to be primarily composed of 79% nitrogen and 21% oxygen by volume for analysis purposes; that is, for each kilomole of oxygen, there are 79/21 = 3.76 kilomole of nitrogen. The reaction of methane with air is then written as

$$CH_4 + 2O_2 + 2(3.76)\,N_2 \rightarrow CO_2 + 2H_2O + 7.52N_2. \tag{3.79}$$

In this reaction, nitrogen is assumed as inert and does not undergo any chemical reaction. Nitrogen thus appears on both sides of the equation and simply effects the product temperature by absorbing part of the released chemical energy and raising its own internal energy.

However, in some high-temperature and high-pressure reactions, nitrogen may undergo reaction and form air pollutants such as nitrous oxide, N_2O, nitrogen dioxide, NO_2, or nitric oxide, NO. Also, in this reaction, air is supplied as 100% *theoretical air* or *stoichiometric air* that supplies sufficient amount of oxygen for complete combustion of all elements that are capable of being oxidized. In a complete combustion, all carbon oxidizes to form CO_2, all hydrogen oxidizes to form H_2O, and sulfur oxidizes to form SO_2. In an incomplete combustion reaction, the product may contain some fuel as unburned fuel, some carbon in the form of CO, and even as carbon particles. Incomplete combustion is caused by insufficient supply of oxygen as well as inadequate mixing of fuel and air in the mixture. In a real reaction process, air is supplied in excess to achieve complete combustion. A combustion reaction with 50% excess air, that is, 150% theoretical air or stoichiometric air, is represented as follows:

$$CH_4 + 2(1.5)O_2 + 2(3.76)\,(1.5)\,N_2 \rightarrow CO_2 + 2H_2O + O_2 + 11.28N_2. \quad (3.80)$$

One of the main reasons of supplying excess air is to avoid any incomplete reaction caused by the lack of air or oxygen and other contributing factors such as inadequate mixing and turbulence. In an incomplete combustion reaction, some carbon forms carbon monoxide (CO) instead of carbon dioxide (CO_2). For example, a combustion process with 105% theoretical air may lead to an incomplete reaction as follows:

$$CH_4 + 2(1.05)O_2 + 2(3.76)(1.05)N_2 \rightarrow 0.95CO_2 + 0.05CO$$

$$+ 2H_2O + 0.175O_2 + 7.896N_2. \quad (3.81)$$

Carbon monoxide is poisonous and not desirable in any combustion exhaust or any reformed gas mixture. In fuel cell applications, a reformed fuel gas mixture containing hydrogen and carbon monoxide is subjected to a gas-shift reaction process to convert carbon monoxide into carbon dioxide before supplying the hydrogen-rich gas mixture to the fuel cell. In a gas-shift reaction process using steam or water, carbon monoxide is converted into carbon dioxide as follows:

$$CO + H_2O \rightarrow CO_2 + H_2. \quad (3.82)$$

Example 3.4

Determine the molar and mass composition of the product for the combustion of octane (C_8H_{18}) with 200% stoichiometric air and determine the dew point temperature of the mixture at an operating pressure of 0.1 MPa.

Solution

The combustion equation is give as

$$C_8H_{18} + (12.5 \times 2.0)O_2 + (12.5 \times 2.0 \times 3.76)N_2$$

$$\rightarrow 8CO_2 + 9H_2O + (12.5) \times O_2 + (94.0)N_2.$$

The total number of moles in the product is

$$\sum_{i=1}^{N} n_i = n_{CO_2} + n_{H_2O} + n_{O_2} + n_{N_2} = 8 + 9 + 12.5 + 94.0 = 123.5.$$

Molar composition is given as

$$y_{CO_2} = \frac{n_{CO_2}}{\sum_{i=1}^{N} n_i} = \frac{8}{123.5} = 0.0647$$

$$y_{H_2O} = \frac{n_{H_2O}}{\sum_{i=1}^{N} n_i} = \frac{9}{123.5} = 0.0728$$

$$y_{O_2} = \frac{n_{O_2}}{\sum_{i=1}^{N} n_i} = \frac{12.5}{123.5} = 0.1012$$

$$y_{N_2} = \frac{n_{H_2O}}{\sum_{i=1}^{N} n_i} = \frac{94}{123.5} = 0.7611.$$

Total mass of the product per kilomole of the mixture is

$$\sum_{i=1}^{N} m_i = m_{CO_2} + m_{H_2O} + m_{O_2} + m_{N_2}$$

$$= M_{CO_2} \times c_{CO_2} + M_{H_2O} \times c_{H_2O} + M_{O_2} \times c_{O_2} + M_{N_2} \times c_{N_2}$$

$$= 44 \times 0.0647 + 18.016 \times 0.0728 + 32 \times 0.1012 + 28.01 \times 0.7612$$

$$= 2.8468 + 1.3115 + 3.2384 + 21.3184.$$

$$\sum_{i=1}^{N} m_i = 28.7151 \text{ kg/kmol of mixture}$$

Mass fraction is given as

$$x_{CO_2} = \frac{m_{CO_2}}{\sum_{i=1}^{N} m_i} = \frac{2.8468}{28.7151} = 0.0991$$

$$x_{H_2O} = \frac{m_{H_2O}}{\sum_{i=1}^{N} m_i} = \frac{1.3115}{28.7151} = 0.0456$$

$$x_{O_2} = \frac{m_{O_2}}{\sum_{i=1}^{N} m_i} = \frac{3.2384}{28.7151} = 0.1127$$

$$x_{N_2} = \frac{m_{N_2}}{\sum_{i=1}^{N} m_i} = \frac{21.3184}{28.7151} = 0.7424.$$

Partial pressure of water in the mixture product

$$P_{H_2O} = y_{H_2O} \times P = 0.0728 \times 0.1 = 7.28 \text{ kPa}.$$

The dew point temperature of the mixture product is the saturation temperature of water at the partial pressure of water in the mixture and given as

$$T_{dp} = T_{Sat}(P_{H_2O}) = 39.7°C$$

based on saturation thermodynamic properties of water given in Table C.5.

This indicates that if the mixture temperature is cooled below the dew point temperature of T_{dp} = 39.7°C, water vapor in the mixture will condense to liquid water.

3.13 Enthalpy of Formation $\left(\bar{h}_f^0\right)$

Enthalpy of formation of the product in a chemical reaction is the difference in enthalpy of the product and sum of enthalpy of all reactants. In order to demonstrate this, let us consider the combustion process under an SSSF process. The first law of thermodynamics for a reaction process in a control volume and assuming SSSF process is given as

$$Q_{CV} + \sum_{R=1}^{N} n_R h_R = \sum_{P=1}^{N} n_P h_P \tag{3.83a}$$

or

$$Q_{CV} + H_R = H_P \tag{3.83b}$$

$$Q_{CV} = H_P - H_R. \tag{3.83c}$$

Considering the enthalpy of the reactants as zero for a reference state of 25°C, 0.1 MPa, the enthalpy of the product at the reference state is then given by the net heat transfer, and this is termed as the enthalpy of formation of product

$$\bar{h}_f^0 = H_P = Q_{CV}, \tag{3.84}$$

where

$\bar{h}_f^0$ = enthalpy of formation at the reference state of 25°C and 0.1 Mpa.

The enthalpy of formation or the heat transfer quantity can be determined by experiment but are typically determined by statistical thermodynamics for different compounds. Enthalpy of formation for some of the common elements and compounds are given in Table C.6.

Enthalpy of formation of the components and compounds at any other states relative to the reference base states is estimated by adding the change in enthalpy between the given state and the reference state of 25°C, 0.1 MPa, as

$$\bar{h}_{T,P} = \left(\bar{h}_f^0\right)_{298°C,0.1\,MPa} + \overline{\Delta h}_{298°C,\,0.1\,MPa \to T,P}, \tag{3.85}$$

where

$$\overline{\Delta h}_{298°C,\,0.1\,MPa \to T,P} = \text{Change of enthalpy between a state and the reference state}$$

$$= \int_{298°C,\,0.1\,MPa}^{T,P} C_P(T)\,dT. \tag{3.86}$$

3.14 First Law for Reacting Systems

The first law of thermodynamics for a reacting system under SSSF process with negligible changes in kinetic energy and potential energy is given as

$$Q_{CV} + \sum_{R=1} n_R h_R = \sum_{P=1} n_P h_P + W_{CV} \tag{3.87a}$$

or

$$Q_{CV} + \sum_{R=1} n_R \left(\bar{h}_f^0 + \overline{\Delta h}\right)_R = \sum_{P=1} n_P \left(\bar{h}_f^0 + \overline{\Delta h}\right)_P + W_{CV}, \tag{3.87b}$$

where
$\overline{\Delta h} = \int_{25°C,\,0.1\,MPa \to T,P} C_P\, dT$ = Change in enthalpy between the state of the component and the reference state as described in Section 3.10.

3.15 Enthalpy of Combustion (h_{RP})

The *enthalpy of combustion* is the difference in total enthalpy of the products and total enthalpy of the reactants at a given temperature and pressure and given as

$$H_{RP} = H_P - H_R \tag{3.88a}$$

or

$$H_{RP} = \sum_{P=1} n_P \left(\bar{h}_f^0 + \overline{\Delta h}\right)_P - \sum_{R=1} n_R \left(\bar{h}_f^0 + \overline{\Delta h}\right)_R \tag{3.88b}$$

Separating terms for the enthalpy of formation and change of enthalpy, Equation 3.88b is written as

$$H_{RP} = \sum_{P=1} n_P \bar{h}_{f\,P}^0 - \sum_{R=1} n_R h_{f\,R}^0 + \sum_{P=1} n_P \overline{\Delta h_P} - \sum_{R=1} n_R \overline{\Delta h_R}. \tag{3.89}$$

Comparing Equation 3.82 and the first law equation for reacting system with no work, we can see that enthalpy of combustion is equivalent to the heat transfer rate across the control volume surrounding the combustion chamber.
Similarly, the internal energy of combustion is given as the difference in internal energy of the products and the internal energy of reactants

$$U_{RP} = U_P - U_R. \tag{3.90}$$

Writing in terms of enthalpy,

$$U_{RP} = \sum_{P=1} n_P \left(\bar{h}_f^0 + \overline{\Delta h} - p\bar{v}\right)_P - \sum_{R=1} n_R \left(\bar{h}_f^0 + \overline{\Delta h} - p\bar{v}\right)_R \tag{3.91}$$

Another frequently used term relating the enthalpy of combustion and internal energy of combustion is the *heating value,* which is equal to the *negative of enthalpy of combustion* for a *constant pressure process* and *negative of internal energy of combustion* for a *constant volume process.* The heating value for a combustion process also differs depending on whether the product contains liquid water or vapor water. The *higher heating value (hhv)* is referred to the combustion process with liquid water, $H_2O(l)$, in the products and *lower heating value* with vapor water, $H_2O(v)$, in the products. The enthalpy of combustion of some of the common fuels at standard temperature and pressure is given in Table C.8.

3.16 Temperature of Product of Combustion

One of the key variables in a combustion process is the temperature of the product of combustion at the exhaust. This temperature depends on the reacting components, heat of combustion of the reaction, any associated heat transfer and work done, and the amount of excess air used. The maximum temperature that a mixture can reach is for the case with no heat losses such as in an adiabatic process, involving no work done, and for using 100% theoretical air. Use of any additional excess air or heat loss and work results in a lower temperature of the mixture product. This maximum temperature is referred to as *adiabatic flame temperature.* The temperature of product of combustion or the adiabatic flame temperature is determined by the application of the first law of thermodynamics for the reacting system as

$$Q_{CV} + \sum_{R=1}^{N_R} n_R h_R = \sum_{P=1}^{N_P} n_P h_P. \tag{3.92a}$$

Considering Q_{CV} as negative for heat loss from the control volume, Equation 3.86a can be rewritten as

$$\sum_{P=1}^{N_P} n_P h_P = \sum_{R=1}^{N_R} n_R h_R - Q_{CV} \tag{3.92b}$$

or

$$\sum_{P=1}^{N_P} n_P \left(\bar{h}_f^0 + \overline{\Delta h}\right)_P = \sum_{R=1}^{N_R} n_R \left(\bar{h}_f^0 + \overline{\Delta h}\right)_R - Q_{CV}. \tag{3.93}$$

For adiabatic flame temperature, $Q_{CV} = 0$ and Equation 3.93 reduces to

$$\sum_{P=1}^{N_P} n_P \left(\bar{h}_f^0 + \overline{\Delta h}\right)_P = \sum_{R=1}^{N_R} n_R \left(\bar{h}_f^0 + \overline{\Delta h}\right)_R. \tag{3.94}$$

The right-hand side (RHS) of Equation 3.94 is estimated based on the temperature of the reactants. The mixture product temperature is then estimated using an iterative process until the total enthalpy of the products given by the left-hand side (LHS) of Equation 3.94 matches with the total enthalpy of the reactants given by the RHS. It can be noted here that adiabatic flame temperature represents the maximum possible temperature of the product of combustion. In reality, there will be positive heat loss from the combustion chamber to outside across the chamber wall geometry and will result in a lower temperature of the product given by Equation 3.94.

Example 3.5

Consider combustion of ethanol (C_2H_5OH) with 120% stoichiometric air in an SSSF process. The reactants enter the combustion process at 25°C and 0.1 MPa. Determine (a) the composition of the product of combustion and (b) the temperature of the product of combustion assuming no loss of heat from the combustion chamber and (c) the dew point temperature of the mixture.

Solution

The corresponding reaction equation for the combustion process is

$$C_2H_5OH + (1.2 \times 3)O_2 + (1.2 \times 3 \times 3.76)N_2$$

$$\rightarrow 2CO_2 + 3H_2O + (0.6) \times O_2 + (13.536)N_2$$

or

$$C_2H_5OH + (3.6)O_2 + (13.536)N_2$$

$$\rightarrow 2CO_2 + 3H_2O + (0.6) \times O_2 + (13.536)N_2.$$

The total number of moles in the product is

$$n_{\mathrm{p}} = n_{\mathrm{CO_2}} + n_{\mathrm{H_2O}} + n_{\mathrm{O_2}} + n_{\mathrm{N_2}} = 2 + 3 + 0.6 + 13.536 = 19.136$$

$$n_{\mathrm{p}} = 19.136$$

Molar composition of the product is given as

$$y_{\mathrm{CO_2}} = \frac{n_{\mathrm{CO_2}}}{\sum_{i=1}^{N} n_i} = \frac{2}{19.136} = 0.105$$

$$y_{H_2O} = \frac{n_{H_2O}}{\sum_{i=1}^{N} n_i} = \frac{3}{19.136} = 0.157$$

$$y_{O_2} = \frac{n_{O_2}}{\sum_{i=1}^{N} n_i} = \frac{0.6}{19.136} = 0.031$$

$$y_{N_2} = \frac{n_{H_2O}}{\sum_{i=1}^{N} n_i} = \frac{13.536}{19.136} = 0.707$$

Adiabatic flame temperature is the temperature of the product assuming no loss of heat energy from the combustion chamber and this is estimated based on the first law equation (Equation 3.87) for a reacting system or Equation 3.94 for adiabatic flame temperature as

$$\sum_{i=1}^{N_P} n_P \left(\bar{h}_f^0 + \overline{\Delta h}\right)_P = \sum_{R=1}^{N_R} n_R \left(\bar{h}_f^0 + \overline{\Delta h}\right)_R$$

Expanding this equation for the associated reaction process

$$n_{C_2H_5OH}\left(\bar{h}_f^0 + \overline{\Delta h}\right)_{C_2H_5OH} + n_{O_2}\left(\bar{h}_f^0 + \overline{\Delta h}\right)_{O_2} + n_{N_2}\left(\bar{h}_f^0 + \overline{\Delta h}\right)_{N_2}$$

$$= n_{CO_2}\left(\bar{h}_f^0 + \overline{\Delta h}\right)_{CO_2} + n_{H_2O}\left(\bar{h}_f^0 + \overline{\Delta h}\right)_{H_2O} + n_{O_2}\left(\bar{h}_f^0 + \overline{\Delta h}\right)_{O_2} + n_{N_2}\left(\bar{h}_f^0 + \overline{\Delta h}\right)_{N_2}$$

For reactants entering the combustion chamber at the reference temperature, $\Delta h_{C_2H_5OH} = 0, \Delta h_{O_2} = 0$, and $\Delta h_{N_2} = 0$. Also, considering the enthalpy of formations for oxygen and nitrogen as zero, we get

$$n_{C_2H_5OH}\left(\bar{h}_f^0\right)_{C_2H_5OH} = n_{CO_2}\left(\bar{h}_f^0 + \overline{\Delta h}\right)_{CO_2} + n_{H_2O}\left(\bar{h}_f^0 + \overline{\Delta h}\right)_{H_2O} + n_{O_2}\left(\overline{\Delta h}\right)_{O_2} + n_{N_2}\left(\overline{\Delta h}\right)_{N_2} \tag{3.95}$$

Enthalpy of formation for ethanol, carbon dioxide, and water are obtained from Table C.6 as

$$\left(\bar{h}_f^0\right)_{C_2H_5OH} = -235{,}000 \text{ kj/kmol}, \qquad \left(\bar{h}_f^0\right)_{CO_2} = -393{,}522 \text{ kj/kmol, and}$$

$$\left(\bar{h}_f^0\right)_{H_2O} = -241{,}826 \text{ kj/kmol}$$

Substituting in Equation 3.95, we have

$$1(-235,000)_{C_2H_5OH}$$
$$= 2\left(787,044 + \overline{\Delta h}\right)_{CO_2} + 3\left(725,478 + \overline{\Delta h}\right)_{H_2O} + 0.6\left(\overline{\Delta h}\right)_{O_2} + 13.536\left(\overline{\Delta h}\right)_{N_2}$$

or

$$2\left(\overline{\Delta h}\right)_{CO_2} + 3\left(\overline{\Delta h}\right)_{H_2O} + 0.6\left(\overline{\Delta h}\right)_{O_2} + 13.536\left(\overline{\Delta h}\right)_{N_2} = 1,277,522 \quad (3.96)$$

Temperature of the product can be estimated based on a trial-and-error basis. A correct guess temperature will satisfy Equation 3.96:

Trial 1: Guess $T_p = 1000$ K

Corresponding enthalpy values from Table C.7:

$$\Delta\bar{h}_{CO_2} = 33,397 \text{ kj/kmol}, \quad \Delta\bar{h}_{H_2O} = 26,000 \text{ kj/kmol},$$
$$\Delta\bar{h}_{O_2} = 22,703 \text{ kj/kmol}, \quad \Delta\bar{h}_{N_2} = 21,463 \text{ kj/kmol}$$

Substituting, we get the left-hand side of the equations as

LHS = 448,939 kj/kmol.

Trial 2: $T_p = 1500$ K

Corresponding enthalpy values from Table C.7:

$$\Delta\bar{h}_{CO_2} = 61,705 \text{ kj/kmol}, \quad \Delta\bar{h}_{H_2O} = 48,149 \text{ kj/kmol},$$
$$\Delta\bar{h}_{O_2} = 40,600 \text{ kj/kmol}, \quad \Delta\bar{h}_{N_2} = 38,405 \text{ kj/kmol}$$

Substituting, we get the left-hand side of the equations as

LHS = 448,939 kj/kmol.

Trial 3: $T_p = 2000$ K

Corresponding enthalpy values from Table C.7:

$$\Delta\bar{h}_{CO_2} = 91,439 \text{ kj/kmol}, \quad \Delta\bar{h}_{H_2O} = 72,788 \text{ kj/kmol},$$
$$\Delta\bar{h}_{O_2} = 59,176 \text{ kj/kmol}, \quad \Delta\bar{h}_{N_2} = 56,137 \text{ kj/kmol}$$

Substituting, we get the left-hand side of the equations as

LHS = 1,196,618 kj/kmol.

Trial 4: $T_p = 2400$ K

Corresponding enthalpy values from Table C.7:

$$\Delta\bar{h}_{CO_2} = 115{,}779 \text{ kj/kmol}, \quad \Delta\bar{h}_{H_2O} = 93{,}741 \text{ kj/kmol},$$

$$\Delta\bar{h}_{O_2} = 74{,}453 \text{ kj/kmol}, \quad \Delta\bar{h}_{N_2} = 70{,}640 \text{ kj/kmol}$$

Substituting, the left-hand side of the equations is computed as

$$\text{LHS} = 1{,}513{,}636 \text{ kj/kmol}.$$

Since the LHS value has exceeded the RHS value of Equation 3.88 as trial temperature is changed from 2000 K to 2400 K in trials 4 to 5, the correct temperature can be approximated by a linear interpolation of the last two trial values as follows:

$$T_p = 2000 + \frac{1{,}277{,}522 - 1{,}196{,}618}{1{,}513{,}636 - 1{,}196{,}618} \times (2400 - 2000)$$

or

$$T_p = 2102 \text{ K or } 1829°\text{C}.$$

3.17 Absolute Entropy $\left(\bar{s}_f^0\right)$

The absolute entropy of substance at any given temperature and pressure is given as

$$\bar{s}_{T,P} = \bar{s}_T^0 - \bar{R}\ln\frac{P}{P^0} + \left(\bar{s}_{T,P} - \bar{s}_{T,P}^*\right). \tag{3.97}$$

The *first term* on the RHS, $\bar{s}_T^0$, is the absolute entropy of substance at any temperature and it is computed in reference to the base reference temperature of T_0 as

$$\bar{s}_T^0 = \bar{s}_f^0 + \int_{T_0}^{T} \frac{\bar{C}_P}{T} dT, \tag{3.98}$$

where $\bar{s}_f^0$ is the absolute entropy of a substance at a reference temperature of $T_0 = 25°\text{C}$ and a reference pressure of $P_0 = 0.1$ MPa. This absolute entropy at the reference temperature is measured from reference entropy values at a base temperature of absolute zero by experimental methods and statistical thermodynamics. According to the third law of thermodynamics, the entropy of a perfect substance like a perfect crystal is zero at the absolute zero temperature. Any real substance with imperfect crystalline structure that is associated with a certain degree of randomness has a finite entropy value at the absolute temperature. The entropy of some common gases like oxygen, nitrogen, hydrogen, carbon dioxide, carbon monoxide, and nitrogen oxide is zero at the absolute zero temperature. The absolute entropy values for some of the common substances are given in Table C.6.

The *second term* on the RHS is the ideal gas term taking into account the difference in pressure from the reference pressure of $P_0 = 0.1$ MPa to pressure, P.

The *third term* on the RHS is a contribution due to the deviation of ideal gas behavior from real gas.

The mixture entropy is computed on the basis of the absolute entropy of the component gases as

$$\bar{s}_{\text{mix}} = \sum_{i=1} y_i \bar{s}_i, \tag{3.99}$$

where the absolute entropy of each ideal gas component is given by

$$\bar{s}_i = \left(\bar{s}_{\text{T}}^0 + \int_{T_0}^{T} \frac{\bar{C}_{\text{P}}}{T} \right)_i - \bar{R} \ln \frac{y_i P}{P^0}. \tag{3.100}$$

The *change in entropy of a reacting system* is given as

$$\Delta S = \sum_{P=1}^{N_{\text{P}}} n_{\text{P}} \bar{s}_{\text{P}} - \sum_{R=1}^{N_{\text{R}}} n_{\text{R}} \bar{s}_{\text{R}}. \tag{3.101}$$

3.18 Gibbs Function of Formation $\left(\bar{g}_{\text{f}}^0\right)$

The Gibbs function of formation is defined in the same manner as in the case of enthalpy of formation. The enthalpy of formations for all basic substances in its most stable form is assumed to be zero at the reference state of 25°C, 0.1 MPa. Gibbs function formations for those basic element substances at any other states and the Gibbs function for all other substances that are formed from the basic element substances are then estimated on the basis of the element substances and their variation from the reference state.

The Gibbs free energy of formation per unit mass is given as

$$g_f = h_f - Ts_f \text{ on mass basis} \tag{3.102a}$$

and

$$\bar{g}_f = \bar{h}_f - T\bar{s}_f \text{ on molar basis.} \tag{3.102b}$$

The Gibbs free energy of formation ΔG_f or $\bar{g}_f^0$, which is defined as the difference in the Gibbs free energy of the product and the reactants, is given as

$$\Delta G_f = \text{Gibbs free energy of formation} = (G_f)_{\text{products}} - (G_f)_{\text{reactants}} \tag{3.103a}$$

$$\begin{aligned}\Delta G_f &= \sum_P G_f - \sum_R G_f \\ &= \sum_P n_p \bar{g}_{fp} - \sum_R n_R \bar{g}_{fR} \\ \Delta G_f &= \sum_P n_P \left(\bar{g}_f^0 + \overline{\Delta g}\right)_P - \sum_R n_R \left(\bar{g}_f^0 + \overline{\Delta g}\right)_R\end{aligned} \tag{3.103b}$$

Note that if both the reactants and the product are at a reference state of 25°C and 0.1 MPa pressure, then the sensible components Δg vanishes and the Gibbs free energy of formation is given as

$$\Delta G_f = \sum_P n_P \left(\bar{g}_f^0\right)_P - \sum_R n_R \left(\bar{g}_f^0\right)_R. \tag{3.104}$$

The change in Gibbs function values can also be determined from the enthalpy of formation and the absolute entropy values. Starting from the definition of Gibbs function as $G = H - TS$, we can write the change in Gibbs function of the reacting system as follows:

$$\Delta G = \Delta h - T\Delta s, \tag{3.105}$$

where Δh and Δs are the change in enthalpy and entropy of the combustion as given by Equations 3.88 and 3.101, respectively.

Equation 3.105 can be further expressed in terms of reactant and product components as follows:

$$\Delta G = \sum_{P=1}^{N_P} n_P \bar{g}_P - \sum_{R=1}^{N_R} n_R \bar{g}_R \tag{3.106a}$$

or

$$\Delta G = \sum_{P=1}^{N_P} n_P \bar{h}_f^0 - \sum_{R=1}^{N_R} n_R \bar{h}_f^0 - T\left(\sum_{P=1}^{N_P} n_P \bar{s}_f^0 - \sum_{R=1}^{N_R} n_R \bar{s}_f^0 \right). \tag{3.106b}$$

Example 3.6

Consider a process for combustion of methanol (CH_3OH) with oxygen. Determine the change in Gibbs energy for this reaction at a standard state of 25°C and 0.1 MPa.

Solution

Combustion reaction of methanol with oxygen is given by

$$2CH_3OH + 3O_2 \rightarrow 2CO_2 + 4H_2O.$$

The change in Gibbs free energy is given by

$$\Delta G = \sum_{P} n_p \bar{g}_{fp} - \sum_{R} n_R \bar{g}_{fR}$$

$$\Delta G = \sum_{P} n_p \left(\bar{g}_f^0 + \overline{\Delta g}\right)_P - \sum_{R} n_R \left(\bar{g}_f^0 + \overline{\Delta g}\right)_R.$$

Setting $\overline{\Delta g_R} = 0$ and $\overline{\Delta g_P} = 0$ for operation at the standard state, we have

$$\Delta G = \sum_{P} n_p \left(\bar{g}_f^0\right)_P - \sum_{R} n_R \left(\bar{g}_f^0\right)_R$$

$$\Delta G = \left[n_{CO_2} \left(\bar{g}_f^0\right)_{CO_2} + n_{H_2O} \left(\bar{g}_f^0\right)_{H_2O} \right] - \left[n_{CH_3OH} \left(\bar{g}_f^0\right)_{CH_3OH} - n_{O_2} \left(\bar{g}_f^0\right)_{O_2} \right]$$

$$\text{From Table C.6, } (\bar{g}_f)_{CH_3OH} = -162,551\frac{kJ}{kmol}, (\bar{g}_f)_{O_2} = 0\frac{kJ}{kmol}$$

$$(\bar{g}_f)_{CO_2} = -394,389\frac{kJ}{kmol} \text{ and } (\bar{g}_f)_{H_2O} = -228,582\frac{kJ}{kmol}$$

$$\Delta G = [2(-394,389) + 4(-228,582)] - [2(-162,551) - 3(0)]$$

$$\Delta G = -1,378.004\frac{kJ}{kmol}.$$

PROBLEMS

1. Oxygen gas stream is heated from 400 K to 900 K with pressure dropping from 200 kPa to 150 kPa. Calculate the change in enthalpy, change in entropy, and change in Gibbs function of the gas stream based on ideal gas relations.
2. The volume composition of a gas mixture is given as H_2: 30%, CO_2: 20%, N_2: 48% and H_2O: 2%. Determine (a) the mass composition of the mixture, (b) the gas constant of the mixture, (c) the constant pressure specific heat of the mixture, and (d) how much heat is needed to raise the temperature of this mixture heat transfer rate to heat the mixture from 25°C to 100°C.
3. Consider combustion of ethanol (C_2H_5OH) with 120% stoichiometric air in a SSSF process. The reactants enter the combustion process at 100°C and 0.1 MPa. Determine (a) the composition of the product of combustion and (b) the temperature of the product of combustion assuming no loss of heat from the combustion chamber and (c) the dew point temperature of the mixture.
4. Methanol (CH_3OH) is burned with 150% stoichiometric air. Methanol enters the combustion chamber at the reference temperature and pressure and air enters at 50°C and atmospheric pressure. Determine (a) the mass composition of the mixture product, (b) the temperature of the mixture product assuming adiabatic and SSSF process.
5. A mixture of 80% liquid octane and 20% ethanol by volume is burned with 100% theoretical air with fuel and air entering at the reference temperature and pressure. Determine the product composition and the heating value of this fuel mixture.
6. Natural gas with methane as the primary component is burned with 250% stoichiometric air with both gases entering at the reference temperature and pressure. What is the adiabatic flame temperature of the product of combustion?

7. In a catalytic reactor, a certain bio-mass stack is converted into a bio-fuel gas mixture that contains 50% methane, 45% carbon dioxide, and 5% hydrogen by volume. Determine the lower heating value of this bio-fuel gas mixture.
8. A gaseous fuel is composed primarily of methane (CH_4) and propane (C_3H_{10}). Calculate the viscosity and thermal conductivity of the gas mixture at 300 K and 2 atm pressure.
9. Compute the viscosities of hydrogen and oxygen at 350 K and 1000 K.
10. Compute the viscosity of a gas mixture composed of 64% N_2, 16% O_2 and 20% H_2O at 353 K and 1 atm.

References

Bird, R. B., W. E. Stewart and E. N. Lightfoot. *Transport Phenomena*. Wiley, New York, 1960.

Cengel, Y. A. and M. A. Boles. *Thermodynamics*, 5th Edition. McGraw Hill, 2006.

Mills, A. F. *Mass Transfer*. Prentice Hall, New Jersey, 2001.

Moran, M. J. and H. N. Shapiro. *Fundamentals of Engineering Thermodynamics*, 6th Edition, Wiley, New York, 2008.

Sonntag, R. E., C. Borgnake and G. J. van Wylen. *Fundamentals of Thermodynamics*, 6th Edition, Wiley, New York, 2003.

Wilke, C. R. A viscosity equation for gas mixtures. *Journal of Chemical Physics* 18: 517–519, 1950.

4

Thermodynamics of Fuel Cells

Thermodynamics and electrochemical kinetics describe the energy conversion process, performance, and ratings of a fuel cell. As we have described in Chapter 2, the kinetics associated with the electrochemical reactions are critical in describing the real performance of an electrochemical fuel cell. While the rate of a single-step electrochemical reaction is described by the electrode kinetics, thermodynamics, however, establishes the maximum theoretical limit for the performance of an electrochemical fuel cell like in any other energy conversion devices including the heat engine. In an open circuit fuel cell, the chemical equilibrium condition at the electrodes establishes the maximum theoretical voltage defined by the thermodynamics. But the open circuit fuel cell would not produce any power because there is no net flow of electrons between the electrodes. As current starts flowing through the external load circuit and electrical power is produced, operating voltage drops owing to number voltage losses caused by the irreversibilities associated with the electrochemical reactions, charge transport, and reactant species transport and depletion.

In this chapter, the thermodynamic principles of fuel cells are presented to estimate the reversible open circuit voltage. Additionally, the procedure for estimating the required reactant gas flow rates, heat generation, and water production rates based on thermodynamic principles is presented.

4.1 Conventional Power Generation—Heat Engine

Conventional thermal power generations are based on heat engine principles, which were derived from the Kelvin–Planck statement of the second law of thermodynamics as demonstrated in Figure 4.1. A major objective of a power-generating heat engine is to convert the chemical energy content of a fuel into electrical energy. Energy in the form of heat (Q_H) is added to the heat engine from a high-temperature heat source (T_H), and a fraction of this heat is converted into work (W) and the rest is rejected as heat energy (Q_L) to a low-temperature heat sink (T_L).

A major requirement for a heat engine is that it operates between two temperature limits: a high-temperature heat source from which heat is added and a low-temperature heat sink at which a fraction of the heat is rejected. This requirement limits the maximum efficiency of a heat engine to less than 100%.

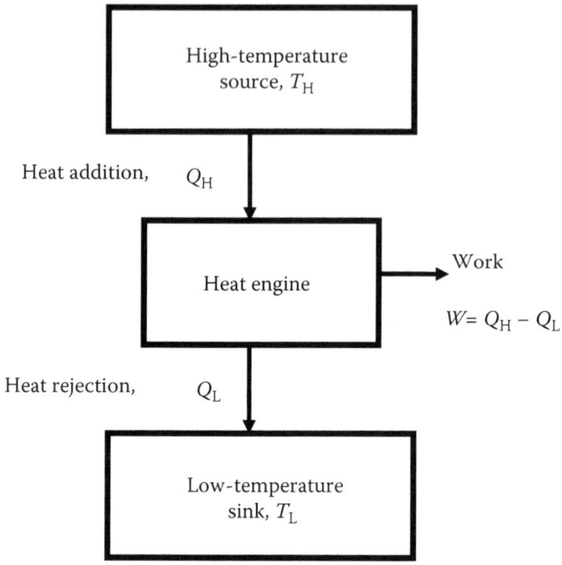

FIGURE 4.1
Energy conversion process in heat engines.

The maximum possible performance of a heat engine set by that given by a reversible heat engine operating on a Carnot cycle, which involves four reversible processes: (i) reversible isothermal heat addition, Q_H, (ii) reversible adiabatic expansion (work), W, (iii) reversible isothermal heat rejection, Q_L, and (iv) reversible adiabatic compression. Thermal efficiency of the heat engine is given by

$$\eta_{th} = \frac{W}{Q_H} = \frac{Q_H - Q_L}{Q_H} = 1 - \frac{Q_L}{Q_H} \tag{4.1}$$

For a reversible heat engine operating on a Carnot cycle, the ratio of reversible isothermal heat addition, Q_H, at the high-temperature source and the reversible isothermal heat rejection, Q_L, at the lower-temperature sink is given as

$$\frac{Q_H}{Q_L} = \frac{T_H}{T_L} \tag{4.2}$$

and the maximum thermal efficiency of a reversible heat engine is given as

$$\eta_{th} = 1 - \frac{T_L}{T_H}, \tag{4.3}$$

where T_H and T_L are temperature in absolute temperature scale.

A real power-generating heat engine differs significantly from the reversible heat engine. For example, a thermal steam power plant operates on a vapor power Rankine cycle, which involves heat addition by evaporation or boiling of the working fluid in a boiler, work output by expansion of high-pressure and high-temperature vapor in a turbine, and heat rejection by condensation of the vapor in a condenser. Figure 4.2 shows a typical thermal power-generating plant that operates on a Rankine vapor power cycle. A vapor power cycle uses a working fluid that is alternately vaporized and condensed. The fuel at first is burnt in a combustion process in a furnace or in a combustion chamber to convert the chemical energy contained in it into heat energy. The heat energy is then transferred from the product of combustion to the working fluid in the boiler that represents the high-temperature heat source of the heat engine cycle. The thermal heat energy is then converted into useful mechanical energy through the expansion of the working fluid in a rotating turbine or in a piston-cylinder arrangement. Finally, the mechanical energy is converted into electrical energy in a generator unit. A major portion of the energy is rejected through the flue gas exhaust through the stack and by heat rejection in the condenser to a coolant medium, which represents the low-temperature heat sink of the heat engine cycle.

Another category of heat engine is based on a gas-powered system that includes *internal combustion engines* of the *spark ignition* and *compression–ignition types*, and *gas turbine*. All these systems are referred to as the internal combustion engines as combustion takes place inside the system in contrast to vapor power systems where combustion takes place outside the system. In all these systems, there is a change in the composition of the working fluid

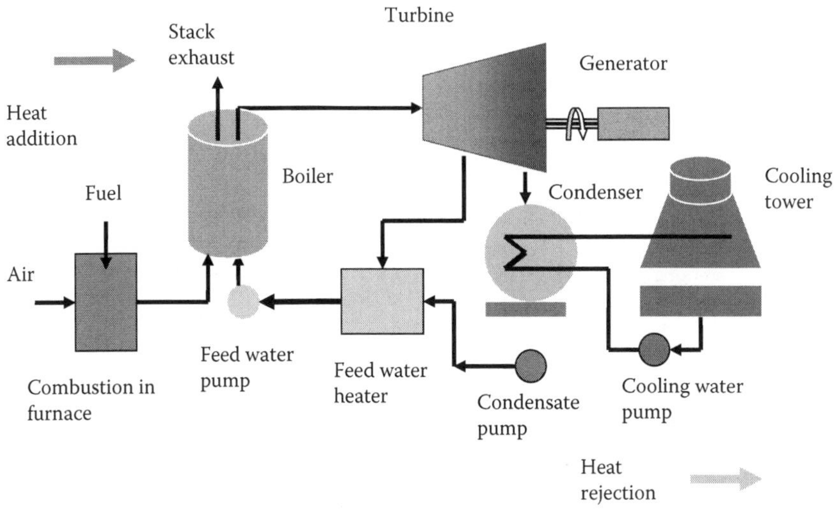

FIGURE 4.2
Rankine vapor cycle external combustion heat engine.

from air to the product gas mixture after the combustion. A typical internal combustion heat engine cycle involves four processes as depicted in Figure 4.3. These processes are (i) air intake and compression, (ii) combustion that represents the high-temperature heat addition, (iii) expansion and work, and (iv) heat rejection in an exhaust process to the environment that represents the low-temperature heat sink.

A simple gas turbine operating on a Brayton cycle is shown in Figure 4.4. This also involves four processes: (i) air intake and compression in a rotary air compressor, (ii) combustion of fuel with the incoming compressed air in a combustion chamber that represents the high-temperature heat addition, (iii) expansion and work in a rotary gas turbine, and (iv) heat rejection in an exhaust process to the environment. The rotary compressor and the turbine are connected by a common shaft so that a fraction of the work produced by the gas turbine is supplied to the air compressor, and the rest results in the net work output.

The thermal efficiency of a heat engine is defined as the ratio of net work output to the total heat energy added at the high-temperature source and expressed as

$$\eta_{th} = \frac{W_{net}}{Q_H} = \frac{Q_H - Q_L}{Q_H} = 1 - \frac{Q_L}{Q_H}. \tag{4.4}$$

The amount of heat addition (Q_H), heat rejection (Q_L), and turbine work output (W_{net}) in a heat engine cycle is estimated by applying the first law of thermodynamics. The thermal efficiency of a real heat engine cycle is less than the reversible Carnot cycle efficiency given by Equation 4.3.

In general, the lower-temperature reservoir in a heat engine power cycle is limited by the ambient condition like the ambient air temperature in an air-cooled condenser or the water temperature in a water-cooled condenser.

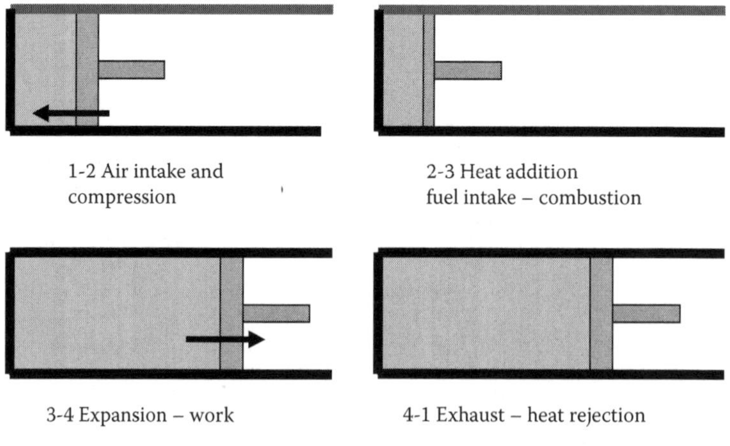

FIGURE 4.3
Internal combustion cylinder-piston heat engine.

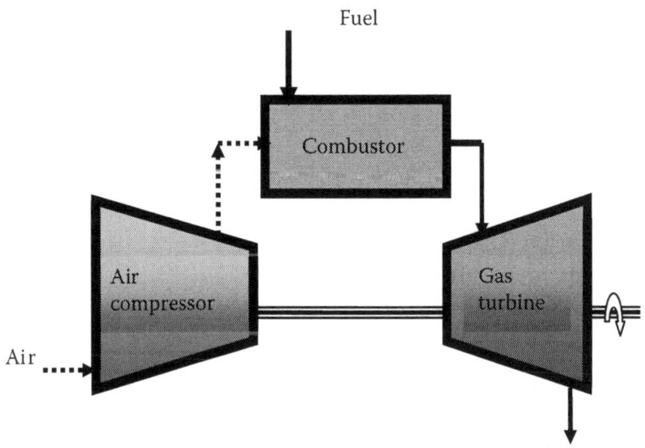

FIGURE 4.4
Gas turbine power generation system.

The high temperature is limited by the condition of the vapor in the boiler in a vapor power cycle or the temperature of the product of combustion in the internal combustion engine.

4.2 Energy Conversion in Fuel Cells

A fuel cell is a machine that takes fuels like hydrogen and oxygen as inputs and converts the chemical energy content of the fuel directly into electrical energy through electrochemical reactions, producing water and heat as the by-products as shown in Figure 4.5.

Direct hydrogen and oxygen contact and combustion are avoided in a fuel cell. Hydrogen and oxygen are separated by an *electrolyte medium.* A direct hydrogen–oxygen combustion reaction is replaced by two electrochemical half reactions at two *electrode–electrolyte interfaces.* The electrochemical reactions at the anode and the cathode sides take place simultaneously, producing electricity, and water and heat as the only by-products when hydrogen is used as the fuel. Heat is produced due to the irreversibilities associated with

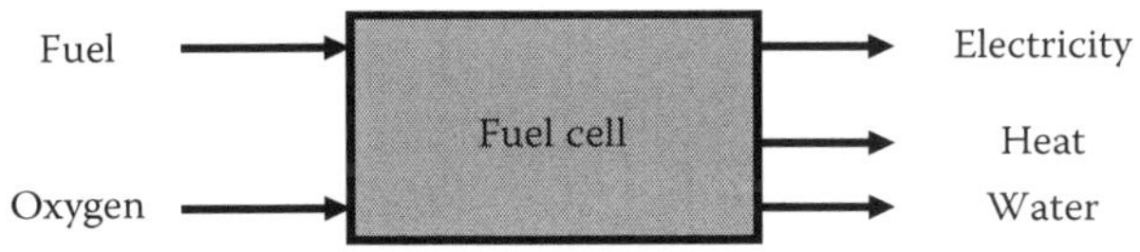

FIGURE 4.5
Energy conversion process in a fuel cell.

the electrochemical reactions at the electrodes and conducting resistances of electrodes and electrolytes to electrons and ions, respectively.

There are different types of fuel cells classified by the different electrolyte materials used as described in Chapter 1. The anode and cathode reaction characteristics are therefore different in each fuel cell type. Table 4.1 shows a

TABLE 4.1

Common Fuel Cell Types

Fuel Cell Type	Electrolyte Type	Charge Carrier	Electrochemical Reaction
Alkaline FC (AFC)	Sodium hydroxide, NaOH, or potassium hydroxide (KOH)	Hydroxyl ion OH^+	Anode: $H_2 + 2OH^- \rightarrow 2H_2O + 2e^-$ Cathode: $\frac{1}{2}O_2 + 2e^- + H_2O \rightarrow 2OH^-$ Overall: $H_2 + \frac{1}{2}O_2 \rightarrow H_2O$
Phosphoric acid FC (PAFC)	Phosphoric acid (H_3PO_4)	Proton or hydrogen ion H^+	Anode: $H_2 \rightarrow 2H^+ + 2e^-$ Cathode: $\frac{1}{2}O_2 + 2H^+ + 2e^- \rightarrow H_2O$ Overall: $H_2 + \frac{1}{2}O_2 \rightarrow H_2O$
Proton exchange membrane FC (PEMFC)	Solid polymer	Proton or hydrogen ion H^+	Anode: $H_2 \rightarrow 2H^+ + 2e^-$ Cathode: $\frac{1}{2}O_2 + 2H^+ + +2e^- \rightarrow H_2O$ Overall: $H_2 + \frac{1}{2}O_2 \rightarrow H_2O$
Direct methanol fuel cell (DMFC)	Solid polymer	Proton or hydrogen ion H^+	Anode: $CH_3OH + H_2O \rightarrow CO_2 + 6H^+ + 6e^-$ Cathode: $\frac{3}{2}O_2 + 6H^+ + 6e^- \rightarrow 3H_2O$ Overall: $CH_3OH + \frac{3}{2}O_2 \rightarrow 2H_2O + CO_2$
Molten carbonate FC (MCFC)	Lithium carbonate Li_2CO_3 or potassium carbonate K_2CO_3	Carbonate ion CO_3^{2-}	Anode: $H_2 + CO_3^{2-} \rightarrow CO_2 + H_2O + 2e^-$ Cathode: $\frac{1}{2}O_2 + CO_2 + 2e^- \rightarrow CO_3^{2-}$ Overall: $H_2 + 1/2O_2 \rightarrow H_2O$
Solid oxide FC (SOFC)	Yttria stabilized zirconia (YSZ)	Oxygen ion O^{2-}	Anode: $H_2 + O^{2-} \rightarrow H_2O + 2e^-$ Cathode: $\frac{1}{2}O_2 + 2e^- \rightarrow O^{2-}$ Overall: $H_2 + \frac{1}{2}O_2 \rightarrow H_2O$

list of common fuel cell types along with the associated electrolyte type, the charge carrier ion, and the electrochemical reactions.

A detailed description of the energy conversion processes through electrochemical reactions in a fuel cell is given here by considering a hydrogen–oxygen *polymer electrolyte membrane fuel cell (PEMFC)* and a *solid oxide fuel cell (SOFC)* with basic cell components as shown in Figures 4.6 and 4.7, respectively. Similar descriptions for other types of fuel cells will be presented in Chapter 9.

In a PEMFC, a proton-conducting polymer membrane electrolyte is sandwiched between two porous electrically conducting electrodes.

The hydrogen gas enters the anode side, transports through the porous anode electrode, and undergoes an electrochemical reaction at the electrode–membrane interface in the presence of a catalyst layer, and ionized releasing *electrons* (e^-) and hydrogen ions (H^+) or *protons*. The proton transports through the membrane toward the cathode interface. The electrons move through the electrically conductive anode electrode toward external load or the electrical circuit, performing the electrical work and producing electrical power. The oxygen gas enters the cathode side, transports though the porous cathode electrode, and reacts electrochemically with the returning electrons (e^-) from the anode electrode and proton (H^+) from the electrolyte to form water at the cathode electrode–membrane interface.

The electrochemical reactions at the anode and cathode sides take place simultaneously, producing electricity, water, and some heat owing to the irreversibilities associated with the electrode reactions, and charge conducting

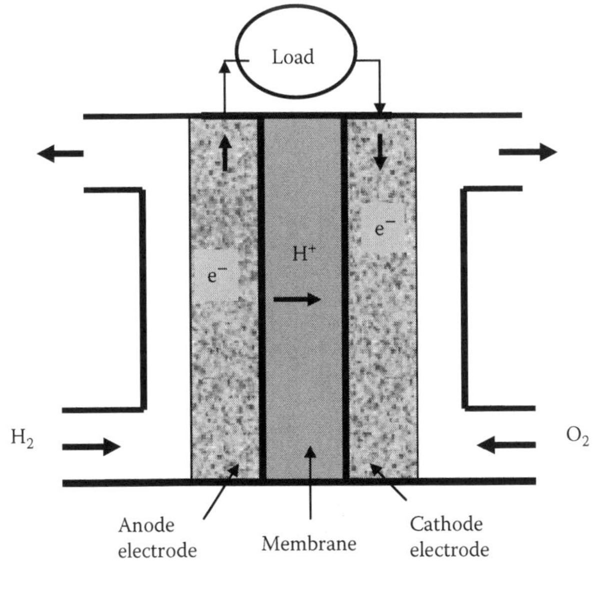

FIGURE 4.6
Electrochemical reactions and energy conversion process in a PEMFC.

resistances of electrodes and proton membrane. It is also required that the electrolyte material allows only the proton migration and not the electrons or hydrogen. Any migration of electrons through the electrolyte will cause reduced electron flow through the external circuit, causing voltage loss, and any fuel crossover through the membrane will also cause reduced performance.

The reactions in a PEMFC are summarized as follows:

$$\begin{aligned}
&\text{Anode reaction:} && H_2 \rightarrow 2H^+ + 2e^- && \text{(a)}\\
&\text{Cathode reaction:} && \frac{1}{2}O_2 + 2H^+ + 2e^- \rightarrow H_2O && \text{(b)}\\
&\text{Overall reaction:} && H_2 + \frac{1}{2}O_2 \rightarrow H_2O && \text{(c)}
\end{aligned} \qquad (4.5)$$

The basic components and the overall reaction are similar in an SOFC with the exception of the electrochemical reactions at the anode and cathode electrodes as demonstrated in Figure 4.7.

At the cathode electrode, the reduction of oxygen takes place with the formation of a negatively charged *oxygen ion*. The oxygen ion transports through the solid oxide ion conducting membrane electrolyte toward the anode where it combines with the hydrogen gas producing water and electrons that travels to the cathode side through the external electrical circuit.

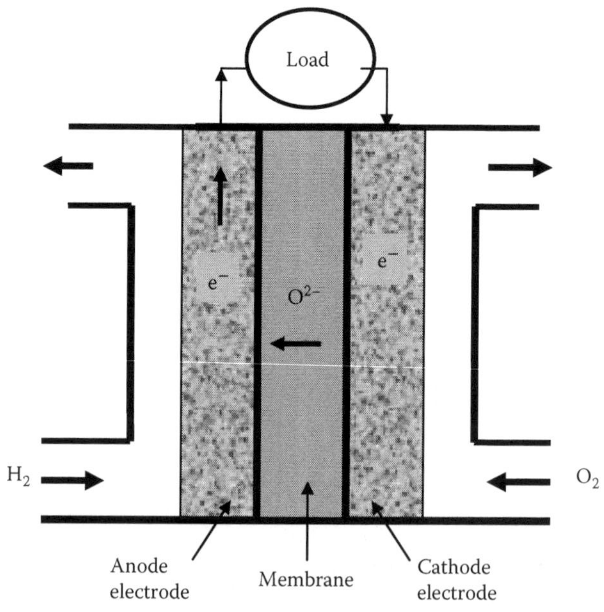

FIGURE 4.7
Electrochemical reactions and energy conversion process in an SOFC.

The reactions in an SOFC are summarized as follows:

$$\begin{aligned} &\text{Anode reaction:} && H_2 + O^{2-} \rightarrow H_2O + 2e^- && \text{(a)} \\ &\text{Cathode reaction:} && \frac{1}{2}O_2 + 2e^- \rightarrow O^{2-} && \text{(b)} \\ &\text{Overall reaction:} && H_2 + \frac{1}{2}O_2 \rightarrow H_2O && \text{(c)} \end{aligned} \tag{4.6}$$

In these reactions, we can say 1 mol of hydrogen combines with half a mole of oxygen and produces 1 mol of water. Also, two electrons are transmitted through the external circuit performing the electrical work. The chemical energy content of hydrogen is converted into electrical work or energy by oxidizing hydrogen with oxygen and producing water.

If carbon monoxide is also supplied as one of the species of a reformed fuel gas mixture to the anode of an SOFC, then the reactions are given as follows:

$$\begin{aligned} &\text{Anode reaction:} && CO + O^{2-} \rightarrow CO_2 + 2e^- && \text{(a)} \\ &\text{Cathode reaction:} && \frac{1}{2}O_2 + 2e^- \rightarrow O^{2-} && \text{(b)} \\ &\text{Overall reaction:} && CO + \frac{1}{2}O_2 \rightarrow CO_2 && \text{(c)} \end{aligned} \tag{4.7}$$

In this reaction, 1 mol of carbon monoxide combines with half a mole of oxygen and produces 1 mol of carbon dioxide. At the cathode, oxygen is reduced to oxide ions by combining with incoming electrons. At the anode, carbon monoxide is oxidized to carbon dioxide and releases two electrons. The electrons are transmitted through the external circuit performing the electrical work. The chemical energy content of carbon monoxide is converted into electrical work or energy after being oxidized, producing water and heat.

When a reformed fuel gas stream consisting of a mixture of hydrogen and carbon monoxide is used as a fuel to the SOFC, the reactions can be written in a general form as

Anode reaction:

$$\begin{aligned} &(x_{H_2})H_2 + (x_{CO})CO + O^{2-} \\ &\rightarrow (x_{H_2})H_2O + (x_{CO})CO_2 + 2e^- \qquad \text{(a)} \end{aligned}$$

Cathode reaction:

$$\frac{1}{2}O_2 + 2e^- \rightarrow O^{2-} \qquad \text{(b)}$$

Overall reaction:

$$(x_{H_2})H_2 + (x_{CO})CO + \frac{1}{2}O_2 \rightarrow (x_{H_2O})H_2O + (x_{CO_2})CO_2 \quad \text{(c)} \tag{4.8}$$

where x_{H_2} and x_{CO_2} are the mole fractions of hydrogen and carbon monoxide in the fuel gas mixture.

As we can see that in the operation of a fuel cell, the electrical work is done by moving electrons through an external circuit. The maximum possible electrical work is equivalent to the change in Gibbs free energy ($\bar{g}_f$). This is the energy available to do external work, neglecting energy change by temperature and change in entropy. Considering Equation 3.19b for the Gibbs energy, we can derive the expression for change in Gibbs energy as

$$dG = \delta Q - \delta W + pdV + Vdp - Tds - sdT \tag{4.9}$$

Using $\delta Q = Tds$ for a reversible process and assuming constant pressure and temperature processes, Equation 4.9 reduces to

$$dG = -\delta W + pdV. \tag{4.10}$$

Now, for a system that involves only electrical work, $\delta W = \delta W_e$, and no additional work because of change in volume as in a simple compressible substance, we get

$$dG = -\delta W_e. \tag{4.11a}$$

Using molar change in Gibbs free energy, the electrical work for the process is written as

$$W_e = -\Delta \bar{G}, \tag{4.11b}$$

where the symbol Δ represents the change in Gibbs free energy of the reactants and the products in the process.

4.2.1 Electrical Work in Fuel Cells

In an electrochemical cell, the electrical work is done owing to the flow of electrons through an electrical load circuit under the effect of a potential difference between the two electrodes or the voltage of the cell. When N_e electrical charges flow through an external load circuit owing to the potential difference E, the *fuel cell electrical work* is given as

$$W_{cell} = \text{Charge} \times \text{Cell Voltage}$$

or

$$W_{cell} = N_e E \tag{4.12}$$

If n_e is the number of electrons per molecule of reactant transferred, then $n_e N_a$ is the total number of electrons that pass through the external circuit, where N_a is the *Avogadro number*, defined as the number of molecules per mole of substance and is a constant value, $N_a = 6.022 \times 10^{23}$. If e is the charge of one electron, then the total charge that flows through the circuit is given by

$$N_e = n_e N_a e = n_e F, \text{ coulombs (C)} \tag{4.13}$$

where $F = N_a e$ is *Faraday's constant* defined as the charge carried by a mole of electrons. Considering the value of the charge as $e = 1.602 \times 10^{-19}$ C, *Faraday's constant* is given as

$$F = N_a e = 6.022 \times 10^{23} \times 1.602 \times 10^{-19}$$

or

$$F = 96{,}485 \text{ C/mol.}$$

Substituting Equation 4.13 into Equation 4.12, the electrical work is given as

$$W_{cell} = n_e F\, E \text{ joules.} \tag{4.14}$$

Equating Equations 4.11 and 4.14, we get

$$W_{cell} = n_e F\, E = -\Delta \overline{g_f} \tag{4.15}$$

where $\Delta \overline{g_f}$ represents change in Gibbs energy per unit mass. *Cell voltage* is given as

$$E = \frac{-\Delta \overline{g_f}}{n_e F}. \tag{4.16}$$

4.2.2 Reversible Cell Voltage

If the total chemical energy content of the fuel, that is, the enthalpy of formation or the heating value, can be converted into electrical energy, then the maximum possible voltage is given as

$$E_{max} = \frac{-\Delta h_f}{n_e F}. \tag{4.17}$$

For a hydrogen fuel cell at the standard conditions, this is estimated as

$$E_{\text{max}} = \frac{-\Delta h_{\text{f}}}{2F}. \tag{4.18}$$

This is the voltage that would be achievable in a 100% efficient fuel cell. However, as we have mentioned, it is the Gibbs free energy that is available for conversion into electrical work in a fuel cell.

The *maximum electrical work* in a fuel cell is obtained when all reactions are reversible with no losses and is equal to the change in the Gibbs free energy of formation *at the reference standard temperature and pressure (STP)*, and it is given as

$$W_{\text{cell,max}} = \Delta \bar{g}_{\text{f}}^{0}. \tag{4.19}$$

Equating Equations 4.15 and 4.19,

$$-n_{\text{e}}FE_{\text{rev}}^{0} = \Delta \bar{g}_{\text{f}}^{0} \tag{4.20}$$

and the reversible open circuit voltage of the cell or the electromotive force (EMF) is given as

$$E_{\text{rev}}^{0} = \frac{-\Delta \bar{g}_{\text{f}}^{0}}{n_{\text{e}}F}. \tag{4.21}$$

In reality, the cell operating voltage would be lower than the value given by Equation 4.21 because of voltage losses caused by a number cell irreversibilities associated with the electrochemical reactions and transport processes that will be discussed in later chapters.

4.2.3 Cell Power

Electrical power, produced as current, is drawn from fuel cell against a constant cell voltage. The electrical power produced by the cell is given as the product of the cell voltage and the total current drawn from the cell through the external circuit and expressed as

$$P = E\,I, \tag{4.22}$$

where I is the total current flowing through the electrical circuit, defined as the rate of charge (coulomb) transfer per unit time and expressed as

$$I = n_{\text{e}}F\dot{n}, \tag{4.23a}$$

where $\dot{n}$ is the consumption rate of reactant (mol/s) and can be defined based on the operating current flow as

$$\dot{n} = \frac{I}{n_e F}. \tag{4.23b}$$

Another form of fuel cell power is given as *power density*

$$p = E\,j, \tag{4.24}$$

where j is the current density, defined as the current per unit area of the fuel cell as

$$j = \frac{I}{A_{cell}}, \tag{4.25}$$

where A_{cell} is the cell reaction surface area.

A higher current output form the cell to the electrical circuit requires a higher reactant gas consumption rate, and theoretically this should result in an increased cell power output. However, as the fuel consumption rate and current output increases, the cell output voltage drops from the maximum reversible voltage due to a number of irreversibility factors associated with the electrochemical reactions and transport processes.

4.3 Changes in Gibbs Free Energy

A detailed description of Gibbs function or Gibbs free energy and the procedure to estimate change in Gibbs free energy are given in Chapter 3. In this section, we are considering the change in Gibbs free energy for a fuel cell. The Gibbs free energy is defined as the enthalpy minus the energy transfer due to entropy change; that is,

$$G = H - TS. \tag{4.26}$$

Similarly, the Gibbs free energy of formation per unit mass is given as

$$g_f = h_f - Ts, \text{ on a mass basis} \tag{4.27a}$$

and

$$\bar{g}_f = \bar{h}_f - T\bar{s}, \text{ on a molar basis.} \tag{4.27b}$$

In the case of energy transfer in a fuel cell, it is the Gibbs free energy of formation ΔG_f or $\Delta \bar{g}_f^0$, which is defined as the difference in the Gibbs free energy of the product and the reactants, and is given as

$$\Delta G_f = \text{Gibbs free energy of formation} = (G_f)_{\text{products}} - (G_f)_{\text{reactants}}$$

$$\Delta G_f = \sum_P G_f - \sum_R G_f$$

$$= \sum_P n_P \overline{g_{f_P}} - \sum n_R \overline{g_{f_R}}$$

$$\Delta G_f = \sum_P n_P(\bar{g}_f^0 + \overline{\Delta g})_P - \sum_R n_R(\bar{g}_f^0 + \overline{\Delta g})_R. \tag{4.28}$$

Note that if the reactants and the products are at the standard reference state of 25°C and 0.1 MPa pressure, then the sensible components $\overline{\Delta g}$ vanish and the difference between the Gibbs function of formation is given as

$$\Delta G_f = \sum_P n_P \left(\bar{g}_f^0\right)_P - \sum_R n_R \left(\bar{g}_f^0\right)_R. \tag{4.29}$$

For the overall reaction of a hydrogen–oxygen fuel cell given by Equation 4.7c or 4.8c, $n_{H_2O} = 1$, $n_{H_2} = 1$ and $n_{O_2} = \frac{1}{2}$, and the change in Gibbs function is given as

$$\Delta \overline{g_f} = (\overline{g_f})_{H_2O} - (\overline{g_f})_{H_2} - \frac{1}{2}(\overline{g_f})_{O_2} \tag{4.30a}$$

or

$$\Delta \overline{g_f} = \left(\bar{g}_f^0 + \overline{\Delta g}\right)_{H_2O} - \left(\bar{g}_f^0 + \overline{\Delta g}\right)_{H_2} - \frac{1}{2}\left(\bar{g}_f^0 + \overline{\Delta g}\right)_{O_2} \tag{4.30b}$$

Alternatively, we can also estimate the difference in Gibbs function from enthalpy and entropy changes as

$$\Delta \bar{g}_f = \Delta \bar{h}_f - T\Delta \overline{s_f}, \tag{4.31}$$

where

$$\Delta \bar{h}_f = \text{Change in enthalpy of formation} = (h_f)_{\text{products}} - (h_f)_{\text{reactants}}$$

$$\Delta\bar{h}_{\mathrm{f}} = \sum_{\mathrm{P}} n_{\mathrm{P}}\left(\bar{h}_{\mathrm{f}}^{0} + \overline{\Delta h}\right) - \sum_{\mathrm{R}} n_{\mathrm{R}}\left(\bar{h}_{\mathrm{f}}^{0} + \overline{\Delta h}\right) \tag{4.32}$$

and

$$\Delta S_{\mathrm{f}} = \text{change in absolute entropy} = (S_{\mathrm{f}})_{\text{products}} - (S_{\mathrm{f}})_{\text{reactants}}$$

$$\Delta\bar{s}_{\mathrm{f}} = \sum_{\mathrm{P}} n_{\mathrm{P}}\left(\bar{s}_{\mathrm{T}}^{0}\right) - \sum_{\mathrm{R}} n_{\mathrm{R}}\left(\bar{s}_{\mathrm{T}}^{0}\right). \tag{4.33}$$

The procedure for estimating the enthalpy of formation and absolute entropy is reviewed in Chapter 3. For the hydrogen–oxygen fuel cell, Equations 4.32 and 4.33 can be expressed as

$$\Delta\bar{h}_{\mathrm{f}} = \left(\bar{h}_{\mathrm{f}}^{0} + \overline{\Delta h}\right)_{\mathrm{H_2O}} - \left(\bar{h}_{\mathrm{f}}^{0} + \overline{\Delta h}\right)_{\mathrm{H_2}} - \frac{1}{2}\left(\bar{h}_{\mathrm{f}}^{0} + \overline{\Delta h}\right)_{\mathrm{O_2}} \tag{4.34}$$

and

$$\Delta\bar{s}_{\mathrm{f}} = \left(\bar{s}_{\mathrm{T}}^{0}\right)_{\mathrm{H_2O}} - \left(\bar{s}_{\mathrm{T}}^{0}\right)_{\mathrm{H_2}} - \frac{1}{2}\left(\bar{s}_{\mathrm{T}}^{0}\right)_{\mathrm{O_2}}. \tag{4.35}$$

Note that the difference in enthalpy of formation, absolute entropy, and Gibbs function of formation is the functions of temperature. Once $\Delta\bar{h}_{\mathrm{f}}$ and $\Delta\bar{s}_{\mathrm{f}}$ are estimated as functions of temperature, then $\Delta\bar{g}_{\mathrm{f}}$ can also be estimated through substitution in Equation 4.31.

Table 4.2 shows the values of enthalpy of formation and absolute entropy of the basic elements of a hydrogen–oxygen fuel cell with reference state considered as 298 K and 0.1 MPa for the enthalpy and 0 K and 0.1 MPa for the absolute entropy. This is reproduced from Table C.7.

TABLE 4.2

Enthalpy of Formation and Absolute Entropy Data for Hydrogen–Oxygen Fuel Cell

Elements	$\bar{h}_{\mathrm{f}}^{0}$ (J/kmol)	$\bar{s}_{\mathrm{T}}^{0}$	$\bar{g}_{\mathrm{f}}^{0}$	$\overline{\Delta h}$
Hydrogen	0	130.678		0
Oxygen	0	205.148		0
Water (v)	−241,826	188.835	−228,582	
Water (l)	−285,830	69.950	−237,141	

Let us consider two separate cases: (i) water existing in liquid form and (ii) water existing in vapor form.

i. Water in liquid form

The enthalpy of formation is given as

$$\Delta\bar{h}_f = \left(\bar{h}_f^0 + \overline{\Delta h}\right)_{H_2O(l)} - \left(\bar{h}_f^0 + \overline{\Delta h}\right)_{H_2} - \frac{1}{2}\left(\bar{h}_f^0 + \overline{\Delta h}\right)_{O_2}.$$

Substituting data from Table 4.2

$$\Delta\bar{h}_f = (-285,830) - (0+0) - \frac{1}{2}(0+0)$$

$$\Delta\bar{h}_f = -285,830 \text{ J/mol with water as liquid.}$$

The change in absolute entropy is given as

$$\Delta\bar{s}_f = \left(\bar{s}_T^0\right)_{H_2O} - \left(\bar{s}_T^0\right)_{H_2} - \frac{1}{2}\left(\bar{s}_T^0\right)_{O_2}$$

or

$$\Delta\bar{s}_f = (69.950) - (130.678) - \frac{1}{2}(205.148)$$

$$= -163.302 \text{ J/mol K.}$$

The Gibbs free energy of formation is given as

$$\Delta\bar{g}_f = \Delta\bar{h}_f - T\Delta\bar{s}_f$$

$$= -285,830 - 298.15(-163.302)$$

$$= -285,830 + 48688.4913$$

$$= -237,141.51 \text{ J/mol}$$

$$\Delta\bar{g}_f = -237,141.51 \text{ J/mol.}$$

ii. Water in vapor form

Substituting data from Table 4.2, the enthalpy of formation is given as

$\overline{\Delta h_{\mathrm{f}}} = -241,826$ J/mol with water as in vapor phase.

The change in absolute entropy is given as

$$\Delta\overline{s_{\mathrm{f}}} = \left(\overline{s}_{\mathrm{T}}^{0}\right)_{\mathrm{H_2O}} - \left(\overline{s}_{\mathrm{T}}^{0}\right)_{\mathrm{H_2}} - \frac{1}{2}\left(\overline{s}_{\mathrm{T}}^{0}\right)_{\mathrm{O_2}}$$

or

$$\Delta\overline{s_{\mathrm{f}}} = (188.835) - (130.678) - \frac{1}{2}(205.148)$$

$$= -44.417 \text{ J/mol K.}$$

The Gibbs free energy of formation is given as

$$\begin{aligned}\Delta\overline{g}_{\mathrm{f}} &= \Delta\overline{h}_{\mathrm{f}} - T\Delta\overline{s}_{\mathrm{f}} \\ &= -241{,}826 - 298.15\,(-44.417) \\ &= -228{,}583.07 \text{ J/mol}\end{aligned}$$

$\Delta\overline{g}_{\mathrm{f}} = -228,583.07$ J/mol with water in the vapor form.

The reversible cell voltage is given by Equation 4.21 as

$$E_{\mathrm{rev}}^{0} = \frac{-\Delta\overline{g_{\mathrm{f}}}}{n_{\mathrm{e}}F} = \frac{-(-237,141.5087)}{(2)\,(96,485)}$$

$E_{\mathrm{rev}}^{0} = 1.2289$ V with water in the liquid form

and

$$E_{\mathrm{rev}}^{0} = \frac{-\Delta\overline{g_{\mathrm{f}}}}{n_{\mathrm{e}}F} = \frac{-(-228,583.07)}{(2)(96,485)}$$

$E_{\mathrm{rev}}^{0} = 1.18455\ V$ with water in the vapor form.

Note that if all the energy from the fuel, that is, the higher heating value or the enthalpy of formation, is converted into electrical energy, then the maximum possible voltage is given as

$$E^0_{max} = \frac{-\Delta h_f}{2F} = \frac{-(-285,830)}{(2)(96,485)}$$

$E^0_{max} = 1.4812$ V for water as liquid

and

$$E^0_{max} = \frac{-\Delta h_f}{2F} = \frac{-(-241,826)}{(2)(96,485)}$$

$E^0_{max} = 1.253$ V for water as vapor.

This is the voltage that would be achievable in a 100% efficient fuel cell.

Example 4.1

Consider a hydrogen–oxygen fuel cell operating at a constant temperature of 400 K and a pressure of 0.1 MPa and assuming water as vapor in the product. Estimate (a) the enthalpy of formation, (b) the entropy of formation, (c) the Gibbs free energy of formation, and (d) the reversible fuel cell voltage for a hydrogen–oxygen fuel cell.

Solution

For the overall reaction of hydrogen–oxygen fuel cell given by Equation 4.7c, $n_{H_2O} = 1$, $n_{H_2} = 1$, and $n_{O_2} = \frac{1}{2}$, and the enthalpy of formation and entropy are given as

$$\Delta\bar{h}_f = \left(\bar{h}^0_f + \overline{\Delta h}\right)_{H_2O} - \left(\bar{h}^0_f + \overline{\Delta h}\right)_{H_2} - \frac{1}{2}\left(\bar{h}^0_f + \overline{\Delta h}\right)_{O_2}$$

and

$$\Delta\bar{s}_f = \left(\bar{s}^0_T\right)_{H_2O} - \left(\bar{s}^0_T\right)_{H_2} - \frac{1}{2}\left(\bar{s}^0_T\right)_{O_2}.$$

From Table C.7, we get following data for the basic elements of the reaction

Elements	$\bar{h}^0_f$	$\bar{s}^0_T$	$\overline{\Delta h}$
Hydrogen	0	139.219	2961
Oxygen	0	213.873	3027
Water (v)	–241,826	198.787	3450

The enthalpy of formation is given as

$$\Delta\overline{h_{\rm f}} = \left(\overline{h}_{\rm f}^{0} + \overline{\Delta h}\right)_{\rm H_2O} - \left(\overline{h}_{\rm f}^{0} + \overline{\Delta h}\right)_{\rm H_2} - \frac{1}{2}\left(\overline{h}_{\rm f}^{0} + \overline{\Delta h}\right)_{\rm O_2}$$

or

$$\Delta\overline{h_{\rm f}} = (-241{,}826 + 3450) - (0 + 2961) - \frac{1}{2}(0 + 3027)$$

$$\Delta\overline{h_{\rm f}} = -246{,}300.5 \text{ kJ/kmol.}$$

The change in absolute entropy is given as

$$\Delta\overline{s_{\rm f}} = \left(\overline{s}_{\rm T}^{0}\right)_{\rm H_2O} - \left(\overline{s}_{\rm T}^{0}\right)_{\rm H_2} - \frac{1}{2}\left(\overline{s}_{\rm T}^{0}\right)_{\rm O_2}$$

or

$$\Delta\overline{s_{\rm f}} = (198.787) - (139.219) - \frac{1}{2}(213.873)$$

$$= -47.3685 \text{ kJ/kmol K.}$$

The Gibbs free energy of formation is given as

$$\Delta\overline{g}_{\rm f} = \Delta\overline{h_{\rm f}} - T\Delta\overline{s_{\rm f}}$$

$$= -246{,}300.5 - 400(-47.3685) = 227{,}353.1 \text{ kJ/kmol}$$

$$\Delta\overline{g}_{\rm f} = -227{,}353.1 \text{ kJ/kmol.}$$

The reversible cell voltage is given as

$$E_{\rm rev} = \frac{-\Delta\overline{g}_{\rm f}}{n_{\rm e}F} = \frac{-(-227{,}353.1)}{(2)(96{,}485)}$$

$$E_{\rm rev} = 1.178 \text{ V.}$$

4.4 Effect of Operating Conditions on Reversible Voltage

As thermodynamic properties like enthalpy, entropy, and Gibbs function vary with operating conditions such as temperature, pressure, and gas concentration, the reversible open circuit voltage also varies with these operating conditions. Let us consider here the effect of these variables on the reversible cell voltage.

4.4.1 Effect of Variation of Temperature

The effect of temperature on Gibbs free energy of formation and reversible voltage is given by Equations 4.31 and 4.21 as

$$\Delta\overline{g}_{\mathrm{f}} = \Delta\overline{h}_{\mathrm{f}} - T\Delta\overline{s}_{\mathrm{f}} \tag{4.31}$$

$$E_{\mathrm{rev}} = \frac{-\Delta\overline{g}_{\mathrm{f}}}{n_{\mathrm{e}}F}. \tag{4.21}$$

As we can see, the temperature not only has a direct effect but also indirectly affects the Gibbs free energy through the functional variation of enthalpy of formation and entropy with temperature.

The procedure outlined in Example 4.1 is used to determine the enthalpy of formation, entropy of formation, Gibbs free energy of formation, and the reversible cell voltage of a hydrogen–oxygen fuel cell over a range of temperature, 298–1473 K, and results are summarized in Table 4.3 and plotted in Figure 4.8.

Data presented in Table 4.3 shows variations in enthalpy of reaction and entropy of reaction with increase in temperature. The change in Gibbs function is estimated based on Equation 4.30a, and results show a decrease in negative of Gibbs energy of formation. Thus, the reversible work of a fuel cell also decreases with increase in temperature. This is in contrast to the reversible work of a Carnot heat engine where the net reversible work and, hence, the thermal efficiency increase with increase in high-temperature heat source.

For example, in a hydrogen fuel cell with hydrogen and oxygen as reactants to produce water vapor, the change in Gibbs energy of reaction decreases as the temperature increases, and so the maximum work output from the fuel cell also decreases. Although these ideal calculations show that a lower operating cell temperature results in higher reversible voltage, the voltage losses for some real fuel cell like SOFC, however, decreases at higher temperature due to enhanced ionic conductivity. Hence, in practice, the operating fuel cell voltage is usually higher at higher operating temperature for these types of fuel cell.

4.4.2 Effect of Pressure on Gibbs Function and Reversible Voltage

In order to show the dependence of Gibbs function on pressure, let us consider Equation 3.44d

TABLE 4.3
Thermodynamic Data of Hydrogen-Oxygen Fuel Cell over a Range of Temperature

Temp (K)	Temp (C)	Enthalpy of Reaction, kJ/mol	Entropy of Reaction, J/mol.K	Gibbs Energy of Reaction, J/mol	Voltage (V)
273	0	−286.6225	−166.1743	−241260	1.2513
283	10	−286.3007	−164.9659	−239620	1.2428
293	20	−285.9788	−163.8298	−237980	1.2343
298	25	−285.8185	−163.2828	−237160	1.2301
303	30	−285.6587	−162.7477	−236350	1.2259
313	40	−285.3411	−161.7087	−234730	1.2175
323	50	−285.0261	−160.7072	−233120	1.2091
333	60	−284.7132	−159.741	−231520	1.2008
343	70	−284.4012	−158.8087	−229930	1.1926
353	80	−284.0885	−157.9078	−228350	1.1844
363	90	−283.7729	−157.0327	−226770	1.1762
373	100	−283.4518	−156.1722	−225200	1.168
473	200	−243.5864	−49.0335	−220390	1.1431
573	300	−244.5529	−50.8943	−215390	1.1172
673	400	−245.4524	−52.3414	−210230	1.0904
773	500	−246.2767	−53.479	−204940	1.063
873	600	−247.0255	−54.387	−199550	1.035
973	700	−247.6976	−55.1215	−194060	1.0066
1073	800	−248.2841	−55.7155	−188500	0.9777
1173	900	−248.7601	−56.1796	−182860	0.9485
1273	1000	−249.0776	−56.5022	−177150	0.9188
1373	1100	−249.1571	−56.6503	−171380	0.8889
1473	1200	−248.8803	−56.5701	−165550	0.8587

$$dG = VdP - SdT. \tag{3.44d}$$

For an isothermal process, the change in the Gibbs function expression reduces to

$$dG = VdP. \tag{4.36}$$

Now, considering the ideal gas equation of state and substituting $V = \frac{n\bar{R}T}{P}$ into Equation 4.36, we get

$$dG = n\bar{R}T\frac{dP}{P}. \tag{4.37}$$

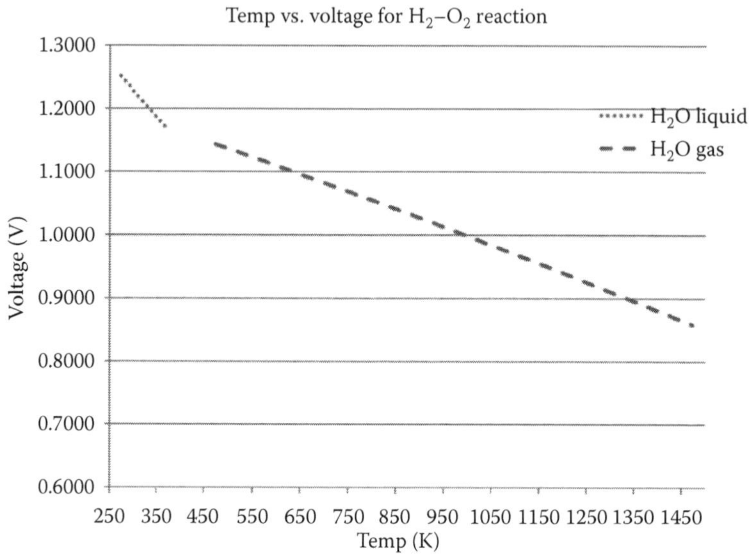

FIGURE 4.8
Variation of reversible voltage with temperature for a hydrogen–oxygen fuel cell.

Integrating from standard reference state pressure P_0 to any arbitrary pressure P, we get

$$G = G^0 + n\bar{R}T \ln\left(\frac{P}{P_0}\right). \tag{4.38}$$

Dividing both sides of the equation by the number of moles, we get the expression on a molar basis as

$$\bar{g} = \bar{g}^0 + \bar{R}T \ln\left(\frac{P}{P_0}\right). \tag{4.39}$$

A general expression for Gibbs function at a given temperature and pressure can be written with respect to the Gibbs function at a temperature T and the reference pressure ($P_0 = 0.1$ MPa), as

$$\bar{g}_i(T, P_i) = \bar{g}^0(T) + \bar{R}T \ln\left(\frac{P_i}{P_0}\right). \tag{4.40}$$

4.4.3 Effect of Gas Concentration—The Nernst Equation

In this section, we will consider the effect of species concentration in reactant and product gas mixtures on the change of Gibbs function reaction and reversible voltage. As was discussed in Chapter 3, the species concentrations in a gas mixture can also be represented in terms of the species partial pressures. Let us consider a generic chemical reaction involving two reactants A and B and producing two products M and N as below:

$$n_A R_A + n_B R_B \rightarrow n_M R_M + n_N R_N, \tag{4.41}$$

where n_A, n_B, n_M, and n_N are the number of moles for each component in reactant and product gas mixtures. The change on Gibbs function is written as

$$\Delta G = \sum G_P - \sum G_R \tag{4.42}$$

or

$$\Delta G = G_M + G_N - G_A - G_B. \tag{4.43}$$

Using Equation 4.39, we get

$$\text{or } \Delta G = n_M\left[\bar{g}_M^0 + \bar{R}T\ln\left(\frac{P_M}{P_0}\right)\right] + n_N\left[\bar{g}_N^0 + \bar{R}T\ln\left(\frac{P_N}{P_0}\right)\right]$$

$$-n_A\left[\bar{g}_A^0 + \bar{R}T\ln\left(\frac{P_A}{P_0}\right)\right] - n_B\left[\bar{g}_B^0 + \bar{R}T\ln\left(\frac{P_B}{P_0}\right)\right]. \tag{4.44}$$

All terms representing the standard Gibbs function terms can be grouped as a change of standard Gibbs function term as

$$\Delta G^0 = n_M\left[\bar{g}_M^0\right] + n_N\left[\bar{g}_N^0\right] - n_A\left[\bar{g}_A^0\right] - n_B\left[\bar{g}_B^0\right]. \tag{4.45}$$

Combining Equations 4.44 and 4.45, we get the change in Gibbs function of reaction

$$\Delta G = \Delta G^0 + \bar{R}T\ln\frac{\left(\frac{P_M}{P_0}\right)^{n_M}\left(\frac{P_N}{P_0}\right)^{n_N}}{\left(\frac{P_A}{P_0}\right)^{n_A}\left(\frac{P_B}{P_0}\right)^{n_B}}, \tag{4.46}$$

where P_M and P_N are the partial pressures of the components M and N, respectively, in the products; P_A and P_B are the partial pressure of the reactant components; and P_0 is the standard pressure. For the standard reference pressure of 1 atm, the equation can be written in a simplified form as

$$\Delta G = \Delta G^0 + \bar{R}T \ln \frac{P_{\mathrm{M}}^{n_{\mathrm{M}}} P_{\mathrm{N}}^{n_{\mathrm{N}}}}{P_{\mathrm{A}}^{n_{\mathrm{A}}} P_{\mathrm{B}}^{n_{\mathrm{B}}}}. \tag{4.47}$$

Now, representing the change of Gibbs function in terms of cell voltage given by Equations 4.16 and 4.21, we get

$$-n_{\mathrm{e}}FE = -n_{\mathrm{e}}FE_{\mathrm{rev}}^0 + \bar{R}T \ln \frac{P_{\mathrm{M}}^{n_{\mathrm{M}}} P_{\mathrm{N}}^{n_{\mathrm{N}}}}{P_{\mathrm{A}}^{n_{\mathrm{A}}} P_{\mathrm{B}}^{n_{\mathrm{B}}}}$$

or

$$E = E_{\mathrm{rev}}^0 - \frac{\bar{R}T}{n_{\mathrm{e}}F} \ln \frac{P_{\mathrm{M}}^{n_{\mathrm{M}}} P_{\mathrm{N}}^{n_{\mathrm{N}}}}{P_{\mathrm{A}}^{n_{\mathrm{A}}} P_{\mathrm{B}}^{n_{\mathrm{B}}}}. \tag{4.48a}$$

This equation can also be generalized for a reaction with arbitrary number of reactants and products as

$$E = E_{\mathrm{rev}}^0 - \frac{\bar{R}T}{n_{\mathrm{e}}F} \ln \frac{\prod_{i=1}^{N} P_{P_i}^{n_i}}{\prod_{i=1}^{N} P_{R_i}^{n_i}}, \tag{4.48b}$$

where the symbol $\prod$ represents the product term.

Since partial pressure of the gas species in reactant and product gas mixtures are represented by the gas concentration or mole fraction through the relation given by $y_i = \frac{P_i}{P}$, Equation 4.48a is also a representation of the effect of gas concentration on reversible voltage. In terms of gas mole fraction with operating pressure $P = 1$ atm, Equation 4.48a can also be expressed as

$$E = E_{\mathrm{rev}}^0 - \frac{\bar{R}T}{n_{\mathrm{e}}F} \ln \frac{y_{\mathrm{M}}^{n_{\mathrm{M}}} y_{\mathrm{N}}^{n_{\mathrm{N}}}}{y_{\mathrm{A}}^{n_{\mathrm{A}}} y_{\mathrm{B}}^{n_{\mathrm{B}}}} \tag{4.49a}$$

and in a general form as

$$E = E_{\mathrm{rev}}^0 - \frac{\bar{R}T}{n_{\mathrm{e}}F} \ln \frac{\prod_{i=1}^{N} y_{P_i}^{n_i}}{\prod_{i=1}^{N} y_{R_i}^{n_i}}. \tag{4.49b}$$

Equations 4.48 and 4.49 are referred to as the *Nernst equation*, which is used to determine the deviation of cell voltage from reversible cell voltage due to variation of gas composition in the anode and cathode gas mixtures. As we can see, the real electrochemical voltage (E) generated in the cell will be modified from the ideal reversible voltage (E^0_{rev}) by the term given by $\frac{\bar{R}T}{n_e F} \ln \frac{P_M^{n_M} P_N^{n_N}}{P_A^{n_A} P_B^{n_B}}$ or $\frac{\bar{R}T}{n_e F} \ln \frac{\prod_{i=1}^{N} P_{Pi}^{n_i}}{\prod_{i=1}^{N} P_{Ri}^{n_i}}$. This term is often referred to as the *Nernst loss*.

As an example, let us consider a fuel cell reaction involving hydrogen and oxygen as

$$H_2 + \frac{1}{2} O_2 \rightarrow H_2O$$

with the number of electrons transferred in this reaction as $n_e = 2$. For this reaction, we have $R_A \approx H_2$, $R_B \approx O_2$, $n_A = n_{H_2} = 1$, $n_B = n_{O_2} = 1/2$ for the reactants and $M \approx H_2O$, $n_M = 1$, $n_N = 0$ for the product. Thus, for this electrochemical reaction with direct hydrogen and oxygen, we can write the expression for the change in Gibbs energy as

$$\Delta G = \Delta G^0 + \frac{\bar{R}T}{2F} \ln \frac{P_{H_2O}}{P_{H_2} P_{O_2}^{1/2}} \tag{4.50}$$

and the *Nernst equation* (Equation 4.48) reduces to

$$E = E^0_{rev} - \frac{\bar{R}T}{2F} \ln \frac{P_{H_2O}}{P_{H_2} P_{O_2}^{1/2}}. \tag{4.51}$$

As we can see, the actual electrochemical voltage (E) generated in the hydrogen–oxygen fuel cell will be less than the ideal voltage (E^0_{rev}) by the term $\frac{\bar{R}T}{2F} \ln \frac{P_{H_2O}}{P_{H_2} P_{O_2}^{1/2}}$. We can also see that the lower the value of the term $P_{H_2} P_{O_2}^{1/2}$ compared to P_{H_2O}, that is, the product of partial pressures of the reactants is less than the partial pressure of the products, the lower is the cell voltage. When hydrogen is supplied as a gas mixture, like the mixture of hydrogen and carbon dioxide in a reformed hydrocarbon gas mixture, the mole fraction or the partial pressure of hydrogen will be lowered and results in a lower cell voltage. Similarly when oxygen is supplied as air, the mole fraction and the partial pressure of oxygen are also reduced. Considering air as a mixture of 21% oxygen and 79% nitrogen, the overall reaction in a hydrogen fuel cell is written as

$$H_2 + 0.5O_2 + (0.5)(3.76)N_2 = H_2O + (0.5)(3.76)N_2.$$

For this reaction, the mole fractions and the partial pressures are given as

$$\text{Hydrogen: } x_{H_2} = 1 \text{ and } P_{H_2} = 0.1\,\text{MPa}$$

$$\text{Oxygen: } Y_{O_2} = \frac{n_{O_2}}{n_{O_2} + n_{N_2}} = \frac{0.5}{0.5 + (0.5)(3.76)} = 0.21 \text{ and}$$

$$P_{O_2} = 0.21 \times 0.1\,\text{MPa} = 0.021\,\text{MPa}$$

$$\text{Water: } Y_{H_2O} = \frac{n_{H_2O}}{n_{H_2O} + n_{N_2}} = \frac{1}{1 + 0.5 \times 3.76} = 0.347 \text{ and}$$

$$P_{H_2O} = 0.347 \times 0.1\,\text{MPa} = 0.0347\,\text{MPa}$$

The reversible voltage for this fuel cell is given as

$$E = E^0 - \frac{\bar{R}T}{2F} \ln \frac{P_{H_2O}}{P_{H_2} P_{O_2}^{1/2}}$$

$$E = 1.2289 - \frac{8.314 \times 298}{(2)(96,485)} \ln \frac{0.0347}{(0.1)(0.021)}$$

$$E = 1.2289\text{ V} - 0.03601129$$

$$E = 1.1929\text{ V.}$$

4.4.3.1 Effect of Hydrogen Partial Pressure

In order to show the effect of variation of hydrogen partial pressure in a hydrogen–oxygen fuel cell, let us rearrange Equation 4.51 in the following manner:

$$E = E_{rev}^0 - \frac{\bar{R}T}{2F} \ln \frac{P_{H_2O}}{P_{O_2}^{1/2}} - \frac{\bar{R}T}{2F} \ln\left(P_{H_2}\right). \tag{4.51}$$

Hence, change in voltage owing to a change in hydrogen partial pressure from $P_{H_{2_1}}$ and $P_{H_{2_2}}$ while keeping P_{H_2O} and P_{O_2} unchanged is expressed as

$$\Delta E = \frac{\bar{R}T}{2F} \ln\left(P_{H_{2_2}}\right) - \frac{\bar{R}T}{2F} \ln\left(P_{H_{2_1}}\right) = \frac{\bar{R}T}{2F} \ln\left(\frac{P_{H_2}}{P_{H_1}}\right). \tag{4.52}$$

4.4.3.2 *Effect of Oxygen Partial Pressure*

In a similar manner, we could isolate the contribution from oxygen as

$$E = E^0 - \frac{\bar{R}T}{2F}\ln\frac{P_{H_2O}}{P_{H_2}} - \frac{\bar{R}T}{2F}\ln\left(P_{O_2}^{1/2}\right). \tag{4.53}$$

The change in voltage owing to a change in oxygen partial pressure from $P_{O_{2_1}}$ and $P_{O_{2_2}}$ while keeping P_{H_2O} and P_{O_2} unchanged is expressed as

$$\Delta E = \frac{\bar{R}T}{2F}\ln\left(\frac{P_{O_{2_2}}}{P_{O_{2_1}}}\right)^{1/2}. \tag{4.54}$$

Example 4.2

A direct methanol fuel cell (DMFC) uses liquid methanol (CH_3OH) as fuel. The electrochemical half and overall reactions in a DMFC is summarized in Table 4.1. The overall reaction in a DMFC is given as

$$CH_3OH + \frac{3}{2}O_2 \rightarrow 2H_2O + CO_2$$

with six free electrons moving from the anode electrode to the cathode electrode for each molecule of methanol fuel. Determine the following:

a. Change in molar Gibbs energy for this reaction and the reversible cell voltage at standard reference temperature and pressure.
b. Estimate the cell voltage as operation temperature changes from reference temperature to 400 K.
c. Estimate the change in voltage caused by changes in partial pressure of the oxygen as we change from pure oxygen to air as a source of oxygen and operating at 400 K.
d. Estimate the change in voltage caused by change in system pressure operating at 400 K.

Solution

a. The change in Gibbs function of formation is given by Equation 4.28 as

$$\Delta G_f = \sum_P n_P\left(\bar{g}_f^0\right)_P - \sum_R n_R\left(\bar{g}_f^0\right)_R.$$

For the direct methanol reaction

$$\Delta G_f = n_{H_2O}(\bar{g}_f)_{H_2O} + n_{CO_2}(\bar{g}_f)_{CO_2} - n_{O_2}(\bar{g}_f)_{O_2} - n_{CH_3OH}(\bar{g})_{CH_3OH}$$

$$= 2\left(\bar{h}_f - T\bar{s}_f\right)_{H_2O} + \left(\bar{h}_f - T\bar{s}_f\right)_{CO_2} - \left(\bar{h}_f - T\bar{s}_f\right)_{CH_3OH} - \frac{3}{2}\left(\bar{h}_f - T\bar{s}_f\right)_{O_2}$$

$$= 2(-241{,}826 - 298 \times 188.835) + (-393{,}522 - 298 \times 213.794)$$

$$-(-239{,}220 - 298 \times 126.809) - \frac{3}{2}(0 - 298 \times 205.148)$$

$$\Delta G_f = -688{,}996.05 \frac{\text{kJ}}{\text{kmol}}.$$

The reversible cell voltage at STP is given as

$$E^0_{rev} = \frac{-\Delta G_f}{n_e F} = \frac{688{,}996}{6 \times 96{,}485} = 1.19 \text{ V}.$$

The open circuit cell voltage based on the Nernst equation is given by Equation 4.48b as

$$E = E^0_{rev} - \frac{\bar{R}T}{n_e F} \ln \frac{y_M^{n_M} y_N^{n_N}}{y_A^{n_A} y_B^{n_B}}.$$

For this electrochemical reaction, the number of moles are $n_A = n_{CH_3OH} = 1$, $n_B = n_{O_2} = 1$, $n_M = n_{H_2O} = 2$, and $n_N = n_{CO_2} = 1$, and the mole fractions are given as $y_A = y_{CH_3OH} = \frac{2}{5}$, $y_B = y_{O_2} = \frac{3}{5}$, $y_M = y_{H_2O} = \frac{2}{3}$, and $y_N = y_{CO_2} = \frac{1}{3}$.

Substituting, we get

$$E = E^0_{rev} - \frac{\bar{R}T}{n_e F} \ln\left[\frac{y^2_{H_2O} y^1_{CO_2}}{y^1_{CH_3OH} y^{3/2}_{O_2}}\right]$$

$$= 1.19 - \frac{8.314 \times 298}{6 \times 96{,}485} \ln\left[\frac{\left(\frac{2}{3}\right)^2 \left(\frac{1}{2}\right)^1}{\left(\frac{2}{5}\right)^1 \left(\frac{3}{5}\right)^{3/2}}\right]$$

$$E = 1.19 - 0.0007634$$

$$E = 1.189 \text{ V}.$$

b. For the cell voltage at an operating temperature of $T = 400$ K, we use the open circuit cell voltage based on the Nernst equation as

$$E = E_{rev}(T) - \frac{\bar{R}T}{n_e F} \ln \frac{y_M^{n_M} y_N^{n_N}}{y_A^{n_A} y_B^{n_B}},$$

where

$$E_{rev}(T) = \frac{-\overline{\Delta G_f(T)}}{n_e F} \text{ and } \Delta G_f = \Delta H_f - T\Delta S_f.$$

For change of enthalpy of formation

$$\Delta H_i = n_{H_2O}\left(\overline{h_f} + \Delta h\right)_{H_2O} + n_{CO_2}\left(\overline{h_f} + \Delta h\right)_{CO_2}$$

$$-n_{O_2}\left(\overline{h_f} + \Delta h\right)_{O_2} - n_{CH_3OH}\left(\overline{h_f} + \Delta h\right)_{CH_3OH}.$$

For $T = 400$ K:

Enthalpy change caused by temperature change from $T_1 = 298$ K to $T_2 = 400$ K:

$$\Delta h_{O_2} = 3027 \text{ kJ/kmol}, \ \Delta h_{CO_2} = 4003 \text{ kJ/kmol} \ \text{ and } \ \Delta h_{H_2O} = 3450 \text{ kJ/kmol}$$

$$\bar{s}^0_{TO_2} = 213.873 \text{ kJ/kmol}\cdot\text{K}, \ \bar{s}^0_{TCO_2} = 225.314 \text{ kJ/kmol}\cdot\text{K and}$$
$$\bar{s}^0_{TH_2O} = 198.787 \text{ kJ/kmol}\cdot\text{K}.$$

For methanol, we can use Equations 3.48a and 3.53b assuming constant specific heat at reference temperature:

$$\Delta h_{CH_3OH} = C_{poCH_3OH}(T_2 - T_1)$$

$$= 1.405 \times (400 - 298)$$

$$\Delta h_{CH_3OH} = 143.21 \text{ kJ/kmol}$$

$$\Delta H_f = n_{H_2O}\left(\overline{h_f} + \Delta h\right)_{H_2O} + n_{CO_2}\left(\overline{h_f} + \Delta h\right)_{CO_2}$$

$$-n_{O_2}\left(\overline{h_f} + \Delta h\right)_{O_2} - n_{CH_3OH}\left(\overline{h_f} + \Delta h\right)_{CH_3OH}$$

$$= 2(-241{,}826 + 3450) + (-393{,}522 + 4003)$$

$$-\frac{3}{2}(0 + 3027) - (-201{,}200 + 143.21)$$

$$\Delta H_f = -672{,}599.31 \frac{\text{kJ}}{\text{kmol}}.$$

For entropy change of methanol

$$\Delta s_{CH_3OH} = C_{poCH_3OH} \ln\frac{T_2}{T_1} - R_{CH_3OH} \ln\frac{P_2}{P_1}$$

$$= 1.405 \ln\frac{T_2}{T_1} - 0.2595 \ln\frac{P_2}{P_1}$$

$$\Delta s_{CH_3OH} = 1.405 \ln\frac{400}{298} - 0.2595 \ln\frac{0.1}{0.1}$$

$$\Delta s_{CH_3OH} = 0.4135 \frac{\text{kJ}}{\text{kmol}}.$$

For entropy change, let us use Equation 3.99

$$\Delta\overline{s}_f = \sum_P n_P\left(\overline{s}_f^0 + \overline{\Delta s}\right) - \sum_R n_R\left(\overline{s}_f^0 + \overline{\Delta s}\right).$$

For methanol reaction:

$$\Delta S_f = n_{H_2O}(\overline{s_T})_{H_2O} + n_{CO_2}(\overline{s_T})_{CO_2}$$

$$-n_{O_2}(\overline{s_T})_{O_2} - n_{CH_3OH}\left(\overline{s}_f + \overline{\Delta s}\right)_{CH_3OH}$$

$$\Delta S_f = 2 \times (198.787) + 1 \times (225.314)$$

$$-\frac{3}{2}(213.873) - 1 \times (126.809 + 0.4135)$$

$$\Delta \bar{s}_f = 174.856 \frac{\text{kJ}}{\text{kmol}}.$$

The Gibbs free energy of formation is given as

$$\Delta G_f = \Delta H_f - T\Delta S_f$$

$$\Delta G_f = -672{,}599.31 - 400 \times 174.856$$

$$\Delta G_f = -742{,}541.71 \frac{\text{kJ}}{\text{kmol}}.$$

Reversible cell voltage

$$E_{\text{rev}}(T) = \frac{-\overline{\Delta G_f(T)}}{n_e F} = \frac{672{,}599.31}{6 \times 96{,}485}$$

$$E_{\text{rev}} = 1.16 \text{ V}$$

$$E = E_{\text{rev}}(T) - \frac{\bar{R}T}{n_e F} \ln \frac{Y_M^{n_M} Y_N^{n_N}}{Y_A^{n_A} Y_B^{n_B}}$$

$$E = E_{\text{rev}}(T) - \frac{\bar{R}T}{n_e F} \ln \left[\frac{Y_{H_2O}^2 Y_{CO_2}^1}{Y_{CH_3OH}^1 Y_{O_2}^{3/2}} \right]$$

$$= 1.16 - \frac{8.314 \times 400}{6 \times 96{,}485} \ln \left[\frac{\left(\frac{2}{3}\right)^2 \left(\frac{1}{2}\right)^1}{\left(\frac{2}{5}\right)^1 \left(\frac{3}{5}\right)^{3/2}} \right]$$

$$E = 1.158 \text{ V}.$$

4.5 Fuel Cell Efficiency

A fuel cell performance may be expressed by considering different quantities such as thermodynamic efficiency based on energy forms; voltage efficiency based on operating voltage and all irreversible losses, and current efficiency based on excess fuel supplied.

4.5.1 Thermodynamic Efficiency

The thermodynamic efficiency of a fuel cell is defined based on the ratio of electrical energy or work produced and the available energy of the fuel. As we have discussed before, the available chemical energy content of a fuel is expressed as the enthalpy of formation (Δh_f). However, the maximum amount of energy that can be converted into electrical work in a fuel cell is limited by the Gibbs free energy $\Delta \bar{g}_f$. Thus, one way of defining the fuel cell efficiency is the ratio of electrical work produced to the available Gibbs free energy as

$$\eta_{fc} = \frac{W_{cell}}{-\Delta \bar{g}_f}. \tag{4.55}$$

Since the maximum possible electrical work is equivalent to the change in Gibbs free energy ($\Delta \bar{g}_f$) considering no losses or irreversibilities, the maximum theoretical limit for the efficiency is 100% in an ideal fuel cell.

However, this may not be the best representation of fuel cell efficiency since the available chemical energy content of a fuel for power conversion is the *enthalpy of formation* or the *heating values* of the fuel. Thus, a more practical definition of fuel cell efficiency is given as the ratio of electrical work produced to the enthalpy of formation of the fuel

$$\eta_{fc} = \frac{W_{cell}}{-\Delta \bar{h}_f} = \frac{n_e F E}{-\Delta \bar{h}_f}. \tag{4.56}$$

Since the maximum electrical work in a fuel cell is limited by the available Gibbs free energy, the *maximum thermodynamic* or *reversible efficiency* of a fuel cell is defined as the ratio of Gibbs free energy change for conversion into electrical energy to the net fuel energy available in the form of enthalpy of formation as

$$\eta_{fc,max} = \eta_{fc,rev} = \frac{\Delta \bar{g}_f}{\Delta \bar{h}_f}. \tag{4.57}$$

This also corresponds to the condition of open circuit reversible voltage, E^0_{rev}, with no current flowing through the external circuit. Such a condition leads to the maximum electrical energy conversion and Equation 4.57 for reversible thermodynamic efficiency of a fuel cell can also be expressed as

$$\eta_{fc,rev} = \frac{\Delta \bar{g}_f}{\Delta \bar{h}_f} = \frac{n_e F E^0_{rev}}{-\Delta \bar{h}_f}. \tag{4.58}$$

We can see that for a reversible fuel cell, not all energy available would be converted into electrical work as the change in Gibbs free energy would be less than the enthalpy of formation of the reaction. Hence, the ideal or reversible efficiency of a fuel cell is less than 100%, limited by the Gibbs free energy and not by the Carnot cycle efficiency that sets the maximum theoretical efficiency of heat engine based on the high and low temperature limits.

The enthalpy of formation in a hydrogen–oxygen fuel cell can be estimated by assuming the product water as liquid or as vapor and referred to as higher heating value (*hhv*) or lower heating value (*lhv*), respectively. The fuel cell efficiency can therefore be defined on the basis of using either one of these heating values depending on the state of water in the product. Considering the product water as liquid, the enthalpy of formation for the hydrogen–oxygen fuel cell reaction is estimated as $\Delta\bar{h}_f = -285.835$ kJ/mol, which is the higher heating value. Similarly, for product water as vapor, the enthalpy of formation for this reaction is estimated as $\Delta\bar{h}_{f_1} = -241.83$ kJ/mol, which is referred to as the lower heating value. Since the higher heating value represents the upper bounds of available energy for conversion, it is more appropriate to use the higher heating value while computing the reversible efficiency of a fuel cell.

For example, for a hydrogen–oxygen fuel cell operating at STP conditions, the higher heating value is $\Delta\bar{h}_f = -285.835$ kJ/mol and the Gibbs free energy is estimated as $\Delta\bar{g}_f = -237{,}206$ J/mol. The reversible thermodynamic efficiency of the fuel cell is then given as

$$\eta_{\text{cell,rev}} = \frac{-237{,}206}{-285{,}837} \times 100 = 83.07\%.$$

Figure 4.9 shows the variation of difference of Gibbs free energy and reversible thermodynamic efficiency of a hydrogen–oxygen fuel cell as a function of temperature and at a standard pressure. The figure also includes Carnot cycle efficiency of a reversible heat engine with the low temperature heat sink given by the standard reference temperature of 25°C and with increasing temperature values for the high-temperature heat source. In order to match this fuel cell efficiency, a conventional heat engine needs to operate in a cycle with a maximum temperature of 1150 K and a low temperature of 25°C.

Note that Gibbs function decreases with increase in temperature. Thus, the reversible work and thermodynamic efficiency of a fuel cell decrease with increase in temperature. This is in contrast to the reversible thermodynamic efficiency of a Carnot heat engine where the efficiency or reversible work increases with increase in temperature.

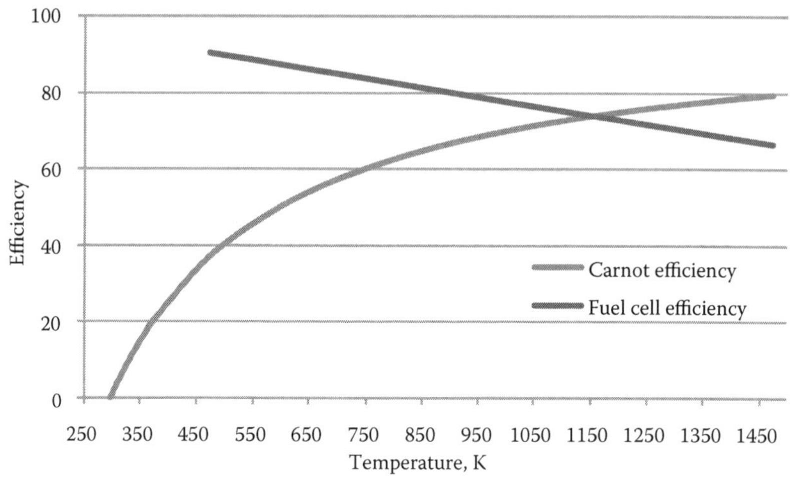

FIGURE 4.9
Variation of fuel cell and Carnot cycle efficiencies with temperature.

4.5.2 Voltage Efficiency

As we have discussed, the reversible fuel cell operation and reversible fuel cell voltage are referred to as the condition of fuel cell when no current is flowing through the external circuit. However, as the current starts flowing through the external circuit, a number of irreversible losses take place and fuel cell voltage and efficiency drop from the reversible values given the current–voltage polarization curve. The voltage drops caused by these irreversibilities are also referred to as *cell polarizations* or *cell overpotentials*. The three major factors that cause these irreversible losses are (i) *activation loss* caused by kinetics of the electrochemical reaction at the electrodes, (ii) *ohmic losses* caused by electrical resistances to the flow of ions and electrons, and (iii) *concentration* or *mass transfer losses* caused by the depletion of reactants at the electrode active reaction sites and over-accumulation of water that blocks reactants in reaching the reaction side. A detailed description of these irreversible losses and the polarization curves that depict the variation of cell voltage with increase in current are given in Chapter 5. In this section, we will consider these voltage losses and define the fuel cell efficiency as the ratio of actual work to reversible work.

$$\eta_{fc} = \frac{W_{fc,real}}{W_{fc,rev}}. \tag{4.59}$$

Substituting the expressions for actual work and reversible work in terms of voltages, Equation 4.59 leads to the definition of cell *voltage efficiency* as the ratio of actual voltage and the ideal or reversible voltage

$$\eta_{\text{fc,v}} = \frac{E}{E^0_{\text{rev}}}. \tag{4.60}$$

For example, the *reversible voltage* for a hydrogen–oxygen cell at the STP is $E^0_{\text{rev}} = 1.229$ V with water in liquid form. Considering a fuel cell with an operating cell voltage of 0.7 V, the cell voltage efficiency can be estimated as

$$\eta_{\text{fc,v}} = \frac{E}{E^0_{\text{rev}}} = \frac{0.7}{1.229} \times 100 = 56.95\%.$$

Similarly, considering the *reversible voltage* for a hydrogen–oxygen cell at the STP as $E^0_{\text{rev}} = 1.18455$ V with water in vapor form, the voltage efficiency is given as

$$\eta_{\text{fc,v}} = \frac{E}{E^0_{\text{rev}}} = \frac{0.7}{1.185} \times 100 = 59.07\%.$$

4.5.3 Current or Fuel Utilization Efficiency

Another important factor that needs to be considered while considering fuel cell efficiency is the fact that excess fuel is generally supplied in order to offset for any unwanted consumptions such as fuel crossover loss through electrolyte, incomplete and undesirable reactions, and leakage loss through cell components and to sustain the electrochemical reaction across the entire active surface area. Any unconsumed fuel will exit the cell as an element exhaust gas mixture. The fuel *utilization factor* or *stoichiometric factor* is defined as a measure of the excess fuel supplied as

$$\xi_{\text{u}} = \frac{\text{Fuel supplied at inlet to the cell}}{\text{Fuel consumed in reaction}} = \frac{\dot{n}_{\text{f,in}}}{\dot{n}_{\text{i}}}. \tag{4.61}$$

For a given fuel *stoichiometric factor*, ξ_{f}, the fuel supply rate at inlet is given as

$$\dot{n}_{\text{f,in}} = \xi_{\text{f}}\dot{n}_{\text{f}}, \tag{4.62}$$

where $\dot{n}_{\text{f}}$ is the consumption rate of reactant (mol/s) given by Equation 4.23b as

$$\dot{n}_f = \frac{I}{n_e F}. \tag{4.23b}$$

The *current efficiency*, η_I, is then defined as the ratio of the mass of fuel consumed in the reaction to the mass of fuel supplied to the cell and expressed as

$$\eta_I = \frac{I/n_e F}{\dot{n}_{f,in}}. \tag{4.63}$$

Substituting Equation 4.23b into Equation 4.62, the *current* or *fuel utilization efficiency* can be expressed as a function of the stoichiometric factor

$$\eta_I = \eta_{fc,fuel} = \frac{1}{\xi_f}. \tag{4.64}$$

Equation 4.64 shows that for a fuel stoichiometric factor greater than one, the current efficiency of the fuel cell is less than 100%.

Physically, a current or fuel utilization efficiency value represents the fraction of the fuel converted into current. The remaining fraction of the fuel leaves the cell without reacting or without being consumed for the production of the current. The excess fuel that exits the cell may be recycled back into the cell or may be burnt to produce heat for other system use.

4.5.4 Overall Efficiency

The overall efficiency of a fuel cell is then represented by a product of all three above-mentioned fuel cell efficiencies, that is, thermal efficiency, voltage efficiency, and current of fuel utilization efficiency, as

$$\eta_{fc} = \eta_{fc,rev} \times \eta_{fc,v} \times \eta_{fc,fuel}. \tag{4.65}$$

4.6 Fuel Consumption and Supply Rates

4.6.1 Oxygen Consumption and Supply Rates

Oxygen gas can be supplied directly for the cathodic reaction or it can be supplied in the form of air. Let us consider these two cases separately.

4.6.1.1 Direct Oxygen Consumption

Considering the number of electron charges as n_e for each mole of oxygen in a cathodic reaction, the total charge or current I is given as

$$I = n_e F \times \dot{n}_{O_2}. \tag{4.66}$$

The consumption of oxygen for a single fuel cell is then given as

$$\dot{n}_{O_2} = \frac{I}{n_e F}\,(\text{mol/s}) \quad \text{for a single cell.} \tag{4.67}$$

It can be noted here that in a hydrogen–oxygen fuel cell, the number of electron charge transferred for each mole of oxygen is $n_e = 4$.

For a stack of N_c number of cells, the oxygen consumption is

$$\dot{n}_{O_2} = \frac{I\,N_c}{n_e F}\,(\text{mol/s}) \quad \text{for a stack} \tag{4.68}$$

and in terms of mass flow rate of oxygen,

$$\dot{m}_{O_2} = \frac{I\,N_c M_{O_2}}{n_e F}\,(\text{kg/s}). \tag{4.69}$$

The gas consumption rate can also be expressed in terms of total power consumption rate for a stack. The power consumption in a single cell and in a stack is expressed as

$$P_c = V_c\,I. \tag{4.70}$$

and the total power of a fuel cell stack is expressed as

$$P_t = V_c\,I\,N_c, \tag{4.71}$$

where P_c and P_t are the power output of a single cell and total power output of fuel cell stack, respectively.

Rearranging, we get the expression for the average current in a cell as

$$I = \frac{P_t}{V_c\,N_c}. \tag{4.72}$$

Substituting Equation 4.72 for the current into Equation 4.68, we get the expression for *molar oxygen consumption* in terms of power as

$$\dot{n}_{O_2} = \frac{P_t}{n_e F V_c} \text{ (mol/s)} \quad \text{for the stack} \tag{4.73a}$$

and

$$\dot{n}_{O_2} = \frac{P_t}{n_e F V_c N_c} \text{ (mol/s)} \quad \text{for a cell} \tag{4.73b}$$

and in terms of *mass oxygen consumption rate* as

$$\dot{m}_{O_2} = \frac{M_{O_2} P_t}{n_e F V_c} \text{ (kg/s)} \quad \text{for the stack} \tag{4.74a}$$

and

$$\dot{m}_{O_2} = \frac{M_{O_2} P_t}{n_e F V_c N_c} \text{ (kg/s)} \quad \text{for a single cell.} \tag{4.74b}$$

Note that in a hydrogen–oxygen fuel cell, we can use $n_e = 4$ for the number of electron charge transferred for each mole of oxygen and $M_{O_2} = 32 \times 10^{-3}\,\text{kg/mol}$ in Equation 4.74.

4.6.1.2 Oxygen Consumption as Air

Considering air with oxygen mole fraction as Y_{O_2}, the number of moles of oxygen per kilogram of air is

$$n_{O_2} = \frac{Y_{O_2}}{M_{air}}. \tag{4.75}$$

Consumption of air for a cathodic reaction is then given as

$$\dot{m}_{air} = \frac{M_{air} P_t}{Y_{O_2} n_e F V_c} \text{ (kg/s).} \tag{4.76}$$

Usually, an excess amount of oxygen is supplied for sustaining complete reaction at the entire cathode active reaction area. Defining the excess air supply in terms of the stoichiometric factor, ξ_{air}, the air supply rate at inlet is given as

$$\dot{m}_{air} = \frac{\xi_{air} M_{air} P_t}{Y_{O_2} n_e F V_c} \text{ (kg/s)} \quad \text{for a stack} \tag{4.77a}$$

and

$$\dot{m}_{\text{air}} = \frac{\xi_{\text{air}} M_{\text{air}} P_{\text{t}}}{Y_{O_2} n_{\text{e}} F V_{\text{c}} N_{\text{c}}} \text{ (kg/s)} \quad \text{for a single cell.} \tag{4.77b}$$

In terms of total current, the air supply rate is given as

$$\dot{m}_{\text{air}} = \frac{\xi_{\text{air}} M_{\text{air}} I}{Y_{O_2} n_{\text{e}} F} \text{ (kg/s)} \quad \text{for a stack} \tag{4.78a}$$

and

$$\dot{m}_{\text{air}} = \frac{\xi_{\text{air}} M_{\text{air}} I}{Y_{O_2} n_{\text{e}} F N_{\text{c}}} \text{ (kg/s)} \quad \text{for a cell.} \tag{4.78b}$$

Considering air as a mixture of 21% oxygen and 79% nitrogen, we can use the mole fraction of oxygen as $Y_{O_2} = 0.21$ along with $M_{\text{air}} = 28.97 \times 10^{-3}$ kg/mol in Equation 4.78.

The *exit air flow rate* is determined as the difference between inlet air flow rate and the oxygen consumption:

$$\dot{m}_{\text{o,air}} = \dot{m}_{\text{i,air}} - \dot{m}_{O_2}. \tag{4.79}$$

Substituting Equations 4.74 and 4.78b, the exit air flow rate is given as

$$\dot{m}_{\text{o,air}} = \frac{\xi_{\text{air}} M_{\text{air}} P_{\text{t}}}{Y_{O_2} n_{\text{e}} F V_{\text{c}}} - \frac{M_{O_2} P_{\text{t}}}{n_{\text{e}} F V_{\text{c}}}. \tag{4.80}$$

4.6.2 Hydrogen Consumption and Supply Rates

In a similar manner, we can estimate hydrogen consumption in an anodic reaction. Considering the number of electron charges as n_{e} for each mole of hydrogen, in an anodic reaction, the mole consumption of hydrogen is given as

$$\dot{n}_{H_2} = \frac{I}{n_{\text{e}} F} \text{ (mol/s)} \quad \text{for a single cell} \tag{4.81a}$$

and

$$\dot{n}_{H_2} = \frac{I\,N_c}{n_e F}\text{(mol/s)} \quad \text{for a stack.} \tag{4.81b}$$

In terms of total fuel cell power, the *hydrogen mole consumption rate* is given as

$$\dot{n}_{H_2} = \frac{P_t}{n_e F V_c}\text{(mol/s)} \quad \text{for a stack} \tag{4.82a}$$

and

$$\dot{n}_{H_2} = \frac{P_t}{n_e F V_c N_c}\text{(mol/s)} \quad \text{for a cell.} \tag{4.82b}$$

The *mass hydrogen consumption rate* is given as

$$\dot{m}_{H_2} = \frac{M_{H_2} P_t}{n_e F V_c}\text{(kg/s)} \quad \text{for a stack} \tag{4.83a}$$

and

$$\dot{m}_{H_2} = \frac{M_{H_2} P_t}{n_e F V_c N_c}\text{(kg/s)} \quad \text{for a cell.} \tag{4.83b}$$

Note that in a hydrogen–oxygen fuel cell, we can use $n_e = 2$ for the number of electron charge transferred for each mole of hydrogen and $M_{H_2} = 2.02\times 10^{-3}$ kg/mol in Equations 4.82 and 4.83.

4.7 Water Production Rate

In a hydrogen–oxygen fuel cell, 1 mol of water is produced for every two electron charges. Thus, the water production rate is given as

$$\dot{n}_{H_2O} = \frac{I\,N_c}{n_e F}\text{(mol/s)} \tag{4.84a}$$

in terms of average current and

$$\dot{n}_{H_2O} = \frac{P_t}{n_e F V_c}\text{(mol/s)} \quad \text{for a stack} \tag{4.84b}$$

in terms of power and cell voltage.

The water mass production rate is expressed as

$$\dot{m}_{H_2O} = \frac{M_{H_2O} P_t}{n_e F V_c} \text{ (kg/s)} \quad \text{for a stack} \tag{4.85a}$$

and

$$\dot{m}_{H_2O} = \frac{M_{H_2O} P_t}{n_e F V_c N_c} \text{ (kg/s)} \quad \text{for a cell.} \tag{4.85b}$$

In Equations 4.84 and 4.85, the number of electron charge transferred for each mole of water produced is $n_e = 2$ for a hydrogen–oxygen fuel cell.

4.8 Heat Generation in a Fuel Cell

As we have discussed before, the operating voltage (E) of a fuel cell is less than the open circuit maximum possible voltage as current is drawn from the cell. This is due to the fact that only a fraction of fuel energy is available in the form of Gibbs energy for conversion, and the rest represents lost work. As we have discussed in an earlier section, the energy generated in the hydrogen fuel cell reaction is the enthalpy change, ΔH, of the hydrogen oxidation. The maximum available energy for conversion to electrical work is the change in Gibbs free energy expressed as $\Delta G = \Delta H - T\Delta S$. The difference ($-T\Delta S$) is the energy released in the form of heat owing to entropy change. This heat release is referred to as the *reversible heat generation*, Q_{rev}. Additionally, a part of the available energy is also lost owing to the number of irreversibilities associated with activation losses, mass transfer losses, and ohmic losses for resistances to ion and electron flows. A detailed description of these irreversibilities will be discussed in more detail in Chapters 5 through 7. As a consequence of these irreversibilities, a fraction of the energy is converted into heat within the fuel cell. This component of heat generation is referred to as the *irreversible heat generation*, Q_{irrev}. This heat energy results in a temperature distribution within the fuel cell and affects the cell's operating conditions. This waste heat has to be removed continuously in order to ensure a continuous isothermal operation of the fuel cell. Estimation of the waste heat generation in a fuel cell is important to determine the cooling requirement, to employ an appropriate cooling system to transfer heat from the cell, and to consider a thermal management system for better overall efficiency of the fuel cell power generation system. In this section, we primarily focus on the estimation of heat generation rate during a steady-state fuel cell operation.

For a given fuel supply rate ($\dot{n}_f$), a fraction (ϕ_f) of the fuel takes part in the electrochemical reaction, producing electric power and the by-products water and heat. The rest of it may either react chemically to produce the product water or perform a side chemical reaction to produce other products and hence produce additional heat energy. We can classify the following three consumption rates of fuel molecules:

i. Fraction of fuel participated in electrochemical reaction producing product: $\phi_f \dot{n}_f$
ii. Number of moles of fuel participated in non-electrochemical reaction producing product: $\varphi(1-\phi_f)\,\dot{n}_f$, where φ is the fraction of non-electrochemical reaction producing product
iii. Number of moles of fuel participated in non-electrochemical reaction producing another product: $(1-\varphi)(1-\phi_f)\,\dot{n}_f$.

Each one of these reaction terms will generate heat in the fuel cell. For simplicity, the consumption of fuel molecules by the side reactions that produce other products may be assumed as negligible as a first approximation.

4.8.1 Heat Generation owing to Electrochemical Reaction

The reversible heat generation owing to electrochemical reaction is due to the difference in the energy available in the form of enthalpy of formation and the energy available for electrical work; that is, the change in Gibbs free energy and can be estimated as follows:

$$Q_{\text{rev}} = \phi_f\, \dot{n}_f(\Delta H - \Delta G). \tag{4.86}$$

Using the thermodynamic relation equation, we can also express this reversible heat generation in terms of entropy change as

$$Q_{\text{rev}} = \phi_f\, \dot{n}_f(-T\Delta S) \tag{4.87a}$$

and for $\phi_f = 1$ as

$$Q_{\text{rev}} = \dot{n}_f(-T\Delta S), \tag{4.87b}$$

where the entropy change of the chemical reaction can be estimated directly on the basis of the procedure outlined in Chapter 3 and Section 4.3.

Noting that the fuel consumption rate can be expressed in terms of operating current as $\dot{n}_f = \frac{I}{n_e F}$, Equation 4.87 can be written as

$$Q_{gen,rev} = \frac{I}{n_e F}(-T\Delta S). \tag{4.87c}$$

It can be mentioned here that reversible heat generation can be computed for each of the half electrochemical electrode reactions separately and defined as follows:

$$Q_{gen,reva} = \frac{I}{n_e F}(-T\Delta S_a) \quad \text{for an } \textit{anode electrode} \tag{4.88a}$$

and

$$Q_{gen,revc} = \frac{I}{n_e F}(-T\Delta S_c) \quad \text{for a } \textit{cathode electrode} \tag{4.88b}$$

Estimation of entropy change for a half electrochemical reaction at an individual electrode is complicated, and therefore reversible heat generation in a fuel cell is often computed as a single term on the basis of the entropy change of the overall hydrogen oxidation reaction. For additional discussion on the procedure for the estimation of entropy change and the associated individual electrode, heat generation can be found in the literature (Fischer and Seume, 2006; Forland and Ratkje, 1980; Rajkje and Tomil, 1993).

Other major factors that contribute to the generation of heat are the irreversibility associated with the electrochemical reaction and the irreversible voltage drop owing to the resistance of ion and electron transport through the fuel cell components. With an operating fuel cell voltage V_c that is less than the reversible voltage E_{rev}, the heat generation rate owing to these irreversibilities and the associated lost electrical work is expressed as

$$Q_{gen,irrev} = n_e F(E_{rev} - V_c)\phi_f \dot{n}_f, \tag{4.89a}$$

where E_{rev} is the reversible open circuit potential and V_c is the actual terminal cell voltage.

Noting that the fuel consumption rate can be expressed in terms of operating current as $\dot{n}_f = \frac{I}{n_e F}$, the heat equation (Equation 4.89a) can be written as

$$Q_{\text{gen,irrev}} = I(E_{\text{rev}} - \dot{E}_{\text{c}})\phi_{\text{f}} \quad \text{for a cell} \tag{4.89b}$$

and

$$Q_{\text{gen,irrev}} = I(E_{\text{rev}} - \dot{V}_{\text{c}})\phi_{\text{f}} N_{\text{c}} \quad \text{for a stack.} \tag{4.89c}$$

An approximation to Equation 4.89 can be derived by assuming that the major deviation of operating voltage from reversible voltage is dominated by the irreversible voltage drop caused by the resistance of fuel cell components to electron and ion transport. Expressing such a voltage drop as $(E_{\text{rev}} - V_c) \approx \eta_{\text{ohm}} = I\,R_c$, Equation 4.89 can be approximated as

$$Q_{\text{gen,ohm}} = I\eta_{\text{ohm}} = I^2\dot{R}_{\text{c}}\,\phi_{\text{f}} \quad \text{for a cell} \tag{4.90a}$$

and

$$Q_{\text{gen,ohm}} = I^2 R_{\text{c}}\,\phi_{\text{f}} N_{\text{c}} \quad \text{for a stack.} \tag{4.90b}$$

This irreversible heat generation term is also referred to as the *ohmic heating of the fuel cell.*

Another approximation to Equation 4.89 is derived by assuming that the irreversible voltage drop is primarily caused by the activation overpotential (η_{act}) as

$$Q_{\text{gen,act}} = I\eta_{\text{act}}\,\phi_{\text{f}} \quad \text{for a cell} \tag{4.91a}$$

and

$$Q_{\text{gen,act}} = I\eta_{\text{act}}\,\phi_{\text{f}} N_{\text{c}} \quad \text{for a stack.} \tag{4.91b}$$

Again, a more accurate representation of this component irreversible heat generation can be given in terms of anode and cathode activation overpotentials as

$$Q_{\text{gen,acta}} = I\eta_{\text{act,a}}\,\phi_{\text{f}} \quad \text{for an anode reaction} \tag{4.92a}$$

and

$$Q_{\text{gen,actc}} = I\eta_{\text{act,c}}\,\phi_{\text{f}} \quad \text{for a cathode reaction.} \tag{4.92b}$$

The total heat generation owing to the electrochemical reaction part of the fuel cell operation is given as

$$Q_{\text{elec,gen}} = \phi_{\text{f}} \left[\dot{n}_{\text{f}}(-T\Delta S) + I(E_{\text{rev}} - V_{\text{c}}) \right] \quad \text{for a cell} \tag{4.93a}$$

or

$$Q_{\text{elec,gen}} = \phi_{\text{f}} \left[\dot{n}_{\text{f}}(\Delta H - \Delta G) + I(E_{\text{rev}} - V_{\text{c}}) \right] \quad \text{for a cell.} \tag{4.93b}$$

Considering ohmic heating as the primary contributor to heat generation owing to voltage loss, Equation 4.93a can be approximated as

$$Q_{\text{elec,gen}} = \phi_{\text{f}} \left[\dot{n}_{\text{f}}(-T\Delta S) + I^2 R_{\text{c}} \right] \quad \text{for a cell} \tag{4.94a}$$

and considering the activation overpotential as the primary contributor to heat generation owing to voltage loss, Equation 4.93a can be approximated as

$$Q_{\text{elec,gen}} = \phi_{\text{f}} \left[\dot{n}_{\text{f}}(-T\Delta S) + I\eta_{\text{act}} \right] \quad \text{for a cell} \tag{4.94b}$$

and substituting for $\dot{n}_{\text{f}} = \dfrac{I}{n_{\text{e}}F}$ as

$$Q_{\text{elec,gen}} = \phi_{\text{f}} \left[\frac{I}{n_{\text{e}}F}(-T\Delta S) + I\eta_{\text{act}} \right]. \tag{4.94c}$$

Heat generation at the electrode–electrolyte interface owing to electrochemical reaction can be expressed in terms of dominant cathode reaction and negligible anode reaction as

$$Q_{\text{gen}} = \left[\frac{T(-\Delta S)}{n_{\text{e}}F} + \eta_{\text{act,c}} \right] I. \tag{4.94d}$$

4.8.2 Heat Generation owing to Non-electrochemical Reaction

The second and third terms that represent the fraction of the reaction taking place in a direct chemical reaction are completely irreversible and produce heat and product water or a different product. The heat generation rate owing to these reaction terms is given by

$$Q_{\text{nelec-gen}} = (1-\phi_{\text{f}})(-\Delta H_{\text{cr}})\dot{n}_{\text{f}}, \tag{4.95}$$

where ΔH_{cr} is the enthalpy of formation for the chemical reaction.

4.8.3 Total Heat Generation in a Fuel Cell

Combining Equations 4.93a and 4.95, the total heat generation per cell is given as

$$Q_{\text{gen}} = \phi_{\text{f}}\left[\dot{n}_{\text{f}}(-T\Delta S) + I(E_{\text{rev}} - V_{\text{c}})\right] + (1-\phi_{\text{f}})(-\Delta H_{\text{cr}})\dot{n}_{\text{f}} \tag{4.96a}$$

or

$$Q_{\text{gen}} = \phi_{\text{f}}\left[\dot{n}_{\text{f}}(-T\Delta S) + I^2 R_{\text{c}}\right] + (1-\phi_{\text{f}})(-\Delta H_{\text{cr}})\dot{n}_{\text{f}} \tag{4.96b}$$

or

$$Q_{\text{gen}} = \phi_{\text{f}}\left[\frac{I(-T\Delta S)}{n_{\text{e}}F} + I^2 R_{\text{c}}\right] + (1-\phi_{\text{f}})(-\Delta H_{\text{cr}})\frac{I}{n_{\text{e}}F}. \tag{4.96c}$$

For $\phi_{\text{f}} = 1$, that is, all fuel moles going through electrochemical reaction, the equation reduces to

$$Q_{\text{gen}} = I\left[\frac{(-T\Delta S)}{n_{\text{e}}F} + (E_{\text{rev}} - V_{\text{c}})\right] \quad \text{for a cell} \tag{4.97a}$$

and

$$Q_{\text{gen}} = N_{\text{c}}I\left[\frac{(-T\Delta S)}{n_{\text{e}}F} + (E_{\text{rev}} - V_{\text{c}})\right] \quad \text{for a stack.} \tag{4.97b}$$

Another simple form of heat generation term can be estimated on the basis of the difference between the maximum possible voltage $E_{\text{max}} = \frac{-\Delta \overline{h}_{\text{f}}}{n_{\text{e}}F}$ and the real operating voltage V_{c} as

$$Q_{\text{gen}} = N_{\text{c}}I(E_{\text{max}} - V_{\text{c}}). \tag{4.98}$$

For a total fuel cell power of P_t, the total heat generation rate is given as

$$Q_{gen} = P_t \left[\frac{(-T\Delta S)}{n_e F V_c} + \left(\frac{E_{rev}}{V_c} - 1 \right) \right]. \tag{4.99}$$

Substituting

$$Q_{gen} = P_t \left[\frac{(\Delta H - \Delta G)}{n_e F V_c} + \left(\frac{E_{rev}}{V_c} - 1 \right) \right]$$

or

$$Q_{gen} = P_t \left[\left(\frac{E_{max} - E_{rev}}{V_c} \right) + \left(\frac{E_{rev}}{V_c} - 1 \right) \right] \quad \text{for a stack.} \tag{4.100}$$

Equation 4.100 is also equivalent to the heat generation on the basis of the assumption that net heat generation is due to the difference between the maximum possible cell voltage based on the enthalpy of reaction and the operating voltage.

$$Q_{gen} = P_t \left[\left(\frac{E_{max}}{V_c} - 1 \right) \right] \quad \text{for a stack} \tag{4.101}$$

If we consider the ohmic loss as the dominating factor, then Equation 4.94a can be reduced to

$$Q_{gen} = \left[\frac{I(-T\Delta S)}{n_e F} + I^2 R_c \right] \quad \text{per cell} \tag{4.102a}$$

and

$$Q_{gen} = N_c \left[\frac{I(-T\Delta S)}{n_e F} + I^2 R_c \right] \quad \text{for a stack.} \tag{4.102b}$$

Another simplified estimate of heat generation term is given by neglecting the ohmic heating for a first approximation as

$$Q_{gen} = \mathrm{I} \left[\frac{(-T\Delta S)}{n_e F} \right] \quad \text{for a cell} \tag{4.103a}$$

and for a stack.

$$Q_{\text{gen}} = N_{\text{c}} I \left[\frac{(-T\Delta S)}{n_{\text{e}} F} \right] \quad \text{for a stack} \tag{4.103b}$$

and

$$Q_{\text{gen}} = P_{\text{t}} \left[\frac{(-T\Delta S)}{n_{\text{e}} F V_{\text{c}}} \right] \quad \text{for a stack} \tag{4.103c}$$

The heat release must be removed from the cell to maintain isothermal operation, either by conduction heat dissipation through the fuel cell components or by convection from the external surface, heat convection by reactant gas flows, or use of a separate cooling stream through the system.

Example 4.3

Consider a hydrogen–oxygen fuel cell operating at 80°C and with a rated power output of P_{t} = 80 kW and assuming the cell voltage efficiency as $\eta_{\text{fc,v}}$= 50%. Estimate the following: (a) hydrogen mass flow supply rate, (b) oxygen mass flow supply rate, (c) air supply rate if oxygen is supplied as air, (d) water mass production rate, (e) the mass and volume of water produced in 1 h of operation of the fuel cell, and (f) the rate of heat generation.

Solution

The average cell operating voltage is computed from Equation 4.60 as

$$E_{\text{c}} = \eta_{\text{fc,v}} \times E_{\text{rev}}.$$

For the hydrogen–oxygen fuel cell operating at a temperature of 80°C, the reversible cell voltage is E_{rev} = 1.18 (see Table 4.2).

$$V_{\text{c}} = 1.18 \times 0.6 = 0.708 \text{ V}$$

Hydrogen supply rate:
The hydrogen supply rate is given by Equation 4.83a:

$$\dot{m}_{\text{H}_2} = \frac{M_{\text{H}_2} P_{\text{t}}}{n_{\text{e}} F V_{\text{c}}}.$$

Using the molecular weight of hydrogen gas as 2.02×10^{-3} kg/mol, we get

$$\dot{m}_{\text{H}_2} = \frac{2.02 \times 10^{-3} \times 80 \times 1000}{2 \times 96{,}485 \times 0.708}$$

$$m_{H_2} = 0.00118 \text{ kg/s} \ \ (2.11 \text{ cfm})$$

Oxygen supply rate:
The oxygen supply rate is given by Equation 4.74a:

$$\dot{m}_{O_2} = \frac{M_{O_2} P_t}{n_e F V_c}$$

$$\dot{m}_{O_2} = \frac{32 \times 10^{-3} \times 80 \times 1000}{4 \times 96,485 \times 0.708}$$

$$m_{O_2} = 0.009368 \text{ kg/s} \ \ (16.82 \text{ cfm})$$

Air supply rate:
The air supply rate is given by Equation 4.76:

$$\dot{m}_{air} = \frac{M_{air} P_t}{Y_{O_2} n_e F V_c}.$$

Using the molecular weight of air as 28.97 × 10^{-3} kg/mol, we get

$$m_{air} = \frac{28.97 \times 10^{-3} \times 80 \times 1000}{0.21 \times 4 \times 96,485 \times 0.708}$$

$$m_{air} = 0.0404 \text{ kg/s} \ (72.5 \text{ cfm})$$

The water production rate is given by Equation 4.85a:

$$\dot{m}_{H_2O} = \frac{M_{H_2O} P_t}{n_e F V_c}$$

$$m_{H_2O} = \frac{18.02 \times 10^{-3} \times 80 \times 1000}{2 \times 96,485 \times 0.708}$$

$$m_{H_2O} = 0.01055 \text{ kg/s}.$$

The amount of water produced in an hour is given as

$$m_{H_2O} = 0.01055 \times 3600 = 37.98 \text{ kg}.$$

Assuming the density of water as $\rho_{H_2O} = 1.0\,\text{g/cm}^3$, the volume of water generated is $\forall = 37{,}980\ \text{cm}^3 = 37.98$ liters for every 80 kWh of power generation.

Heat generation rate:
Assuming that the heat generation is due to the difference between the operating voltage and the maximum possible output voltage, the heat generation rate is given by Equation 4.101:

$$Q_{gen} = P_t\left(\frac{E_{max}}{V_c} - 1\right),$$

where

$$E_{max} = \frac{-\Delta\bar{h}_f}{n_e F} = \frac{284{,}088.5}{2 \times 96{,}485} = 1.472\ \text{V}.$$

Substituting, we get the heat generation rate as

$$= 80\left(\frac{1.472}{0.708} - 1\right)$$
$$= 86.32\ \text{kW}.$$

Let us now estimate heat generation from Equation 4.99:

$$Q_{gen} = P_t\left[\frac{(-T\Delta S)}{n_e F V_c} + \left(\frac{E_{rev}}{V_c} - 1\right)\right].$$

For $T = 80°\text{C}$, $\Delta S = -157.9078\ \text{kJ/kg} \cdot \text{K}$

$$Q_{gen} = 80\left[\frac{((80 + 273.2) \times 157.9078)}{2 \times 96{,}485 \times 0.708} + \left(\frac{1.18}{0.708} - 1\right)\right]$$

$$Q_{gen} = 80[0.408 + 0.6666] = 86\ \text{kW}.$$

4.9 Summary

Reversible cell voltage, $E^0_{rev} = \dfrac{-\Delta\bar{g}^0_f}{n_e F}$

Fuel Cell Efficiency

$$\text{Reversible efficiency, } \mu_{\text{fc,rev}} = \frac{\Delta\bar{g}_{\text{f}}}{\Delta\bar{h}_{\text{f}}}$$

$$\text{Voltage efficiency, } \eta_{\text{fc,v}} = \frac{E}{E^0_{\text{rev}}}$$

$$\text{Current or fuel utilization efficiency, } \eta_{\text{I}} = \eta_{\text{fc,fuel}} = \frac{1}{\xi_{\text{f}}}$$

$$\text{Overall efficiency, } \eta_{\text{fc}} = \eta_{\text{fc,rev}} \times \eta_{\text{fc,v}} \times \eta_{\text{fc,fuel}}$$

Reactant Gas Consumption and Supply Rate

$$\text{Hydrogen mass consumption rate, } \dot{m}_{\text{H}_2} = \frac{M_{\text{H}_2} P_{\text{t}}}{n_{\text{e}} F V_{\text{c}}}$$

$$\text{Oxygen mass consumption rate, } \dot{m}_{\text{O}_2} = \frac{M_{\text{O}_2} P_{\text{t}}}{n_{\text{e}} F V_{\text{c}}}$$

$$\text{Air consumption rate, } \dot{m}_{\text{air}} = \frac{M_{\text{air}} P_{\text{t}}}{Y_{\text{O}_2} n_{\text{e}} F V_{\text{c}}}$$

$$\text{Water production rate, } \dot{m}_{\text{H}_2\text{O}} = \frac{M_{\text{H}_2\text{O}} P_{\text{t}}}{n_{\text{e}} F V_{\text{c}}}$$

Heat Generation Rate

$$Q_{\text{gen}} = P_{\text{t}} \left[\frac{(-T\Delta S)}{n_{\text{e}} F V_{\text{c}}} + \left(\frac{E_{\text{rev}}}{V_{\text{c}}} - 1 \right) \right]$$

PROBLEMS

1. For a PEMFC with hydrogen and oxygen as reactants operating at 80°C, estimate (a) the cell reversible voltage, (b) the open circuit voltage based on the Nernst equation, (c) the reversible cell voltage and open circuit voltage when oxygen is supplied as air, and (d) the change in voltage caused by change in system operating pressure to 2 atm.
2. Consider a SOFC fuel cell operating at a temperature of 800°C with hydrogen gas stream consisting of 90% H_2 and 10% H_2O and air as a mixture of 21% oxygen and 79% nitrogen. Estimate (a) the cell reversible voltage and (b) the open circuit voltage based on the Nernst equation.

3. For a direct alcohol fuel cell (DAFC) with ethanol (CH_3OH) as fuel and operating at 100°C, estimate (a) the cell reversible voltage (b) the open circuit voltage based on the Nernst equation, and (c) the reversible cell voltage and open circuit voltage when oxygen is supplied as air.
4. Consider a hydrogen–oxygen fuel cell operating at 900°C and with a rated power output of P_t = 1.0 MW and assuming the cell voltage efficiency as $\eta_{fc,v}$= 50%. Estimate (a) hydrogen mass flow supply rate, (b) oxygen mass flow supply rate, (c) air supply rate if oxygen is supplied as air, (d) water mass production rate, (e) the mass and volume of water produced in 1 h of operation of the fuel cell, and (f) the rate of heat generation.

References

Atkins, P. W. *Physical Chemistry*, 6th Edition. Oxford University Press, 1998.

Breiter, M. W. *Electrochemical Processes in Fuel Cell*. Springer-Verlag, Heidelberg, 1969.

Chen, E. L. and P. I. Chen. Integration of fuel cell technology into engineering thermodynamics textbooks. Proceedings of the ASME 2001 IMECE Vol. 3 (CD-ROM), New York, November 11–16, 2001, ASME Paper AES-23647.

Fischer, K. and J. R. Seume. Location and magnitude of heat sources in solid oxide fuel cells. Proceedings of the 4th International Conference on Fuel Cell Science, Engineering and Technology, FUELCELL2006-97167, 2006.

Forland, T. and S. K. Ratkje. Entropy production by heat, mass, charge transfer and specific chemical reactions. *Electrochemical Acta* 25: 157–163, 1980.

Hamann, C. H., A. Hamnett and W. Vielstich. *Electrochemistry*. Wiley-VCH, New York, 1998.

Hart, A. B. and G. J. Womack. *Fuel Cells—Theory and Womack*. Chapman and Hall, London, 1967.

Hoogers, G. Editor. *Fuel Cell Technology Handbook*. CRC Press, Boca Raton, FL, 2003.

Ito, Y., H. Kaiya, S. Yoshizawa, S. K. Ratkje and T. Forland. Electrode heat balances of electrochemical cells. *Journal of the Electrochemical Society* 131: 2504–2509, 1984.

Larminie, J. and A. Dicks. *Fuel Cell System Explained*, 2nd Edition. Wiley & Sons, West Sussex, UK, 2003.

Lide, D. R. *CRC Handbook of Chemistry and Physics*, 76th Edition. CRC Press, Boca Raton, FL, 1995, pp. 5-63–5-69.

Newman, J. and K. E. Thomas-Alyea. *Electrochemical Systems*, 3rd Edition. Wiley Interscience, Hoboken, NJ, 2004.

O'Hayre, R. O., S.-W. Cha, W. Colella, and F. B. Prinz. *Fuel Cell Fundamentals*. John Wiley & Sons, Hoboken, NJ, 2006.

Ruka, R. J., J. E. Bauerle and L. Dykstra. Seeback coefficient of a $(ZrO_2)_{0.85}(CaO)_{0.15}$ electrolyte thermcell. *Journal of the Electrochemical Society* 115: 497–501, 1968.

Xianguo, L. *Principles of Fuel Cells*. Taylor & Francis, New York, 2006.

5

Electrochemical Kinetics

The electrochemical kinetics study involves the study of electrochemical reaction rates and the key factors that determine whether a reaction will be fast or slow, and how the reaction rate may be changed. The study of kinetics is important in the design and operation of a fuel cell. The rate of electron transfer at the electrodes or the current produced by the fuel cell depends on the rate of electrochemical reaction. The key factors that affect the electron transfer are ionic and electronic resistances in electrolyte and in electrodes and the rate of mass transport through the electrodes. In order to understand how these factors affect reaction rates, phenomena at molecular level during a chemical reaction need to be understood. The processes at the electrode and electrolyte interface are illustrated in Figure 5.1.

The processes that govern the electrode reaction rates are the mass transfer between the bulk solution and electrode surface, the electron transfer at the electrode, and the chemical reactions involving electron transfer. These processes are heterogeneous reactions between electrode and electrolyte and are characterized by both chemical and electrical changes. Several steps are involved in these reactions. For electron transfer to the electrodes, first electroactive species must be transported to the electrode surface by migration or diffusion. At the electrode, adsorption of electroactive material may be involved both before and after the electron transfer step. In the whole sequence of reactions, the slowest step determines the overall rate of the electrochemical process. In this chapter, we discuss electrochemical kinetics that governs the reaction rate and hence the rate of electrochemical energy output from a fuel cell.

5.1 Electrical Double Layer

When an electrode is immersed in an electrolyte, a potential is set up at the electrode–electrolyte interface, where the electronic charge on the electrode attracts ions with opposite charge and orients the solvent dipoles. There exist two layers of charge, one in the electrode and another in the electrolyte. This separation of charge set up is commonly known as the *electrical double layer.*

There are several reasons for the electrical double layer at the electrode. One reason is occurrence charge separation during the electron transfer

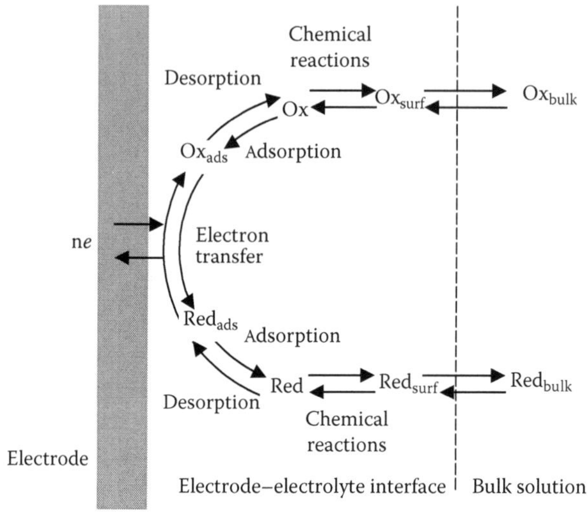

FIGURE 5.1
Processes at the electrode and electrolyte interface.

across the interface. Other reasons for the occurrence of potential differences are due to surface-active groups in the ionizable media and orientation of permanent or induced dipoles. The double layer at the interface has complex structures with electrical, compositional, and structural features. The electrical and compositional features are the excess charge densities on each phase (electrode and electrolyte) and the structural features are the distribution of the constituents such as ions, electrons, dipoles, and neutral molecules in the two phases and the interfacial region.

Many models have been put forward to explain the electrical, compositional, and structural aspects relevant to the electrochemical reactions that occur in fuel cells. Here, we introduce the evolution of the theoretical aspects that have been used to explain the effects occurring in this region.

The model put forward in the 1850s by *Helmholtz* is analogous to an electrical capacitor that has two plates of charge separated by some distance with the potential drop occurring in a linear manner between the two plates. In this model, no electron transfer reaction is assumed to occur at the electrode and the solution is composed only of electrolyte. Since the interface is to remain neutral, the charge held on the electrode is balanced by the redistribution of ions close to the electrode surface. The attracted ions are assumed to approach the electrode surface and form a layer balancing the electrode charge, the distance of approach is assumed to be limited to the radius of the ion and a single sphere of solvation round each ion. The overall result is two layers of charge (the double layer). The locus of the electrical centers of these solvated ions is called the outer Helmholtz plane (OHP) and a potential drop is confined to OHP in solution. The potential variation with distance under this model is shown in Figure 5.2. The potential drop across the interface is linear and the capacitance (CH) of the double layer, as in the case of a parallel-plate condenser, is given by

$$C_{H} = \frac{\varepsilon\varepsilon_0}{d}, \tag{5.1}$$

where ε is the dielectric constant in the medium between the plates, ε_0 is the permittivity of free space (8.85419×10^{-12} C^2 N^{-1} m^{-2} or Farad/m), and d is the separation between charges. Assuming $\varepsilon = 6$ and $d = 3$ Å, the value of C_H is approximately 17.7 μF cm^{-2}. The model of Helmholtz does not account for many factors such as diffusion/mixing in solution, the possibility of absorption onto the surface, and the interaction between solvent dipole moments and the electrode.

The second model is the *Gouy–Chapman model* developed in 1910 (Gouy, 1903, 1906; Chapman, 1913). In this model, the double layer is not as compact as in the Helmholtz rigid layer. The ions are assumed to be able to move in solution owing to thermal forces and thus the electrostatic interactions are in competition with Brownian motion. Figure 5.3 shows the charge distribution and potential from the electrode surface. The solvated ions interact with

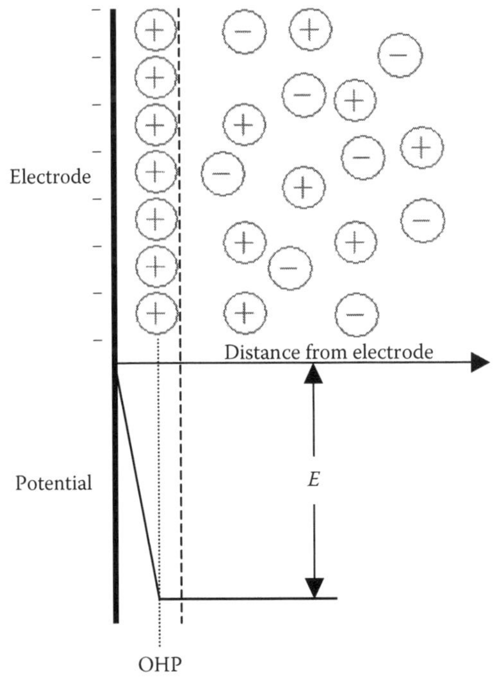

FIGURE 5.2
Helmholtz model of double layer and potential distribution.

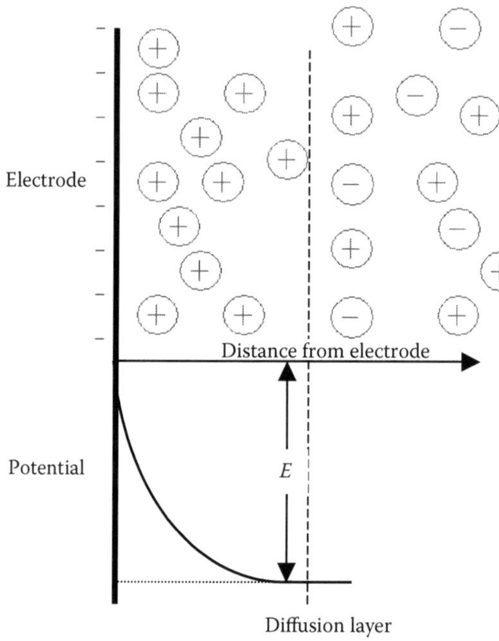

FIGURE 5.3
Gouy–Chapman model of double layer and potential distribution.

electrode with long-range electrostatic forces so that their interactions are independent of the chemical properties of the ions. These ions are distributed in the three-dimensional region called the diffuse layer that extends into the bulk of the solution. The double layer represents a compromise between electrical forces (tending to maintain the ordering) and thermal forces (tending to make the arrangement random).

The differential capacitance that accounts the capacitance for the diffuse layer, called *Gouy–Chapman capacitance,* is given as

$$C_G = \left(\frac{2N\varepsilon\varepsilon_0 z^2 e^2}{k_B T}\right)^{1/2} \cosh\frac{zeV}{2k_B T}, \tag{5.2}$$

where N is the number of ions of positive and negative sign per unit volume in the bulk of the electrolyte, z is the number of units of electronic charge, k_B is Boltzmann's constant, T is absolute temperature, e is charge of electron, and V is the potential drop from the electrode to the bulk of the electrolyte. For dilute aqueous solutions at 25°C, Equation 5.2 can be written in terms of bulk electrolyte concentration C^* (mol/L) as

$$C_G = 228\, zC^{*1/2} \cosh(10.5\, zV)\ (\mu F/cm^2). \tag{5.3}$$

Thus, for $V = 0$ V, $z = 1$, and $C^* = 1$ mol/L, the diffuse layer capacitance is 228 μF/cm².

The *Stern model* (1924) essentially combines the Helmholtz and Gouy–Chapman models as shown in Figure 5.4. Thus, the Stern model has two parts of double layer: (a) compact layer ("rigid layer") of ions at the distance of closest approach (OHP) and (b) diffuse layer. The concentration of ions and the potential distribution from the electrode vary as shown in Figures 5.3 and 5.4. The potential has a sharp drop between the electrode and OHP beyond which the potential gradually falls to a value characteristic of bulk electrolyte.

On the basis of the Stern model, the total capacitance for the double layer is

$$C = \left(\frac{1}{C_H} + \frac{1}{C_G} \right)^{-1}, \tag{5.4}$$

where C_H is given by Equation 5.1 and C_G is given by Equation 5.2. For concentrated electrolyte $C_G \gg C_H$, $C = C_H$; that is, the model is very similar to that of Helmholtz. For very dilute $C_G \ll C_H$, therefore, $C = C_G$. Though the Stern model shows reasonable values of C versus V relations for electrolytes with nonadsorbable ions such as Na^+ or F^-, it does not take into account the role

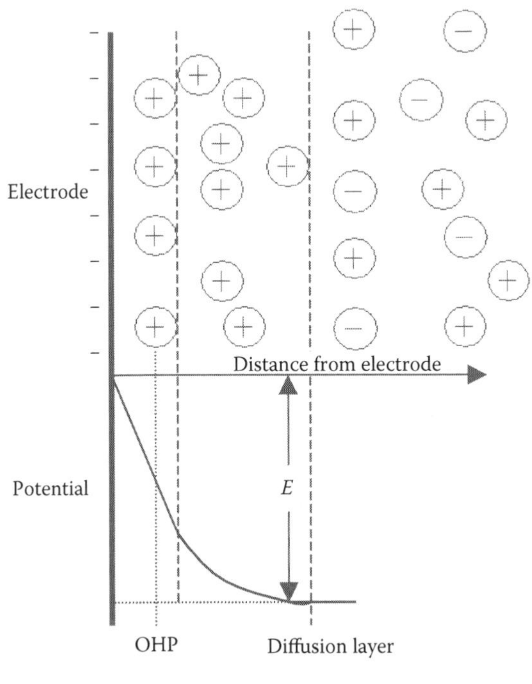

FIGURE 5.4
Stern model of double layer and potential distribution.

of the solvent as related to the hydration of the ions and its influence on the structure of the double layer.

Example 5.1

Using the Stern model, calculate the total capacitance of the double layer on an electrode with an applied potential of 0.1 V for a dilute electrolyte with a charge concentration of (i) 10^{-4} mol/L solution and (ii) 10^{-2} mol/L. Consider a single electron transfer reaction and dielectric constant of the media to be 10 and the separation between charges to be 10 Å.

Answer

The total capacitance of the double layer from the Stern model is given by Equation 5.4

$$C = \left(\frac{1}{C_H} + \frac{1}{C_G}\right)^{-1},$$

where C_H and C_G are given by Equations 5.1 and 5.3, respectively, as

$$C_H = \frac{\varepsilon\varepsilon_0}{d},$$

$C_G = 228\, zC^{*1/2} \cosh\,(10.5\, zV)$ (μF/cm^2).

Now, we have $\varepsilon = 10$, $d = 10$ Å, $z = 1$, $\varepsilon_0 = 8.85419 \times 10^{-12}$ F/m.

The Helmholtz capacitance is calculated as

$$C_H = \frac{\varepsilon\varepsilon_0}{d} = \frac{10 \times 8.85419 \times 10^{-12}\ (\text{F/m})}{10 \times 10^{10}\ (\text{m})} = 8.85 \times 10^{-2}\ \text{F/m}^2 = 8.85\ \mu\text{F/cm}^2.$$

The Gouy–Chapman capacitance C_G is calculated as

i. For concentration $C^* = 10^{-4}$ mol/L:

$$\begin{aligned} C_G &= 228\, zC^{*1/2} \cosh(10.5\, zV) \\ &= 228 \times 1 \times (10^{-4})^{1/2} \cosh(10.5 \times 1 \times 0.1\ \text{V})\ \mu\text{F/cm}^2 \\ C_G &= 3.66\ \mu\text{F/cm}^2 \end{aligned}$$

ii. For concentration $C^* = 10^{-2}$ mol/L:

$$\begin{aligned} C_G &= 228\, zC^{*1/2} \cosh(10.5\, zV) \\ &= 228 \times 1 \times (10^{-2})^{1/2} \cosh(10.5 \times 1 \times 0.1\ \text{V})\ \mu\text{F/cm}^2 \\ C_G &= 36.62\ \mu\text{F/cm}^2 \end{aligned}$$

Thus, the total capacitance is

i. For concentration $C^* = 10^{-4}$ mol/L:

$$C = \left(\frac{1}{C_H} + \frac{1}{C_G}\right)^{-1} = (1/8.85 + 1/3.66)^{-1}\ \mu F/cm^2 = 2.59\ \mu F/cm^2$$

ii. For concentration $C^* = 10^{-2}$ mol/L:

$$C = \left(\frac{1}{C_H} + \frac{1}{C_G}\right)^{-1} = (1/8.85 + 1/36.62)^{-1}\ \mu F/cm^2 = 7.130\ \mu F/cm^2$$

A fourth model proposed, the *Grahame model* (Grahame, 1951), which is referred to as the *triple-layer model*, takes into consideration that ions could be dehydrated in the direction of the electrode and specifically adsorbed on the electrode. Thus, an *inner layer* between the electrode surface and the Helmholtz layer further modifies the structure of the double layer. The locus of electrical centers of unhydrated ions strongly attached to the electrode is called *inner Helmholtz plane* (IHP). Figure 5.5

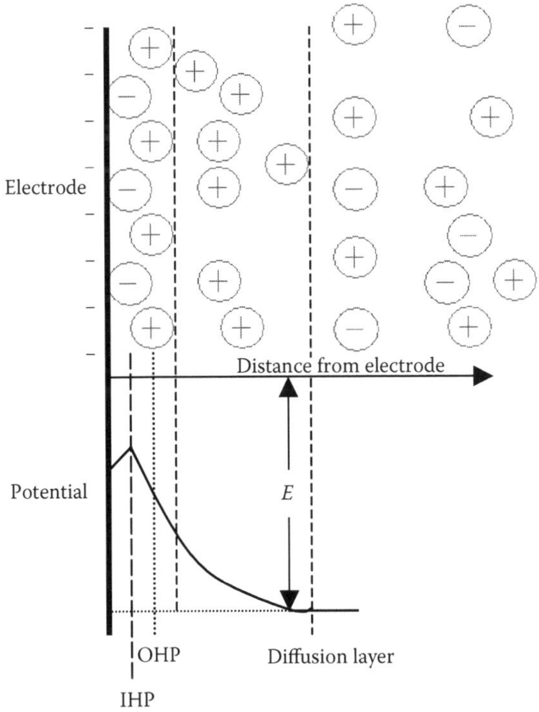

FIGURE 5.5
Triple-layer models of double layer and potential distribution.

shows the triple-layer model and the potential from the electrode surface. For this model, the capacitance is given as

$$C = \left\{ \frac{1}{C_I} + \left(\frac{1}{C_H} + \frac{1}{C_G} \right) \frac{dq_1}{dq_E} \right\}^{-1}, \tag{5.5}$$

where C_I is the capacitance of the space between the electrode and the IHP and can be calculated with an equation similar to Equation 5.1, and dq_1/dq_E represents the rate of change of the specifically adsorbed charge with charge on the electrode.

5.2 Electrode Kinetics

As seen from Figure 5.1, the electron transfer occurs at the interface between the electrode and the electrolyte. In this section, we present relationships between the heterogeneous rate constants for electron transfer and the voltage and introduce the reversible and irreversible processes in the context of electrolysis reactions. Since the potential (voltage) is a measure of electron energy, changing the applied voltage can vary the rate of the electron transfer reaction at the electrode surface. Since V (volt) = J (joule)/e (coulomb), a volt is simply the energy (J) required to move charge (c). The voltage applied to an electrode supplies electrical energy to move the electron. The transfer involves quantum-mechanical tunneling of electrons between the electrode and the electroactive species. Since electrons possess charge, an applied voltage can alter the "energy" of the electrons within a metal electrode.

This can be explained by band theory in which the behavior of an electron motion in the field of atomic nuclei and other electrons is treated. The available energy states for electrons in the materials form bands instead of having discrete energies as in the case of free atoms. The electrons that are free to move form an energy band called *conduction band* and the valence electron energy form a *valence band*. Crucial to the conduction process is whether or not there are electrons in the conduction band. In insulators, the electrons in the valence band are separated by a large gap from the conduction band; in conductors like metals, the valence band overlaps the conduction band, and in semiconductors, there is a small enough gap between the valence and conduction bands that thermal or other excitations can bridge the gap. An important parameter in the band theory is the *Fermi level* (E_F), the highest available electron energy level at low temperatures. In an electrolytic cell, the electrode potential is controlled. In Figure 5.6, the energy of electrons (Fermi levels) in the electrode and the lowest unoccupied orbital of the species in the electrolyte are shown. If they were at the

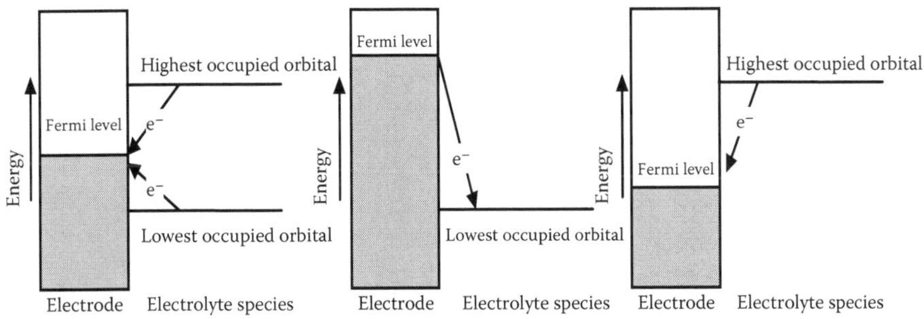

FIGURE 5.6
Fermi level in electrode for different applied potential.

same energy, electron transfer would be taking place but at equal rates in both directions. Thus, at equilibrium electrode potential, the Fermi levels are equal and there is no net electron transfer between the electrode and the electrolyte. Thus, the oxidation and reduction processes are balanced. The greater the difference, the faster the rate of reduction of the species in the electrolyte.

By increasing the electrode potential (voltage) in the negative direction, the electron energy in the electrode is increased. It is therefore thermodynamically favorable for an electron to jump from the electrode to the electrolyte and the reduction process occurs. If the electrode potential is increased in the positive direction, then the electron energy in the electrode is lower than that in the electrolyte. It is thermodynamically unfavorable for the electron transfer to occur from electrolyte to electrode leading to oxidation. Thus, using electrode potential, oxidation and reduction processes can be controlled. Though the electron transfer is possible depending on the applied potential to the electrode, the actual occurring of the electron transfer is limited and it depends upon the rate (kinetics) of the electron transfer reaction. The reaction rates are limited because of the energy barrier called an *activation energy*, which slows the reaction rates (Bard and Faulkner, 1980; Bockris and Srinivasan, 1969).

For the oxidation process, even though the electrons are energetically downhill, they have to overcome a hump, an activation energy barrier, before transferring to the electrolyte. This is illustrated in Figure 5.7, where the reaction proceeds if the free energy can overcome the activation energy hill. The free energy of activation determines the probability of electron transfer.

Consider an electrochemical reaction occurring at an electrode between reduction (Red) and oxidation (Ox) forms of a chemical species

$$Ox^{+} + ne^{-} \underset{k_b}{\overset{k_f}{\rightleftharpoons}} Red. \tag{5.6}$$

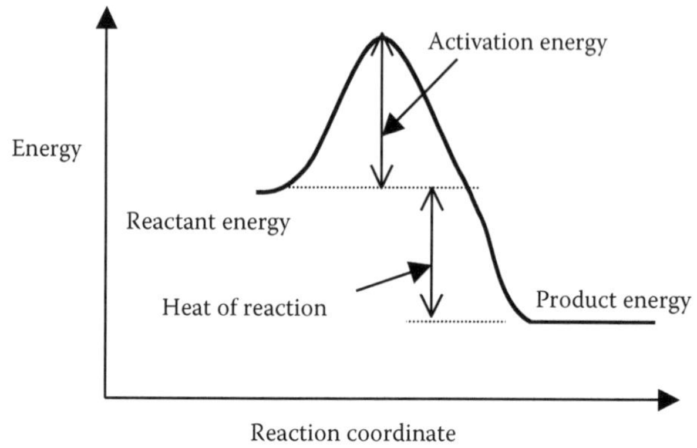

FIGURE 5.7
An activation energy barrier for conversion of reactant to products.

Here, the rates of reactions (mol/s) for forward and backward reactions are dependent on the reaction rate constants (#/s) k_f and k_b, respectively.

As shown in Figure 5.8, the rate constants k_f and k_b depend on the free energies of activation ΔG_f and ΔG_b for forward and backward reactions. The reaction rate constant in general is given by the Arrhenius expression

$$k = fe^{-\frac{G}{RT}}, \tag{5.7}$$

where ΔG is the free energy of activation, f is the frequency factor for the reaction constant, R is the universal gas constant, and T is the absolute

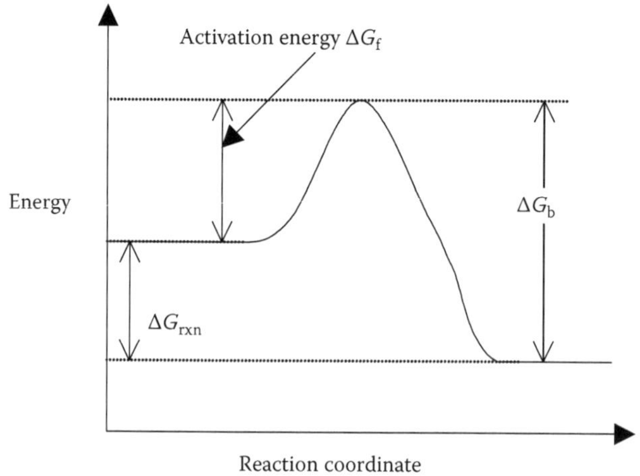

FIGURE 5.8
Free-energy changes during a reaction.

temperature. From statistical mechanics, the frequency factor f is given as $k_B T/h$, where k_B is Boltzmann constant and h is Planck's constant.

The reaction rate v is the product of the reaction rate constant and the reactant concentration. For reduction process, the reaction rate v_f is written as

$$v_f = C_{Ox,surf} k_f = C_{Ox,surf} \frac{k_B T}{h} e^{-\frac{G_f}{RT}}, \tag{5.8}$$

where $C_{Ox,surf}$ is the concentration of oxidant species. Here, the reaction rate for reduction process is

$$k_f = \frac{k_B T}{h} e^{-\frac{G_f}{RT}}. \tag{5.9}$$

Similarly, the reaction rate for oxidation reaction is given as

$$v_b = C_{Red,surf} k_b = C_{Red,surf} \frac{k_B T}{h} e^{-\frac{G_b}{RT}}, \tag{5.10}$$

where $C_{Ox,surf}$ is the concentration of reductant species. And the rate constant for oxidation reaction is

$$k_b = \frac{k_B T}{h} e^{-\frac{G_b}{RT}}. \tag{5.11}$$

The overall rate of electrochemical reaction is the difference between rate of forward and backward reaction rates

$$v_{net} = C_{Ox,sufr} k_f - C_{Red,surf} k_b. \tag{5.12}$$

For positive net reaction ($v_{net} > 0$), the reduction reaction dominates, and for negative net reaction ($v_{net} < 0$), the oxidation reaction dominates. Note that by definition, the reductive current is negative and the oxidative current is positive; the difference in sign simply tells us that the current flows in opposite direction across the interface depending upon an oxidation or reduction. The free energies of activation for the electrode reaction are related to both the chemical properties of the reactants/transition state and the response of both to potential.

Now, consider oxidation and reduction reactions taking place in an electrochemical cell such as the fuel cell with anode and cathode electrodes. Since the reaction rate is proportional to the current, for a given forward (reduction) reaction rate, v_f, the cathodic current is

$$i_c = nFAv_f, \tag{5.13}$$

where A is the electrode surface area, F is the Faraday constant, and n is the number of electrons transferred. Similarly, the anodic current is given as

$$i_a = nFAv_b. \tag{5.14}$$

Now, the total current flowing i is the difference between the reductive i_c and oxidative i_a currents

$$i = i_c - i_a. \tag{5.15}$$

From Equations 5.13 and 5.14, the total current is given as

$$i = nFA\,\{C_{Ox,surf}k_f - C_{Red,surf}k_b\}. \tag{5.16}$$

It should be noted here that the concentrations in the rate expression are always the electrode surface concentrations. The surface concentration may differ from the bulk concentrations. As will be seen in later sections, the electrode kinetic behavior is strongly influenced by the interfacial potential difference.

5.3 Single- and Multistep Electrode Reactions

In an electrode reaction, there is at least one step that involves transfer of electron between electrode and electrolyte in the overall electrochemical reaction. In the reaction, there may be a single electron transfer step or multiple steps. The simplest reaction at the electrode is single electron transfer. For example,

$$Fe(H_2O)_6^{3+} + e^- \rightarrow Fe(H_2O)_6^{2+} \tag{5.17}$$

$$Fe(CN)_6^{3-} + e^- \rightarrow Fe(CN)_6^{4-}. \tag{5.18}$$

In these reactions, the electron is transferred to and from the electrode. Similarly, in a secondary lithium ion battery, lithium electrodeposition/dissolution is a single-step reaction

$$Li^+ + e^- \rightarrow Li. \tag{5.19}$$

For single-step reactions, the kinetics reactions are relatively simple.

Even though the reaction is represented as a single electron transfer step, there are other steps that take place. For example, the electrodeposition step is followed by nucleation, surface diffusion of lithium, and crystal growth. In general, the reaction has to overcome several energy barriers unlike that shown in Figure 5.5.

In a successive reaction, two or more intermediate steps occur in series, that is, an intermediate produced in the first step is reacted in the second. If more than two intermediate steps are involved, then the species produced in the second step reacts in the third step. An example of this successive reaction of relevance to fuel cells is the electroreduction of oxygen to water reaction:

$$O_2 + 4H_3O^+ + 4\,e^- \rightarrow 6H_2O.$$

This reaction involves four electron transfers on a platinum electrode in acid medium with successive reactions:

$$\left.\begin{aligned} &O_2 + M + H_3O^+ + e^- \rightarrow MHO_2 + H_2O \\ &MHO_2 + M \rightarrow MO + MOH \\ &MO + H_3O^+ + e^- \rightarrow MOH + H_2O \\ &2MOH + 2H_3O^+ + 2e^- \rightarrow 2M^+ + 4H_2O \end{aligned}\right\} \quad (5.20)$$

Here, M represents the electronically conducting electrode material (e.g., Pt) that is not involved in the overall reaction and plays the role of an electrocatalyst for the reaction. The last intermediate step occurs in two identical consecutive steps since electron transfer occurs by quantum mechanical tunneling, which involves only one electron transfer at a time. When multistep reactions take place, there is the possibility of parallel-intermediate steps. The parallel-step reactions could lead to the same final product or to different products. Direct electro-oxidation of organic fuels, such as hydrocarbons or alcohols, in a fuel cell exhibits this behavior. For instance, in the case of methanol, a six-electron transfer, complete oxidation to carbon dioxide can occur consecutively in six or more consecutive steps. In addition, partially oxidized reaction products could arise, producing formaldehyde and formic acid in parallel reactions. These, in turn, could then be oxidized to methanol.

In multistep reactions, the rate-determining step is referred to as the step in reactions that proceed in two or more intermediate stages, either consecutively or in parallel. The reaction does not take place in one smooth process over a single energy barrier as shown in Figure 5.5, but with multiple energy barriers as shown in Figure 5.9. For a consecutive reaction with about five intermediate steps, one can show from a plot of the free energy versus distance along a reaction coordinate that the step exhibiting the highest energy state with respect to the initial or final state controls the rate of the reaction.

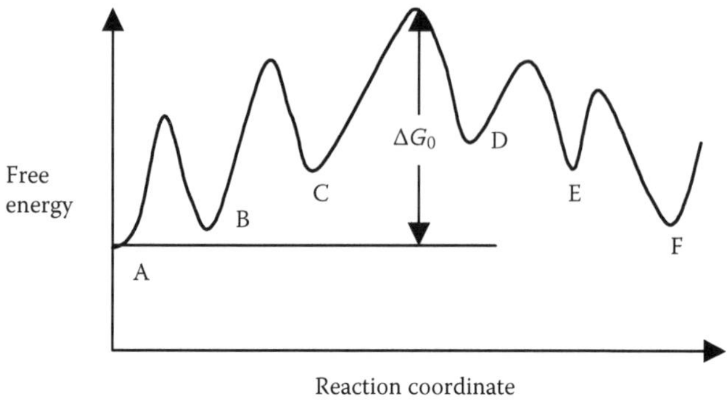

FIGURE 5.9
Typical free energy versus distance along reaction coordinate plot.

For the chemical reaction represented in Figure 5.9, the rate of the forward reaction is determined by the highest energy barrier

$$v = k_{C\to D}\, C_R, \tag{5.21}$$

where $k_{C\to D}$ is the rate constant for forward reaction and C_R is the concentration of reactant.

To visualize the concept of the rate-determining step, electrical circuit analogues are useful. For a consecutive reaction, an electrical circuit with a series of two or more resistances and power can be used, and for a parallel reaction, parallel resistance and power as shown in Figure 5.10 can be used. For consecutive reactions with three reaction steps, the electric circuit has three resistors, R_1, R_2, and R_3, in series; in addition, the power source (a fuel cell or a battery) has an internal resistance, R_i. The current (I) through the electrical circuit is given by the expression

$$I = \frac{E}{R_1 + R_2 + R_3 + R_i}. \tag{5.22}$$

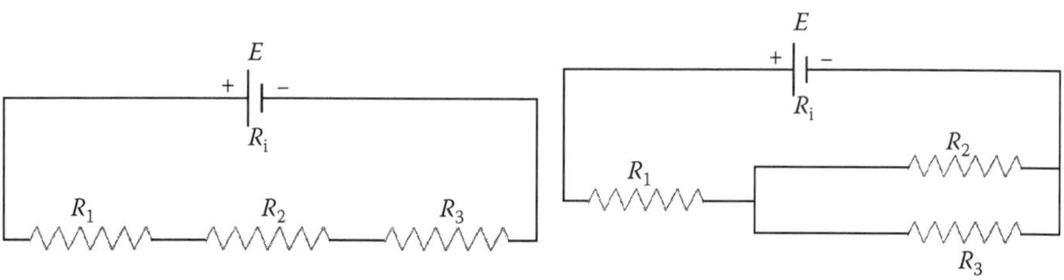

FIGURE 5.10
Circuit analogues of energy barrier for multistep reaction.

If we assume R_1, R_2, and R_i are small resistances, then the current is given by

$$I = \frac{E}{R_2}. \tag{5.23}$$

Thus, the reaction corresponding to the resistance R_2 is the rate-determining reaction.

For a consecutive reaction and two parallel reactions, the circuit has a series resistance followed by two parallel resistors and the current is given by

$$I = \frac{E}{R_1 + R_i + \dfrac{R_2 R_3}{R_2 + R_3}}. \tag{5.24}$$

Again, if R_i and R_1 are much less than R_2 or R_3 and that $R_2 \ll R_3$, I will approximate to $I = E/R_2$.

In terms of a chemical reaction, one may consider the sequence

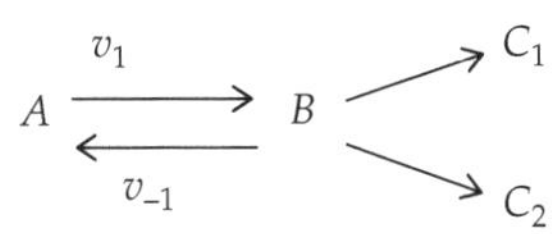

(5.25)

For simplicity, the rates (v) of the intermediate steps C_1 to B and C_2 to B may be considered negligible. Thus,

$$v_1 - v_{-1} = v_2 + v_3. \tag{5.26}$$

If $v_2 > v_3$, then

$$v_1 - v_{-1} = v_2. \tag{5.27}$$

Further, if $v_1 \gg v_2$, then

$$v_1 = v_{-1}. \tag{5.28}$$

Thus, the step $A \rightarrow B$ is virtually in equilibrium and the step $B \rightarrow C$ controls the rate of the overall reaction.

For a single-step reaction, the rate of reaction is given by Equation 5.11. For multistep reactions, calculations of potential energy versus reaction coordinate are much more complex and sophisticated since the interaction energies between multiple atoms need to be considered.

The rate constants for the forward and backward reactions, k_f, k_b, can then be modified to

$$k_\mathrm{f} = \frac{k_\mathrm{B}T}{h} e^{-\frac{\Delta G_{i\to g}}{RT}} \tag{5.29}$$

and

$$k_\mathrm{b} = \frac{k_\mathrm{B}T}{h} e^{-\frac{\Delta G_{n\to g}}{RT}} \tag{5.30}$$

for a reaction that occurs in n-steps with the gth step being the rate-controlling step.

As mentioned earlier on the energetic reactions, the reactions rates can be modified by changing the potential difference across the electrode and the electrolyte. Consider the electron transfer reaction

$$\mathrm{Ox}^+ + n\mathrm{e}^- \underset{k_\mathrm{Ox}}{\overset{k_\mathrm{Red}}{\rightleftharpoons}} \mathrm{Red}. \tag{5.31}$$

The rate constants k_{Ox} and k_{Red} can be influenced by the applied voltage according to the transition state theory from chemical kinetics. When a potential is applied to the electrode, the potential barrier changes as shown in Figure 5.11. The height of the potential barrier depends on the electrode

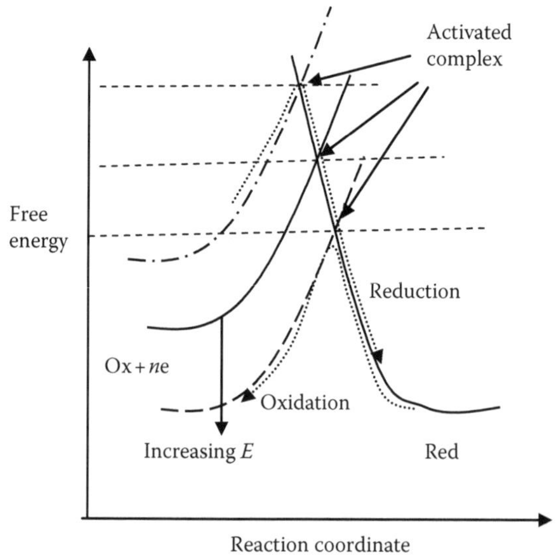

FIGURE 5.11
Effect of applied electrode potential on oxidation and reduction reactions.

potential E. The electrode potential E is shifted from the equilibrium value (solid curve) to a more positive value (dashed curve below solid curve). An increase in E favors the anodic direction (oxidation) and disfavors the cathodic direction (reduction). In other words, anodic reactions require a positive value for the overpotential to increase the rate of reaction, whereas for cathodic reactions, a negative value for the overpotential is required to increase the rate of reaction.

Thus, for applied voltage, the free-energy profiles are modified and forward or reverse reactions are thermodynamically favored as the overall barrier height (i.e., activation energy) is altered as a function of the applied voltage. In general, rate constants for the forward and reverse reactions will be altered by the applied voltage. The relative energy of the electron resident on the electrode changes by $-nF(E - E)^0$ for change in potential by E.

In order to formulate a model, we will assume that the effect of voltage on the free-energy change will follow a linear relationship (this is undoubtedly an oversimplification). Using this linear relationship, the activation free energies for reduction and oxidation will vary as a function of the applied voltage (E). Now, from Figure 5.12, the barrier for oxidation $\Delta G_a(E)$ is reduced by a fraction β of the total energy change $-nF(E - E^0)$. The free energy of activation for oxidation owing to potential E is

$$\Delta G_a(E) = \Delta G_{0a} - \beta F(E - E^0), \tag{5.32}$$

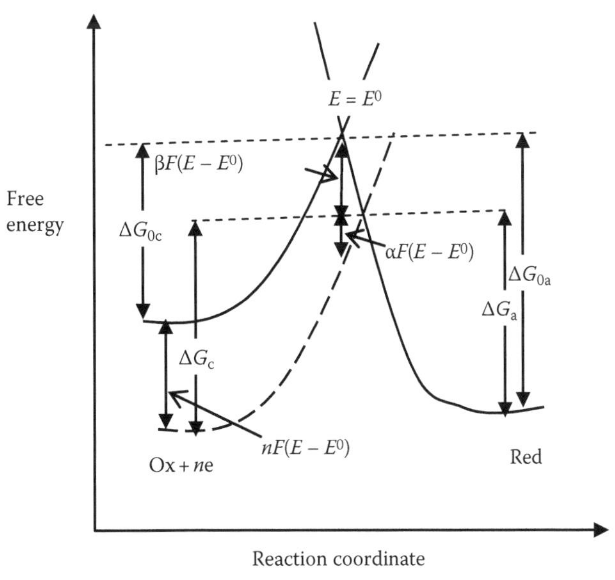

FIGURE 5.12
Effect of potential on the free energies of activation for oxidation and reduction.

where ΔG_{0a} is the free energy at reversible potential $E = E^0$. The potential E^0 corresponds to reversible potential at which chemical species at the electrode surface are equal for forward and reverse reactions.

Again, from Figure 5.9, the barrier for cathodic reaction at the same electrode the free energy of activation increases by $\alpha F(E - E^0)$. Thus,

$$\Delta G_c(E) = \Delta G_{0c} + \alpha F(E - E^0). \tag{5.33}$$

The dimensional parameters α and β are called the *transfer coefficients* for reduction and oxidation, respectively. They reflect the sensitivity of the transition state to the applied voltage. Since n is the total number of electrons transferred, the transfer coefficients are related by the following equation

$$\alpha + \beta = n. \tag{5.34}$$

If α and β = 0, then the transition state shows no potential dependence. Physically, it provides an insight into the way the transition state is influenced by the voltage. For a single-step reaction involving a single electron transfer ($n = 1$),

$$\alpha + \beta = 1. \tag{5.35}$$

Then,

$$\beta = 1 - \alpha. \tag{5.36}$$

The transfer coefficient has values between zero to unity, and most cases of reactions on a metallic surface, it is around 0.5.

The free energy on the right-hand side of both of the above equations can be considered as the chemical component of the activation free-energy change; that is, it is only dependent upon the chemical species and not the applied voltage. We can now substitute the activation free energy terms above into the expressions for the oxidation and reduction rate constants, which give

$$k_{\text{Ox}} = \frac{k_B T}{h} e^{-\Delta G_a(E)/RT} \tag{5.37}$$

$$k_{\text{Red}} = \frac{k_B T}{h} e^{-\Delta G_c(E)/RT}. \tag{5.38}$$

These rate equations now can be written using Equations 5.32 and 5.33 as

$$k_{\text{Ox}} = \frac{k_B T}{h} e^{-\Delta G_{0a}/RT} e^{\beta F(E-E^0)/RT} \tag{5.39}$$

$$k_{\text{Red}} = \frac{k_B T}{h} e^{-\Delta G_{0c}/RT} e^{-\alpha F(E-E^0)/RT}. \tag{5.40}$$

On the right-hand side of Equations 5.39 and 5.40, the first term is independent of potential. By defining the rate constants $k_{0\text{Red}}$ and $k_{0\text{Ox}}$ (cm/s) as the measure of reaction rate when $E = E^0$, the rate constants can be written as

$$k_{\text{Ox}} = k_{0\text{Ox}} e^{\beta F(E-E^0)/RT} \tag{5.41}$$

$$k_{\text{Red}} = k_{0\text{Red}} e^{-\alpha F(E-E^0)/RT}, \tag{5.42}$$

where $k_{0\text{Red}}$ and $k_{0\text{Ox}}$are given as

$$k_{0\text{Ox}} = \frac{k_B T}{h} e^{-\Delta G_{0a}/RT} \tag{5.43}$$

$$k_{\text{Red}} = \frac{k_B T}{h} e^{-\Delta G_{0c}/RT}. \tag{5.44}$$

These results show that rate constants for the electron transfer steps are proportional to the exponential of the applied voltage. Thus, the rate of reaction can be changed simply by varying the applied voltage. It can be seen that with an increase in applied voltage E, ΔG_a reduces. Hence, anionic reaction is faster and electrons are more easily transferred from the solution to the metal. For a decrease in E, ΔG_c decreases and the cathodic reaction is faster. Going from E^0 to $E > E^0$, the Gibbs free energy of electrons in the metal is lowered, which makes electron transfer to the metal more likely.

5.4 Electrode Reaction in Equilibrium—Exchange Current Density

The rate of an electrochemical reaction at an electrode/electrolyte interface is expressed as a current density (A/cm^2 or mA/cm^2) and is measured at constant temperature. Determination of the variation of the current density as a function of the potential is one of the most important diagnostic criteria in elucidating the mechanism of an electrochemical reaction, that is, the reaction path, intermediate steps, and the rate-determining step. In any reaction, the overall reaction is the net difference between the forward and reverse reactions. Consider the single-step electron transfer reaction under no overpotential at the electrode.

$$\textit{Reduction reaction: } \mathrm{Ox} + ne^- \rightarrow \mathrm{Red} \quad (5.45)$$

$$\textit{Oxidation reaction: } \mathrm{Ox} + ne^- \Downarrow \mathrm{Red} \quad (5.46)$$

If reaction rates for the forward (reduction) reaction is v_1 and that for the backward (oxidation) reaction is v_2, then the net reaction rate is

$$v = v_1 - v_2. \quad (5.47)$$

In general, the reverse (i.e., oxidation) and forward (reduction) reactions rates may not be equal. We have the reaction rates

$$v_1 = C_{\mathrm{Ox,surf}}(k_B T/h)e^{-\Delta G_1/RT - \alpha F(E-E^0)/RT} \quad (5.48)$$

$$v_2 = C_{\mathrm{Red,surf}}(k_B T/h)e^{-\Delta G_2/RT + \beta F(E-E^0)/RT}, \quad (5.49)$$

where $C_{\mathrm{Ox,surf}}$ is the concentration of the reactant surface of the electrode and $C_{\mathrm{Red,surf}}$ is the concentration of the product at the surface of the electrode, ΔG_1 is the activation barrier for the forward reaction, and ΔG_2 is the activation barrier for the reverse reaction. From Figure 5.6, we see that the forward activation barrier ΔG_1 and reverse activation barrier ΔG_2 are related to the change in free energy for reaction ΔG_{rxn} given as

$$\Delta G_{\mathrm{rxn}} = \Delta G_1 - \Delta G_2. \quad (5.50)$$

Here, the reaction is assumed as first order with respect to the concentrations of reactants and products. For fuel cells, current or current density is of interest; hence, the reaction rates are expressed in terms of current density.

The current density is related to the reaction rate as

$$j = nFv. \quad (5.51)$$

Now, the forward current density is given as

$$j_1 = nFC_{\mathrm{Ox,surf}}(k_B T/h)e^{-\Delta G_1/RT - \alpha F(E-E^0)/RT}, \quad (5.52)$$

and for the reverse direction, the current density is

$$j_2 = nFC_{\mathrm{Red,surf}}(k_B T/h)e^{-\Delta G_2/RT + \beta F(E-E^0)/RT}. \quad (5.53)$$

For the case when there is a thermodynamic equilibrium, the forward and reverse reactions must balance each other so that the net current density is zero.

$$j_1 - j_2 = 0. \tag{5.54}$$

Thus, for a thermodynamic equilibrium condition,

$$j_1 = j_2 = j_0. \tag{5.55}$$

j_0 is called the *exchange current density* for the reaction. At equilibrium, the net reaction rate is zero, though forward and reverse reactions take place.

The exchange current density is now:

$$j_0 = nFC_{\text{Ox,surf}} k_{0\text{Red}} e^{-\alpha F(E_{\text{eq}}-E^0)/RT} = nFC_{\text{Red,surf}} k_{0\text{Ox}} e^{\beta F(E_{\text{eq}}-E^0)/RT}, \tag{5.56}$$

where E_{eq} is the equilibrium potential. The rate constants $k_{0\text{Red}}$ and $k_{0\text{Ox}}$ are defined as

$$k_{0\text{Ox}} = (k_B T/h) e^{-\Delta G_1/RT} \tag{5.57}$$

$$k_{0\text{Red}} = (k_B T/h) e^{-\Delta G_2/RT}. \tag{5.58}$$

For equilibrium condition, the rate constants are equal.

$$k_{0\text{Red}} = k_{0\text{Ox}} = k_0. \tag{5.59}$$

At equilibrium, the bulk concentrations are also found at the electrode surface

$$C_{\text{Red,surf}} = C_{\text{Red,bulk}} \text{ and } C_{\text{Ox,surf}} = C_{\text{Ox,bulk}}. \tag{5.60}$$

Hence, from Equations 5.56 and 5.59, and noting that $\alpha + \beta = n$,

$$e^{nF(E_{\text{eq}}-E^0)/RT} = C_{\text{Ox,bulk}}/C_{\text{Red,bulk}}. \tag{5.61}$$

Also,

$$e^{nF(E_{\text{eq}}-E^0)/RT} = C_{\text{Ox,surf}}/C_{\text{Red,surf}}. \tag{5.62}$$

Solving for E_{eq},

$$E_{\text{eq}} = E^0 + RT/nF \ln(C_{\text{Ox,bulk}}/C_{\text{Red,bulk}}), \tag{5.63}$$

and also we have

$$E_{eq} = E^0 + RT/nF \ln(C_{Ox,surf}/C_{Red,surf}), \tag{5.64}$$

which is the *Nernst relation*. The first term on the right-hand side represents the standard equilibrium potential for conditions of unit activities of reactants and products. The second term reflects the change in the reversible potential with the change in concentrations of reactants and products.

At net current zero, the forward and reverse reactions are in equilibrium and the expected maximum voltage is determined form the *Nernst equation*. Now, the exchange current density is

$$j_0 = nFC_{Ox,surf}k_0 e^{-\alpha F(E_{eq}-E^0)/RT} \tag{5.65}$$

or

$$j_0 = nFC_{Red,surf}k_0 e^{\beta F(E_{eq}-E^0)/RT}. \tag{5.66}$$

Raising the power of Equation 5.61 by $-\alpha$, we have

$$e^{-\alpha F(E_{eq}-E^0)/RT} = \left(\frac{C_{Ox,surf}}{C_{Red,surf}}\right)^{-\alpha}. \tag{5.67}$$

Substituting this into the exchange current density equation,

$$j_0 = nFk_0 C^{\beta}_{Ox,bulk} C^{-\alpha}_{Red,bulk}. \tag{5.68}$$

For a special case of $C_{Ox,bulk} = C_{Red,bulk} = C$, the exchange current density becomes

$$j_0 = nF\, k_0\, C. \tag{5.69}$$

5.5 Equation for Current Density—The Butler–Volmer Equation

When a potential is applied to the electrode, there is deviation from the equilibrium. From Equations 5.52 and 5.53, the forward reduction reaction and backward oxidation reaction current densities are

$$j_1 = nFC_{\text{Ox,surf}} k_{0\text{Ox}} e^{-\alpha F(E-E^0)/RT} \tag{5.70}$$

$$j_2 = nFC_{\text{Red,surf}} k_{0\text{Red}} e^{-\beta F(E-E^0)/RT}. \tag{5.71}$$

The net current density is

$$j = j_1 - j_2 = nFC_{\text{Ox,surf}} k_{0\text{Ox}} e^{-\alpha F(E-E^0)/RT} - nFC_{\text{Red,surf}} k_{0\text{Red}} e^{\beta F(E-E^0)/RT}. \tag{5.72}$$

Thus, we have the expression for current density

$$j = nFe^{-\alpha F(E-E^0)/RT} \left\{ C_{\text{Ox,surf}} k_{0\text{Ox}} - C_{\text{Red,surf}} k_{0\text{Red}} e^{nF(E-E^0)/RT} \right\}. \tag{5.73}$$

Using the expressions for exchange current density given by Equations 5.65 and 5.66, the net current density is written as

$$j = j_0 \left\{ \frac{C_{\text{Ox,surf}}}{C_{\text{Ox,bulk}}} e^{-\alpha nF(E-E^0)/RT} - \frac{C_{\text{Red,surf}}}{C_{\text{Red,bulk}}} e^{\beta F(E-E^0)/RT} \right\}. \tag{5.74}$$

This equation, which relates the current density to applied potential for half-cell, is called *Butler–Volmer (BV) equation*. The current density is dependent on the exchange current density and the transfer coefficient parameter α. The equation for current density can be written in terms of activation overpotential or voltage loss, $\eta = E - E_{eq}$, deviation from equilibrium potential as

$$j = j_0 \left\{ \frac{C_{\text{Ox,surf}}}{C_{\text{Ox,bulk}}} e^{-\alpha F\eta/RT} - \frac{C_{\text{Red,surf}}}{C_{\text{Red,bulk}}} e^{\beta F\eta/RT} \right\}. \tag{5.75}$$

For $n = 1$, this can be written as

$$j = j_0 \left\{ \frac{C_{\text{Ox,surf}}}{C_{\text{Ox,bulk}}} e^{-\alpha F\eta/RT} - \frac{C_{\text{Red,surf}}}{C_{\text{Red,bulk}}} e^{(1-\alpha)F\eta/RT} \right\}. \tag{5.76}$$

The Butler–Volmer states that the current generated by an electrochemical reaction varies exponentially with activation overpotential. The exponential terms are controlled by the concentration ratios $C_{\text{Ox,surf}}/C_{\text{Ox,bulk}}$ and $C_{\text{Red,surf}}/C_{\text{Red,bulk}}$. In fact, the electrode surface concentrations of reactant and product determine the limiting current density of the fuel cell.

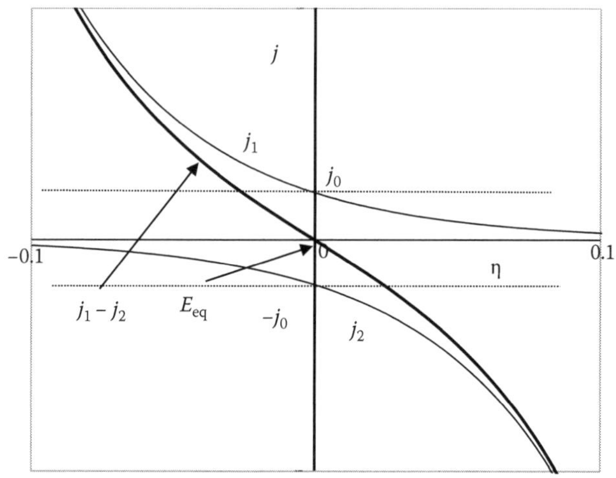

FIGURE 5.13
The variation of *i* versus η as the given BV equation.

Figure 5.13 shows the functional relation between current density and activation overpotential. The forward current density is positive and reverse current densities are shown in a thin line and a thick line represents the net current density. For $\eta = 0$, the current density is j_0. By identifying the forward current density with the anode and the reverse current density to the cathode, we see that for a large negative overpotential, the cathode current density is negligible and the net current density is essentially the anode current density. For a large positive overpotential, the anodic current density is small and the net current density is the same as the cathodic current density. The variation of current density with overpotential shows a linear region near $\eta = 0$, and for higher η values (positive or negative), it has an exponential region. For very large values of η, the current density levels off, and in these regions, the current density is limited by mass transfer rather than by heterogeneous kinetics.

5.6 Activation Overpotential and Controlling Factors

For a well-mixed solution, that is, when the electrode surface and bulk concentrations are identical, the effects of concentration can be neglected. This implies that the concentrations of reactant and product species at the electrode are not affected by the net reaction rate. In this case, the BV equation simplifies to

$$j = j_0\left(e^{-\alpha F\eta/RT} - e^{\beta F\eta/RT}\right). \tag{5.77}$$

In the case of $n = 1$,

$$j = j_0\left(e^{-\alpha F\eta/RT} - e^{(1-\alpha)F\eta/RT}\right). \tag{5.78}$$

It is easier to see the effects of various parameters such as exchange current density and transfer coefficient α on the current density using the simple form of the BV equation. Without mass transfer effects, the overpotential associated with the current density solely controls the activation energy. In Figure 5.14, the effect of exchange current density on the relation between activation overpotential and the current density is illustrated. The figure is shown for different exchange current densities, 10^{-2} A/cm^2 (curve a), 10^{-6} A/cm^2 (curve b), 10^{-8} A/cm^2 (curve c), and for $n = 1$, $\alpha = 0.5$, and $T = 298.15$ K.

For $\eta = 0$, each curve shows no current flow since the system is in total equilibrium. However, as a voltage different to that of equilibrium is applied, then different responses are observed depending upon the value of j_0. The negative value of overpotential drives the cathode current and hence it corresponds to cathode loss. Similarly, the positive overpotential corresponds to anode reactions. For large value of j_0 (curve a), a small change in η results in a large current change. Thus, there is little or negligible activation barrier to either of the reactions. For this case, the electrode reaction is said to be reversible since both k_{red} and k_{ox} are large. At the other extreme, when j_0, is very "small" (curve c), then a large value of η is needed to alter the current. This implies the fact that there is now a high barrier to activation and so the

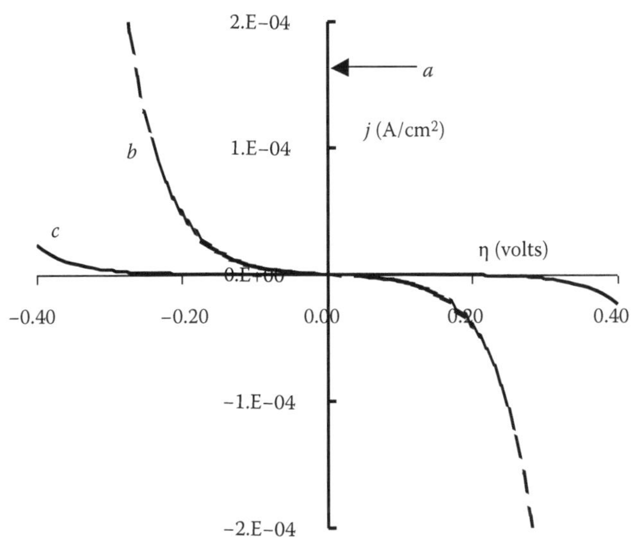

FIGURE 5.14
The effect of exchange current density on the current density and overpotential relation.

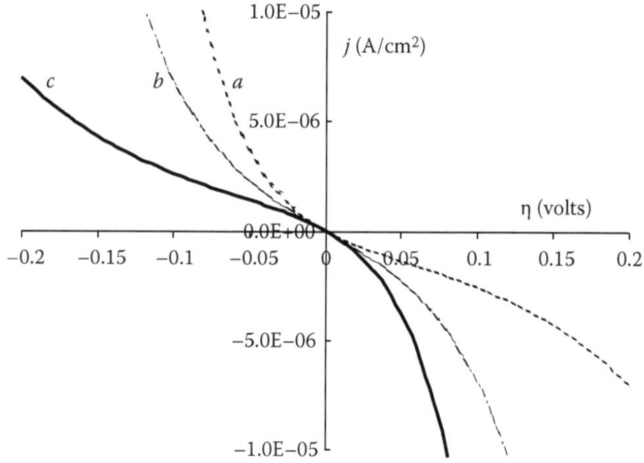

FIGURE 5.15
The effect of charge transfer coefficient on the current density and overpotential relation.

rates of the reduction and oxidation processes become slow. Electrode reactions of this type are termed irreversible. Intermediate behavior is generally referred to as quasi-reversible (curve b).

In Figure 5.15, the effect of the charge transfer coefficient on the current density and overpotential relation is shown for different values of the charge transfer coefficient $\alpha = 0.75$ (curve a), $\alpha = 0.5$ (curve b), $\alpha = 0.25$ (curve c), $n = 1$, and $j_0 = 10^{-6}$ A/cm^2. The charge transfer coefficient represents the fraction of additional energy that goes toward the reaction at the electrode. It is also considered as the symmetry coefficient of the electrode reaction. For $\alpha = 0.5$ (curve b), the current density curve is symmetric with respect to positive and negative overpotential. This indicates that the overpotential affects equally the anodic and cathodic reactions at an electrode. However, for α different than 0.5, the current density and overpotential curve is nonsymmetrical with respect to current density axis. For $\alpha = 0.25$, the overpotential required for the cathodic reduction reaction is larger than the overpotential required for anodic oxidation. On the other hand, for $\alpha = 0.75$, the required overpotential for cathodic reduction reactions are smaller than the anodic oxidation reaction. Thus, for small α, the cathodic losses (polarizations) are higher; hence, it is ideal to have α greater than 0.5.

5.7 Tafel Equation—Simplified Activation Kinetics

We have seen that the BV equation is dependent on several parameters, and when applying it to fuel cell reaction kinetics, some simplifications are useful. These simplifications are *two limiting cases* of the BV equation: (1) low overpotential region (called "polarization resistance"), where η is *very small,* and (2) high overpotential region, where η is *very large.*

For small values of η (typically less than 15 mV at room temperature), the exponential term e^x can be approximated using Taylor series expansion as $e^x \approx 1 + x$, for small x. For small η, then Equation 5.77 can be written as

$$j = j_0\left(\frac{-nF\eta}{RT}\right). \tag{5.79}$$

For small η, the net current density varies linearly with the overpotential in a narrow range of potential near equilibrium potential E_{eq}. It should be noted that the current density is independent of the charge transfer coefficient α for small values of overpotential. The ratio $-\eta/j$ has the dimensions of resistance and is called the *charge transfer resistance*, R_{ct}, and is given as

$$R_{ct} = \frac{RT}{nFj_0}. \tag{5.80}$$

Theoretically, the exchange current density j_0 can be obtained by measuring i versus η for a low range of η. Unfortunately, this measurement is not practical because of large experimental errors introduced by other fuel cell losses arising from ohmic resistances, mass transport effects, and reactant and product crossover effects. These losses are discussed in the next section.

Example 5.2

Determine for what values of η the linear form of the BV equation is a good approximation.

Answer

For this, we use the simple form of the BV equation and compare it with the linear form of the BV equation given by Equations 5.78 and 5.79

$$j = j_0\left(e^{-\alpha F\eta/RT} - e^{(1-\alpha)F\eta/RT}\right),\ j = j_0\left(\frac{-nF\eta}{RT}\right).$$

First, writing the equation in nondimensional form and denoting them as ratios R_{BV} and R_{lin}, respectively, we have

$$R_{BV} = \frac{j}{j_0} = e^{-\alpha F\eta/RT} - e^{(1-\alpha)F\eta/RT},\ R_{lin} = \frac{j}{j_0} = \frac{-nF\eta}{RT}.$$

Assuming $n = 1$, $T = 353$ K (80°C), and $\alpha = 0.5$, the ratios of the current density and exchange current density, R_{lin} and R_{BV}, are calculated for negative η values and are compared in Table 5.1. In this table, the relative percent difference between the two are also listed.

From Table 5.1, we can see that the error between the linear approximation and the full BV equations is less than 1% for η less than 14 mV. Thus, the linear form of the BV equation is a good approximation for η less than 14 mV at a temperature of 80°C.

For *large values* of η either positive or negative (greater than 140 mV at 80°C), the second exponential term becomes negligible. For example, at large negative η,

$$e^{-\alpha F\eta/RT} \gg e^{\beta F\eta/RT}. \tag{5.81}$$

The *BV equation* now simplifies to

$$j = j_0 e^{-\alpha F\eta/RT}. \tag{5.82}$$

TABLE 5.1

Error in Linear Approximation of BV Equation for Small η

η (mV)	R_{lin}	R_{BV}	$\left\lvert \frac{R_{BV} - R_{lin}}{R_{BV}} \right\rvert \times 100$
0	0.000	0.000	0.000
1	0.033	0.033	0.004
2	0.066	0.066	0.018
3	0.099	0.099	0.040
4	0.131	0.131	0.072
5	0.164	0.164	0.112
6	0.197	0.197	0.162
7	0.230	0.230	0.220
8	0.263	0.264	0.287
9	0.296	0.297	0.363
10	0.328	0.330	0.448
11	0.361	0.363	0.542
12	0.394	0.397	0.644
13	0.427	0.430	0.756
14	0.460	0.464	0.876
15	0.493	0.498	1.004
16	0.526	0.532	1.142
17	0.558	0.566	1.287
18	0.591	0.600	1.442
19	0.624	0.634	1.605
20	0.657	0.669	1.776
21	0.690	0.704	1.955

Solving for η, we have

$$\eta = \frac{RT}{\alpha F}\ln j_0 - \frac{RT}{\alpha F}\ln j. \tag{5.83}$$

This form of equation is known as the *Tafel equation,* which was derived empirically. The Tafel equation is generally written in the following form:

$$\eta_c = a - b\log(j) \text{ for a } \textit{cathodic reaction,} \tag{5.84}$$

$$\eta_a = a + b\log(j) \text{ for an } \textit{anodic reaction,} \tag{5.85}$$

where a and b are called *Tafel constants,* and at 80°C, they are for cathodic reaction

$$a = RT/\alpha F \ln j_0 = (2.303\, RT/\alpha F)\log j_0 = (0.070/\alpha)\log j_{0c} \tag{5.86}$$

$$b = 2.303\, RT/nF = 0.070/\alpha n_c \tag{5.87}$$

and for anodic reaction

$$a = 0.070/\beta \log j_{0a} \tag{5.88}$$

$$b = 0.070/\beta n_a \tag{5.89}$$

where j_{0c} and j_{0a} are exchange current densities for the cathode and anode, respectively, and n_c and n_a are the number of electrons for anodic and cathodic elementary charge transfer steps, respectively. The values of the n_c and n_a need not be an integer if experimentally derived. This is because there can be more than one charge transfer reaction in parallel.

Example 5.3

Show for what values of η the Tafel equation is a good representation of the BV equation.

Answer

Similar to Example 5.2, we use the simple form of the BV equation and compare it with the Tafel equation given by Equations 5.78 and 5.82:

$$j = j_0\left(e^{-\alpha F\eta/RT} - e^{(1-\alpha)F\eta/RT}\right),\; j = j_0 e^{-\alpha F\eta/RT}.$$

Writing the equation in nondimensional form and denoting them as ratios R_{BV} and R_{Tafel}, respectively, we have

$$R_{BV} = \frac{j}{j_0} = e^{-\alpha F\eta/RT} - e^{(1-\alpha)F\eta/RT}, \; R_{lin} = \frac{j}{j_0} = \frac{-nF\eta}{RT} e^{-\alpha F\eta/RT}.$$

Assuming $n = 1$, $T = 298$ K (25°C), and $\alpha = 0.5$, the ratios of the current density and exchange current density, R_{Tafel} and R_{BV}, are calculated for negative η values and are compared in Table 5.2 along with the relative percent difference between the two ratios.
Table 5.2

From Table 5.2, we can see that the error between the Tafel approximation of BV and the full BV equations is less than 1% for η larger than 118 mV. Thus, the Tafel form of the BV equation is a good approximation for η larger than 118 mV at 25°C.

TABLE 5.2

Error in Tafel Approximation of BV Equation for Large η

η (mV)	R_{Tafel}	R_{BV}	$\left\lvert \frac{R_{BV} - R_{Tafel}}{R_{BV}} \right\rvert \times 100$
0	1.000	0.000	—
10	1.215	0.392	210.244
20	1.476	0.798	84.925
30	1.793	1.235	45.183
40	2.178	1.718	26.727
50	2.645	2.267	16.676
60	3.213	2.902	10.724
70	3.903	3.647	7.025
80	4.742	4.531	4.655
90	5.760	5.586	3.108
100	6.997	6.854	2.085
110	8.499	8.382	1.404
112	8.837	8.723	1.297
114	9.187	9.078	1.199
116	9.552	9.447	1.108
117	9.739	9.637	1.065
118	9.931	9.830	1.024
119	10.126	10.027	0.985
120	10.325	10.228	0.947
130	12.542	12.462	0.640
140	15.236	15.170	0.433
150	18.508	18.454	0.293

The plot of η versus *j*, called the *Tafel plot*, is a straight line with intersection *a* and slope *b* (Tafel slope). The ratio of Tafel slope and current density has units of resistance. In the Tafel kinetic region, the charge transfer resistance (R_{ct}) is defined by

$$R_{ct} = \frac{d\eta}{dj} = \frac{b}{j}. \tag{5.90}$$

Thus, in the Tafel kinetic region, R_{ct} decreases inversely with *j* or, in other words, η increases with an increase of *j* for a cathodic reaction.

Thus, experimentally, j_0 and α can be determined using a Tafel plot. The Tafel form of equation holds if the reverse reaction contributes to less than 1% of the current density or

$$e^{-\alpha F\eta/RT}/e^{\beta F\eta/RT} = e^{-nF\eta/RT} \leq 0.01. \tag{5.91}$$

At 25°C, this corresponds to

$$|\eta| > 0.118/n \text{ (volts)}. \tag{5.92}$$

This result was also seen from Example 5.3. For large η, the forward reaction dominates and thus the reaction process is completely irreversible. Though the Tafel equation predicts the forward reaction for large η, it does not account the mass transfer limited current at high η. If electrode kinetics are fairly fast, then mass transfer limited currents are easily reached at high η. For such cases, the Tafel equation does not apply well. On the other hand, when the electrode kinetics is slow, then the significant overpotential is required and the Tafel relationship holds good.

For a special case of α = β, a case of symmetric charge transfer coefficient for anodic and cathodic reactions, Equation 5.77 becomes

$$j = j_0\left(e^{-\alpha F\eta/RT} - e^{\alpha F\eta/RT}\right). \tag{5.93}$$

From trigonometry,

$$\sinh(x) = \frac{(e^x - e^{-x})}{2}. \tag{5.94}$$

Hence, current density for α = β is given as

$$\begin{aligned} j &= 2j_0\frac{1}{2}\left(e^{(-\alpha F\eta/RT)} - e^{-(-\alpha F\eta/RT)}\right) \\ &= 2j_0\sinh\left(\frac{-\alpha F\eta}{RT}\right). \end{aligned} \tag{5.95}$$

Example 5.4

Calculate decrease in cathode activation loss in a hypothetical cell reaction for a current density of 0.5 A/cm^2 when the reaction temperature is raised form 25°C to 80°C. The electrode exchange current density at 25°C is $j_0 = 10^{-4}$ A/cm^{-2} and $\alpha = 1$, and at 80°C, effective exchange current density is $j_0 = 10^{-3}$ A/cm^{-2} while $\alpha = 1$ remains the same.

Answer

First, we calculate the activation loss at 25°C using the Tafel equation given by Equation 5.83:

$$\eta = \frac{RT}{\alpha F}\ln\frac{j}{j_0}.$$

At 25°C, the activation loss is

$$\begin{aligned}\eta &= \frac{RT}{\alpha F}\ln\frac{j}{j_0}\\ &= \frac{8.314\,(\mathrm{J/mol\,K})\times 298\,(\mathrm{K})}{1\times 96{,}485\,(\mathrm{C/mol})}\ln\left(\frac{0.5\,(\mathrm{A/cm^2})}{10^{-4}\,(\mathrm{A/cm^2})}\right) = 0.219\ \mathrm{V}.\end{aligned}$$

Since the activation loss is 219 mV, the use of the Tafel equation is justified.

Now, at 80°C, the activation loss is

$$\eta = \frac{8.314\,(\mathrm{J/mol\,K})\times 353\,(\mathrm{K})}{1\times 96{,}485\,(\mathrm{C/mol})}\ln\left(\frac{0.5\,(\mathrm{A/cm^2})}{10^{-3}\,(\mathrm{A/cm^2})}\right) = 0.189\ \mathrm{V}.$$

The net decrease in activation loss is = 219 mV − 189mV = 30 mV.

5.8 Relationship of Activation Overpotential with Current Density—Tafel Plots

The Tafel plot, a plot of log j versus η, is useful for evaluation of kinetic parameters for reactions. Now, using constant a and b from Equations 5.88 and 5.89, we have the Tafel equation for cathodic reaction at 25°C as

$$\eta = (0.0591/\alpha)\,(\log i_0 - \log i). \tag{5.96}$$

The plot of the Tafel equation is shown in Figure 5.16 for a cathodic reaction for $\alpha = 0.5$, $n = 1$, and $j_0 = 10^{-9}$A/cm^2. The Tafel line shows a straight line with a slope of $0.0591/\alpha n$. From Equation 5.36 for $\eta = 0$, $j = j_0$, in the figure, the line intercepts at $\eta = 0$ for $j = j_0$. In fact, it corresponds to the value of j_0 used for making this plot.

For low voltage, the Tafel line deviates from the Butler–Volmer relation. A method given by Allen and Hickling (1957) allows plotting of i versus η, even at low voltages. The simple form of the BV equation (Equation 5.77) can be cast into the following form:

$$j = j_0 e^{-\alpha\eta F/RT} (1 - e^{-\alpha\eta F/RT}). \tag{5.97}$$

By taking logarithm of terms in this equation, it can be written as

$$\log \frac{j}{1 - e^{n\eta F/RT}} = \log j_0 - \frac{\alpha\eta F}{2.3RT}. \tag{5.98}$$

This equation shows that if $\log [j/(1 - e^{n\eta F/RT})]$ versus η is plotted, then the intercept of the line at $\eta = 0$ gives $\log j_0$ and the slope gives $-\alpha F\eta/RT$.

The Tafel plot (the plot of $\log j$ versus η) and the plot of $\log [j/(1 - e^{n\eta F/RT})]$ versus η is shown in Figure 5.17 for both anodic and cathodic reactions for $n = 1$, $\alpha = 0.5$, $T = 298$ K, and $j_0 = 10^{-7}$ A/cm^2. The anodic branch has a slope of $(1 - \alpha)F/2.3RT$ and the cathodic branch has a slope of $-\alpha F/2.3RT$. Equation 5.98 is useful as it is applicable to electrode reactions that are not totally irreversible and for the cases where the mass transfer effects are not important.

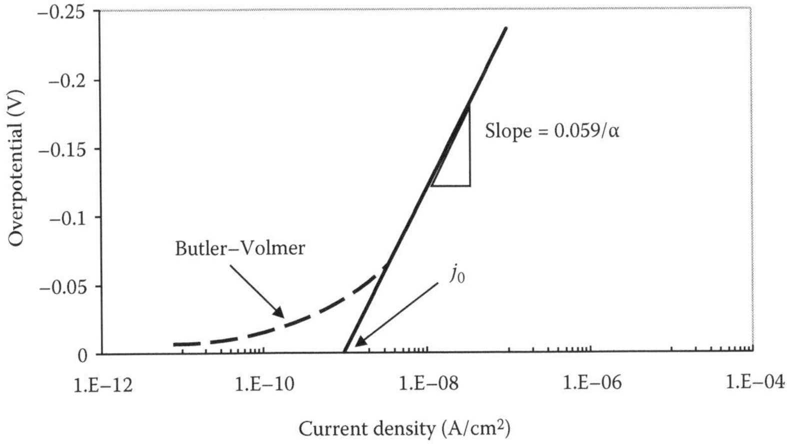

FIGURE 5.16
Tafel plot for cathodic reaction at 25°C for $\alpha = 0.5$, $n = 1$, and $j_0 = 10^{-9}$ A/cm^2.

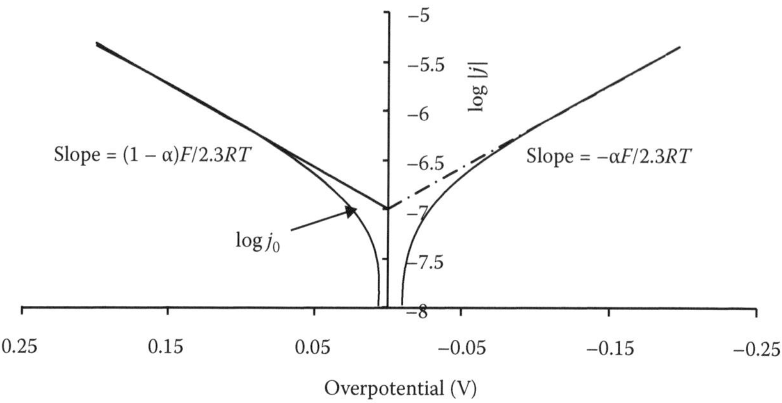

FIGURE 5.17
Tafel plot and the plot of log $[j/(1-e^{n\eta F/RT})]$ versus η, for hydrogen oxidation and reduction reaction, $n = 1$, $\alpha = 0.5$, $T = 298$ K, and $j_0 = 10^{-7}$ A/cm^2.

5.9 Fuel Cell Kinetics

In the previous sections, the electrode kinetics were developed for the half-cell (single electrode). Here, we use Butler–Volmer kinetic equations for the fuel cell anode and cathode and develop the net current density and overpotential relationships (Kordesch and Simander, 2000; Larminie and Dicks, 1999; Mench, 2008; O'Hayre et al., 2006; Vielstich et al., 2003). Consider the charge transfer reaction

$$\mathrm{Ox}^+ + ne^- \underset{k_{\mathrm{Ox}}}{\overset{k_{\mathrm{Red}}}{\rightleftharpoons}} \mathrm{Red}. \tag{5.99}$$

From Equation 5.75, the net current density is written in terms of charge transfer coefficients for reduction and oxidation processes α_{Red} and α_{Ox}, respectively, as

$$j = j_0\left\{\left(\frac{C_{\mathrm{Ox,surf}}}{C_{\mathrm{Ox,bulk}}}\right)e^{-\alpha_{\mathrm{Red}}F\eta/RT} - \left(\frac{C_{\mathrm{Red,surf}}}{C_{\mathrm{Red,bulk}}}\right)e^{\alpha_{\mathrm{Ox}}F\eta/RT}\right\}. \tag{5.100}$$

Using Equation 5.68, the exchange current density is now written as

$$j_0 = nFk_0 C_{\mathrm{Ox,bulk}}^{\alpha_{\mathrm{Ox}}} C_{\mathrm{Red,bulk}}^{-\alpha_{\mathrm{Red}}} \tag{5.101}$$

$$j_0 = nFC_{\mathrm{Red,bulk}}\, k_{0\mathrm{Red}}\, e^{-\alpha F\eta/RT}. \tag{5.102}$$

Since the BV equation is valid for fuel cell reactions in the anode and cathode, the anode and cathode net current densities are given as

$$j_a = j_{0a}\left\{\left(\frac{C_{Ox,surf}}{C_{Ox,bulk}}\right)_a e^{-\alpha_{Red,a}F\eta_a/RT} - \left(\frac{C_{Red,surf}}{C_{Red,bulk}}\right)_a e^{\alpha_{Ox,a}F\eta_a/RT}\right\} \tag{5.103}$$

and

$$j_c = j_{0c}\left\{\left(\frac{C_{Ox,surf}}{C_{Ox,bulk}}\right)_c e^{-\alpha_{Red,c}F\eta_c/RT} - \left(\frac{C_{Red,surf}}{C_{Red,bulk}}\right)_c e^{\alpha_{Ox,c}F\eta_c/RT}\right\}. \tag{5.104}$$

Here, j_{0a} and j_{0c} are the exchange current densities for anode and cathode, respectively. Since reactions are different in anode and cathode, $j_{0a} \neq j_{0c}$. The forward (reduction) reaction charge transfer coefficients for anode and cathode are $\alpha_{Red,a}$ and $\alpha_{Red,c}$, respectively. Similarly, backward (oxidation) reaction charge transfer coefficients for anode and cathode are $\alpha_{Ox,a}$ and $\alpha_{Ox,c}$, respectively. The oxidation and reduction transfer coefficients are related by the following expressions:

$$\alpha_{Red,a} + \alpha_{Ox,a} = n \tag{5.105}$$

$$\alpha_{Red,c} + \alpha_{Ox,c} = n. \tag{5.106}$$

Since charge is conserved, the anode and cathode current densities are equal; that is, fuel cell current density,

$$j = j_c = j_a \tag{5.107}$$

If mass transfer effects are negligible, the current density equation for the fuel cell,

$$j = j_{0a}\left\{e^{-\alpha_{Red,a}F\eta_a/RT} - e^{\alpha_{Ox,a}F\eta_a/RT}\right\} = j_{0c}\left\{e^{-\alpha_{Red,c}F\eta_c/RT} - e^{\alpha_{Ox,c}F\eta_c/RT}\right\}. \tag{5.108}$$

Now, for small values of overpotential (low electrode loss region), the linear kinetic region applies. Using the cathode current density equation we have for small η_c,

$$j = j_{0c}\left(\frac{-nF\eta_c}{RT}\right) \tag{5.109}$$

or in terms of overpotential

$$\eta_c = -\frac{j}{j_{0c}} \frac{RT}{nF}. \tag{5.110}$$

For the fuel cell, the overpotential on the anode is positive; this makes the first term on the right-hand side of the anode current density equation (Equation 5.103) negligible compared to the second term for moderate values of overpotential. Thus, the anode current density is reduced to

$$j = -j_{0a} e^{\alpha_{Ox,a} F \eta_a / RT}. \tag{5.111}$$

Thus, at anode, oxidation current is dominant. The negative sign indicates net oxidation wherein the electrons are leaving the electrode. For cathode, the overpotential is negative; hence, the second term on the right-hand side of the cathode current density equation (Equation 5.108) is negligible in comparison with the first term. The cathode current density reduces to

$$j = j_{0c} e^{-\alpha_{Red,c} F \eta_c / RT}. \tag{5.112}$$

For large values of η_a and η_c (corresponding high electrode loss region), the anode and cathode overpotentials can be written in Tafel kinetic form

$$\eta_a = \frac{RT}{\alpha_{Ox,a} F} \ln j_{0a} - \frac{RT}{\alpha_{Ox,a} F} \ln j \tag{5.113}$$

$$\eta_c = \frac{RT}{\alpha_{Red,c} F} \ln j_{0c} - \frac{RT}{\alpha_{Red,c} F} \ln j. \tag{5.114}$$

In Figure 5.18, the cathode overpotential as a function of current density is shown for $n = 1$, $\alpha_{Red,c} = 0.5$, $j_{0c} = 10^{-5}$ A/cm^2 and at 298.15 K. The linear kinetic and Tafel kinetic regions are represented by Equations 5.111 and 5.114, respectively. Here, the negative value of the overpotential corresponds to the cathode polarization or loss. For low current densities, the overpotential increases linearly with j. As η increases, the Tafel kinetic region shows logarithmic dependence. At high current density, the mass transfer effects further affect the overpotential. In this figure, the effect of mass transfer is not shown, which limits the current density at high overpotential.

From the figure, the Tafel slope $b = 2.303\ RT/\alpha_{Red,c}$ is approximately 120 mV per decade of j. A higher Tafel slope results in higher overpotentials at fuel cell operating current density. Hence, it is desirable to minimize the Tafel slope so as to achieve high voltages at high operating current densities. Tafel slope is smaller with large values of charge transfer coefficient. Typically, Tafel slopes

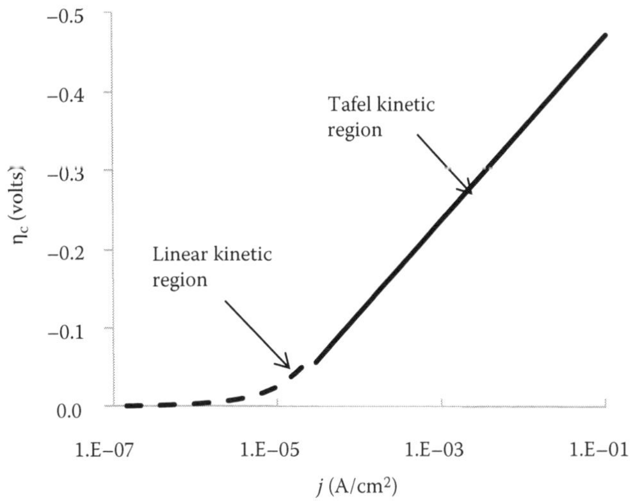

FIGURE 5.18
Linear and Tafel kinetic regions for fuel cell electrode.

are dependent on the mechanism and cannot be easily changed or reduced. The Tafel slope for oxygen reduction on high area platinum catalyst is on the order of 65–90 mV/decade, but it is around 120 mV/decade on bulk platinum.

In the case of hydrogen–oxygen fuel cells, the anode exchange current density j_{0a} (~10^{-4} A/cm^2 on Pt at 1 atm and 25°C) is several orders of magnitude larger than the cathode exchange current density j_{0c} (~10^{-9}A/cm^2) and the overpotential on cathode η_c is larger than the overpotential on anode η_a. The exchange current densities for hydrogen oxidation and oxygen reduction reactions for various electrode material surface and electrolyte combinations are listed in tables in Appendix C.

5.10 Fuel Cell Irreversibilities—Voltage Losses

In the previous section, it was discussed that for large exchange current densities, the electrode reaction is reversible since both $k_{Red,c}$ and $k_{Ox,c}$ are large. For small exchange current densities, the rates of the reduction and oxidation processes are slow and the electrode reactions are irreversible. Thus, electrode kinetics plays a vital role in determining the performance of fuel cells. Higher required overpotential implies higher cell voltage loss. In Figure 5.19, the effect of activation overpotential on the fuel cell current density is shown for different exchange current densities, $j_{0c} = 10^{-3}$ A/cm^{-2}, $j_{0c} = 10^{-5}$ A/cm^2, $j_{0c} = 10^{-5}$ A/cm^{-2}, $n = 1$, $\alpha = 0.5$, and at 298.15 K. Here, the cell

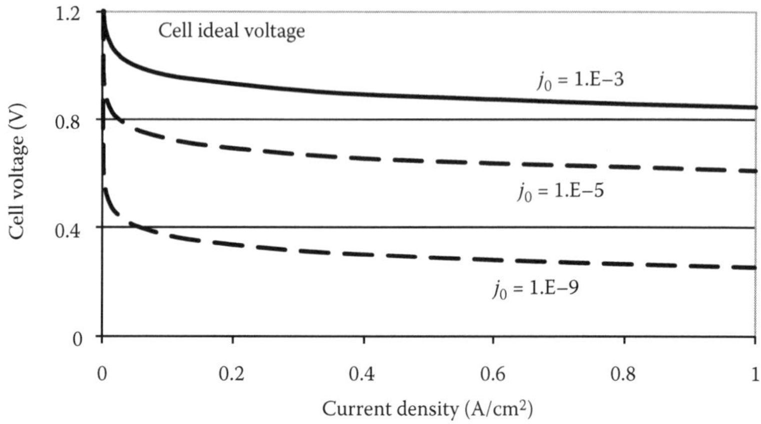

FIGURE 5.19
Cell voltage characteristics for different exchange current densities.

voltage is obtained by subtracting the activation overpotential from the cell theoretical EMF of 1.2 V.

In fact, there are other losses in cell voltage. Besides activation losses, the cell has ohmic losses, mass transfer losses, and the losses owing to fuel and oxidant crossover and short circuits in the cell. The mass transfer losses primarily affect the cell potential at large current loading. At lower current loads, the activation and the ohmic losses predominate. The electrolyte has generally the largest ohmic loss. Figure 5.20 shows the losses owing to anode and cathode activation losses and ohmic loss because of electrolyte.

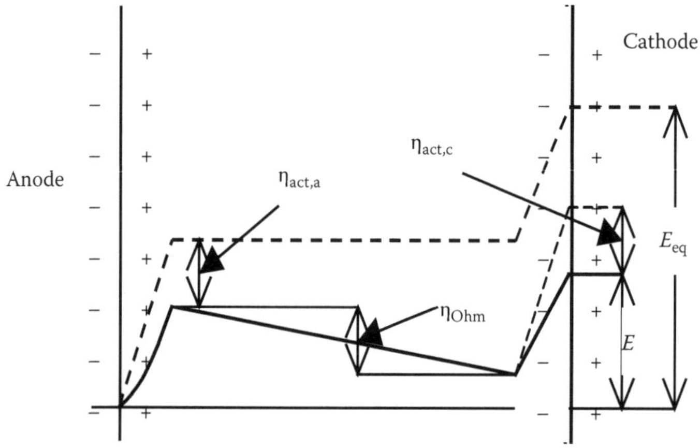

FIGURE 5.20
Potential loss owing to activation and ohmic losses.

In the fuel cell, there are a series of steps involved in electrode reactions:

a. Dissolution of the reactant gases in the electrolyte
b. Diffusion of the dissolved reactant gases to the active sites in the electrode
c. Chemisorption of the reactant that involves adsorption of reactants or intermediate species formed by dissociative adsorption on the electrode from the electrolyte
d. Charge transfer from reactant in electrolyte or from the above chemisorbed species to the electrode
e. Diffusion of species away from the electrode
f. Transfer of conducting ions from one electrode to the other through the electrolyte
g. Transfer of electrons from one electrode to the other through the external load

In each of these steps, there are losses and the single-cell overpotential during operation of a fuel cell is diminished by the losses in overpotential at the anode, at the cathode, and in the electrolyte. The common word for voltage losses or overpotential is *polarization* in electrochemistry. In Figure 5.21, the cell overpotential or polarization curve is shown where regions I–IV show predominant losses in the fuel cell. It should be noted that the losses owing to activation, ohmic resistance, concentration, and fuel and oxidant crossover and short circuits in cell happen at different degrees for the entire range of current loading.

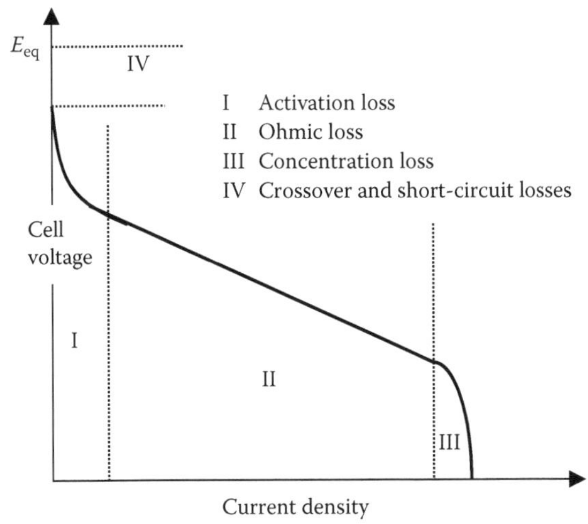

FIGURE 5.21
Fuel cell overpotential or polarization curve showing the activation, ohmic, concentration, and fuel and oxidant crossover and short circuit losses.

Thus, one may express the overpotential as the sum of all the losses *through* an equation:

$$\eta = -\eta_{act,a} - \eta_{act,c} - \eta_{Ohm} - \eta_{mt,a} - \eta_{mt,c} - \eta_{other}. \tag{5.115}$$

The variables $\eta_{act,a}$ and $\eta_{act,c}$ are the activation overpotentials at the anode and cathode, respectively, η_{Ohm} is the ohmic resistance losses in the fuel cell, $\eta_{mt,a}$ and $\eta_{mt,c}$ are the mass transfer losses at the anode and cathode, respectively. These are discussed here.

5.10.1 Activation Losses

The activation losses are nonlinear with current as seen from earlier discussions. Typically, the activation losses introduce a sharp initial drop in the cell open circuit EMF with current load. Losses are different at each electrode, cathode, or anode as the double layer configuration is different. The activation loss is directly related to the energy barrier (resistances) for oxidation and reduction at the electrodes. This energy barrier depends on several parameters as seen from the BV equation.

The activation losses for anode and cathode are given by Equations 5.113 and 5.114 as

$$\eta_{act,a} = -\frac{RT}{\alpha_{Ox,a}F}\ln\frac{j}{j_{0,a}} \tag{5.116}$$

$$\eta_{act,c} = -\frac{RT}{\alpha_{Red,c}F}\ln\frac{j}{j_{0c}}. \tag{5.117}$$

As noted earlier, for a hydrogen–oxygen fuel cell, the cathode activation losses $\eta_{act,c}$ are dominant and hence anode activation losses $\eta_{act,a}$ are neglected. Various processes and parameters influence the activation losses. The reaction mechanisms, operating conditions, type and structure of catalyst, species concentration, impurities, service history, and age determine the activation losses in a fuel cell. Depending on the complexity of reaction mechanism, a higher overpotential may be required to overcome energy barrier. For example, methanol oxidation requires a higher overpotential than hydrogen oxidation. The operating parameters such as pressure and temperature affect the reaction rates.

From the previous section, it was seen that the activation losses depend primarily on the exchange current density, j_0, and the charge transfer coefficient α. For a reaction with larger exchange current density, the required overpotential is smaller and hence there are lower electrode losses. Thus, increasing j_0 enhances the electrode kinetic performance. In order to understand how j_0 can be increased, we look at the definition of exchange current density. It is given by

$$j_0 = nFC_{Ox,surf}\,k_{0Red}e^{-\alpha F\eta/RT}. \tag{5.118}$$

From this equation, we can see that there are a few control parameters that will enable an increase in j_0. The parameters *n*, *F*, and *R* cannot be changed for a given reaction. The decrease of energy barrier $\{\alpha F\eta/RT\}$, increase in temperature *T*, increase of reactant concentration C_R, and increasing reaction area lead to a higher value of j_0. The activation energy barrier has a strong influence on j_0. Since it appears as an exponent, a slight decrease in the activation energy barrier can induce substantial increase in j_0.

The most effective way to decrease the activation energy barrier is to use highly catalytic electrodes. The catalytic electrode changes the shape of the free-energy surface of the reaction. For example, in hydrogen charge transfer reaction, the free-energy curve for metal–hydrogen (M–H) bond is greatly altered such that the net energy is decreased and favors the electron transfer reaction as shown in Figure 5.22. The reaction is faster when the catalyst is present and the activation energy is lower with the catalyst present. It should be noted that the energies of reactants and products have not changed. The heat of reaction is the same. The relative amounts of reactants and products stay the same. The catalyst only allows the reaction to reach equilibrium faster. The intermediate strength bond for M–H provides the best catalytic effect for hydrogen oxidation reaction. The platinum group metals such as Pt, Pd, Ir, and Rh have high catalytic activity and hence are preferred material in low-temperature fuel cells. High-temperature fuel cells like SOFC and MCFC have low activation losses compared to low-temperature fuel cells like DMFC, PEMFC, and PAFC.

From Equation 5.118, we see that an increase in temperature increases i_0. The reactant has higher thermal activity—higher intensity thermal vibrations with increase in temperature. The increased thermal activity enhances the possibility of reactant energy to reach the activation energy level and

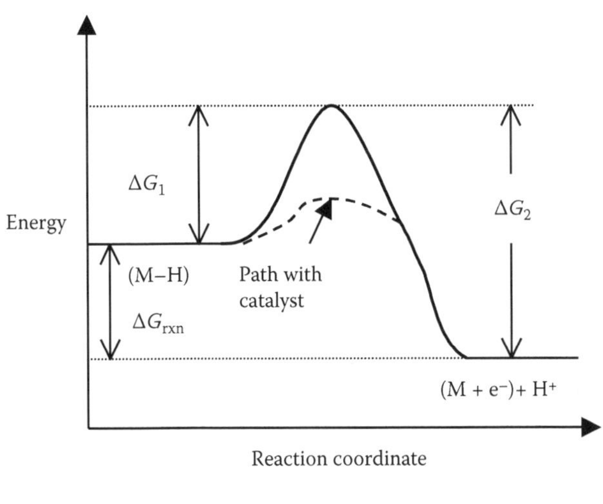

FIGURE 5.22
Reaction plane with catalysts for hydrogen oxidation.

thus increase reaction rate. The changing of the temperature has an exponential effect on j_0.

The reactant concentration has a direct linear effect on exchange current density j_0. Increasing the reactant concentration increases the reaction kinetics. One way to increase the concentration of the gaseous reactant is to increase the operating pressure. It is important to remember that the concentration of the reactant near the electrode is what counts in the reaction. As will be seen later on mass transport loss, a higher reaction at the electrode depletes the reactant concentration and thereby limits the higher reaction rates.

For an electrochemical reaction, availability of surface area is very important as it determines the amount charge transfer in a reaction. Higher electrode surface area is most desired in fuel cell. The electrode is generally made with extremely rough surface so that the effective reaction sites are increased by several orders of magnitude than on a smooth surface electrode. The effective exchange current density for a rough surface electrode will be greater than that for a smooth surface electrode. For fuel cell electrodes, the catalysts are therefore made in nanosized particles and are embedded on a rough surface electrode to provide large effective surface area.

5.10.2 Ohmic Losses

As shown in Figure 5.21, the fuel cell overpotential curve shows a linear region where the fuel cell internal ohmic resistance dominates. The ohmic losses arise owing to resistance to the charge transport in the fuel cell. There are two types of charged species, electrons and ions, which are transported in electrodes and electrolytes, respectively. When charges are transported, a voltage drop exists and this voltage drop for a given current flow is governed by Ohm's law given from Equation 2.24 as

$$V = IR. \tag{5.119}$$

Here, the resistance R is related to *conductivity* σ or *resistivity* ρ as

$$R = \frac{\rho L}{A} = L/\sigma A, \tag{5.120}$$

where L represents the thickness of the electrode or the electrolyte and A is the area of charge flow.

In terms of the cell current density j, the overpotential for ohmic losses can be written from Equation 5.119 as

$$\eta_{\text{Ohm}} = jr, \tag{5.121}$$

where r is called the fuel cell *area specific resistance* with unit Ω cm^2 and is given as

$$r = L\rho = L/\sigma. \tag{5.122}$$

From Equation 5.120, the area specific resistance can be written in terms of the fuel cell area A,

$$r = RA. \tag{5.123}$$

The ohmic overpotential from Equations 5.121 and 5.122 in terms of resistivity or conductivity can be written as

$$\eta_{\mathrm{Ohm}} = jL\rho = jL/\sigma. \tag{5.124}$$

The voltage drop occurs at both electrodes (anode and cathode), at various layers such as gas diffusion layer and catalysts, at the electrolyte, at bipolar plates, at the contact between the electrodes and electrolyte (owing to contact resistances), and at any interconnectors that have current path. Since the current flows serially in all these components of the cell, the total ohmic resistance is the sum of individual resistance contributions.

Thus, the total cell resistance R can be written as

$$R = \sum R_i, i = \text{ionic, electronic, and contact.} \tag{5.125}$$

Typically, for well-designed and well-built fuel cell, the dominating ohmic loss is from the electrolyte. If electrodes including bipolar plates are made of solid metal or graphite, their electrical resistances are negligible. However, if there are passive films formed on the electrodes or if the electrodes are ceramic materials such as in SOFC, these could have higher electronic resistances.

The electro-catalyst layers, which are porous, conduct both ions and electrons to facilitate oxidation and reduction reactions. For example, the electro-catalyst layers in most PEMFCs are 5–30 μm thick. The ion conductivity of these layers varies from 1 to 5 S/m and hence electro-catalyst layer area specific resistance values vary from 0.01 Ω cm^2 to 0.03 Ω cm^2 (or 10–300 mΩ cm^2). The Nafion electrolyte in PEMFC has a conductivity of 10 S/m when hydrated. Hence, for electrolyte thickness of 50–200 μm, the area specific resistance varies from 50 mΩ cm^2 to 200 mΩ cm^2. Thus, it can be seen that the electro-catalyst layer contribution to ohmic resistance is significant. In Table 5.3, typical thickness and area-specific resistance values for selected fuel components are listed.

Material conductivity and thickness play a major role in determining the internal resistance of the cell. Ways to reduce the cell resistance include (i) use of electrodes with the highest possible conductivity, (ii) use of thin

TABLE 5.3

Typical Thickness and Area-Specific Resistance Ranges for Fuel Cell Components

Component	Thickness (L)	r (mΩ cm^2)
PEMFC electro-catalyst layer	5–30 μm	10–300
PEMFC electrolyte	50–200 μm	50–200
PEMFC gas diffusion layer	100–300 μm	0.1–0.3
PEMFC graphite bipolar plate	2–4 mm	1–8
Cell contact resistance	—	30
AFC, PAFC, and MCF electrolytes	0.5–2 mm	50–20,000
SOFC electrolyte	10–300 μm	10–3000
Total cell (average)	—	100–200

electrolytes, and (iii) design of fuel cell with good contact conductance materials for plates and connectors. The electrolyte cannot be too thin as it needs to support catalyst layers and often it is the support onto which the electrodes are built. Also, it should be sufficiently thick to withstand pressure differential between anode and cathode spaces and to prevent any shorting of one electrode to another through the electrolyte.

Example 5.5

Consider a 100 cm^2 PEMFC with anode and cathode catalyst layers each of thickness 20 μm and conductivity of 4 S/m, on either side of the Nafion electrolyte of thickness 100 μm and conductivity of 10 S/m. The total cell specific contact resistance is 50 mΩ cm^2. Calculate (i) the cell total resistance and (ii) the net ohmic loss if the current density is 1 A/cm^2.

Answer

i. First, we calculate the resistances from anode and cathode catalyst layers, electrolyte, and the contact resistance.

   From Equation 5.120, we have the resistance of the anode and cathode catalyst layers

$$R_{\text{anode,or cathode catalyst layer}} = \frac{L}{\sigma A} = \frac{20\times 10^{-6}(\text{m})}{4(\text{S/m})\times 100(\text{cm}^2)(1\ \text{m}^2/10{,}000\ \text{cm}^2)} = 5\times 10^{-4}\,\Omega.$$

   Anode and cathode catalyst layers' total resistance = 0.5 mΩ + 0.5 mΩ = 1 mΩ.

   The electrolyte resistance

$$R_{\text{electrolyte}} = \frac{100\times 10^{-6}(\text{m})}{10(\text{S/m})\times 100(\text{cm}^2)(1\ \text{m}^2/10{,}000\ \text{cm}^2)} = 1\ \text{m}\Omega.$$

The contact resistance

$$R_{contact} = \frac{r}{A} = \frac{50 \times 10^{-3}(\Omega cm^2)}{100(cm^2)} = 0.50 \text{ m}\Omega$$

Cell total resistances from Equation 5.125 is written as

$$R = \sum R_i, \ i = \text{catalyst layers, electrolyte, contact} = 1 \text{ m}\Omega + 1 \text{ m}\Omega + 0.5 \text{ m}\Omega = 2.5 \text{ m}\Omega.$$

ii. The ohmic overpotential is given by Equation 5.121

$$\eta_{Ohm} = jr = jA\sum R_i.$$

For $j = 1$ A/cm^2, the ohmic loss is

$$\eta_{Ohm} = 1 \text{ (A/cm}^2) \times 100 \text{ (cm}^2) \times 2.5 \times 10^{-3} \text{ }(\Omega) = 0.25 \text{ V}.$$

5.10.3 Mass Transport Loss

The charge-transfer reactions occur at interfaces of electrodes with electrolytes, and this depends on the availability of sufficient concentration of reactant species at the interface. At large reaction rates at the interface, that is, at high current density, the concentration of the reactant at the interface depletes at a higher rate. Hence, at high current densities, the concentrations of reactants are low at the interface particularly for gases with very low solubility in the electrolyte (e.g., hydrogen or oxygen). With increase in overpotential, more reactions are driven, and this results in more depletion of the reactant at the interface. Ultimately, the effective concentration of the reactant species reaches zero at the interface at large current densities. Thus, maximum reaction rates and, hence, the maximum current limit are reached at high overpotential owing to *concentration* or *mass transfer* effects.

The balance between the rate of transport of species and the rate of consumption at the interface determines the maximum current. The key transport processes are *convection*, *diffusion*, and *migration*. Migration refers to the transport of ionic species toward or away from the electrode owing to the effect of the electric field. Higher electric field gives higher migration rates. One can overcome most limitations caused by *migration* by using supporting electrolytes. Diffusion refers to the transport of the reactant or product species because of gradient in concentration. At low concentrations, the diffusion process mainly governs the transport of the species. The diffusion limits electrochemical reactions in the fuel cell owing to the slowness of transport of these species from the bulk to the OHP of the double layer where the charge transfer occurs. In general, depletion of the reactant affects the fuel cell performance by increasing the cell overpotential in two ways: (i) according to the Nernst equation, the fuel cell reversible potential decreases when the reactant concentration at the interface is lower than the bulk concentration; (ii) the reaction activation loss increases with decrease in the reactant concentration at the interface relative to the bulk concentration.

Convection refers to the transport of the reactant or product species by bulk fluid motion driven by natural or applied mechanical forces. The natural convection limitations are due to convective transport caused by differences in densities as a result of temperature or concentration. The species transport to the interface can also be limited by the fuel cell flow structures and their conditions. For example, in PEMFC, blockage of flow channels or pore structures in diffusion or electrode-catalyst layers owing to the liquid phase can restrict the supply of the reactant to the interface. Accumulation of inert gases that do not participate in chemical reaction will limit the partial pressure of the reactant at the interface. This results to decreased reactions at the interface. The accumulation of chemical impurities at the reaction sites will prevent adsorption of desired reactant species. For example, in PEMFC, the presence of carbon monoxide degrades the platinum catalyst because the platinum preferentially adsorbs carbon monoxide, leaving few reaction sites for hydrogen adsorption and oxidation. This leads to high anodic overpotential.

Consider a fuel cell electrode–electrolyte interface. The reactants are supplied through the flow field channels to the interface. The electrode at the interface is porous and, depending on the type of fuel cell, the interface may have single or more porous layers. For example, in PEMFC, the reactants are supplied through flow field channels to the gas diffusion and electrocatalyst layers where reaction takes place. As the reaction is taking place at the interface, the reactant species concentration and, hence, its partial pressure decrease from bulk value at flow channels to a lower value at the electrocatalyst layer. This space being porous with micro-sized pores, the transport of the reactants is governed by the diffusion process. The consumption of the reactant determines the gradient in the concentration. According to the Nernst and Merriam model, the concentration gradient is considered in a layer called the *diffusion layer* near the electrode across which the concentration of the reactant species varies linearly with distance. Typical values of the diffusion layer thickness are approximately 100–300 μm.

Depending on the reaction rates or the current load, the concentration gradient varies as shown in Figure 5.23. Here, C_R is the concentration of the reactant. For a cathodic electron transfer reaction, the concentration refers to the oxidant species concentration and the reactant concentration can be written as C_{Ox}.

Consider a single-step electron transfer reaction at the interface of the electrode

$$\mathrm{Ox}^{n+} + ne \rightarrow \mathrm{Red}. \tag{5.126}$$

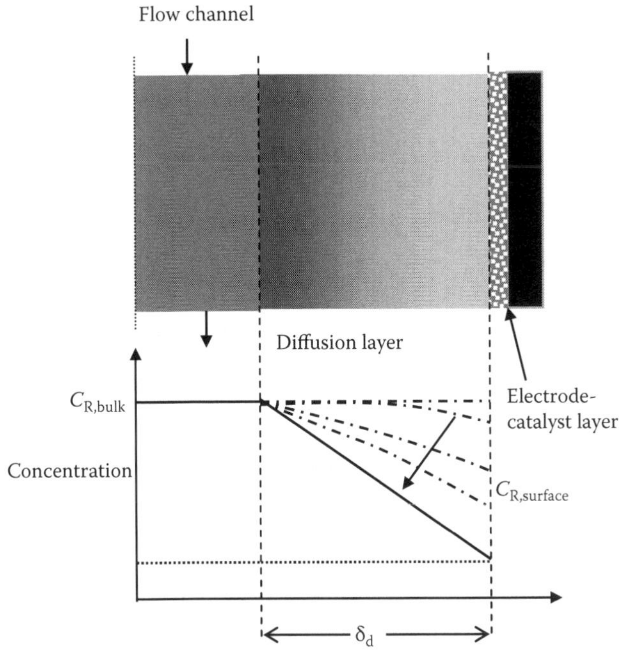

FIGURE 5.23
Reactant concentration profile in the diffusion layer.

Assuming the charge transfer to be in equilibrium condition, that is, the net current density is zero, the Nernst equation can be used to relate the cell reversible potential to the concentration of reactant through the equation

$$E_{eq} = E^0 + RT/nF \ln(C_{Ox,bulk}). \tag{5.127}$$

At equilibrium, the interface concentration is the same as the bulk concentration. Now, if we consider the potential of the electrode E at which the net current density is not zero, that is, $j_{net} = j \neq 0$, then the interfacial concentration is lower than the bulk concentration and the Nernst equation is written as

$$E = E^0 + RT/nF \ln(C_{Ox,surface}). \tag{5.128}$$

The current density $j \neq 0$ results in the departure of potential E from the equilibrium potential E_{eq}. This potential E occurs owing to the difference between concentrations at the electrode surface and at the bulk concentration.

Now, taking the difference between Equations 5.127 and 5.128, we have

$$E - E_{eq} = RT/nF \ln(C_{Ox,surface}/C_{Ox,bulk}). \tag{5.129}$$

By defining the concentration overpotential owing to mass transfer at cathode $\eta_{mt,c}$ as the difference between the electrode potential at j to the electrode potential at equilibrium,

$$\eta_{mt,c} = E - E_{eq}. \tag{5.130}$$

The mass transfer overpotential for cathode is written as

$$\eta_{mt,c} = RT/nF \ln(C_{Ox,surface}/C_{Ox,bulk}). \tag{5.131}$$

We have now related the reactant concentrations at the electrode surface and the bulk concentration to the resultant overpotential owing to concentration effects. The bulk concentration is larger than the surface concentration, and hence, $\eta_{mt,c}$ is negative.

Now, let us look at the current density generated because of the net charge transfer at the interface. Now, we know that there is a concentration gradient at the diffusion layer, which varies linearly with the distance from electrode surface. If the rate of diffusion of the reactant or the diffusion flux is J_D, then we can relate J_D to the current density j as

$$J_D = j/nF. \tag{5.132}$$

The diffusion flux is governed by the concentration gradient in the diffusion layer. Fick's law of diffusion relates the diffusion flux to the concentration gradient as

$$J_D = -D(dC/dx)_{x=surface}, \tag{5.133}$$

where D is the diffusion coefficient of the reactant species.

Using the Nernst and Merriam model of the diffusion layer, where the concentration varies linearly from surface value at the electrode to the bulk value at a distance of diffusion layer thickness δ, the concentration gradient is given as

$$(dC/dx)_{x=surface} = (C_{Ox,bulk} - C_{Ox,surface})/\delta. \tag{5.134}$$

Now, from Equation 5.133, we can write the current density as

$$i = -DnF(C_{Ox,bulk} - C_{Ox,surface})/\delta. \tag{5.135}$$

The maximum diffusion flux and, hence, the possible maximum current density correspond to the maximum gradient in the concentration. And this is possible if the surface concentration gradient $C_{\mathrm{Ox,surface}} = 0$. Using this limit on Equation 5.133, the maximum concentration gradients is

$$\lim_{C_{\mathrm{Ox,surface}}\to 0}(\mathrm{d}C/\mathrm{d}x)_{x=\mathrm{surface}} = \lim_{C_{\mathrm{Ox,surface}}\to 0}(C_{\mathrm{Ox,bulk}} - C_{\mathrm{Ox,surface}})/\delta = C_{\mathrm{Ox,bulk}}/\delta \qquad (5.136)$$

The maximum concentration gradient gives the maximum current density called *limiting current density* j_{L} owing to concentration or mass transfer effects

$$j_{\mathrm{L}} = -DnFC_{\mathrm{Ox,bulk}}/\delta. \qquad (5.137)$$

Typical values of the diffusion coefficients are on the order of 10^{-5} cm^2/s at room temperature for most species undergoing electrochemical reactions in aqueous media. For fuel cell reactants such as hydrogen and oxygen, the solubility at room temperature is on the order of 10^{-4} mol/cm^3. Thus, one can assume the bulk concentration 10^{-4} mol/cm^3. Using the value of diffusion layer thickness of 10^{-2} cm, the theoretical limiting current density for the electro-oxidation of hydrogen or for the electro-reduction of oxygen at planar electrodes is approximately 10^{-4} A/cm^2. This is a small value. In practice, the design of the diffusion layer and electro-catalyst layer structures are made such that the effective diffusion coefficient is increased on the order of 10^{-2} cm^2/s, and effective diffusion layer thickness is decreased.

For this three-dimensional porous gas diffusion, electrodes are used in fuel cells to provide a three-dimensional reaction zone and the diffusion of the reactant species to the electro-active sites by radial diffusion. The pore sizes and particles used are on the order of nanometers, resulting in the effective diffusion layer thickness being several orders of magnitude smaller. Limiting current densities on the order of 1–10 A/cm^2 can be reached with such designs.

Now, from Equation 5.135, we have the bulk concentration as

$$C_{\mathrm{Ox,surface}}/C_{\mathrm{Ox,bulk}} = 1 + j\delta/DnF/C_{\mathrm{Ox,bulk}}. \qquad (5.138)$$

The bulk concentration is given in terms of *limiting current density* as

$$C_{\mathrm{Ox,bulk}} = -j_{\mathrm{L}}\delta/DnF. \qquad (5.139)$$

From Equation 5.139, we have

$$C_{\mathrm{Ox,surface}}/C_{\mathrm{Ox,bulk}} = 1 - j/j_{\mathrm{L}}. \qquad (5.140)$$

Substituting Equation 5.140 into Equation 5.131 and identifying the limiting current density for cathode as j_{Lc}, we have the mass transfer overpotential for cathode as

$$\eta_{mt,c} = \frac{RT}{nF}\ln\left(1-\frac{j}{j_{Lc}}\right). \tag{5.141}$$

From this equation, we see that this expression is valid for current density smaller than the limiting current density. For very small current density compared to the limiting current density ($j \ll j_{Lc}$), the concentration overpotential is negligible.

Similarly for anode, the mass transfer overpotential is written as

$$\eta_{mt,a} = \frac{RT}{nF}\ln\left(1-\frac{j}{j_{La}}\right), \tag{5.142}$$

where j_{La} is the limiting current density for anode.

Using Equation 5.141, the effect of concentration loss on the fuel cell performance is shown in Figure 5.24 for different limit current densities. The plots are shown for $n = 1$, $T = 353$ K, and $j_L = 1, 2, 3$ A/cm². From the figure, it is clear that the fuel cell potential is affected by the concentration loss for large current densities.

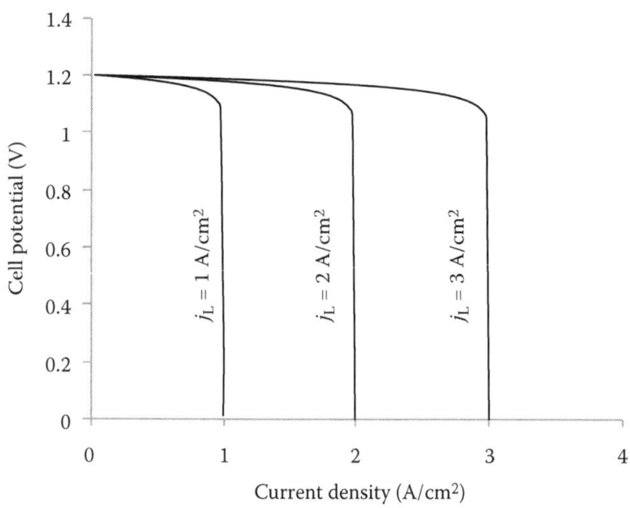

FIGURE 5.24
Fuel cell potential change owing to concentration loss for $n = 1$, $T = 353$ K, and $j_L = 1, 2, 3$ A/cm² and at room temperature.

It should be noted that this model has limitations as it does not account for (i) the inert gases or impurities present with reactants such as nitrogen in air or the presence of carbon monoxide or carbon dioxide when hydrogen is supplied from a fuel reformer, and (ii) the production and removal of reaction products, such as water.

Example 5.6

For fuel cell, the limiting current density for anode is 20 A/cm^2 and that for the cathode is 2 A/cm^2. Assuming single electron transfer reaction steps both at anode and cathode, determine the mass transfer overpotential for anode and cathode if the fuel cell is operating at 80°C with a fuel cell current density of 1.5 A/cm^2.

Answer

The anode and cathode mass transfer losses are given by Equations 5.141 and 5.142.

For the cathode, the mass transfer loss is

$$\eta_{\mathrm{mt,c}} = \frac{RT}{nF}\ln\left(1-\frac{j}{j_{\mathrm{Lc}}}\right) = \frac{8.314(\mathrm{J/mol\,K})\times 353(\mathrm{K})}{1\times 96,485(\mathrm{C/mol})}\ln\left(1-\frac{1.5(\mathrm{A/cm^2})}{2(\mathrm{A/cm^2})}\right) = 42\ \mathrm{mV}.$$

For the anode, the mass transfer loss is

$$\eta_{\mathrm{mt,a}} = \frac{8.314(\mathrm{J/mol\,K})\times 353(\mathrm{K})}{1\times 96,485(\mathrm{C/mol})}\ln\left(1-\frac{1.5(\mathrm{A/cm^2})}{20(\mathrm{A/cm^2})}\right) = 2.4\ \mathrm{mV}.$$

5.10.4 Reactant Crossover and Internal Currents

In the previous sections, we looked into the losses in the fuel cell potential contributed by the resistance to the reaction kinetics at the cathode and anode (activation losses), resistance to ion or electron transport (ohmic losses), and the mass concentration variation near the electrode (mass transfer losses). In addition to these losses, fuel cells show significant potential losses as a result of a short circuit in the electrolyte and crossover of reactants through the electrolyte.

Although the electrolyte of a fuel cell conducts mainly ions, it is not completely insulated from electrons. It will always be able to support very small amounts of electron conduction. This electron conduction in electrolyte or *internal current* is a net loss of current to external load. In a practical fuel cell, some reactants will diffuse from one electrode to another through the electrolyte where it will react without external electron transfer.

Consider a fuel cell with anodic and cathodic reactions,

$$A \leftrightarrow A^{n+} + ne^- \quad (5.143)$$

$$A^{n+} + B + ne^- \leftrightarrow C \quad (5.144)$$

and the net reaction

$$A + B \leftrightarrow C. \quad (5.145)$$

In Figure 5.25, a schematic of the various processes that occur is shown where both the fuel (A) and oxidant (B) are considered to diffuse to the other side of the electrolyte and some electrons transfer from the anode to the cathode through electrolyte. Even though there is no external current density (i_{ext} = 0) under open-circuit conditions, there are internal short-circuiting currents because of (1) the small electronic conductivity of the electrolyte membrane, the electrical-short-circuit current, and (2) the permeating fuel (A) and oxidant (B) across the membrane that cause small local crossover currents at the cathode and the anode, respectively. This leads to a net potential loss even under open-circuit conditions.

For example, in PEMFC, hydrogen from the anode can diffuse through its electrolyte to the cathode and will undergo oxidation on the Pt catalyst with oxygen electrochemically. The crossing over of one hydrogen molecule from

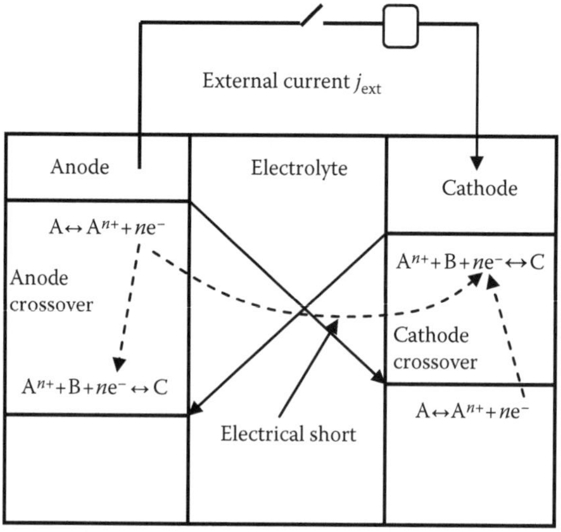

FIGURE 5.25
Schematic of oxidation and reduction reactions and the resulting external current and internal loss currents. Note that the internal loss current is the sum of the current caused by anode crossover, cathode crossover, and electrical short.

the anode to cathode results in a waste of two electrons and amounts to the crossing of two electrons from the anode to cathode internally, rather than as an external current. Thus, the electrochemical hydrogen oxidation provides electrons, or a crossover current at the cathode, for the oxygen reduction at the cathode even under open-circuit conditions. Similarly, there is some oxygen reduction occurring at the anode because of the small amount of oxygen diffusing through the electrolyte from the cathode to the anode. This picks up electrons from the hydrogen oxidation occurring at the anode, thus resulting in a crossover current at anode.

The net flow of fuel and electrons through the electrolyte is small, typically the equivalent of only a few mA/cm^2. However, for the low-temperature fuel cell PEMFC, though the Nernst equation predicted that the open-circuit voltage is 1.2 V, the measured values are at 1 V. The voltage loss 0.2 V represents significant efficiency loss. The electrolyte supports a very small amount of electronic conductivity, so that small short-circuiting currents are possible.

The fuels crossover and internal currents are equivalent; that is, they both contribute voltage loss owing to a small equivalent cell current. However, fuel crossover and the internal currents have a different physical effect on fuel cell. In the internal current, the oxidation reaction has already taken place and the electrons are short-circuited through electrolyte. In case of fuel crossover such as hydrogen permeation from the anode to the cathode, first the fuel crosses over from the anode to the cathode and then oxidation and reduction reactions occur near the cathode. With reactant crossover and internal currents, a small amount of current is lost. In both cases, the current losses are similar to activation losses, and hence as an approximation, the current and potential behavior can be represented by the Tafel law.

The total electric current is the sum of the external current and the current caused by fuel crossover and internal currents. In terms of cell current density, i,

$$j = j_{ext} + j_{loss}. \tag{5.146}$$

Using the Tafel equation (Equation 5.114), the overpotential caused by fuel crossover and internal currents is given as

$$\eta_{other} = \frac{RT}{\alpha F}\ln\left(\frac{j}{j_0}\right). \tag{5.147}$$

From Equation 5.146, and at $T = 353$ K and $\alpha = 0.5$, Equation 5.147 reduces to

$$\eta_{other} = 0.06088\ln\left(\frac{j_{ext} + j_{loss}}{j_0}\right)(\text{V}). \tag{5.148}$$

The coefficient on the right-hand side of the equation is ~61 mV. Now, assuming no external current is drawn, $i_{ext} = 0$, and a cathode exchange current density of 10^{-5} A/cm^2, the overpotential caused by current losses versus the internal current density is shown in Figure 5.26. For an internal current density of 1 mA/cm^2, the cell overpotential is 0.28 V. From the figure, we see a steep increase in overpotential at small internal current density. Thus, even if the cell current density to the load is zero, the open-circuit voltage of the cell is 0.92 V for an internal current density of 1 mA/cm^2.

In the case of PEMFC, the effect of hydrogen permeation through the electrolyte can be neglected once the current density reaches operational ranges. This is because the hydrogen concentration in the catalyst layer decreases at high current densities and hence the permeation probability decreases. In the case of DMFC, the fuel crossover effects are extreme, since the liquid methanol solution used as a fuel has a higher molecular concentration that results in large crossover. Typically, in DMFC, the measured OVC is approximately 0.7 V compared to the theoretical 1.2 V caused by these large crossover effects. The methanol or hydrogen crossover is a function of concentration (partial pressure), membrane permeability, and membrane thickness.

There are several approaches to reduce or limit the internal current losses. Choosing high ionic conductivity and low electronic conductivity electrolyte reduces the electron transfer through the electrolyte. In order to reduce the reactant crossover, the following approaches have been used. (i) *Use of thicker electrolyte to increase the diffusion length.* This approach has been used in DMFC to reduce methanol solution diffusion. Often the electrolyte thickness also increases ohmic losses; thus, this approach is limited to low-power applications. (ii) *Changing porosity and structure of the electrolyte material.* Different PEMFC electrolytes with different hydrogen diffusion rates have

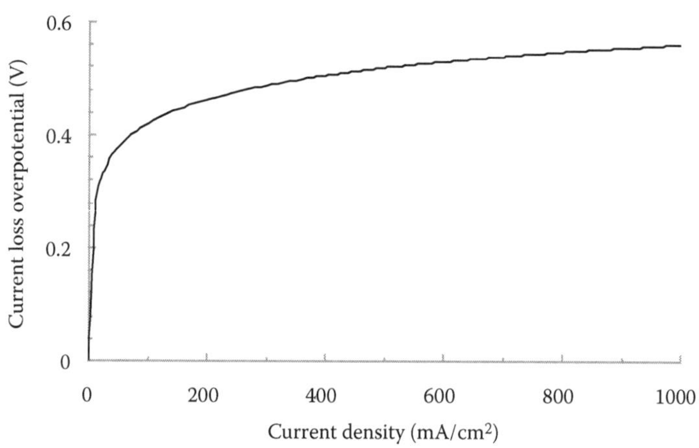

FIGURE 5.26
The internal short circuit current open-circuit condition with a cathode exchange current density of 10^{-5} A/cm^2.

been used. (iii) *Providing a diffusion barrier between reactant and electro-catalyst layer.* This technique is used for DMFC where a limited amount of the methanol solution as needed for electro-oxidation reaction is made available to the electro-catalyst layer with a diffusion barrier layer.

5.11 Fuel Cell Polarization Curve

In the previous section, each of the fuel cell losses, cathode and anode activation losses, ohmic losses, mass transfer losses, and losses owing to short circuit and reactant crossover was discussed, and expression for each loses overpotential or the polarizations were obtained. Now, we have net fuel cell overpotential from Equation 5.149

$$\eta = \eta_{\mathrm{act,a}} + \eta_{\mathrm{act,c}} + \eta_{\mathrm{Ohm}} + \eta_{\mathrm{mt,a}} + \eta_{\mathrm{mt,c}} + \eta_{\mathrm{other}}. \tag{5.149}$$

Note that the overpotential owing to short circuit and crossover η_{other} reduces the cell potential lower than the reversible potential even when the external current is zero. By using the expressions for the losses owing to activation, ohmic resistances, and mass transfer effects, the cell overpotential is now written as

$$\eta = -\frac{RT}{\alpha_{\mathrm{Ox,a}}F}\ln\frac{j}{j_{0\mathrm{a}}} - \frac{RT}{\alpha_{\mathrm{Red,c}}F}\ln\frac{j}{j_{0\mathrm{c}}} - jr + \frac{RT}{nF}\ln\left(1-\frac{j}{j_{\mathrm{La}}}\right) + \frac{RT}{nF}\ln\left(1-\frac{j}{j_{\mathrm{Lc}}}\right), \tag{5.150}$$

where $j = j_{\mathrm{ext}} + j_{\mathrm{loss}}$.

The cell-specific resistance r can be written as

$$r = r_{\mathrm{i}} + r_{\mathrm{e}} + r_{\mathrm{con}}, \tag{5.151}$$

where the RHS terms are ionic, electronic, and contact resistances, respectively. Electronic and contact resistances impose losses owing to the flow of external current density j_{ext}, whereas the ionic resistance imposes losses owing to j_{ext} and j_{loss}.

Therefore, the ohmic resistance can be written as

$$jr = (r_{\mathrm{e}} + r_{\mathrm{c}})\, j_{\mathrm{ext}} + r_{\mathrm{i}}\,(j_{\mathrm{ext}} + j_{\mathrm{loss}}). \tag{5.152}$$

Typically, the ionic resistance is large compared to the electronic and contact resistances. The overpotential is now written as

$$\eta = -\frac{RT}{\alpha_{\mathrm{Ox,a}}F}\ln\frac{j_{\mathrm{ext}}+j_{\mathrm{loss}}}{j_{0\mathrm{a}}} - \frac{RT}{\alpha_{\mathrm{Red,c}}F}\ln\frac{j_{\mathrm{ext}}+j_{\mathrm{loss}}}{j_{0\mathrm{c}}} - (r_{\mathrm{e}}+r_{\mathrm{con}})j_{\mathrm{ext}} - r_{\mathrm{i}}(j_{\mathrm{ext}}+j_{\mathrm{loss}})$$
$$-\frac{RT}{nF}\ln\left(\frac{j_{\mathrm{La}}}{j_{\mathrm{La}}-j_{\mathrm{ext}}-j_{\mathrm{loss}}}\right) + \frac{RT}{nF}\ln\left(\frac{j_{\mathrm{Lc}}}{j_{\mathrm{Lc}}-j_{\mathrm{ext}}-j_{\mathrm{loss}}}\right). \quad (5.153)$$

The kinetic losses, ohmic losses, mass transfer losses, and short circuit and crossover losses are illustrated in Figure 5.27. The dominant losses are typically activation losses and ohmic losses. At high current density, the mass transfer or concentration losses dominate. In the illustration, the internal losses are shown as constant; however, the internal losses also depend on the current density.

The fuel cell voltage E is thus the difference between open-circuit voltage and total overpotential

$$E = E_{\mathrm{eq}} - \eta. \quad (5.154)$$

The cell performance or the polarization curves are illustrated in Figure 5.28 where each of the losses is deducted from the cell open-circuit voltage in steps beginning with internal losses, followed by activation losses, and then ohmic losses, and finally concentration losses.

If the anode losses are negligible compared with cathode losses, then the overpotential can be simplified as

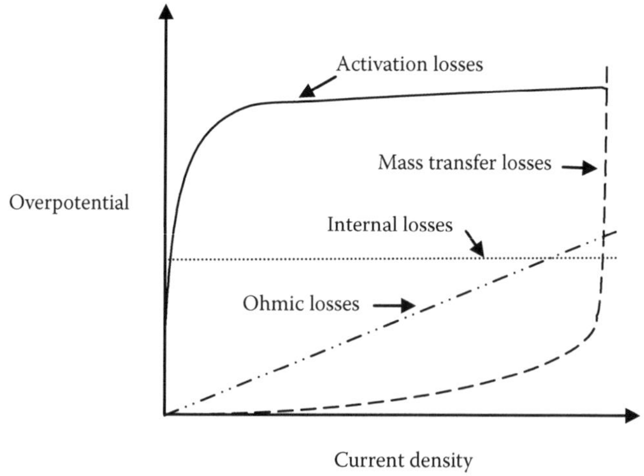

FIGURE 5.27
Contribution of activation losses, ohmic losses, and mass transfer losses and internal losses owing to short circuit and reactant crossover.

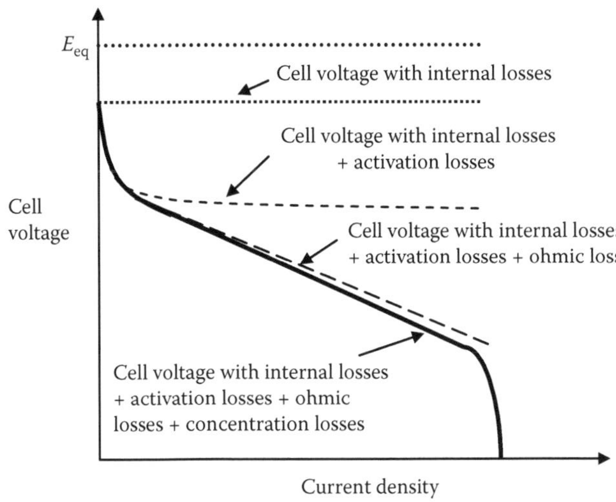

FIGURE 5.28
Fuel cell polarization curves owing to internal losses, activation losses, ohmic losses, and concentration losses. Note that the actual magnitude of each loss is different for each fuel cell design and construction.

$$\eta = -\frac{RT}{\alpha_{\text{Red,c}}F}\ln\frac{j_{\text{ext}}+j_{\text{loss}}}{j_{0c}} - (j_{\text{ext}}+j_{\text{loss}})r - \frac{RT}{nF}\ln\frac{j_{\text{Lc}}}{j_{\text{Lc}}-j_{\text{ext}}-j_{\text{loss}}}. \tag{5.155}$$

As seen from Equation 5.155, the activation and mass transfer losses are similar in the anode and cathode though often the cathode losses dominate. The fuel cell polarization can be written in approximate form as

$$\eta = -A\ln\frac{j}{j_0} - jr - B\ln\frac{j_{\text{L}}}{j_{\text{L}}-j}. \tag{5.156}$$

Here, the coefficients A and B are given as

$$A = -\frac{RT}{\alpha F}, \quad B = \frac{RT}{nF}. \tag{5.157}$$

In Equation 5.156, the values of A and B, current density i_0, internal specific resistance r, and limiting current density j_{L} depend on the type and design of the fuel cell. In Table 5.4, representative values of these parameters are listed for low-temperature (PEMFC) and high-temperature fuel cells (SOFC). For SOFC, the current density is large compared to the PEFC. The activation loss for the SOFC should be calculated from the full BV equation. In Figure 5.29, the cell voltages for PEMFC and SOFC are plotted using these parameters.

It is often easy to develop a fuel cell performance curve using experiential data. One can write the fuel cell voltage in the following forms:

$$E = E_{OCV} - A' \ln j - jr - B \ln \frac{j_L}{j_L - j} \tag{5.158}$$

$$E = E_{cov} - A' \ln j - jr - C \ln nj, \tag{5.159}$$

where A', r, B, C, n, and i_L are obtained from curve fits of experimental data. The fuel cell open-circuit voltage E_{OCV} can be obtained from experiments or theoretically calculated.

TABLE 5.4
Typical Parameters for Low-Temperature PEMFC and High-Temperature SOFC

Parameters	PEMFC	SOFC
Open-circuit voltage (V)	1.22	1.06
i_0 (A/cm^2)	1×10^{-4}	0.1
i_L (A/cm^2)	1.5	1.5
i_{loss} (A/cm^2)	0.002	0.002
r (Ω cm^2)	0.03	0.09
A (V)	0.05	0.03
B (V)	0.06	0.08

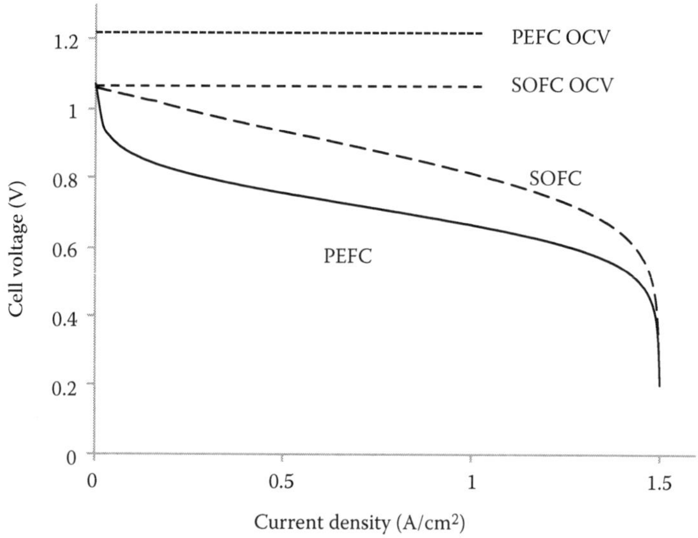

FIGURE 5.29
Polarization curves for PEMFC and SOFC.

5.12 Summary

The response of the fuel cell is determined by the electrochemical processes and associated kinetics at the electrode and electrode interface. The electrochemical processes depend on the mass and charge transfer between the bulk electrolyte solution and electrode surface. The rates at which these transfers occur are determined by the number of localized phenomena and largely depend on the materials involved. These processes are presented in this chapter and the relations between the fuel cell potential and current density are given in terms of BV and Tafel equations. The key losses in the fuel cell include the activation losses, ohmic losses, mass transport losses, and losses owing to reactant crossover and internal currents that are discussed in this chapter. The fuel cell polarization curve is presented and is discussed for low-temperature and high-temperature fuel cells such as PEMFC and SOFC, respectively.

PROBLEMS

1. Several processes happen at the electrode–electrolyte interface in an electrochemical cell. Identify the transfer process involving mass and charge at the electrode–electrolyte interface.
2. Various models have been proposed for the electric double layer at an electrode–electrolyte interface. Briefly explain the structure of the electric double layer starting from the Helmholtz model to the triple-layer model and then identify the key features of each model.
3. Find the values of overpotential as a function of temperature from 25°C to 800°C for which the linear approximation and Tafel approximation of BV equations are valid. (Hint: see Examples 5.2 and 5.3.)
4. Calculate anodic and cathodic activation overpotentials at current densities of 0.01 A/cm^2, 0.1 A/cm^2, and 0.5 A/cm^2 for the following conditions:

Temperature	353 K
j_{0a}	1.0 A/cm^2
j_{0c}	10^{-3} A/cm^2
$\alpha_{Ox,a}$	0.5
$\alpha_{Ox,c}$	0.5

References

Allen, P. L. and A. Hickling. Electrochemistry of sulphur. Part 1. Overpotential in the discharge of the sulphide ion. *Transactions of the Faraday Society* 53: 1626, 1957.

Bard, J. and L. R. Faulkner. *Electrochemical Methods Fundamentals and Applications*. John Wiley & Sons, New York, 1980.

Bockris, J. O'M. and S. Srinivasan. *Fuel Cells: Their Electrochemistry*. McGraw Hill Publishing Company, New York, 1969.

Chapman, D. L. A contribution to the theory of electrocapillarity. *Philosophical Medicine* 25: 475, 1913.

Gouy, G. Sur la fonction electrocapillaire (Electrocapillarity). Part I. *Annales des Chimie et des Physique* 7(29): 145, 1903; Part II 8(8): 291, 1906; Part III, 8(9): 75, 1906.

Grahame, D. C. The role of the cation in the electrical double layer. *Journal of Electrochemical Society* 98: 343, 1951.

Kordesch, K. and G. Simander. *Fuel Cells and Their Applications*. John Wiley & Sons Ltd., New York, 2000.

Larminie, J. and A. Dicks, Editors. *Fuel Cell Systems Explained*. John Wiley & Sons Ltd., UK, 1999.

Mench, M. M. *Fuel Cell Engines*. John Wiley & Sons, New Jersey, 2008.

O'Hayre, R., S. W. Cha, W. Colella and F. B. Prinz. *Fuel Cell Fundamentals*. John Wiley & Sons, New York, 2006.

Stern, O. The theory of the electrolytic double-layer. *Zeitschrift fur Elektrochemie* 30: 508, 1924.

Vielstich, W., A. Lamm and H. Gasteiger, Editors. *Handbook of Fuel Cells-Fundamentals, Technology and Applications*. John Wiley & Sons, New York, 2003.

6

Heat and Mass Transfer in Fuel Cells

Operation of a fuel cell depends on a number of transport processes such as flow of reactant gases through the gas flow channels, mass transport of reactant gas species from gas flow channels and through the porous electrodes, ion transport through the membranes, and electron transport through electrodes and interconnects. Figure 6.1 shows transport of gas flow in flow channels as well heat and mass transport through the channels and electrode–membrane tri-layers of the fuel cell.

As fuel and oxygen/air are supplied continuously at the gas flow channels, the reactant gas species transport through the gas supply channels and through electrodes and are continuously consumed at the electrode–membrane interfaces through electrochemical reactions. Water and heat generated in the fuel cell also transport through the membrane and electrodes and are transferred to the flowing gases in the channels. Poor transport of mass and heat contributes significantly to the fuel cell losses and performance. While charge transport contributes to the *ohmic losses,* the mass transport of reactant gases influences the *mass transfer losses.* Poor mass transport may lead to insufficient supply of reactant gases to the electrode–membrane interface for electrochemical reactions and results in the so-called concentration overpotential or mass transfer loss.

In this chapter, we will primarily focus on fluid flow, heat, and mass transport through gas flow channels and in solid porous electrodes, and its effect on the *mass transfer loss.* Solid-phase diffusion, charge transport in electrolyte membrane, and *ohmic loss* will be discussed in Chapter 7. Water transport will also be discussed in Chapter 7.

6.1 Fluid Flow

Fluid flow and pressure variation in a fuel cell play a critical role in the distribution of reactant gas concentration at electrochemical reaction sites and, hence, in the distribution of local current densities and cause *mass transfer loss.* The governing equations for reactant gas flows in gas flow channels and in porous electrode–gas diffusion layers are given by conservation of mass and momentum equations. Solutions to these equations result in the distribution of pressure, P, and velocity field, which is also referred to as the bulk motion in the gas flow channels and porous electrode–gas diffusion layers.

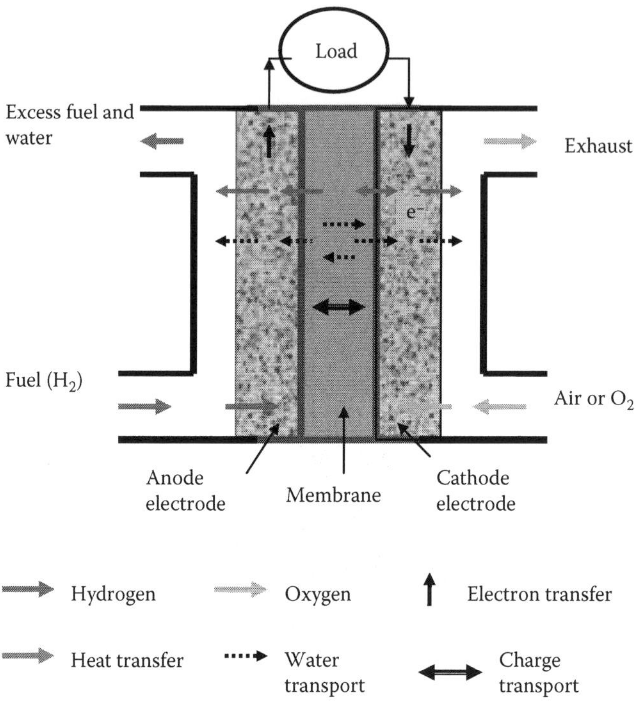

FIGURE 6.1
Fluid, heat, and mass species transport in a tri-layer fuel cell.

Before considering the effect of fluid flow on the fuel cell, we will briefly review some of the basic principles and relations of fluid flow, heat, and mass transport.

6.1.1 External Flow

Consider flow of fluid with uniform upstream velocity u_∞ over a flat surface as shown in Figure 6.2.

Flow of fluid over the surface is characterized by the formation of a *hydrodynamic* or *velocity boundary layer*, which is defined as the thin layer of fluid over which the velocity of the fluid varies from no-slip zero velocity at the surface to the outer stream velocity over the thickness of the boundary layer $y = \delta$. Because of the effect of viscosity, fluid flow slows down near the stationary solid surface and maintains the no-slip fluid–solid interface boundary conditions. The flow is assumed to be viscous within the boundary layer and inviscid outside the boundary layer. The boundary layer thickness increases in the downstream x-direction and results in a varying x-component velocity profile $u(y)$ as shown in the figure.

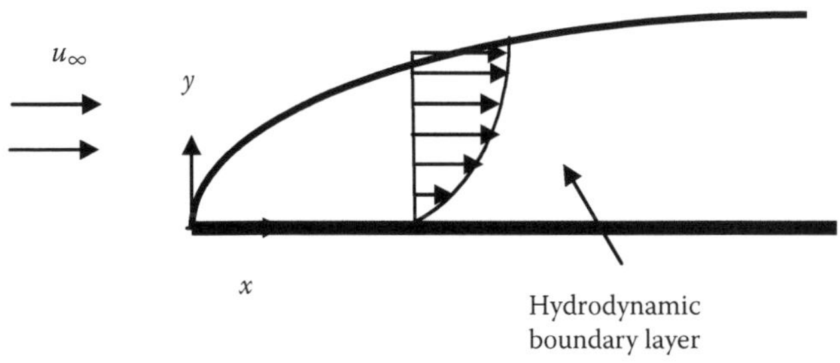

FIGURE 6.2
Hydrodynamic boundary layer for flow over a flat plate.

The fluid shear stress at the wall is given by Newton's law as

$$\tau_w = \mu \left.\frac{\partial u}{\partial y}\right|_{y=0} \tag{6.1}$$

and in the form of a dimensionless parameter called *wall skin friction* as

$$C_f = \frac{\tau_w}{\frac{1}{2}\rho u_\infty^2} = \frac{\mu \left.\frac{\partial u}{\partial y}\right|_{y=0}}{\frac{1}{2}\rho u_\infty^2}. \tag{6.2}$$

The skin friction decreases in the flow direction as the boundary layer thickness increases in the downstream x-direction. The wall shear stress and hence the skin friction can be obtained from the known velocity field, which is defined by the continuity and momentum equations of fluid motion. The skin frictions are generally expressed in the form of a correlation as a function of characteristics flow Reynolds number as

$$C_{fx} = f(Re) = CRe_x^m, \tag{6.3}$$

where the Reynolds number for the external flow of a flat plate is given as

$$Re_x = \frac{\rho u_\infty x}{\mu}. \tag{6.4}$$

External flow is characterized as laminar or turbulent on the basis of the critical Reynolds number:

$$Re_{crit} \approx 5 \times 10^5 \tag{6.5}$$

The flow is considered to be laminar for $Re_x < Re_{crit}$ and turbulent for $Re_x \geq Re_{crit}$.

6.1.2 Internal Flows

Internal flows include flow in conduits like pipes, tubes, channels, and enclosures. As flow enters the channel, boundary layers develop and grow on both top and bottom surfaces. The flow slows down within the boundary layer owing to the effect of viscosity with no-slip conditions at the wall and it accelerates in the center core region to satisfy mass continuity as shown in the Figure 6.3.

At some distance away from the entrance, the boundary layers meet and flow is assumed as viscous over the entire cross section of the channel. The internal flow is categorized into two distinct regions: (i) *hydrodynamic entrance region* where velocity profile varies with the axial length of the channel and (ii) *hydrodynamic fully developed region* where velocity profile remains invariable with the longitudinal distance along the channel, or becomes fully developed.

Hydrodynamic Entry Length ($L_{e,h}$): The length required for the velocity profile to become fully developed.

Internal flow is characterized as laminar or turbulent on the basis of the critical Reynolds number: $Re_{crit} \approx 2300$. The flow is assumed as laminar for $Re_D < 2300$ and as turbulent for $Re_D \geq 2300$.

The Reynolds number for internal flows is defined as

$$Re_D = \frac{\rho U_{av} D}{\mu} \tag{6.6}$$

The diameter D can be replaced by the hydraulic diameter $D_H = \dfrac{4A}{P}$ for no-circular flow geometry.

Like external flows, the wall shear stress and the skin friction can also be expressed by Equations 6.1 and 6.2, respectively. For internal flows, the wall viscous shear stress causes a pressure drop in the channel and this is expressed as

$$\Delta P = f \frac{L}{D_H} \frac{V^2}{2} \tag{6.7}$$

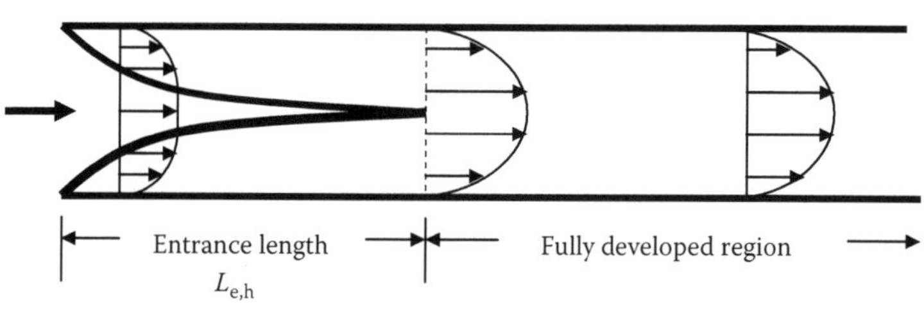

FIGURE 6.3
Hydrodynamic flow development for internal flow.

where f is defined as the friction factor, which is related to the wall shear stress and the skin friction as

$$f = \frac{C_f}{4}. \tag{6.8}$$

A list of correlations for skin friction or pressure drop for fully developed flow is given in Table 6.1.

TABLE 6.1

Fully Developed Flow Correlations for Heat Transfer Coefficients

Geometry	b/a	$q_s'' =$ Const	$T_s =$ Const	fRe_D
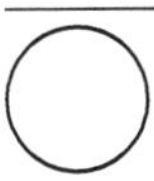		4.36	3.66	64
 a b	1.0	3.61	2.98	57
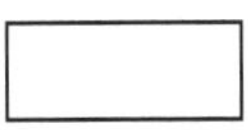 a b	2.0	4.12	3.39	62
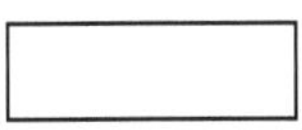	4.0	5.33	4.44	73
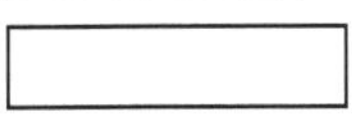	8.0	6.49	5.60	82
	∞	8.23	7.54	96
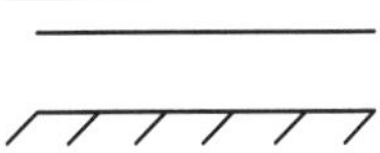	∞	5.38	4.861	96
		3.11	2.47	53
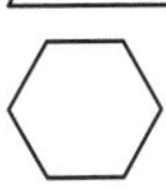		4.02	3.35	60.25
Octagon		4.20	3.46	61.52

Source: Incropera, F. P. et al.: *Fundamentals of Heat and Mass Transfer*, 6th Edition. 2007. Copyright Wiley-VCH Verlag GmbH & Co. KGaA. Reproduced with permission; Shah, R. K. and London, A. L. *Laminar Flow Forced Convection in Ducts*. Academic Press, New York, 1978; Asako, Y. et al. *International Journal of Heat and Mass Transfer*, 31, 2590–2593, 1988.

Note: $q_s'' =$ Const represents the case with constant surface heat flux and $T_s =$ Const represents the case with constant surface temperature.

The friction for laminar flow in a circular channel is given by $f = \dfrac{64}{Re_D}$. The friction factor for a rectangular channel depends on the aspect ratio, b/a. For example, it varies from $fRe_{D_H} = 57$ for an aspect ratio of $b/a = 1$ to $fRe_{D_H} = 96$ for an aspect ratio of $b/a = \infty$. A correlation of $fRe_{D_H}(b/a)$ for laminar flow in a rectangular channel is given as

$$fRe_{D_H} = 24\left[1 - 1.3553\left(\frac{b}{a}\right) + 1.9467\left(\frac{b}{a}\right)^2 - 1.7012\left(\frac{b}{a}\right)^3 + 0.9564\left(\frac{b}{a}\right)^4 - 0.2537\left(\frac{b}{a}\right)^5\right]. \tag{6.9}$$

As we have mentioned before, the pressure drop in the reactant gas flow channels plays a critical role in the operation and performance of a fuel cell. The higher the pressure drop, the higher is the decrease in the reactant gas pressures in the bulk fluid flow, and this affects the gas distribution in the electrode layer in cases where pressure-driven advection flow is important. Additionally, a higher pressure drop in the gas flow channels results in higher pumping or parasitic power requirement of a fuel cell.

A more detailed discussion of gas flow channel analysis and design is considered in Chapter 10.

6.1.3 Gas Flow Channels

6.1.3.1 Conservation of Mass

The conservation of mass or continuity equation is given as

$$\frac{\partial \rho}{\partial t} + \nabla \cdot (\rho \vec{V}) = 0 \tag{6.10a}$$

and in Cartesian coordinates as

$$\frac{\partial \rho}{\partial t} + \frac{\partial(\rho u)}{\partial x} + \frac{\partial(\rho v)}{\partial y} + \frac{\partial(\rho w)}{\partial z} = 0, \tag{6.10b}$$

with velocity field given as $\vec{V} = \hat{i}u + \hat{j}v + \hat{k}w$.

For incompressible flow, density ρ is constant and Equation 6.10a reduces to

$$\nabla \cdot \vec{V} = 0 \tag{6.11a}$$

and in Cartesian coordinate as

$$\frac{\partial u}{\partial x} + \frac{\partial v}{\partial y} + \frac{\partial w}{\partial z} = 0. \tag{6.11b}$$

6.1.3.2 Conservation of Momentum

The conservation of momentum equation is derived from Newton's second law of motion, which expresses proportionality between applied force and resulting acceleration of a particle. Momentum equation for Newtonian viscous fluid is given by the *Navier–Stokes equation*

$$\rho\frac{D\vec{V}}{Dt} = \rho\vec{g} - \nabla P + \frac{\partial}{\partial x_j}\left[\mu\left(\frac{\partial u_i}{\partial x_j} + \frac{\partial u_j}{\partial x_i}\right) - \frac{2}{3}\mu(\nabla\cdot\vec{V})\right]. \tag{6.12}$$

For Cartesian coordinate, the *Navier–Stokes equation* is expressed as follows:

x-momentum:

$$\begin{aligned}\rho\frac{Du}{Dt} &= \rho g_x - \frac{\partial p}{\partial x} + \frac{\partial}{\partial x}\left[\mu\left(2\frac{\partial u}{\partial x} - \frac{2}{3}\nabla\cdot\vec{V}\right)\right] \\ &\quad + \frac{\partial}{\partial y}\left[\mu\left(\frac{\partial u}{\partial y} + \frac{\partial v}{\partial x}\right)\right] + \frac{\partial}{\partial z}\left[\mu\left(\frac{\partial w}{\partial x} + \frac{\partial u}{\partial z}\right)\right]\end{aligned} \tag{6.13a}$$

y-momentum:

$$\begin{aligned}\rho\frac{Dv}{Dt} &= \rho g_y - \frac{\partial p}{\partial y} + \frac{\partial}{\partial x}\left[\mu\left(\frac{\partial u}{\partial y} + \frac{\partial v}{\partial x}\right)\right] + \frac{\partial}{\partial y}\left[\mu\left(2\frac{\partial v}{\partial y} - \frac{2}{3}\nabla\cdot\vec{V}\right)\right] \\ &\quad + \frac{\partial}{\partial z}\left[\mu\left(\frac{\partial v}{\partial z} + \frac{\partial w}{\partial y}\right)\right]\end{aligned} \tag{6.13b}$$

z-momentum:

$$\begin{aligned}\rho\frac{Dw}{Dt} &= \rho g_z - \frac{\partial p}{\partial z} + \frac{\partial}{\partial x}\left[\mu\left(\frac{\partial w}{\partial x} + \frac{\partial u}{\partial z}\right)\right] + \frac{\partial}{\partial y}\left[\mu\left(\frac{\partial v}{\partial z} + \frac{\partial w}{\partial y}\right)\right] \\ &\quad + \frac{\partial}{\partial z}\left[\mu\left(2\frac{\partial w}{\partial z} - \frac{2}{3}\nabla\cdot\vec{V}\right)\right].\end{aligned} \tag{6.13c}$$

For incompressible fluid flow and for constant fluid viscosity (μ), the Navier–Stokes equation reduces to

$$\rho\frac{D\vec{V}}{dt} = \rho\vec{g} - \nabla P + \mu\nabla^2\vec{V} \tag{6.14a}$$

or

$$\rho\left[\frac{\partial \vec{V}}{\partial t}+(\vec{V}\cdot\nabla)\vec{V}\right]=\rho\vec{g}-\nabla P+\mu\nabla^2\vec{V} \tag{6.14b}$$

and in three-component Cartesian coordinates as follows:

x-momentum:

$$\rho\left(\frac{\partial u}{\partial t}+u\frac{\partial u}{\partial x}+v\frac{\partial u}{\partial y}+w\frac{\partial u}{\partial z}\right)=\rho g_x-\frac{\partial p}{\partial x}+\mu\left(\frac{\partial^2 u}{\partial x^2}+\frac{\partial^2 u}{\partial y^2}+\frac{\partial^2 u}{\partial z^2}\right) \tag{6.15a}$$

y-momentum:

$$\rho\left(\frac{\partial v}{\partial t}+u\frac{\partial v}{\partial x}+v\frac{\partial v}{\partial y}+w\frac{\partial v}{\partial z}\right)=\rho g_y-\frac{\partial p}{\partial y}+\mu\left(\frac{\partial^2 v}{\partial x^2}+\frac{\partial^2 v}{\partial y^2}+\frac{\partial^2 v}{\partial z^2}\right) \tag{6.15b}$$

z-momentum:

$$\rho\left(\frac{\partial w}{\partial t}+u\frac{\partial w}{\partial x}+v\frac{\partial w}{\partial y}+w\frac{\partial w}{\partial z}\right)=\rho g_z-\frac{\partial p}{\partial z}+\mu\left(\frac{\partial^2 w}{\partial x^2}+\frac{\partial^2 w}{\partial y^2}+\frac{\partial^2 w}{\partial z^2}\right) \tag{6.15c}$$

6.1.4 Fluid Flow in Porous Electrodes

The electrodes of a fuel cell are made of a gas diffusion layer (GDL) and a catalyst layer (CL) for some fuel cells like PEMFC as shown in Figure 6.4.

Reaction gases flow though the pores in GDL toward the reaction sites at the electrode–membrane interface. Simultaneous gas flow and reaction takes place in the region of the CL. Water produced at the electrode–membrane interface migrates through the porous layers to the gas flow channels.

The mass continuity and momentum equations for the bulk fluid flow in a channel are significantly altered in a porous media owing to the presence of complex flow geometries.

6.1.4.1 Mass Continuity in Porous Media

The conservation of mass or the mass continuity equation for flow in a porous media is given as

$$\frac{\partial \rho}{\partial t}+\nabla\cdot\rho\vec{V}=S_i, \tag{6.16}$$

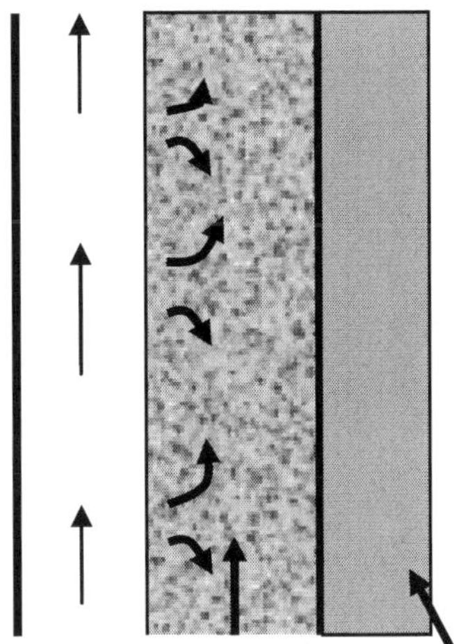

FIGURE 6.4
Fluid flow in porous media.

where S_i represents the reactant gas consumption rates for hydrogen and oxygen, and mass source term for the water transport. For the GDL with no volume reaction zone, the source term is dropped and included as a boundary condition at the electrode–membrane interfaces.

6.1.4.2 Momentum Equation in Porous Media

A number of different approaches are proposed and used in modeling flow through porous media. Some of the most popular approaches include (i) Darcy's law, (ii) Brinkman equation, and (iii) a modified Navier–Stokes equation. In the absence of the bulk fluid motion or advection transport, the reaction gas species can only transport through the GDL and CL by the diffusion mechanisms, which we will discuss in a later section.

6.1.4.2.1 *Darcy's Law*

As we know, the fluid flow takes place under the influence of different body and surface forces. Darcy's law defines the fluid flow in a porous media under the influence of pressure gradient force only, and this is expressed as

$$\vec{V} = \frac{\kappa}{\mu}\nabla P \tag{6.17}$$

where κ is the permeability of the porous media and μ is the dynamic viscosity of the fluid.

In this formulation, the presence of inertia and viscous forces is neglected and the region is assumed as homogeneous porous media characterized by the permeability.

6.1.4.2.2 Brinkman's Equation

One of the difficulties in using Darcy's law for relatively slower bulk flow in a porous media involving multiple pore size distribution is the matching of boundary conditions to the adjacent free flow such as in gas flow in channels as shown in Figure 6.4.

In order to match the solutions of Navier–Stokes equation to the solution of Darcy's equation at the channel–porous media interface, Darcy's equation is modified to include a viscous force term in the momentum equation and this is given by the *Brinkman's equation* (Martys, 2001; Martys et al., 1994) as

$$\nabla P = -\frac{\mu}{\kappa}\vec{V} + \mu_e \nabla^2 \vec{V}, \tag{6.18}$$

where μ_e is the effective viscosity to be used for the fluid in the porous media.

Notice that Brinkman's equation includes both the pressure force and the viscous force terms. The effective viscosity for the slower-moving fluid in the porous media is selected such that continuity in shear stress is maintained at the interface between the faster-moving gas flow in the channel and the slower-moving gas flow in porous electrode. The continuity in shear stress at the interface is given as

$$\mu\frac{\partial u}{\partial z} = \mu_e \frac{\partial u}{\partial z}. \tag{6.19}$$

6.1.4.2.3 Modified Navier–Stokes Equation

The Navier–Stokes equation used for flow through channels needs to be modified to describe the flow in a porous media with complex flow geometry involving fluid and solid regions. The approach used to characterize the flow in a porous media is based on a volume average technique outlined by Whitaker (1969) and Slattery (1969). The modified Navier–Stokes or momentum equation based on this averaging technique is given as

$$\rho\left[\frac{\partial\langle\vec{V}\rangle}{\partial t} + \left(\langle\vec{V}\rangle\cdot\nabla\right)\langle\vec{V}\rangle\right] = \rho\vec{g} - \nabla\langle P\rangle + \mu\nabla^2\langle\vec{V}\rangle - \frac{\mu}{\kappa}\varepsilon\langle\vec{V}\rangle \tag{6.20}$$

where density and viscosity, μ, are for the fluid, and κ is the permeability of the porous media. The symbol $\langle\rangle$ represents a volume average quantity given as

$$\langle\varphi\rangle = \frac{1}{\forall}\int_{\forall}\varphi d\forall \tag{6.21}$$

where φ represents flow variables such as velocity and pressure and $\forall$ represents the volume.

For simplicity, the modified Navier–Stokes equation is written without the volume average symbol as

$$\rho\left[\frac{\partial\vec{V}}{\partial t}+\left(\vec{V}\cdot\nabla\right)\vec{V}\right]=\rho\vec{g}-\nabla P+\mu\nabla^2\vec{V}-\frac{\mu}{\kappa}\varepsilon\vec{V}. \tag{6.22}$$

The modified Navier–Stokes equation reduces to Brinkman's equation when the inertia force term is dropped and it reduces to Darcy's law when both inertia and viscous force terms are dropped.

A selection of appropriate momentum equation for the porous GDL and CL will depend on the operating conditions and the design of GDL and CL regions for a fuel cell.

6.1.5 Inlet and Boundary Conditions

6.1.5.1 Inlet Conditions

Inlet conditions for pressure and velocity at the inlet to the flow channels are given in terms of total constant inlet pressure and average inlet reactant gas velocity, respectively:

$$P_{\mathrm{i}} = P_{\mathrm{i,in}} \tag{6.23a}$$

$$u_{\mathrm{i}} = U_{\mathrm{i,in}} \tag{6.23b}$$

The average inlet gas velocity is computed from the reactant gas consumption rate as

$$U_{\mathrm{i,in}} = \frac{\dot{m}_{\mathrm{i}}}{\rho_{\mathrm{i}}A_{\mathrm{ch}}} \tag{6.24}$$

where $\dot{m}_i$ is the mass flow rate of the reactant gas at the inlet to the channel. The index i refers to anode and cathode gas flow channels.

In a fuel cell, the reactant gas mass flow rate at the inlet to the gas flow channels is assumed to be ideally equal to the gas consumption rate at the electrode–membrane interface, which depends on the operating cell current density and given as

$$\dot{m}_{\mathrm{i}} = \frac{\xi_{\mathrm{i}}IM_{\mathrm{i}}}{n_{\mathrm{e}}F}\,(\mathrm{kg/s}), \tag{6.25}$$

where ξ_{i} is the stoichiometric factor that represents the amount of excess reactant gas.

6.1.5.2 Boundary Conditions

At all impermeable solid surfaces, a no-slip condition, that is, a zero velocity boundary condition, is assumed. Boundary conditions at the interface of the fluid flow channel and porous media are given on the basis of the assumption of continuity in the solutions of pressure and normal component velocity for the two adjacent regions.

$$P_{ch} = P_{pm} \tag{6.26a}$$

and

$$v_{ch} = v_{pm}. \tag{6.26b}$$

For Brinkman's equation, additional boundary condition is given in terms of continuity in shear stress as

$$\mu \frac{\partial u_{ch}}{\partial z} = \mu_e \frac{\partial u_{pm}}{\partial z}. \tag{6.27}$$

6.2 Heat Transfer in Fuel Cells

As we have discussed in Chapter 4, heat is generated in a fuel cell owing to the irreversibilities associated with the electrochemical reactions at the electrolyte–electrode interface and conducting resistances of electrodes and proton membrane to electrons and ion flows. While the ohmic heating or the joule heating owing to the charge transport is volumetric over the entire regions of membrane and the electrodes, the heat generation owing to reaction irreversibilities can be assumed as surface heat generation at the electrode–membrane interface where electrochemical reaction takes place. However, for some fuel cell MEAs with three-phase reaction region that includes electrode, electrolyte, and catalyst materials, this heat generation is treated as volumetric.

As we have discussed in Chapter 4, the total electrochemical heat generation includes a reversible heat generation component as well as an irreversible heat generation component, and this is expressed as

$$Q_{gen,elec} = \left[\dot{n}_f(-T\Delta S) + I(E_{rev} - V_c)\right] \quad \text{for a cell,} \tag{6.28a}$$

where $\dot{n}_f = \dfrac{I}{n_e F}$ is the fuel consumption rate.

The entropy change of the chemical reaction can be estimated directly on the basis of the procedure outlined in Chapter 3 and Section 4.3. Noting that the fuel consumption rate can be expressed in terms of the operating current as $\dot{n}_f = \frac{I}{n_e F}$, the reversible heat generation of Equation 6.28a can be written as

$$Q_{\text{gen,rev}} = \frac{I}{n_e F}(-T\Delta S_a) + \frac{I}{n_e F}(-T\Delta S_c). \tag{6.28b}$$

As we have discussed in Chapter 4, the reversible heat generation can be computed for each of the half electrochemical electrode reactions separately. However, the reversible heat generation in fuel cell is often computed as a single term on the basis of entropy change of the overall hydrogen oxidation reaction.

The irreversible heat generation owing to the ohmic heat is given as

$$Q_{\text{gen,ohm}} = I^2 \dot{R}_c \text{ or a cell.} \tag{6.29a}$$

and the irreversible heat generation caused by the activation overpotential as

$$Q_{\text{gen,act}} = I\eta_{\text{act}} \tag{6.29b}$$

The generated heat dissipates through the electrodes, membranes, and bipolar plates by conduction and is carried away by gas and coolant streams by convection. Figure 6.5 shows the different modes of heat transport across the fuel cell.

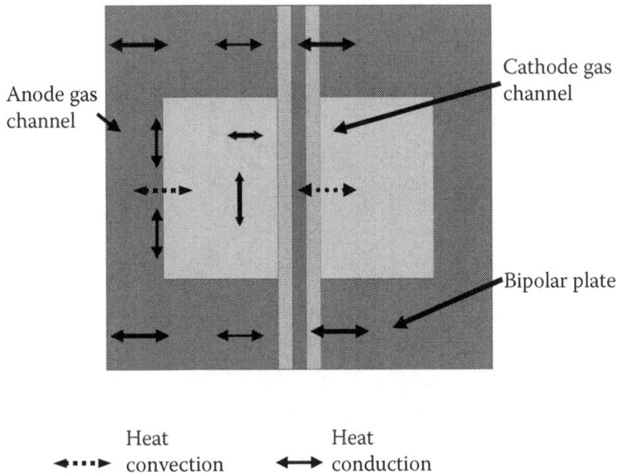

FIGURE 6.5
Heat transport in a fuel cell.

The heat generation and subsequent dissipation result in a temperature distribution within the fuel cell and influence the cell's performance. Heat transfer plays a very important role in the analysis, design, and operation of a fuel cell, and overall thermal management of the fuel cell power generation system.

In the following section, a brief discussion on the mechanism of heat transfer as applicable to fuel cell is given.

6.2.1 Heat Transfer Modes and Rate Equations

Heat transfer is defined as the energy transfer owing to the presence of spatial temperature variation. There are three basic modes of heat transfer: conduction, convection, and radiation.

6.2.1.1 Conduction Heat Transfer

This mode is primarily important for heat transfer in solids such as the electrodes and membranes, and in stationary fluid such as electrolytes and liquids in a porous membrane. Figure 6.6 demonstrates heat transfer by conduction in a plane slab representing a solid or stationary fluid layer.

The conduction rate equation is governed by Fourier's law, which states the heat flow rate per unit area or heat flux as

$$\vec{q} = -k\nabla T, \tag{6.30}$$

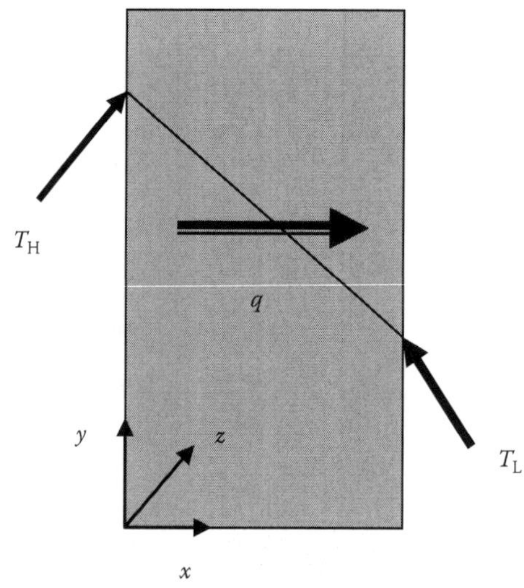

FIGURE 6.6
Heat transfer by conduction in a solid or stationary fluid layer.

where $q'' = \frac{q}{A}$ is the heat flow per unit area per unit time or heat flux, and k is the thermal conductivity of the material.

The heat flux vector in the Cartesian coordinate system is written as

$$\vec{q} = -\left(\hat{i}k_x \frac{\partial T}{\partial x} + \hat{j}k_y \frac{\partial T}{\partial y} + \hat{k}k_z \frac{\partial T}{\partial z} \right), \tag{6.31}$$

where the heat flux components are

$$q''_x = -k_x \frac{\partial T}{\partial x}, \; q''_y = -k_y \frac{\partial T}{\partial y}, \; \text{and } q''_z = -k \frac{\partial T}{\partial z}.$$

6.2.1.2 Convection Heat Transfer

Convection heat transfer is the transfer of heat energy owing to the combined effect of molecule motion or diffusion plus energy transfer by bulk fluid motion, which is also referred to as advection. The convection heat transfer occurs between a moving fluid and an exposed solid surface. Let us consider the fluid flow over a solid surface at a temperature T_S as shown in Figure 6.7. The fluid upstream temperature and velocity are T_∞ and u_∞, respectively.

As we have discussed earlier, because of the effect of viscosity or no-slip condition, there is a development of a thin fluid region, known as the hydrodynamic boundary layer, inside which velocity varies from the solid surface velocity to the outer stream velocity, u_∞. Similarly, there is a development of a thermal boundary layer inside which fluid temperature changes from solid surface temperature T_S to outer fluid temperature T_∞.

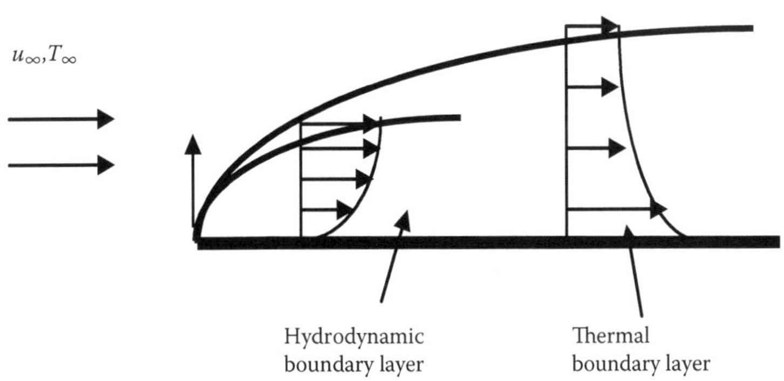

FIGURE 6.7
Hydrodynamic and thermal boundary layers for flow over a solid surface.

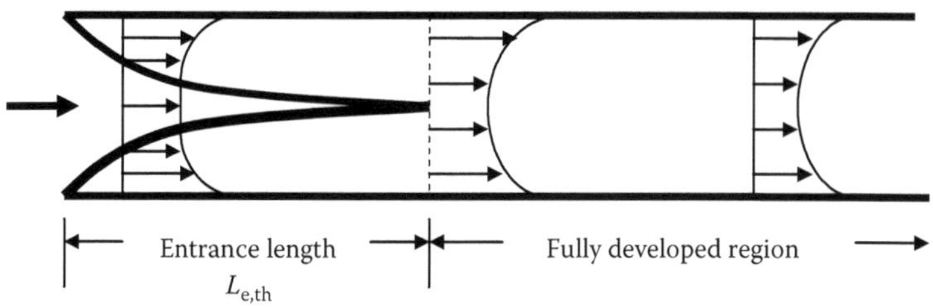

FIGURE 6.8
Thermal entry length and thermally fully developed region.

For internal flows, thermal boundary layers develop from both top and bottom surfaces and develop into two regions: *thermal entry length* and *thermal fully developed* regions similar to hydrodynamic internal flow as shown in Figure 6.8.

Thermal fully developed region: The region where the dimensionless temperature profile remains invariable along the longitudinal length of the channel.

Thermal entry length ($L_{e,th}$): The length required for the dimensionless temperature profile to become fully developed.

Criterion for entry length: Criterion for hydrodynamic entry length for laminar flow:

$$\frac{L_{e,th}}{D} \approx 0.06\, Re_D. \tag{6.32}$$

For example, for a maximum laminar flow Reynolds number in a circular channel given by the critical Reynolds number, $Re_{Crit} = 2300$,

$$\frac{L_{e,th}}{D} \approx 0.06 \times 2300$$

or

$$L_{e,th} = 138\, D.$$

For example, if we consider a circular channel of 1 mm diameter, then the hydrodynamic entrance length is approximately 13.8 cm.

Criterion for *thermal entry length* for laminar flow:

$$\frac{L_{e,th}}{D} \approx 0.06\, Re_D Pr. \tag{6.33}$$

Criterion for hydrodynamic and thermal entry length for turbulent flow:

$$20 \approx \frac{L_e}{D} \approx 40. \tag{6.34}$$

Since the fluid is stationary at the solid surface, the heat is transferred by conduction through the stationary fluid layer normal to the surface owing to molecular motion or diffusion, and this is expressed by the conduction rate equation (Equation 6.30) as

$$q_s'' = -k_f \left.\frac{\partial T}{\partial y}\right|_{y=0}, \tag{6.35}$$

where k_f is the thermal conductivity of the fluid.

The heat transferred by conduction from the surface is carried away by the bulk motion of the fluid. For a velocity field given by $\vec{V} = \hat{i}u + \hat{j}v + \hat{k}w$, the convective heat transfer by fluid flow is given as

$$q_{\text{conv}}'' = \rho c \vec{V} T \tag{6.36}$$

and in scalar form in a Cartesian coordinate as

$$q_{\text{conv},x}'' = \rho c u T,\ q_{\text{conv},y}'' = \rho c v T,\ \text{and}\ q_{\text{conv},z}'' = \rho c w T \tag{6.37}$$

In order to determine the heat transfer rate by convection, the temperature distribution in the thermal boundary layer needs to be known. This temperature distribution depends on the nature of the fluid motion or the velocity field, and this is determined by solving the energy equation along with the mass and momentum equations for specific flow geometry.

6.2.2 Convection Modes and Heat Transfer Coefficient

On the basis of the nature of the flow field, the convection heat transfer is classified as *forced convection, free or natural convection*, or *phase change heat transfer* such as in condensation and boiling. In *forced convection*, the flow field is induced by some external forces generated by pumps, fans, or winds. On the other hand, for *free or natural convection*, the flow is induced by natural forces such as buoyancy or Marangoni forces. In both forced and free convections, energy being transferred is in the form of sensible energy of the fluid. On the other hand, in *phase change heat transfer*, the energy transfer is in the form of latent heat of the fluid, and the flow field is created because of the formation of vapor bubbles as in boiling heat transfer or because of the condensation of vapor on a solid surface as in condensation heat transfer.

Irrespective of this classification of convection heat transfer modes, the overall effect is given by a *convection rate equation* governed by *Newton's law of cooling* expressed as

$$q_c'' = h_c(T_S - T_\infty), \tag{6.38}$$

where h_c is the *convection heat transfer coefficient* or *film coefficient*.

Combining Equations 6.35 and 6.38, we have the defining equation for convection heat transfer coefficient

$$h_c = \frac{-k_f \left.\dfrac{\partial T}{\partial y}\right|_{y=0}}{(T_S - T_\infty)}. \tag{6.39}$$

The convection heat transfer coefficient depends on a number of factors such as surface geometry, flow field, and the thermophysical and transport properties of fluid. In order to determine the convection heat transfer coefficient and hence the convection heat transfer, it is necessary to solve the energy equation for the temperature distribution along with the equation of motion for the velocity field.

Convection heat transfer coefficients are derived for many flow conditions in the form of a correlation. For forced convections, the correlations are of the form

$$Nu = f(Re, Pr), \tag{6.40}$$

where

$$Nu = \text{Nusselt number} = \frac{h_c L_c}{k} \tag{6.41a}$$

$$Re = \text{Reynolds number} = \frac{\rho U_c L_c}{\mu} \tag{6.41b}$$

$$Pr = \text{Prandtl number} = \frac{\mu c_P}{k} = \frac{v}{\alpha} \tag{6.41c}$$

L_c and U_c represent the characteristic length and velocity in the problem.

For free convection, the heat transfer correlations are given in terms of the Grashof number, *Gr*, and Prandtl number as

$$Nu = f(Gr, Pr) = f(Ra), \tag{6.42}$$

where

$$Gr = \text{Grashof number} = \frac{g\beta(T_S - T_\infty)L_c^{-3}}{\nu^2} \tag{6.43a}$$

$$Ra = \text{Rayleigh number} = Gr\ Pr \tag{6.43b}$$

For simplicity, such correlations are used as convective boundary conditions for many heat conduction problems taking into account convection heat transfer from/to solid surfaces instead of solving a complete set of differential equations for flow field and convection heat equations.

Some of the widely used correlations that are applicable to the flow channels in a fuel cell are outlined here. However, before selecting an appropriate heat transfer correlation, the hydrodynamic and thermal entry lengths of flow geometry are estimated based on criteria outlined by Equations 6.33 and 6.35.

6.2.2.1 *Fully Developed Correlations*

For both hydrodynamic and thermally fully developed flow, convection heat transfer coefficients and friction coefficients are constant since the velocity and the dimensionless temperature profiles do not vary along the channel length in the fully developed regions.

6.2.2.2 *Thermal Entry Length*

For thermal entry length solution, it is assumed that the velocity field is fully developed and temperature field is developing. The thermal entry length correlation is given by Housen (Incropera et al., 2007):

$$N\bar{u}_D = 3.66 + \frac{0.0668\left(\frac{D}{L}\right)Re_D Pr}{1+0.04\left[\left(\frac{D}{L}\right)Re_D Pr\right]^{2/3}} \quad \text{for constant surface temperature.} \tag{6.44}$$

This is also valid for combined entry length with $Pr \geq 5$ for which velocity field develops at a faster rate than thermal field.

6.2.2.3 *Combined Entry Length*

For both velocity and temperature fields as developing, the heat transfer correlations for gases with lower values of Prandtl number ($Pr < 5$), *Sieder and Tate's correlation* (Incropera et al., 2007) is applicable:

$$N\bar{u}_D = 1.86\left(\frac{Re_D Pr}{L/D}\right)^{1/3}\left(\frac{\mu}{\mu_s}\right)^{0.14}.$$

For $T_S = \text{Const}$, $0.60 \leq Pr \leq 5$ and

$$0.0044 \leq \frac{\mu}{\mu_s} \leq 0.75. \tag{6.45}$$

This equation is applicable for fuel cell reactant gases like hydrogen and air with $Pr \sim 0.7$, and for heating and cooling fluids.

6.2.3 Conservation of Energy and Heat Equation

In order to estimate the heat transfer rates in different parts of the fuel cell, it is necessary to determine the temperature distribution or temperature field in the medium resulting from the heat generations and thermal boundary conditions. The temperature field is determined by solving the heat equation, which is a statement of conservation of energy or the first law of thermodynamics.

A simplified heat model applicable to many convection problems is derived from conservation of energy based on the following assumptions: (i) constant thermal conductivity, k; (ii) negligible viscous dissipation, Φ; (iii) negligible compressibility effect; and (iv) negligible radiation heat transfer rate. The energy equation for such a model is derived as

$$\rho c_p \left[\frac{\partial T}{\partial t} + (\vec{V} \cdot \nabla) T \right] = \nabla \cdot (k \nabla T) + \dot{Q} \tag{6.46}$$

and in Cartesian coordinates as

$$\begin{aligned} &\rho c_p \left(\frac{\partial T}{\partial t} + u \frac{\partial T}{\partial x} + \upsilon \frac{\partial T}{\partial y} + w \frac{\partial T}{\partial z} \right) \\ &= \frac{\partial}{\partial x}\left(k \frac{\partial T}{\partial x} \right) + \frac{\partial}{\partial x}\left(k \frac{\partial T}{\partial x} \right) + \frac{\partial}{\partial x}\left(k \frac{\partial T}{\partial x} \right) + \dot{Q}, \end{aligned} \tag{6.47}$$

where $\dot{Q}$ is the volumetric heat generation.

6.2.3.1 Gas Flow Channel

The heat equation in the anode and cathode gas channels involves convection and conduction heat transfer modes and no heat generation. The equation is expressed as

$$\begin{aligned} &\rho_i c_{p_i} \left(\frac{\partial T}{\partial t} + u \frac{\partial T}{\partial x} + \upsilon \frac{\partial T}{\partial y} + w \frac{\partial T}{\partial z} \right) \\ &= \frac{\partial}{\partial x}\left(k_i \frac{\partial T}{\partial x} \right) + \frac{\partial}{\partial x}\left(k_i \frac{\partial T}{\partial x} \right) + \frac{\partial}{\partial x}\left(k_i \frac{\partial T}{\partial x} \right), \end{aligned} \tag{6.48}$$

where index i represents the anode and cathode gas stream. The velocity components u, v, and w are given by the solution of the Navier-Stokes equation.

6.2.3.2 Electrode–Gas Diffusion Layer

For the porous electrode–gas diffusion layer, the heat equation involves primarily heat generation caused by ohmic heating and heat transfer by conduction and convection. The heat equation is expressed as

$$\rho_i c_{p_i}\left(\frac{\partial T}{\partial t}+u\frac{\partial T}{\partial x}+v\frac{\partial T}{\partial y}+w\frac{\partial T}{\partial z}\right) = \frac{\partial}{\partial x}\left(k_i\frac{\partial T}{\partial x}\right)+\frac{\partial}{\partial x}\left(k_i\frac{\partial T}{\partial x}\right)+\frac{\partial}{\partial x}\left(k_i\frac{\partial T}{\partial x}\right)+\dot{Q}_i, \tag{6.49}$$

where index i represents the anode and cathode electrode–gas diffusion layer. The velocity components u, v, and w are given by Darcy's equation or Brinkman's equation. $\dot{Q}_i$ represents the heat generation in electrodes by ohmic heating owing to electron flow.

6.2.3.3 Electrolyte Membrane

For the solid electrolyte membrane, the heat equation involves ohmic heat generation and conduction heat dissipation, and is expressed as

$$\rho_e c_{p_e}\frac{\partial T}{\partial t} = \frac{\partial}{\partial x}\left(k_e\frac{\partial T}{\partial x}\right)+\frac{\partial}{\partial x}\left(k_e\frac{\partial T}{\partial x}\right)+\frac{\partial}{\partial x}\left(k_e\frac{\partial T}{\partial x}\right)+\dot{Q}_e, \tag{6.50}$$

where $\dot{Q}_e$ represents the heat generation in electrolyte membrane by ohmic heating caused by migration of ions.

6.2.4 Inlet and Boundary Conditions

6.2.4.1 Boundary Conditions

6.2.4.1.1 *Adiabatic or Symmetric Surface*

A zero net heat flux condition is used at all *adiabatic* or *symmetric surfaces.* This condition is given as

$$n \cdot q = 0, \tag{6.51a}$$

where

$$q = -k\nabla T + \mathbf{u}C_p T. \tag{6.51b}$$

6.2.4.1.2 Interfaces

At the channel and gas diffusion layer interfaces, a continuity condition is applied as

$$n \cdot (q_1 - q_2) = 0, \tag{6.52a}$$

where

$$q_i = -k_i \nabla T_i + T_i C_p \mathbf{u}. \tag{6.52b}$$

The heat flux discontinuity condition is used at the *electrode–membrane interface*

$$-n \cdot (q_1 - q_2) = Q_s, \tag{6.53}$$

where Q_s is the surface heat generation at membrane and electrode interfaces given by Equations 6.28 due to the reversible and irreversible component of electrochemical reaction.

6.2.4.2 Channel Inlet Conditions

A known gas stream inlet temperature is assigned as

$$T_i = T_{in}. \tag{6.54}$$

A typical temperature distribution in a tri-layer fuel cell with adjacent gas flow channels is shown in Figure 6.9.

It can be seen from this plot that the heat generated within the tri-layer fuel cell dissipates by conduction through the tri-layers of electrodes and

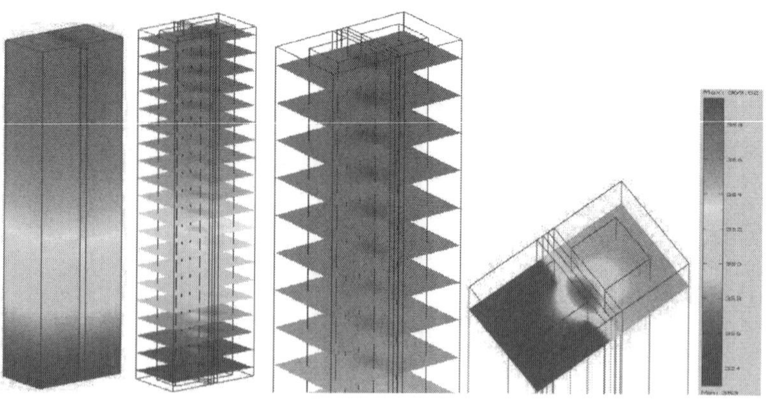

FIGURE 6.9
Typical temperature distribution in a fuel cell.

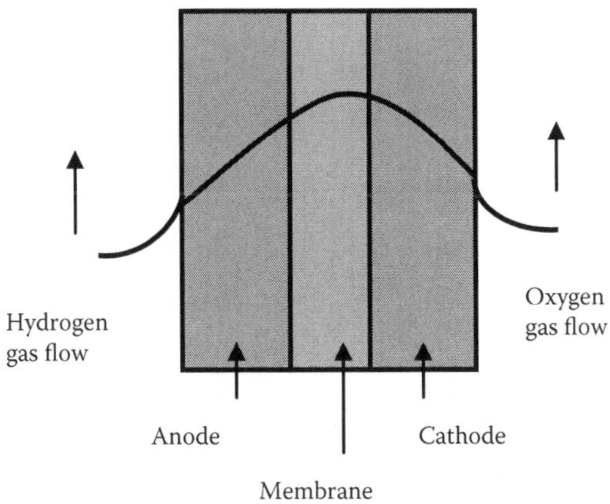

FIGURE 6.10
Typical temperature distribution in a tri-layer cell with adjacent gas flow stream.

electrolytes and by convection to the adjacent reactant gas flow streams. It can be seen that temperature in the cell is increasing from the inlet to the outlet section, indicating that the gas stream is effective in carrying away the heat generated in the cell. Results also show that the anode and cathode gas flows are more effective in transferring heat away from the cell at the inlet section because of the bigger temperature difference and entry length effect. The average temperature level also tends to increase with higher operating current density and in such a case more effective cooling will be required to maintain the cell at a desirable operating temperature level.

Figure 6.10 shows a typical temperature distribution in the electrode–gas diffusion layers, membrane, and gas channels at a given cross section.

6.3 Mass Transfer in Fuel Cells

Mass species transport describes the motion of species in a mixture as well as in base fluids and solids. There are two basic modes of mass transfer: *mass transfer by diffusion* owing to the presence of species concentration gradient and *mass transfer by convection or advection* owing to the bulk fluid motion. Figure 6.11 shows different modes of mass transport across the fuel cell.

There is a close similarity between heat and mass transfer in terms of transport rate equation and transport conservation equation. The diffusion and convective mass transfer modes are similar to the conduction and convection modes of heat transfer. Both diffusion and convection mass transfer play a significant role in the transport of reactant gas species through the gas flow channels and gas diffusion layers/electrodes.

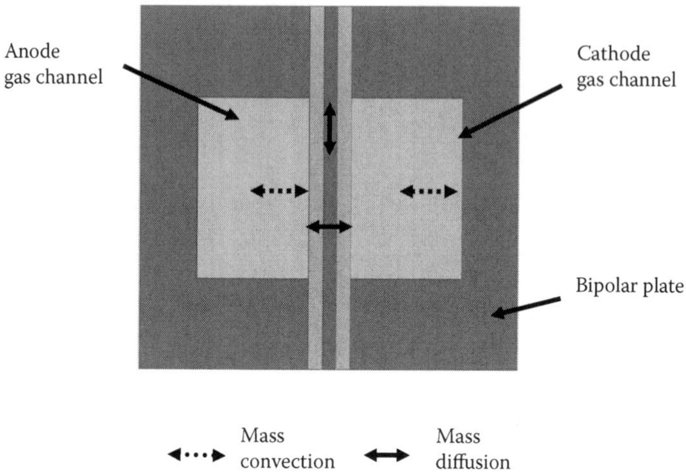

FIGURE 6.11
Mass transports in a fuel cell.

A brief description of these basic mass transport modes is given along with the mass transport rate and species mass conservation equations in the following sections.

6.3.1 Basic Modes and Transport Rate Equation

6.3.1.1 Diffusion Mass Transfer

This mode is primarily important for mass transfer in a stationary medium such as in a solid and stationary fluid.

Diffusion mass transfer, also referred to as *molecular diffusion* or *ordinary diffusion,* is defined as the transport of species owing to the random molecular motion and collisions in the presence of a concentration gradient. Species migrate from the region of high concentration to the region of low concentration. Molecular diffusion is also caused by the presence of temperature gradient and pressure gradient. A temperature difference may establish a concentration gradient and hence may cause mass transfer. Such temperature-driven diffusion process is referred to as the *Soret effect or thermal diffusion.* The presence of a pressure gradient may cause bulk fluid motion and hence convective mass species transport.

Let us consider the diffusion of species i in a mixture of species i and j as shown in Figure 6.12. Initially, the species are separated by a membrane in the middle with a higher concentration of i species on the left side and a higher concentration of j on the right side.

As we remove the separating membrane, mass diffusion takes place in the direction of decrease in concentrations. Hence, species i diffuses from the

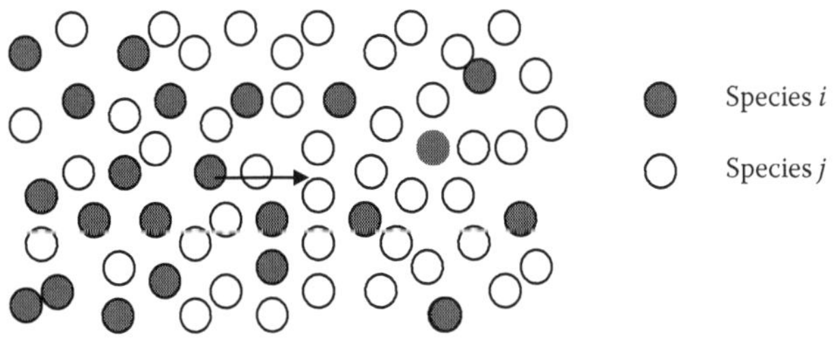

FIGURE 6.12
Diffusion mass transfer.

high-concentration region (left) to the low-concentration region (right) and species *j* diffuses from right to left. Finally, uniform concentrations of both species *i* and *j* are reached, and there is no diffusion transport of species.

The *diffusion rate equation* is given by *Fick's law of diffusion*, which expresses the transfer of a species *i* in a mixture of *i* and *j*:

$$j_i'' = -D_{ij}\nabla C_i \tag{6.55a}$$

or

$$j_i'' = -CD_{ij}\nabla y_i \text{ in terms of mole fraction} \tag{6.55b}$$

where
- j_i'' = molar diffusion flux of species *i* (kmole/m^2·s)
- C_i = molar concentration or molar density of species *i* (kmole/m^3)
- C = total molar concentration or molar density (kmole/m^3)
- y_i = mole fraction of *i*
- D_{ij} = diffusion coefficient or mass diffusivity of species *i* of the mixture. different species in the mixture
- i,j = indices representing component species in the mixture

In scalar form, Equation 6.55a can be written in three components and, for example, in the Cartesian coordinate system as

$$j_{x_i}'' = -D_{ij}\frac{\partial C_i}{\partial x},\ j_{y_i}'' = -D_{ij}\frac{\partial C_i}{\partial y},\ \text{and } j_{z_i}'' = -D_{ij}\frac{\partial C_i}{\partial z}. \tag{6.55c}$$

Equation 6.55a can also be written as

$$m_i'' = -D_{ij}\nabla \rho_i \text{ in terms of mass density} \tag{6.56a}$$

or

$$m_i'' = -\rho D_{ij}\nabla x_i \text{ in terms of mass fraction} \tag{6.56b}$$

where

m_i'' = mass diffusion flux of species i (kg/m^2 · s)
ρ_i = mass concentration or density of species i (kg/m^3)
ρ = total mass density (kg/m^3)
x_i = mass fraction of i

Diffusion mass transfer exists in all phases, that is, in gases, liquids, and solids, with a higher order of magnitude in gases than in liquids and a higher order of magnitude in liquids than in solids.

6.3.1.1.1 *Binary Diffusion*

For *binary diffusion* with only two species a and b present in a mixture, the diffusion coefficient is expressed as D_{ab}. For example, if a humidified hydrogen gas is supplied as fuel, then the binary diffusion involves two species, hydrogen and water vapor, and we deal with the diffusion coefficient D_{H_2,H_2O} to describe the hydrogen transport by diffusion.

6.3.1.1.2 *Multicomponent Diffusion*

For diffusion in a multicomponent system, the gas mixture involves three or more species. For example, the gas supply in the anode side may involve multicomponent diffusion of both hydrogen and water species. If oxygen is supplied as air in the cathode side of the fuel cell, then it involves multicomponent diffusion of three components: oxygen, nitrogen, and water vapor. The diffusion coefficient D_{aj} of species a depends not only on the concentration gradient of species a but also on the flux components of the other species in the mixture.

The *Chapman–Enskog kinetic theory of gases* (Hirschfelder et al., 1964) is used to describe the multicomponent diffusion flux of species i in a mixture of n gas species and expressed as the *Stefan–Maxwell equation* (Bird et al., 2002). The diffusion flux of species i is given as

$$m_{i,d}'' = -\rho x_i \sum_{j=1'}^{n} D_{ij} \left\{ \frac{M}{M_j} \left(\nabla x_j + x_j \frac{\nabla M}{M} \right) + \left(y_j - x_j \right) \frac{\nabla P}{P} \right\} + D_i^T \frac{\nabla T}{T} \quad (6.57)$$

where

x_i = mass fraction of species i
y_j = mole fraction of species j
D_{ij} = binary component diffusivity of the multicomponent diffusivity matrix (m^2/s)
D_i^T = thermal diffusion coefficient
ρ = mixture gas density = $\dfrac{P}{R_{mix}T}$

The second term is included for the molar flux owing to pressure gradient force. The third term on the right-hand side represents the diffusion flux component owing to temperature gradient and this is known as the *diffusion thermo* or *Dufour effect.*

For lower temperature gradient and negligible diffusion thermal effect, a simplified form of the *Stefan–Maxwell equation* is written as

$$m''_{i,d} = -\rho x_i \sum_{j=1'}^{n} D_{ij}\left(\nabla y_j + (y_j - x_j)\frac{\nabla P}{P}\right) \tag{6.58a}$$

and for negligible pressure gradient, the diffusion flux is given as

$$m''_{i,d} = -\rho x_i \sum_{j=1'}^{n} D_{ij} \nabla y_j. \tag{6.58b}$$

6.3.1.2 Convection Mass Transfer

The convection mass transport of species i may also take place if there exists a bulk fluid motion. The convection mass transfer is analogous to convection heat transfer and occurs between a moving mixture of fluid species and an exposed solid surface. Like hydrodynamic and thermal boundary layers, a concentration boundary layer forms over the surface if the free stream concentration of a species i, $C_{i\infty}$, differs from species concentration at the surface, C_{iS}, in an external flow over a solid surface as demonstrated in Figure 6.13.

For internal flows, concentration boundary layers develop from both top and bottom surfaces and develop into two regions as shown in the figure: *concentration entry length* and *concentration fully developed* regions similar to hydrodynamic internal flow as shown in Figure 6.14.

These regions are defined as follows:

Concentration fully developed region: The region where the dimensionless concentration profile remains invariable along the longitudinal length of the channel.

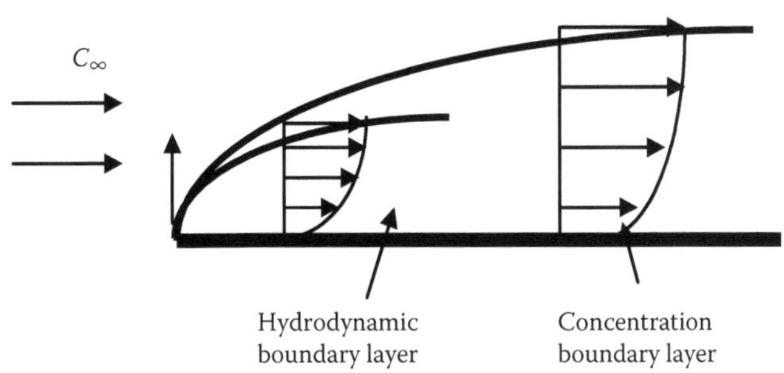

FIGURE 6.13
Concentration boundary layer for flow over a solid surface.

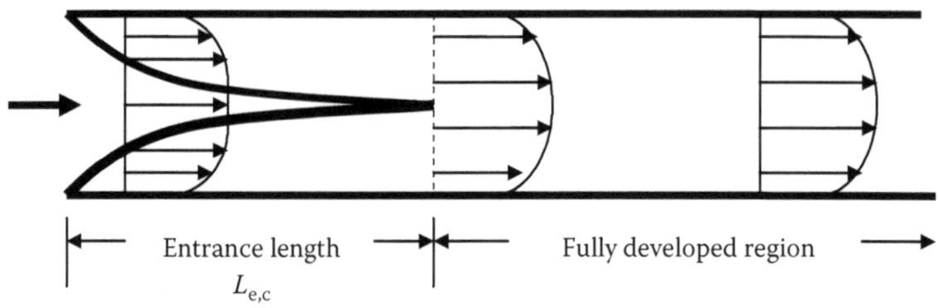

FIGURE 6.14
Concentration entry length and fully developed region for internal flow in a channel.

Concentration entry length ($L_{e,c}$): The length required for the dimensionless concentration profile to become fully developed.

Since the fluid is stationary at the solid surface, the mass of species i is transferred by molecular diffusion normal to the surface and is expressed by the diffusion rate equation (Equation 1.78) as

$$j''_{is} = -D_{ij} \left. \frac{\partial C_i}{\partial y} \right|_{y=0} \tag{6.59}$$

The mass transfer by molecular diffusion from the surface is carried away by the bulk motion of the fluid. For a given velocity field $\vec{V} = \hat{i}u + \hat{j}v + \hat{k}w$, the convective mass transfer by fluid flow is given as

$$\vec{j}''_{\text{conv}} = C_i \vec{V} \tag{6.60a}$$

and in scalar form in a Cartesian coordinate as

$$j''_{\text{conv},x} = C_i u,\ j''_{\text{conv},y} = C_i v,\ \text{and}\ q''_{\text{conv},z} = C_i w. \tag{6.60b}$$

In a unidirectional flow field, with bulk fluid velocity, V, the convective mass transport is given as

$$j''_{\text{conv}} = C_i V. \tag{6.60c}$$

In order to determine the mass species transport by convection or fluid motion, the species concentration distributions need to be known. Like in heat transfer, the concentration distribution depends on the nature of the fluid motion and determined by solving the mass species transport equation along with the mass and momentum equations for the bulk fluid in the media.

6.3.1.3 Combined Diffusion and Convection Mass Transport

The total or absolute transport is the sum of diffusive transport and convective transport

$$N_i'' = j_i'' + C_i V \tag{6.61a}$$

or

$$N_i'' = -D_{ij}\frac{dC_i}{dx} + C_i V \tag{6.61b}$$

where the bulk fluid velocity V represents the molar average velocity of the mixture and expressed for binary mixture as

$$CV = N'' = \sum_{j=1}^{n} N_j'' \tag{6.62}$$

The bulk fluid velocity is given as

$$V = \frac{1}{C}\sum_{j=1}^{n} N_j'' \tag{6.63}$$

Substituting Equation 6.63 into equation, we arrive at an alternative form of *the total molar transport flux of species i*

$$N_i'' = j_i'' + C_i\left(\frac{1}{C}\sum_{j=1}^{n} N_j\right) \tag{6.64a}$$

or

$$N_i'' = -CD_{ij}\nabla y_i + y_i\sum_{j=1}^{n} N_j \tag{6.64b}$$

and *in terms of mass flux* as

$$m_i'' = -\rho D_{ij}\nabla x_i + x_i\sum_{j=1}^{n} N_j \tag{6.65}$$

Note that the diffusive molar flux given by Ficks's law can also be expressed in terms of the diffusion velocity of the species relative to the mixture molar average velocity.

Mass transport owing to combined multicomponent diffusion and convection is written by combining Equations 6.57 and 6.60 as

$$m_i'' = -\rho x_i \sum_{j=1'}^{n} D_{ij} \left\{ \frac{M}{M_j} \left(\nabla x_j + x_j \frac{\nabla M}{M} \right) + (y_j - x_j) \frac{\nabla P}{P} \right\} + D_i^T \frac{\nabla T}{T} + x_i \rho \vec{V}. \tag{6.66}$$

For lower temperature gradient and negligible diffusion thermal effect, a simplified form is written as

$$m_i'' = -\rho x_i \sum_{j=1'}^{n} D_{ij} \left(\nabla y_j + (y_j - x_j) \frac{\nabla P}{P} \right) + x_i \rho \vec{V} \tag{6.67a}$$

and in terms of mole fraction as

$$j_i'' = C y_i \sum_{j=1'}^{n} D_{ij} \left(\nabla y_j + (y_j - x_j) \frac{\nabla P}{P} \right) + C \vec{V} y_i \tag{6.67b}$$

6.3.2 Mass Species Transport in Fuel Cells

In order to determine the reactant gas transport rates in the gas channels and in electrode/gas diffusion layers and the consumption rates at electrode–membrane interfaces, it is necessary to determine the gas concentration distributions from mass species conservation equation. Conservation of mass species over a differential element, considering mass diffusion and convection, leads to

$$\left[\frac{\partial C_i}{\partial t} + (\vec{V} \cdot \nabla) C_i \right] = \nabla \cdot (J_i'') + S, \tag{6.68}$$

where

S_i = the volumetric species molar consumption (kmole/m^3·s).

6.3.2.1 Mass Species Transport Equation in Gas Flow Channels

The mass species transport equation in the anode and cathode gas channels involves both diffusion and convection and no species consumption in the absence of any kind of reaction or fuel reforming. Considering the diffusion flux given by *Fick's law of diffusion*, the equation is expressed in terms of mole concentration as

$$\left(\frac{\partial C_i}{\partial t}+u\frac{\partial C_i}{\partial x}+v\frac{\partial C_i}{\partial y}+w\frac{\partial C_i}{\partial z}\right)$$
$$=\frac{\partial}{\partial x}\left(D_{ij}\frac{\partial C_i}{\partial x}\right)+\frac{\partial}{\partial x}\left(D_{ij}\frac{\partial C_i}{\partial x}\right)+\frac{\partial}{\partial x}\left(D_{ij}\frac{\partial C_i}{\partial x}\right), \quad (6.69)$$

where the velocity components u, v, and w are computed based on the Navier–Stokes equation given by Equation 6.15.

6.3.2.2 Mass Species Transport Equation in Electrodes

For the porous electrode–gas diffusion layer with bulk fluid motion given by Darcy or Brinkman's equation, the governing species transport is given as

$$\frac{\partial}{\partial t}(\rho x_i)=\nabla\cdot j_i''+S_i. \quad (6.70)$$

Substituting Equation 6.66 for the net mass flux owing to the mass species transport equation for electrodes is given as

$$\frac{\partial}{\partial t}(\rho x_i)+\nabla\cdot(\rho\vec{V}x_i)$$
$$=\nabla\cdot\left[\rho x_i\sum_{j=1'}^{n}D_{ij}\left\{\frac{M}{M_j}\right\}\left(\nabla x_j+x_j\frac{\nabla M}{M}\right)-(y_j-x_j)\frac{\nabla P}{P}-D_i^T\frac{\nabla T}{T}\right]+S_i. \quad (6.71)$$

and substituting Equation 6.67a as

$$\frac{\partial}{\partial t}(\rho x_i)+\nabla\cdot\left(\rho\vec{V}x_i\right)=\nabla\cdot\left[-\rho x_i\sum_{j=1'}^{n}D_{ij}\left(\nabla y_j+(y_j-x_j)\frac{\nabla P}{P}\right)\right]+S_i. \quad (6.72)$$

In Equation 6.72, the first term represents the unsteady accumulation term; S_i is the volumetric source or sink term given by the three-phase electrochemical reaction region in the active layer of the electrodes. The volumetric reaction term is neglected from the mass species transport equation for the electrode if the active layer is assumed as the electrode–membrane interface with surface reaction and this is taken into account in the assigned boundary condition at the interface.

For steady-state analysis, a simplified model with negligible thermal diffusion and assuming surface reaction, the equation reduces to

$$\nabla\cdot\left(\rho\vec{V}x_i\right)=\nabla\cdot\left[\rho x_i\sum_{j=1'}^{n}D_{ij}\left\{\frac{M}{M_j}\right\}\left(\nabla x_j+x_j\frac{\nabla M}{M}\right)-(y_j-x_j)\frac{\nabla P}{P}\right]. \quad (6.73)$$

For the anode gas stream mixture consisting of hydrogen and water, the mass species transport equations are written as

$$\nabla\cdot\left(\rho\vec{V}x_{H_2}\right)=\nabla\cdot\left[\rho x_{H_2}\sum_{j=1'}^{n}D_{ij}\left\{\frac{M}{M_j}\right\}\left(\nabla x_j+x_j\frac{\nabla M}{M}\right)-(y_j-x_j)\frac{\nabla P}{P}\right] \quad (6.74a)$$

and

$$\nabla\cdot\left(\rho\vec{V}x_{H_2O}\right)=\nabla\cdot\left[\rho x_{H_2O}\sum_{j=1'}^{n}D_{ij}\left\{\frac{M}{M_j}\right\}\left(\nabla x_j+x_j\frac{\nabla M}{M}\right)-(y_j-x_j)\frac{\nabla P}{P}\right]. \quad (6.74b)$$

Note also that mass fraction of water in the gas stream can also simply be computed from

$$x_{H_2O}=1-x_{H_2}. \quad (6.74c)$$

For the cathode gas stream mixture consisting of oxygen, nitrogen, and water vapor, the mass transport is given by the following three equations:

Oxygen:

$$\nabla\cdot\left(\rho\vec{V}x_{O_2}\right)=\nabla\cdot\left[\rho x_{O_2}\sum_{j=1'}^{n}D_{O_2j}\nabla x_j-(y_j-x_j)\frac{\nabla P}{P}\right] \quad (6.75a)$$

Water:

$$\nabla\cdot\left(\rho\vec{V}x_{H_2O}\right)=\nabla\cdot\left[\rho x_{H_2}\sum_{j=1'}^{n}D_{H_2O,j}\nabla x_j-(y_j-x_j)\frac{\nabla P}{P}\right] \quad (6.75b)$$

Nitrogen:

$$x_{N_2}=1-x_{H_2}-x_{H_2O} \quad (6.75c)$$

A simplified model for gas species transport can be given based on *Fick's law of diffusion* for hydrogen in the anode side and oxygen in the cathode side without considering the binary multicomponent diffusion and based on mole concentration as follows:

$$\left(\frac{\partial C_i}{\partial t}+u\frac{\partial C_i}{\partial x}+v\frac{\partial C_i}{\partial y}+w\frac{\partial C_i}{\partial z}\right)$$
$$=\frac{\partial}{\partial x}\left(D_{ij}^{\text{eff}}\frac{\partial C_i}{\partial x}\right)+\frac{\partial}{\partial x}\left(D_{ij}^{\text{eff}}\frac{\partial C_i}{\partial x}\right)+\frac{\partial}{\partial x}\left(D_{ij}^{\text{eff}}\frac{\partial C_i}{\partial x}\right)+S_i \tag{6.76}$$

where the index i represents hydrogen on the anode side and oxygen on the cathode side.

S_i represents the gas consumption rate for hydrogen and oxygen and the generation rate for water.

6.3.2.3 Boundary Conditions for Concentration

6.3.2.3.1 *Symmetric Surface*

A zero net flux is used at all at symmetric surfaces. This condition is given as

$$n \cdot J_i = 0, J_i = -D_i \nabla C_i + C_i \mathbf{u}. \tag{6.77}$$

6.3.2.3.2 *Interfaces*

At the channel and gas diffusion layer interfaces, a continuity condition is applied:

$$n \cdot (J_1 - J_2) = 0, J_i = -D_i \nabla C_i + C_i \mathbf{u}, \tag{6.78}$$

where 1 and 2 represents gas species in the gas channel and in the electrode, respectively.

The mass flux discontinuity condition is used at the electrode–membrane interface

$$-n \cdot (J_1 - J_2) = S_i, \tag{6.79}$$

where S_i is a sink term representing species consumption rate at the membrane and electrode interfaces for cathode and anode electrodes.

6.3.2.4 Channel Inlet Conditions

$$T_i = T_{i,in}, C_i = C_{i,in}, \text{ and } P_i = P_{i,in} \tag{6.80}$$

Figure 6.15 shows a typical reactant gas concentration distribution in the electrode–gas diffusion layers and gas flow channels of a tri-layer fuel cell for a typical operating current density.

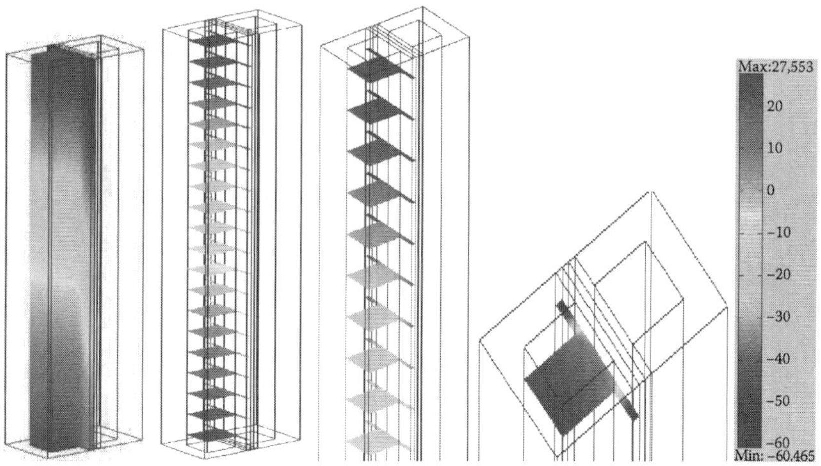

(a)

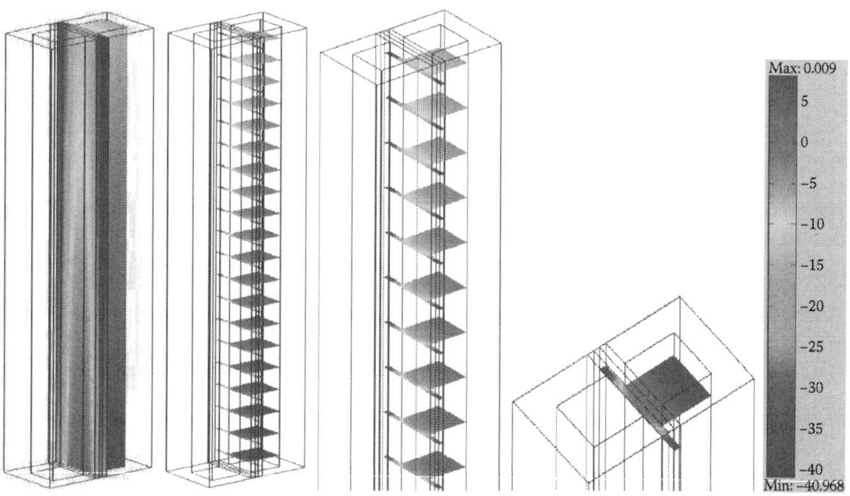

(b)

FIGURE 6.15
Typical gas concentration distribution in a tri-layer fuel cell with adjacent gas flow channels. (a) Hydrogen gas concentration distribution. (b) Oxygen gas concentration distribution.

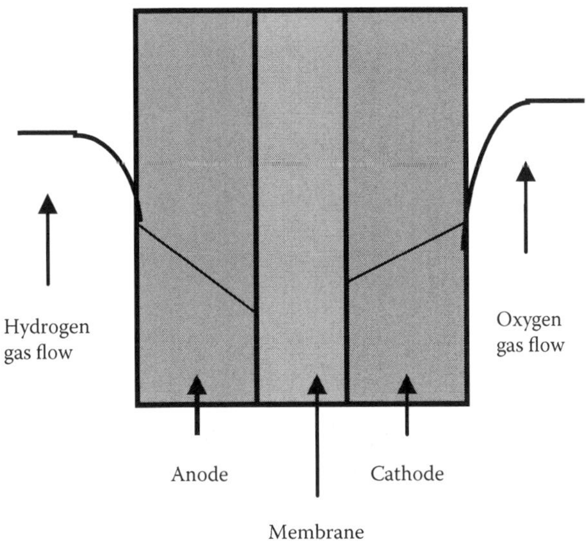

FIGURE 6.16
Typical reactant gas distributions at a given cross section of a tri-layer cell.

Results in contour plots show decrease in hydrogen and oxygen gas concentrations in the gas diffusion layer and in the channel down the length of the channel. Also, as expected, there are nonuniform gas distributions in the diffusion layer with lower gas concentration near the land areas compared to the areas in contact with the channel.

Figure 6.16 shows typical hydrogen and oxygen gas concentration profiles in the electrodes and in the adjacent gas channels at a given cross section of the cell.

As we can see, gas concentration drops sharply from the bulk gas stream to the electrode surface within the concentration boundary layer by convection and then linearly decreases by diffusion to the electrode–electrolyte reaction interface.

6.3.3 Convection Mass Transfer Coefficient

In Equation 6.59, the concentration gradient at the fluid–solid interface is obtained from the solution of the fluid equation of motion and mass species transport equation. Determination of convection mass transfer from a solid surface through the solution of flow field and mass species transport equation could be quite complex and time-consuming depending on the flow and surface geometry under consideration. This procedure is often simplified with the introduction of the convection mass transfer coefficient similar to the convection heat transfer coefficient.

Like in the convection heat transfer process, the convective mass transfer of species *i* from the moving bulk fluid to the solid surface is given by a convection rate equation analogous to *Newton's law of cooling* as

$$j''_{is} = h_m (C_{i\infty} - C_{is}), \tag{6.81}$$

where h_m is called the *convection mass transfer coefficient* or *mass film coefficient*. The defining equation for the convection mass transfer coefficients is obtained by comparing Equations 6.59 and 6.81 as

$$h_m = \frac{j''_{is}}{(C_{i\infty} - C_{is})} \tag{6.82a}$$

or

$$h_m = \frac{-D_{ij} \left.\dfrac{\partial C_i}{\partial y}\right|_{y=0}}{(C_{i\infty} - C_{is})}. \tag{6.82b}$$

Like the convection heat transfer coefficient expressed as the Nusselt number, the convection mass transfer coefficient is also expressed in the form a dimensionless number known as the *Sherwood number*, which is defined as

$$Sh = \frac{h_m L_c}{D_{ij}}. \tag{6.83}$$

The correlations for the convective mass transfer coefficient is given in a similar manner as in convection heat transfer, but in terms of mass transfer parameters.

For forced convections, the correlations are of the form

$$Sh = f(Re, Sc), \tag{6.84}$$

where

$$Sh = \text{Sherwood number} = \frac{h_m L_c}{D_{ij}} \tag{6.85a}$$

$$Re = \text{Reynolds number} = \frac{\rho U_c L_c}{\mu} \tag{6.85b}$$

$$Sc = \text{Schmidt number} = \frac{\mu}{\rho D_{ij}} = \frac{v}{D_{ij}} \tag{6.85c}$$

The mass transfer parameters *Sherwood number, Sh,* and *Schimdt number, Sc,* are analogous to the Nusselt number, *Nu,* and the Prandtl number, *Pr,* respectively, in convection heat transfer. The Sherwood and Prandtl numbers are related by the *Lewis number* as

$$Le = \frac{Sc}{Pr}. \tag{6.86}$$

For dilute mixture and low mass transfer rates, mass transfer is quite analogous to the heat transfer rate with $Le \approx 1$, and mass transfer correlations are derived from that of heat transfer correlations by simply replacing the Nusselt number, *Nu,* with the Sherwood number, *Sh,* and replacing the Prandtl number with the Schmidt number. However, one important difference will be the case where only one side of the channel is exposed to the gas diffusion layer and permeable and the rest of the three surfaces are impermeable to species flux.

For example, the convective mass transfer correlation for a gas flow channel with square cross-sectional area is given as

$$Sh_j = \frac{h_m D_h}{D} = 3.66 \text{ for uniform surface flux, } J_s'' = \text{Constant}$$

$$Sh_c = \frac{h_m D_h}{D} = 4.36 \text{ for uniform surface concentration, } C_s = \text{Constant}$$

where Sh_j and Sh_c represent cases with the Sherwood number for constant surface flux and constant surface concentration.

In fuel cell applications, however, gas flow channels consist of only one surface that is adjacent to the permeable porous electrode. In a square channel with only one surface permeable and with constant surface flux, the Sherwood number for the fully developed laminar flow is 2.71 for constant surface flux and 2.44 for constant surface concentration.

A list of selected mass transfer correlations for fully developed laminar flow is given in Table 6.2.

Convection mass transfer coefficients are often used as convective boundary conditions for gas diffusion in a stationary media. However, while applying mass transfer correlations to describe mass species transport from the electrode–gas diffusion layer to gas flow stream in the channel, it is assumed that species mass transport rate at the wall is small and does not alter the hydrodynamic, thermal, and concentration boundary layers like in boundary layers with wall suction or blowing.

TABLE 6.2

Sherwood Number for Fully Developed Laminar Flow in Ducts of Different Cross Sections

Geometry	b/a	$j_s'' =$ Const	$C_s =$ Const
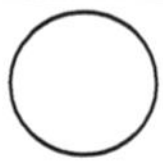		4.36	3.66
	0.2	5.74	4.80
		4.41	3.91
	0.5	4.12	3.38
		3.54	3.19
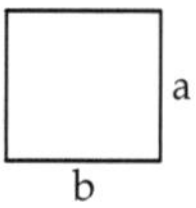	1.0	3.61	2.97
		2.71	2.44
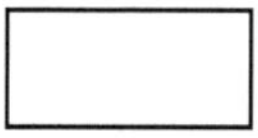	2.0	4.12	3.39
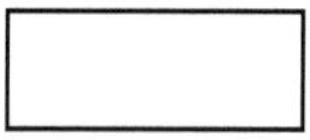	4.0	5.33	4.44
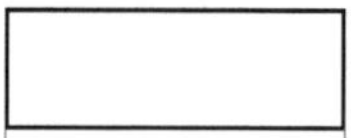	5.0	5.74	4.80
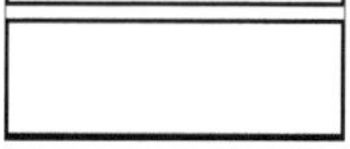		0.96	0.83

Let us demonstrate this by considering a problem with reactant gas flow adjacent to an electrode as shown in Figure 6.1. Considering one-dimensional diffusion in the electrode of thickness, a_i, the governing equation and boundary conditions for this problem is given as

Mass species equation:

$$\frac{\mathrm{d}}{\mathrm{d}x}\left(D_{ij}\frac{\mathrm{d}C_i}{\mathrm{d}x}\right)=0 \tag{6.87}$$

Boundary conditions:

$$\text{BC 1: at } x=0,\ -D_{ij}\frac{\mathrm{d}C_i}{\mathrm{d}x}=h_{\mathrm{m}}(C_{i\infty}-C_i) \tag{6.88a}$$

$$\text{BC 2: at } x=a_i,\ -D_{ij}\frac{\mathrm{d}C_i}{\mathrm{d}x}=S_i. \tag{6.88b}$$

Here, boundary condition (2) states the continuity in the mass flux given by the diffusion of hydrogen gas stream to the electrode–membrane interface with the consumption through electrochemical reaction.

6.3.3.1 Mass Transfer Resistances

In the transport of reactant gases from the gas flow channel through an electrode composed of a GDL and CL, there are a number of mass transfer resistances that influence the mass transport as shown in Figure 6.17. These resistances are (i) convective mass transfer resistance in the gas flow channel, $R_{\mathrm{conv,m}}$; (ii) diffusion mass resistance in the gas diffusion layer, $R_{\mathrm{diff,m}}$; and (iii) diffusion and reaction resistance owing to reaction kinetics. Figure 6.17 shows a typical reactant gas concentration distribution across the gas channel, the gas diffusion layer, and the active CL.

While the gas concentration distribution in the gas channel is obtained by the solution of the Navier–Stokes equation along with the governing equation for mass species transport, the overall resistance for convective mass transport and convection mass transfer rate from bulk gas stream to the adjacent electrode surface is often given by the convection mass transfer coefficient. For such a case, the convective mass transfer rate equation over a surface of area A can be written as

$$j_i = h_{\mathrm{m}}A(C_{i,\mathrm{ch}} - C_{i,\mathrm{s}}). \tag{6.89}$$

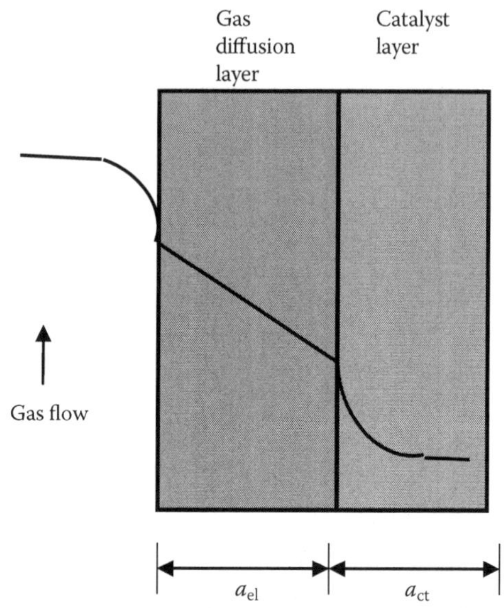

FIGURE 6.17
Gas concentration distributions in gas flow channel, gas diffusion, and active catalyst layers.

Equation 6.89 can be recast to define the convective mass transfer resistance as

$$R_{conv,m} = \frac{1}{h_m A}. \tag{6.90}$$

As we can see, the higher the convective mass transfer coefficient, the lower the convective mass transfer resistance and that leads to a smaller concentration drop between the bulk fluid and the solid surface.

The gas concentration distribution in the gas diffusion layer is simply obtained by the solution of the diffusion equation and boundary conditions given below:

Mass species equation:

$$\frac{d}{dx}\left(D_{ij}\frac{dC_i}{dx}\right) = 0 \tag{6.91}$$

Boundary conditions:

$$\text{BC 1: at } x = 0, \quad C_i(0) = C_{i,s} \tag{6.92a}$$

$$\text{BC 2: } x = a_{el}, \quad C_i(a) = C_{i,el} \tag{6.92b}$$

Solution to Equations 6.91 and 6.92 leads to the linear concentration distribution in a gas diffusion layer given as

$$C_i(x) = C_{i,\mathrm{s}} + \frac{C_{i,\mathrm{el}} - C_{i,\mathrm{s}}}{a_{\mathrm{el}}} x \tag{6.93}$$

and the diffusion mass transfer rate based on Darcy's equation is given as

$$j_i = -D_{ij} A \frac{\mathrm{d}C_i}{\mathrm{d}x} = \frac{D_{ij} A}{a_{\mathrm{el}}} (C_{i,\mathrm{s}} - C_{i,\mathrm{el}}). \tag{6.94}$$

This solution also defines the diffusion mass transfer resistance as

$$R_{\mathrm{diff,m}} = \frac{a_{\mathrm{el}}}{D_{ij} A}. \tag{6.95}$$

For the rate of reaction at the electrode–electrolyte interface where the reactant species disappears by a first-order reaction, the rate of mass transfer rate at the surface is equal to the rate of reactant consumption and is given as

$$j = k_1'' A C_{i,\mathrm{ct}} \tag{6.96}$$

and the mass transfer reaction resistance as

$$R_{\mathrm{m,r}} = \frac{1}{k_1'' A}, \tag{6.97}$$

where k_1'' is the reaction rate constant.

Combining Equations 6.92, 6.94, and 6.96, we can define the net mass transfer rate and the combined mass transfer resistance as follows:

$$j_i = \frac{C_{i,\mathrm{ch}} - 0}{\sum R_{\mathrm{m}}}, \tag{6.98a}$$

where the sum of all resistance is given as

$$\sum R_{\mathrm{m}} = R_{\mathrm{m,conv}} + R_{\mathrm{m,diff}} + R_{\mathrm{m,r}} = \frac{1}{h_{\mathrm{m}} A} + \frac{a_{\mathrm{el}}}{D_{ij}^{\mathrm{eff}} A} + \frac{1}{k_1'' A} \tag{6.98b}$$

Substituting Equation 6.98b into Equation 6.98a, the mass transfer rate is given as

$$\dot{j}_i = \frac{C_{i,\mathrm{ch}}}{\left(\dfrac{1}{h_{\mathrm{m}}A} + \dfrac{a_{\mathrm{el}}}{D_{ij}^{\mathrm{eff}}A} + \dfrac{1}{k_1''A}\right)} \tag{6.99}$$

and the net mass flux is given as

$$j_i'' = \frac{\dot{j}}{A} = \frac{C_{i,\mathrm{ch}}}{\left(\dfrac{1}{h_{\mathrm{m}}} + \dfrac{a_{\mathrm{el}}}{D_{ij}^{\mathrm{eff}}} + \dfrac{1}{k_1''}\right)}. \tag{6.100}$$

For negligible *mass transfer reaction resistance*, we can define the net mass transfer rate and the combined mass transfer resistance as follows:

$$\dot{j}_i = \frac{C_{i,\mathrm{ch}} - C_{i,\mathrm{el}}}{\left(\dfrac{1}{h_{\mathrm{m}}A} + \dfrac{a_{\mathrm{el}}}{D_{ij}^{\mathrm{eff}}A}\right)} \tag{6.101}$$

and the net mass flux is given as

$$j_i'' = \frac{\dot{j}}{A} = \frac{C_{i,\mathrm{ch}} - C_{i,\mathrm{el}}}{\left(\dfrac{1}{h_{\mathrm{m}}} + \dfrac{a_{\mathrm{el}}}{D_{ij}^{\mathrm{eff}}}\right)}, \tag{6.102}$$

There are two limiting cases that we can encounter. In the first limiting case $\dfrac{a_{\mathrm{el}}k_1''}{D_{ij}} \ll 1$, the diffusion resistance is much smaller than the reaction resistance and this leads to $C_{i,s} = C_{i,\mathrm{ct}}$, and the mass flux is given as $j_i''(r) = -k_1''C_{i,s}$. This is a case where the diffusion resistance can be neglected and the surface reaction rate is controlled or limited by the reaction rate constant k_1'' or reaction resistance $\dfrac{1}{k_1''}$. For the second limiting case $\dfrac{a_{\mathrm{el}}k_1''}{D_{ij}} \gg 1$, the diffusion resistance is much larger than the reaction resistance and this leads to $C_{i,\mathrm{el}} = C_{i,\mathrm{ct}}$, and the mass flux is given as $j_i''(r) = \dfrac{D_{ij}(C_{i,s} - C_{i,\mathrm{el}})}{L}$. In this case, the reaction is controlled or limited by diffusion resistance.

6.3.3.2 Concentration Distribution in the Active Reaction Layer

In the active layer, gas diffuses and depleted or consumed owing to reaction kinetics and the concentration distribution is given by the following equation and boundary:

$$\frac{d}{dx}\left(D_{ij}\frac{dC_i}{dx}\right) - S_i = 0, \tag{6.103}$$

where S_i is the volumetric gas consumption in the active layer. Assuming a given by a first-order homogeneous reaction given as

$$S_i = k_1'''C_i, \tag{6.104}$$

where k_1''' is the volumetric reaction constant.

Substituting Equation 6.104 into Equation 6.103 and assuming constant diffusivity, we have

$$D_{ij}\frac{d^2C_i}{dx^2} - k_1'''C_i = 0. \tag{6.105}$$

Boundary conditions at the ends of the active layer are

1. $x = 0,\ C_i = C_{i,r}$ (6.106a)

2. $x = a_{ct},\ \frac{dC_i}{dx} = 0$ (6.106b)

The solution to Equations 6.98 and 6.99 is given as

$$\frac{C_i}{C_{i,r}} = \frac{\cosh m\left(1 - \frac{x}{L}\right)}{\cosh m}, \tag{6.107}$$

where

$$m = \sqrt{k_1'''a_{ct}^2 / D_{ij}}.$$

6.4 Diffusion Coefficient

6.4.1 Diffusion Coefficient for Binary Gas Mixture

As we can see, *Fick's law* introduces the *binary diffusion coefficient* as a material transport property. This is independent of the concentration and it is a property of the binary gas pair. The binary diffusion coefficients for gases are estimated using theoretical as well as empirical formulas.

A commonly used theoretical formula for the binary diffusion coefficient is derived based on the *Chapman–Enskog kinetic theory* (Bird et al., 2002; Hirschfelder et al., 1964; Sherwood et al., 1975) for low-pressure gas mixture and is given as

$$D_{ab} = \frac{0.001858T^{1.5}\left(\dfrac{1}{M_a}+\dfrac{1}{M_b}\right)^{1/2}}{P\sigma_{ab}^2\Omega_D}, \tag{6.108}$$

where D_{ab} = diffusion coefficient (cm^2/s), T = absolute temperature (K), P = pressure (atm), M = molecular weight, and Ω_D = dimensionless collision integral parameter given as a function of $f(kT/\varepsilon_{ab})$.

The constants σ_{ab} and ε_{ab} are the constants for intermolecular interactions given based on *Lennard-Jones potential* given as

$$\phi(r) = 4\varepsilon_{ab}\left[\left(\frac{\sigma_{ab}}{r}\right)^{12} - \left(\frac{\sigma_{ab}}{r}\right)^{6}\right] \tag{6.109}$$

Lennard-Jones potential parameters such as collision diameter, σ_{ab}, and energy of interaction, ε_{ab}, for nonpolar and nonreacting molecule pairs can be computed from the corresponding values of pure component species based on the following rules:

$$\sigma_{ab} = \frac{1}{2}(\sigma_a + \sigma_b) \tag{6.110a}$$

and

$$\frac{\varepsilon_{ab}}{k_B} = \left(\frac{\varepsilon_a}{k_B}\frac{\varepsilon_b}{k_B}\right)^{1/2}. \tag{6.110b}$$

The Lennard-Jones potential parameters σ and ε for gases are listed in Table 6.3.

The dimensionless collision parameter Ω_D depends on the interaction between two species based on the Lennard-Jones potential. The parameter depends on the temperature and energy interaction parameter and expressed as $\Omega_D = f(kT/\varepsilon_{ab})$. The calculation of this quantity depends on the integration of the interaction between two gas pairs. Values of the collision parameter as a function of temperature and the energy interaction parameter for different gas pairs are given in Table 6.4 (Bird et al., 2002; Sherwood et al., 1975).

TABLE 6.3

Critical Properties and Lennard-Jones Intermolecular Force Parameters

		Lennard-Jones Parameters		Critical Constants	
Substance	**Molecular Weight**	**σ (Å)**	**$\frac{\varepsilon}{k}$(K)**	**T_c (K)**	**P_c(atm)**
H_2	2.016	2.915	38.0	33.3	12.80
O_2	32.0	3.433	113.0	154.4	49.7
N_2	28.02	3.681	91.5	126.2	33.5
CO	28.01	3.590	110	133	34.5
CO_2	44.01	3.996	190	304.2	72.9
Air	28.97	3.617	97.0	132	36.4
H_2O	18	2.641	809.1	647.3	217.5

Source: Bird, R. et al.: *Transport Phenomena,* 2nd Edition. Copyright Wiley-VCH Verlag GmbH & Co. KGaA. Reproduced with permission.

TABLE 6.4

Th e Collision Integral Parameter $\Omega_D = f(kT/\varepsilon_{ij})$

kT/ε	Ω_D
1.0	1.439
1.5	1.198
2.0	1.075
2.5	0.9996
3.0	0.9490
3.5	0.9120
4.0	0.8836
4.5	0.8610
5.0	0.8422
6.0	0.8124
7.0	0.7896
8.0	0.7712
9	0.7556
10	0.7424
20	0.6640
30	0.6232
40	0.5960
50	0.5756

Source: Reproduced from Bird, R. et al.: *Transport Phenomena,* 2nd Edition. Copyright Wiley-VCH Verlag GmbH & Co. KGaA. Reproduced with permission.

The *empirical correlations* for the binary diffusion coefficient are derived based on kinetic theory and using experimental data. A widely used correlation is given as (Fuller et al., 1966)

$$D_{ab} = \frac{0.00100T^{1.75}\left(\dfrac{1}{M_a} + \dfrac{1}{M_b}\right)^{1/2}}{P\left[\left(\sum v\right)_a^{1/3} + \left(\sum v\right)_b^{1/3}\right]^2}, \tag{6.111}$$

where M_a and M_b are the molecular weights of species a and b, respectively, and the quantity $\left(\sum v\right)$ represents the summation of atomic-diffusion volumes for each species of the binary mixture. A list of data for atomic-diffusion volume for some common gas species is given in Table 6.5 (Cussler, 1984; Sherwood et al., 1975).

Another frequently used equation for binary gas diffusion coefficients is derived based on kinetic theory and experimental data (Bird et al., 2002; Slattery and Bird, 1958) as

$$\frac{PD_{ab}}{\left(P_{ca}\,P_{cb}\right)^{1/3}\left(T_{ca}T_{cb}\right)^{5/12}\left(\dfrac{1}{M_a} + \dfrac{1}{M_b}\right)^{1/2}} = a_0\left(\frac{T}{\sqrt{T_{ca}T_{cb}}}\right)^{b_0}, \tag{6.112}$$

where D_{ab} = diffusion coefficient of binary gas mixture (cm^2/s), P = pressure (atm), T = temperature (K), and a_0, b_0 are empirical constants give as (Bird et al., 2002) follows:

TABLE 6.5

Atomic and Molecular-Diffusion Volume at Ambient Temperature (25°C) and Atmospheric Pressure

Element	Atomic Diffusion Volume (v)	Molecular Diffusion Volume (v)
Air		20.1
Carbon (C)	16.5	
Carbon dioxide (CO_2)		26.9
Carbon monoxide (CO)		18.9
Hydrogen (H, H_2)	1.98	7.07
Water (H_2O)		12.7
Oxygen (O, O_2)	5.48	16.6
Nitrogen (N, N_2)	5.69	17.9
Sulfur (S)	17.0	
Sulfur dioxide (SO_2)		41.1

Source: Reproduced from Sherwood, T. K. et al. *Mass Transfer*, Internal Student Edition. McGraw-Hill, Kogakusha, Tokyo, 1975; Cussler, E. L. *Diffusion—Mass Transfer in Fluid Systems*. Cambridge University Press, 1984.

For a pair of nonpolar gases such as H_2, O_2, and N_2:

$$a_0 = 2.745 \times 10^{-4} \text{ and } b_0 = 1.823$$

For a pair of polar (H_2O) and nonpolar gas:

$$a_0 = 3.640 \times 10^{-4} \text{ and } b_0 = 2.334$$

Table 6.6 shows typical binary coefficient data for common gas species.

Example 6.1

Estimate the diffusion coefficient of oxygen in nitrogen at 80°C and 2 atm using Chapman–Enskog's theoretical formula and by Fuller's empirical correlations.

Solution

First, estimate the Lennard-Jones potential parameters from Equation 6.110

$$\sigma_{ab} = \frac{1}{2}(\sigma_a + \sigma_b)$$

TABLE 6.6

Experimental Values of Binary Diffusion Coefficient for at Near Ambient Temperature and Atmospheric Pressure

Gas Pair	Temperature (K)	Diffusion Coefficient (cm^2/s)
Air–H_2	273.0	0.611
Air–O_2	273.0	0.177
CH_4–H_2	298.0	0.726
CH_4–H_2O	307.7	0.292
CO–N_2	295.8	0.212
CO–H_2	295.6	0.743
CO_2–H_2	298.0	0.646
CO_2–N_2	298.0	0.165
CO_2–O_2	293.2	0.160
CO_2–CO	296.1	0.152
H_2–N_2	297.2	0.779
H_2–O_2	273.2	0.697
H_2–H_2O	307.1	0.915
H_2–SO_2	285.5	0.525
O_2–N_2	273.2	0.181
O_2–H_2O	308.1	0.282

Source: Reproduced from Cussler, E. L. *Diffusion—Mass Transfer in Fluid Systems*. Cambridge University Press, 1984.

and

$$\frac{\varepsilon_{ab}}{k_B} = \left(\frac{\varepsilon_a}{k_B}\frac{\varepsilon_b}{k_B}\right)^{1/2}$$

From Table 6.3, $\sigma_a = 3.443$, $\frac{\varepsilon_a}{k_B} = 113$ for O_2 and $\sigma_b = 3.681$, $\frac{\varepsilon_b}{k_B} = 91.5$ for N_2. Substituting,

$$\sigma_{ab} = \frac{1}{2}\left(\sigma_{O_2} + \sigma_{N_2}\right) = \frac{1}{2}(3.443 + 3.681) = 3.562\ \text{Å}$$

and

$$\frac{\varepsilon_{ab}}{k_B} = \sqrt{\frac{\varepsilon_{O_2}}{k_B}\frac{\varepsilon_{N_2}}{k_B}} = (113 \times 91.5)^{1/2} = 101.68332$$

$$\frac{\varepsilon_{ab}}{k_B T} = 101.68332 / (80 + 273) = 0.288054$$

$$\frac{k_B T}{\varepsilon_{ab}} = 3.47156 \approx 3.5$$

From Table 6.2: at $\frac{k_B T}{\varepsilon_{ab}} = 3.5$, we get $\Omega_D = 0.912$.

The binary diffusion coefficient is based on the *Chapman-Enskog kinetic theory* and is given by Equation 6.108:

$$D_{ab} = \frac{0.001858T^{1.5}\left(\dfrac{1}{M_{O_2}} + \dfrac{1}{M_{N_2}}\right)^{1/2}}{P\sigma^2_{O_2N_2}\Omega_D} = \frac{0.001858T^{1.5}\left(\dfrac{1}{32} + \dfrac{1}{28.02}\right)^{1/2}}{2\sigma^2_{ij}\Omega_D}$$

$$D_{ab} = \frac{0.001858T^{1.5}\left(\dfrac{1}{32} + \dfrac{1}{28.02}\right)^{1/2}}{2(3.5)^2 \times 0.912}$$

$$D_{O_2,N_2} = 0.13776346\ \text{cm}^2/\text{s}$$

For estimation based on the *Fuller empirical correlation* equation (Equation 6.111), the appropriate molecular weight and diffusion volume are given from Table 6.5 as

$$\text{For oxygen: } M_{O_2} = 32, \left(\sum v\right)_{O_2} = 16.6$$

and

$$\text{For nitrogen: } M_{N_2} = 28.02, \left(\sum v\right)_{N_2} = 17.9$$

Substituting into Equation 6.111

$$D_{O_2N_2} = \frac{0.00100(80+273)^{1.75}\left(\dfrac{1}{32}+\dfrac{1}{28.02}\right)^{1/2}}{2\left[(16.6)^{1/3}+(17.9)^{1/3}\right]^2}$$

$$D_{O_2,N_2} = \frac{7.45780}{53.2922}$$

$$D_{O_2,N_2} = 0.13930 \text{ cm}^2/\text{s}.$$

Example 6.2

Estimate the binary diffusion coefficient of H_2 in H_2O at 25°C and 1.0 atm empirical correlations by Fuller and by Equation 6.63.

Solution

Molecular weights and critical properties of the gas pairs are as follows:

$$M_{H_2O} = 18.05,\ M_{H_2} = 2.016,\ T_{c,H_2} = 33.3 \text{ K},\ P_{c,H_2} = 12.8 \text{ atm},\ T_{c,H_2O} = 647.3 \text{ K},\ P_{c,H_2O} = 21.5 \text{ atm}$$

and for a pair of polar (H_2O) and a nonpolar H_2:

$$a = 3.640 \times 10^{-4} \text{ and } b = 2.334.$$

Substituting in Equation 6.112,

$$D_{ij} = \left[a\left(\frac{T}{\sqrt{T_{ci}T_{cj}}}\right)^b (P_{ci}\,P_{cj})^{1/3}(T_{ci}T_{cj})^{5/12}\left(\frac{1}{M_i}+\frac{1}{M_j}\right)^{1/2}\right]\Big/P$$

$$D_{H_2,H_2O} = \left[3.64\times10^{-4}\left(\frac{25+273.3}{\sqrt{33.3\times647.3}}\right)^{2.334}(12.8\times21.5)^{1/3}\right.$$

$$\left.(33.3\times647.3)^{5/12}\left(\frac{1}{2.016}+\frac{1}{18.05}\right)^{1/2}\right]\Big/1$$

$$D_{H_2,H_2O} = 0.58\ \text{cm}^2/\text{s}.$$

6.4.2 Diffusion in Liquids

Diffusion in liquids is much smaller than that in gases because of the higher density of liquids. Diffusion in liquid is generally described by the hydrodynamic theory based on the *Nernst–Einstein equation*, which is derived based on the assumption of motion of a slowly moving rigid solid spherical particle in a continuum medium of solvent and thus limited to an infinitely dilute solution. Basically, we assume that the solute particles are large relative to that of the base liquid and without considering the molecular motion. This is in contrast to the assumption of molecular motion in the kinetic theory of gases. According to this equation, the diffusion of a species *a* in a stationary liquid medium *b* is given by

$$D_{ab} = k_B T\frac{U_a}{F_b}, \tag{6.113}$$

where U_a is the velocity of the particle under the action of a hydrodynamic viscous force, F_b, exerted by the fluid medium, and k_B is Boltzmann's constant (1.38×10^{-6}).

The velocity can be determined from the solution of the hydrodynamic flow field. For a no-slip condition at the interface of the diffusing spherical particles and the liquid, the force over the spherical particle is given by *Stoke's law* as

$$F_b = 6\pi\mu_b U_a R_a, \tag{6.114}$$

where μ_b is the dynamic viscosity of the liquid and R_a is the radius of the diffusing spherical particle, which can be assumed as half of the collision diameter.

Combination of Equations 6.113 and 6.114 leads to the *Stokes–Einstein equation* as

$$D_{ab} = \frac{k_B T}{6\pi\mu_b R_a} \tag{6.115}$$

For slip condition at the interface of the diffusing particles and the liquid medium, the force over the particles is given as

$$F_b = 4\pi\mu_b U_a R_a, \tag{6.116}$$

and the corresponding expression for diffusivity is given as

$$D_{ab} = \frac{k_B T}{4\pi\mu_b R_a}. \tag{6.117}$$

Another empirical correlation based on the Stokes–Einstein equation for low concentration of species *i* in liquid medium *j* is given by Wilke and Chang (Sherwood et al., 1975),

$$D_{ab} = \frac{7.4\times 10^{-8}(\psi_b M_b)^{1/2} T}{\mu_b v_a^{0.6}} \tag{6.118}$$

where

v_a = molar volume of solute at normal boiling point (cm^3/mol); molar volumes at normal boiling point are listed in Table 6.5
μ_b = viscosity of liquid (centipoise, cP)
T = absolute temperature (K)
ψ_b = association parameter of the solvant liquid

Some suggested values of the association parameter ψ_b are as follows: 2.6 for water, 1.9 for methanol, and 1.5 for ethanol (Sherwood et al., 1975).

Example 6.3

Estimate oxygen diffusion in water using the Stokes–Einstein equation and Wilke's empirical correlation at 25°C.

Solution

Estimate the oxygen molecule diameter from collision diameter as

$$R_a = \frac{\sigma_a}{2} = \frac{1}{2}\times 3.433 = 1.716\ \text{Å} = 1.716\times 10^{-6}\ \text{cm}.$$

Substituting into the *Stokes – Einstein equation,*

$$D_{ab} = \frac{k_B T}{6\pi\mu_b R_a} = \frac{(1.38\times 10^{-16})\times 298\ \text{K}}{(6\pi\times 0.01\ \text{g/cm s})(1.716\times 10^{-6}\ \text{cm})}$$

$$D_{O_2,H_2O} = 1.272\times 10^{-7}\ \text{cm}^2/\text{s}.$$

Estimation based on Wilke and Chang's empirical equation (Equation 6.118) is

$$D_{ab} = \frac{7.4\times10^{-8}\left(\psi_{H_2O}M_{H_2O}\right)^{1/2}T}{\mu_{H_2O}v_{O_2}^{0.6}}$$

$$D_{ab} = \frac{7.4\times10^{-8}(2.6\times18)^{1/2}(303\ \text{K})}{(1\ \text{cP})(25.6\ \text{cm}^3/\text{g mol})^{0.6}}$$

$$D_{O_2,H_2O} = 2.192\times10^{-5}\ \text{cm}^2/\text{s}.$$

6.4.3 Diffusion in Porous Solids

The reactant gas species transport to reaction sites through the porous electrodes based on the concept of gas diffusion in porous media. In porous media, the diffusion mechanism can be of three different types: *ordinary diffusion, Knudsen diffusion,* and *surface diffusion.* If the pores are much larger than the mean free path length, then the molecules collide with each other more frequently than with the pore walls, and ordinary diffusion is assumed to be the dominant diffusion mechanism. *Knudsen diffusion* is encountered in smaller pores or at lower pressure or density. In this case, molecules collide more frequently with the walls than with other gas molecules. The Knudsen diffusion coefficient given is based on kinetic theory as

$$D_{kd} = \frac{2}{3}a\bar{v}_1, \tag{6.119}$$

where a is the effective pore radius in meters and $\bar{v}_1$ is the average molecular speed of species, which is given as

$$\bar{v}_1 = \left(\frac{8\bar{R}T}{\pi M_1}\right)^{1/2}. \tag{6.120}$$

With the substitution of the value for the gas constant and combining Equations 6.119 and 6.120, the expression for Knudsen diffusion is given as

$$D_{kd} = \frac{2}{3}a\left(\frac{8\bar{R}T}{\pi M_1}\right)^{1/2} \tag{6.121a}$$

or

$$D_{\mathrm{kd}} = 97a\left(\frac{T}{M_1}\right)^{1/2} \ (\mathrm{m}^2/\mathrm{s}), \tag{6.121b}$$

where T is temperature in Kelvin and M_1 is the molecular weight.

In an intermediate range of pore sizes, both ordinary and Knudsen diffusion contribute to the transfer of the species in the media. In this range, the combined ordinary and Knudsen diffusion can be represented by assuming parallel resistances and expressed as

$$D_i^{\mathrm{eff}} = \left(\frac{1}{D_{\mathrm{md}}} + \frac{1}{D_{\mathrm{kd}}}\right)^{-1} \tag{6.122}$$

Surface diffusion takes place along the surface of the solid in parallel to the regular diffusion in the pore. This surface diffusion involves multiple physical processes: (i) rapid adsorption of solute species on pore surface, (ii) diffusion of species in the surface liquid layer owing to the presence of the surface concentration gradient, and (iii) rapid desorption from the surface.

$$j_i = -D_{\mathrm{sd}}\nabla(c_{\mathrm{ad}}), \tag{6.123}$$

where D_{sd} is the surface diffusion coefficient and $c_{\mathrm{ad}} = s_g\rho_p c_s$ is the density of adsorbed species (mass or moles per unit volume of porous mass) given by the adsorption isotherm. c_s is the surface concentration of adsorbed diffusing species.

The gas phase and liquid surface diffusion takes place in parallel and the net mass flux can be assumed as additive and given as

$$j_i = -D_{ij}\nabla c_i - D_{\mathrm{sd}}\nabla c_{\mathrm{sd}}. \tag{6.124}$$

Other important factors that have to be considered for porous media are the presence of the tortuous path and changes and reduction in the cross-sectional area of pore channels. The effective diffusivity is expressed as

$$D_{ij}^{\mathrm{eff}} = \frac{\varepsilon}{\tau} D_{ij}, \tag{6.125}$$

where ε is the *porosity* of the media that accounts for the reduction of the free area for diffusion owing to the presence of the solid phase and τ is the *tortuosity factor* that accounts for the increase in the diffusion path owing to the tortuous path of the pores and has to be determined by experiments. Typical tortuosity factor values fall in range of 2–6.

Another alternative form of effective mass diffusion coefficient for porous media that takes into account the effect of porosity and the tortuous morphology of the porous structure or the tortuous path is given by using the Bruggemann empirical correction formula (De La Rue and Tobias, 1959; Springer et al., 1991) as

$$D_{ij}^{eff} = D_{ij}\varepsilon^{1.5}. \tag{6.126}$$

This Bruggemann empirical formula is restricted to a low tortuosity factor and porosity range of 0.4–0.5.

6.5 Mass Transfer Resistance in Fuel Cells

As we have mentioned earlier, one of the fuel cell voltage losses is the mass transfer loss or concentration loss caused by lower reactant gas concentration distribution at the reaction sites. Mass transport establishes reactant gas concentration distributions in gas supply channels and in the electrodes of a fuel cell, and hence in the distribution of local current densities. The gas supply rates to the anode–membrane and cathode–membrane interface must be sufficient enough to meet the gas consumption rate given by the electrochemical reaction rates. Any insufficient supply of gas to reaction sites may cause sluggishness in the reactions and cause *mass transfer loss* and reduction in fuel cell output voltage.

Mass transfer loss primarily occurs at high current densities with increased demand for electrochemical reactions and gas consumption rates. At the higher current densities, the fuel supply may not be sufficient enough to maintain reactant gas concentration at the electrode CL at a positive level to sustain the reaction. At this point, all the reactant gas supply is completely consumed by the reaction and produces the maximum current density, which is also referred to as the *limiting current density*, j_1. At the limiting current density, the net output cell voltage is zero as the reversible cell voltage is completely balanced by the cell voltage losses such as activation, ohmic, and, primarily, mass transfer loss.

Major factors that contribute to the mass transfer loss are as follows:

1. Convective mass transfer resistance in the gas flow channel
2. Diffusive mass transfer resistance in the electrode–gas diffusion layer

3. Diffusion–reaction resistance in the CL
4. Transport of different species such reaction gas, electrons, and ions to/from the electrode–electrolyte interface
5. Removal of reaction product such as water from the reaction sites
6. Pressure drop in the gas flow channel

6.5.1 Estimation of Limiting Current Density

In order to derive the limiting current density, let us equate the net mass transfer rate of a reactant gas given by Equation 6.102 and the gas consumption rate based on the current density as follows:

$$j_i'' = \frac{C_{i,\mathrm{ch}} - C_{i,\mathrm{el}}}{\left(\dfrac{1}{h_\mathrm{m}} + \dfrac{a_\mathrm{el}}{D_{ij}^\mathrm{eff}}\right)} = \frac{j}{n_\mathrm{e}F} \tag{6.127}$$

Equation 6.127 can also be used to estimate the reactant gas concentration at the electrode reaction surface as

$$C_{i,\mathrm{el}} = C_{i,\mathrm{ch}} - \frac{j}{n_\mathrm{e}F}\left(\frac{1}{h_\mathrm{m}} + \frac{a_\mathrm{el}}{D_{ij}^\mathrm{eff}}\right) \tag{6.128a}$$

Notice that in situations where net mass transfer resistance is controlled by the diffusion resistance in the electrode–gas diffusion layer only, the reactant gas concentration at the reaction surface is given as

$$C_{i,\mathrm{el}} = C_{i,\mathrm{ch}} - \frac{j}{n_\mathrm{e}F}\left(\frac{a_\mathrm{el}}{D_{ij}^\mathrm{eff}}\right). \tag{6.128b}$$

Equating 6.127 for the current density as a function of reactant gas concentration

$$j = n_\mathrm{e}F\frac{C_{i,\mathrm{ch}} - C_{i,\mathrm{el}}}{\left(\dfrac{1}{h_\mathrm{m}} + \dfrac{a_\mathrm{el}}{D_{ij}^\mathrm{eff}}\right)}. \tag{6.129}$$

Equation 6.80 shows that the current density depends directly on the reactant gas concentration in the channel as well as the reactant gas concentration at the electrode–electrolyte interface. The current density increases with higher gas concentration, $C_{i,\mathrm{ch}}$, in the channel and lower concentration, $C_{i,\mathrm{el}}$, at the reaction surface. For a given electrode–gas diffusion layer and gas channel design with a fixed $C_{i,\mathrm{ch}}$ value, the maximum current density or the *limiting current density* is reached when the gas concentration at the electrode–electrolyte interface becomes zero or $C_{i,\mathrm{el}} = 0$, and this is expressed as

$$j = n_\mathrm{e}F\frac{C_{i,\mathrm{ch}}}{\left(\dfrac{1}{h_\mathrm{m}}+\dfrac{a_\mathrm{el}}{D_{ij}^\mathrm{eff}}\right)}. \tag{6.130}$$

In situations where net mass transfer resistance is controlled by the diffusion resistance in the electrode–gas diffusion layer only, the *limiting current density* can be approximated as

$$j_\mathrm{l} = n_\mathrm{e}FD_{ij}^\mathrm{eff}\left(\frac{C_{i,\mathrm{ch}}}{a_\mathrm{el}}\right). \tag{6.131a}$$

In situations where net mass transfer resistance is controlled by the mass convection resistance in the electrode–gas diffusion layer only, the *limiting current density* can be approximated as

$$i_\mathrm{l} = n_\mathrm{e}Fh_\mathrm{m}C_{i,\mathrm{ch}}. \tag{6.131b}$$

Basically, the limiting current density represents the limiting condition for mass transport in a fuel cell and the highest current density that a fuel cell can operate.

6.5.2 Mass Transfer or Concentration Loss

While the limiting current density defines the limit of the operating condition of a fuel cell, the mass transfer loss or concentration loss represents the cell voltage loss caused by the difference in the reactant and the product gas concentrations between the reaction surface and the bulk gas flow in channels. The variation of gas concentration at the reaction surface causes fuel cell voltage loss in two ways: (1) decrease in thermodynamic fuel cell voltage given by *Nernst voltage* and (2) increase in activation or the electrochemical reaction loss.

1. Decrease in Nernst voltage owing to concentration

   Let us consider the Nernst equation (Equation 4.49) for real thermodynamic fuel cell voltage given by Equation 4.49.

$$E = E_\mathrm{rev}^0 - \frac{\bar{R}T}{n_\mathrm{e}F}\ln\frac{\prod_{i=1}^{N} C_{\mathrm{P}i}^{n_i}}{\prod_{i=1}^{N} C_{\mathrm{R}i}^{n_i}} \tag{6.132}$$

In this equation, the Nernst voltage loss term represents the voltage loss owing to the variation in reactant and product gas concentrations in the supply gas stream. If we just have only single reactant specie for one electrode side, the Nernst voltage is given as

$$E^0_{\text{Nernst}} = \frac{\bar{R}T}{n_e F}\ln\frac{1}{C_{\text{R}i}}. \tag{6.133}$$

Concentration overpotential or loss is estimated as the change in Nernst voltage loss owing to the variation of reactant gas concentration from the bulk gas flow stream to the gas concentration at the electrode reaction surface as follows:

$$\eta_{\text{mass,Nernst}} = \frac{\bar{R}T}{n_e F}\ln\frac{1}{C_{i,\text{el}}} - \frac{\bar{R}T}{n_e F}\ln\frac{1}{C_{i,\text{ch}}}$$

or

$$\eta_{\text{mass,Nernst}} = \frac{\bar{R}T}{n_e F}\ln\frac{C_{i,\text{ch}}}{C_{i,\text{el}}}. \tag{6.134}$$

From Equation 6.131a for the limiting current density with mass transfer limited by diffusion resistance, the reactant concentration in the bulk gas flow in the channels is written as

$$C_{i,\text{ch}} = \frac{j_l a_{\text{el}}}{n_e F D_{ij}^{\text{eff}}}. \tag{6.135}$$

Substituting Equation 6.135 into Equation 6.128b, we have

$$C_{i,\text{el}} = \frac{j_l a_{\text{el}}}{n_e F D_{ij}^{\text{eff}}} - \frac{j}{n_e F}\frac{a_{\text{el}}}{D_{ij}^{\text{eff}}} \tag{6.136}$$

Combining Equations 6.135 and 6.136,

$$\frac{C_{i,\text{ch}}}{C_{i,\text{el}}} = \frac{j_l}{j_l - j}. \tag{6.137}$$

Now, substituting Equation 6.137 into Equation 6.134, we have the expression for the *mass transfer* or *concentration overpotential* as

$$\eta_{\text{mass,Nernst}} = \frac{\bar{R}T}{n_e F}\ln\frac{j_l}{j_l - j}. \tag{6.138}$$

6.5.3 Effect of Concentration on Activation Loss

As we have discussed in Chapter 4, electrochemical reaction kinetics is given by the Butler–Volmer equation (Equation 5.76)

$$j = j_0 \left\{ \frac{C_{\mathrm{R,r}}}{C_{\mathrm{R,ch}}} e^{\alpha n_e F \eta_{\mathrm{act}}/(RT)} - \frac{C_{\mathrm{P,r}}}{C_{\mathrm{P,ch}}} e^{(1-\alpha) n_e F \eta_{\mathrm{act}}/(RT)} \right\}. \tag{6.139}$$

At higher current density where mass transfer loss is predominant, the second term that represents the reduction reaction or product becomes insignificant and the equation can be simplified by dropping this term as

$$j = j_0 \left\{ \frac{C_{\mathrm{R,r}}}{C_{\mathrm{R,ch}}} e^{\alpha n_e F \eta_{\mathrm{act}}/(RT)} \right\}. \tag{6.140}$$

Solving for the activation overpotential

$$\eta_{\mathrm{act}} = \frac{RT}{\alpha n_e F} \ln \frac{j C_{\mathrm{R,ch}}}{j_0 C_{\mathrm{R,r}}} \tag{6.141}$$

Equation 6.141 represents the activation overpotential on the basis of the reactant gas concentration in bulk gas flow. The mass transfer loss can be estimated on the basis of the changes in activation overpotential owing to the variation in reactant gas concentration from the bulk flow to the reaction surface as follows:

$$\eta_{\mathrm{mass,act}} = \eta_{\mathrm{act}}(C_{\mathrm{R,r}}) - \eta_{\mathrm{act}}(C_{\mathrm{R,ch}})$$

$$= \frac{RT}{\alpha n_e F} \ln \frac{j C_{\mathrm{R,ch}}}{j_0 C_{\mathrm{R,r}}} - \frac{RT}{\alpha n_e F} \ln \frac{j C_{\mathrm{R,ch}}}{j_0 C_{\mathrm{R,ch}}}$$

$$\eta_{\mathrm{mass,act}} = \frac{RT}{\alpha n_e F} \ln \frac{C_{\mathrm{R,ch}}}{C_{\mathrm{R,r}}}. \tag{6.142}$$

Substituting Equation 6.137 for the ratio $\frac{C_{i,\mathrm{ch}}}{C_{i,\mathrm{r}}} = \frac{j_l}{j_l - j}$, we get

$$\eta_{\mathrm{mass,act}} = \frac{RT}{\alpha n_e F} \ln \frac{j_l}{j_l - j}. \tag{6.143}$$

Equation 6.143 represents the mass transfer loss owing to the activation reaction.

Combining Equations 6.138 and 6.143, we get the net mass transfer loss

$$\eta_{\text{mass}} = \frac{RT}{n_e F}\left(1+\frac{1}{\alpha}\right)\ \ln\left(\frac{i_l}{i_l - i}\right). \tag{6.144}$$

Considering that the mass transfer loss is dominated by the oxygen concentration in cathode site, the limiting current density is estimated based on Equation 6.131a as

$$j_l = n_e F D_{O_2}^{\text{eff}}\left[\frac{C_{O_2,\text{ch}}}{a_c}\right], \tag{6.145}$$

where

$C_{O_{2,\text{ch}}}$ = reactant concentration in the cathode gas channel
$D_{O_2}^{\text{eff}}$ = oxygen diffusion coefficient in cathode electrode
a_c = thickness of cathode electrode

It can be noticed here that the expression is only valid for $j < j_l$ and the mass transfer loss is very small or insignificant for very low current densities, $j_i \ll j_l$. As the operating current density approaches the limiting current density value, the mass transfer loss sharply increases.

6.6 Summary

Limiting current density:

$$j_l = nFD_{O_2}^{\text{eff}}\left[\frac{C_{O_{2,\text{ch}}}}{a_c}\right]$$

Mass transfer loss:

$$\eta_{\text{mass}} = \frac{RT}{n_e F}\left(1+\frac{1}{\alpha}\right)\ \ln\left(\frac{j_l}{j_l - j}\right)$$

Activation overpotential based on bulk gas flow concentration:

$$\eta_{\text{act}} = \frac{RT}{\alpha n_e F}\ln\frac{jC_{\text{R,ch}}}{j_0 C_{\text{R,r}}}$$

PROBLEMS

1. Estimate the binary diffusion coefficient of hydrogen in water vapor at 100°C and 1 atm using Chapman–Enskog's theoretical formula and Fuller's empirical correlations.
2. Estimate the binary diffusion coefficient of methane in hydrogen at 200°C and 2 atm pressure using (a) Chapman–Enskog's theoretical formula, (b) Fuller's empirical correlations, and (c) Bird's correlation.

References

Asako, Y., H. Nakamura and M. Fagri. Developing laminar flow and heat transfer in the entrance region of regular polygonal ducts. *International Journal of Heat and Mass Transfer* 31: 2590–2593, 1988.

Bird, R., W. Stewart and E. Lightfoot. *Transport Phenomena*, 2nd Edition. John Wiley & Sons, New York, 2002.

Cussler, E. L. *Diffusion—Mass Transfer in Fluid Systems*. Cambridge University Press, 1984.

De La Rue, R. E. and C. W. Tobias. On the conductivity of dispersions. *Journal of the Electrochemical Society* 106: 827–833, 1959.

Fuller, E. N., P. D. Schettler and J. C. Gliddings. *Industrial and Engineering Chemistry* 58(5): 19, 1966.

Hirschfelder, J. O., C. P. Curtiss and R. B. Bird. *Molecular Theory of Gasses and Liquids*. Wiley, New York, 1964.

Incropera, F. P., D. P. Dewitt, T. L. Bergman and A. Lavine. *Fundamentals of Heat and Mass Transfer*, 6th Edition. John Wiley & Sons, New York, 2007.

Martys, N. Improved approximation of the Brinkman equation using lattice Boltzmann method. *Physics of Fluids* 13(6): 1807–1810, 2001.

Martys, N., D. P. Bentz and E. J. Garboczi. Computer simulation study of the effective viscosity in Brinkman's equation. *Physics of Fluids* 6(4): 1434–1438, 1994.

Shah, R. K. and A. L. London. *Laminar Flow Forced Convection in Ducts*. Academic Press, New York, 1978.

Sherwood, T. K., R. L. Pigford and C. R. Wilke. *Mass Transfer*, Internal Student Edition. McGraw-Hill, Kogakusha, Tokyo, 1975.

Slattery, J. C. Single-phase flow through porous media. *AIChE Journal* 15: 866–872, 1969.

Slattery, J. C. and R. B. Bird. Calculation of the diffusion co-efficient of dilute gases and of the self-diffusion co-efficient of dense gases. *AIChE Journal* 4: 137–142, 1958.

Springer, T. E., T. A. Zawodzinski and S. Gottesfeld. Polymer electrolyte fuel cell model. *Journal of the Electrochemical Society* 138(8): 2334–2342, 1991.

Whitaker, S. Fluid motion in porous media. *Industrial and Engineering Chemistry* 1: 14–28, 1969.

Further Reading

Li, X. *Principles of Fuel Cells*. Taylor & Francis Publishers, 2006.

Majumdar, P. *Computational Methods for Heat and Mass Transfer*. Taylor and Francis Publishers, New York, 2005.

Mangar, Y. N. and R. M. Manglik. Modeling of convective heat and mass transfer characteristics of anode-supported planar solid oxide fuel cells. *Journal of Fuel Cell Science and Technology* 4: 185–193, 2007.

Mills, A. F. *Mass Transfer*. Prentice Hall, New Jersey, 2001.

O'Hayre, R., S.-W. Cha, W. Cotella and F. B. Prinz. *Fuel Cell Fundamentals*. John Wiley & Sons, Inc, New York, 2006.

Venjata, P. P., M. A. Jog and R. M. Manglik. Computational modeling of planar SOFC: Effects of volatile species/oxidant mass flow rate and electrochemical reaction rate on convective heat transfer. Proceedings of the 2008 ASME International Mechanical Engineering Congress and Exposition, Boston, Massachusetts, IMECE2008-69249, 2008.

Wesselingh, J. A. and R. Krishna. *Mass Transfer in Multicomponent Mixtures*. Delft University Press, 2000.

Wilke, C. L. *Chemical Engineering Progress* 45: 218–223, 1949.

Woods, L. C. *An Introduction to the Kinetic Theory of Gasses and Magnetoplasmas*. Oxford Press, Oxford, 1993.

Xianguo Li, *Principles of Fuel Cells* Taylor & Francis, New York, 2006.

7

Charge and Water Transport in Fuel Cells

In this chapter, we will be considering the fundamental principles and mechanism of charge transport in a fuel cell. As we have discussed before, in the operation of a fuel cell, charges such as *electrons* and *ions* are produced and consumed in two electrochemical reactions at the anode–electrolyte and cathode–electrolyte interfaces. Electrons transport through the electrodes and interconnect to the external electrical circuit. Ions transport through the electrolyte from the electrode where it is produced to the electrode where it is consumed. Ohmic voltage loss is caused by the resistances to the motion of ions through the electrolytes as well as electrons through the electrodes, interconnect materials, and contact interfaces. Additionally, ion transport plays a critical role in the transport of water in a PEM fuel cell. We will also include our discussion on water transport in this chapter.

7.1 Charge Transport

As with any transport phenomena, charge transport through a medium takes place under the influence of some kind of force. In a fuel cell, the electrons and ion charges are generated and consumed at the two electrode–electrolyte interfaces through electrochemical half reactions. These reactions result in an *electrical voltage* or *potential difference* between the electrodes, which acts as a driving force for the *transport of electrons* from the anode side to the cathode side. This potential difference is also a measure of the cell voltage that produces electrical power. For *ion transport,* these reactions result in the accumulation of ions at one electrode and reduction at another electrode. This process results in an electrical potential gradient as well as a concentration gradient of the ions across the electrolyte. The ions transport through the fuel cell electrolytes under the influence of both electrical potential gradient and concentration gradient as the driving forces. Also, a pressure gradient across the electrolyte between the anode and cathode sides acts as an additional driving force that establishes the convection mode for the migration of ions. The *ion transport* is governed by the combined effect of all three transport modes: (i) *migration:* charge transport under the influence of electrical potential gradient, (ii) *diffusion:* charge transport under the influence of concentration gradient, and (iii) *convection:* charge transport under the influence of flow velocity.

7.1.1 Charge Transport Modes and Rate Equations

A description of different charge transport modes and the associated rate equations for the charge flux are given here.

7.1.1.1 Charge Transport by Diffusion

The mass transport of charge flux owing to concentration gradient is given based on *Fick's law*

$$J_i = -D_i \, \nabla C_i \tag{7.1a}$$

and in a one-dimensional Cartesian coordinate as

$$J_i = -D_i \frac{dC_i}{dx}, \tag{7.1b}$$

where D_i is the diffusion coefficient of the ion in the electrolyte and C_i is the charge concentration.

In terms of molar charge flux, Equation 7.1 is given as

$$J_i = -(Z_i F) \, D_i \, \nabla C_i \tag{7.2a}$$

and for a one-dimensional Cartesian coordinate as

$$J_i = -(Z_i F) D_i \frac{dC_i}{dx}, \tag{7.2b}$$

where Z_i is the charge number of the charge carrier. For example, $Z_i = +1$ for H^+ and $Z_i = -2$ for O^{2-}.

7.1.1.2 Charge Transport by Convection

The mass charge flux owing to convection mode is given as

$$J = C_i \vec{V} \tag{7.3a}$$

and for a one-dimensional problem with unidirectional flow as

$$J = uC_i. \tag{7.3b}$$

The convection velocity, u, in the porous electrolyte structure is given either by Darcy's law equation (Equation 6.17) or by Brinkman's equation (Equation 6.18) as discussed in Chapter 6.

Considering the convection velocity given by Darcy's law, the pressure-driven mass flux is given as

$$J = C_i \frac{K_\mathrm{p}}{\mu} \nabla P, \tag{7.4}$$

where K_p is the hydraulic permeability and μ is the viscosity of the fluid.

The net mass transport by combined diffusion and convection is given as

$$J_i = -D_i \nabla C_i + C_i \frac{K_\mathrm{p}}{\mu} \nabla P \tag{7.5a}$$

and for a one-dimensional Cartesian coordinate as

$$J_i = -D_i \frac{\mathrm{d}C_i}{\mathrm{d}x} + C_i \frac{K_\mathrm{p}}{\mu} \frac{\mathrm{d}P}{\mathrm{d}x}. \tag{7.5b}$$

7.1.1.3 Charge Transport by Electrical Potential Gradient

Charge transport take place because of the presence of electrical potential gradient in the electrolyte, electrodes, and interconnect materials. The ionic flux or ionic current flow takes place because of the presence of electric potential gradient in the electrolyte. Similarly, the electron transport takes place because of the presence of an electrical potential gradient through the electrodes and interconnect materials. Since the electrolyte layer is sandwiched between two electrodes with an electrical potential difference, the ions in the electrolyte move in the direction of the electrical potential gradient. The charge transport flux owing to migration under the influence of electrical potential is given as

$$J_{i,\mathrm{m}} = -z_i \frac{F}{RT} D_i C_i \nabla \Phi, \tag{7.6a}$$

where Z_i is the charge number for a charge carrier.

The charge flux can also be expressed based on *Ohm's law* assuming a direct proportionality between charge flux and the electrical potential gradient as

$$j = -\sigma_\mathrm{c} \nabla \Phi \tag{7.6b}$$

and in a one-dimensional Cartesian coordinate as

$$j = -\sigma_\mathrm{c} \frac{\mathrm{d}\Phi}{\mathrm{d}x}, \tag{7.6c}$$

where $\frac{d\Phi}{dx}$ is the electrical potential gradient and σ_c is the *charge conductivity*, an electrical transport property of the ions in the electrolyte and of the electrons in electron conducting electrode and interconnect layers. The common unit of charge conductivity is $S/m = (\Omega \cdot m)^{-1}$ or $S/cm = (\Omega \cdot cm)^{-1}$.

In terms of *mass charge flux*, Equation 7.6c can be written as

$$J = -\frac{\sigma_c}{z_i F}\frac{d\Phi}{dx}, \tag{7.6d}$$

where

z_i = charge number of the carrier
F = Faraday constant

7.1.1.4 Nernst–Planck's Equation

The net charge mass transport as a result of the combined effect of all three modes: migration owing to potential gradient, diffusion, and convection is given as

$$J_i = -z_i \frac{F}{RT} D_i C_i \nabla\Phi - D_i \nabla C_i + C_i \vec{V}_i \tag{7.7a}$$

or as

$$J_i = -\sigma_c \nabla\Phi - D_i \nabla C_i + C_i \vec{V}_i \tag{7.7b}$$

and in a one-dimensional case as

$$J_i = -\sigma_c \frac{d\Phi}{dx} - D_i \frac{dC_i}{dx} + C_i u_i \tag{7.8a}$$

or

$$J_i = -\sigma_c \frac{d\Phi}{dx} - D_i \frac{dC_i}{dx} + C_i \frac{K_p}{\mu}\frac{dP}{dx}. \tag{7.8b}$$

Equation 7.9 is referred to as the *Nernst–Planck's equation* (Bernardi and Verbrugge, 1991; Verbrugge and Hill, 1990).

The relationship between the charge transport and the current density is given as

$$i = F\sum_{i=1} z_i J_i \tag{7.9}$$

where summation is used for the number of different charges that transport through the media.

7.1.1.5 Schlogl's Equation

The fluid dynamics and the net convection velocity developed under the influence of electrical potential and pressure gradient are described by *Schlogl's equation* of motion as (Bernardi and Verbrugge, 1991)

$$\vec{V} = \frac{K_\phi}{\mu} z_f C_f \nabla\phi - \frac{K_p}{\mu}\nabla P \tag{7.10}$$

where K_ϕ is the electrokinetic permeability and K_p is the membrane hydraulic permeability owing to pressure gradient.

Note that in order to estimate charge transport or electrical current in the electrode and electrolyte layers, it is necessary to solve for the electrical potential field in the electrode and electrolyte layers.

The electrical potential can be written from the *Nernst–Planck equation* (Equation 7.7a) and Equation 7.9

$$\sum_{i=1} \frac{(z_i F)^2}{RT} D_i C_i \nabla\Phi = -i - F\sum_{i=1} z_i D_i \nabla C_i + F\sum_{i=1} \frac{F}{RT} C_i \vec{V}. \tag{7.11a}$$

Considering charge conductivity $\sigma = \sum_{i=1} \frac{(z_i F)^2}{RT} C_i D_i$, Equation 7.11a can be written for charge potential as

$$\nabla\phi = -\frac{i}{\sigma} - \frac{F}{\sigma}\left(\sum_i z_i D_i \nabla C_i\right) + \frac{F}{\sigma}\left(\sum_i z_i C_i\right)\vec{V} \tag{7.11b}$$

7.1.2 Charge Transport and Electrical Potential Equation

The relationship of charge transport and electrical potential field is analogous to the solution of heat transfer on the basis of the temperature field. In order to estimate the charge transport or electrical current in the electrode and electrolyte layers, it is necessary to solve the electrical potential equation. This is analogous to the solution of the temperature field from the heat equation before estimating the heat transfer rate through a medium based on the temperature gradient.

As we have discussed in Chapter 5, the electrical double layer, as depicted in Figure 7.1, plays a critical role in the distribution of electrical potential at the electrode–electrolyte interfaces and to the ion transport through the electrolyte from the anode side to the cathode side. Figure 7.1 shows the comprehensive details of the electrical double layer structure, which is composed of an inner Helmholtz plane (IHP), an outer Helmholtz plane (OHP), and the diffusion layer.

At the anode electrode–electrolyte interface, there is an increase in the electrical potential owing to the formation and accumulation of charge species in the electrical double layer that spans over the anode–electrolyte interface.

The sign of the charges along an electrode surface depends on the electrode types. For example, in a hydrogen fuel cell, there is accumulation of negative charges along the anode electrode surface and positive charges in the adjacent electrolyte media. For simplicity, we will consider a simplified

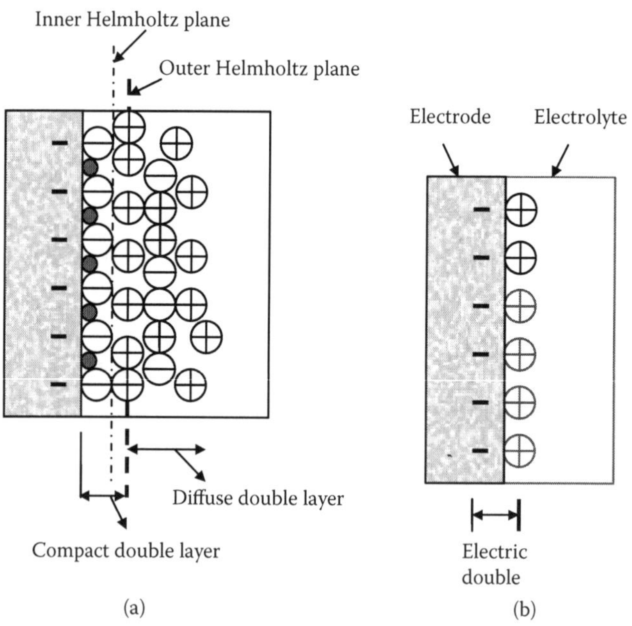

FIGURE 7.1
Accumulation of charges in electric double layer and electrical potential. (a) Comprehensive double layer model. (b) Simplified double layer model.

representation of the electrical double layer across the electrode–electrolyte interface as shown in Figure 7.1b.

The electrical double layer acts like a capacitor with an increase in electrical potential over a very small thickness of the order of nanometers at the anode–electrolyte interface. Similarly, at the cathode–electrolyte interface, there is again an increase in electrical potential because of the formation and accumulation of charged ions over the electrical double layer. This is followed by the drop in electrical potential over the thickness of the cathode electrode because of the resistance of electron transport.

The electrochemical reactions and the charge transfer or current flow at the electrode–electrolyte interfaces are driven by the potential or voltage jump across the double layer and represent the activation overpotential or voltage drop. This buildup of charges and charge transfer through electrochemical reaction is equivalent to the capacitance–resistance model of an electrical circuit. In the electrolyte layer, the electrical potential drops owing to the presence of ohmic voltage loss caused by the resistance of the ionic transport.

Figure 7.2a shows a typical distribution of electrical potential in the anode–electrolyte–cathode layers of a fuel cell with lower potential at the anode and higher potential at the cathode. The reversible potentials for the anode and cathode electrodes are given on the basis of the electrochemical half-reactions and are shown as $\phi_{rev,a}$ and $\phi_{rev,c}$, respectively. The corresponding real potentials are given as ϕ_a and ϕ_c based on positive activation overpotential, $\eta_{act,a}$, at the anode and negative activation overpotential, $\eta_{act,c}$, at the cathode:

$$\phi_a = \phi_{rev,a} + \eta_{act,a} \tag{7.12a}$$

and

$$\phi_c = \phi_{rev,c} + \eta_{act,c}. \tag{7.12b}$$

As we can see, the electrical potential decreases over the thicknesses of the anode and cathode electrodes owing to the ohmic resistance of the electrode gas diffusion layers to electron transport. These losses are referred to as the anode and cathode *ohmic overpotentials* $\eta_{ohm,a}$ and $\eta_{ohm,c}$, respectively.

Because of the presence of these double layers, there is a drop in the electric potential over the thickness of the electrolyte, and this causes the positively charged ions to transport from the anode side to the cathode side and complete the electrochemical reactions. This drop in the electrical potential is referred to as the *electrolyte ohmic overpotential*, $\eta_{ohm,e}$.

Since the thicknesses of the electrical double layer are in the ranges of nanometers, which is significantly smaller compared to the electrode and electrolyte thicknesses on the order of microns, the linear variation in electrical potential over the electric double layer is generally approximated with

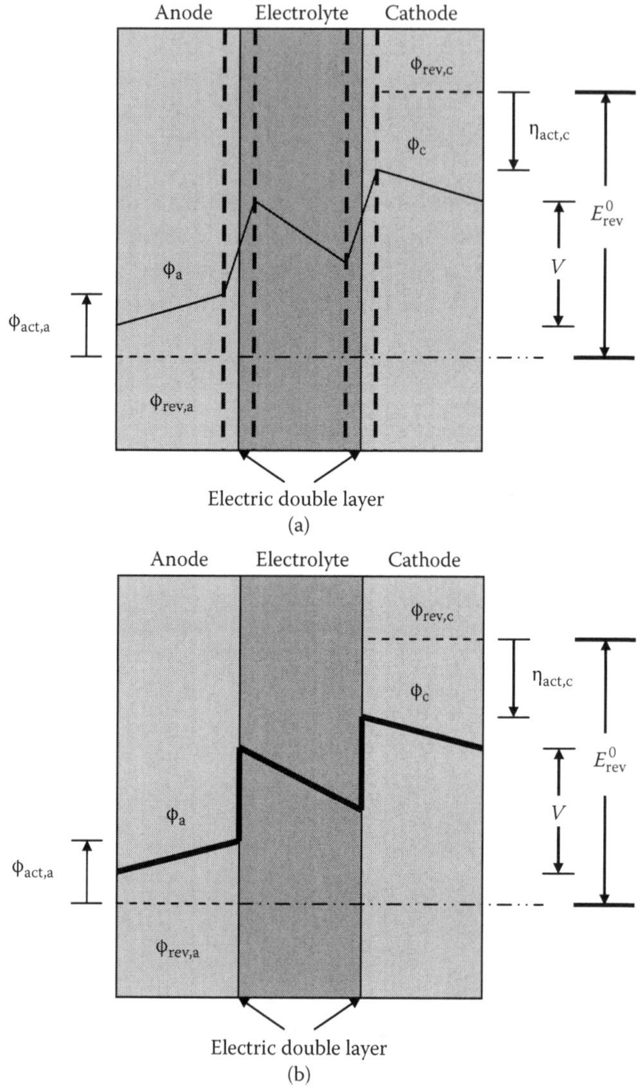

FIGURE 7.2
Distribution of electrical potential in the anode–electrolyte–cathode layers of a fuel cell. (a) With variation in electrical potential within electrical double layer. (b) With negligible variation in electrical potential within electrical double layer.

a sharp increase or jump in electric potential at the electrode–electrolyte interfaces without showing the presence of the electric double layers as shown in Figure 7.2b.

The total potential drop in the tri-layer cell represents the total voltage drop, which is the sum of anode and cathode activation overpotentials, $\eta_{act,a}$

and $\eta_{act,c}$, and the ohmic losses in the electrolyte and electrode layers. The cell operating voltage is then given as

$$V = E^{o}_{rev} - \eta_{act,a} - \eta_{act,c} - \eta_{ohm,a} - \eta_{ohm,c} - \eta_{ohm,e}, \tag{7.13}$$

where

E^{o}_{rev} = cell reversible voltage
$\eta_{act,a}$ = anode activation overpotential
$\eta_{act,c}$ = cathode activation overpotential
$\eta_{ohm,a}$ = anode ohmic over potential
$\eta_{ohm,c}$ = cathode ohmic overpotential
$\eta_{ohm,e}$ = electrolyte ohmic overpotential

Equation 7.13 is written in terms of net activation overpotential and net ohmic loss as

$$V = E^{o}_{rev} - \eta_{act} - \eta_{ohm}. \tag{7.14}$$

It can be noticed that even though there is a net increase in electrical potential or voltage from the anode side to the cathode side, there are potential drops in anode, cathode, and electrolyte layers, and hence there is charge transport in the direction of decrease in electrical potentials.

7.1.2.1 Charge Transport Equations

The charge transport equations in electrolyte, electrodes, and current collector or bipolar plates are derived based on charge balance.

The current conservation equation is given as

$$\nabla \cdot j = 0. \tag{7.15}$$

Considering current flow owing to potential gradient only and neglecting the diffusion and convection terms, the charge transport equations are given as follows:

7.1.2.1.1 Electrolyte

The charge transport equation in an electrolyte with solid or stationary immobilized liquid electrolyte can be derived on the basis of charge balance and assuming steady-state diffusion of charge particles based on ohm's law as

$$\nabla(-\sigma_e \nabla \phi_e) = 0 \text{ for the electrolyte layer.} \tag{7.16}$$

7.1.2.1.2 Electrodes

In the anode and cathode electrodes, electrons transfer from the electrode–electrolyte interfaces to the current collector plate. Considering electronic diffusion and current source in the active region, the charge transport equations in the electrode are given as

Anode Electrode

$$\nabla(-\sigma_a \nabla \phi_a) = j_a \text{ in the anode active layer} \tag{7.17a}$$

$$\nabla(-\sigma_a \nabla \phi_a) = 0 \text{ in the anode gas diffusion layer} \tag{7.17b}$$

Cathode Electrode

$$\nabla(-\sigma_c \nabla \phi_c) = j_c \text{ in the cathode active layer} \tag{7.18a}$$

$$\nabla(-\sigma_c \nabla \phi_c) = 0 \text{ in the cathode gas diffusion layer} \tag{7.18b}$$

where ϕ_a, ϕ_e and ϕ_c are the potential function in anode, electrolyte, and cathode layers respectively. σ_a and σ_c are the electronic conductivities in anode and cathode electrodes, and σ_e is the ionic conductivity in the electrolyte membrane.

In Equations 7.17a and 7.18a, charged transfer current densities i_a and i_c are given on the basis of Butler–Volmer charge transfer kinetics described in Chapter 5 as follows:

$$j = j_{o,i}\left(\frac{C_R}{C_{R,o}} e^{\alpha_i n_{e,i} F \eta_{act,i}/(RT)} - \frac{C_P}{C_{P,o}} e^{-(1-\alpha_i) n_{e,i} F \eta_{act,i}/(RT)}\right) \tag{7.19a}$$

where

$j_{o,i}$ = exchange current density that represents the current density at equilibrium

α_i = transfer coefficient that represents the symmetry of the activation barriers associated with forward and backward reactions

$\eta_{act,i}$ = activation voltage loss that modifies the forward and backward activation barrier

$n_{e,i}$ = number of electrons transferred in the electrochemical reaction

C_R, C_P = actual reaction surface concentrations of the reactant and product

$C_{R,o}$, $C_{P,o}$ = reference reaction surface concentrations of the reactant and product

For the negligible effect of reaction rates on the concentrations of the reactant and product at the electrode, that is, assuming $C_R \approx C_{R,o}$ and $C_P \approx C_{P,o}$, the equation reduces to

$$j = j_{o,i}\left(e^{\alpha_i n_{e,i} F \eta_{act,i}/(RT)} - e^{-(1-\alpha_i) n_{e,i} F \eta_{act}/(RT)}\right) \tag{7.19b}$$

For anode and cathode electrodes, Equation 7.19b can be written as

$$\text{Anode: } j_a = j_{o,a}\left(e^{\alpha_a n_a F \eta_{act,a}/(RT)} - e^{-(1-\alpha_a) n_a F \eta_{act,a}/(RT)}\right) \tag{7.20a}$$

$$\text{Cathode: } j_c = j_{o,c}\left(e^{\alpha_c n_c F \eta_{act,c}/(RT)} - e^{-(1-\alpha_c) n_c F \eta_{act,c}/(RT)}\right). \tag{7.20b}$$

7.1.2.2 Boundary Conditions for Electrical Potential

In the solution of electrical potential equations for the electrode gas diffusion layer, the active layer is considered as the boundary and the charge transfer current densities given by Butler–Volmer reaction kinetics are considered as the boundary conditions. At the inlet surfaces of the electrodes, a constant voltage boundary condition can be assigned. For example, at the left side of the anode surface zero voltage and at the right side of the cathode, a cell voltage can be assigned in the following manner:

$$\phi_a = 0 \text{ at the left surface of the anode} \tag{7.21a}$$

and

$$\phi_c = V_c \text{ at the right surface of the cathode} \tag{7.21b}$$

In the solution of electrical potential equations for the electrodes and electrolyte, continuity conditions for electrical potential and current can generally be applied at the interfaces. In the remaining electrolyte surfaces, zero potential gradient or insulation boundary conditions are applied.

In the solution of electrolyte potential equation with *electrode active layer* as the boundary, constant ionic current densities j_a and j_c at the anode and cathode active layer boundaries are specified as boundary conditions on the basis of an agglomerate model as

$$j_a = a_{al}(1 - \varepsilon_{al}) j_{agg,a} \tag{7.22a}$$

and

$$j_c = a_{cl}(1 - \varepsilon_{cl}) j_{agg,c}, \tag{7.22b}$$

where a_{al} and a_{cl} are the thicknesses of anode and cathode active layers and $j_{agg,a}$ and $j_{agg,c}$ represent the current densities in the anode and cathode active layers, which are given on the basis of the *agglomerate model* discussed below.

7.1.3 Agglomerate Model for the Active Catalyst Layer

The distribution of current density within the active layer depends on the distribution of reactant gases within the active catalyst layers. The catalyst layer is a complex porous composite structure that involves the complex interaction of couple transport and reaction of reactant gases and charge species transport. The agglomerate model for such complex composite structure was developed on the basis of the proposed agglomerate structure obtained through microscopic images (Ridge et al., 1989). The electrode active catalyst layer was assumed to be a porous structure composed of a number of agglomerates of either cylindrical or spherical electrolytes. The agglomerates are made of a number of electrically conducting particles coated with catalyst particles and embedded in the layers of electrolyte materials. For example, in a PEM fuel cell, the agglomerates of the electrode catalyst layer are composed of carbon-supported catalyst particles distributed in the proton conduction polymer layer as shown in Figure 7.3.

The agglomerate model presented here is based on diffusion and electrochemical reaction kinetics of hydrogen and oxygen species in the active

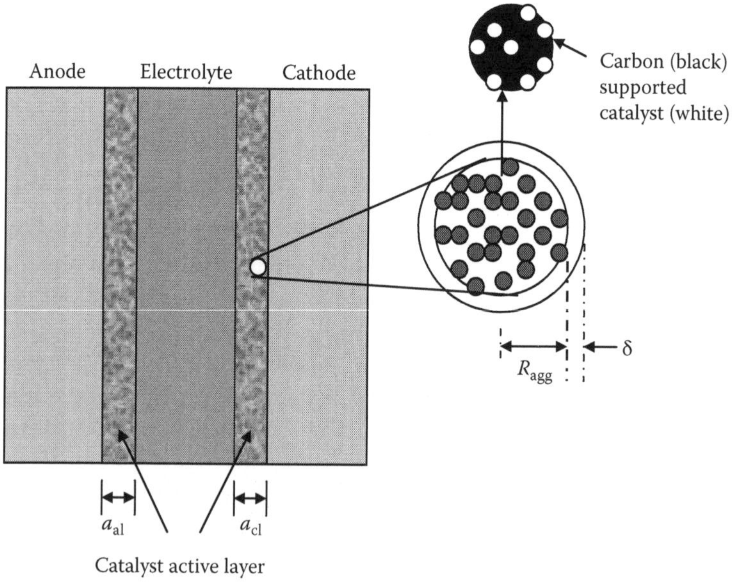

FIGURE 7.3
Active catalyst layers of electrodes.

layers composed of spherical agglomerates. Reactant gas diffuses through the intraparticle pores of the diffusion layer and through the pores between agglomerates filled with polymer electrolyte. Inside the agglomerate particles, the gas diffuses through the electrolyte to the surface carbon-supported catalyst particles and reacts electrochemically in the presence of the catalyst particles.

The gas transport equations and associated boundary conditions in the active layers are given as

Agglomerate layer

$$\frac{1}{r^2}\frac{\partial}{\partial r}\left(r^2 D_{\text{agg}}^{\text{eff}}\frac{\partial C_{\text{agg},i}}{\partial r}\right)=\frac{A_c i(1-\varepsilon_{\text{agg}})i}{n_e F}=0,\ 0<r<R_{\text{agg}} \tag{7.23}$$

Electrolyte film

$$\frac{1}{r^2}\frac{\partial}{\partial r}\left(r^2 D_{\text{f}}^{\text{eff}}\frac{\partial C_{\text{f},i}}{\partial r}\right)=0,\ R_{\text{agg}}<r<(R_{\text{agg}}+\delta) \tag{7.24}$$

Boundary conditions

At $r = 0$,

$$\frac{\partial C_{\text{agg},i}}{\partial r}=0 \tag{7.25a}$$

At $r = R_{\text{agg}}$

$$C_{\text{agg},i}(R) = C_{\text{f},i}(R_{\text{agg}}) \tag{7.25b}$$

and

$$D_{\text{agg}}^{\text{eff}}\frac{\partial C_{\text{agg},i}}{\partial r}=D_{\text{f}}^{\text{eff}}\frac{\partial C_{\text{f},i}}{\partial r} \tag{7.25c}$$

At $(R_{\text{agg}} + \delta)$,

$$C_{\text{agg},i}=\frac{C^*}{C_{i,s}^*}C_{i,s}=HC_{i,s}, \tag{7.25d}$$

where i is an index representing anode and cathode; R_{agg} is the radius of agglomerate particles; $D_{agg,i}$ is the gas diffusivity of agglomerate active layers; ε_{agg} is the porosity of the active layer; $C_{agg,i}$ and $C_{f,i}$ are the gas concentration within the agglomerate particles and in the electrolyte film, respectively; and A_c is the catalyst surface area per unit volume.

On the basis of the distribution of gas concentrations within the active layer, the current densities within the anode and cathode active layers are given. Number of agglomerate models and corresponding solutions for the current density distribution are available (Jaouen et al., 2002; Kamarajugadda and Mazumder, 2008).

On the basis of an analytical solution of gas concentration–agglomerate model given by Equations 7.23 through 7.25, Jaouen et al. (2002) presented the volumetric current density distribution within the active layers as

$$j_{agg,i} = -A_c(1-\varepsilon_{agg})j_0^{ref}\left(\frac{C^*}{C_i^{ref}}\right)^{1-\alpha_i/n}\exp\left(-\frac{\alpha_i F}{RT}\eta_i\right)\mathrm{Eff}_1\mathrm{Eff}_2(1-\varepsilon_{agg})$$
$$\times\frac{R_{agg}^3}{(R_{agg}+\delta)^3}\frac{C_{i,s}}{C_{i,s}^*} \tag{7.26a}$$

where

C^* = gas concentration in the polymer
C^*_{ref} = standard reference concentration of the reactant gas
$C_{i,s}$ = concentration at the interface of gas pore–electrolyte coating
$C^*_{i,s}$ = concentration of the bulk reactant at the inlet
A_c = catalyst surface area for reaction per unit volume of the active layer
ε_{agg} = porosity of the active layer

Factors Eff_1 and Eff_2 are correction factors for pure kinetic current density owing to diffusion limitation in the agglomerate and in the electrolyte layers. These factors are expressed as

$$\mathrm{Eff}_1 = \frac{3}{qr_{agg}}\left(\frac{1}{\tanh\left(qR_{agg}\right)}-\frac{1}{qR_{agg}}\right) \tag{7.26b}$$

and

$$\mathrm{Eff}_2 = \frac{1}{1+\frac{1}{3}\frac{R_{agg}^2\delta}{(R_{agg}+\delta)}q^2\varepsilon_{agg}^{1.5}\mathrm{Eff}_1}, \tag{7.26c}$$

where

$$q^2 = \frac{A_c(1-\varepsilon_{agg})\, i_0^{ref}\left(\dfrac{C^*}{C_i^{ref}}\right)\exp\left(-\dfrac{\alpha_i F}{RT}\eta\right)}{n_e F D_{i,j} C^*\, \varepsilon_{agg}^{1.5}}. \tag{7.26d}$$

The total catalyst surface area, A_c, is computed based on the platinum mass loading (mg/cm^3), active layer thickness, particle size, and platinum-to-carbon support mass (Pt|C) as

$$A_c = A_o \frac{m_{Pt}}{L_{al}}, \tag{7.27}$$

where
m_{Pt} = platinum catalyst loading (mg/cm^3)
L_{al} = thickness of the active layer
A_o = catalyst surface area per unit mass of catalyst particles

$$= \frac{6}{d_{Pt}\rho_{Pt}}$$

d_{Pt} = platinum particle diameter
ρ_{Pt} = platinum density

Platinum mass loading in terms of catalyst surface area per unit mass of catalyst particles is obtained from experimental evaluation of active layer structure. A typical empirical correlation for such a structure is given as (Marr and Li, 1999; Secanell et al., 2007)

$$A_o = 2.2779 \times 10^6(\text{Pt}|\text{C})^3 - 1.5857 \times 10^6(\text{Pt}|\text{C})^2 - 2.0153 \times (\text{Pt}|\text{C}) + 1.5950 \times 10^6. \tag{7.28}$$

7.2 Solid-State Diffusion

The mass transport of charge species owing to concentration gradient in a solid electrolyte is governed by solid-state diffusion. In a solid-state diffusion process, atoms and ions transport through lattice of crystalline structures like in ceramic and other solid nonmetals like polymers owing to the presence of a nonuniform concentration distribution of the migrating elements.

Figure 7.4 shows an initial nonuniform distribution of element i in a medium of j. Atoms of species i diffuse from the region of high concentration to the region of low concentration and establish a more uniform concentration distribution of the species. Self-diffusion also takes place in a relatively pure crystalline solid material controlled by a process known as *vacancy mechanism* or *the hopping process*. The ion transport in crystalline electrolyte is controlled by this vacancy diffusion or hopping diffusion mechanism. In this

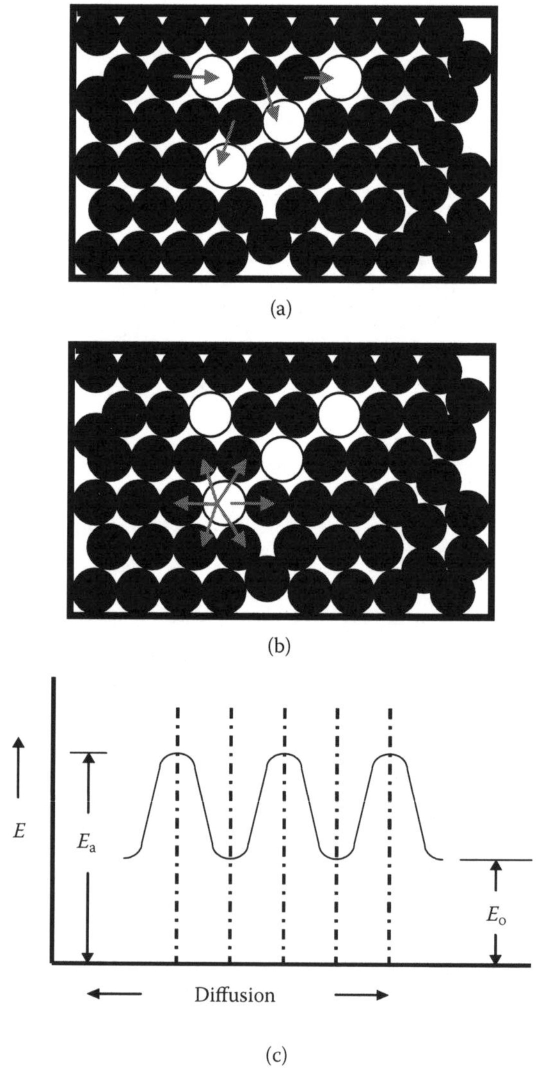

FIGURE 7.4
Solid-state self-diffusion by hopping or vacancy mechanism.

mechanism, an atom of the same species may exchange position with neighboring atom sites by moving into any adjacent vacant lattice site as shown in Figure 7.4a. Another way of looking at this exchange process is the movement of the vacancies into any of the adjacent atom sites as shown in Figure 7.4b. The vacancy positions can be thought of as a large number of natural point defects in a material. However, in many electronic materials as well in electrolyte materials, the vacancy sites or unoccupied charge sites are created by introducing or doping impurity atoms to enhance electronic or ionic conductivity of the materials through this vacancy diffusion process.

The *vacancy diffusion process* depends strongly on the number of vacant lattice sites in the material and possible jump directions. The probability of the charge element to move to a new vacant site depends on the amplitude and frequency of vibration associated with the element. A diffusing element, however, has to possess enough high energy to overcome any opposing barrier energy level to leave its equilibrium lattice position and jump to the adjacent vacant lattice site or next equilibrium position. The barrier energy level is a measure of the energy associated with the bonds and opposing resistive force of the surround atoms. Figure 7.4c shows the variation in energy (E) possessed by a vibrating and diffusing atom with respect to the energy associated with the equilibrium position (E_0) as it moves from one equilibrium position to a vacant position. For an atom to move to an adjacent vacant side, its associated energy level has to be greater than the energy barrier level, also known as the *activation energy* ($E_a = \Delta G_a$) for the hopping process.

The rate of vacancy diffusion or rate of hopping process is described by the *Arrhenius equation* given in terms of diffusion coefficient as

$$D = D_0 e^{-\frac{\Delta G_a}{RT}} \tag{7.29}$$

where D_0 is the reference diffusion constant, ΔG_a is the barrier energy or activation energy for diffusion or hopping process (kJ/kmol), R is the gas constant (kj/kmol·K), and T is the absolute temperature (K).

7.3 Charge Conductivity

The charge conductivity is influenced by the charge concentration and charge convective velocity and expressed as

$$\sigma_i = |z_i| F C_i u_i, \tag{7.30}$$

where C_i is the molar concentration of the charge carrier, u_i is the charge mobility velocity and $|z_i|$ is the absolute value of the charge number for a charge carrier. The charge number for an electron (e^-) charge is –1 and that for a proton or hydrogen ion (H^+) is +1.

Equation 7.30 shows that charge conductivity depends directly on the charge mobility velocity (u_i) and the charge concentration (C_i) in the base materials such as the electrode and interconnect material for electrons transport and the electrolyte materials for ion transport.

The charge mobility is given by

$$u_i = \frac{|z_i|F}{RT}D. \tag{7.31}$$

Combining Equations 7.30 and 7.31,

$$\sigma_i = \frac{C_i\left(|z_i|F\right)^2}{RT}D. \tag{7.32}$$

We will consider the ionic and electronic conductivities in the following section.

7.3.1 Ionic Conductivity (σ_i)

As mentioned earlier, the *ionic conductivity* is an electrical transport property of the electrolyte materials that represents the coupling coefficient or the constant of proportionality between charge flux and the electrical potential gradient given by Equation 7.7. Another alternate form of the charge molar flux owing to the electrical potential gradient can be derived on the basis of ionic transport velocity using convection mode as

$$J = (z_iF)C_iu_{i,t}. \tag{7.33}$$

In this equation, z_i represents the absolute value of the charge number of the ion. The ionic transport velocity is given as

$$u_{i,t} = -A_i(z_iF)\frac{\mathrm{d}\Phi}{\mathrm{d}x}, \tag{7.34}$$

where A_i is the ion mobility constant.

Substituting Equation 7.34 for the ion transport into Equation 7.31, we get

$$J = -A_i(z_iF)^2C_i\frac{\mathrm{d}\Phi}{\mathrm{d}x}. \tag{7.35}$$

Comparing Equation 7.6b for charge flux given by *Ohm's law* and Equation 7.35, we get the defining equation of ionic conductivity as

$$\sigma = A_i(z_iF)^2C_i. \quad (7.36)$$

Equation 7.36 shows that ionic conductivity is a direct function of ion concentration.

An equilibrium concentration distribution of ions in an electrolyte can be derived by setting the net charge flux owing to combined driving forces caused by electrical potential gradient and concentration gradient to zero and solving as follows:

$$J_i = -(z_iF)D_i\frac{dC_i}{dx} - A_i(z_iF)^2C_i\frac{d\Phi}{dx} = 0. \quad (7.37)$$

Rearranging and integrating,

$$\int_{C_{i,0}}^{C_i}\frac{dC_i}{C_i} = -\int_{\Phi=0}^{\Phi}\frac{A_i(z_iF)}{D_i}d\Phi \quad (7.38)$$

Solving, we get the ion concentration distribution as

$$C_i = C_{i,0}e^{-\left(\frac{A_iz_iF}{D_i}\right)\Phi}, \quad (7.39)$$

where $C_{i,0}$ is the ion concentration for electrical potential $\Phi = 0$.

The ion concentration distribution in electrolyte media under electrical field is also expressed by the Boltzmann distribution as

$$C_i = C_{i,0}e^{-\left(\frac{z_iF}{RT}\right)\Phi} \quad (7.40)$$

Comparing Equations 7.37 and 7.38, we get the expression for the *ion mobility* constant in terms of temperature and ion diffusion coefficient as

$$A_i = \frac{D_i}{RT}. \quad (7.41)$$

The ionic conductivity is then obtained by substituting Equation 7.41 for the mobility constant into Equation 7.36

$$\sigma = \frac{(z_i F)^2}{RT} C_i D_i \tag{7.42}$$

Equation 7.42 shows dependence of ionic conductivity on the ion diffusivity, ion concentration, and temperature.

7.3.1.1 Ionic Conductivity in Solid Electrolytes

The ion transport in crystalline electrolyte is controlled primarily by the vacancy diffusion or hopping diffusion mechanism discussed in Section 7.1. Vacancy spots are created in crystalline base electrolyte material by doping it with an impurity or alloy element. The ion concentration distribution in the electrolyte is controlled by the density of the doping element in the base electrolyte material.

Substituting Equation 7.29 for diffusion coefficient into Equation 7.42 for ionic conductivity, we get

$$\sigma = \frac{C_i}{RT}(z_i F)^2 D_0 e^{-\frac{\Delta G_a}{RT}}. \tag{7.43}$$

The charge conductivity can be expressed in a simplified form as

$$\sigma = \sigma_0 e^{-\frac{\Delta G_a}{RT}} \tag{7.44}$$

where σ_0 is the reference ion conductivity of the material given by

$$\sigma_0 = \frac{C_i}{RT}(z_i F)^2 D_0. \tag{7.45}$$

Equation 7.45 is similar to Equation 7.40 and gives a direct relation between the charge conductivity and the ion concentration. In a crystalline electrolyte, the ion concentrations and the number of vacancy sites are controlled by doping the base material with an impurity element.

7.3.1.2 Ionic Conductivity in Polymer Electrolyte Membrane

The purpose of the polymer electrolyte is to transport the positively charged proton from the anode to the cathode side. One of the most popular polymer membranes is the Nafion-117, which is made of material structure that has a large amount of hydrophilic regions through which proton can migrate. Since proton conductivity depends on water content, it is essential that the membrane is sufficiently and uniformly hydrated from the anode side to the cathode side to maintain effective transport of the proton.

The proton conductivity is then obtained from Equation 7.42 with $|z_i| = 1$ for hydrogen ion (H^+) as

$$\sigma_{H^+} = \frac{F^2}{RT} C_{H^+} D_{H^+}. \tag{7.46}$$

A more detailed description of the construction and material composition of polymer electrolyte membrane will be given in Chapter 9.

7.3.1.3 Ionic Conductivity in Ceramic Electrolyte Membrane

While a detailed description of ceramic solid oxide membrane will be given in Chapter 9, let us briefly describe here the ion conductivity of negatively charged oxygen ions. The state-of-the-art ceramic electrolyte material used in SOFC is the *yttria stabilized zirconia* (YSZ) because YSZ has a higher oxide ion conductivity than any other materials, the lowest electronic conductivity, and the lowest gas permeability to prevent gas crossover losses. The base material in the YSZ is the zirconia (ZrO_2), which is doped with yttria (Y_2O_3) as the dopant element to stabilize the structures of zirconia over the low- to high-temperature range during heating. The substitution of Zr^{4+} at a lattice position with the Y^{3+} ions creates vacancies in the oxygen sublattice and causes oxygen ion conduction in the stabilized zirconia. The electrolyte material is made with 8–10 mol% Y_2O_3-stabilized ZrO_2 with an operating temperature of 800°C–1000°C for good ionic conductivity, reaction kinetics, and lowest electronic conductivity. Typical composition contains 8% yttria (Y_2O_3) mixed with zirconia (ZrO_2). Yttria introduces high concentration of oxygen vacancies into zirconia crystal structure and results in higher ion conductivity.

Table 7.1 shows typical variation of ion conductivity of zirconia YSZ electrolyte with yttria concentration (Fergus et al., 2009).

TABLE 7.1
Variation Ionic Conductivity of YSZ with Volume Fraction of Yttria

Yttria Concentration (%)	Ionic Conductivity (σ_i) at 1000°C, S/cm	Activation Energy (ΔG_a) at 850°C–1000°C, eV (kJ/mol)
3%Y_2O_3	0.049	0.80 (80)
8%Y_2O_3	0.137	0.91 (91)
10%Y_2O_3	0.13	0.83 (83)
12%Y_2O_3	0.068	1.04 (104)

Source: Fergus, J. W. et al., *Solid Oxide Fuel Cells—Materials Properties and Performance*, CRC Press, 2009.

The YSZ conductivity increases with yttria dopant concentration up to 8% owing to increases in oxygen vacancies. With further increase in dopant concentration, the ionic conductivity decreases because of increased interactions between oxygen and yttrium ions.

The conductivity of the YSZ electrolyte is a strong function of temperature and it increases with temperature. The temperature dependence of the electrolyte conductivity is given by the following curve-ft correlations of experimental data function relations

$$\sigma_i = a \exp\left(-\frac{b}{T}\right), \tag{7.47}$$

where the empirical coefficients are given as $a = 3.34 \times 10^4$ $(\Omega\text{-m})^{-1}$ and $b = 1.03 \times 10^4$ K.

Another alternative correlation is given in terms of activation energy in similarity with Equation 7.44 as

$$\sigma_{\text{electrolyte}} = \sigma_0 e^{-\frac{G_a}{RT}}, \tag{7.48}$$

where ΔG_a is the activation energy and σ_0 is the reference conductivity determined empirically for the migrating element. Typical values of activation energy for SOFC based on the YSZ electrolyte is in the range of 80–105 kJ/mol.

A plot of these equations shows strong dependence of ionic conductivity of 8% YSZ with temperature in the range of 800°C–1000°C (Figure 7.5).

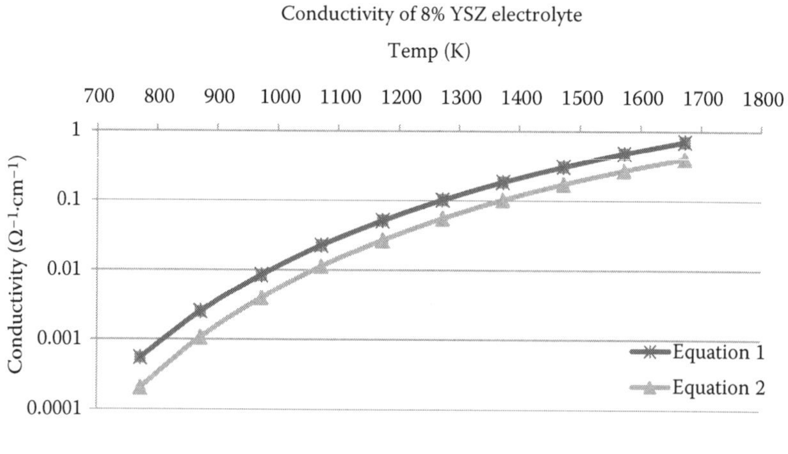

FIGURE 7.5
Variation of ionic conductivity of 8% YSZ with temperature.

Equation 1 is based on the YSZ conductivity given by Equation 7.47. Equation 2 is based on Equation 7.48 with $\sigma_0 = 9 \times 10^7$ K/(Ω.m) and $\Delta G_a = 100$ kJ/mol. The conductivity value decreases by a factor of 10 as temperature is reduced from 1000°C to 800°C.

7.3.1.4 Ionic Conductivity in Liquid Electrolyte

Ions transport in a liquid electrolyte under the influence of an electrical force induced by an electrical potential field and against an opposing drag force associated with the friction of the fluid over the ion charge particle.

The electrical force is given by

$$F_e = |z_i| e \frac{d\Phi}{dx}, \tag{7.49}$$

where e = unit of charge, $|z_i|$ = absolute value of charge number, and $\frac{d\Phi}{dx}$ = electrical potential gradient.

The friction drag force over a spherical charge particle moving through a liquid electrolyte is given based on Stoke's law as

$$F_d = 6\pi\mu d_i V, \tag{7.50}$$

where μ = liquid viscosity of liquid electrolyte, d_i = charge diameter, and V = ion velocity.

Equating Equations 7.49 and 7.50, and defining the ion mobility velocity as the ratio of ion velocity to the electrical potential gradient, we get the expression for the *mobility velocity of the ion* as

$$u_i = \frac{V}{\frac{d\Phi}{dx}} = \frac{|z_i| e}{6\pi\mu d_i}. \tag{7.51}$$

The ion conductivity is then given on the basis of this mobility velocity from Equation 7.30 as

$$\sigma_i = |z_i| F C_i u_i. \tag{7.30}$$

Ionic mobility velocities are given in terms of ionic equivalent conductance (λ_i^o) through the following relation:

$$u_i = \frac{\lambda_i^o}{|z_i| F^2}.$$

TABLE 7.2

Ion Equivalent Conductance, λ_i^o and Ionic Diffusion Coefficient, D_i, in Aqueous Electrolyte Solutions

Ion Type	λ_i^o (S.cm²)/equiv	D_i (cm²/s)
H^+	349.8	9.312 × 105
K^+	73.52	1.957×10^{-5}
Li^+	38.69	1.030×10^{-5}
Na^+	50.11	1.334×10^{-5}
OH^-	197.6	5.260×10^{-5}

Source: Newman, J. and Thomas-Alyea, K. E.: *Electrochemical Systems*, 3rd Edition. 2004. Copyright Wiley Interscience. Reproduced with permission.

The ionic diffusion coefficient is then calculated from the Nernst–Einstein equation

$$D_i = \frac{RT\lambda_i^o}{|z_i| F^2}.$$

Table 7.2 shows typical estimates of ionic conductance and ionic diffusion coefficient of selected ions with infinite dilution in aqueous solution.

7.3.2 Electronic Conductivity (σ_e)

Electronic conductivity is given in terms of material electrical property known as resistivity as

$$\sigma_e = \frac{1}{\rho_e}, \tag{7.52}$$

where resistivity is given as

$$\rho_e = R\frac{L}{A}. \tag{7.53}$$

The ohmic resistance owing to motion of electrons through electronic conductors such as electrodes and interconnects is given in terms of *resistivity* (ρ) as

$$R_{\text{ohm,e}} = \rho_e \frac{L}{A} \tag{7.54}$$

TABLE 7.3

Electrical Conductivity of Liquid Electrolytes

Electrolyte	Temperature (°C)	Concentration (mol/l)	Conductivity (S/cm)
KOH in water (AFC)	18	5	0.528
		10	0.393
NaOH in water	18	5	0.345
		10	0.205
			0.110
	50	5	0.670
		10	0.575
		15	0.440
	100	5	1.24
		10	1.41
		15	1.33
H_3PO_4 in water (PAFC)	18	6	0.625
		11	0.151
H_3PO_4 (PAFC)	200	Concentrated	0.6
L_2CO_3 or K_2CO_3	650		0.3

Source: McDougall, A., *Fuel Cells*, MacMillan, 1976; and Li, X., *Principles of Fuel Cells*, Taylor & Francis, 2006.

and in terms of electronic conductivity as

$$R_{\text{ohm,e}} = \frac{L}{\sigma_e A}, \tag{7.55}$$

where

$$\rho_e = \text{resistivity } (\Omega\text{-cm}).$$

Electrical conductivities of some typical electrolytes are given in Table 7.3.

7.4 Ohmic Loss in Fuel Cells

As we have discussed before, the charge transport through different components of the fuel cell contributes to voltage loss in the fuel cell following Ohm's law given by Equation 7.6. For example, ion transport through the electrolyte causes voltage loss between the electrodes owing to the ionic resistivity of the electrolyte materials. Similarly, electron transport causes voltage loss owing to the electronic resistivity of electrode and interconnect materials.

This voltage loss owing to charge transport is referred to as ohmic loss, η_{ohmic}, in a fuel cell. In order to determine the expression for the ohmic loss, let us consider charge transport in a plane conductor film as shown in Figure 7.6.

The charge transport across a conductor of cross-sectional area, A, and thickness, L, is given by Ohm's law and Equation 7.6b as

$$I = -\sigma_c A \frac{d\Phi}{dx}. \tag{7.56}$$

Integrating Equation 7.56 across a charge conductor of thickness L and cross-sectional area A, we get total voltage drop across the electrolyte or the ohmic loss owing to charge transport as

$$I = \sigma_c A \frac{\phi_0 - \phi_L}{L}. \tag{7.57}$$

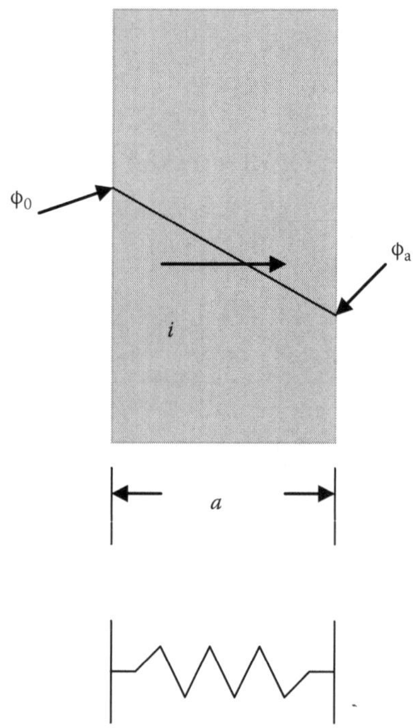

$$R_{ohm} = \frac{L}{\sigma A}$$

FIGURE 7.6
Charge transport and ohmic resistance in a charge conductor.

Rearranging,

$$I = \frac{\phi_0 - \phi_L}{\dfrac{L}{\sigma_c A}}. \tag{7.58}$$

On the basis of the electrical circuit analogy, we can define the ohmic resistance owing to charge transport through the charge conductor given as

$$R_{ohm,i} = \frac{L}{\sigma_c A}, \tag{7.59}$$

where
σ_c = charge conductivity $(\Omega\text{-cm})^{-1}$ or (S-cm^{-1}).

Another alternative expression of ohmic resistance is given in terms of *resistivity* (ρ) as

$$R_{ohm,i} = \rho_c \frac{L}{A}, \tag{7.60}$$

where
ρ_c = resistivity (Ω-cm).

As we can see, the ohmic resistance increases with higher thickness of the conductor and decreases with higher charge conductivity and higher cross-sectional area of the conductor.

From Equation 7.58, we can also express the ohmic voltage drop

$$\eta_{ohmic} = \phi_o - \phi_L = IR. \tag{7.61}$$

Equations 7.60 and 7.61 are the general representation of the ohmic resistance and ohmic voltage loss, respectively. The total ohmic voltage loss is the sum of all ohmic loss components owing to electronic resistances in interconnects and electrodes, and ionic conductivity in the electrolyte as shown in Figure 7.7.

The total potential drop or voltage drop in a fuel cell is given as the sum of all components of the ohmic losses in the electrolyte, electrode layers, and interconnect materials.

$$\eta_{ohm} = \eta_{ohm,i} + \eta_{ohm,a} + \eta_{ohm,c} + \eta_{ohm,int}, \tag{7.62}$$

where
$\eta_{ohm,i}$ = ohmic overpotential owing to the ionic conductivity of electrolyte
$\eta_{ohm,a}$ = ohmic overpotential owing to the electronic conductivity of anode

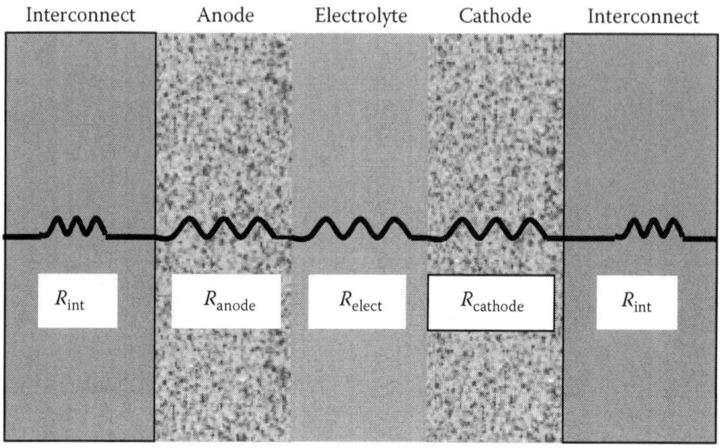

FIGURE 7.7
Ohmic resistance in a fuel cell.

$\eta_{ohm,c}$ = ohmic overpotential owing to the electronic conductivity of cathode
$\eta_{ohm,int}$ = ohmic overpotential owing to the electronic conductivity of interconnect

Additional ohmic resistance takes place because of contact resistances at the interfaces of electrodes, electrolytes, and interconnects.

Combining all components of electronic ohmic overpotential by a single component, Equation 7.62 is written in a simplified form as

$$\eta_{ohm} = \eta_{ohm,i} + \eta_{ohm,e}. \tag{7.63}$$

Following Equation 7.63, the ionic and electronic components of ohmic overpotentials are given as

$$\eta_{ohmi,i} = IR_{ohm,i} \tag{7.64}$$

and

$$\eta_{ohmi,e} = IR_{ohm,e}, \tag{7.65}$$

where the ohmic resistance owing to ion transport through the electrolyte layer is given as

$$R_{ohmi,i} = \frac{L_{electrolyte}}{A\sigma_i} \tag{7.66}$$

and the ohmic resistance owing to electron transport through the electrode and interconnect is given as

$$R_{\text{ohmi,e}} = \frac{L_e}{A\sigma_e}, \tag{7.67}$$

where σ_i is the ionic electrolyte material and σ_e is the electronic conductivities of electrode and interconnect materials. Note that unit of ohmic resistance is ohm (Ω) or siemens ($S = 1/\Omega$).

Generally, in a fuel cell, the *electrolyte ohmic overpotential* is the dominant component of the ohmic overpotential owing to the lower ionic conductivity value as compared to the electronic conductivity of electrodes and interconnect materials. Research effort to improve ohmic loss in a fuel cell is, therefore, focused on the improvement of the electrolyte in terms of higher ionic conductivity and lower thickness. Use of a thinner electrolyte is limited by a number of factors such as structural integrity, manufacturability and defects, increased parasitic loss owing to fuel crossover loss, and dielectric limit of the electrolyte.

7.5 Water Transport Rate Equation

Water produced by electrochemical reactions in a fuel cell needs to be removed for efficient operation of the cell. Water transport in electrode–gas diffusion layers, electrolyte, and gas flow channels plays a critical role in the design of a fuel cell. Figure 7.8 shows water generation at the electrode–electrolyte interface and mechanisms of water transport in a fuel cell.

In a *cation transfer fuel cell* such as PAFC and PEMFC with positively charged ion, water is produced by the electrochemical reaction at the electrolyte–cathode interface. On the other hand, in an *anion transfer fuel cell* such as AFC, MCFC, and SOFC with negatively charged ion, water is produced at the anode–electrolyte interface.

Water produced at the electrode–electrolyte interfaces transports through the electrode–gas diffusion layers by diffusion and convection toward the gas flow channels where it may be transferred to the reactant gas flow streams by convection and diffusion if the gas streams are sufficiently dry. It is essential that water is removed from the electrolyte–electrode interface either by the flowing gas streams or by some external water collection system in order to prevent any accumulation or flooding of the electrode–electrolyte interface regions that blocks the pores of the electrode–gas diffusion layer and prevents reactant gas to reach reaction sites causing cell *concentration polarization* or *mass transfer loss*. Water flooding issue and mass transfer loss are

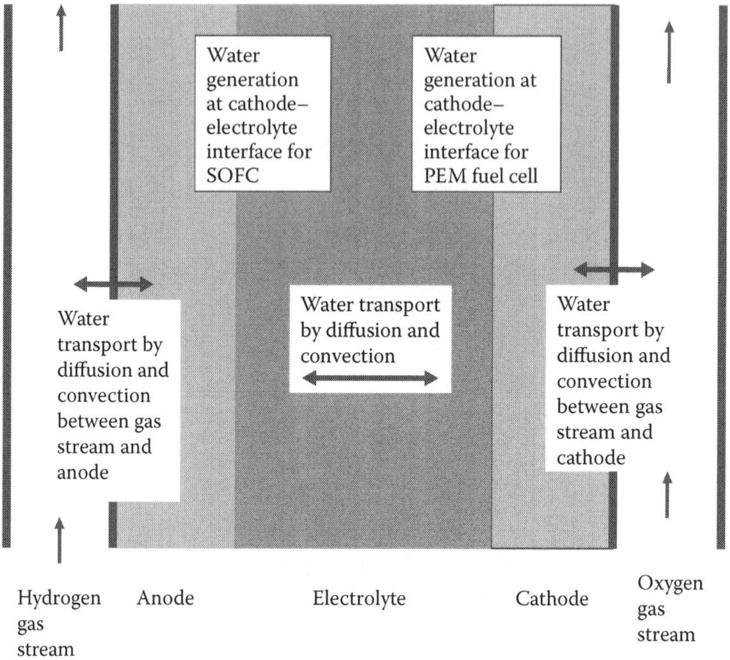

FIGURE 7.8
Water generation and transport in a fuel cell.

more critical in a *cation transfer fuel* cell because of significantly lower oxygen diffusivity in water at the cathode side as compared to hydrogen diffusivity in water at the anode side of the *anion transfer fuel cell.* A proper balance of water or water management is, therefore, needed for efficient operation of the fuel cell. Water management issue is more critical for a PEM fuel cell, which requires a proper level of hydration of the polymer membrane for higher proton conductivity.

7.5.1 Water Transport in Electrolyte Membranes

Like the transport of reactant gas species in anode and cathode electrodes, the major driving forces for the transport of water across the membrane between the electrodes are (i) diffusion owing to water concentration difference and (ii) convection flow driven by a pressure gradient force. In addition, in a polymer electrolyte membrane, there is an additional driving force known as *electro-osmotic drag force* with the transport of ions. For example, in a PEM fuel cell, the polymer membrane like *Nafion* must contain sufficient amount of water for the mobility of protons H^+ from the anode electrode to the cathode electrode. As the protons transport through the hydrated electrolyte membrane, they drag a number of water molecules along with them. The number of water molecules dragged by each proton is given by the *electro-osmotic drag coefficient*, $n_{\text{drag}} = n_{\text{H}_2\text{O}} / \text{H}^+$. A more detailed discussion of the composition and construction of a Nafion polymer electrolyte membrane will be given in Chapter 9.

Water transport flux owing to electro-osmotic drag in a cell with operating current density i is given as

$$j_{\text{H}_2\text{O,es}} = n_{\text{drag}} \frac{i}{n_{\text{e}} F}, \tag{7.68}$$

where

$\frac{i}{n_{\text{e}} F}$ = proton flux, which is two times the hydrogen flux.

The major components of water flux in the membrane are shown in Figure 7.9. Diffusion of water owing to concentration gradient is given as

$$j_{\text{H}_2\text{O,d}} = -D_{\text{H}_2\text{O}} \nabla c_{\text{H}_2\text{O}}. \tag{7.69}$$

Water flux owing to the pressure-driven convection flow is given as

$$j_{\text{H}_2\text{O,conv}} = c_{\text{H}_2\text{O}} \overrightarrow{v_{\text{H}_2\text{O}}}, \tag{7.70}$$

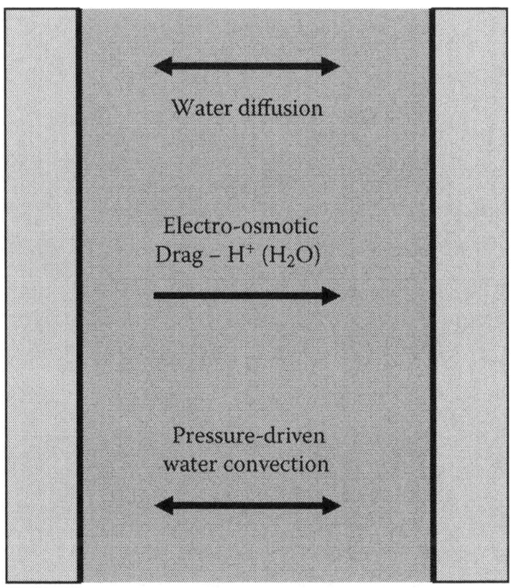

FIGURE 7.9
Water transport mechanisms in a polymer membrane.

where the velocity vector is given by Darcy's law as

$$\overrightarrow{V_{H_2O}} = -\frac{K_{H_2O}}{\mu_{H_2O}} \nabla P, \tag{7.71}$$

where
K_{H_2O} = hydraulic permeability coefficient of water
μ_{H_2O} = dynamic viscosity of water

Substituting Equation 7.71 into Equation 7.70, we have the expression for water flux owing to convection as

$$j_{H_2O} = -C_{H_2O} \frac{K_{H_2O}}{\mu_{H_2O}} \nabla P. \tag{7.72}$$

A combination of all these effects leads to the net water flux as

$$j_{H_2O} = 2n_{\text{drag}} \frac{i}{n_e F} - D_{H_2O} \nabla c_{H_2O} - c_{H_2O} \frac{K_{H_2O}}{\mu_{H_2O}} \nabla P, \tag{7.73}$$

and in one dimension along the thickness of the membrane as

$$j_{H_2O} = 2n_{\text{drag}} \frac{i}{n_e F} - D_{H_2O} \frac{dc_{H_2O}}{dz} \nabla c_{H_2O} - c_{H_2O} \frac{K_{H_2O}}{\mu_{H_2O}} \frac{dP}{dz}, \tag{7.74}$$

Equation 7.74 is referred to as the generalized form of the *Nernst–Planck's equation.*

For the case of a negligible diffusion term, Equation 7.74 reduces to the so-called *Schlogl equation* as

$$j_{H_2O} = n_{\text{drag}} \frac{i}{n_e F} - c_{H_2O} \frac{K_{H_2O}}{m_{H_2O}} \nabla P. \tag{7.75}$$

If pressure-driven convection is neglected, then the equation reduces to

$$j_{H_2O} = n_{\text{drag}} \frac{i}{n_e F} - D_{H_2O} \nabla c_{H_2O}. \tag{7.76}$$

For water transport in *Nafion* polymer membrane in a PEM fuel cell, the equation is given as

$$J_{H_2O} = 2n_{drag}^{sat}\frac{\lambda_m}{22}\frac{i}{n_e F} - \frac{\rho_{dry}D_{\lambda_m}(\lambda_m)}{M_m}\nabla\lambda_m - \lambda_m\frac{\rho_{air}}{M_m}\frac{K_{H_2O}}{\mu_{H_2O}}\nabla P, \quad (7.77)$$

where

ρ_{dry} = density of the dry membrane
$D_{\lambda_m}(\lambda_m)$ = diffusion coefficient of water in the polymer membrane
M_m = equivalent weight of the polymer membrane
λ_m = water content in the membrane given in terms of number of water molecule per unit sulfonate charge site $\left(H_2O/SO_3^-\right)$

Note that the relation between water concentration, C_{H_2O}, and water content, λ_m, is given by

$$C_{H_2O} = \lambda_m\frac{\rho_{dry}}{M_m}. \quad (7.78)$$

For a *Nafion polymer membrane*, the Nernst–Planck's equation without the pressure-driven convection term is written in terms of water content in the membrane as

$$J_{H_2O} = n_{drag}\frac{i}{n_e F} - \frac{\rho_{dry}D_{\lambda_m}(\lambda_m)}{M_m}(\nabla.\lambda_m). \quad (7.79)$$

As we can see, distribution water in the membrane is established by the combined effect of three competing effects. Since water is generated at the cathode–electrolyte interface by the electrochemical reaction, the water concentration tends to be higher at the cathode side and water migrates from the cathode side to the anode side by back diffusion. This is in opposite direction to the water transport caused by the electro-osmotic drag of protons from the anode side to the cathode side. However, electro-osmotic drag in a PEM fuel cell often dominates over the back diffusion and causes drying of the membrane at the anode side and results in the higher accumulation of water or the so-called flooding of the membrane at the cathode side. While drying of the membrane significantly reduces the proton transport in the membrane, flooding of the membrane at the cathode side prevents oxygen reactant gas from reaching the cathode–electrolyte interface for the electrochemical reaction, and this dramatically reduces the cell performance. In order to overcome this situation, PEM fuel cell is often maintained at higher pressure at the cathode side compared to the anode side by supplying air or oxygen gas stream at the cathode side at a higher pressure than the hydrogen gas stream. Additionally, hydrogen is supplied as a humidified gas stream with 5%–10% moisture to sufficiently keep the membrane hydrated. The level of cathode side pressurization and anode side humidification of hydrogen gas stream is selected based on analysis and design evaluation of the fuel cell.

7.5.2 Water Transport Equation

The water transport equation in gas flow channels and in electrode gas diffusion layer is similar to the gas species transport equations presented in Chapter 6, and it is given as follows:

Gas channels:

$$\nabla \cdot \left(\rho u C_{H_2O}\right) = \nabla \cdot \left(D_{ij} \nabla C_{H_2O}\right) \tag{7.80}$$

Electrode–gas diffusion layer:

$$\nabla(\rho u C_i) = \nabla \cdot (D_{eff} \nabla C_i) \tag{7.81}$$

Electrolyte membrane:

$$\nabla \cdot J_{H_2O} = 0, \tag{7.82}$$

where water flux, J_{H_2O}, is given by Equation 7.74.

Figure 7.10 shows a typical water concentration distribution in a PEM fuel cell with adjacent reactant gas supply channels. Results show a higher level of

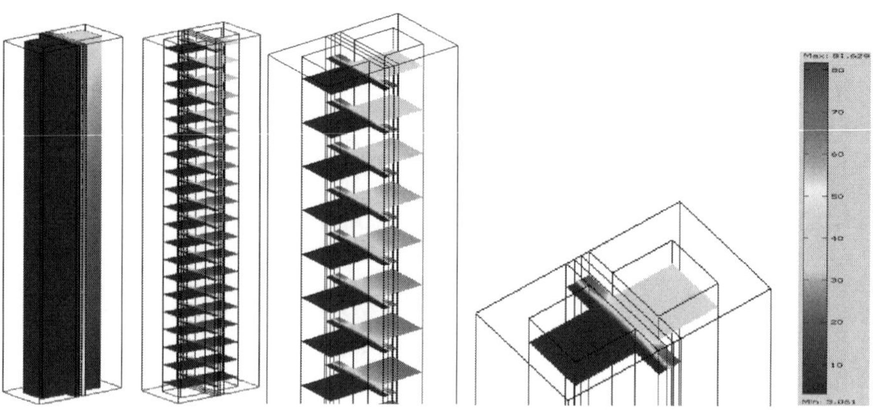

FIGURE 7.10
A typical water concentration distribution across a PEM fuel cell.

water content in the cathode side compared to the anode side for the selected operating current density. Gas flows in the cathode side carry away a higher amount of water as evidenced by the increased level of water concentration at the exit section. Water generation as well as level of water content increases with higher operating current densities.

It is evident that the pickup of water is higher on the cathode side channel compared to the anode side channel.

7.6 Summary

Nernst–Planck's Equation for charge transport:

$$J_i = -z_i \frac{F}{RT} D_i C_i \nabla \Phi - D_i \nabla C_i + C_i \vec{V}_i$$

Schlogl's equation of motion:

$$\vec{V} = \frac{K_\phi}{\mu} z_f C_f \nabla \phi - \frac{K_P}{\mu} \nabla P$$

Ionic conductivity:

$$\sigma_i = \frac{C_i\left(|z_i| F\right)^2}{RT} D$$

Ohmic loss:

$$\eta_{ohm} = \eta_{ohm,i} + \eta_{ohm,e}$$

$$\eta_{ohmi,i} = IR_{ohm,i}$$

and

$$\eta_{ohmi,e} = IR_{ohm,e}$$

Ohmic resistances:

$$R_{\text{ohmi,i}} = \frac{L_{\text{electrolyte}}}{A\sigma_{\text{i}}}$$

and

$$R_{\text{ohmi,e}} = \frac{L_{\text{e}}}{A\sigma_{\text{e}}}$$

Nernst–Planck's equation for water transport:

$$j_{\text{H}_2\text{O}} = 2n_{\text{drag}}\frac{i}{n_{\text{e}}F} - D_{\text{H}_2\text{O}}\frac{\text{d}c_{\text{H}_2\text{O}}}{\text{d}z}\nabla c_{\text{H}_2\text{O}} - c_{\text{H}_2\text{O}}\frac{K_{\text{H}_2\text{O}}}{\mu_{\text{H}_2\text{O}}}\frac{\text{d}P}{\text{d}z}$$

Schlogl equation for water transport:

$$j_{\text{H}_2\text{O}} = n_{\text{drag}}\frac{i}{n_{\text{e}}F} - -c_{\text{H}_2\text{O}}\frac{K_{\text{H}_2\text{O}}}{\mu_{\text{H}_2\text{O}}}\nabla p.$$

References

Bernardi, D. M. and M. W. Verbrugge. Mathematical model of a gas diffusion electrode bonded to polymer electrolyte. *AIChE Journal* 37: 1151–1163, 1991.

Fergus, J. W., R. Hui, X. Li, W. P. Wilkinson and J. Zhang. *Solid Oxide Fuel Cells—Materials Properties and Performance*. CRC Press, 2009.

Jaouen, F., G. Linbergh and G. Sundholm. Investigation of mass transport limitations in the solid polymer fuel cell cathode, mathematical model. *Journal of the Electrochemical Society* 149(4): A437–A447, 2002.

Kamarajugadda, S. and S. Mazumder. Computational modeling of the cathode catalyst layer of a PEMFC. Proceedings of 2008 ASME Summer Heat Transfer Conference, HT2008-56020, Jacksonville, FL, USA, 2008.

Li, X. *Principles of Fuel Cells*. Taylor & Francis, New York, 2006.

Marr, C. and X. Li. Composition and performance modeling of catalyst layer in a proton exchange membrane fuel cell. *Journal of Power Sources* 77: 17–27, 1999.

McDougall, A. *Fuel Cells*. MacMillan, 1976.

Newman, J. and K. E. Thomas-Alyea. *Electrochemical Systems*, 3rd Edition. Wiley Interscience, 2004.

Ridge, S. J., R. E. White, Y. Tsou, R. N. Beaver and G. A. Eisman. Oxygen reduction in a proton exchange membrane test cell. *Journal of Electrochemical Society* 136(7): 1902–1909, 1989.

Secanell, M., K. Karan, A. Suleman and N. Djilali. Multi-varible optimization of PEMFC cathodes using an agglomerate model. *Electrochimica Acta* 52: 6318–6337, 2007.

Verbrugge, M. W. and R. Hill. Transport phenomena in perfluoro sulfonic acid membranes during the passage of current. *Journal of the Electrochemical Society* 137(4): 1131–1138, 1990.

Further Reading

Badwal, S. P. S. Zirconia-based solid electrolytes: Microstructure, stability and ionic conductivity. *Solid State Ionics* 52: 23–32, 1992.

Bard, A. J. and L. R. Faulkner. *Electrochemical Methods—Fundamentals and Applications*. Wiley & Sons, 1980.

Bird, R., W. Stewart and E. Lightfoot. *Transport Phenomena*. Wiley Publishers, New York, 2002.

Bockris, J. O'M. and S. Srinivasan. *Fuel Cells: Their Electrochemistry*. McGraw-Hill, 1069.

Breiter, M. W. *Electrochemical Processes in Fuel Cell*. Springer-Verlag, Heidelberg, 1969.

Broka, K. Characterization of the proton exchange membrane fuel cell. Techn. Lic Thesis, Royal Institute of Technology, Stockholm, 1995.

Cussler, E. L. *Diffusion Mass Transfer in Fluid Systems*. Cambridge University Press, Cambridge, UK, 1984.

De La Rue, R. E. and C. W. Tobias. On the conductivity of dispersions. *Journal of the Electrochemical Society* 106: 827–836, 1959.

Fergus, J. W. Electrolytes for solid oxide fuel cells. *Journal of Power Sources* 162: 30–40, 2006.

Fuller, E. N., P. D. Schettler and J. C. Gliddings. New method for prediction of binary gas-phase diffusion co-efficients. *Industrial and Engineering Chemistry* 58(5): 18–27, 1966.

Futerko, P. and I. M. Hsing. Two-dimensional finite element method study of the resistance of membrane in polymer electrolyte fuel cells. *Electrochimica Acta* 45: 1741–1751, 2000.

Hamann, C. H., A. Hamnett and W. Vielstich. *Electrochemistry*. Wiley-VCH, New York, 1998.

Hartvigsen, J., S. Elangovan and A. Kandkar. *Science and Technology of Zirconia V*, Editors S. P. S. Badwai, M. J. Bannister and R. H. J. Hannink. Technomic Publishing Company Inc., PA, 1993, p. 682.

He, W., J. S. Yi and T. V. Nguyen. Two phase flow model of the cathode of PEM fuel cells using interdigitated flow fields. *AIChE Journal* 46: 2053–2063, 2000.

Herbin, R., J. M. Fiard and J. R. Ferguson. *First European Solid Oxide Fuel Cell Forum Proceedings*, Editor U. Bossel. Lucern, Switzerland, 2004, p. 317.

Hoogers, G., Editor. *Fuel Cell Technology Handbook*. CRC Press, Boca Raton, FL, 2003.

Kordesch, H. A. and G. Samader. *Fuel Cells and Their Applications*. VCH, New York, 1996.

Liebhafsky, H. A. and E. J. Cairns. *Fuel Cells and Fuel Batteries*. Wiley, New York, 1968.

Mench, M. W. *Fuel Cell Engines*. John Wiley & Sons, Inc., New Jersey, 2008.

O'Hayre, R., S.-W. Cha, W. Cotella and F. B. Prinz. *Fuel Cell Fundamentals*. John Wiley & Sons, Inc., New York, 2006.

Springer, T. E., T. A. Zawodzinski and S. Gottesfeld. Polymer electrolyte fuel cell model. *Journal of Electrochemical Society* 138(8): 2334–2342, 1991.

Thornton, P. A. and C. J. Colangelo. *Fundamentals of Engineering Materials*. Prentice Hall, New Jersey, 1985.

Wilke, C. R. *Chemical Engineering Progress* 45: 218–223, 1949.

8

Fuel Cell Characterization

The characterization of a fuel cell or its component refers to the process of testing and assessment of overall fuel cell performance or individual component performance. As seen in Chapter 5, fuel cell performance depends on various losses, namely, activation losses, ohmic losses, concentration losses, and fuel crossover and short-circuit losses. Each of these losses depends not only on the operating conditions but also on the materials, design, and construction of fuel cells and their components. Even though two fuel cells may have the same design, the uncertainty in manufacturing may lead to slightly different fuel cell polarization curves. The fuel cell has several components that are physically coupled, and their design and construction can vary from cell to cell. Hence, each fuel cell needs to be characterized for its performance and to find the best or optimum operating conditions for practical applications.

The characterization of a fuel cell involves finding the best operating conditions and identifying and quantifying various losses under different operating conditions such as normal or design current density, high and low current density, different reactant pressure, temperature, and reactant concentrations, and in transients such as startup and shut-down operations. By the process of characterization, one can determine the best or optimal operation condition for a given fuel cell at the required current density.

A number of diagnostic techniques are available to characterize a fuel cell and its components. These techniques include electrochemical, electrical, optical, and structural methods. In this chapter, we will focus on the characterization techniques for a fuel cell and its components. The description of the techniques used in characterizing a parameter or component and the characteristics are discussed in this chapter.

8.1 Characterization of Fuel Cells and Fuel Cell Components

The overall performance of the fuel cell is characterized by the polarization curve. From the polarization curve, one can discriminate the key losses in the fuel cell, namely, kinetic, ohmic, concentration, and other parasitic losses. The polarization curve and with it the power density curve provide information on the highest voltage delivered by the fuel cell at the required current

TABLE 8.1

Parameters or Phenomena to Be Characterized to Evaluate Fuel Cell Performance

Component	Phenomena	Characteristic Parameters
Fuel cell	Losses	E_{eq}, j–E curve, j–P curve
Fuel cell	Kinetics	j_0, η_{act}, α, A_{active}
Fuel cell	Ohmic resistance	η_{ohmic}, R_i, i = electrolyte, electrode, contact, interconnect
Fuel cell	Mass transport	η_{conc}, j_L, ΔP, D, concentration distribution
Fuel cell	Reactant crossover, short-circuit	η_{other}
Fuel cell	Heat generation, losses	j–E curve, q''_{loss}
Electrocatalyst	Activity, conductivity, structure	L, ε, loading, particle size, $A_{electro\text{-}active}$, σ_{ionic}, $\sigma_{electronic}$
Electrode	Conductivity, structure	$\varepsilon_{electrode}$, $\sigma_{electronic}$
Flow field plates	Structure	ΔP, concentration distribution
Gas diffusion layer	Structure	ΔP, L, ε, $\sigma_{electronic}$
Electrolyte layer	Ion transfer	$j_{short\text{-}circuit}$, $m_{crossover}$, σ_{ionic}, $\sigma_{electronic}$

TABLE 8.2

Diagnostic Methods or Techniques for Characterization of Fuel Cells or Components and Parameters Measured

Electrochemical Methods	Optical/Radiation Methods	Physical Methods
Potentiostatic Galvanostatic -j–E measurement (polarization curve) Power density curve	Microscopy-optical (OM) Transmission electron (TEM) Scanning electron (SEM) Atomic force (AFM) -Microstructure (pore size and distribution, particle or grain size, crystal structure) of catalysts	Brunauer–Emmett–Teller (BET) method -Surface area measurement
Current interrupt measurement -Ohmic resistance	X-ray diffraction (XRD) -Chemical identification -Crystal structure	Volume infiltration method, mercury porosimetry -Porosity
Electrochemical impedance spectroscopy -Kinetic parameters	Nuclear magnetic resonance (NMR) -Chemical identification	Flow and pressure drop measurement -Gas permeability
Cyclic voltammetry -Electrode kinetics, electrode and catalysts surface area	Neutron radiography -Microstructure	
	Spectroscopy-Auger Electron (AES), x-ray photoelectron (XPS), secondary ion mass spectroscopy (SIMS) -Chemical identification	

density. The components of the fuel cell such as electrolyte membrane or matrix, flow field plates, gas diffusion layers, electrocatalyst, and so on, have unique electrochemical and structural characteristics that determine the overall fuel cell performance.

The characteristics of each component or components in essence determine the losses in the fuel cell. For example, a fuel cell with abnormally high ohmic resistance may have an electrolyte with high resistance, or electrical contacts may have high resistance. The cell and component characterization helps determine these losses and provide guidance to focus on components for improvement.

The characterization processes of cell and component should provide key parameters that will enable to identify and quantify the various losses and performance of each component. In Table 8.1, the parameters or the components to be characterized to fully understand the fuel cell performance are listed. In Table 8.2, various diagnostic techniques employed in characterizing the fuel cell and its components are listed.

8.2 Electrochemical Characterization Techniques

As listed in Table 8.2, there are a variety of techniques to study electrochemical system behavior. These include in situ and ex situ measurements, steady and transient techniques, and AC and DC methods. Each of (or a combination of) these methods is employed to characterize the electrochemical behavior of fuel cells and their components. In the following sections, key electrochemical techniques are discussed.

8.2.1 Current–Voltage Measurement

The overall fuel cell performance and power density are primarily determined by the current–voltage response of the fuel cell. The j–E curve helps identify the better-performing fuel cell. The measurement of current and voltage is the basic one and the effects of other operational conditions such as pressure, temperature, and gas flow rate on the fuel cell performance are determined through the j–E curve. It should be noted that the testing procedures affect the j–E curve. For example, if the fuel cell was shut down for some period and restarted, the j–E curve may differ from the curve obtained from a cell that was shut down and restarted immediately. Often, there are hysteresis effects if the directions of j–E sweeps are reversed. Thus, when two fuel cells' performances are to be compared, it is important that the j–E curves are obtained under identical operating conditions and identical testing procedures are used.

The open circuit voltage indicates the maximum possible voltage of the fuel cell for no current flow. For a given chemical reaction, the Gibbs free energy of formation gives the theoretical equilibrium cell voltage at a given temperature. Typically, the open-circuit voltage of most low-temperature fuel cells operating with air and hydrogen is in the range of 0.95–1.0 V. Any lower value of voltage is an indication of a voltage loss, either a crossover or electronic short circuit through the membrane. However, there may be other problems such as poisoning of the catalyst or the electrolyte, or in the case of PEM fuel cells, it can indicate total dehydration of the membrane.

The cell voltage as a function of current density can be obtained by controlling the voltage and measuring the current or controlling the current and measuring the voltage, referred to as the potentiostatic method or galvanostatic method, respectively. Under steady-state conditions, either potentiostatic or galvanostatic measurement methods can be used to obtain the *j–E* curve since they give the same *j–E* curve. For unsteady-state conditions, potentiostatic or galvanostatic measurements may give a different *j–E* curve. This is because the response of the fuel cell varies in each measurement method and the system may not have enough time to relax to its steady state during a short period.

The steady-state measurement of the *j–E* curve requires the cell to be at its steady-state condition. Typically, the data are obtained by incrementally increasing the current density starting from a low value, for example, 5 mA/cm^2, to the maximum desired value, for example, 1 A/cm^2; 5 to 7 data points are collected per decade of current density. It is important to stay at each current value for several minutes (except in the case of quick screening tests), to allow the voltage to reach a stable value. Small incremental increases in current and adequate dwell time at each point ensure good water equilibration within the cell and provide stable performance data. Small cells (<1 kW) may take few to several minutes to reach steady state, while for a large fuel cell (>5 kW), it may require more than 30 minutes to reach steady state after an abrupt change in current or voltage. Thus, scan rate for a small fuel cell and a large fuel cell will be quite different. For a small fuel cell, one can find the required scan rate by taking the measurements at different scan speeds and find the scan rate at which the *j–E* curve does not change for further decrease in scan speed.

The *j–E* curve can be used not only to quantitatively describe the overall fuel cell performance but also to identify and quantify the activation loss, ohmic loss, and the mass transfer limited current density. At low current density, the ohmic loss is negligible and hence the activation loss can be directly obtained from the *j–E* curve at low current density. The semi-log plot of the *j–E* curve is linear for low current density and it can be fit to a Tafel equation (Equation 5.83) as shown in Figure 8.1 at low current density. Using the line fit to the Tafel equation,

$$\eta = \frac{RT}{\alpha F}\ln j_0 - \frac{RT}{\alpha F}\ln j, \tag{8.1}$$

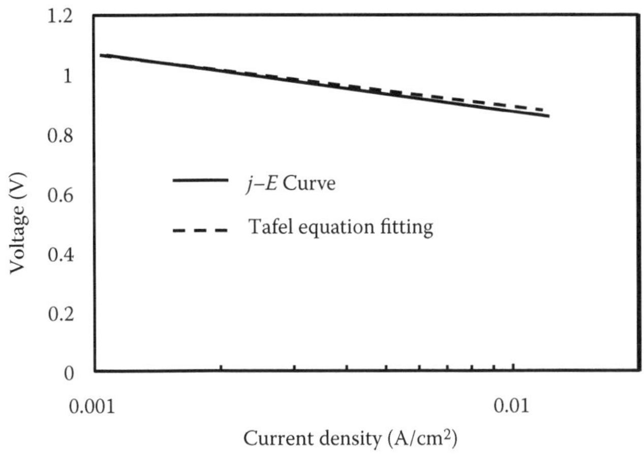

FIGURE 8.1
The linear nature of the j–E curve at low current density on a semi-log plot and the Tafel equation fitting.

the charge transfer coefficient α and the exchange current density j_0 can be determined. Using the activation loss equation (Equation 8.1), the activation loss at current densities can be approximated and calculated for the entire j–E curve. The complete plot of the j–E curve shown in Figure 8.2 then enables the quantification of the activation and ohmic losses.

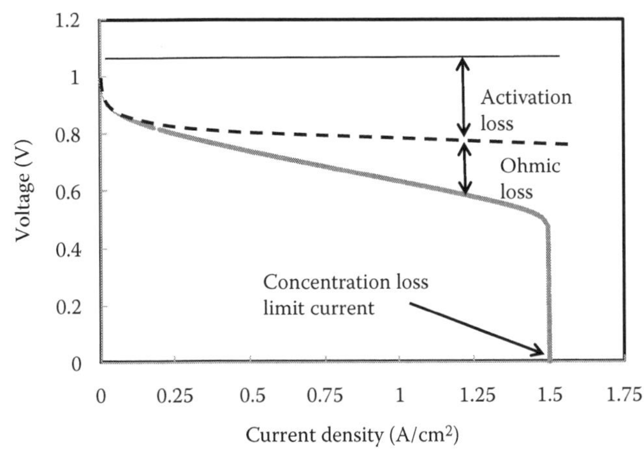

FIGURE 8.2
Fuel cell j–E curve, with activation loss line identifying ohmic loss and activation loss.

8.2.2 Electrochemical Impedance Spectroscopy

It is well known that electrical resistance is the ability of a circuit element to resist the flow of electrical current. For a flow of DC current i to an applied DC potential E, the resistance is given by Ohm's law as

$$R = \frac{E}{i}. \tag{8.2}$$

However, in the real world, circuit elements exhibit much more complex behavior; the simple concept of resistance cannot be used and in its place, *impedance*, a more general circuit parameter, is used. Like resistance, impedance is the ability of the system to impede the flow of electrical current through it. Though it is similar to resistance, impedance is not time independent; it is a time- or frequency-dependent parameter. Similar to resistance, impedance is defined as the ratio of the time-dependent current to the time-dependent potential,

$$Z = \frac{E(t)}{i(t)}. \tag{8.3}$$

Electrochemical impedance is usually measured by applying an AC potential to an electrochemical cell and then measuring the current through the cell (Barsoukov and Macdonald, 2005, Ivers-Tiffée et al., 2003, Orazem and Tribollet, 2008, Springer et al., 1996). The response to this potential is an AC current signal. According to *ASTM G-15*, the definition of *electrochemical impedance* is the frequency-dependent, complex valued proportionality factor, $\Delta E/\Delta i$, between the applied potential (or current) and the response current (or) potential in an electrochemical cell. This factor becomes the impedance when the perturbation and response are related linearly (the factor value is independent of the perturbation magnitude) and the response is caused only by the perturbation.

The magnitude of the excitation signal is small. This is done so that the cell's response is pseudo-linear. In normal electrochemical impedance spectroscopy (EIS) practice, a small (1 to 10 mV) AC signal is applied to the cell. The signal is small enough such that a pseudo-linear segment of the cell's current versus voltage curve is used as shown in Figure 8.3. The cell's nonlinear response to the DC potential is not measured because in EIS, the cell current at the excitation frequency is measured. If the system is nonlinear, the current response will contain harmonics of the excitation frequency.

Measuring an EIS spectrum often takes many hours. The system being measured must be at a steady state throughout the time required to measure the EIS spectrum. In a linear (or pseudo-linear) system, the current response to a sinusoidal potential will be a sinusoid at the same frequency but shifted in phase as shown in Figure 8.4.

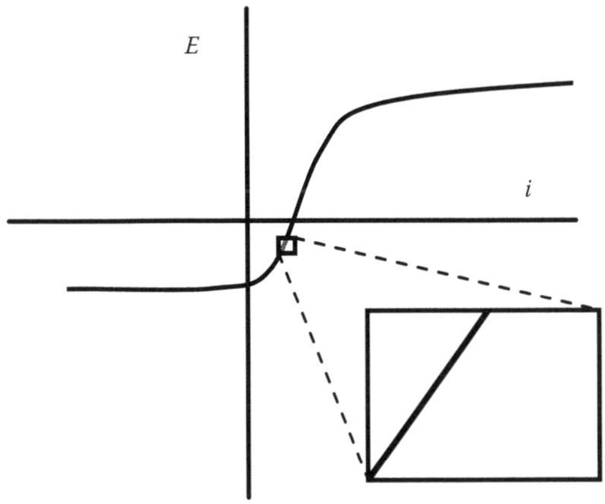

FIGURE 8.3
E–i curve showing pseudo-linearity.

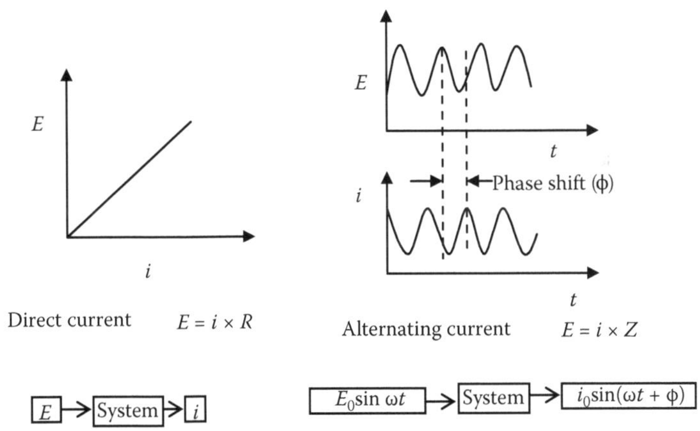

FIGURE 8.4
A DC voltage results in current i showing system resistance R, whereas a sinusoidal potential results in a sinusoidal current showing system impedance.

For example, an application of sinusoidal potential can be given as

$$E(t) = E_0 \sin(\omega t), \tag{8.4}$$

where E_0 is the voltage amplitude and ω is the radial frequency (radians per second). The radial frequency is expressed in terms of frequency f (hertz) as

$$\omega = 2\pi f. \tag{8.5}$$

For a linear system, the current response is shifted in phase (ϕ) and has a different amplitude, i_0, given as

$$i(t) = i_0 \sin(\omega t + \phi). \tag{8.6}$$

From Equation 8.3, an expression analogous to resistance, the impedance of the system is given as

$$Z = \frac{E_0 \sin \omega t}{i_0 \sin(\omega t + \phi)} = Z_0 \frac{\sin \omega t}{\sin(\omega t + \phi)} \tag{8.7}$$

Thus, impedance is expressed in terms of a magnitude, Z_0, and a phase shift ϕ.

$$e^{j\varphi} = \cos \phi + j \sin \phi, \tag{8.8}$$

Using Euler's relationship, the impedance is expressed as a complex function. The potential and current responses are described with imaginary and real components as

$$E = E_0 e^{j\omega t} \tag{8.9}$$

$$i = i_0 e^{j(\omega t - \phi)}. \tag{8.10}$$

The impedance is then represented as a complex number

$$Z = Z_0 e^{j\omega t} = Z_0(\cos \phi + j \sin \phi). \tag{8.11}$$

A sinusoidal current or voltage can be drawn as a rotating vector with a rotation speed equal to ω radians per second as shown in Figure 8.5. The in-phase or real component shown defines the observed voltage or current. It becomes the real component of the rotating vector. The out-of-phase or imaginary component shown defines the non-observed voltage or current. Assuming the voltage is forcing the current, if current is in-phase with voltage, the vectors are coincident and rotate together. When voltage and current are out of phase, they rotate at the same frequency ω but are separated by a constant angle shift ϕ. In EIS measurements, one vector is viewed using the other as a frame of reference. Thus, the reference point rotates and the time dependence of the signals (ωt) is not viewed. Both the current and voltage vectors are referred to the same reference frame.

If the real part of the impedance Z is plotted on the x-axis and the imaginary part is plotted on the y-axis of a chart, a "Nyquist plot" is obtained. For an electric circuit containing resistance and a capacitor in parallel, the impedance Z in Nyquist plot is shown in Figure 8.6. Note that in this plot, the

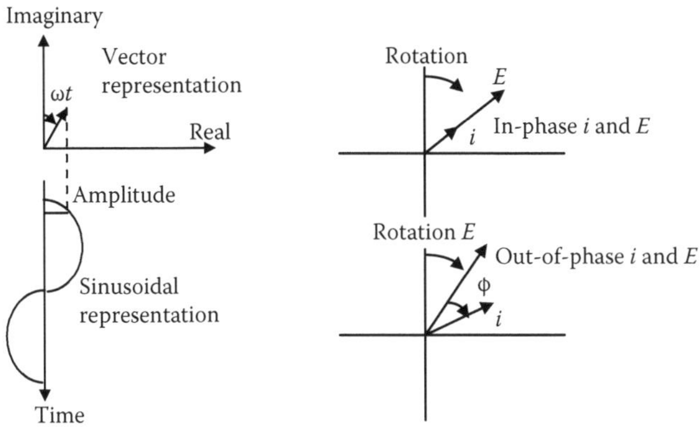

FIGURE 8.5
Vector and sinusoidal representation of current and voltage.

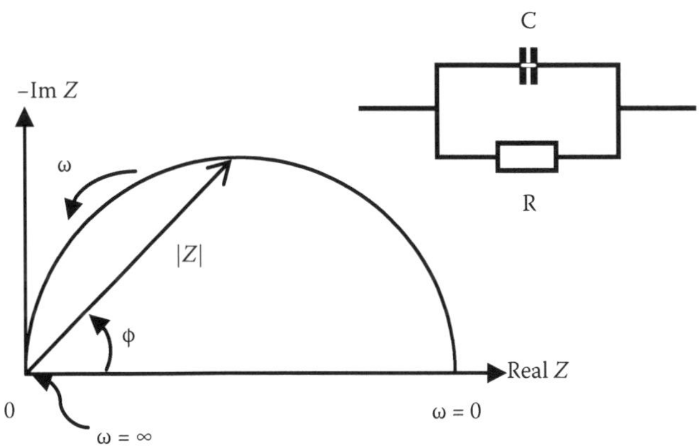

FIGURE 8.6
Nyquist plot of the impedance vector of parallel RC circuit.

y-axis is negative and that each point on the Nyquist plot is the impedance at one frequency. The low-frequency data are on the right side of the plot and higher frequencies are on the left.

8.2.2.1 Equivalent Circuit Models

EIS data are commonly analyzed by fitting them to an equivalent electrical circuit model corresponding to a fuel cell component or components. Most of the circuit elements in the model are common electrical elements such as resistors, capacitors, and inductors. As an example, the electrolyte ohmic resistance can be represented with a resistor. Very few electrochemical cells can be modeled using a single equivalent circuit element. Instead, EIS models usually consist of a number of elements in a network. Both serial and parallel combinations of elements occur. Consider the example of an *electrical double layer*. It exists at the interface between an electrode and its surrounding electrolyte. This double layer is formed as ions from the solution "stick on" the electrode surface. Charges in the electrode are separated from the charges of these ions. The separation is very small, on the order of angstroms. Charges separated by an insulator form a capacitor. And there are various resistances or impedances that are coupled to the electrode–electrolyte interface. These include polarization resistance, charge transfer resistance, and diffusion impedance or *Warburg impedance*.

8.2.2.2 Constant Phase Element

Capacitors in EIS experiments often do not behave ideally. Instead, they act like a constant phase element (CPE) as defined below for the impedance of a capacitor:

$$Z = A(j\omega)^{-\alpha}. \tag{8.12}$$

When this equation describes a capacitor, the constant $A = 1/C$ (the inverse of the capacitance) and the exponent $\alpha = 1$. For a CPE, the exponent α is less than one. The "double layer capacitor" on real cells often behaves like a CPE instead of a capacitor.

8.2.2.3 Polarization Resistance

Whenever the potential of an electrode is forced away from its value at open circuit, it is referred to as polarizing the electrode. When an electrode is polarized, it can cause current to flow via electrochemical reactions that occur at the electrode surface. The amount of current is controlled by the kinetics of the reactions and the diffusion of reactants both toward and away from the electrode. The open circuit potential is controlled by the equilibrium between two different electrochemical reactions. One of the reactions generates cathodic current and the other generates anodic current. The open circuit potential ends up at the potential where the cathodic and anodic currents are equal.

For kinetically controlled reactions occurring, the potential of the cell is related to the current by the following (known as the Butler–Volmer equation, Equation 5.77):

$$j = j_0(e^{-\alpha F\eta/RT} - e^{\beta F\eta/RT}). \tag{8.13}$$

This equation can be written as

$$j = j_0(e^{-\eta/\beta_a} - e^{\eta/\beta_c}), \tag{8.14}$$

where β_a is the anodic beta coefficient in volts/decade and β_c is the cathodic beta coefficient in volts/decade. If a small signal approximation (η is small) is applied to Equation 8.14, we get the following:

$$j_0 - \frac{\beta_a \beta_c}{\beta_a + \beta_c}\left(\frac{1}{R_p}\right), \tag{8.15}$$

where the new parameter, R_p, is the polarization resistance. The polarization resistance behaves like a resistor.

8.2.2.4 Charge Transfer Resistance

When the polarization depends only on the charge transfer kinetics, the Butler–Volmer equation is given as (Equation 5.78)

$$j = j_0(e^{-\alpha F\eta/RT} - e^{(1-\alpha)F\eta/RT}). \tag{8.16}$$

When the overpotential, η, is very small and the electrochemical system is at equilibrium, the expression for the charge transfer resistance changes into (Equation 5.80)

$$R_{ct} = \frac{RT}{nFj_0} \tag{8.17}$$

From this equation, the exchange current density can be calculated when R_{ct} is known.

8.2.2.5 Warburg Impedance

The Warburg impedance is a result of mass transport owing to diffusion. The Warburg impedance is significant at small species concentration owing to slow diffusion, and for large species concentration, the impedance is negligible. This impedance depends on the frequency of the potential perturbation. At high frequencies, the Warburg impedance is small since diffusing reactants do not have to move very far. At low frequencies, the reactants have to diffuse farther, thereby increasing the Warburg impedance. The equation for the "infinite" Warburg impedance is

$$Z = \frac{\sigma}{\sqrt{\omega}}(1 - j) \tag{8.18}$$

On a Nyquist plot, the infinite Warburg impedance appears as a diagonal line with a slope of 1. On a Bode plot, the Warburg impedance exhibits a phase shift of 45°.

In Equation 8.18, σ is the *Warburg coefficient* defined as

$$\sigma = \frac{RT}{n^2F^2A\sqrt{2}}\left(\frac{1}{C_{\mathrm{Ox,bulk}}\sqrt{D_{\mathrm{Ox}}}}\frac{1}{C_{\mathrm{Red,bulk}}\sqrt{D_{\mathrm{Red}}}}\right) \tag{8.19}$$

where ω is the radial frequency, n is the number of electrons transferred, D_{Ox} is the diffusion coefficient of the oxidant, D_{Red} is the diffusion coefficient of the reductant, A is the surface area of the electrode, and $C_{\mathrm{Ox,bulk}}$ and $C_{\mathrm{Ox,bulk}}$ are the bulk concentration of the oxidant diffusing species and bulk concentration of the reductant diffusing species, respectively. This form of the Warburg impedance is only valid if the diffusion layer has an infinite thickness. Quite often, this is not the case. If the diffusion layer is bounded, the impedance at lower frequencies no longer obeys the equation above. Instead, it has a more general form of equation called the *"finite" Warburg equation*:

$$Z_0 = \frac{\sigma}{\sqrt{\omega}}(1-j)\tanh(\delta(j\omega/D)^{1/2}) \tag{8.20}$$

where δ is Nernst diffusion layer thickness and D is an average value of the diffusion coefficients of the diffusing species. For high frequencies or for an infinite thickness of the diffusion layer, Equation 8.20 simplifies to the *infinite Warburg impedance* (Equation 8.18).

In Table 8.3, the elements used in common equivalent circuit models along with equations for admittance and impedance are given for each element.

The impedance of a resistor is independent of frequency and has only a real component. Because there is no imaginary impedance, the current through a resistor is always in phase with the voltage. The impedance of an inductor increases as frequency increases. Inductors have only an imaginary

TABLE 8.3

Circuit Elements Used in the Models

Equivalent Element	Admittance	Impedance
R	$1/R$	R
C	$j\omega C$	$1/(j\omega C)$
L	$1/(j\omega L)$	$j\omega L$
W (infinite Warburg)	$\frac{\sqrt{\omega}}{\sigma(1-j)}$	$\frac{\sigma}{\sqrt{\omega}}(1-j)$
O (finite Warburg)	$\frac{\sqrt{\omega}}{\sigma(1-j)}\coth(\delta(j\omega/D)^{1/2})$	$\frac{\sigma}{\sqrt{\omega}}(1-j)\tanh(\delta(j\omega/D)^{1/2})$
Q (CPE)	$(j\omega)^{\alpha}/A$	$A(j\omega)^{-\alpha}$

impedance component. As a result, an inductor's current is phase shifted 90° with respect to the voltage. The impedance versus frequency behavior of a capacitor is opposite to that of an inductor. A capacitor's impedance decreases as the frequency is raised. Capacitors also have only an imaginary impedance component. The current through a capacitor is phase shifted –90° with respect to the voltage.

The common equivalent circuit models used to interpret simple EIS data are shown in Table 8.4 along with fuel cell representative components and Nyquist plots.

Since for a purely resistor the imaginary components of the resistance is zero, the Nyquist plot for a resistor is a single point on the real axis with a value R. A purely capacitive coating can be represented by a series of a capacitor and a resistor. The Nyquist plot for a series RC circuit is a vertical line where the intercept of the line with the real axis gives an estimate of the resistance value. The imaginary component of the impedance (contributed by the capacitor) dominates the response of the circuit. One limitation of this Nyquist plot is that the value of the capacitance cannot be determined from the plot. It can be determined by a curve fit or from an examination of the data points. Also, the plot does not indicate which frequency was used to take each data point.

The impedance at the electrochemical reaction interface where reaction occurs can be represented as a parallel combination of charge transfer resistance and a double-layer capacitance. The Nyquist plot for the parallel RC circuit is a characteristic semicircle where the high-frequency intercept of the impedance semicircle is zero and the low-frequency intercept of the semicircle is resistance R_{CT}. The diameter of the semicircle R_{CT} provides information on the reaction kinetics of the electrochemical reaction interface. A large-diameter semicircle (large R_{CT}) indicates sluggish reaction kinetics while a small-diameter semicircle indicates facile reaction kinetics.

The Randle cell represents the combination of electrolyte resistance, a double-layer capacitance, and a charge transfer or polarization resistance. The Nyquist plot for a Randle cell is also a semicircle; however, the high-frequency intercept of the impedance semicircle is electrolyte resistance R_{EL}. Thus, electrolyte resistance can be found by reading the real axis value at the high-frequency intercept. This is the intercept near the origin of the plot. The real axis value at the other (low frequency) intercept is the sum of the polarization resistance and the electrolyte resistance. The diameter of the semicircle is therefore equal to the polarization resistance.

For a semi-infinite diffusion process at cathode represented by Warburg impedance, the Nyquist plot appears as a straight line with a slope of 45°. The impedance increases linearly with decreasing frequency. The infinite diffusion model is only valid for infinitely thick diffusion layer. For finite diffusion layer thickness, the finite Warburg impedance converges to infinite Warburg impedance at high frequency. At low frequencies or for small

TABLE 8.4

Common Electrical Elements

Component	Circuit, Nyquist Plot
Resistor—polarization or charge transfer resistance	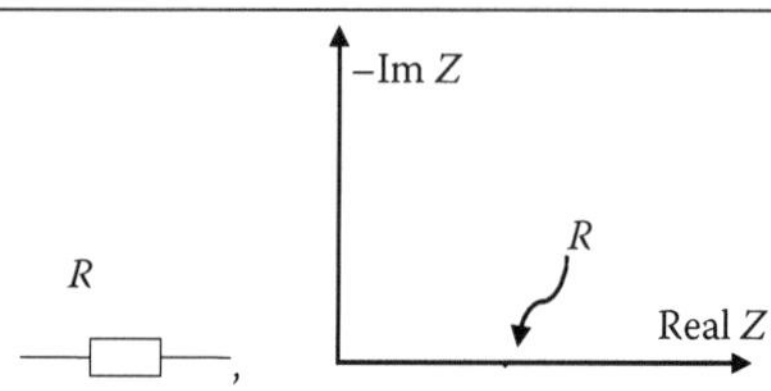
Purely capacitive coating—a series capacitor and resistor	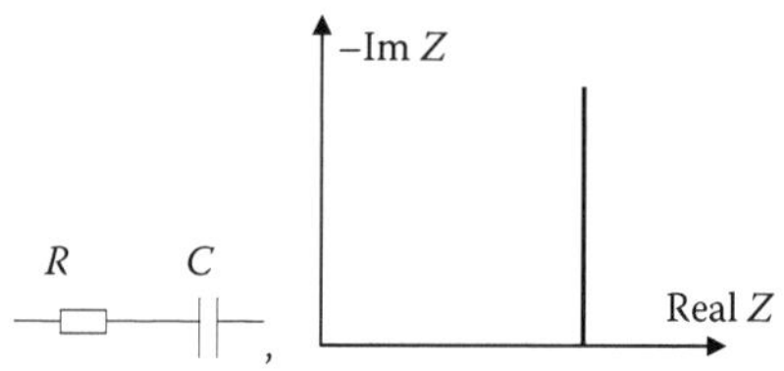
Electrochemical reaction interface—parallel double-layer capacitor and charge transfer resistor	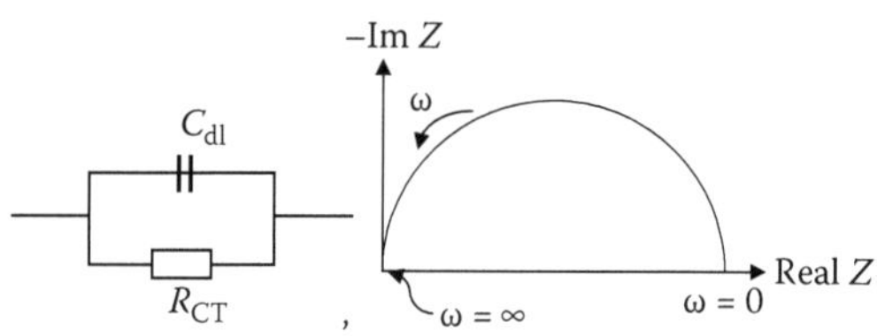
Randles cell—electrolyte resistance, a double-layer capacitance and a charge transfer or polarization resistance	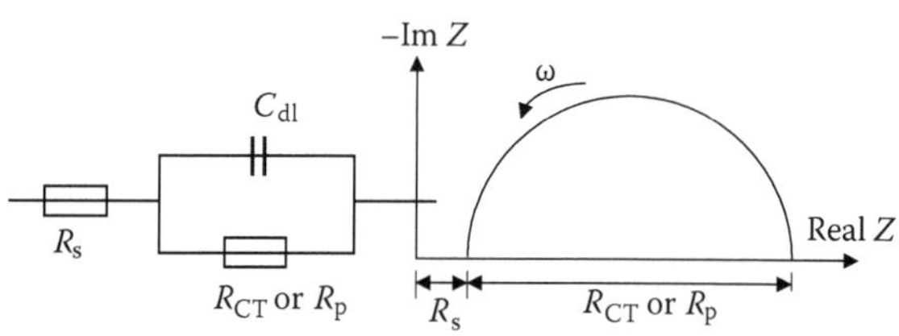
Infinite Warburg—diffusion is the rate determining step with infinite diffusion layer thickness, as series cell impedance	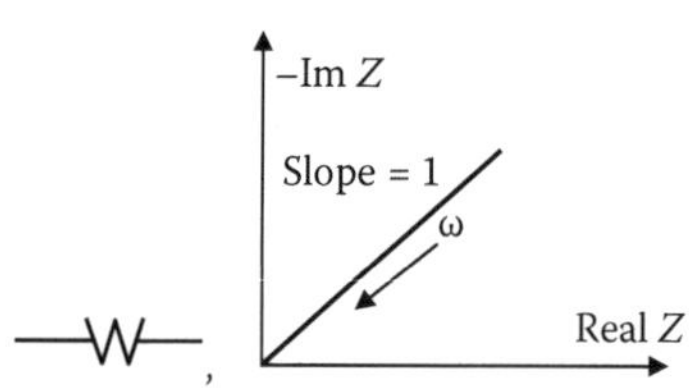
Finite Warburg—for a cathode with fixed diffusion layer thickness (porous bounded Warburg)	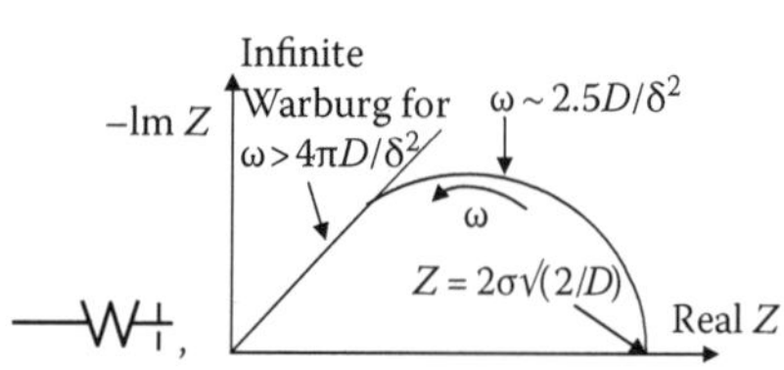

(continued)

TABLE 8.4 (Continued)

Common Electrical Elements

Component	Circuit, Nyquist Plot
Mixed kinetic and charge transfer control at cathode of fuel cell—series electrolyte resistor and parallel RC representing cathode activation kinetics and mass transfer effects through infinite Warburg impedance	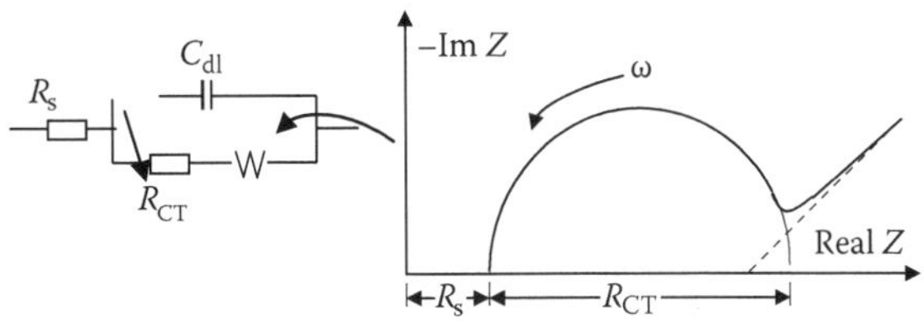

diffusion layer thickness, the finite Warburg impedance returns toward real impedance axis.

The fuel cell cathode can be represented by a series electrolyte resistance, a parallel double-layer capacitance, a charge transfer impedance, and finite Warburg impedance for diffusion process. This circuit model polarization is due to a combination of kinetic and diffusion processes. The Nyquist plot for this shows a semicircle with a 45° straight line.

8.2.2.6 Fuel Cell Equivalent Circuit Modeling

Equivalent circuit modeling of EIS data is used to extract physically meaningful properties of the fuel cell by modeling the impedance data in terms of an electrical circuit composed of ideal resistors (*R*), capacitors (*C*), and inductors (*L*). The real systems do not necessarily behave ideally with processes that occur distributed in time and space; hence, often specialized circuit elements are used. These include the generalized CPE and Warburg element. The Warburg element is used to represent the diffusion or mass transport impedances of the cell as described previously. A generalized equivalent circuit element for a single cell fuel cell is shown in Figure 8.7 along with a physical picture of the fuel cell reactant conversion and transport, as well as an equivalent circuit for key impedances that represent ohmic losses, anode and cathode activation losses, and mass transfer effects.

In the equivalent circuit analog, resistors represent conductive pathways for ion and electron transfer. As such, they represent the bulk resistance of a material to charge transport such as the resistance of the electrolyte to ion transport or the resistance of a conductor to electron transport. Resistors are also used to represent the resistance to the charge-transfer process at the electrode surface. Capacitors and inductors are associated with space-charge polarization regions, such as the electrochemical double layer, and adsorption/desorption processes at an electrode, respectively.

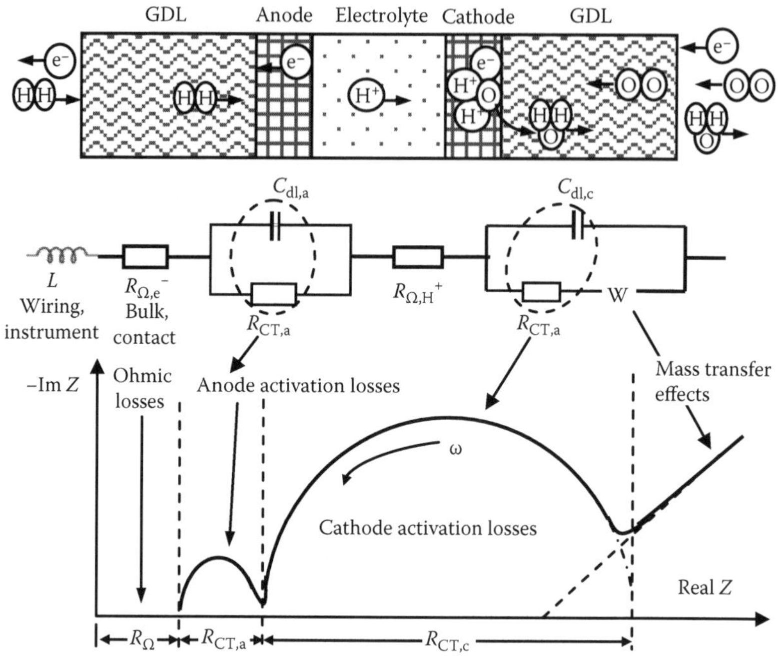

FIGURE 8.7
The PEMFC physical picture, equivalent circuit, and the Nyquist plot using the impedance model.

8.2.2.7 Time and Frequency Domains

In EIS, the data are represented in two domains, the time domain and the frequency domain. In the time domain, signals are represented as signal amplitude versus time. In the frequency domain, the data are plotted as amplitude versus frequency. The Fourier transform and inverse Fourier transform are used to switch between the domains. In EIS systems, lower frequency data are usually measured in the time domain. An FFT (fast Fourier transform as done in a digital computer) is used to convert the current signal into the frequency domain.

Modern EIS analysis uses a computer to find the model parameters that cause the best agreement between a model's impedance spectrum and a measured spectrum. For most EIS data analysis software, a nonlinear least-squares fitting (NLLS) *Levenberg–Marquardt* algorithm is used. NLLS starts with initial estimates for all the model's parameters, which must be provided by the user. Starting from this initial point, the algorithm makes changes in several or all of the parameter values and evaluates the resulting fit. If the change improves the fit, the new parameter value is accepted. If the change worsens the fit, the old parameter value is retained. Next, a different parameter value is changed and the test is repeated. Each trial with new values is called an iteration. Iterations continue until the goodness of fit exceeds an acceptance criterion, or until the number of iterations reaches a limit.

A common cause of problems in EIS measurements and their analysis is drift in the system being measured. In practice, a steady state can be difficult to achieve. The cell can change through adsorption of solution impurities, growth of an oxide layer, buildup of reaction products in solution, coating degradation, and temperature changes, to list just a few factors.

8.2.3 Current Interrupt Measurement

The current interrupt technique is the most widely used method of ohmic drop and ohmic resistance evaluation of various electrochemical systems including fuel cells. The principle behind the current interrupt method is the performance of the voltage response of the fuel cell for a given step change of current flow.

An interruption to current can be accomplished through either a fast switch or a superimposed square wave. A simple circuit shown in Figure 8.8 with a fuel cell, switch, load, and an oscilloscope or digital computer voltage recorder can be used to perform current interrupt test. First, the switch is closed and the load resistor is adjusted until the desired test current and voltage are established. Then, the load current is then switched off. The transient voltage data are recorded with the computer. In the case of the oscilloscope, triggering will need to be set so that the oscilloscope moves into store mode. A schematic of the potential–time response for a current interrupt is shown in Figure 8.9.

When the load (current) is abruptly changed, voltage recovery (or decay) as a function of time occurs mainly because of three components: (i) ohmic drop, (ii) activation overpotential, and (iii) concentration overpotential. On current interruption, the first component manifests as a jump or abrupt rise, since the ohmic drop is passive. When the current is set equal to zero (as

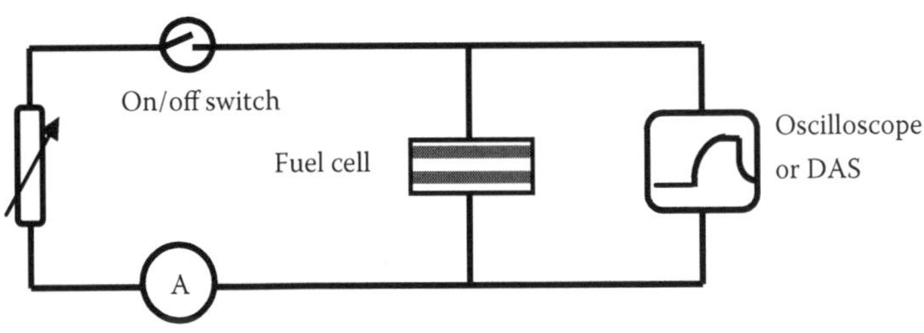

FIGURE 8.8
Fuel cell circuit for current interrupt test consisting of load, on/off switch, and a transient recording device such as oscilloscope of data acquisition system (DAS).

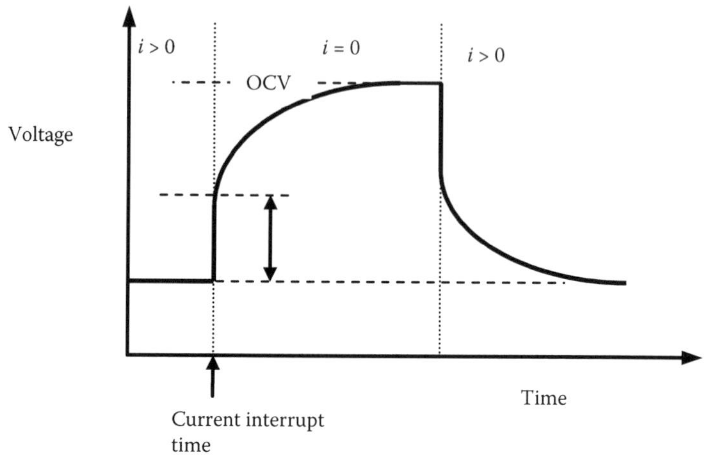

FIGURE 8.9
Schematic of the current interrupt and corresponding voltage waveform. Current interruption voltage suddenly rises corresponding to the ohmic loss recovery and slowly rises to open circuit voltage (OCV).

during the current interrupt), the ohmic voltage drop becomes equal to zero. At that instant, the cell recovers a voltage value equivalent to the ohmic drop instantaneously, that is, at $t = 0$. The second and third components of the voltage recovery take place in an exponential manner.

The recovery time depends on the impedance at the electrode–electrolyte interface. Depending on the time constant associated with the resistance and capacitance (RC constant) of the interface, the voltage exponentially recovers. By measuring the jump (or drop) at zero time (or realistically within 10 s), one can obtain the value of the ohmic resistance of the cell. The ohmic resistance of the cell R_{ohm} (Ω-cm^2) is determined as the quotient of the instantaneous change in voltage and the cell current density i (A cm^{-2}) just prior to the interrupt event, $R_{ohm} = \delta(V)/i$. If the cell is operating far below mass transfer limits, then the voltage recovery corresponds to the activation loss in the cell.

The advantages of this method include a single data value that is easily interpreted. Furthermore, there is no requirement for additional equipment because the interrupt is brought about by the load. The primary disadvantage of this method is that it imposes a significant perturbation on the cell, if only for a short duration (i.e., tens of microseconds). It should be noted that in this method, the data are degraded when long cell cables are used because of stray capacitances and inductances. Under some circumstances for electrochemical systems with porous electrodes, the interrupter method may overestimate the ohmic voltage change and therefore overestimate the ohmic resistance of the cell.

8.2.4 Cyclic Voltammetry

The cyclic voltammetry technique is one of the most commonly used electroanalytical techniques for the study of electroactive species and electrode surfaces. Cyclic voltammetry was introduced in Chapter 2. In a cyclic voltammetry experiment, the working electrode potential is ramped linearly versus time to a set potential. When cyclic voltammetry reaches a set potential, the working electrode's potential ramp is inverted. This inversion can happen multiple times during a single experiment.

Typically, the equipment required to perform cyclic voltammetry consists of a conventional three-electrode potentiostat connected to three electrodes: working, reference, and auxiliary electrodes. The potentiostat applies and maintains the potential between the working and reference electrodes while at the same time measuring the current at the working electrode. During the experiment, charge flows between the working electrode and the auxiliary electrode. A recording device such as a computer or plotter is used to record the resulting cyclic voltammogram as a graph of current versus potential. Figure 8.10 depicts a generic cyclic voltammogram. The potential is graphed along the x-axis with more positive (or oxidizing) potentials plotted to the left and more negative (or reducing) potentials plotted to the right. The current is plotted on the y-axis of the voltammogram, with cathodic (i.e., reducing) currents plotted up along the positive direction, and anodic (i.e., oxidizing) currents plotted down in the negative direction.

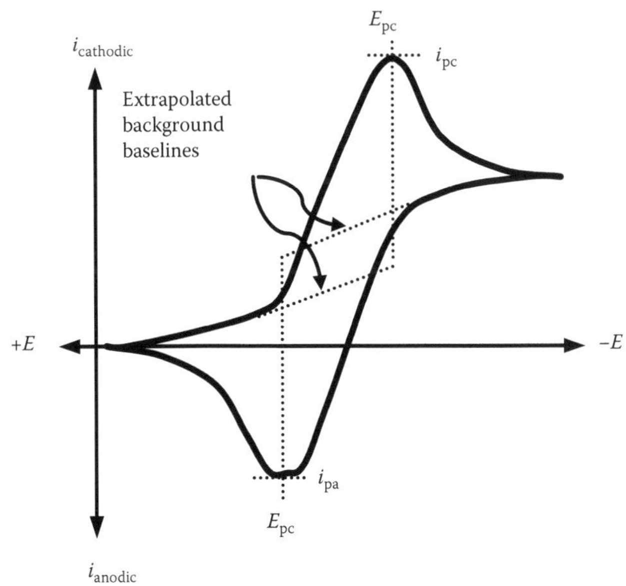

FIGURE 8.10
A typical cyclic voltammogram.

8.3 Characterization of Electrodes and Electrocatalysts

The overall electronic resistance of the electrode is an important quantity (Cooper and Smith, 2006, Hack et al., 1990). A four-probe resistivity method is employed to avoid contact resistance problems. The resistance can be measured using a variety of commercial LCR (inductance/capacitance/resistance) meters, and the resistivity values can be calculated with the knowledge of the geometry of the probe employed. As shown in Figure 8.11, in the four-probe resistivity method, the electrode layer of thickness *t* resistance is measured by passing a current through two outer probes, and measuring the voltage through the inner probes allows the measurement of the substrate resistivity. A current is passed through the outer probes and induces a voltage in the inner voltage probes.

The sheet resistance ρ is given as

$$\rho = \frac{\pi}{\ln 2} \frac{E}{I} \quad (\Omega), \tag{8.21}$$

where E is the measured voltage (volts) and I is the source current (amperes). The bulk or volume resistance is calculated as

$$\rho = \frac{\pi}{\ln 2} \frac{E}{I} t \quad (\Omega\text{-cm}) \tag{8.22}$$

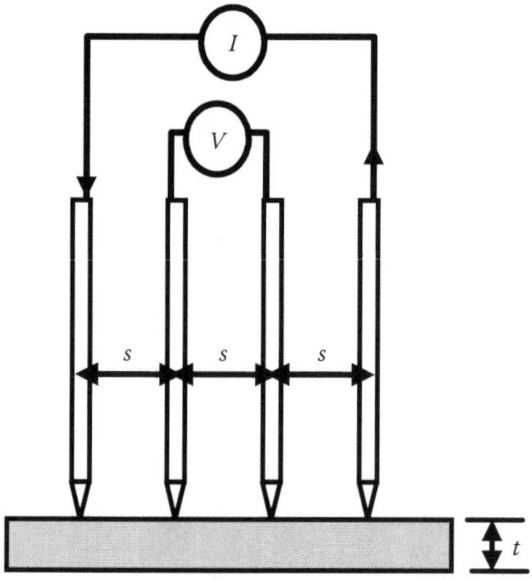

FIGURE 8.11
A four-point probe method to measure resistivity of a layer.

when the wafer thickness is less than half the probe spacing ($t < s/2$). For thicker layer ($t \geq s/2$), the bulk resistivity is calculated as

$$\rho = \frac{\pi t}{\ln\left(\dfrac{\sinh(t/s)}{\sinh(t/2s)}\right)} \frac{E}{I} \quad (\Omega\text{-cm}). \tag{8.23}$$

It is important to know as much about the structure of the electrode as possible to determine how to improve its efficiency for carrying out the electrochemical reaction. As listed in Table 8.2 electrochemical, optical, radiation, and physical methods are employed to characterize electrodes and electrocatalysts. Various techniques including transmission electron microscopy (TEM), x-ray diffraction (XRD), neutron radiography, spectroscopy-auger electron (AES), x-ray photoelectron (XPS), secondary ion mass spectrometry (SIMS), small angle x-ray scattering (SAXS), and scanning electron microscopy (SEM) are employed to investigate the catalyst chemical composition, nanoparticles, and electrode structures including surface area. Most of the physical tests have to be conducted ex situ and with a small sample. For example, for SEM, the electrode samples can be used as such, but for TEM, a thin slice of the electrode material is often required to get good quality images. From the TEM images, one can study the catalyst structure in detail and estimate the surface area by counting the particles in a given grid and estimating the particle size.

On the electrochemical methods, EIS and CV methods give a variety of parameters for electrode and electrocatalysts as discussed in previous sections. Notable is the electrocatalyst active surface area, which is indicative of fuel cell reaction kinetics. It was shown in Chapter 2 that parameters obtained with cyclic voltammogram can be used to obtain information on the redox potential and evaluation of electron transfer kinetics. Catalyst surface area and poisoning effects can also be estimated using a cyclic voltammogram. Catalyst utilization and electrochemical surface area are important parameters in the performance of catalyst and membrane electrode assembly. The technique for determining the electrochemical surface area (ECSA) of fuel cell electrodes involves cycling the electrode of interest over a voltage range where charge transfer reactions are adsorption-limited at the activation sites. The applied electrode potential is such that the number of reactive surface sites can be obtained by recording the total charge required for monolayer adsorption/desorption.

Fuel cell electrodes can be examined for their electrocatalytic behavior by ex situ or in situ voltammetry tests. In the case of ex situ tests, also known as half-cell tests, the properties of the electrode are evaluated using a standard three-electrode cell where an aqueous solution (e.g., perchloric acid) simulates the proton-conducting electrolyte in a PEMFC. Half-cell tests are a convenient and relatively fast method of screening electrocatalysts; however, they are not suitable for assessment of fuel cell electrodes under operating conditions.

For characterization of low-temperature fuel cell with platinum electrodes, the following reactions are generally used: the hydrogen adsorption/desorption (HAD),

$$\mathrm{Pt-H_{ads}} \underset{\text{reverse}}{\overset{\text{forward}}{\leftrightarrow}} \mathrm{Pt+H^{+}+e^{-}}, \tag{8.24}$$

and the oxidative stripping of adsorbed carbon monoxide,

$$\mathrm{Pt-CO_{ads}+H_2O \rightarrow Pt+CO_2+2H^{+}+2e^{-}}. \tag{8.25}$$

The electro-reduction of protons and adsorption of hydrogen on the catalyst surface, that is, reaction in the reverse direction, are the processes considered in CV. The hydrogen adsorption charge density (q_{Pt} in coulombs/cm^2) caused by this reaction determined from the CV test is used to calculate the ECSA of the fuel cell electrode. For in situ experiments, a two-electrode configuration is used in which one of the electrodes of the fuel cell serves as both a counter electrode and a pseudo-reference electrode. Typically, the electrochemical activity of the fuel cell cathode is of most interest because of the sluggish kinetics of the oxygen reduction reaction. Therefore, the cathode is often chosen to be the working electrode. The fuel cell anode is used as the reference electrode with the inherent assumption that polarization of this electrode is small relative to the polarization imposed on the fuel cell cathode, the working electrode. The current densities obtained in the ECSA tests are relatively small and justify this assumption.

The fuel cell electrode of interest (working electrode) is filled with water or slowly purged with a non-reactive gas such as nitrogen or argon, while hydrogen is fed to the other electrode (reference electrode). Anodic and cathodic currents occur at the electrode. A typical CV voltammogram of the HAD reaction is shown in Figure 8.12. The voltammogram exhibits multiple peaks associated with both the oxidation and reduction reactions. Each peak is indicative of the adsorption onto or desorption from a particular crystal index of platinum, for example, 100 and 110 indices. Integration of the hydrogen desorption/adsorption peaks that result as a consequence of the forward and reverse scans, respectively, is used to estimate the ECSA of the electrocatalyst. The shaded area in Figure 8.13 represents the charge density Q_{Pt} arising from hydrogen adsorption on the Pt catalyst during the reverse sweep.

The ECSA of the Pt catalyst is calculated using the following equation where

$$\mathrm{ECSA}\left(\mathrm{cm}^2_{\mathrm{Pt}}/\mathrm{g}_{\mathrm{Pt}}\right)=\frac{Q_{\mathrm{Pt}}}{\Gamma L}, \tag{8.26}$$

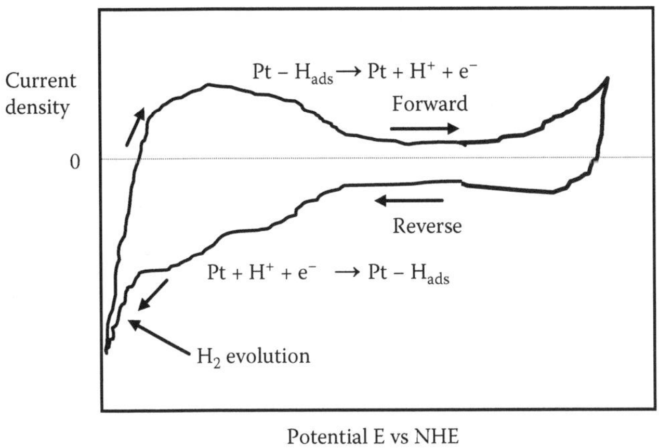

FIGURE 8.12
Cyclic voltammogram of PEM fuel cell catalyst layer for ECSA analysis by hydrogen adsorption/desorption.

where Q_{Pt} is charge density integrated over the shaded area. Γ is the charge required to reduce a monolayer of protons on Pt; typical for smooth electrodes, $\Gamma = 210\ \mu C/cm^2_{pt}$, and $L(g_{pt}/cm^2_{electrode})$ is Pt content or loading in the electrode. The baseline current shown in gray is the sum of the capacitive current caused by charging/discharging of the electrical double layer (positive on the forward scan and negative on the reverse scan) and the transport

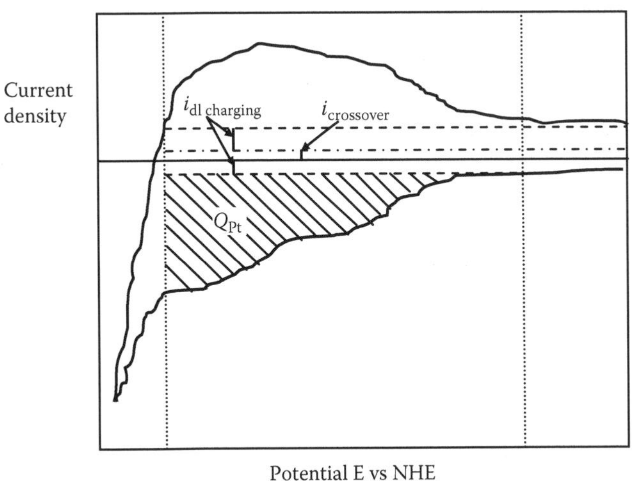

FIGURE 8.13
Cyclic voltammogram of a PEMFC catalyst layer highlighting the region of interest. The shaded area is the charge density owing to H adsorption during the reverse scan and is used in the ECSA calculation.

limited H_2 crossover current. The capacitive or non-faradic current arises owing to the double layer at the electrode–electrolyte interface. This current involves charge accumulation but not chemical reactions or charge transfer. This current is directly proportional to the scan rate, v (V/s),

$$i_{\mathrm{dl}} = C\frac{\mathrm{d}E}{\mathrm{d}t} = C_{\mathrm{dl}} \cdot v, \tag{8.27}$$

where C_{dl} (farad/$\mathrm{cm}^2_{\mathrm{electrode}}$) is the specific capacitance of the electrode double layer. The electrical charge associated with double-layer charging and fuel crossover must be accounted for in the analysis to avoid overestimating the charge attributed to the electrocatalytic activity. The capacitance of the electrode C_{dl} can be determined from the double-layer charging current density i_{dl}. It should be noted that both the platinum and carbon support contribute to the electrode's capacitance,

$$C_{\mathrm{dl}} = A_{\mathrm{Pt}} \cdot C^*_{\mathrm{dl}} + A_{\mathrm{carbon}} \cdot C^*_{\mathrm{dl,carbon}}, \tag{8.28}$$

where $C^*_{\mathrm{dl,Pt}}$ and $C^*_{\mathrm{dl,cardon}}$ are the specific double-layer capacitance (in F/cm^2) for Pt and carbon, respectively, and A_{Pt} and A_{carbon} are the area of platinum and carbon in the electrode ($\mathrm{cm}^2/\mathrm{cm}^2_{\mathrm{electrode}}$), respectively. The specific area of platinum A_{Pt} is determined from the ECSA and the catalyst loading L,

$$A_{\mathrm{Pt}}\left(\mathrm{cm}^2_{\mathrm{Pt}}/\mathrm{cm}^2_{\mathrm{electrode}}\right) = \mathrm{ECSA} \times L. \tag{8.29}$$

For CO stripping test, pure CO or a small concentration of CO in inert gas such as argon is bubbled into the electrolyte for a period depending on CO concentration and then its adsorption on the electrode is driven under a constant potential control for several minutes. The electrolyte is purged for several minutes with argon, keeping electrode potential at open circuit potential to eliminate CO being reversibly adsorbed on the surface. Then two to four cyclic voltammmetry sweeps are recorded. The first anodic sweep is performed to electro-oxidize the irreversibly adsorbed CO and the subsequent voltammetries are performed in order to verify the completeness of the CO oxidation. Figure 8.14 shows typical two cyclic voltammograms obtained on Pt/C with a CO adsorbed ad-layer. To calculate CO stripping charge, the area under the peak has to be integrated and the charge owing to double-layer charging and oxide formation has to be subtracted. The simplest approach is to consider double-layer charging and oxide formation the same as in the absence of CO, that is, to use the second cycle in the same experiment for baseline subtraction and to contribute the difference between the first and the second cycle only to CO oxidation. The calculated peak charge Q_{CO}, is related to the reaction in Equation 8.22.

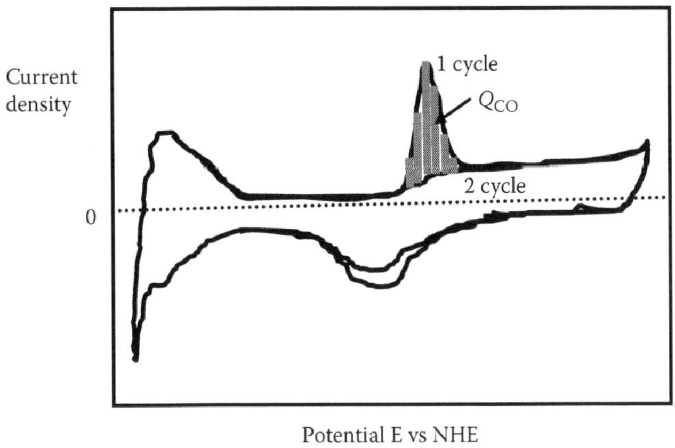

FIGURE 8.14
CO stripping cyclic voltammograms for Pt MEA with (first cycle) and without (second cycle) a CO adsorbed ad-layer. The shaded area represents the charge related to the CO oxidation reaction.

The active surface of the catalyst is calculated by means of the following equation:

$$\mathrm{ECSA}_{CO} = \frac{Q_{CO}}{\tau_{CO} L}, \tag{8.30}$$

where τ_{CO} is charge density required to oxidize a monolayer of CO on bright Pt and is given as $\tau_{CO} = 00.484\ \mathrm{mC/cm^2}$.

Not all of the catalysts used to make the fuel cell electrode is accessible to reactants (e.g., protons and oxygen at the cathode) or is in electrical contact and thus is not able to participate in the electrochemical reaction. The fraction of the catalyst that is available to participate in electrode reactions is given by the ratio of ECSA to the specific area of the catalyst obtained by the catalyst manufacturer using chemisorption or other ex situ techniques (also in $\mathrm{cm^2_{Pt}/g_{Pt}}$). This ratio is referred to as utilization. Utilization is an idealized condition because the very low reaction rate used during the ECSA measurement results in negligible transport limitations. In a fuel cell operating at a practical current density, oxygen and proton transport resistances could decrease the amount of catalyst that participates in the cathode reaction, effectively decreasing utilization.

8.4 Characterization of Membrane Electrode Assembly

The MEA of the PEFC consists of proton conducting electrolyte membrane sandwiched between electrode catalyst layers. Some fuel cell MEAs have gas diffusion layers on the electrode catalyst layer. The characterization of the MEA includes the polarization characteristics and individual component characterization (Barbir, 2005, Laraminie and Dicks, 2000, Springer et al., 1993, Springer et al., 1991). The MEAs are characterized for their voltage–current performance (polarization curve), proton resistance, and electrochemical active surface areas for both the anode and cathode using impedance spectra described earlier. The impedance spectra (analyzed in the form of Nyquist plots) can be divided into clearly separated domains, as follows: a high-frequency feature describing the impedance of the cell and membrane, and the intermediate- to low-frequency range related to the oxidation reaction and the so-called CO poisoning of the catalysts.

Permeation of reactant from one electrode to the other through the PEM degrades fuel cell performance, efficiency, and durability. In addition, severe crossover autocatalytically accelerates membrane degradation and pinhole formation via locally generated heat leading to proximate membrane thinning, which further accelerates the crossover process. Although crossover of both fuel (e.g., hydrogen or methanol) and oxidant (oxygen) occurs, the latter generally occurs at a lower rate and thus most often fuel crossover is the property of interest. To experimentally determine the fuel crossover, a suitable inert gas such as nitrogen is used to purge the fuel cell cathode while hydrogen is passed through the fuel cell anode. The potential of the fuel cell cathode (i.e., the working electrode) is swept by means of a linear potential scan to potentials at which any hydrogen gas present at the fuel cell cathode is instantaneously oxidized under mass transfer limited conditions. Such experiments are referred to as linear sweep voltammetry (Bard and Faulkner, 2001). Methanol crossover can also be determined using this basic voltammetric method, with the hydrogen being replaced by the liquid fuel (Ren et al., 2000).

The output of working electrode current density (current normalized by the active area of the working electrode) versus potential is used to determine the hydrogen crossover flux (mol/cm^2/s) from Faraday's law,

$$J_{\text{crossover},H_2} = \frac{i_{\text{lim}}}{n \cdot F}, \tag{8.31}$$

where i_{lim} is the transport limiting current density (A/cm^2), n is the number of electrons taking part in the reaction (electron-mole/mole), and F is the Faraday constant (96,485 C/electron-mole). The rate of H_2 crossover from the anode to the cathode through the membrane is proportional to the mass transport limited current density. The current typically attains either a constant or linearly increasing value with increasing electrode potential. A constant, electrode potential–independent current is indicative of a fuel cell with a very high (infinite) electrical resistance (i.e., no internal shorting) whereas a linearly increasing current indicates that the cell has a finite resistance caused by internal shorting. The electrical resistance of the cell can be estimated from the slope of the voltage versus current plot.

Advanced instruments such as electron microscopy techniques (SEM and TEM) are used to characterize the microstructure of MEA components at various resolutions. Digital mapping of the morphology/catalyst clusters can be used to perform quantitative analysis of the microstructure.

The electrolyte membrane is characterized separately for its mechanical properties, hydrogen permeability, conductivity, and water retention and transport properties (Dunbar and Masel, 2007, Fuller and Newman, 1992, Trabold et al., 2006). The hydrogen permeability of the electrolyte membranes is measured by using a forced convection drying oven, consisting of two compartments separated by a vertical membrane. The contents of the compartments are kept under constant agitation where gas concentrations are measured by gas chromatography. The permeability coefficient of any gas through a membrane is directly related to the size and thickness of the membrane. A low-hydrogen permeability membrane should be better for H_2/O_2 fuel cell applications. The hydrogen permeability tests are used to assess membrane hydrogen permeability coefficient.

We learned from the previous section that electrochemical active surface area can be determined from the CV method. The accurate measurement of surface area of the electrolyte membrane or electrodes is done by a technique known as the Brunauer–Emmett–Teller (BET) method (Brunauer et al., 1938). It is based on the physical adsorption of gas molecules on a solid surface. It is assumed that gas molecules physically adsorb on a solid in layers infinitely and there is no interaction between each adsorption layer. The *BET equation* is expressed as

$$v = \frac{v_{\mathrm{m}} c p}{(p_{\mathrm{o}} - p)[1 + (c-1)(p/p_{\mathrm{o}})]}, \tag{8.32}$$

where v is the adsorbed gas quantity (e.g., in volume units), v_{m} is the monolayer adsorbed gas quantity, p and p_{o} are the equilibrium and the saturation pressure of adsorbates at the temperature of adsorption, and the constant c is the BET constant, which is expressed as

$$c = \exp\left(\frac{E_1 - E_{\mathrm{L}}}{RT}\right), \tag{8.33}$$

where E_1 is the heat of adsorption for the first layer and E_{L} is that for the second and higher layers and is equal to the heat of liquefaction. In a typical experiment, a first, dry sample is placed under high vacuum to evacuate all the gas. It is then cooled to liquid nitrogen temperature (77 K) and exposed to inert gas, such as a nitrogen, argon, or krypton environment. A layer of inert gas will physically adhere to the sample, lowering the pressure in the analysis chamber. The amount adsorbed can be determined by either volumetric or gravimetric methods. Equation 8.33 can be written as

$$\frac{p}{v(p_o - p)} = \frac{1}{v_m c} + \frac{c-1}{v_m c}\frac{p}{p_o}. \tag{8.34}$$

Equation 8.34 is an adsorption isotherm and can be plotted as a straight line with $1/v[(p_0/p)-1]$ on the y-axis and (p/p_o) on the x-axis. This plot is called a BET plot, which is linear in the range of $0.05 < p/p_o < 0.35$. The values of

TABLE 8.5

Electrolyte Membrane Characterization Methods

Method	Characteristic Parameter	Remarks
X-ray diffraction (XRD)	Identification of elements and crystalline nature of the materials	
Stress–strain curves with three-point bending method	Mechanical strength of membrane	Measurement of flexural elastic moduli as a function of temperature
Nitrogen gas sorption analyzer	Identification of membrane pores	Measurement of N_2 adsorption–desorption isotherms
The Fourier transform infrared (FT-IR) spectra		
Thermogravimetric analysis (TGA) and differential thermal analysis (DTA)	Thermal degradation process and stability of the membrane	Measurements are typically carried out under dry air with a certain heating rate (5–10°C/min)
SEM scanning electron microscopy (SEM)	Surface morphology-phase separation at the surfaces of membranes	
X-ray energy dispersive spectrometer	Crystalline structure and elements	
Transmission electron microscopy (TEM)	Nanostructure, segregation of the crystals at nanoscale	
Nuclear magnetic resonance (NMR)	Spatial distribution of water in membrane	Also, measurement of chemical shift for a particular chemical
Neutron radiography	Spatial distribution of water in membrane	
AC impedance	Proton conductivity	Calculated from the electrolyte resistance (R) obtained from the Nyquist plot
Fourier transform infrared (FT-IR) spectra	Proton conductivity of the membrane	Presence of water and protonated water

the slope A and the y-intercept I of the line are used to calculate the monolayer adsorbed gas quantity v_m and the BET constant c with the following equations:

$$v_m = 1/(A + I); \quad c = 1 + A/I. \tag{8.35}$$

A total surface area S_{total} and a specific surface area S are calculated by the following equations:

$$S_{total} = \frac{v_m Ns}{V}; \quad S = \frac{S_{total}}{a} \tag{8.36}$$

where v_m is in units of volume, which is also the unit of the molar volume of the adsorbate gas; N is Avogadro's number; s is the adsorption cross section of the adsorbing species; V is the molar volume of adsorbate gas; and a is the mass of adsorbent (in grams).

Some of the advanced techniques used in characterizing the electrolyte membrane are listed in Table 8.5 along with characteristics measured (Lakshminarayana and Nogami, 2009).

8.5 Characterization of Bipolar Plates

For PEMFC, graphite composites are considered the standard material for bipolar plates because of their low surface contact resistance and high corrosion resistance. Unfortunately, graphite and graphite composites are brittle and permeable to gases with poor cost effectiveness for high volume manufacturing processes relative to metals such as aluminum, stainless steel, nickel, titanium, and so on. Metallic bipolar plates seem more appropriate than graphite composite bipolar plates (Tawfik et al., 2007). Metallic plates have higher mechanical strength, easier manufacturability, better durability to shocks and vibration, no permeability, and lower interface contact resistance than graphite composite plates. Recently, metallic bipolar plates have been considered for their particular suitability to transportation applications owing to higher mechanical strength, better durability to shocks and vibration, no permeability, and much superior manufacturability and cost effectiveness when compared to carbon-based materials.

However, the main drawback with metals is the lack of ability to combat corrosion in the harsh acidic and humid environment inside the PEM fuel cell without forming oxidants, passive layers, and metal ions that cause considerable power degradation. Various coatings are applied to improve the corrosion resistance of the metals used without sacrificing surface contact resistance and maintaining cost-effectiveness. Thus, key parameters for graphite bipolar plate characterization are gas permeability and electrical resistivity (Ghouse et al., 1998), and that for metallic bipolar plate is to assess the coatings and their interaction with other materials.

To measure the gas permeability of the graphite bipolar plate samples, a nitrogen gas is passed through the specimen to produce a differential pressure across the specimen DP and measure the rate of nitrogen flow, Q. The permeability is calculated using the formula

$$K = \frac{Q \times W}{\Delta P \times A} \tag{8.37}$$

where K is permeability (cm²/s), Q is flow rate (atm cm³/s), W is sample length (cm), A is sample area (cm²), and ΔP is absolute pressure drop across the sample (atm).

The electrical resistivity of the graphite bipolar plates can be measured either by passing DC current and measuring the voltage drop across the plate sample or by direct resistance measurement using a milliohmmeter. Using the measured electrical resistance R, the electrical resistivity is calculated as

$$\rho = R\frac{A}{d} = \frac{\Delta E}{I}\frac{A}{d}, \tag{8.38}$$

where ρ is electrical resistivity (Ω cm), R is resistance (Ω), A is the cross-sectional area for current (cm²), ΔE is the voltage drop across the sample (mV), I is current (mA), and d is the distance between the voltage terminals (cm).

Often the metal bipolar plate develops a passive film layer, which can increase the contact resistance. This resistance can be measured using the contact electric resistance technique (Kim et al., 2002). In this technique, the two sample surfaces are brought into contact and then separated repeatedly with a chosen frequency as depicted in Figure 8.15. The passive

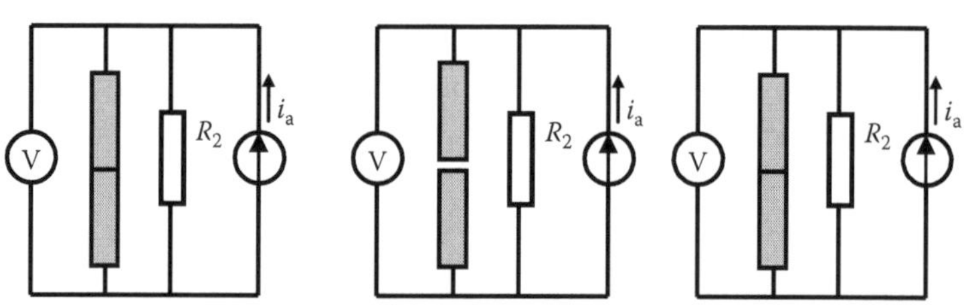

FIGURE 8.15
Contact electrical resistance measurement.

film is formed and grown on the surfaces when the two surfaces are disconnected and exposed to the environment. When the two surfaces are in contact and a direct current is passed through the samples, the DC electrical resistance of the system is determined using Ohm's law by measuring the voltage drops. Since the bulk metal has high conductivity, the voltage drop is mainly due to the passive film or interface of the metal and passive film.

The contact resistance is calculated as

$$R_c = \frac{E_c}{E_s - E_c} R_2, \tag{8.39}$$

where E_c is the contact voltage, E_s is the separation voltage, and R_2 is the shunt resistance.

Characterization studies on coated bipolar plates include examination for any possible chemical changes in the composition of the coating or the membrane electrode assembly (MEA) that may affect the stability of the coating, substrate, or the ionic conductivity of the cell. Analyses using scanning electron microscope (SEM) and energy dispersive x-ray (EDX) are performed on the land and valley surfaces of the reactant flow fields at both the anode and the cathode. The EXD measurements are used to identify dissociation and the dissolution of the coating binders. Samples scraped from the anode and cathode electrodes of the MEA from fuel cells with bipolar plates with 1000 h of operation are analyzed using x-ray diffraction (XRD) analysis. These analyses provide information on crystal growth that could result in power degradation (Hung et al., 2009).

8.6 Characterization of Porous Structures of Electrodes and Membranes

The high porosity in the electrode and diffusion layer structure is required for efficient transport of species between electrolyte membrane and gas flow fields. The porosity of the material is defined as the fraction of the void space in the volume of the material. Thus, porosity

$$\phi = \frac{V_P}{V_T}. \tag{8.40}$$

It can also be expressed in terms of the density of the porous sample (ρ_p) and the bulk density (ρ_b) of the material used to make the porous sample.

$$\phi = 1 - \frac{\rho_p}{\rho_b}. \tag{8.41}$$

However, in an electrode or gas diffusion layer, the pores may not be connected to one another and to the surface of the layer to effectively influence the flow of reactants and products. Hence, effective porosity is defined where the pores that are interconnected and open to surface are counted toward porosity of the layer. The effective porosity can be measured using various techniques including the imbibition method, gas expansion method, and porosimetry methods.

In the imbibition method, the porous sample is immersed, under vacuum, in a fluid that preferentially wets the pores (Dullien, 1992). The pore volume is determined by the difference between the total volume of the water and the volume of water left after soaking. The total volume of the porous sample is measured by immersion of the sample in a liquid that does not enter the pores such as mercury at low pressure. In the gas expansion method, a sample of known bulk volume is enclosed in a container of known volume. It is connected to another container with a known volume that is evacuated (i.e., near vacuum pressure). When a valve connecting the two containers is opened, gas passes from the first container to the second until a uniform pressure distribution is attained. Using ideal gas law, the volume of the pores is calculated as

$$V_p = V_T - V_a - V_b \frac{P_2}{P_1} \tag{8.42}$$

where V_p is the effective volume of the pores, V_T is the bulk volume of the sample, V_a is the volume of the container containing the sample, V_b is the volume of the evacuated container, P_1 is the initial pressure in volume V_a and V_p, and P_2 is the final pressure present in the entire system.

For the detailed information, pore structure porosimetry techniques are used. These methods enable measurement of pore diameter, pore shape, pore volume, and pore distribution in the electrode catalyst and gas diffusion layers. However, for PEMFC, these layers have hydrophobic and hydrophilic pores and there is no suitable technique available for characterization of such complex pore structures. Combination of multiple porosimetry techniques are employed to characterize layers with both hydrophobic and hydrophilic pores. The pore structure characterization techniques include capillary flow porosimetry, water intrusion porosimetry, and mercury intrusion porosimetry (Jena and Gupta, 2002). In water intrusion porosimetry, the sample is immersed in water and pressure is increased on water to force it into the hydrophobic pores. Since water can spontaneously enter the hydrophilic pores of the sample but cannot spontaneously enter the hydrophobic pores, a differential pressure on water is applied so that water is forced into the hydrophobic pores. The measured intrusion volume of water gives the volume of hydrophobic pores and the differential pressure on water gives the pore diameter with Washburn's equation as

$$P_L - P_G = 4\sigma \frac{\cos\theta}{d_p}, \tag{8.43}$$

where d_p is pore diameter, σ is the surface tension of water, θ is the contact angle of water, P_L is the pressure of liquid, and P_G is the pressure of gas.

In mercury intrusion porosimetry, mercury surrounds the sample and application of differential pressure on mercury forces it into the pores. Mercury does not wet hydrophilic and hydrophobic pores and cannot enter these pores spontaneously owing to a small contact angle. Application of pressure on mercury can force it into the pores. The measured intrusion volume is equal to the pore volume and the differential intrusion pressure is related to pore diameter as given in Equation 8.43, where σ and θ are the surface tension and contact angle of mercury, respectively. Mercury porosimetry is valuable in determining the pore structure of the catalyst layer, especially for gas diffusion electrodes, where the distribution of gas and liquid phase pores is essential for the optimization of performance.

A typical mercury intrusion porosimetry test involves placing a sample into a container, evacuating the container to remove contaminant gases and vapors (usually water) and, while still evacuated, allowing mercury to fill the container. This creates an environment consisting of a solid, a nonwetting liquid (mercury), and mercury vapor. Next, pressure is increased toward ambient while the volume of mercury entering larger openings in the sample bulk is monitored. When pressure has returned to ambient, pores of diameters down to approximately 12 mm have been filled. The sample container is then placed in a pressure vessel for the remainder of the test. A maximum pressure of approximately 60,000 psia (414 MPa) is typical for commercial instruments and this pressure will force mercury into pores down to approximately 0.003 mm in diameter. The volume of mercury that intrudes into the sample owing to an increase in pressure from P_i to P_{i+1} is equal to the volume of the pores in the associated size range r_i to r_{i+1}, sizes being determined by substituting pressure values into Washburn's equation.

The measurement of the volume of mercury moving into the sample may be accomplished in various ways. A common method that provides high sensitivity is attaching a capillary tube to the sample cup and allowing the capillary tube to be the reservoir for mercury during the experiment. Only a small volume of mercury is required to produce a long "string" of mercury in a small capillary. When external pressure changes the variation in the length of the mercury column in the capillary, it indicates the volume passing into or out of the sample cup. However, electronic means of detecting the rise and fall of mercury within the capillary are much more sensitive, providing even greater volume sensitivity down to less than a microliter.

In the extrusion flow porometry, a wetting liquid, such as galwick, is used to fill the hydrophilic and the hydrophobic pores of the sample. The inert gas is used to remove the liquid from pores and permit gas flow. The airflow rate and differential air pressures are measured using dry and wet samples. The differential pressure is related to pore diameter as

$$d = 4\sigma \frac{\cos\theta}{\Delta P}, \tag{8.44}$$

where d is pore diameter; γ and θ are surface tension and contact angle, respectively, of the wetting liquid; and Δp is differential pressure.

Mercury porosimetry can distort the pore size owing to the elastic nature of the carbon-PTFE composite; also, for thin electrodes and for electrodes consisting of two or three layers of different porosity, this method is of limited application (Abell et al., 1999)

Advanced techniques such as the scanning technique have been used to directly image the porous microstructures and pore size distributions within porous electrodes (Quinzio et al., 2002). In this technique, the measurements use a piezoelectric-driven scanning probe similar to that used in scanning tunneling microscopy. The probe can measure and map the conductivity through cross sections of porous electrodes. The information on the poor conducting pore can be obtained. In addition, by fitting the probe with a force gauge, the topography of the surface can be simultaneously measured. This technique can provide high-resolution (to 0.1 micron) images of the cross-section, or alternatively by scanning large numbers of pores, it can provide pore size distributions in localized regions of the electrode structure.

8.7 Fuel Cell Test Facility

There are varieties of fuel cell test stations available depending on the nature of tests considered. Table 8.6 lists key tests and equipment associated with fuel cell testing facility.

TABLE 8.6
Fuel Cell Test Facility Components

Test Parameter/Supply	Instrument/Equipment/Material
Power testing	Stationary power (e.g., 120 VAC, split phase, 60 Hz) Auxiliary power units Controllable solid-state power supplies Online uninterruptible power supply
Reactant supply	Hydrogen, natural gas methane, air, oxygen Blowers, compressors
Loads	AC and DC load banks, load profile controller
Continuous power measurement	Auxiliary input: real power, power factor, frequency AC: real power, power factor, frequency Revenue quality meter (kW-h) Solid state metering (watt/var/pf/freq) DC: power
Continuous fuel measurement	High accuracy flow meters Online gas chromatograph for fuel energy
Power quality monitoring	Voltage sags/swells Total harmonic distortion
Exhaust gas analysis	Online high-resolution gas analysis
Sensors	Pressure, temperature, flow measurement
Basic facilities	Deionization water system Vent hood Humidification system Heat exchanger and cooling flows
Data and system control	Computer-based data acquisition system Control system for valve, pressure, and other safety equipment
Purge gas	Nitrogen purge system
Safety instrumentation	Hydrogen, oxygen, and other fuel gas leak detectors and alarms

For fundamental science and engineering analyses, such as evaluating new materials and components, electrode designs, and so on, a system for analytical capability is required. Important features of such a system include the performance (accuracy, stability, precision) of the control and measurement system. In fuel cell testing, this includes environmental and reactant control (flow rate, dew point/humidity, cell temperature, pressure, etc.) and electrical measurements (voltage and current). In addition, the instrumentation should have the capability for state-of-the-art analytical methods, such as continuous, real-time cell resistance measurement by current interrupt and high-frequency resistance, EIS, and controlled voltage/current/power scanning, cycling, among others. On the test data, verifying the integrity and significance of the data is required. The quality of the data is only as good as the quality (accuracy, precision, etc.) of the equipment used to make the measurement.

Experimental conditions such as humidifier water quality, repeatable and accurate humidity streams, cell fixture design and cell assembly, and other external conditions can affect the performance of the cell and the validity and quality of the research data. For highly analytical testing, it is very important to consider and control all parameters of the testing setup. For example, when using electronic load, the closer the load rating is to the expected performance of the cell or stack under test, the more accurate the results will be. Mass flow control and humidification systems may also be subject to unacceptable accuracy errors if not sized correctly. This is why it is difficult to perform testing on both low-power and high-power applications with one test system so it is desirable to use two test systems each properly sized for the desired application. Testing that focuses on more functional parameters of fuel cell assemblies (most stack testing) has different needs that should be considered. Stack testing may require monitoring the electrical performance of each cell as well as multiple temperatures.

8.8 Summary

The characterization of the fuel cell or its component is an important component in the use and operation of the fuel cell. The losses in the fuel cell, namely, activation losses, ohmic losses, concentration losses, and fuel crossover and short-circuit losses, should be characterized to find its optimum operating condition for practical applications. There are a number of diagnostic methods to characterize fuel cell and its components. The basic methods of measurement include potentiostatic, galvanostatic, current interrupt measurement, electrochemical impedance spectroscopy, and cyclic voltammetry. In addition to current–voltage, pressure, and temperature measurements, various tools used in characterizing the fuel cell and its components include optical microscopy, transmission electron microscopy, scanning electron microscopy, atomic force microscopy, the Brunauer–Emmett–Teller method, x-ray diffraction, mercury porosimetry, nuclear magnetic resonance, neutron radiography and Auger electron spectroscopy, x-ray photoelectron spectroscopy, and secondary ion mass spectroscopy. Key parameters of interest to characterize the fuel cell and its components that can be measured from these tools include microstructure of catalysts such as porosity, pore size and distribution, particle or grain size, and crystal structure; surface area of catalysts and electrodes; gas permeation; and chemical composition. The principle of measurement in each of these tools and the methods of measurement are presented in this chapter. There are a number of fuel cell testing facilities currently available from various vendors that enable one to characterize operational parameters for the fuel cell.

References

Abell, A. B., K. L. Willis and D. A. Lange. Mercury intrusion porosimetry and image analysis of cement-based materials. *Journal of Colloid and Interface Science* 211: 39–44, 1999.

Barbir, F. *PEM Fuel Cells: Theory and Practice*. Elsevier Academic Press, Burlington, MA, 2005.

Bard, A. J. and L. Faulkner. *Electrochemical Methods: Fundamentals and Applications*. John Wiley & Sons, New York, 2001.

Barsoukov, E. and J. R. Macdonald, Editors. *Impedance Spectroscopy—Theory, Experiment, and Applications*. Wiley-Interscience, Hoboken, New Jersey, 2005.

Brunauer, S., P. H. Emmett and E. Teller. Adsorption of gases in multimolecular layers. *Journal of the American Chemical Society* 60: 309–319, 1938.

Cooper, K. R. and M. Smith. Electrical test methods for on-line fuel cell ohmic resistance measurement. *Journal of Power Sources* 160(2): 1088–1095, 2006.

Dullien, F. A. L. Porous Media. *Fluid Transport and Pore Structure*. Academic Press, San Diego, CA, 1992.

Dunbar, Z. and R. I. Masel. Quantitative MRI study of water distribution during operation of a PEM fuel cell using Teflon® flow fields. *Journal of Power Sources* 171(2): 678–687, 2007.

Fuller, T. F. and J. Newman. Experimental determination of the transport number of water in nafion 117 membrane. *J Electrochem Soc* 139(5): 1332–1337, 1992.

Ghouse, M., H. Abaoud, A. Al-Boeiz and S. Al-Zaharani. Fabrication and characterization of the graphite bi-polar plates used in a 0.25 kW PAFC stack. *International Journal of Hydrogen Energy* 23(8): 721–730, 1998.

Hack, H. P., P. J. Moran and J. R. Scully. In: *The Measurement and Correction of Electrolyte Resistance in Electrochemical Tests—ASTM STP 1056*, Editors L. L. Scribner and S. R. Taylor. American Society for Testing and Materials, Philadelphia, PA, pp. 5–26, 1990.

Hung, Y., H. Tawfik and D. Mahajan. Durability and characterization studies of polymer electrolyte membrane fuel cell's coated aluminum bipolar plates and membrane electrode assembly, *Journal of Power Sources* 186: 123–127, 2009.

Ivers-Tiffée, E., A. Weber and H. Schichlein. *Chapter 17—Electrochemical Impedance Spectroscopy in Handbook of Fuel Cells*, Vol. 2. John Wiley & Sons, Hoboken, New Jersey, 2003.

Jena, A. K. and K. Gupta. Characterization of pore structure of filter media. *Fluid/Particle Separation Journal* 14: 227–241, 2002.

Kim, J. S., W. H. A. Peelen, K. Hemmes and R. C. Makkus. Effect of alloying elements on the contact resistance and the passivation behavior of stainless steel. *Corrosion Science* 44: 635–655, 2002.

Lakshminarayana, G. and M. Nogami. Synthesis and characterization of proton conducting inorganic-organic hybrid nanocomposite films from mixed phosphotungstic acid/phosphomolybdic acid/tetramethoxysilane/3-glycidoxypropyltrimethoxysilane/phosphoric acid for H_2/O_2 fuel cells. *Journal of Renewable and Sustainable Energy* 1(063106): 1–18, 2009.

Laraminie, J. and A. Dicks. *Fuel Cells Systems Explained*. John Wiley & Sons, New York, 2000.

Orazem, M. E. and B. Tribollet. *Electrochemical Impedance Spectroscopy*. John Wiley & Sons, Hoboken, New Jersey, 2008.

Quinzio, M. V., G. To and A. H. Zimmerman. Scanning porosimetry for characterization of porous electrode structures Battery, Conference on Applications and Advances, 2002, The Seventeenth Annual, pp. 291–295, 2002.

Ren, X., T. E. Springer, T. A. Zawodzinski and S. Gottesfeld. Methanol transport through nafion membranes—Electro-osmotic drag effects on potential step measurements. *Journal of the Electrochemical Society* 147: 466–474, 2000.

Springer, T. E., M. S. Wilson and S. Gottesfeld. Modeling and experimental diagnostics in polymer electrolyte fuel cells. *Journal of the Electrochemical Society* 140(12): 3513–3526, 1993.

Springer, T. E., T. A. Zawodzinski and S. Gottesfeld. Polymer electrolyte fuel cell model. *Journal of the Electrochemical Society* 138(8): 2334–2342, 1991.

Springer, T. E., T. A. Zawodzinski, M. S. Wilson and S. Gottesfeld. Characterization of polymer electrolyte fuel cells using AC impedance spectroscopy. *Journal of the Electrochemical Society* 143: 587–599, 1996.

Tawfik, H., Y. Hung and D. Mahajan. Metal bipolar plates for PEM fuel cell—A review. *Journal of Power Sources* 163: 755–767, 2007.

Trabold, T. A., J. P. Owejan, D. L. Jacobson, M. Arif and P. R. Huffman. In situ investigation of water transport in an operating PEM fuel cell using neutron radiography: Part 1—Experimental method and serpentine flow field results. *International Journal of Heat and Mass Transfer* 49: 4712–4720, 2006.

9

Fuel Cell Components and Design

As we have discussed in Chapter 1, a number of different fuel cells have been under development and are in use for a wide range of applications. These fuel cells are generally classified by the type of electrolyte used, type of ion transferred, and the range of applicable temperature level. The most commonly used fuel cells are alkaline fuel cell (AFC), phosphoric acid fuel cell (PAFC), polymer electrolyte membrane fuel cell (PEMFC), molten carbonate fuel cell (MCFC), and solid oxide fuel cell (SOFC). Among these, PAFC and PEMFC are referred to as *cation fuel cells* with positively charged proton (H^+) as transporting ion. Fuel cells such as AFC, MCFC, and SOFC are termed *anion fuel cells*, which involve negatively charged ions transporting through the electrolytes such as hydroxyl ion (OH^-), carbonate ion $\left(CO_3^{2-}\right)$, and oxide ion (O^{2-}), respectively. In this chapter, descriptions of these fuel cells are given with regard to the design, major components, materials, operation, and technical challenges.

9.1 Alkaline Fuel Cell

AFC has alkali (NaOH or KOH) as electrolyte as a mobile liquid or in immobilized form in a porous matrix. This fuel cell can use a variety of non-precious metals as a catalyst at the anode and cathode. The AFC has excellent performance compared to other candidate fuel cells and also has flexibility to use a wide range of electrocatalysts. Francis T. Bacon developed AFC in the 1940s and 1950s, referred to as Bacon fuel cell. In the early 1960s, aircraft engine manufacturer Pratt & Whitney licensed the Bacon patents and won the National Aeronautics and Space Administration (NASA) contract to power the Apollo spacecraft with alkali cells. The AFC was developed for space application by UTC fuel cells, where it provided on-board electric power. Union Carbide Corp. (UCC) developed AFCs for terrestrial mobile applications starting in the late 1950s, lasting until the early 1970s. UCC fuel cell systems were used in the US Army and the US Navy, an alkaline direct hydrazine powered motorcycle, and the "Electrovan" of General Motors. Professor Karl V. Kordesch built his Austin A-40 car, fitted with UCC fuel cells with lead acid batteries as hybrid (Kordesch and Hacker, 2003; Cifrain and Kordesch, 2003). It was demonstrated on public roads for three years. However, the large commercialization of this fuel cell did not catch up with other competing fuel cells such as MCFC or SOFC because of its inherent issue of carbonate formation with electrolyte from carbon dioxide in the oxidant stream.

9.1.1 AFC Basic Principles and Operations

In an alkaline electrolyte fuel cell, hydroxyl (OH^-) ions are available and mobile. At the anode, these react with hydrogen, releasing energy and electrons, and producing water.

Figure 9.1 shows the operating configuration of the AFC.

$$H_2 + 2OH^- \rightarrow 2H_2O + 2e^- \text{ (anode)} \tag{9.1}$$

At the cathode, oxygen reacts with electrons taken from the electrode, and water in the electrolyte, forming new OH ions.

$$\frac{1}{2}O_2 + 2^{e-} + H_2O \rightarrow 2OH^- \text{ cathode} \tag{9.2}$$

$$H_2 \frac{1}{2}O_2 \rightarrow H_2O + \text{electric energy} + \text{beat (cell)} \tag{9.3}$$

For these reactions to proceed continuously, the OH^- ions must be able to pass through the electrolyte and there must be an electrical circuit for the electrons to go from the anode to the cathode. Since KOH has the highest conductance among the alkaline hydroxides, it is the preferred electrolyte. The KOH solution molarity is typically between 30% and 80%, depending on the operating

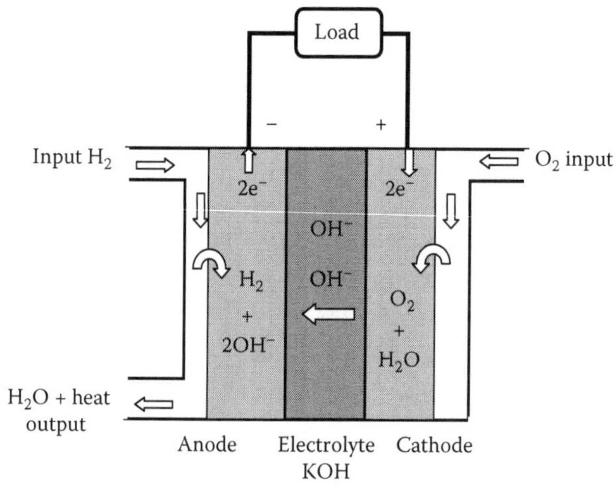

FIGURE 9.1
Principles of operation of AFCs.

temperature. A higher molarity reduces the vapor pressure of the solution, and thus high-temperature systems require a high electrolyte concentration. Note that although water is consumed at the cathode, it is created twice as fast at the anode. This leads to a water management challenge on the AFC.

9.1.2 AFC Components and Configurations

The fuel cell developed by Bacon operated at relatively high temperature (200 to 240°C) and pressure (40 to 55 atm) to keep electrolyte (45% KOH) from boiling. The Bacon cell performance was very good with cell voltage of 0.8 V at a current density of 1000 mA/cm^2. Figure 9.2 shows the configuration of the Bacon fuel cell. The anode consisted of a two-layer structure of nickel with porous Ni of 16 μm maximum pore diameter on the electrolyte side and 30 μm pore diameter on the gas side, and the cathode consisted of a porous structure of lithiated NiO. The high temperature of Bacon's AFC enabled the use of non-noble metal nickel catalysts. The electrolyte (liquid), electrode/catalyst (solid), and reactant (gas) phase boundary in the porous electrodes was maintained by a differential gas pressure across the electrode. In many cell designs, the electrolyte is circulated (mobile electrolyte) so that heat can be removed and water eliminated by evaporation.

The AFC can be categorized into three main configurations, static electrolyte, mobile electrolyte, and charged electrolyte systems. The Bacon fuel cell is an example of a mobile electrolyte system. In this system, the electrolyte is

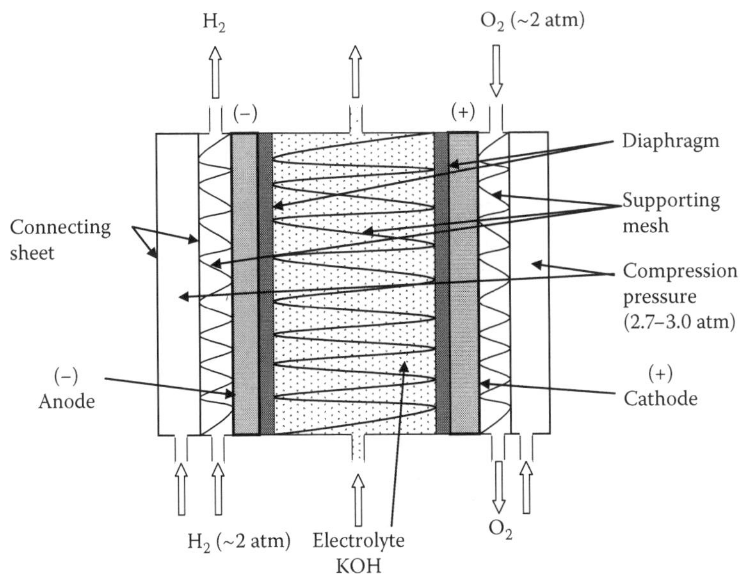

FIGURE 9.2
Configuration of a Bacon fuel cell.

pumped from the stack into an electrolyte reservoir. The mobile electrolyte is constrained within the porous electrode structure by either asbestos (as in the case of Bacon cell) or other porous separation layer between the electrode and the mobile electrolyte or by careful control of the differential pressure in the anode and cathode and the surface tension in the porous electrode structure.

The KOH solution is pumped around the fuel cell. Hydrogen is supplied to the anode, but must be circulated, as it is at the anode that the water is produced. The hydrogen will evaporate the water, which is then condensed out at the cooling unit that the hydrogen is circulated through. The hydrogen comes from a compressed gas cylinder, and the circulation is achieved using an ejector circulator. The system uses air, rather than oxygen.

One of the main issues with AFC is that of electrolyte and electrode degradation caused by the formation of carbonate/bicarbonate $\left(CO_3^{2-}/HCO_3^-\right)$ in the liquid alkaline electrolyte on reaction of OH^- ions with carbon dioxide contamination in the oxidant stream. Unless pure oxygen is used, CO_2 has to be scrubbed from air and electrolyte needs to be replaced often.

The carbonates are formed as

$$CO_2 + 2OH^- \rightarrow CO_3^{2-} + H_2O \tag{9.4}$$

$$CO_2 + OH^- \rightarrow HCO_3^-. \tag{9.5}$$

The major cause of the degradation is that the carbonate/hydrogen carbonate $\left(CO_3^{2-}/HCO_3^-\right)$ precipitate into large solid metal carbonate crystals (Na_2CO_3 or K_2CO_3) and fill the electrolyte-filled pores of electrodes blocking pores and mechanically disrupting and destroying the active layers. The potassium hydroxide is thus gradually changed to potassium carbonate. The effect of this is that the concentration of OH^- ions reduces as they are replaced with carbonate CO_3^- ions, which greatly affects the performance of the cell.

The circulating electrolyte system has several advantages over an immobilized system: (i) no humidification of the reactant is required because the water content of the caustic electrolyte remains uniform everywhere inside the stack, (ii) heat can be managed by a heat exchanger compartment in the stack where the recirculating electrolyte itself works as a cooling liquid inside each cell, (iii) any accumulated impurities in the circulating stream can easily be removed, (iv) the OH^- concentration is maintained fairly constant with minimum gradient, and (v) the electrolyte prevents the buildup of gas bubbles between electrodes and electrolyte as they are washed away continuously.

One disadvantage of the mobile electrolyte is it requires extra equipment. A pump is needed to pump the corrosive fluid. The extra pipework means more possibilities for leaks, and the surface tension of the KOH solution makes for a fluid that is prone to finding its way through the smallest of gaps. Also, it becomes harder to design a system that will work in any orientation.

An alternative to a "free" electrolyte, which circulates, is for each cell in the stack to have its own separate electrolyte that is held in a matrix material between the electrodes.

The KOH solution can be held in a matrix material, such as asbestos or other porous matrix. Though asbestos has excellent porosity, strength, and corrosion resistance, it has public safety problems. The system uses pure oxygen at the cathode, though that is not obligatory for a matrix-held electrolyte. The hydrogen is circulated as with the previous system, in order to remove the product water. In spacecraft systems, this product water is used for drinking, cooking, and cabin humidification. However, a cooling system with cooling water, or other fluid, such as a glycol–water mixture, will also be needed. In the Orbiter systems, the cooling fluid was a fluorinated hydrocarbon dielectric liquid. This matrix-held electrolyte system was essentially like the PEMFC—the electrolyte is like solid and can be in any orientation. A major advantage is of course the electrolyte does not need to be pumped around and there is also no problem of the internal "short circuit," which can be the result of a pumped electrolyte. However, there is the problem of water management where water is produced at the anode and water is used at the cathode. The fuel cell must be designed so that the water content of the cathode region is kept sufficiently high by diffusion from the anode. For one thing, the saturated vapor pressure of KOH solution does not rise so quickly with temperature as it does with pure water. Besides the water management problem, carbon dioxide contamination of the electrolyte needs to be addressed by renewal of the electrolyte or rebuilding a complete fuel cell.

Another configuration is fuel-charged AFC. Fuel-charged systems have been used in a number of successful fuel cell demonstrators. The principle is shown in Figure 9.3.

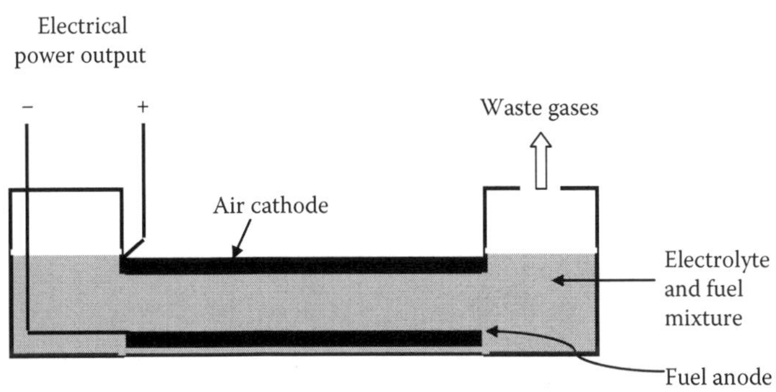

FIGURE 9.3
Fuel-charged AFC.

The electrolyte is KOH solution, with a fuel, such as hydrazine or ammonia, mixed with it. The fuel anode is a platinum catalyst. The fuel is also fully in contact with the cathode. The strong smell of ammonia is an advantage since it indicates any leakage in the system immediately. Ammonia poisoning is medically completely reversible. AFC is not very sensitive to ammonia in the fuel gas since the electrolyte rejects ammonia and the residual ammonia in the hydrogen can be recirculated through the dissociator catalytic heating unit. No shift converter, selective oxidizer, or further co-reactants like water are required. This results in a compact lightweight dissociator.

This dissolved fuel type of cell can be used with other liquid fuels such as methanol (Muller et al., 2000; Sinor J. E. Consultants, 1997). An advantage of a liquid fuel like methanol is its high theoretical capacity per volume and weight. According to

$$CH_3OH + 6OH^- \rightarrow 5H_2O + 6e^- + CO_2, \quad (9.6)$$

the capacity of 1 kg of methanol is 5025 Ah. With a density of $\rho = 0.79$ kg m^{-3}, the capacity per volume is approximately 4000 Ah l^{-1}. The air electrode is the positive part of the cell. In an alkaline solution, the oxygen combines with the reaction water to reform most of the OH^- ions consumed in Equation 9.1.

$$3/2O_2 + 3H_2O + 6e^- \rightarrow 6OH^-. \quad (9.7)$$

Therefore, an alkaline methanol cell shows the following overall reaction:

$$CH_3OH + 3/2O_2 \rightarrow CO_2 + 2H_2O. \quad (9.8)$$

In this process, not only oxygen and methanol are consumed but also two OH^- ions per molecule of methanol. Therefore, the molarity of the OH^- ions should be twice that of methanol to reach a complete conversion of the fuel.

Though there is a "fuel crossover" problem, in this case it does not matter greatly, as the cathode catalyst is not platinum, and so the rate of reaction of the fuel on the cathode is very low. The cell is refueled simply by adding more fuel to the electrolyte. However, since carbon dioxide is produced, it reacts with the KOH solution, converting it to carbonate. This effect makes the cell impractical for use as a power source. Indeed, since the electrolyte is "used up" by the cell reaction, it could be argued that the system is not a true fuel cell.

9.1.3 AFC Electrolyte, Electrode, and Catalyst

AFCs can be operated at a wide range of temperatures and pressures. It is also the case that their range of applications is quite restricted. The result of this is that there is no standard type of electrode for the AFC, and different approaches are taken depending on performance requirements, cost limits,

TABLE 9.1

Typical Operational Characteristics of AFC

AFC System Year	Temperature (°C)	Pressure (kPa)	Electrolyte Configuration	Catalyst Anode–Cathode	Performance
Bacon cell/1940–1950	200	4500	Recirculating	Ni–NiO	0.8 V at 1 A/cm^2
Apollo, 1960s	230	340	Static	Ni–NiO	1.5 kWe/109 kg
Space Shuttle orbiter, 1980s	93	410	Static	Pt/Pd–Au/Pt	12 kWe/93 kg
Siemens, 1986	80	220	Recirculating	Ni–Ag	0.8 V at 0.4 A/cm^2
Russian photon system, 1993	100	120	Static	Pt–Pt	Efficiency 65%–75%

Source: Adapted from M. Warshay and P. R. Prokopius, *Journal Power Sources*, 29, 193–200, 1990.

operating temperature, and pressure. A significant cost advantage of AFCs is that both anode and cathode reactions can be effectively catalyzed with nonprecious, relatively inexpensive metals. The AFC has been developed and operated with a variety of catalysts over a very broad range of temperature, pressure, and electrolyte solution concentration. Table 9.1 shows some of the operational parameters for a selection of AFC systems built and tested throughout the years.

9.1.3.1 Electrolyte

The electrolyte in the AFC is concentrated (85 wt%) KOH in cells designed for operation at high temperature (~260°C) or less concentrated (35–50 wt%) KOH for lower temperature (<120°C) operation. AFC electrolyte development has been restricted to KOH water solutions with normalities ranging from 6 to 8. For immobilized cells, the electrolyte is retained in a matrix. Previously, asbestos was used as matrix material. However, because of its health hazards, material like potassium hexatitanate bonded with polytetrafluoroethylene (PTFE) have been considered as matrix material.

Depending on the operating temperature, the pressure of the cell has to be maintained such that the electrode is kept under subcooled conditions. The AFC used in the US Apollo Space Program utilized pure H_2 and O_2 and concentrated electrolyte (85% KOH) to permit cell operation at a lower pressure (410 kPa reactant gas pressure) without electrolyte boiling. The typical performance of this AFC cell was 0.85 V at 150 mA/cm^2, comparing favorably to the performance of the Bacon cell operating at approximately 10 times higher pressure. The AFCs in the Space Shuttle Orbiter operate in the same pressure range as the Apollo program cells, but at a lower temperature (80 to 90°C) and a higher current density (470 mA/cm^2 at 0.86 V).

9.1.3.2 Electrodes and Catalysts

Because of small kinetic losses, a wide range of electrocatalysts can be used; these include Ni, Ag, metal oxides, spinels, and noble metals. The AFCs developed by Bacon utilized non-noble sintered nickel metal catalysts. The high electrical conductivity of these porous electrodes permits use of current collection from monopolar stack plates. Bacon-designed nickel electrodes for AFC were made porous by fabricating them from powdered nickel, which was then sintered to make a rigid structure. To enable a good three-phase contact between the reactant gas, the liquid electrolyte, and the solid electrode, the nickel electrode was made in two layers using two sizes of nickel powders. This gave a wetted fine pore structure for the liquid side, and more open pores for the gas side. This electrode structure was also used in the Apollo mission fuel cells.

Raney metals have been used to achieve a very active and porous form of a metal in the AFC from the 1960s through to the present. The Raney metals are prepared by mixing the active metal (e.g., nickel) with an inactive metal, usually aluminum. The mixing is done in such a way that separate regions of aluminum and the host metal are maintained—it is not a true alloy. The mixture is then treated with a strong alkali, which dissolves out the aluminum. This leaves a highly connected porous material, with very high surface area. The required pore sizes and their distribution can be obtained by mixing the two metals of different grain sizes.

Recently developed electrodes have carbon-supported catalysts, mixed with PTFE, which are then rolled out onto a material such as nickel mesh. The PTFE acts as a binder, and also its hydrophobic properties stop the electrode flooding and provide for controlled permeation of the electrode by the liquid electrolyte. A thin layer of PTFE will often be put over the surface of the electrode to further control the porosity and to prevent the electrolyte passing through the electrode, without the need to pressurize the reactant gases, as has to be done with the porous metal electrodes. Carbon fiber is sometimes added to the mix to increase strength, conductivity, and roughness. The manufacturing process is low cost and can be done using a modified paper making machine. The electrodes are similar to PEFC electrode structures with a porous carbon cloth consisting of a carbon support material with fine metal catalysts, interdispersed with PTFE for hydrophobicity and pressed onto a nickel mesh to improve in-plane conductivity. The PTFE acts as a binder, and its hydrophobic properties also stop the electrode from flooding and provide for controlled permeation of the electrode by the liquid electrolyte. A thin layer of PTFE is put over the surface of the electrode to further control the porosity and to prevent the electrolyte passing through the electrode, without the need to pressurize the reactant gases, as has to be done with the porous metal electrodes. Carbon fiber is sometimes added to the mix to increase strength, conductivity, and roughness.

In a typical design, the electrodes can contain high loadings of noble metals for low-temperature operation (80°C to 90°C). For example, electrodes made of 80% Pt–20% Pd anodes loaded at 10 mg/cm^2 on Ag-plated Ni screen and 90% Au–10% Pt cathodes loaded at 20 mg/cm^2 on Ag-plated Ni screen and then bonded with PTFE to achieve high performance at the lower temperature have been used. A wide variety of materials (e.g., potassium titanate, ceria, asbestos, zirconium phosphate gel) have been used in the microporous separators for AFCs.

9.1.3.3 Stack Configuration

The stack design of the AFC is similar to the PEFC stack where the individual cells in bipolar plate stacks are typically connected in series, with current collection across the entire electrode surface along the interface between the bipolar plate landings and the separator plate. The flow fields in AFCs are similar to those used in other fuel cells, and various parallel and serpentine configurations are used to optimize mass, heat, and reactant/product transport. A subset of stacks is designed using monopolar plates. Monopolar plates are used for many AFC applications. In this design, there is a PTFE sheet between the electrode and the flow field to prevent the liquid electrolyte from passing through the electrode into the channel, which can be by static forces or by weeping, which is caused by electro-osmotic pressure-induced motion resulting from current flow. PTFE is an insulator, and so it is difficult to make an electrical connection to the face of the electrode. The connections are normally made to the edge of the electrode. Simple wires connect the positive of one cell to the negative of another. This gives certain flexibility: it is not necessary to connect the positive of one cell to the negative of the adjacent cell, as must occur with bipolar plates. Instead, they are connected in a series-parallel arrangement to optimize power, compactness, and durability. A monopolar arrangement allows the unique advantage of isolating single cells in the event of replacement or damage is also realized.

9.1.4 AFC Recent Advances

Once development of alkaline cells was underway for space application, terrestrial applications began to be investigated. A significant cost advantage of AFCs is that both anode and cathode reactions can be effectively catalyzed with nonprecious, relatively inexpensive metals. Most low-cost catalyst development work has been directed toward Raney nickel powders for anodes and silver-based powders for cathodes. The essential characteristics of the catalyst structure are high electronic conductivity and stability (mechanical, chemical, and electrochemical). Electrode development is concentrated on multi-layered structures with porosity characteristics optimized for flow of liquid electrolytes and gases (H_2 and O_2). Both metallic (typically hydrophobic) and carbon-based (typically hydrophilic) electrode structures are being investigated. Development of low-cost manufacturing processes including powder mixing and pressing of carbon-based electrodes, sedimentation and spraying, and high-temperature sintering operations continues.

The focus on AFC stack development is directed toward reducing space, weight, and cost. Epoxy resins, polysulfone, and ABS (acrylonytril-butadiene-styrene) have been under investigation. Framing techniques under continuing development include injection molding, filter pressing, or welding.

As discussed above, one of the main issues with AFC is that of electrolyte and electrode degradation caused by the formation of carbonates with carbon dioxide contamination in the oxidant stream. In order to avoid this problem, alkaline anion exchange solid membranes similar to the PEFC membrane are proposed (Varcoe and Slade, 2005). Recently, development of stable solid-state alkaline polymer electrolyte has been carried out. Most of these membranes contain either trimethylammonium or *N*-methyl pyridium groups. The trimethylammonium groups have been found to be most stable in hot alkaline solutions. The membranes such as quaternary ammonium polysulphone are found to be stable up to 120°C (Lu et al., 2008).

9.2 Phosphoric Acid Fuel Cell

PAFC, as its name implies, is a fuel cell that uses phosphoric acid as its electrolyte. It is also the most advanced type of fuel cell available and is the closest to commercialization. There are hundreds of MW demonstration PAFC units worldwide that have been tested, are being tested, or are being fabricated. The plants are in the 50 to 200 kW capacity range, but large plants of 1 MW and 5 MW have been built and operated. The largest plant operated to date has been that built by International Fuel Cells and Toshiba for Tokyo Electric Power. This has achieved 11 MW of grid quality AC power. Major efforts in the United States and Japan are now concentrated on improving PAFCs for stationary dispersed power and on-site cogeneration (CHP) plants. The major industrial participants are International Fuel Cells Corporation in the United States and Fuji Electric Corporation, Toshiba Corporation, and Mitsubishi Electric Corporation in Japan. PAFCs have been developed for electric utilities and for on-site cogeneration and vehicular applications, where PAFCs' unique characteristics make them preferable to the stationary dispersed power plants and on-site cogeneration power plants.

9.2.1 PAFC Basic Principles and Operations

The PAFC works in a similar fashion to the PEMFC described in Section 9.2. The electrolyte is an inorganic acid; concentrated phosphoric acid (100%), which, like the membranes in the PEM cells, will conduct protons and the reactions occurring on the anode and cathode, is similar to the PEFC. At the anode, the hydrogen gas ionizes, releasing electrons and creating H^+ ions (or protons). This reaction releases energy. At the cathode, oxygen reacts with electrons taken from the electrode, and H^+ ions from the electrolyte, to form water.

The electrochemical reactions occurring in PAFCs are

$$\text{At the anode:}\quad H_2 \rightarrow 2H^+ + 2e^- \tag{9.9}$$

$$\text{At the cathode: } \frac{1}{2}O + 2H^+ + 2e^- \rightarrow 2H_2O \tag{9.10}$$

$$\text{Overall cell reaction}\quad H_2 + O \rightarrow H_2O. \tag{9.11}$$

In the PAFC, the electrochemical reactions take place on highly dispersed electrocatalyst particles supported on carbon black. As with the PEMFCs, platinum (Pt) or Pt alloys are used as the catalyst at both electrodes. The PAFC operates at an elevated temperature compared to the PEFC, at 160°C–220°C. The higher-temperature PAFC was chosen for development for terrestrial applications. Since the PAFC has a range of operating temperature (16°C–220°C), operating pressure ranges from 1 atm to several atmospheres. Typical current densities are 150–350 mA/cm^2 with output voltage per cell of 0.5 to 0.8 V. The reactant flow stoichiometries are similar to other fuel cells.

Higher operating pressures require larger parasitic losses and ancillary component costs, so most installed systems now operate at near atmospheric pressure. The main advantage of the PAFC is that the higher temperature eliminates or reduces two major problems with the PEFC, CO poisoning sensitivity and water management. The PAFC cannot accomplish internal reformation like the MCFC or SOFC, but because of the elevated temperature of 200°C compared to 80°C for the PEFC, the anode in the PAFC can tolerate a 1%–2% CO in the feed stream. This allows operation on reformed natural gas and other fuel feedstock with minimal CO filtering, greatly reducing reformer size and control requirements. Because of the high-temperature operation and electrolyte behavior, water management in the PAFC is not a major concern. The electrolyte is a highly concentrated acid solution that is conductive without water and has a very low vapor pressure at high concentration. The electrolyte concentration varies between 90% and 100% during operation depending on the flow rate, current density, and operating temperature. During operation, water is generated at the cathode, which is readily evaporated into the flow stream or absorbed into the electrolyte. If the water is absorbed into the electrolyte, dilution increases the vapor pressure and drives off the water at a faster rate. Therefore, water management in the electrolyte is self-regulating. The generated water leaves the system as steam, which can be used for the steam reformation process in the fuel-processing subsystem or to provide thermal energy for cogeneration application. Another major advantage of the electrolyte system is that control over freeze damage can be accomplished by dilution of the electrolyte to lower concentrations, which can reduce the freeze point of the electrolyte to below −40°C before shipping. Once in operation, the PAFC will self-regulate the acid concentration when the operating temperature reaches a high normal value through the vapor pressure dependency discussed. Compared to the SOFC, the PAFC does not suffer the major material compatibility or manufacturing difficulties.

9.2.2 PAFC Components and Configurations

The basic cell structure consists of an electrolyte (phosphoric acid) contained in a matrix sandwiched between the anode and the cathode as shown in Figure 9.4. A key matrix material is silicon carbide (SiC). The electrode is composed of a catalyst layer where the electrochemical reaction takes place and the substrate that mechanically supports the thin catalyst layer. The purpose of the matrix is not only to hold phosphoric acid as an integral part of the cell structure but also to prevent the crossover of reactant gases to the opposite electrode. The matrix containing the electrolyte is ionically conductive but electronically non-conductive. A separator is used to prevent the mixing of reactant gases. The key material involved is again impregnated graphite.

As to the cooling system, there are three cooling media: water, air, and oil. Water cooling is superior to other methods from the point of view of performance and is appropriate for relatively larger stacks. Air cooling systems are very simple compared to water cooling ones and are generally appropriate for relatively small stacks. Dielectric liquid (oil) cooling systems are

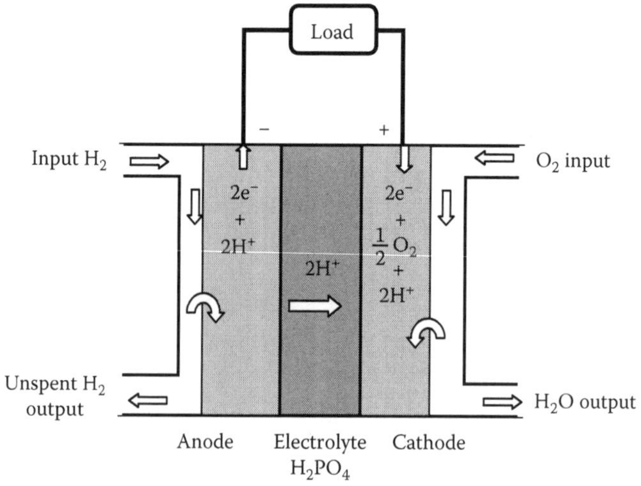

FIGURE 9.4
Principles of operation of PAFC.

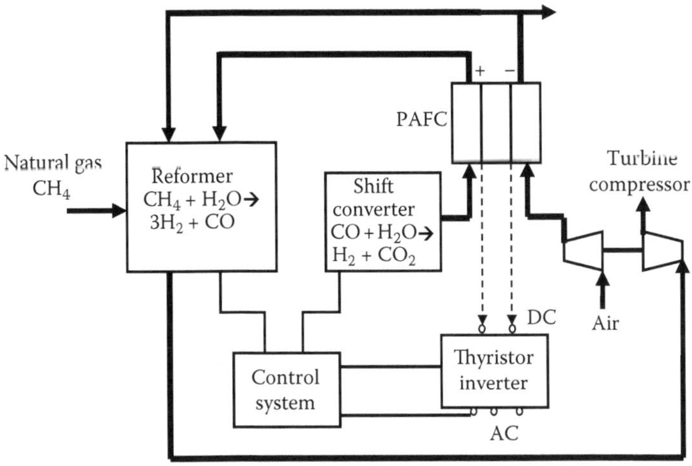

FIGURE 9.5
PAFC power system block diagram.

used for vehicular applications. Their performance is in between those of air and water cooling systems. Such systems are non-pressurized and corrosion resistant.

PAFCs can operate at temperatures up to 220°C and, consequently, are less sensitive to carbon monoxide poisoning compared to PEMFCs. This simplifies fuel processing, that is, the conversion of hydrocarbon fuels such as natural gas to hydrogen-rich fuel, although sulfur still must be removed from the fuel. The operating temperature range is low enough to allow use of moderately priced high-temperature materials for packaging and the balance of plant. A typical PAFC plant contains four key components: the fuel cell stack, fuel processing system (reformer), inverter, and control system as shown in Figure 9.5. On-site cogeneration is recognized to be the most effective application of PAFC plants. Plants can be installed directly at the center of demand and supply both heat and electricity to customers.

9.2.3 PAFC Electrolyte, Electrode, and Catalyst

Because of its commercialization potential, PAFC component development has progressed well from 1965 to the present. Table 9.2 summarizes the evaluation of PAFC components. In the mid-1960s, the conventional porous electrodes were PTFE-bonded Pt black, and the loadings were approximately 9 mg Pt/cm^2. During the past two decades, Pt supported on carbon black has replaced Pt black in porous PTFE-bonded electrode structures as the electrocatalyst. A dramatic reduction in Pt loading has also occurred; the loadings are currently approximately 0.10 mg Pt/cm^2 in the anode and approximately

TABLE 9.2

PAFC Component Evaluation

Component	Year 1965	Year 1975	Current
Anode	PTFE-bonded Pt black 9 mg Pt cm^{-2}	PTFE-bonded Pt/C Vulcan XC-72a 0.25 mg Pt cm^{-2}	PTFE-bonded Pt/C 0.25 mg Pt cm^{-2}
Cathode	PTFE-bonded Pt black 9 mg/cm^2	PTFE-bonded Pt/C Vulcan XC-72a 0.5 mg Pt cm^{-2}	PTFE-bonded Pt/C 0.5 mg Pt cm^{-2}
Electrode support	Ta mesh screen	Graphite structure	Graphite structure
Electrolyte support	Glass fiber paper	PTFE-bonded SiC	PTFE-bonded SiC
Electrolyte	85% H_3PO_4	95% H_3PO_4	100% H_3PO_4
Electrolyte matrix		Porous graphite plate	Porous graphite plate
Cooling channels			1 per ~7 cells; imbedded (*SS*) tubes in graphite plate

Source: Adapted from *Fuel Cell Hand Book*, 7th edition, Department of Energy, 2004.

0.50 mg Pt/cm^2 in the cathode. The operating temperatures and acid concentrations of PAFCs have increased to achieve higher cell performance; temperatures of approximately 200°C (392°F) and acid concentrations of 100% H_3PO_4 are commonly used today. Although the present practice is to operate at atmospheric pressure, the operating pressure of PAFCs surpassed 8 atm in the 11 MW electric utility demonstration plant, confirming an increase in power plant efficiency.

9.2.3.1 Electrolyte

The electrolyte phosphoric acid (H_3PO_4) is a colorless viscous hygroscopic liquid. This inorganic acid has well enough thermal stability, chemical and electrochemical stability and low enough volatility above approximately 150°C to be considered as an electrolyte for fuel cells. Moreover, phosphoric acid is tolerant to CO, in the fuel and oxidant, unlike the AFC. In the PAFC, it is contained by capillary action (it has a contact angle >90°) within the pores of a matrix made of silicon carbide particles held together with a small amount of PTFE. The pure 100% phosphoric acid, used in fuel cells since the early 1980s, has a freezing point of 42°C, so to avoid stresses developing owing to freezing and rethawing, PAFC stacks are usually maintained above this temperature once they have been commissioned. Although the vapor pressure is low, some acid is lost during normal fuel cell operation over long periods at high temperature; it is therefore necessary to replenish electrolyte during operation, or ensure that there is sufficient reserve of acid in the matrix at the start of operation to last the projected lifetime. The SiC matrix comprising particles of approximately 1 mm, is 0.1–0.2 mm thick, which is thin enough to allow reasonably low ohmic losses (i.e., high cell voltages) while having sufficient mechanical strength and the ability to prevent crossover of reactant gases from one side of the cell to the other. Under some conditions, the pressure difference between anode and cathode can rise considerably, depending on the design of the system. The SiC matrix presently used is not robust enough to stand pressure differences greater than 100–200 mbar.

9.2.3.2 Electrodes and Catalysts

Like the PEMFC, the PAFC uses gas diffusion electrodes. In the mid-1960s, the porous electrodes used in the PAFC were PTFE-bonded Pt black, and the loadings were approximately 9 mg Pt cm^{-2} on each electrode. Since then, Pt supported on carbon has replaced Pt black as the electrocatalyst. The carbon is bonded with PTFE (approximately 30–50 wt%) to form an electrode support structure. The carbon has important functions: (i) to disperse the PI catalyst to ensure good utilization of the catalytic metal, (ii) to provide micropores in the electrode for maximum gas diffusion to the catalyst and electrode/electrolyte interface, and (iii) to increase the electrical conductivity of the catalyst.

By using carbon to disperse the platinum, a dramatic reduction in Pt loading has also been achieved over the last two decades; the loadings are currently approximately 0.10 mg Pt cm^{-2} in the anode and approximately 0.50 mg Pt cm^{-2} in the cathode. The choice of carbon is important, as is the method of dispersing the platinum. The activity of the PI catalyst depends on the type of catalyst, its crystallite size, and specific surface area. Small crystallites and high surface areas generally lead to high catalyst activity.

9.2.3.3 Stack

The bipolar plates used in early PAFCs consisted of a single piece of graphite with gas channels machined on either side. As large power size PAFCs are developed for stationary applications, the PAFC stack design consists of a repeating arrangement of a ribbed bipolar plate, the anode, electrolyte matrix, and cathode. The ribbed bipolar plate is easier for continuous manufacturing process in large sheets. The bipolar plates separate the individual cells and electrically connect them in series, while providing the gas supply to the anode and cathode, respectively, similar to other flat fuel cells. The flat surfaces between catalyst layer and substrate promote better and uniform gas diffusion to the electrodes. Phosphoric acid can be stored in the substrate, thereby increasing the lifetime of the stack. A typical PAFC stack may contain 50 or more cells connected in series to obtain the practical voltage level required.

PAFC stack has coolant channels to remove heat generated during cell operation, and these channels are located about every fifth cell. Heat is removed by either liquid (two-phase water or a dielectric fluid) or gas (air) coolants that are routed. Liquid cooling requires complex manifolds and connections, but better heat removal is achieved than with air cooling. The advantage of gas cooling is its simplicity, reliability, and relatively low cost. However, the size of the cell is limited, and the air-cooling passages are much larger than the liquid-cooling passages.

9.2.4 PAFC Recent Advances

Various sizes of PAFC are available for stationary applications that meet market specifications and are generally supplied with guarantees. Cell components are being manufactured at scale and in large quantities. Many of these systems have now run for several years, and so there is a good amount of operating experience on the reliability of the stack and the quality of power produced by the systems. The attribute of high power quality and reliability leads to systems being applied to premium power applications, such as in banks, in hospitals, and in computing facilities.

The major challenge to PAFC systems is the high cost of materials; the high platinum content is particularly costly and represents 10% to 15% of the total PAFC system costs. Other material costs and processing costs also contribute to making the first cost of the PAFC system considerably greater than the average fossil-fueled steam turbine power plant.

There is a need to increase the power density of the cells and to reduce costs and optimize system. Technology advances are being made to reduce the cost and to reach commercialization targets. Some of the technology approaches, experience, and development needs were identified in a US Department of Energy–sponsored workshop (Remick et al., 2010).

The UTC Power–developed PureCell 200 fuel cell system is a 200 kW PAFC system and has been installed in over 260 systems across 19 countries on five continents with more than 8.7 million hours of operation and more than 1.4 billion kWh of electricity generation. The longest running system has operated over 64,000 h. The PureCell 200 produces 200 kW of assured power, plus approximately 1.50 MM Btu/h of heat at 140°F (60°C). The PAFC commercialization target includes development of (i) alternative electrolyte, (ii) highly active catalysts, (iii) low-cost manufacturing methodologies, (iv) low-cost heat exchangers, and (v) ammonia-free producing fuel processing system. For example, UTC Power's PAFC electrolyte ionic conductivity target is to reach greater than 0.65 S/cm. Eliminating anion poisoning and using electrolytes with vapor pressure lower than that of phosphoric acid would improve power plant efficiency by 6 percentage points; for example, increase the beginning-of-life efficiency to 48% and reduce cost by 15% to 20%. The electrolyte improvement was identified as one the most important improvements for the PAFC system. PBI-phosphoric acid–based membrane electrode assemblies have been developed by BASF Fuel Cells (BASF). These MEAs operate at temperatures between 120°C and 180°C, tolerate large concentrations of carbon monoxide, have a high sulfur tolerance, and are able to run independently of humidification.

9.3 Polymer Electrolyte Membrane Fuel Cell

The PEMFC has been under development for the last two decades primarily as a potential replacement for internal combustion engines in electric passenger vehicles with power needs of 50–100 kW. However, PEMFCs have also been considered for larger vehicles of few hundred kilowatts for buses and trucks, as auxiliary power units, for small-scale stationary power generations of few kilowatts for combined heat and power of residential buildings, and even in smaller units of few watts for portable power electronics applications (Li, 2008; O'Hayre et al., 2006).

Some of the attractive features of PEMFC compared to the other types of fuel cells are (i) compactness and lightweight owing to the use of solid polymer electrolyte membrane (PEM); (ii) operation at relatively lower temperatures (80°C), making them suitable for operations involving frequent cycles of startups and shutdowns, and more responsive to load variations; (iii) use of thin membrane resulting in lower ohmic resistance and operations at higher current and power densities; and (iv) ease in fabrication of solid polymer in thinner membrane.

9.3.1 PEMFC Operation and Design

A tri-layer PEMFC consisting of a PEM sandwiched between anode and cathode gas diffusion layer electrodes is shown in Figure 9.6.

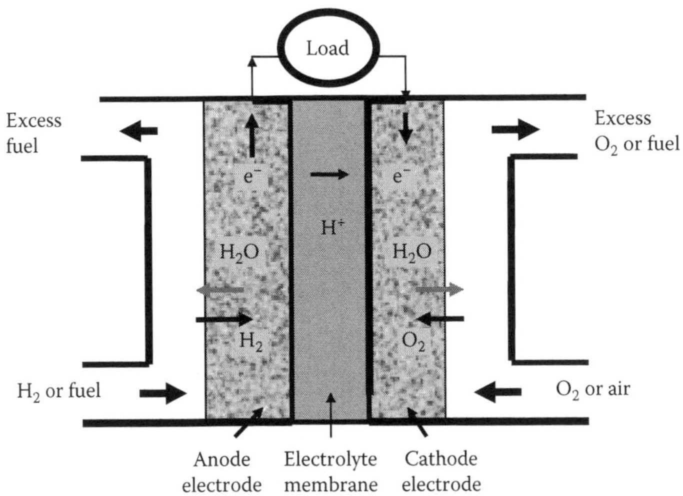

FIGURE 9.6
A three-layer PEMFC.

Humidified hydrogen and oxygen either pure or in the form of air flow through the anode and cathode gas channel, respectively. At the anode side, hydrogen diffuses through the gas diffusion layer toward the catalyst-coated anode electrode–electrolyte interface and undergoes an electrochemical oxidation reaction producing two positively charged hydrogen ions or protons (H^+) and two electrons from each hydrogen molecule. The protons transport through the ion-conducting PEM from the anode side to the cathode side. At the catalyst-coated cathode–electrolyte interface, oxygen undergoes electrochemical reduction reaction by combining with the incoming protons and electrons producing water. The two electrochemical half-reactions and overall reaction in PEMFC with hydrogen as fuel are given as follows:

$$\begin{aligned} &\text{Anode reaction:} \quad H_2O \rightarrow 2H^+ + 2e^- \\ &\text{Cathode reaction:} \quad \frac{1}{2}O_2 + 2H^+ + 2e^- \rightarrow H_2O \\ &\text{Overall reaction:} \quad H_2 + \frac{1}{2}O_2 \rightarrow H_2O \end{aligned} \tag{9.12}$$

The product water tends to migrate across the cell and may cause drying of the membrane near the anode and flooding near the cathode. As we have discussed in Chapter 7, the net distribution of water in the membrane is affected by a number of co-existing transport processes such as electro-osmotic drag of water molecules by the transporting protons, diffusion owing to water concentration gradient, and transport caused by pressure difference across the cell. A proper balance of the water distribution across the cell is essential for effective operation of the polymer membrane. Also, heat generated owing to irreversibilities associated with electrochemical reactions and ohmic heating diffuses through the cells and may require an effective cooling mechanism to maintain the operating cell temperature.

9.3.1.1 Electrode Material and Structure

Electrochemical reactions at the electrode–electrolyte interface are surface phenomena and require large exposed solid surface area as reaction sites. In order to achieve large active surface area and for efficient transport of reactant gases to the reaction sites, the electrodes are made in the form of a highly porous structure. The pore structure typically used in PEMFC is in the form of a macro- or microporous carbon cloth or paper through which reactant gas diffuses toward the interface. The electrodes are characterized by the thickness and pore structure.

Another important aspect of the electrodes is the use of catalyst to accelerate hydrogen oxidation and oxygen reduction reaction at the anode and cathode electrodes. The catalyst loading is characterized by the mass of catalyst (mg) per unit surface area (cm^2), that is, mg/cm^2 of the electrode. In the early design of an electrode in PEMFC, the catalyst layers are applied to the gas diffusion layer or on the Nafion-membrane directly. In such designs, platinum *black* was used as the catalyst in a thin monolayer, which leads to a very high catalyst loading on the order of 1–4 mg/cm^2 and demonstrated excellent long-term performance, but at a very high cost of fuel cell. In the newer designs, the electrodes are divided into two regions: a *catalyst layer* or *active region* adjacent to electrolyte membrane and a *gas diffusion layer* as shown in Figure 9.7.

9.3.1.2 Catalyst Layer

Design and development of a cost-effective and high-performance catalyst layer is an active field of research. Since the use of expensive platinum black as the thin catalyst layer results in a lower surface area and is not very cost-effective, a higher platinum loading per unit area is, therefore, necessary in order to maintain reaction kinetics, and this results in higher cost of the fuel cell. As we have discussed before, the oxygen reduction reaction at the cathode is several orders of magnitude slower than the relative faster hydrogen oxidation reaction at the anode and a major contributor to the fuel cell polarization loss. For example, in a typical value of exchange current density (i_0), a measure of electrode kinetics is around 1 mA/cm^2 for cathode reaction as compared to the 0.001 mA/cm^2 for anode reaction. While platinum black with low Pt loading can be used as a catalyst layer at the anode, the use of the cathode catalyst layer needs significant improvement in terms of lower catalyst loading, faster reaction, and lower polarization loss. Hence, major research effort should be focused on the cathode catalyst layer design to improve electrode reaction kinetics and reduce polarization loss.

In addition to the electrode-bonded and Nafion-bonded catalyst layers, which involve only surface reaction, a design with *Nafion* impregnated into the electrode catalyst layer or *impregnation of electrode* within a thin layer of electrolyte membrane is also used to increase the active surface area as shown in Figure 9.7.

Impregnation of electrode within a thin layer of electrolyte membrane forms a three-phase active reaction zone or *triple-phase boundaries* (*TPBs*). Kim et al. (1995) developed a Nafion-impregnated electrode with a platinum loading of 0.4 mg/cm^2 and a Nafion content of 0.6 mg/cm^2.

In the newer design of the catalyst layer, small catalyst particles are usually supported on relatively larger carbon particles and hence reduce the catalyst loading by a factor of 10 to 0.4 mg/cm^2, leading to a less expansive fuel cell. Figure 9.8 shows a typical catalyst layer with carbon-supported catalyst particles.

Platinum catalyst particles are supported on larger and finely divided carbon particles. A carbon-based power XC-72 is commonly used. The reaction regions are characterized by the active surface area where electrode, electrolyte, and catalyst are present.

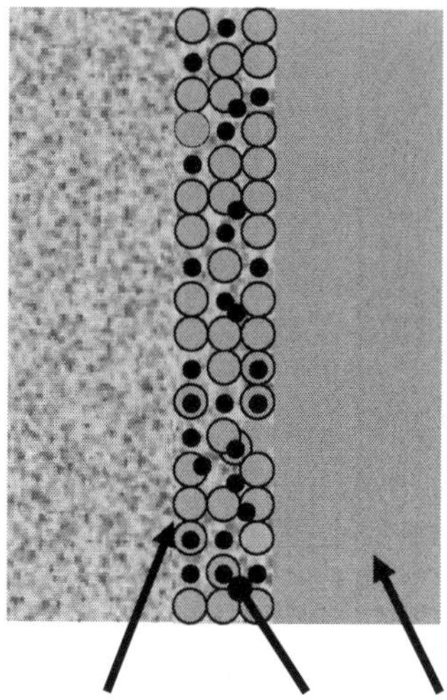

FIGURE 9.7
Nafion-impregnated electrode with dispersed catalyst particles.

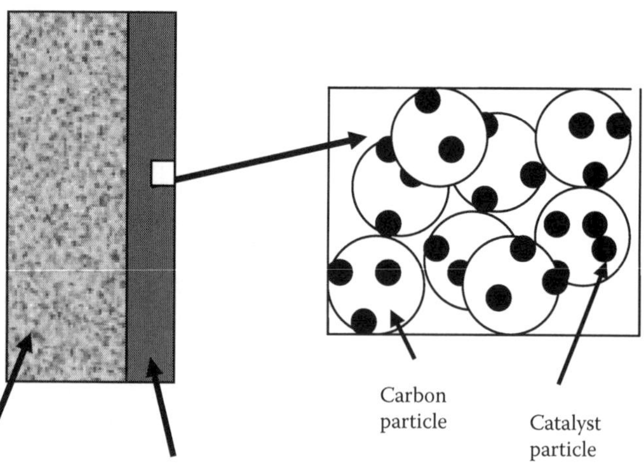

FIGURE 9.8
Catalyst layer with carbon-supported catalyst particles.

This carbon-supported catalyst is then fixed in a thin layer on the electrode surface to enhance the electrochemical reaction and to reduce activation overpotential loss. The most common carbon-based particles being used for catalyst support are VULCAN XC-72R. These carbon particles support the catalyst particles, prevent them from agglomeration, and serve as the conductor to allow transport of electrons through the electrode to external load circuit.

In the PEMFC with an operating temperature range of 70°C–90°C, platinum (Pt) is the most common catalyst used on both anode and cathode electrodes. The use of carbon-supported platinum (Pt/C) forms a higher active surface area and this leads to lowering of the catalyst loading. We have mentioned that one way to improve electrode reaction kinetics, particularly for the cathode oxygen reduction reaction, is to increase the effective surface area of the carbon-supported dispersed platinum catalyst particles in the active region of the electrode. The active electrochemical surface area varies with the ratio Pt/C value, which is generally kept in the range of 10%–60%. For an optimum design in terms of minimum platinum loading that leads to high active area and performance, a number of factors have to be considered. These are ratio of platinum to carbon by weight (Pt/C ratio), platinum loading (mg/cm^2), platinum particle size, and thickness of the active layer.

As we have discussed in Chapter 7, the platinum catalyst loading, m_{pt} is given by Equation 7.27 as

$$m_{pt} = \frac{A_c L_{al}}{A_o}$$

where m_{pt} is platinum catalyst loading (mg/cm^3), L_{al} is the thickness of the active layer (cm), A_o is platinum mass loading given in terms of catalyst surface area per unit mass of catalyst particles, and A_c is the total catalyst surface area.

Platinum mass load is obtained from experimental evaluation of the active layer structure in terms of platinum-to-carbon support mass (Pt|C) given by empirical correlation such as Equation 7.28.

Platinum alloy such Pt/Ru is also used as a catalyst to increase resistance against CO poisoning as a result of PEMFC operating with reformed fuel rather than using pure hydrogen.

Another attractive way of forming the catalyst layer is to form the portion of the electrode as a porous carbon impregnated pore structure coated with catalyst particles. In this design, the catalyst layer coexists with a certain thickness of the electrode material to increase the active surface area. In some of the newer designs, the catalyst layer is impregnated inside the membrane in order to increase the active surface area. The active surface area is the region where electrode, electrolyte, and catalyst co-exist and the rate of electrochemical reactions is highest with the presence of gases, protons, and electrons. This three-phase active area is also referred to as *triple-phase boundaries* (*TPBs*). The thickness of the catalyst layer is generally very small in the range of 10 nm. A thinner catalyst layer is generally preferred because the current is generated in the region close to the electrolyte. Thus, a thicker catalyst layer without any electrolyte impregnation has less catalyst utilization.

One of the major reasons for the higher cost of PEMFC is due to the loading of expensive *Platinum* (Pt) catalyst in the electrodes. Hence the research and development activities are focused on reduced catalyst loading, use of less expensive platinum group metal (PGM) based catalyst, and development of alternate less expensive non-precious metal catalyst (other than platinum) with increased activity and durability.

Approaches to developing higher performance and lower-cost catalysts involve the use of PGM alloys, non-PGM catalysts, and ultralow Pt loading using nano-structured materials such as carbon nanotechnology. Strong research activities are currently underway in an effort to reduce Pt loading below 0.4 mg/cm^2 and the use of decreased PGM to less than 0.125 mg/cm^2. Development of a low-cost nanostructured catalyst for enhanced electrocatalyst activity for cathode reduction is an active field of research. This will help reduce the use of platinum catalyst and lower the cost of PEMFC. The Los Alamos National Laboratory has reported that its highly improved cathode used a PGM-based catalyst such as Pt/Pd-nanostructured particles with Pt catalyst loading significantly lower than the baseline Pt/C loading and demonstrated significantly high catalytic activity.

9.3.1.3 Gas Diffusion Layer

The thin catalyst layer is usually supported by a thicker electrode layer referred to as a gas diffusion layer. A thicker gas diffusion layer provides increased protection and mechanical strength for the catalyst layer and enhanced reactant gas diffusion or transport to reaction sites, but with increased ohmic resistance. The diffusion layer is usually made of carbon paper or carbon cloth coated with a mixture of carbon black and PTFE. The carbon paper and carbon cloth provide structural strength for the electrode and also serve as electron conductor. Since PTFE is hydrophobic, it can prevent the electrode from flooding especially at the cathode where water is produced in PEMFC and force it to move away from the electrode–electrolyte interface toward the gas flow channels. The gas diffusion layer is made of a microporous *carbon cloth* or *paper*. Some popular brands are Toray, CARBEL, and E-TAK. Most gas diffusion layers are made of carbon papers for a thinner design. Carbon cloths are, however, used for higher power densities where water productions rates are higher and carbon cloth provides additional absorption capacity as well as additional structural strength. Some of the important properties of *Toray* carbon diffusion layer are given in Table 9.3.

TABLE 9.3

Properties of Toray Carbon Paper for the Gas Diffusion Layer

Properties	
Thickness (mm)	0.11–0.37
Density (g/cm^3)	0.40–0.45
Porosity (%)	0.8–0.78
Electrical resistivity (mΩcm)	80
Thermal conductivity (W/(m·K))	1.7

9.3.1.4 Electrolyte Membrane

The purpose of the polymer electrolyte is to transport the proton or H^+ from the anode toward the cathode side. Additionally, it provides strong resistance to electron transport so that electrons move away from the membrane toward the electrodes.

A polymer electrolyte membrane (PEM) is used in a PEMFC to transport protons from the anode side to the cathode side and also acts as an electronic insulator that forces electrons to transport through the electrodes to the external electrical circuits. Major requirements are (i) higher ionic conductivity, (ii) lower electronic conductivity, (iii) lower thickness for lower ohmic resistance, (iv) lower fuel crossover, (v) higher structural strength and ease of manufacturability, and (vi) higher stability and durability.

General Electric in 1962 developed the first polymer solid membrane for the Gemini Space Project. This membrane is a polystyrene sulfonic acid membrane that is hydrated for proton transport with an operating temperature of 70°C. Since 1970, *DuPont* has been marketing the polymer membrane under the trade name *Nafion*. The most popular Nafion membranes are Nafion-115 and Nafion-117 of thicknesses 127 and 183 μm, respectively. The corresponding weights are 250 g/m^2 and 360 g/m^2, respectively. The most common polymer membrane is *Nafion*, which is based on a perfluorosulfonic acid/PTFE copolymer, and it is designed to include a large amount of hydrated regions through which protons can migrate efficiently.

Dow Chemical also developed an experimental membrane in 1988 with an equivalent weight of 800. A PEMFC made with this membrane and operated with hydrogen showed improved performance and demonstrated durability over 10,000 hrs.

Current industry standard for PEM is the Nafion, which is limited to operations below 100°C owing to the requirement for hydration. Since proton conductivity in a Nafion membrane depends on water concentration in the membrane, it is essential that the membrane is sufficiently hydrated in order to maintain an effective transport of the ion, and hence maintain the reaction at the desired level. A poor water distribution in the membrane leads to drying at the anode side and flooding at the cathode side. This results in higher ohmic loss because of reduced proton conductivity and an increase in mass transfer losses at high current density owing to poor gas concentration distribution at the reaction surfaces. This limits the operation of the PEMFC to lower current and power densities.

In recent times, there has been considerable research effort to develop PEM membrane for higher operating temperature range. Membrane development effort is primarily concentrated on the development of high-temperature membrane with operation at 120°C or higher, lower relative humidity with less than 10% for operation at 80°C, higher proton conductivity, and a thinner membrane for reduced ohmic loss and improved tolerance for impurities like carbon monoxide.

Recently, there were considerable research efforts to develop PEM membranes for higher operating temperature range. Li et al. (2004) developed a phosphoric acid doped polybenzimidazole (PBI) membrane for operation of PEMFC at higher temperatures up to 200°C. BASF has also developed its Celtec membrane for PEMFC and operation at a higher temperature range of 120°C–180°C using PBI and phosphoric acid. The membrane operates without the need for any humidification and has higher tolerance against carbon monoxide and sulfur, and this makes it suitable for use with reformed hydrocarbon fuel and with simplified purification of gas stream such as natural gas. Gyner Electrochemical Systems (GES) developed the high-temperature and relative dryer membrane using perfluorosulfunic acid with significantly higher conductivity than Nafion and operation at a higher temperature of 120°C and 50% RH.

9.3.1.5 Nafion Membrane Construction

The polymer membrane such as Nafion is designed to include a large amount hydrated regions through which proton or H^+ ions can migrate efficiently. The base structure or backbone of a Nafion membrane is a *polyethylene polymer*, which is modified by replacing the hydrogen with a fluorine and forming a structure known as polytetrafluoroethylene (PTFE), and makes the structure highly resistive chemically, stable, and durable. The electrolyte is usually made by adding a sulphonic acid (HSO_3) side chain to the base PTFE polymer backbone. The sulphonic acid added is in ionic form in which SO_3^- and H^+ ions are held together by the strong ionic attraction as shown in Figure 9.9. This polymeric electrolyte membrane structure with hydrophobic PTFE and hydrophilic sulphonic acid side chain is capable of attracting and absorbing large amounts of water.

So the polymer membrane such as Nafion is characterized as a water-filled cluster interconnected by channels. The surface of the clusters and channels are the sulfonate ions, and the mobile liquid phase is composed of hydrogen ion and water $H^+(H_2O)_n$. When the PEM is well hydrated, the H^+ ion is

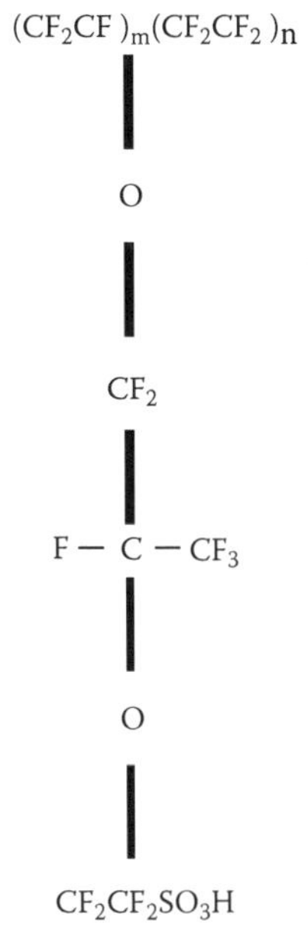

FIGURE 9.9
Polymer structure in Nafion.

relatively weakly attracted to the SO_3^- ion with high electron negativity and can transport easily. This results in a polymer membrane with high proton conductivity and behaves as a good electron insulator as well.

Use of PTFE as the backbone provides high structural strength and enables the polymer electrolyte to be made into a thinner membrane and hence results in low ohmic resistance. The lower limit is set by the structural strength of the membrane.

9.3.1.6 *Major Characteristics of Nafion-117 Membrane*

Verbrugge and Hill (1990) presented the theoretical representation and experimental data to characterize the proton and water in perfluorosulfonic acid membrane. The analysis describes the transport of water molecules carried along with the transport of proton across the membrane. Springer et al. (1991) presented a one-dimensional, isothermal model of PEMFC in which detailed considerations of the Nafion-117 membrane characteristics in terms of water content and water transport properties including water drag coefficient and diffusion coefficient are given. Their results demonstrated increased membrane resistance with current density.

9.3.1.7 Water Content in Nafion—PEM

The water-absorbing capacity of a material is in general expressed in the form of sorption isotherms as a function of humidity condition and temperature. The water content (λ) is defined as the ratio of number of water molecules to number of charged sites $\left(SO_3^-H^+\right)$. Sorption isotherm for Nafion-117 was measured experimentally based on humidity condition at 30°C and the empirical correlation is given by Zawodzinski et al. (1991) (Springer et al., 1991) in the form of water content (λ) as a function of water vapor activity (*a*). The functional form of the sorption isotherm is given as follows:

$$\lambda = 0.043 + 17.81a - 39.85a^2 + 36.0a^3,\ 0 < a \le 1\ (\textit{unsaturated}) \tag{9.13a}$$

and a linear relation is suggested for saturated range as

$$\lambda = 14 + 1.4\,(a - 1),\ 1 \le a \le 3\ (\textit{saturated}) \tag{9.13b}$$

The water content (λ) is given as the ratio of the number of water molecules to the number of charge $\left(SO_3^-H^+\right)$ sites. The humidity condition at the gas diffusion catalyst layer interface is given as the water activity or relative humidity (*a*) and defined as

$$a = \frac{P_{H_2O}}{P_{sat}} \tag{9.14a}$$

or

$$a = \frac{y_{H_2O}P}{P_{sat}}, \tag{9.14b}$$

where y_{H_2O} is the mole fraction of water, P is the total pressure, and P_{sat} is the saturation pressure of water as a function of temperature given in the Table C.5. A curve-fit expression of water saturation data is given by Springer et al. (1991) as

$$\log_{10} P_{sat} = -2.1794 + 0.02953\,T - 9.1837 \times 10^{-5}T^2 + 1.4454 \times 10^{-7}T^3, \tag{9.14c}$$

where temperature T is in degrees Celsius and water vapor P_{sat} is in the unit of *bar*.

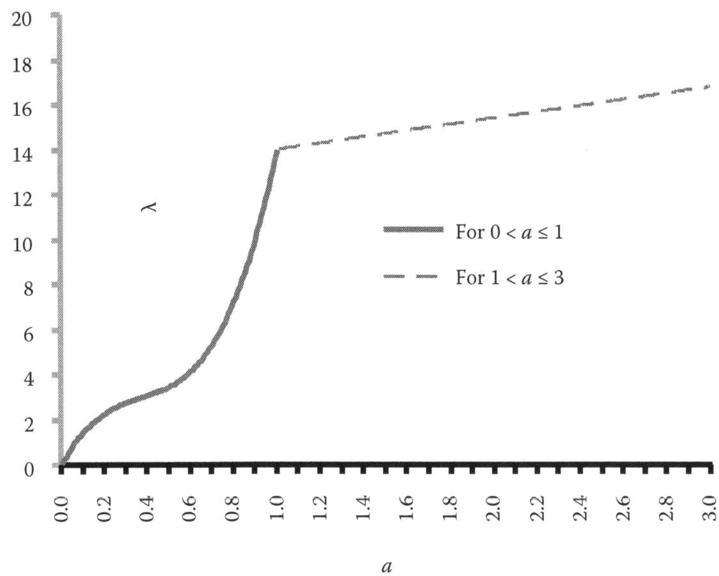

FIGURE 9.10
Water content in Nafion-117 with water activity.

While Equation 9.13a is the fit of the experimental data in the range from dry to equilibrium with saturated water vapor, Equation 9.13b is a linear extrapolation from the equilibrium state with water vapor ($a = 1$) at 30°C and corresponding water content of $\lambda = 14$ to exceeded water saturation ($a = 3$) with corresponding water content of $\lambda = 16.8$.

In the absence of experimental data for extended temperature range, the 30°C sorption data and Equation 9.13 are generally assumed as applicable to membranes operating at a higher temperature of 80°C. The variation of water content in Nafion-117 with water activity is plotted in Figure 9.10.

Example 9.1

Determine water content in the Nafion membrane at the anode–membrane and cathode–membrane interfaces for PEMFC operation at 80°C and operating pressure of 2 atm pressure on both anode and cathode sides. Assume water mole fractions as 0.1 and 0.25 at the anode–membrane and cathode–membrane interfaces, respectively.

Solution

At an operating temperature of 80°C, the corresponding water saturation pressure is $P_{sat} = 47.39$ kPa. Based on this water vapor pressure, the water activity ratios at the anode–membrane and cathode–membrane interfaces are computed an

$$a_{am} = y_{H_2O,a}\frac{P_a}{P_{sat}} = 0.1\times\frac{2\times 101.3}{47.39} = 0.4275$$

And assuming that water is in liquid form at the cathode–membrane interface,

$$a_{cm} = y_{H_2O,c}\frac{P_c}{P_{sat}} = 0.25\times\frac{2\times 101.3}{47.39} = 1.068$$

Water content at the interface is computed using Equation 9.13a for the anode–membrane interface as

$$\begin{aligned}\lambda_{am} &= 0.0043 + 17.81a_{am} - 39.85a_{am}^2 + 36.0a_{am}^3\\ &= 0.0043 + 17.81\times 0.4275 - 39.85\times 0.4275^2 + 36.0\times 0.4275^3\\ &= 0.0043 + 7.613775 - 7.28283 + 2.81261\\ \lambda_{am} &= 3.14786\end{aligned}$$

and using Equation 9.13b as

$$\lambda_{cm} = 14 + 1.4(1.068 - 1) = 14.095.$$

9.3.1.8 Proton Conductivity in Nafion

Proton conductivity in Nafion-117 increases with increase in water content and temperature. The experimental fit of the proton conductivity data $(\Omega - \text{cm})^{-1}$ as a function of temperature and membrane water hydration is given as (Springer et al., 1991)

$$\sigma(T,\lambda) = \sigma_{303K}(\lambda)\exp\left[1268\left(\frac{1}{303} - \frac{1}{T}\right)\right], \tag{9.15a}$$

where $\sigma_{303K}(\lambda)$ is the proton conductivity at a temperature of 30°C and given as

$$\sigma_{303K} = 0.005139\lambda - 0.00326 \text{ for } \lambda > 1. \tag{9.15b}$$

The proton conductivity of Nafion increases exponentially with temperature as given by Equation 9.15a and as shown in Figure 9.11. Nafion conductivity increases strongly with the water content in a linear manner as shown in Figure 9.12.

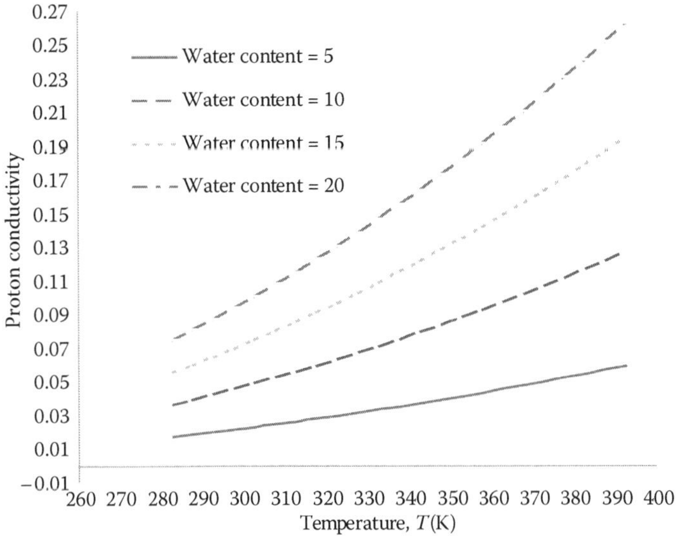

FIGURE 9.11
Variation of proton conductivity of Nafion-117 with temperature for different water content.

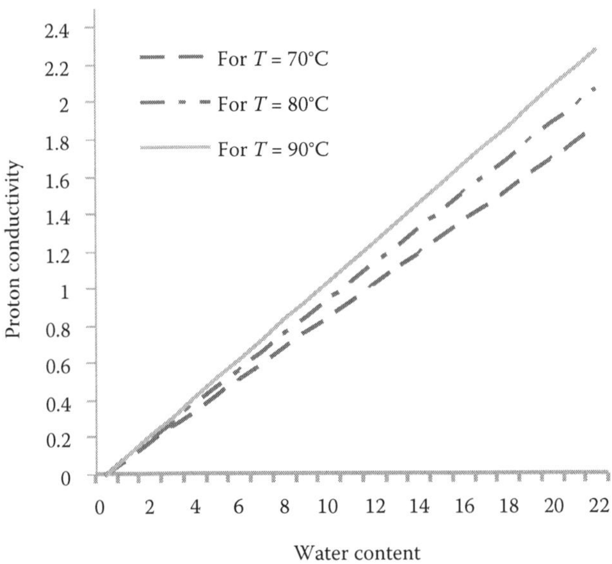

FIGURE 9.12
Variation of proton conductivity of Nafion with water content at 303 K.

Example 9.2

Estimate the proton conductivity for Nafion membrane with 15% moisture in the air stream with an operating pressure and temperature of 1 atm and 70°C, respectively.

Solution

At an operating temperature of 70°C, the corresponding water saturation pressure is P_{sat} = 31.19 kPa. Based on this water vapor pressure, the water activity ratio with 15% moisture content is given as

$$a = y_{H_2O,a}\frac{P_a}{P_{sat}} = 0.15\times\frac{1\times 101.3}{31.19} = 0.4872$$

and water content in the membrane is given by Equation 9.13a

$$\lambda = 0.0043 + 17.81a - 39.85a^2 + 36.0a^3$$
$$\lambda = 0.0043 + 17.81\times 0.4872 - 39.85\times(0.4872)^2 + 36.0\times(0.4872)^3$$
$$\lambda = 3.386$$

The proton conductivity at 30°C and for λ = 3.386 is given based on Equation 9.15b

$$\sigma_{303K}(\lambda) = 0.005139\lambda - 0.00326$$

$$\sigma_{303K}(\lambda) = 0.005139 \times 3.386 - 0.00326$$

$$\sigma_{303K}(\lambda) = 0.01414$$

The proton conductivity at 70°C is now computed from Equation 9.15a

$$\sigma = \sigma_{303K}(\lambda)\exp\left[1268\left(\frac{1}{303} - \frac{1}{343}\right)\right]$$
$$\sigma = 0.01414\exp\left[1268\left(\frac{1}{303} - \frac{1}{343}\right)\right] \quad (9.15a)$$
$$\sigma = 0.0230\ \text{S/cm}$$

9.3.1.9 Membrane Ionic Resistance and Ohmic Loss

Since proton conductivity and hence the membrane electrical resistance vary locally depending on the water content (λ), we can employ two approaches in determining the ionic resistance and ohmic loss in the membrane:

In *approach 1*, the average conductivity is estimated based on the average water content in the membrane as

$$\sigma(T,\bar{\lambda}) = \sigma_{303K}(\bar{\lambda})\exp\left[1268\left(\frac{1}{303} - \frac{1}{T}\right)\right],$$

where the average water content in the membrane is given as

$$\bar{\lambda} = \frac{1}{a_e}\int_0^{a_e} \lambda \, dx.$$

The membrane ohmic resistance is then computed from

$$\eta_{ohm} = i\text{ASR} = i\frac{a_e}{\sigma(\bar{\lambda})}.$$

In approach 2, the membrane ionic resistance is given by integrating over the membrane thickness as

$$R_m = \int_0^{t_a} \frac{dz}{\sigma(\lambda)} \tag{9.15c}$$

and ohmic loss from

$$\eta_{ohm} = iR_m.$$

9.3.1.10 Water Diffusivity in Nafion

Diffusion coefficient of water in Nafion-117 is measured experimentally by Zawodzinski et al. (1991) (Springer et al., 1991) and the empirical correlation is given as

$$D_{H_2O} = D_\lambda \exp\left[2416\left(\frac{1}{303} - \frac{1}{T}\right)\right] (\text{cm}^2/\text{s}), \tag{9.16a}$$

where

$$D_\lambda = (2.563 - 0.33\lambda + 0.0264\lambda^2 - 0.000671\lambda^3) \times 10^{-6} \text{ for } \lambda > 4. \tag{9.16b}$$

The exponential term include the variation of diffusion coefficient with temperature beyond 30°C or 303 K.

Motupally et al. (2000) proposed correlation diffusion coefficient of water in Nafion-115 as

$$D_{H_2O} = 3.10 \times 10^{-3} \lambda(-1 + \exp[0.28\lambda]) \exp\left[2416\left(-\frac{2436}{T}\right)\right] \text{ for } 0 < \lambda < 3 \quad (9.16c)$$

$$D_{H_2O} = 4.17 \times 10^{-4} \lambda(1 + 161 \exp[-\lambda]) \exp\left[-\frac{2426}{T}\right] \text{ for } 3 < \lambda < 10 \quad (9.16d)$$

Example 9.3

Estimate the diffusion coefficient of water in Nafion at 80°C with a membrane water content of $\lambda = 8$.

Solution

For $\lambda = 4$, Equation 9.16b is used to compute D_λ as

$$\begin{aligned} D_\lambda &= \left(2.563 - 0.33 \times (4) + 0.0264 \times (4)^2 - 0.000671 \times (4)^3\right) \times 10^{-6} \\ &= 1.623 \times 10^{-6} \end{aligned}$$

Water diffusivity, D_{H_2O}, is computed from Equation 9.16a as

$$D_{H_2O} = 1.623 \times 10^{-6} \times \exp\left[2416\left(\frac{1}{303} - \frac{1}{353}\right)\right]$$

$$D_{H_2O} = 5.02 \times 10^{-6} \text{cm}^2/\text{s}.$$

9.3.1.11 Electro-Osmotic Drag Coefficient

As we have discussed in Chapter 8, electro-osmotic drag plays a dominant role in the transport of water within the membrane as proton transports from the anode side to the cathode. Electro-osmotic drag coefficient (n_{drag}) is defined as the ratio of number of moles of water dragged per mole of proton transported $\left(n_{H_2O}/H^+\right)$, and it depends on the water content of a membrane. A linear variation electro-osmotic drag coefficient is generally assumed and expressed (Springer et al., 1991) as

$$n_{\text{drag}} = n_{\text{drag}}^{\text{sat}} \frac{\lambda}{22} \text{ for } 0 \leq \lambda \leq 22, \tag{9.17}$$

where $n_{\text{drag}}^{\text{sat}} = 2.5$ is the measured drag coefficient of a fully hydrated Nafion-117 membrane, that is, with a water content of $\lambda = 22$.

Since the polymer membrane needs to be hydrated to conduct protons, the operating temperature of the PEMFC is limited to temperature below the boiling point of water, typically in the range of 70°C–90°C for the Nafion membrane. Another important characteristic of this type of polymer membrane is that as the proton moves from the anode side to the cathode side, it drags with it a number of water molecules. This tends to make the membrane dry at the anode side and a flooding condition at the cathode, making it difficult for oxygen to reach the cathode–electrolyte interface owing to the relatively lower diffusivity of oxygen in water and results in increased mass transfer losses, particularly at a high current density. Since proton conductivity depends on water content, it is essential that the membrane is sufficiently and uniformly hydrated from the anode side to the cathode side to maintain effective transport of the proton, and the desired reaction level.

In order to mitigate such conditions, the anode side hydrogen gas stream is often humidified to some extent and a higher pressure condition is maintained in the cathode side than the anode side. Overall transport and balance of water within the membrane is, therefore, controlled by a number of transport processes as discussed in Chapter 7. As a consequence, water management within the membrane is a critical issue for effective performance of the membrane and for the design of PEMFC. In addition to Nation-115 and -117, DuPont's Nafion membranes are available in different thicknesses such as Nafion-1135 and Nafion-1110 with thicknesses of 89 μm and 254 μm, respectively.

Since PEM such as Nafion needs to be hydrated for proton transport, the operating temperature of PEMFC is limited to temperature below boiling point (100°C) at atmospheric pressure. The typical operating temperature of Nafion is limited to 70°C–90°C.

Research and development activities involving PEMFC also include improved heat and water management within the membrane and PEMFC, increased reliability and durability, and scalability and improved fabrication processes and cost for large-scale production (Berg et al., 2004; Bernardi and Verbrugge, 1991; Dannenberg et al., 2000; Fuller and Newman, 1992; Fuller and Newman, 1993; Gurau et al., 1998; Hu et al., 2004; Motupally et al., 2000; Sone et al., 1996; Sunden and Faghri, 2005; Zawodzinski et al., 1995).

9.4 Molten Carbonate Fuel Cell

Fuel cells are much like a common battery in that they produce direct current electricity through an electrochemical process. Unlike a standard battery, fuel cells use a continuous supply of fuel. As long as fuel is supplied, they produce electricity. The main feature of MCFCs is their high operating temperature around 950 K. This makes them especially suited for co-generation of heat and power. Also, waste heat can be further utilized in a bottoming cycle such as in a previously described hybrid system.

9.4.1 MCFC Basic Principles and Operations

Basically, an MCFC consists of two porous electrodes, separated by a molten electrolyte held in place by a matrix as seen in Figure 9.13. In the three-phase region on the anode side, hydrogen oxidation reaction occurs and hydrogen combines with carbonate ions, producing water and carbon dioxide while releasing an electron.

The overall electrochemical reaction at the anode is seen in Equation 2.1. When using a methane-based fuel, two other reactions occur on the anode side, the reforming reaction and water–gas shift reaction, which are discussed later. In the three-phase region between the cathode, electrolyte, and gaseous oxygen mixed with CO_2, oxygen is reduced to carbonate ions by combining with carbon dioxide and electrons from the external circuit that begins at the anode. The overall electrochemical reaction at the anode is seen

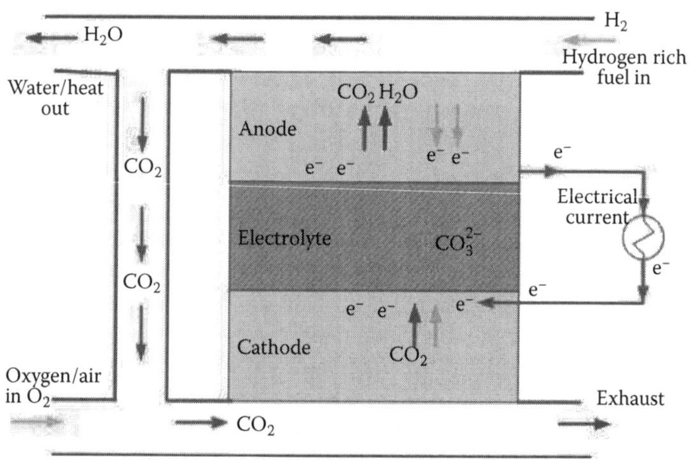

FIGURE 9.13
Basic principle of an MCFC.

in Equation 2.4. The ions produced migrate through the molten carbonate electrolyte matrix in order to complete the circuit. Therefore, the net cell reaction produces water, heat, and electricity (Equation 2.5)

$$H_2 + CO_3^{=} \rightarrow H_2O + CO_2 + 2e^- \quad \text{Anode} \tag{9.18}$$

$$CH_4 + H_2O \rightarrow CO + 3H_2 \quad \text{Reforming} \tag{9.19}$$

$$CO + H_2O \rightarrow CO_2 + H_2 \quad \text{Water-gas shift} \tag{9.20}$$

$$\frac{1}{2}O_2 + CO_2 + 2e^- \rightarrow CO_3^{=} \quad \text{Cathode} \tag{9.21}$$

$$H_2 + \frac{1}{2}O_2 \rightarrow H_2O + \text{Heat} + \text{Electric energy} \quad \text{Net cell reaction} \tag{9.22}$$

$$E_{\text{Nernst}} = +\frac{\Delta G^{\circ}}{nF} + \frac{RT}{2F}\ln\frac{P_{H_{2,a}}P_{O_{2,a}}^{1/2}P_{CO_{2,c}}}{P_{H_2O_{,a}}P_{CO_{2,a}}} \tag{9.23}$$

The Nernst equation (Equation 9.23) relates the operating voltage of a fuel cell to the thermodynamics of the electrochemical reactions. In the Nernst equation, n is the number of electrons transferred in the overall fuel cell reaction, R is the universal gas constant, T is temperature (K), F is Faraday's constant, and P_i are the partial pressures (Pa). The maximum theoretical voltage a fuel cell can produce is often called the "open circuit voltage" and is obtained when no current is produced. It can also be calculated from the first term in the Nernst equation using the change in Gibb's free energy of the overall reaction ΔG°. It can also be seen that if the reacting species cannot be fully utilized or consumed, the voltage will be reduced. This results in fuel cells often operating a fuel utilization of 70% to 85%.

Similar to other fuel cells, MCFC has three main voltage losses: polarization, ohmic, and diffusion. Each loss is associated with different operating ranges of the fuel cell. Polarization losses are a result of slow kinetics at the electrode surface. Although polarization losses can occur at all operating ranges, they are most dominant at low current densities. Ohmic resistance in the cell is directly dependent on current and is translated to a near-linear loss region over the operating range of the fuel cell. Ohmic losses are also highly dependent on temperature. Diffusion is the driving force for transport of the reacting species to the reaction sites. They must travel through porous electrodes that are rather slow, and at high current densities, diffusion is the limiting factor for electrochemical reactions. It is ideal to operate the fuel cell in the ohmic region where relatively large changes in current density will result in small voltage changes. The voltage range of an MCFC is typically 0.75 to 0.9 V and the current density ranges between 100 and 200 mA/cm^2. This requires several cells to be coupled in series to form a stack to increase voltage to a practical value. Also, several stacks can be connected in parallel and series to form a larger power system.

The hydrogen produced at the anode is the main component of the fuel for the MCFC; however, it is evident that water and carbon dioxide are extremely important in a working fuel cell. The water produced by the main anodic reaction (Equation 9.18) drives the water–gas shift reaction forward, which produces even more hydrogen from carbon dioxide. A certain level of water is also required in the anode to deter the formation of carbon particles that can deposit and block fuel gas channels. It is important to also understand that carbon dioxide is one of the oxidants of MCFC. In the half-cell reactions, it is produced at the anode, while being consumed at the cathode. This requires CO_2 from the exhaust gas of the anode to be recirculated or recycled to the cathode, as shown in Figure 9.14. This is typically done in two ways.

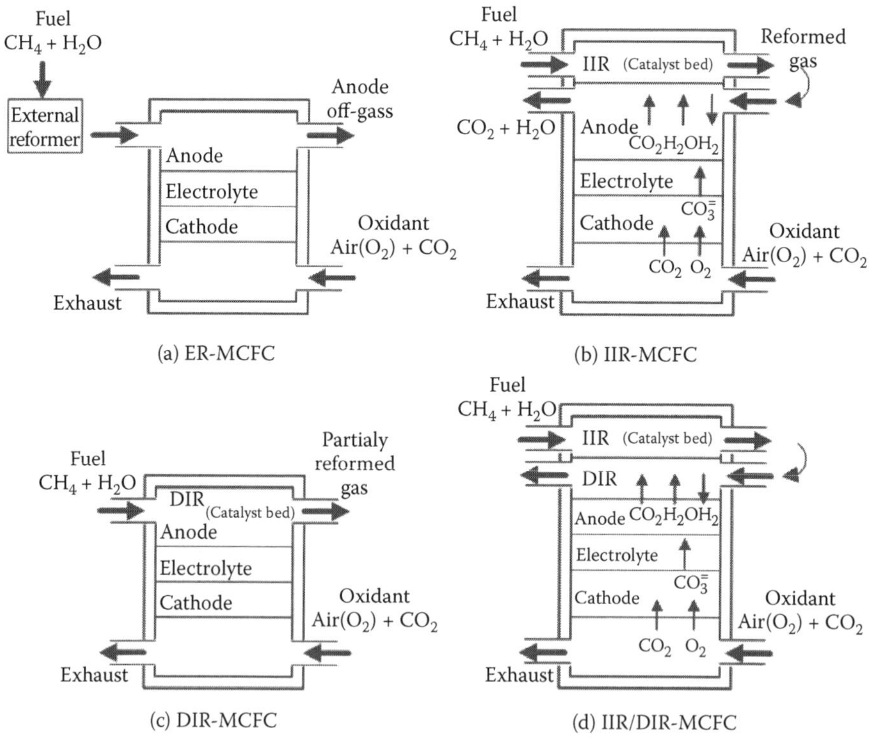

FIGURE 9.14
Schematic representation of placement for reforming in MCFCs.

The first way is to burn the anode exhaust gas with excess air in a combustion chamber and then mix it with cathode inlet gas. Another way is to separate the carbon dioxide from the other components in the anode exhaust by separation methods such as pressure swing absorption techniques or selective membranes.

9.4.2 MCFC Components and Configurations

MCFCs can operate on many different hydrocarbon fuels such as natural gas, gasified coal, and even biomass; however, for reliable operation, fuel supply quality must be regulated. Also, adequate heat transfer and control are essential to the performance and lifetime of the fuel cell system. These requirements can be met using a basic set of equipment as described below.

9.4.2.1 Fuels and Fuel Processing

The MCFC utilizes both H_2 and CO at the anode. Two common carbon-containing fuels converted to be used in a carbonate fuel cell are natural gas and coal. Conventionally, this conversion takes place in a fuel processor that is external to the fuel cell as seen in Figure 9.14a. There are three other designs for handling the reforming and water–gas shift reactions when using natural gas as the fuel. These designs are more thermodynamically advantageous because they utilize the heat from the fuel cell to aid in the reforming process. Figure 9.14b shows an indirect internal reformer (IIR). The IIR consists of a catalyst bed that reforms about half of the natural gas to hydrogen-rich fuel before it enters the anode. The IIR is in close contact with the fuel cell, allowing utilization of the heat released from the exothermic fuel cell reactions. A similar method entails the use of a direct internal reformer (DIR), which allows the reforming reactions to take place directly at the anode compartment. A more recent method is to use a combination of both IIR and DIR as seen in Figure 9.14d. Here, the fuel is partially reformed in the IIR and then enters the DIR where it is further reformed, allowing greater utilization of the methane fuel. In all of these cases, the natural gas must be preheated in order to clean up the gas before entering the fuel cell.

Natural gas contains sulfur compounds that are unfavorable to performance. These compounds must be reduced to levels on the order of parts per billion. High-temperature hydrodesulfurization and the use of absorbents are enough to accomplish such goals. As for coal gas as fuel, many proposed designs have been investigated. Coal gasification leaves toxic and corrosive compounds in the gas, much more so than natural gas. Therefore, much more complicated fuel processing streams must accompany coal gas to remove any contaminants that would be harmful to fuel cell operation. For now, coal and natural gas prices are competitive, but in the future, coal is anticipated to be a cheaper energy source and therefore coal gasification–based carbonate fuel cells may prove to be more economic in the years to come.

9.4.2.2 Combustor

The cathode side of the MCFC requires O_2 and CO_2 to form the ions that complete the fuel cell circuit. CO_2 is not abundant enough in the atmosphere to use just air as the oxidant. Using fuels such as natural gas, however, allows the anode side exhaust to be converted to a CO_2-rich gas by way of an oxidizing unit. A mixture of oxidized gas and air is then fed to the cathode side. An oxidizer can be as simple as a tank in which combustion occurs. In the case of fuel cell systems, an oxidizing catalyst bed can be used to carry out the process. Oxidation system design and performance are well established and can be designed to meet system requirements. Another technique that has been proposed is the use of CO_2 liquid absorption. This method removes the CO_2 from the anode exhaust stream and then recombines it with the cathode entrance stream. This method was projected to raise the efficiency up to 8%, because of better fuel utilization; however, these systems seem to be very large and costly and therefore may be only suitable for large stationary power applications. The oxidizer is best when used with combined heat and power, because it raises the temperature of the gas owing to exothermic oxidation reactions. Therefore, this excess heat needs to be removed before the gas enters the cathode. This heat can be used then to power a turbine or heat a building, which increases the overall efficiency of the system.

9.4.2.3 Cell and Stack Design

Manufacturers of MCFCs take advantage of a simple planar stack design. The fuel cell is much like a layered sandwich of components as seen in Figure 9.15. The main variance in design is whether or not the fuel flow is delivered by an internal or external manifold. FCE uses an external manifold, which is simple but requires gasket seals. Other manufacturers apply an internal manifold system, which has gas channels built into the cell components and uses wet sealing. In either case, separator plates are used to make an electrical connection between cells in a stack. In the case of internal manifold systems, the separator plate also plays the role of supplying fuel and oxidant to the anode and cathode chambers separately, which introduces more design considerations.

9.4.3 MCFC Electrolyte, Electrode, and Catalyst

9.4.3.1 Electrolyte

The purpose of the electrolyte is to attract and selectively diffuse ions from the cathode to the anode, while also preventing gases from diffusing across. The electrolyte of an MCFC is molten and therefore requires a matrix of support to hold it in place. The electrolyte is held in place by capillary forces, which are determined by the pore size of the matrix. $LiAlO_2$ is typically used for the matrix material. The composition of the electrolyte greatly

FIGURE 9.15
Fuel cell structure.

affects the performance and lifetime of the fuel cell. Manufacturers currently prefer Li/K and Li/Na carbonate electrolytes; however, this choice is highly dependent on the operating pressure of the fuel cell. Li/Na is used in pressurized systems because it produces a higher output voltage than Li/K as pressure is increased. Many other factors are used to rate the performance of electrolytes. Electrochemical behavior, creep rate on the metallic surfaces of the cell, wetting ability, gas solubility, and evaporation are all factors that can limit the ability of an electrolyte to perform well.

The electrochemical behavior of electrolytes has been studied extensively. It is believed that the solubility of oxygen within the cathode environment affects the limiting current. The limiting current for molten Li/K is nearly three times that of Li/Na. The polarization losses for Li/Na cells increase as operating temperature decreases below 923 K. This becomes significant in the performance of large cells (up to 1 m^2) where isothermal conditions do not exist. Polarization losses as well as ohmic losses could be reduced as the wetting ability of the electrolyte increases. Recent studies show that Li/Na melts have a higher contact angle than Li/K melts and are more temperature dependent. Better wetting ability may positively affect the electrochemical properties of the cell; however, it is most likely to increase creepage rate on metallic surfaces, partly because of the increase in wetted surface area. This creates a possible optimization problem of contact between the electrolyte and electrode.

In addition to wetting ability, evaporation is a key issue for MCFCs. Potassium is the most volatile species that could be present in the electrolyte. A loss in electrolyte is directly translated to a loss in cell performance.

Therefore, cells that contain Li/K electrolytes are inherently affected. In all cases, electrolyte evaporation could possibly be a cell lifetime limiting issue.

Most often, the electrolyte of an MCFC consists of a mixture of carbonates such as potassium carbonate (K_2CO_3) and lithium carbonate (Li_2CO_3). These carbonates have a melting point of approximately 500°C and are very corrosive. When choosing an electrolyte, it is important to minimize the ohmic resistance of the cell, while maintaining high gas solubility and good reaction kinetics. For example, the resistance of Li_2CO_3 is smaller than that of K_2CO_3, so it would be ideal to use Li_2CO_3 as the electrolyte in order to minimize the ohmic losses. In contrast, the solubility and diffusivity of reactant gases in this Li_2CO_3 are much lower than that of K_2CO_3. Therefore, the optimization of the composition of molten carbonate electrolytes is of much interest when designing a cell.

9.4.3.2 Cathode

Most literature focuses on cathode improvements because there is much room for improvement. Since the cathode side has a large amount of oxidant air in its flow path, there is a need to construct it out of a material that is resistant to the highly corrosive molten carbonate. This allows only a few noble metals to be possible materials for the cathode, while still being cost-effective. Nickel oxide (NiO) is typically chosen as a cathode material. This material, however, partially dissolves owing to the acidity or basicity of the electrolyte. NiO dissolution has a strong impact on the cell lifetime and therefore is a large focus of research activities. The severity of NiO dissolution is regulated by electrolyte composition, gas atmosphere inside the cell, operating pressure, and temperature.

Electrolyte composition has a large effect on the solubility of nickel oxide. Increasing the basicity of the melt greatly reduces the dissolution of NiO. The addition of alkaline earth metals to the eutectic electrolyte mixture greatly reduces the dissolution of NiO; however, above certain concentrations of the metals, the performance of the fuel cell is greatly reduced. An alternative to this would be to use a different cathode material. Lithium ferrite ($LiFeO_2$) and lithium cobaltite ($LiCoO_2$) are much less soluble in the electrolyte matrix; however, $LiFeO_2$ creates high polarization losses that degrade the fuel cell performance. $LiCoO_2$, however, is much more promising, and at atmospheric pressures, it is shown to reduce solubility by an order of magnitude. The doping of the cathode and electrolyte material using alkaline-earth metals is in progress in the area of performance and lifetime; however, these materials increase the price of the fuel cell, which is a big concern in the competitive energy market.

9.4.3.3 Anode

The anode side contains the reaction of hydrogen oxidation by carbonate ions, and therefore there is little free oxygen available to create a corrosive atmosphere. This allows many metals to be used as electrocatalyst for hydrogen oxidation. Nickel, cobalt, and copper can be used in the form of powdered alloys and composites with oxides. Because of the porous structure of these materials, creep and sintering are concerns under the compressive force applied to seal the cell structure. Many materials and additives have been studied to produce stability in the anode with respect to creep and sintering. For example, chromium and aluminum can be added to the anode, forming dispersed oxides, decreasing the amount of creep and sintering, and allowing for longer stable operation of MCFC stacks. Little contribution other than this has been applied to the anode materials because the anode seems to operate smoothly with little degradation over the lifetime goal of 5 years.

9.4.4 MCFC Recent Advances

9.4.4.1 Material Development

Material development has been heavily influenced by the long-term operation test results by the few MCFC developers and manufacturers such as Fuel Cell Energy. The goal of further development is to minimize the voltage decay rate over a lifetime of 40,000 h. Many technological improvements over the last decade have improved the performance and lifetime of MCFCs to the point of commercialization. Further efforts will most likely be focused on new electrolyte mixtures with little dependence on operating temperature and pressure. Also, new cathode materials that do not significantly increase the cost of the fuel cell, while lessening the effects of dissolution and increasing the performance of the fuel cell, will further aid the commercialization effort of MCFCs. Increased efforts and collaboration by the Department of Energy, national laboratories such as the National Renewable Energy Laboratory, and manufacturers is key to the success of MCFC systems and the future outlook is promising.

9.4.4.2 Fuel and Gas Turbine Hybrid Systems

A high-temperature fuel cell can be integrated into a gas turbine cycle both directly and indirectly. When used indirectly as seen in Figure 9.16, the fuel cell exhaust heats the compressed air first, and then the combustion process further heats the gas before reaching the turbine. Part of the turbine exhaust can be used in the combustion process, which is critical for proper operation of the fuel cell. This allows the fuel cell to operate at atmospheric pressure.

An MCFC–Gas Turbine (MCFC/GT) cycle may utilize the exhaust from the gas turbine to create steam for a steam cycle. This cycle, MCFC/GT-ST cycle, can be very efficient; however, it will be more expensive than the MCFC/GT cycle and has not been demonstrated as of yet. A more probable scenario is to directly use the exhaust heat of the MCFC fuel cell to create steam for a

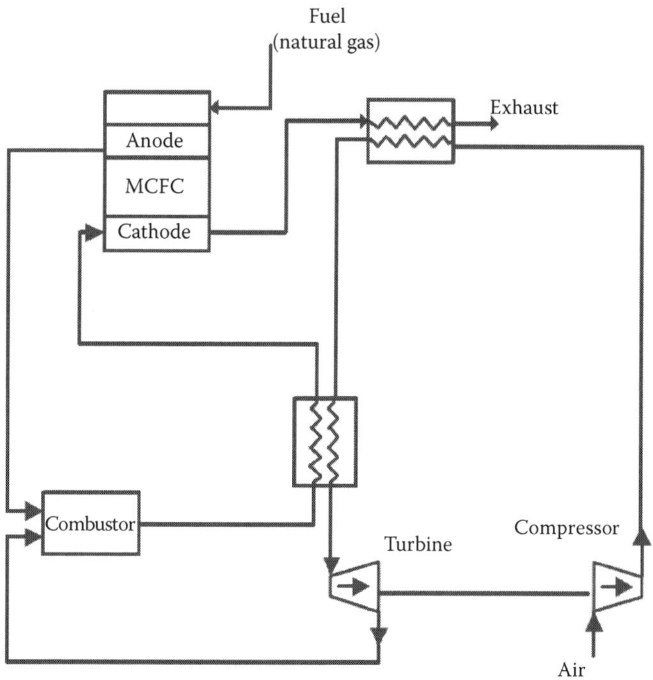

FIGURE 9.16
Indirect turbine/fuel cell layout.

steam cycle. This would be much simpler than the MCFC/GT-ST cycle and would also allow the fuel cell to operate at atmospheric pressure. This cycle, again, would likely be more expensive than the MCFC/GT cycle and be less efficient because of the irreversible losses in heat exchange. In the case of directly creating steam from the fuel cell exhaust, the fuel cell would replace the combustor.

9.5 Solid Oxide Fuel Cell

Some of the attractive features of SOFC are (i) all solid components, (ii) compact, (iii) lower activation losses, (iv) more tolerant to the presence of impurities in the reactant gases, (v) flexible in fuel types, (vi) simpler fuel processing process, (vii) allows internal reforming, and (viii) no corrosion of cell components by liquid. High-temperature operation excludes use of expensive metal catalyst. High-temperature operation also provides a high-quality waste heat for cogeneration, and a better system match resulting in higher overall conversion efficiencies.

In this section, a description of the SOFC is given in terms of its basic principles and operation, different design configurations, components and materials.

9.5.1 Basic Principles and Operation

SOFC is classified as an *anion fuel cell* with negatively charged oxide ion (O^{2-}) migrating through the electrolyte. The basic components and operation of an SOFC are depicted in Figure 9.17.

At the cathode, the reduction of oxygen takes place with the formation of a *negatively charged oxygen ion*. The oxygen ion transports through the solid oxide ion conducting electrolyte toward the anode. At the anode, it combines with hydrogen gas, producing water and electrons that travel to the cathode side through the external electrical circuit.

$$\text{Cathode reaction:} \quad \frac{1}{2}O_2 + 2e^- \rightarrow O^{2-} \tag{9.24}$$

$$\text{Anode reaction:} \quad H_2 + O^{2-} \rightarrow H_2O + 2e^- \tag{9.25}$$

$$\text{Overall reaction:} \quad H_2 + \frac{1}{2}O_2 \rightarrow H_2O \tag{9.26}$$

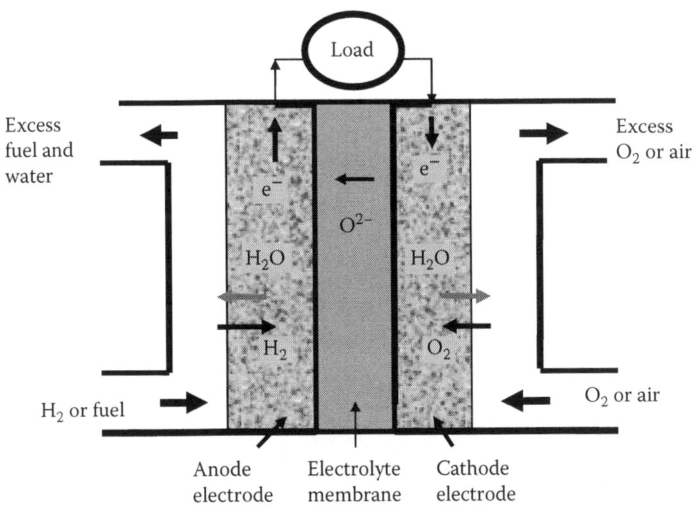

FIGURE 9.17
Operation of SOFC.

Since water is formed at the anode side where hydrogen is supplied as the reactant, the issue of mass transfer loss as a result of water flooding at the anode is relatively low and less critical in SOFC due the higher diffusivity of hydrogen in water. A major characteristic of SOFC is that the oxygen reduction kinetics in the cathode side is relatively fast and results in a lower activation voltage loss, and hence does not require any noble metals as catalyst.

9.5.1.1 SOFC Cell Designs

SOFCs are available in two basic geometrical designs: (i) circular and tubular design and (ii) planar design. The tubular design is primarily considered by Siemens–Westinghouse. Figure 9.18 depicts the tubular design with reactant gas flow configurations using a center manifold. One of the major advantages of this design is the concept of one closed end of the tube, and this

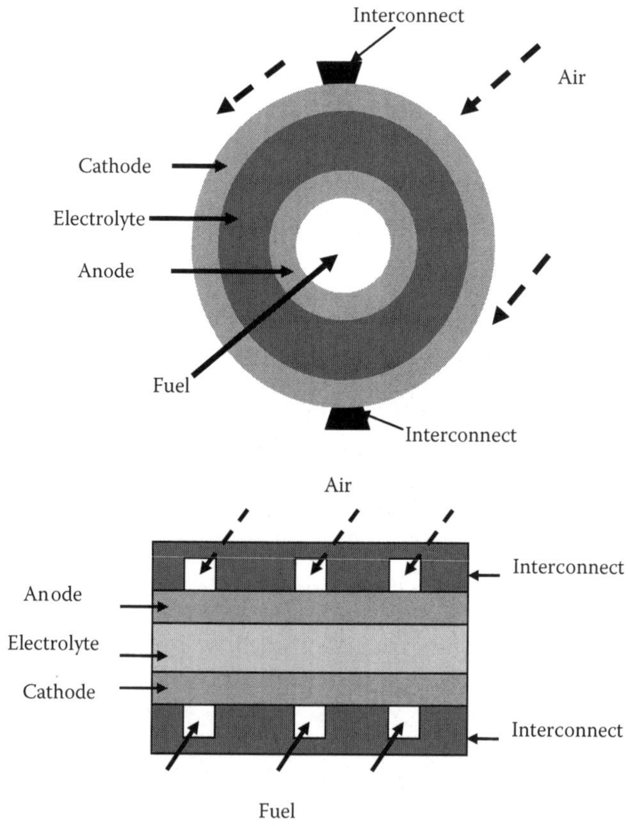

FIGURE 9.18
SOFC designs.

eliminates the need for gas seals between cells and it can provide a robust ceramic structure for the cell. However, it leads to a relatively long current path around the circumferences of the cells, resulting in higher internal resistance. A flat-tube cell design is also considered for a shorter current path and ease of cell stacking.

9.5.1.2 Planar Design

This is the most popular design, which includes integrated planar and sequential cells. Planar SOFC systems have been receiving attention largely because of ease of manufacturing and high performance compared to tubular SOFCs. Such a design allows a number of different gas flow configurations: *cocurrent-flow, counter-flow, and cross-flow* (Mager and Manglik, 2007; Venkata et al., 2008).

The advantages of a planar design are the simplicity in manufacturing of the stacked cell components leading to a highly compact structure with higher power densities. However, a planar design requires sealing to avoid crossover of reactant gases and has increased risk of cell fractures, particularly during thermal cycling.

Planar SOFCs are generally manufactured in three different configurations depending on the structure-supported cell element and operating temperature range as shown in Figure 9.19. These configurations are referred to as (i) *electrolyte-supported cell* with thick electrolyte layer, (ii) *anode-supported cell* with thick anode layer, and (iii) *cathode-supported cell* with thick cathode layer.

The basic SOFC design suitable for operation around 1000°C is an electrolyte-supported cell with a thicker electrolyte of thickness 100 μm that supports thinner anode and cathode electrodes of thickness on the order of 50 μm. Electrolyte conductivity is a strong function of operation temperature of SOFC. For SOFC operating at lower temperatures, the ionic conductivity is lower, and in that case, anode- or cathode-supported cell configurations are preferred. In the electrode-supported cell configuration, the electrolyte is very thin (around 20 μm), and either the anode or cathode is thick enough to support the structure of the cell. The general thickness of the supporting electrode varies between 350 and 1500 μm. Table 9.4 lists the operating temperatures and thicknesses of different planar SOFC configurations.

In order to improve the performance of SOFC, a thinner yttria-stabilized zirconia (YSZ) electrolyte is considered for lower ohmic resistance and for operation in the intermediate temperature range of 500°C–800°C. Ionic conductivity decreases with decrease in temperature and hence the area-specific resistance (ASR) of an electrolyte increases with lower operating temperature. Fabricating the electrolyte in a dense and thinner film reduces the ASR or the resistance to ionic transport, allowing a lower operating temperature. For this purpose, efforts are being made in fabricating SOFC cell on the basis of either a thicker anode-supported or a thicker cathode-supported SOFC

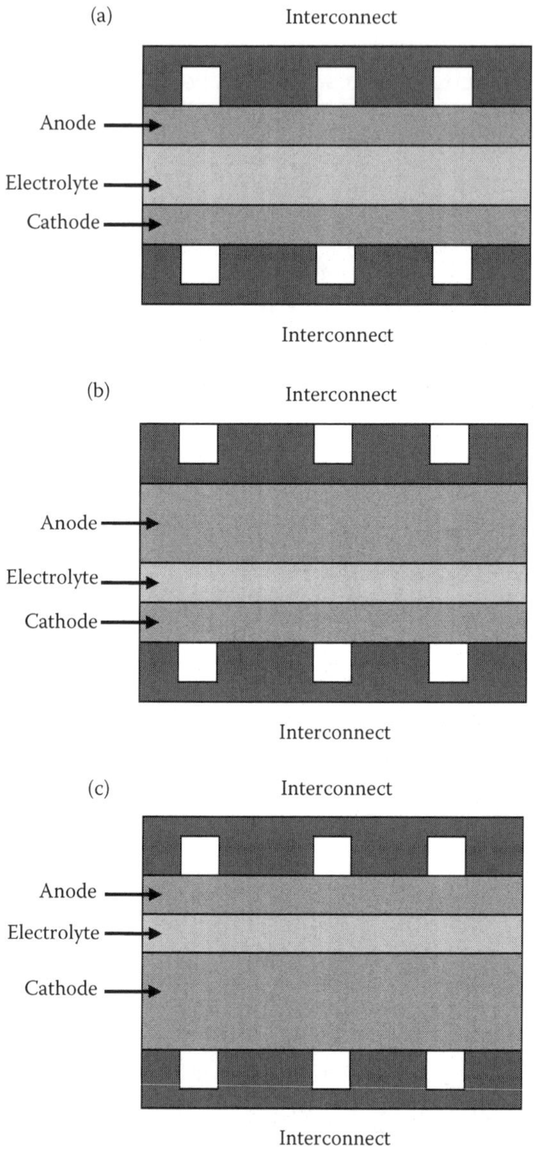

FIGURE 9.19
Different configurations of planar SOFC designs. (a) Electrolyte-supported cell. (b) Anode-supported cell. (c) Cathode-supported cell.

TABLE 9.4

Operating Temperatures and Thicknesses of Different Planar SOFC Configurations

Electrolyte-Supported Cell	Anode-Supported Cell	Cathode-Supported Cell
Temperature: 1000°C Typical thicknesses Anode: ~50 μm Electrolyte: >100 μm Cathode: ~50 μm	Temperature: 600°C–800°C Typical thicknesses Anode: ~300–1500 μm Electrolyte: <20 μm Cathode: ~50 μm	Temperature: 600°C–800°C Typical thicknesses Anode: ~50 μm Electrolyte: <20 μm Cathode: ~300–1500 μm

cell structure. A major challenge in developing an electrode-supported cell with a thin electrolyte layer is to fabricate it free of any defects with no fuel crossover.

Recent development efforts of SOFCs are focused on lowering cell operating temperature for enhanced performance and operating power density, compact stack design, durability, and reduced cost (Kim et al., 2005; Sahibzada et al., 2000). Lowering cell operating temperature below 800°C allows for wider selection of materials for electrolyte, electrode, interconnect, and seals. Kim et al. (2005) fabricated and characterized an anode-supported SOFC cell by dip-coating a thin YSZ film of thickness less than 15 μm over the NiO-YSZ anode support. This design demonstrated a good electrical performance of 0.56 mW cm^{-1} at 850°C and 0.56 mW cm^{-1} at 700°C.

Recent efforts to reduce the operating temperature of SOFCs include the search for novel electrolytes with higher conductivities than the most commonly used YSZ (Badwal and Foger, 1997; Badwal et al., 1998; Inaba and Tagawa, 1996; Steele, 2000; Steele and Heinzel, 2001). Because of the high operating temperature, YSZ-based SOFCs do not require high-cost platinum catalyst for the electrodes like PEMFC. Lowering the operating temperature reduces the reaction kinetics at the electrodes and may require some catalysts such as Ni-based anode electrode. However, the high-temperature gas compositions at the anode and cathode require different materials for the two electrodes as discussed in a subsequent section. High-temperature operation also places additional constraints on the materials for interconnects and seals, which also degrades with long-term operation.

9.5.2 Components of SOFC

The major components of a fuel cell include anode, electrolyte, and cathode. These single cells connected together either in series or parallel with the help of an interconnector to form a fuel cell stack. The electrolyte is an ionic conductor that conducts oxygen ions produced at the cathode–electrolyte interface to the anode–electrolyte interface where these ions combine with hydrogen forming water and electrons. The interconnector serves to conduct electrons through external circuit. A very comprehensive review of materials for SOFC is given in the book edited by Fergus et al. (2009).

9.5.2.1 SOFC Electrolyte

The electrolyte in an SOFC transports negatively charged oxygen ions produced at the cathode–electrolyte interface to the anode–electrolyte interface where these ions combine with hydrogen forming water and electrons. As we have discussed in Chapter 7, the ion transport in crystalline electrolyte is driven by thermally activated vacancy diffusion or hopping diffusion mechanism. The crystal structure must contain enough vacancy sites or unoccupied sites similar to those occupied by oxygen ions. These vacancy sites are also referred to defects, which occur naturally in many oxide crystalline materials. However, in order to achieve sufficient ion conductivity, additional vacancy spots are created in crystalline base electrolyte material by doping it with an impurity or an alloy element.

In addition to the requirement for high ionic conductivity, the electrolyte material must possess the lowest electronic conductivity, negligible reactant fuel migration, compatible coefficient of expansion with adjacent electrode materials, good mechanical properties, stability, and negligible interactions with electrode materials over a range of operating and fabrication conditions. On the basis of these requirements, a number of solid oxide electrolyte materials such a *zirconia* and *ceria fluorites* and lanthanum gallate ($LaGaO_3$)–based pervoskites have been identified and investigated by many researchers.

Among these, *zirconia* (ZrO_2) for a higher temperature range of 800°C–1000°C, lanthanum gallate ($LaGaO_3$) for an intermediate temperature range of 600°C–800°C, and ceria (CeO_2) for a low temperature range of 400°C–600°C have been under development as electrolyte for SOFCs.

In order to create additional oxygen vacancy sites in the materials, the positively charged cation is substituted or doped by another cation with lower valence number. This leads to the creation of vacancies to achieve neutrality. For example, in YSZ, the oxygen vacancy V_o'' is created by replacing the zirconia cation Zr^{4+} with yttria cation Y^{3+}. This is explained by the defect equation written using the *Kroger–Vink notation* as

$$Y_2O_3 \overset{ZrO_2}{\rightarrow} 2Y'_{Zr} + 3O_o^x + V_o'' \tag{9.27}$$

The neutrality condition is given by $2\left[Y'_{Zr}\right] = \left[V_o''\right]$.

The ion conductivity depends on many factors including the difference in the size of dopant and base material ions and degree of interactions between the defect pair and dopant size. Additional factors that influence ionic conductivity are the fabrications and processing methods, aging, and grain boundaries of composite structures. Ceramic electrolytes are also developed with additional co-dopants by adding second or third cations for improving electrochemical kinetics, stability, cost, and processing temperature.

A brief description of these electrolytes is given below:

9.5.2.2 *Zirconia Electrolyte*

Zirconium oxide (ZrO_2) has been considered as the most promising electrolyte material for SOFCs operated at high (800°C–1000°C) and intermediate (600°C–800°C) temperature ranges because of its availability in abundance and low cost. Zirconia-based SOFC systems are being developed using *yttria* and *scandia* as the dopant for enhanced oxygen vacancy sites and for stabilized zirconia structure for phase equilibrium during phase transformations. This is because of their higher ionic conductivity and high mechanical and chemical stability.

However, the state-of-the-art ceramic electrolyte material used in SOFC is the *YSZ* because YSZ has the highest oxide ion conductivity, lowest electronic conductivity, and lowest gas permeability to prevent gas crossover losses. The base material in the YSZ is the zirconia (ZrO_2), which is doped with yttria (Y_2O) as the dopant element to create oxygen vacancy sites and stabilize the structures of zirconia over the low to high temperature range during heating and cooling. The substitution of Zr^{4+} at a lattice position with the Y^{3+} ions creates vacancies in the oxygen sublattice and causes oxygen ion conduction in the stabilized zirconia. The electrolyte material is made with 8–10 mol% yttria (Y_2O_3)-stabilized zirconia (ZrO_2) with operating temperature in the range of 800°C–1000°C for good ionic conductivity, reaction kinetic, and lowest electronic conductivity. A typical composition contains 8% yttria (Y_2O_3) mixed with zirconia (ZrO_2).

Table 9.5 shows typical variation of ion conductivity of zirconia-YSZ electrolyte with yttria concentration (Fergus et al., 2009).

The YSZ conductivity increases with yttria dopant concentration up to 8% owing to increases in oxygen vacancies. With further increase in dopant concentration, the ionic conductivity decreases because of increased interactions between oxygen vacancy and yttrium ions.

The conductivity of the YSZ electrolyte is a strong function of temperature and it increases with temperature. The temperature dependence of the

TABLE 9.5

Variation Ionic Conductivity of YSZ with Volume Fraction of Yttria

Yttria Concentration (%)	Ionic Conductivity (σ_i) at 1000°C, S/cm	Activation Energy (ΔG_a) at 850°C–1000°C, eV (kJ/mol)
3% Y_2O_3	0.049	0.80 (80)
8% Y_2O_3	0.137	0.91 (91)
10% Y_2O_3	0.13	0.83 (83)
12% Y_2O_3	0.068	1.04 (104)

Source: Fergus, J. W., R. Hui, X. Li, D. P. Wilkinson and J. Zhang, *Solid Oxide Fuel Cell*, CRC Press, 2009.

electrolyte conductivity is given by the following curve-ft correlations of experimental data function relations given by Equations 7.48 and 7.49.

$$\sigma_i = ae^{-\frac{b}{T}}, \tag{7.47}$$

where the empirical coefficients are given as $a = 3.34 \times 10^4$ $(\Omega\text{-m})^{-1}$ and $b = 1.03 \times 10^4$ K.

Another alternative correlation is given in terms of activation energy in similarity with Equation 7.44 as

$$\sigma_i = \sigma_0 e^{-\frac{G_a}{RT}}, \tag{7.48}$$

where ΔG_a is the activation energy and σ_0 is the reference conductivity determined empirically for the migrating element. Typical values of activation energy for SOFC based on the YSZ electrolyte are in the range of 80–105 kJ/mol.

As shown in Figure 9.20, a plot of these equations shows strong dependence of ionic conductivity of 8% YSZ with temperature in the range of 800°C–1000°C. Equation 1 is based on the YSZ conductivity given by Equation 7.48. Equation 2 is based on Equation 7.49 with $\sigma_0 = 9 \times 10^7$K/(Ωm) and $\Delta G_a = 100$ kJ/mol. The conductivity value decreases by a factor of 10 as temperature is reduced from 1000°C to 800°C.

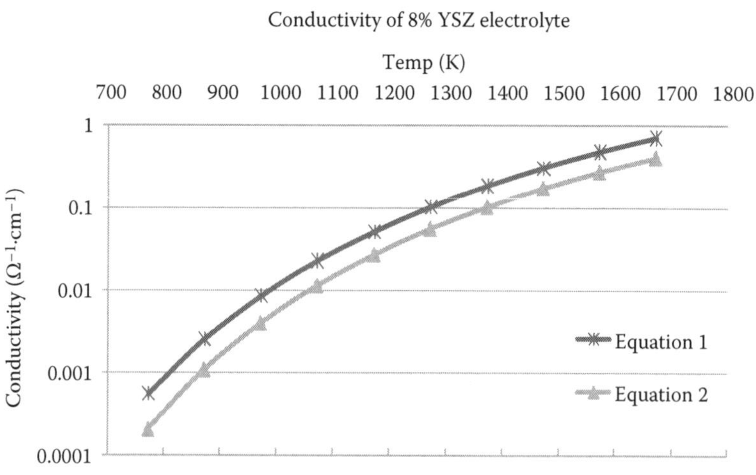

FIGURE 9.20
Variation of ionic conductivity of 8% YSZ with temperature.

9.5.2.3 *Scandia-Stabilized Zirconia (ScSZ)*

Scandia (Sc_2O_3) doped zirconia (ZrO_2) is also of considerable interest as an SOFC electrolyte, particularly in the intermediate temperature range. A higher conductivity is observed for scandia-stabilized zirconia (ScSZ) electrolyte with 8%–10% scandia concentration as compared to YSZ. At an operating temperature of 1000°C, 9% ScSZ has a conductivity of 0.32 S cm^{-1} as compared to 0.164 S cm^{-1} for 8% YSZ. The enhancement is believed to be caused primarily by the smaller difference in the sizes of Zr^{4+} and Sc^{3+} ions as compared to the difference between Zr^{4+} and Y^{3+} ions. Major challenges and efforts in the development of scandia-stabilized zirconia (ScSZ) are to achieve the stability in phase equilibrium through selection of dopant concentrations and fabrication methods.

9.5.2.4 *Ceria Electrolyte*

Another ceramic material that is of considerable interest among researchers of SOFC electrolyte is ceria. The oxygen ion conductivity in ceria is enhanced by doping and co-doping with lower valance number cations. Some of the most popular doped ceria electrolytes are *gadolinia-doped ceria* (GdCeO), *samaria-doped ceria* (SmCeO), and *yttria-doped ceria* (YCeO). These are leading candidates for low-temperature (400°C–600°C) SOFCs and as interlayer for intermediate temperature (600°C–800°C) with YSZ owing to their lower activation energy associated with the ion transport and higher ionic conductivity. Table 9.6 shows the conductivity of various SOFC electrolytes.

9.5.2.5 *Gadolinia-Doped Ceria (GDC or GdCeO)*

Gadolinia-doped ceria (GdCeO) is considered as one of the leading ceria-based SOFC electrolyte for intermediate and low temperature range operations. Like other doped-ceramic electrolyte, the conductivity of $Gd_xCe_{1-x}O_{2-x/2}$ achieves a maximum conductivity value at a certain composition (x) of gadolinia and it is strongly influenced by the temperature and grain boundaries of the composite structure. Typical oxygen ion conductivity of GdCeO falls in the range of 0.016–0.026 S cm^{-1} for the composition range of x = 10%–20% at an operating temperature of 600°C.

TABLE 9.6

Ionic Conductivity of Different Solid Oxide Electrolytes

Electrolyte	Ionic Conductivity, σ_i (S cm^{-1})
YSZ	0.164 at 1000°C 0.03 at 800°C
YDC	0.013 at 700°C
ScSZ	0.1 at 800°C

The ionic conductivity of ceria electrolyte is also expressed by relations given by Equations 7.48 and 7.49.

9.5.2.6 Samaria-Doped Ceria (SmCeO)

The conductivity of samaria-doped Ceria ($Gd_xCe_{1-x}O_{2-x/2}$) varies with samaria concentration and reaches a peak conductivity value in the range of 15%–20%. There is considerable variation in the peak values (0.004–0.023 S cm^{-1}) measured by different researchers.

9.5.2.7 Yttria-Doped Ceria (YDC)

Yttria-doped ceria ($Y_xCe_{1-x}O_{2-x/2}$) shows a peak ionic conductivity for an yttria concentration in the range of x = 10%–15%. The peak conductivity data vary from 0.0025 to 0.0044 at 500°C.

9.5.2.8 SOFC Anode Electrode

The primary function of an anode in SOFC is to allow electrochemical oxidation of fuels, transport electrons to the external circuit, and allow internal reforming and partial oxidation when hydrocarbon fuels are used. The major requirements of an SOFC-anode material include good electrochemical kinetics, high catalytic activity, high porosity for enhanced fuel transport to the electrode–electrolyte interface, high electronic conductivity, and low ionic conductivity. The resistance to electrochemical reactions that takes place at the TPBs depends on the surface catalytic activities for fuel oxidation and reforming as well as on the microstructure and transport properties of the anode. A major challenge is to optimize the charge and mass species transport along the surface, across the interface, and through the bulk electrode. The material should possess a compatible thermal expansion coefficient value that is in good match with other adjacent electrolyte and interconnects. Additional factors include ease of fabrication to obtain desired microstructure properties such as porosity and surface area, and good chemical and thermal stability during fabrication and operation. For an anode-supported cell, it also provides good mechanical strength and maintains stability during manufacturing and operation.

The most common anode material is nickel-yttria stabilized zirconia (Ni-YSZ) Cermet or a mixture of nickel and YSZ. While nickel serves for the required catalytic activity and electron conductivity, YSZ lowers the effective coefficient of expansion to march with adjacent YSZ electrolyte. However, use of YSZ extends the active zone for the anode reaction. Depending on the design, Ni to YSZ volume ratio varies in the range from 30% to 50% in the composite mixture. Electronic conductivity varies in the range of 0.1–1000 S cm^{-1} depending on the porosity and composition of Ni-YSZ composition.

The *typical thickness* of anode varies in the range of 40–100 μm. In recent times, a thicker anode in the range of 350–750 μm is used to develop the so-called anode-supported SOFCs that allow for a thinner electrolyte layer.

9.5.2.9 SOFC Cathode Electrode

In SOFC, the electrochemical reaction with reduction of oxygen into oxygen-ion takes place at the cathode–electrolyte interface as given by Equation 9.24. The cathode electrode in SOFC has to allow oxygen to transport toward the electrode–electrolyte interface where it combines with the incoming electrons to undergo reduction to produce oxygen ion and allow oxygen ion transport away from the reaction regions toward the anode electrode through electrolyte. The requirement of the electrode–electrolyte interface region is to allow the oxygen ion to occupy the vacancy sites and transport through the electrolyte based on vacancy diffusion process. This is demonstrated by the electrochemical reaction written using the Kroger–Vink notation as follows:

$$O_2 + 2V_o'' + 4e^- \rightarrow 2O_o'', \tag{9.28}$$

where the oxygen ion occupies the available oxygen vacancy sites in the YSZ electrolyte.

When the electrode material and electrolyte material possess only electronic and ionic conductivity, respectively, such as Sr-doped $LaMnO_3$ (LSM) electrode and YSZ electrolyte, these criteria are fulfilled in the vicinity of TPB.

The most popular cathode materials are *lanthanum manganite* ($LaMnO_3$), *strontium-doped lanthanum manganite* (LSM), and *lanthanum strontium cobalt ferrite* (LSCF), which are developed to meet major requirements of the cathode electrode such as high reaction activity, high electrical conductivity, low interaction with electrolyte, and thermal and chemical stability. In order to achieve higher electronic conductivity, lanthanum manganite is doped with lower-valence cations such as strontium Sr^{2+} or calcium Ca^{2+}, which leads to the development of *strontium-doped lanthanum manganite,* $La_{1-x}Sr_xMnO_3$ (LSM) and *calcium-doped lanthanum manganite,* $La_{1-x}Ca_xMnO_3$ (LCM).

Strontium-doped lanthanum manganite, $La_{1-x}Sr_xMnO_3$ (LSM) is developed by doping lanthanum manganite with a small fraction (10%–20%) of strontium in order to enhance electronic conductivity of the cathode electrode. *Lanthanum strontium cobalt ferrite* (LSCF) is a mixture of lanthanum oxide, strontium oxide, cobalt oxide, and iron oxide with typical composition given as $La_{0.6}Sr_{0.4}Co_{0.2}Fe_{0.8}O_3$.

While the thickness of a typical cathode electrode in an SOFC is around 50 μm, research is also in progress to develop a cathode-supported SOFC cell that supports a thinner electrolyte deposited over a cathode electrode of thickness on the order of 350–1500 μm. Table 9.7 lists the properties of anode, cathode, and electrolyte materials.

TABLE 9.7

Properties of SOFC Anode, Cathode, and Electrolyte Materials

	Anode	Cathode	Electrolyte
Material used	Ni-YSZ	LSCF	8% YSZ
Density (kg/m_3)	8910	6600	5940
Specific heat (J/g·K)	0.444	0.432	0.471
Thermal conductivity (W/m·K)	90.5	1.95	1.62 (at 1144 K)
Ionic conductivity	–	–	
Electrical conductivity			–
Permeability (m_2)	1.76×10^{-11}	1.76×10^{-11}	–
Porosity	0.5	0.5	–

9.5.2.10 SOFC Interconnect

The identification and fabrication of interconnect materials are a challenge in the development of SOFCs. The primary function of interconnect is to carry the electrical current from the electrochemical cell to the external circuit. Interconnect can be either a metallic or a ceramic material that connects two individual cells. Interconnect must be extremely stable because it is exposed to oxidation and reduction on either side of the material. A generally used interconnect is La(ca)CrO_3. The main disadvantage of this material is it degrades during long-term operation.

9.6 Direct Methanol Fuel Cell

As we discussed in Chapter 1, hydrocarbon fuels like methanol can be used in fuel cells. The general approach is to use an external reforming process to convert the hydrocarbon fuel into a gas mixture before feeding into a fuel cell. However, all fuel cell types with a direct feed of hydrocarbon fuels have also been considered and investigated by many. Among all direct hydrocarbon fuel cells, direct methanol fuel cell (DMFC) is the most popular and attractive because of its high electro-oxidation kinetics at the anode. Also, among all types of acid and alkaline electrolytes, PEM electrolyte in DMFC is the most extensively investigated because of its potential to provide high power density. The DMFC is similar to the design and structure of the PEMFC, which includes a proton-conducting solid PEM with two catalyst-coated electrodes as shown in Figure 9.21.

Basic operation involves supply of a liquid mixture of methanol and water at the anode side. At the anode side, methanol is oxidized in the presence of 1 mol

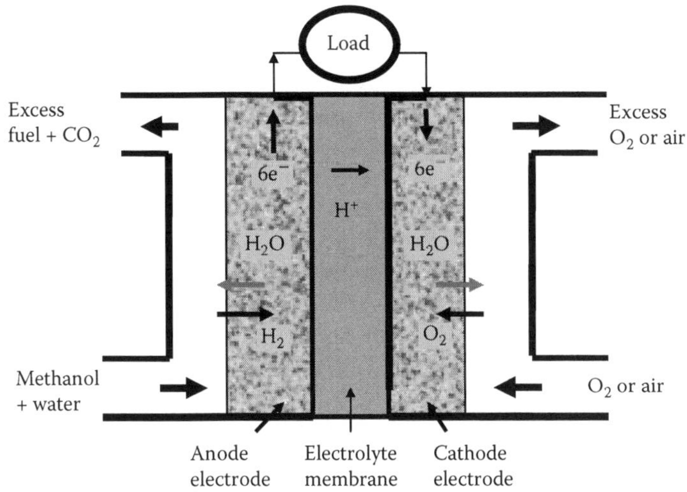

FIGURE 9.21
Operation of DMFC using PEM electrolyte.

of water, producing *1 mol of carbon dioxide, six protons* (H^+), and *six electrons* (e^-) for 1 mol of methanol in the presence of the catalyst. Electron travels through the electrodes and external power circuit toward the cathode. At the cathode, oxygen undergoes reduction reaction by combining with the incoming protons from the membrane and electrons from the external circuit producing 3 mol of water. The electrode half-cell reactions and the overall methanol reactions are as follows:

$$\text{Anode reaction:} \quad CH_3OH + H_2O \rightarrow CO_2 + 6H^+ + 6e^- \tag{9.29a}$$

$$\text{Cathode reaction:} \quad \frac{3}{2}O_2 + 6H^+ + 6e^- \rightarrow 3H_2O \tag{9.29b}$$

$$\text{Overall reaction:} \quad CH_3OH + \frac{3}{2}O_2 \rightarrow 2H_2O + CO_2 \tag{9.29c}$$

As we can see, while methanol oxidation reaction requires water at the anode, three times more water is produced at the cathode with oxygen reduction reaction. Supply of methanol–water mixture at the anode helps in maintaining the polymer membrane well hydrated and avoids drying of the membrane caused by the water migration from anode to cathode owing to electro-osmotic drag as discussed in Chapter 7. A water management system can, therefore, be used to collect water generated at the cathode and supply a fraction of it to the anode.

Since methanol exists as a liquid in the temperature range of –97°C to 64°C at atmospheric pressure, it can be stored, transported, and may be used in liquid form similar to other liquid fuels like gasoline and diesel, and this makes DMFC compact and suitable for portable and mobile applications such as in laptop computers as a battery substitute.

One of the major issues with DMFC is the high mobility or the so-called methanol crossover through the Nafion membrane from the anode side to the cathode side. Methanol crossover causes lower utilization, affects the cathode oxidation reaction, and hence lowers the efficiency of the fuel cell. The crossed-over methanol reacts with oxygen and this lowers the oxygen concentration available for the electrochemical reduction reaction at the cathode. The direct methanol combustion reaction at the cathode generates additional heat, causing local heating as well as aging and poisoning of the catalyst at the cathode–electrolyte interface, and lowers the cathode performance and electric potential.

One major disadvantage of DMFC compared to other types of fuel cell is the lower efficiency caused by the methanol crossover. This is evident from Equation 4.56 for the fuel cell efficiency and Equation 4.93 for heat generation that includes a fuel utilization fraction (ϕ_f).

9.6.1 Gas Diffusion Layer

Like in a PEMFC, the gas diffusion layers in DMFC serve as a distributor of reactant gases to the electrode–electrolyte interface. At the cathode, it distributes oxygen and facilitates in removing any excess water to avoid flooding of the electrode by transferring it to the gas flow channels of the bipolar plate. At the anode, it distributes methanol–water mixture and removes carbon dioxide produced at the anode oxidization reaction (Lindermeir et al., 2004).

9.6.2 Catalyst in DMFC

Like in PEMFC, platinum is the standard catalyst used in the electrode and Nafion polymer electrolyte interface of DMFC for low temperature levels. Catalyst development effort for DMFC involves the search for an alternative to the Pt-Ru catalyst for enhanced anode oxidization reaction.

PROBLEMS

1. Estimate the water content, proton conductivity, water diffusion coefficient, and electro-osmotic drag coefficient for a Nafion membrane under humidity conditions with water activity $a = 0.9$ and an operating temperature of 80°C.
2. Estimate the water content, proton conductivity, water diffusion coefficient, and electro-osmotic drag coefficient for a Nafion membrane with 10% moisture in the air stream with an operating pressure and temperature of 2 bar and 70°C, respectively.

References

Badwal, S. P. S. and K. Foger. Materials for solid oxide fuel cell. *Materials Forum* 21: 183–220, 1997.

Badwal, S. P. S., F. T. Ciacchi, S. Rajendran and J. Drenan. An investigation of conductivity, microstructure and stability of electrolyte compositions in the system 9 mol% (Sc_2O_3–Y_2O_3)–ZrO_2(Al_2O_3). *Solid State Ionics* 109:167–186, 1998.

Berg, P., K. Promislow, J. St Pierre and J. Stumper. Water management in PEM fuel cells. *Journal of the Electrochemical Society* 151(3): A341–A353, 2004.

Bernardi, D. M. and M. W. Verbrugge. Mathematical model of a gas diffusion electrode bonded to polymer electrolyte. *AIChE Journal* 37(8): 1151–1163, 1991.

Cifrain, M. and K. Kordesch. Hydrogen/oxygen (air) fuel cells with alkaline electrolytes. In: *Handbook of Fuel Cells—Fundamentals, Technology and Applications*, Vol. 1, Part 4, Editors W. Vielstich, A. Lamm and H. A. Gasteiger. Wiley, New York, pp. 267–280, 2003.

Dannenberg, K., P. Ekdunge and G. Lindbergh. Mathematical model of the PEMC. *Journal of Applied Electrochemistry* 30: 1377–1387, 2000.

Fergus, J. W., R. Hui, X. Li, D. P. Wilkinson and J. Zhang, Editors. *Solid Oxide Fuel Cell.* CRC Press, Boca Raton, FL, 2009.

Fuel Cell Hand Book, 7th Edition. Department of Energy, 2004.

Fuller, T. F. and J. Newman. Experimental determination of the transport number of water in nafion-117 membrane, *Journal of the Electrochemical Society* 139: 1332–1337, 1992.

Fuller, T. F. and J. Newman. Water and thermal management in solid-polymer-electrolyte fuel cells. *Journal of the Electrochemical Society* 140(5): 1218–1225, 1993.

Gurau, V., S. Kakac and H. Liu. Mathematical model for proton exchange membrane fuel cells. Proceedings of the ASME Advanced Energy Systems Division, pp. 205–214, 1998.

Hu, M., A. Gu, M. Wang, X. Zhu and L. Yu. Three dimensional two phase flow mathematical model for PEM fuel cell: Part I. Model development. *Energy Conversion and Management* 45: 1861–1882, 2004.

Inaba, H. and H. Tagawa. Ceria-based solid electrolytes. *Solid State Ionics* 83(1–2): 1–16, 1996.

Kim, J., S. Lee, S. Srinivasan and C. E. Chamberlin. Modeling of proton exchange membrane fuel cell performance with an empirical equation. *Journal of the Electrochemical Society* 142: 2670–2674, 1995.

Kim, S. D., S. H. Hyun, J. Moon, J.-H. Kim and R. H. Song. Fabrication and characterization of anode-supported electrolyte thin films for intermediate temperature solid oxide fuel cells. *Journal of Power Sources* 139: 67–72, 2005.

Kordesch, K. and V. Hacker. Stack materials and design. In: *Handbook of Fuel Cells—Fundamentals, Technology and Applications*, Vol. 4, Editors W. Vielstich, A. Lamm and H. A. Gasteiger. Wiley, New York, pp. 766–773, 2003.

Li, Q., R. He, J. O. Jensen and N. J. Bjerrum. PBI-based polymer membranes for high temperature fuel cells – preparation, characteristics and fuel cell demonstration. *Fuel Cells* 4(3): 147–159, 2004.

Lindermeir, A., G. Rosenthal, U. Kunz and U. Hoffmann. Improvement of MEAs for direct methanol fuel cells by tuned layer preparation and coating technology. *Fuel Cell* 4(12): 78–85, 2004.

Lu, S., J. Pan, A. Huang, L. Zhuang and J. Lu. Alkaline polymer electrolyte fuel cells completely free from noble metal catalysts. *Proceedings of National Academy of Sciences USA* 105(52): 20611–20614, 2008.

Mager, Y. N. and R. N. Manglik. Modeling of convective heat and mass transfer characteristics of anode-supported planar solid oxide fuel cells. *Journal of Fuel Cell Science and Technology* 4: 185–193, 2007.

Motupally, S., A. J. Becker and J. M. Weider. Diffusion of water in nafion—115 membrane. *Journal of the Electrochemical Society* 147: 3171–3177, 2000.

Muller, J. T., P. M. Urban, W. F. Holderich, K. M. Colbow, J. Zhang and D. P. Wilkinson. Electro-oxidation of dimethyl either in a polymer electrolyte membrane fuel cell. *Journal of the Electrochemical Society*, 147: 4058–4060, 2000.

O'Hayre, R., S.-W. Cha, W. Cotella and F. B. Prinz. *Fuel Cell Fundamentals*. John Wiley & Sons, Inc, New York, 2006.

Remick, R. J., D. Wheeler and P. Singh. MCFC and PAFC R&D Workshop held on November 16, 2009, Palm Spring, Summary Report, January 13, 2010.

Sahibzada, M., B. C. H. Steel, K. Hellgardt, D. Barth, A. Effendi, D. Mantzavinos and I. S. Metcalfe. Intermediate temperature solid oxide fuel cells operated with methanol fuels. *Chemical Engineering Journal* 55: 3077–3083, 2000.

Sinor J. E. Consultants. *Dimethyl Either as a Transportation Fuel Cell: A State-of-the-Art Survey 1997*. U.S. Department of Energy, Washington, DC, 1997.

Sone, Y., P. Ekdunge and D. Somonsson. Proton conductivity of nafion 117 as measured by a four-electrode AC impedance method. *Journal of the Electrochemical Society* 143: 1254–1259, 1996.

Springer, T. E., T. A. Zawodzinski and S. Gottesfeld. Polymer electrolyte fuel cell model. *Journal of Electrochemical Society* 138(8): 2334–2342, 1991.

Steele, B. C. H. Appraisal of $Ce_{1-y}\ Gd_yO_{2-y/2}$ electrolytes for IT-SOFC operation at 500°C. *Solid State Ionics* 129: 95–110, 2000.

Steele, B. C. H. and A. Heinzel. Materials for fuel cell technologies. *Nature* 414: 345–352, 2001.

Sunden, B. and M. Faghri. *Transport Phenomena in Fuel Cells*. WIT Press, Southampton, UK, 2005.

Varcoe, J. R. and R. C. T. Slade. Prospects for problem alkaline anion exchange solid membranes in low temperature fuel cells. *Fuel Cell* 5: 187–200, 2005.

Venkata, P. P., M. A. Jog and R. M. Manglik. Computational modeling of planar SOFC: Effects of volatile species/oxidant mass flow rate and electrochemical reaction rate on convection heat transfer. Proceedings of the 2008 ASME International Mechanical Engineering Congress and Exposition, Boston, 2008.

Verbrugge, M. W. and R. Hill. Transport phenomena in perfluorosulfonic acid membranes during the passage of current. *Journal of the Electrochemical Society* 137(4): 1131–1138, 1990.

Warshay, M. and P. R. Prokopius. The fuel cell in space: Yesterday, today, and tomorrow. *Journal of Power Sources* 29: 193–200, 1990.

Zawodzinski, T. A., M. Neeman, L. Sillerud and S. Gottesfeld. *Journal of Physical Chemistry* 95: 6040–6044, 1991.

Zawodzinski, T. A., J. Davey, J. Valerio and S. Gottesfeld. The water content dependence of electro-osmotic drag in proton-conducting polymer electrolytes. *Electrochemica Acta* 40: 297–302, 1995.

10

Fuel Cell Stack, Bipolar Plate, and Gas Flow Channel

In the previous chapters, we have discussed the basic operating principles of a tri-layer fuel cell or membrane electrode assembly (MEA) that is composed of anode, electrolyte, and cathode, forming the basic building block of a fuel cell stack. In this chapter, we will consider the formation and design of a fuel cell stack using MEAs and interconnect flow-field plates or bipolar plates, and the design and construction of bipolar plates that have integrated flow channels for reactant gases.

The efficiency of a fuel cell depends on the overall kinetics of the electrochemical process and the performance of its components such as MEA and bipolar plate. As we have mentioned before, the success of fuel cells as an alternative power generation system requires reduced cost, reduced volume and weight, and improved performance and durability of both MEA and bipolar plates. Other components that contribute significantly to these requirements in a fuel cell are gaskets or seals, current collector, end plates, and gas flow manifolds.

Research and development effort has been concentrated on the bipolar plate designs to reduce the cost and increase the performance of the fuel cell. Improvements can occur in the performance of a fuel cell through optimization of the channel dimensions and shape in the flow field of bipolar plates. The contact surface area of the reactant gas on the bipolar plates has an effective contribution on the overall reaction of the gases. The reactant gas pressure has an important role in the overall functioning of the fuel cell. Consideration of fluid flow, heat, and mass transfer phenomenon is important while designing the bipolar plate channels.

10.1 Fuel Cell Stack Design

A fuel cell stack is formed by connecting a number of tri-layer single fuel cell units or MEA units separated by interconnect or bipolar plates in series to meet the required power output. The MEAs are placed in good contact on both anode and cathode sides with the electrically conducting plates, often referred to as the *fluid flow-field plates or bipolar plates or interconnect plates,* which have integrated flow channels. Liquid- or gas-phase fuel and oxidant streams are fed through external and internal manifolds and distributed into the

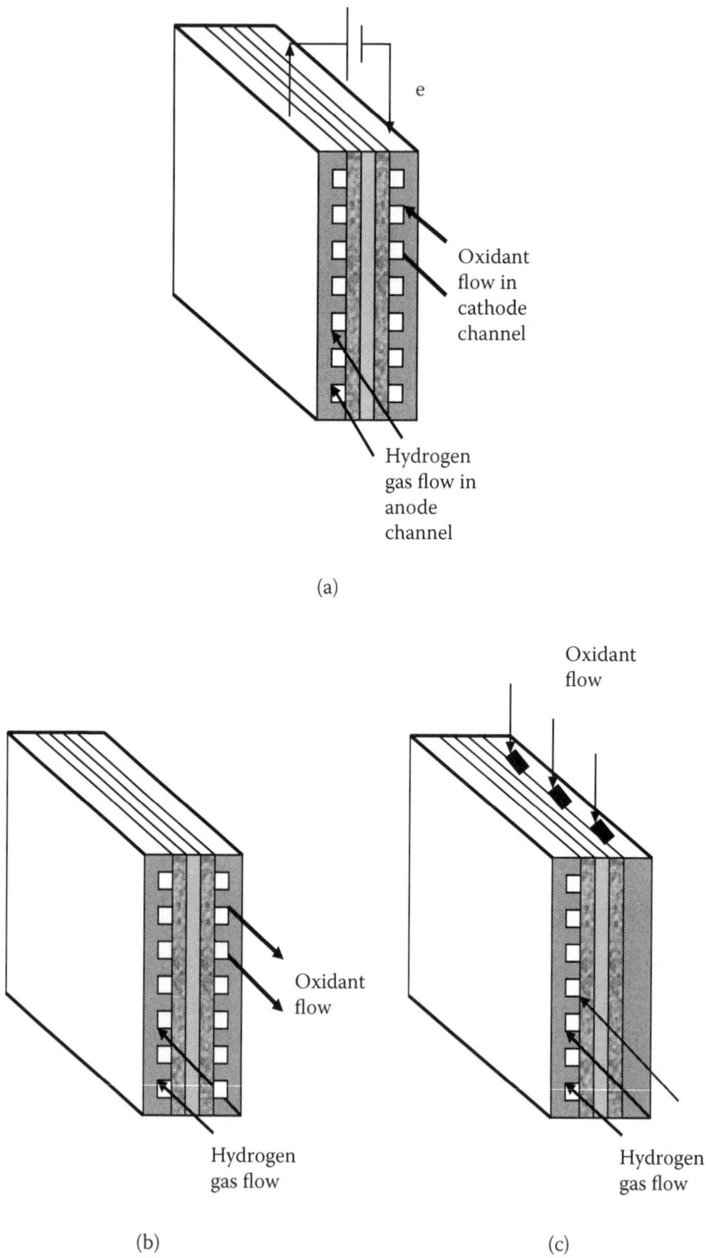

FIGURE 10.1
Gas flow-field plate. (a) Parallel flow-field plate. (b) Counter flow-field plate. (c) Cross flow-field plate.

respective fluid flow-field channels. The flow directions also play a significant role in the effectiveness of the fluid flow-field plates and in the operation of the fuel cell. Fuel and oxidant flow directions through the channels may be in *co-current flow* or in *counter flow* or in *cross flow* directions as shown in Figure 10.1.

In the traditional *fluid flow-field plate*, flow channels are formed on one side of the plate, and so two such plates are needed—one plate on the anode side and another on the cathode side as shown in Figure 10.2a. Such a plate is also

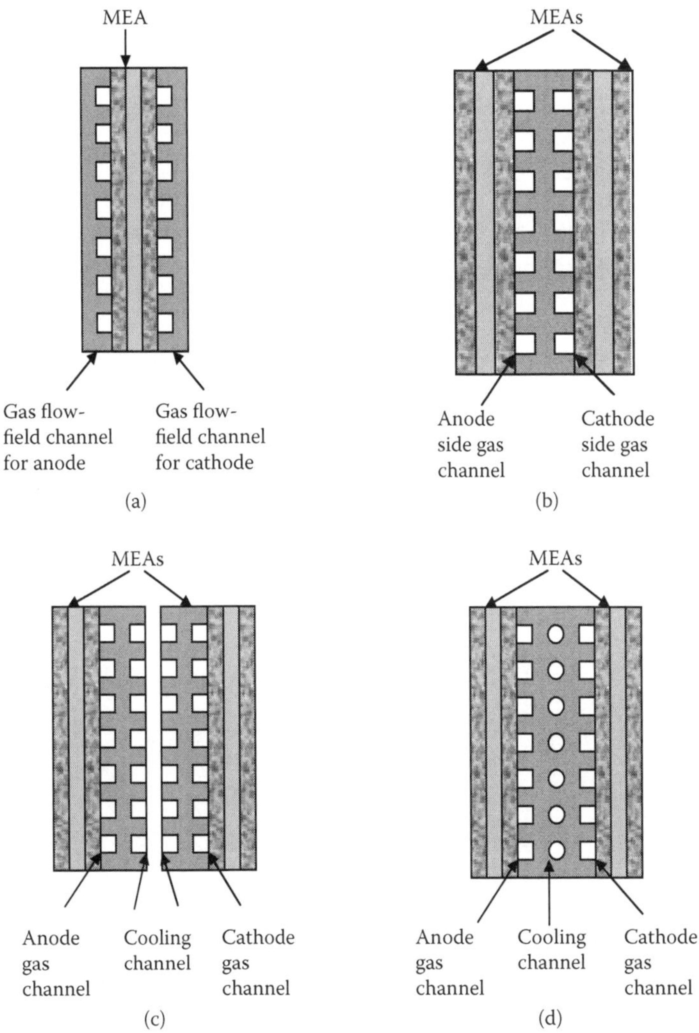

FIGURE 10.2
Gas flow-field plate and bipolar plate with fuel and oxidant flow channels. (a) Gas flow-field plate. (b) Bipolar plate. (c) Cooling channels formed by joining two gas flow-field plates. (d) A bipolar plate with integrated cooling channels.

referred to as *monopolar plate* when used in a single MEA cell. In the so-called *bipolar plate*, the gas flow channels are formed on both sides of the plate—one side is adjacent to the anode side of a MEA and another to the cathode side of the adjacent MEA as shown in Figure 10.2b. Obviously, with the use of bipolar plate, one less plate to build means one less contact surface. This to some extent reduces contact resistance for electron and heat transfer. Often, bipolar plates are also designed to house cooling channels for dissipating heat generated within the cell as shown in Figure 10.2c and d. In Figure 10.2c, the cooling channels are formed by joining two bipolar plates that house grooves for cooling channels on one side and either anode gas channels or cathode gas channels on the other side. Figure 10.2d shows a single bipolar plate that houses both anode gas channels and cathode gas channels on either side and integrated cooling channels at the center of the plate.

A bipolar plate connects the cathode of a tri-layer fuel cell to the anode of the next tri-layer fuel cell and serves a number of purposes: (i) serves as a conductive medium and collector for the electric current generated in the cell, (ii) provides flow channels for an efficient distribution of reactant gases to gas diffusion layers of the electrodes, (iii) effectively removes water and heat generated in the MEAs, and (iv) serves as a structural support to the MEAs. Bipolar plates not only keep the reactant gases such as hydrogen and oxygen separated and prevent them from mixing but also have to be impermeable to the reactant gases, stable, and corrosion resistant to the electrochemical reactions.

A typical cross-sectional view of a fuel cell stack with MEAs along with *fluid flow-field plates* is shown in Figure 10.3 and that with *bipolar plates* is shown in Figure 10.4.

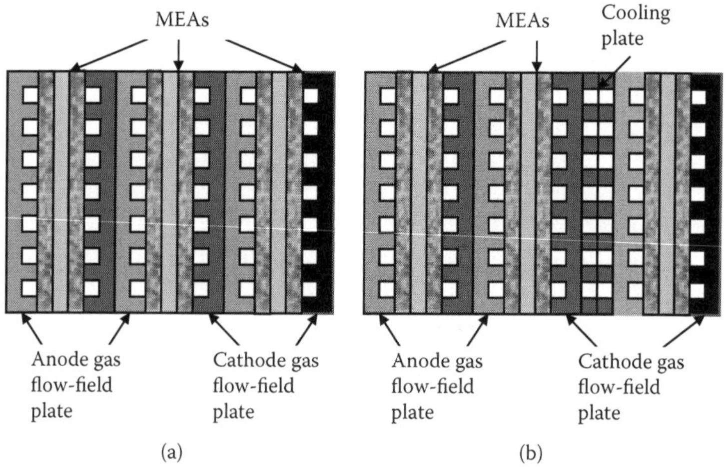

FIGURE 10.3
Fuel cell stack with the series of MEAs and gas flow-field plates. (a) Fuel cell stack with anode and cathode. (b) Fuel cell stack with gas flow-field plates and cooling plates.

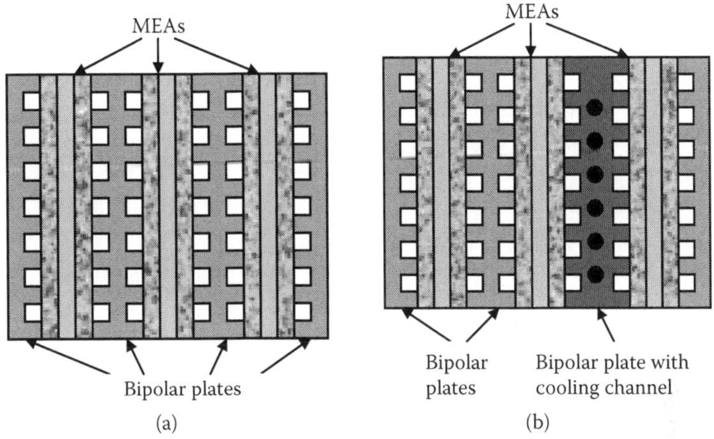

FIGURE 10.4
Fuel cell stack with the series of MEAs and bipolar plates. (a) Bipolar plate without cooling channels. (b) Bipolar plate with and without cooling channels.

10.2 Fuel Cell Stack and Power System

In Figure 10.3a, the fuel cell stack consists of a set of two separate gas flow-field plates for each MEA. The plates have gas flow channels made only on one side of the plate, and with one plate used to supply fuel to the anode side and the other connected to supply oxidant to the cathode side.

In Figure 10.3b, the fuel cell stack consists of bipolar plates that include hydrogen gas supply channels on one side of the plate and half coolant flow-field channels on the other side of the plate. A full cooling channel is formed by joining it to an adjacent bipolar plate that includes air flow channels on the other side of the plate. The presence of such cooling channels in regular intervals along with each MEA is desirable for maintaining uniform temperature across the stack. The cooling plate can be included for each MEA or every alternate MEA depending on the heat dissipation and cell temperature requirements of the stack. Coolant gaskets are also required in between two bipolar plates forming the cooling channels or in between the bipolar plate and the coolant plate.

In Figure 10.4a, the fuel cell stack consists of bipolar plates with a gas flow channel made on both sides of the plate. While the gas flow channel on the left side of the bipolar plate is connected to supply oxidant or air to the cathode side of a MEA, the gas flow channel on the right is used to supply fuel to the anode side of another adjacent MEA. The bipolar plate can also house the integrated coolant channels in a compact design as shown in Figure 10.4b, avoiding the need of a separate coolant plate. Again, the bipolar plate with cooling channels can be used with every MEA or every other MEA.

Figure 10.5 shows an exploded view of a fuel cell stack that includes repeated sets of MEA and bipolar plates.

The stack also includes two half bipolar plates on each side along with end gaskets and end plates. The end plates are used to cover the ends of the stack, house the inlet and outlet ports of the reactant gas streams, interface with the current collectors, and provide the structural support for the stack against any compression loading. The current collector transmits the current from the active area of the MEA to the power cable of the external load circuit.

The scaling up of a fuel cell stack for industrial applications faces a number of technical issues and challenges such as increased ohmic loss owing to series connection of MEAs; gas supply and distribution, through integrated channels; supply and return manifold design; increased water production and hence requirement for water removal and management; and increased

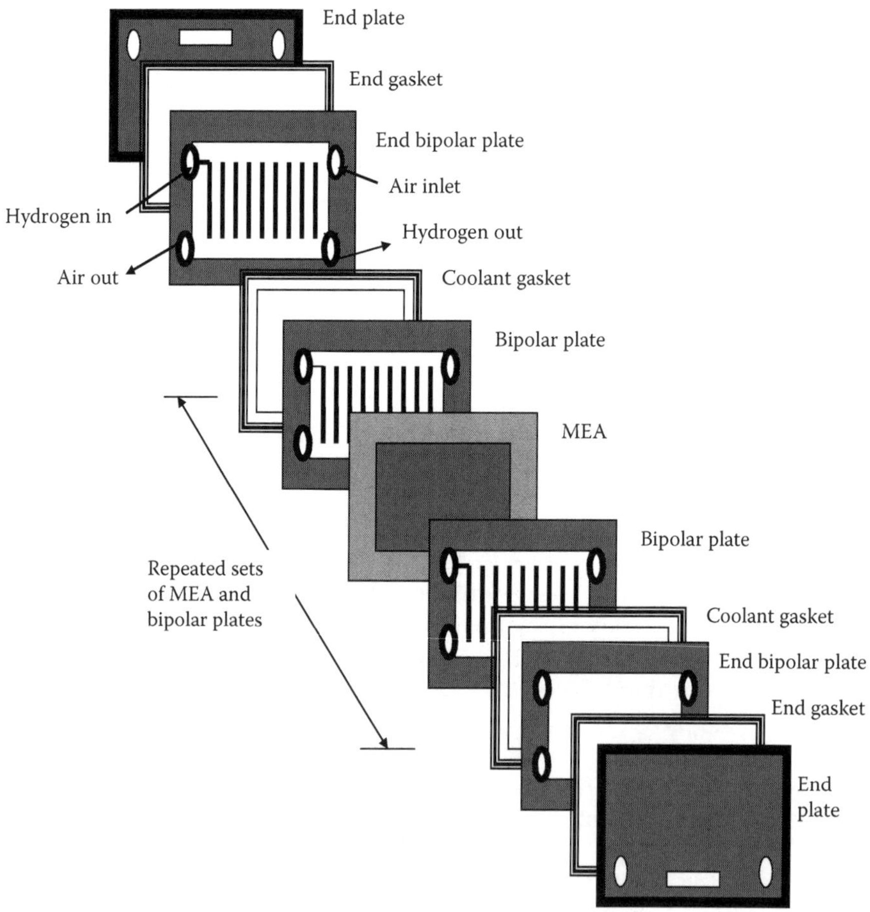

FIGURE 10.5
Exploded view of a fuel cell stack.

heat generation, thermal management for maintaining cell temperature, and utilization of waste heat.

For a larger stack design with high power rating, it is essential to increase the size of the MEA in order to keep the current path short, achieve lower ohmic loss and higher cell performance, and keep the number of cells low. However, scaling up of the fuel cell stack with larger MEA contact area imposes additional challenges for maintaining uniform reactant gas concentration over the entire contact area while effectively removing heat and water products. For example, maintaining a uniform gas distribution as well as removing heat and water product can easily be achieved using a simple gas flow design for a cell size of 5 cm × 5 cm or 10 cm × 10 cm. However, as we scale up MEA size to 20 cm × 20 cm or 30 cm × 30 cm, it is quite a challenge to maintain a uniform gas concentration and current distribution over the entire contact area, as it requires a high-performance gas flow-field design that maintains high heat and mass transfer rates and effectively removes product water from the contact area, avoiding any flooding at the electrode interface.

Because fuel cells will only operate with high efficiency when producing a relatively large fraction of maximum power, the system will need to be designed with multiple fuel cell stacks such that some can be kept idle while others are near maximum loading when demand is well below the overall system maximum. Figure 10.6 shows a large-scale fuel cell power generation system using a number of smaller power fuel cell stack designs. For example, a 250-kW power system can be made up of two 125-kW stack systems as shown in Figure 10.6a; an 8-MW system can be made up of eight 125-kW stack systems as shown in Figure 10.6b; and a 4-MW power system can be made up of an array of four 1-MW stacks shown in Figure 10.6c, to meet the power needs during part load as well as peak load operation.

The figure also shows an external network of return and supply manifolds for gas and coolant, and a return manifold for water. Each fuel cell stack uses an internal manifold to distribute reactant gases in parallel to all MEAs.

Example 10.1

Consider a 1-MW fuel cell stack operating at a temperature of 900°C with an operating voltage and current density of 0.7 V and $0.5\frac{W}{cm^2}$. Estimate the following: (a) mass and volume flow rates of hydrogen, oxygen and flow rate if oxygen is supplied as air, (b) water production rate, (c) heat generation rate, and (d) number of unit fuel cell for a unit cell size of 22 cm × 22 cm.

Solution

The hydrogen supply rate is given by Equation 4.83a

$$\dot{m}_{H_2} = \frac{M_{H_2} P_t}{n_e F V_c}.$$

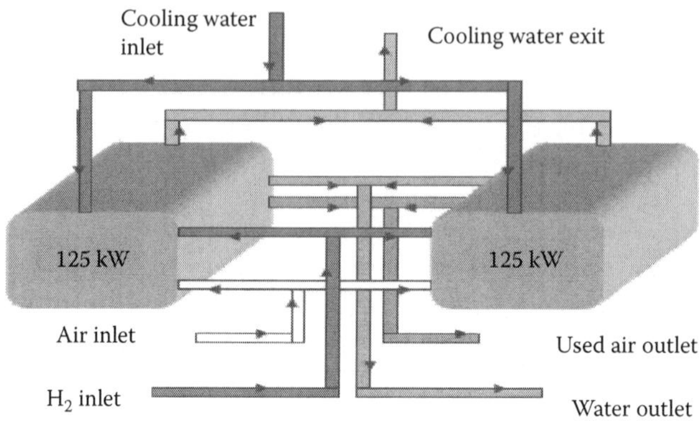

(a)

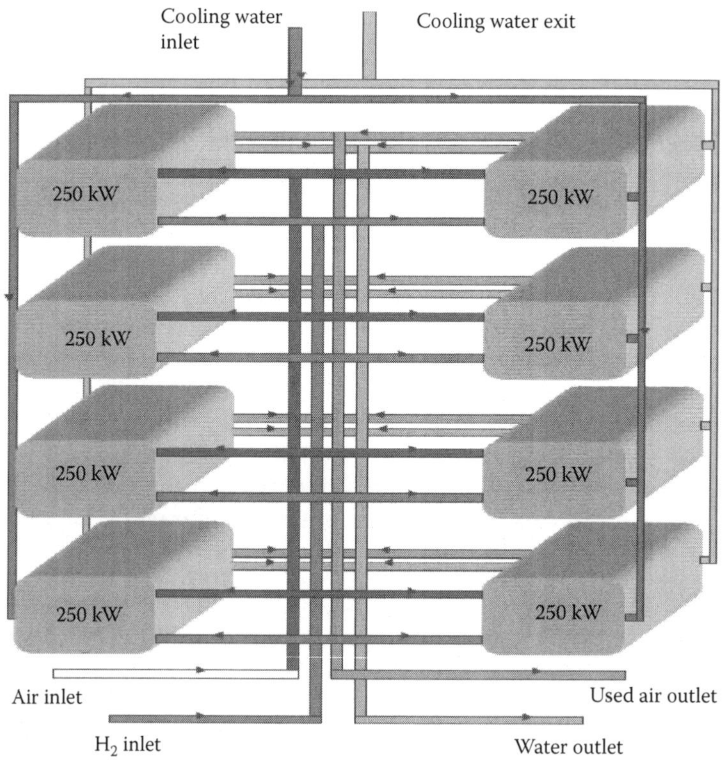

(b)

FIGURE 10.6
Fuel cell power generation system with multiple stack modules with inlet and exit manifolds for gas supply and water outflows. (a) 250 kW system with two 125 kW fuel cell stack module, (b) 2 MW power system with eight 250 kW fuel cell stack module, (c) 4 MW power system with an array of four 1 MW stack module.

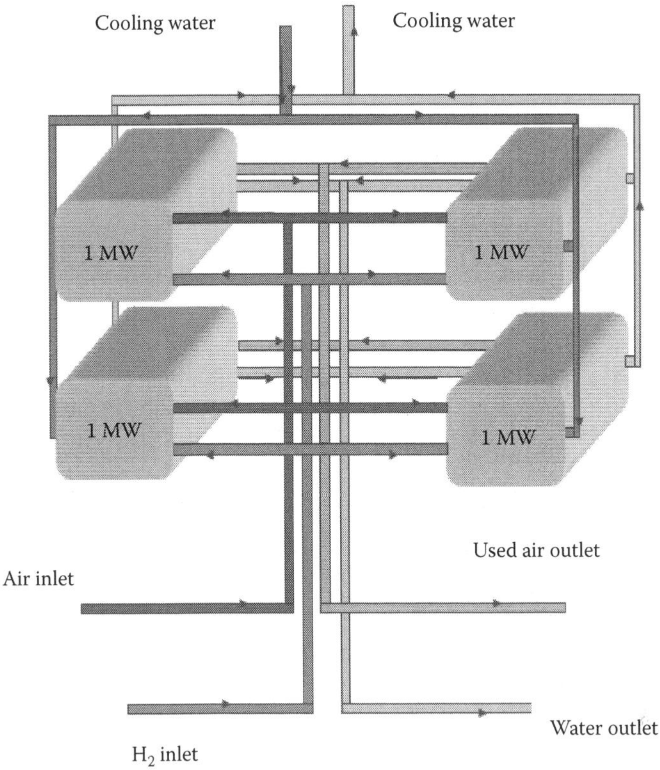

(c)

FIGURE 10.6 (Continued)
Fuel cell power generation system with multiple stack modules with inlet and exit manifolds for gas supply and water outflows. (a) 250 kW system with two 125 kW fuel cell stack module, (b) 2 MW power system with eight 250 kW fuel cell stack module, (c) 4 MW power system with an array of four 1 MW stack module.

Assuming the molecular weight of hydrogen as 2.02×10^{-3} kg/mol, we get

$$\dot{m}_{H_2} = \frac{2.02 \times 10^{-3} \times 1.0 \times 10^6}{2 \times 96{,}485 \times 0.7}$$

$$m_{H_2} = 0.015 \text{ kg/s (26.9 cfm)}$$

$$= 54 \text{ kg/h.}$$

The oxygen supply rate is given by Equation 4.74a

$$\dot{m}_{O_2} = \frac{M_{O_2} P_t}{n_e F V_c}$$

$$m_{O_2} = \frac{32 \times 10^{-3} \times 1.0 \times 10^6}{4 \times 96,485 \times 0.7}$$

$$m_{O_2} = 0.1185 \text{ kg/s}$$

$$= 426.6 \text{ kg/h.}$$

The air supply rate is given by Equation 4.76

$$\dot{m}_{\text{air}} = \frac{M_{\text{air}} P_{\text{t}}}{y_{O_2} n_{\text{e}} F V_{\text{c}}}.$$

Assuming the molecular weight of air as 28.97×10^{-3} kg/mol, we get

$$m_{\text{air}} = \frac{28.97 \times 10^{-3} \times 1.0 \times 10^6}{0.21 \times 4 \times 96,485 \times 0.7}$$

$$m_{\text{air}} = 0.5106 \text{ kg/s}$$

$$= 1838.16 \text{ kg/h}$$

and with 200% stoichiometric:

$$m_{\text{air}} = 1.0212 \text{ kg/s} = 3676.32 \text{ kg/h.}$$

The water production rate is given by Equation 4.85a

$$\dot{m}_{H_2O} = \frac{M_{H_2O} P_{\text{t}}}{n_{\text{e}} F V_{\text{c}}}$$

$$m_{H_2O} = \frac{18.02 \times 10^{-3} \times 1.0 \times 10^6}{2 \times 96,485 \times 0.7}$$

$$m_{H_2O} = 0.1334 \text{ kg/s}$$

$$= 480.24 \text{ kg/h.}$$

The amount of water produced in an hour is given as

$$m_{H_2O} = 0.1334 \times 3600 = 480.24 \text{ kg/h.}$$

Assuming the density of water as $\rho_{H_2O} = 1000 \text{ kg/m}^3$, the volume of water generated in is $\forall = 480 \text{ m}^3\text{/h}$.

Heat generation rate is given in Equation 4.101 for a fuel cell stack as

$$Q_{gen} = P_t\left(\frac{E_{max}}{V_c} - 1\right),$$

where

$$E_{max} = \frac{-\overline{\Delta h_f}}{n_e F} = \frac{248{,}760}{2 \times 96{,}485} = 1.2891 \text{ V}.$$

Substituting, we get the heat generation rate as

$$Q_{gen} = 1.0 \times 10^3\left(\frac{1.2891}{0.7} - 1\right)$$

$$Q_{gen} = 841 \text{ kW}.$$

Power from single cell, $P_{cell} = V_c \times j_c \times A_{cell} = 0.7 \times 0.5 \times 25 \times 25 = 218.75 \text{ W}$.

Number of cell, $N_{cell} = \dfrac{10^6 \text{ W}}{218.75 \text{ W}} = 4571.$

Table 10.1 shows variations in gas consumption rates, water production rate, and heat generation rate for different fuel cell stack power. As we can see, with increased power level, subsystems such as air and fuel delivery system, and water and heat management systems also scale up in size and complexity.

Table 10.2 shows the typical dimension of a 1-MW SOFC stack, estimated on the basis of the assumption of a cell operating voltage and current density of 0.7 V and 0.5 A cm^{-2}, respectively, and an operating temperature of 900°C. Results show that for a cell size of 50 cm × 50 cm and with the thickness of a

TABLE 10.1

Effect of Fuel Cell Stack Power Rating on Gas Consumption Rates, Water Production Rate, and Heat Generation Rate

Power Rating	**10 kW**	**80 kW**	**1 MW**
H_2 gas flow rate (kg/h)	0.54	4.32	54
Air flow rate (kg/s)	4.26	34.12	426.6
Heat generation rate (kJ/s)	8.41	67.28	841
Water production rate (kg/h)	4.80	38.41	480.24
Number of cells	46	366	4571

TABLE 10.2

Dimensions of Fuel Cell Stack versus MEA Size

Dimensions of MEA ($W \times H$) (cm)	25 × 25	33 × 33	50 × 50	75 × 75	100 × 100
Dimensions of unit cell (cm) $W \times H \times L$	25 × 25 × 0.78	33 × 33 × 0.78	50 × 50 × 0.78	75 × 75 × 0.78	100 × 100 × 0.78
Number of cells	4571	2624	1143	508	286
Total length of cell (m)	35.65	20.46	8.91	3.96	2.23

single cell with one MEA and two half bipolar plates on each side as 0.78 cm, the size of a 1-MW SOFC is estimated as 0.5 m × 0.5 m × 8.91 m.

In a fuel cell power system, the stack is integrated with a number of subsystems such as air supply system, fuel supply system, water removal and management; cooling and heat management system; and power supply management and control subsystem as shown in Figure 10.7.

The fuel cell energy system shown in Figure 10.7a includes hydrogen as the fuel. In this system, hot cathode exhaust is used to preheat incoming air. In Figure 10.7b, other fuels such as natural gas can be used along with an external reforming process such as steam reforming. The fuel subsystem

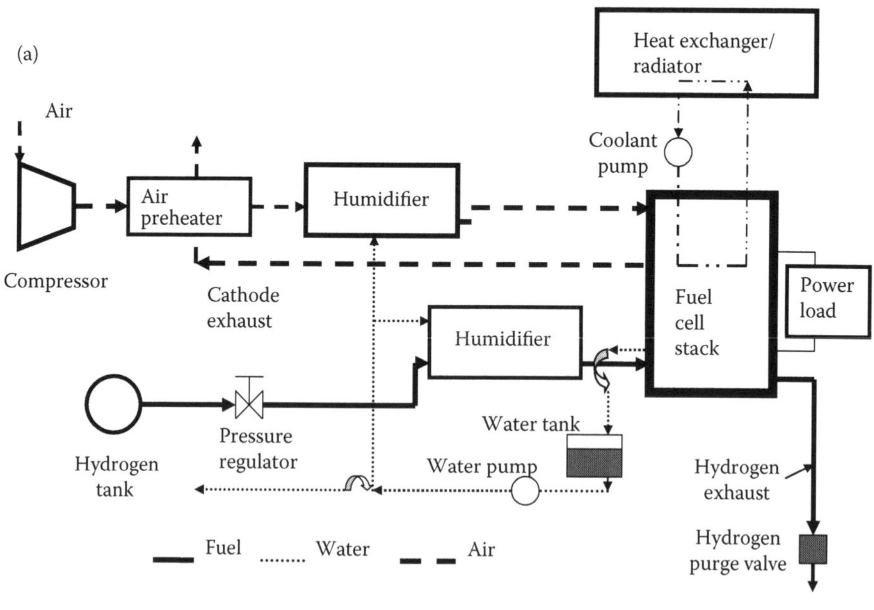

FIGURE 10.7

Fuel cell energy system. (a) Fuel cell energy system with hydrogen as fuel. (b) Fuel cell energy system with fuel reforming.

(b)

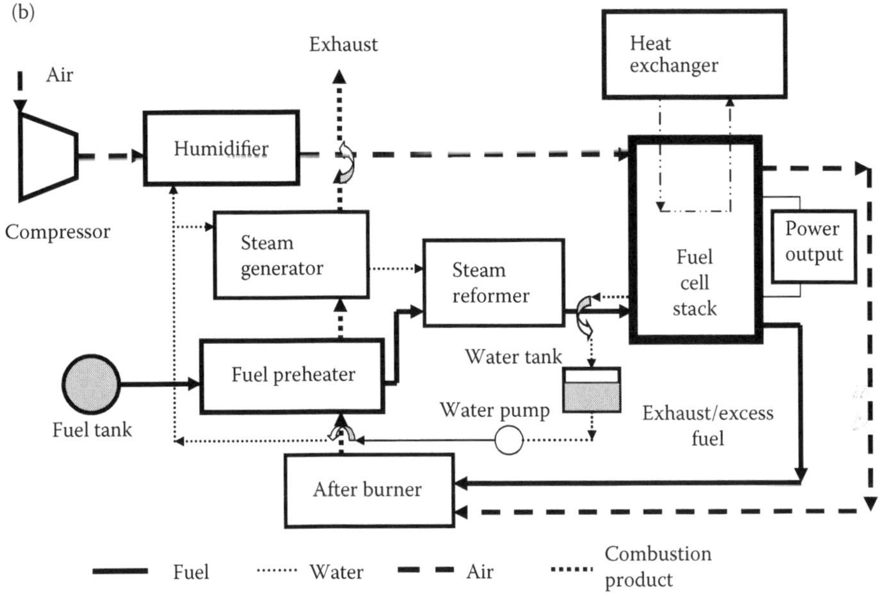

FIGURE 10.7 (Continued)
Fuel cell energy system. (a) Fuel cell energy system with hydrogen as fuel. (b) Fuel cell energy system with fuel reforming.

also includes afterburner to burn excess fuel from the anode exhaust to produce a hot gas stream which is used to preheat incoming fuel stream and to produce steam for the reformer.

10.3 Water Removal and Management

Water has to be collected and removed from the electrode–electrolyte interface, avoiding any liquid water accumulation or flooding of the porous electrode structure that prevents reactant gas transport to reach reaction sites. A water collection and management system for a PEM fuel cell stack is more critical in terms of water removal from the cathode and maintaining enough humidification at the anode. Water removal can be achieved either by collecting water around the edge of the electrode and drain by gravity fed into a water collection tank or by means of the moving gas streams through the gas flow-field channels as shown in Figure 10.8. Water generated at the cathode active layer moves outward through the porous electrode toward cathode flow-field channels where it either moved away by using a fibrous wick

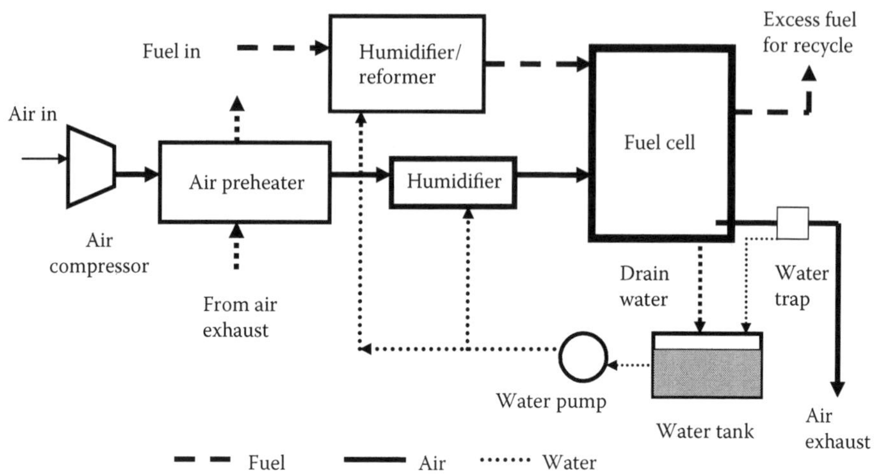

FIGURE 10.8
Water removal by moving gas stream from the cathode electrode surface.

that covers the edges of the electrode and drained off by gravity or moved away by the moving oxidant flow. The oxidant flow carries off the water mainly in vapor form and to some extent as small liquid droplets depending on the pressure and humidity level or the dew point temperature of the oxidant gas stream. The water distribution system shows use of collected water for humidification of gas streams and also for the reformer of the fuel stream.

Choice of water removal scheme by using the wick-drain collection mechanism or by moving gas stream through adjacent flow-field channels or using a combination of both depends on the size and power rating of the stack, gas flow rate and whether pure oxygen or air is used as oxidant. A wick-drain collection system is generally used for a smaller system and is quite sensitive to the orientation and configuration of the cell. A water removal system based on gas flowing stream requires careful consideration of oxidant temperature, humidity level, and gas pressure drop in the gas channel. Gas flow-field design in terms of size, shape, and configuration plays a critical role for effective removal of water by this means.

10.4 Cooling/Heating System for Fuel Cells

Fuel cells require a cooling system to maintain the operation of the cell at essentially a constant temperature to sustain the desired operating condition and avoid condensation and dryout in the fuel cell. Additionally, the system could also be used as a heating system during cold startup and freezing conditions.

A limited cooling may lead to a large temperature gradient in the cell, reduce the fuel cell performance, and may induce high thermal stresses on the cell components and cause failure. The thermal management schemes and the cooling techniques vary with rated power output and size of the fuel cell stack. For a smaller fuel cell stack with a lower power rating, heat dissipation by conduction through the solid, convection through the gas flow channels, and natural convection from the stack surface may be adequate for maintaining cell operating temperature and performance. However, for a larger stack with a high power rating, such a heat management and cooling scheme is not adequate, and requires additional cooling schemes. For example, in a fuel cell stack with a 10-kW power output and with a conversion efficiency of 50%, it is expected that about 10 kW heat will be generated in the stack, and a cooling scheme based on using liquid coolants in separate cooling channels is essential.

Oseen-Senda et al. (2003) gave a good review of a cooling system being used in PEMFC in transportation. Some of the cooling techniques that have been reported in literature include conduction, heat sinks and heat pipes as a passive system and forced air and liquid cooling as active subsystems. Wheeler et al. (2001) utilized a porous bipolar plate design to transport liquid water from the cell to the coolant stream as a passive means for water management in PEM fuel cells. One of the key challenges of the cooling system for PEMFC is that it involves smaller differences in operating temperatures between the cells (80°C–100°C) and ambient compared to internal combustion engines or SOFC fuel cell. Hence, air cooling using heat sinks may not be adequate to keep the system compact and light. It may require a high-performance cooling mechanism involving a higher convection heat transfer coefficient.

High-performance cooling mechanisms such as forced convection liquid cooling and two-phase flow boiling heat transfer have good potential for use in FCs. Some of the key criteria for the selection of the cooling fluid are as follows: the fluid must be non-corrosive, nonfreezing, environmentally safe, and an electrical nonconductor, that is, dielectric to avoid any short circuits.

An integrated cooling/heating system consisting of flow channels is either embedded in the mid-section of the bipolar plate along with gas flow channels as shown in Figure 10.4b or can be housed in a separate cooling plate as shown in Figure 10.3b.

Figure 10.9 shows a combined cooling and heating system for a fuel cell. The cooling loop consists of a circulating pump and radiator/heat exchanger for dissipating the heat. The cooling loop consists of a circulating pump and heater.

The cooling system will be used to remove the heat generated during the operation of the fuel cell to maintain the near-isothermal operating conditions. The heating system will be used for heating the fuel cell during

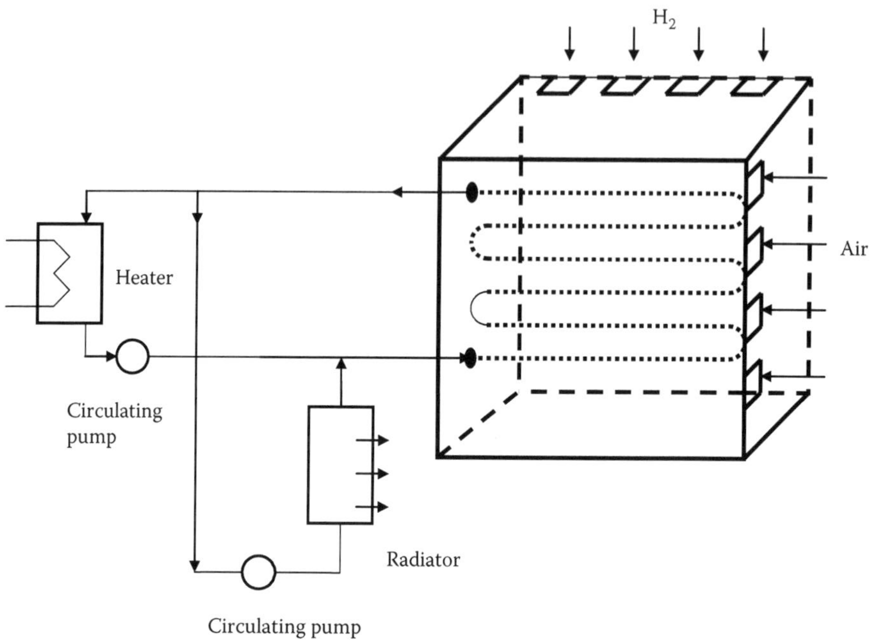

FIGURE 10.9
Schematic of an integrated cooling/heating system.

start-up under cold environment conditions using the same integrated flow channel flow loop.

Example 10.2

Consider an 80-kWe PEM fuel cell stack with an operating power efficiency of 55%. Determine (i) heat generation rate and (ii) mass flow rate of water as coolant assuming 60% of the heat is picked up by the coolant fluid and assuming a coolant temperature rise of $\Delta T_c = 10°C$.

Solution

i. Net heat generation can be estimated from the definition of thermal efficiency as

$$\eta_{th} = \frac{W_e}{W_e + Q_{gen}}.$$

Rearranging and solving heat generation as

$$Q_{gen} = \frac{W_e(1-\eta_p)}{\eta_p}$$

Substituting $W_e = 80$ kW and $\eta_p = 0.55$

$$Q_{gen} = \frac{80(1-0.55)}{0.55}$$

$$Q_{gen} = 65.45 \text{ kW}.$$

ii. Coolant mass flow rate can be estimated on the basis of an energy balance with heat picked up by a coolant flow equal to the sensible heat gain of the single phase liquid water coolant as follows:

$$0.6\, Q_{gen} = \dot{m}_c C_{pc} \Delta T_c$$

Solving for the coolant flow rate:

$$\dot{m}_c = \frac{0.6\, Q_{gen}}{C_{pc}\Delta T_c}$$

Substituting

$$\dot{m}_c = \frac{0.6 \times 65.45 \text{ kJ/s}}{4.197 \dfrac{\text{kJ}}{\text{kg°C}} \times 10\text{°C}}$$

$$\dot{m}_c = 0.9356 \text{ kg/s}.$$

Performance, cost, weight, reliability, and durability of fuel cells are all dependent on the material properties, manufacturing methods, and design of their individual key components, including electrolyte membrane, gas diffusion layer/electrodes, catalyst, bipolar plates, gaskets, and seals. Performance, reliability, and durability of bipolar plates depend on the plate functionality and design requirements in terms of gas flow channel design, plate material selection, cooling/heating mechanism, water removal capability, failure modes, and manufacturing methods. The technical targets for the development of bipolar plates are given by DOE and are shown in Table 10.3.

TABLE 10.3

DOE's Technical Target for Bipolar Plates

Characteristics	2010 Target	2015 Target
Cost (\$/kW)	4–6	3
Weight (kg/kW)	<1	0.4
H_2 permeation rate ($cm^3/s{\cdot}cm^2$)	$<2 \times 10^{-6}$	$<2 \times 10^{-6}$
Corrosion ($\mu A/cm^2$)	<1	<1
Electrical conductivity (S/cm)	>100	>100
Resistivity (Ω/cm^2) (area specific resistance)	0.02	0.02
Flexural strength (MPa)	>25	>25
Flexibility (% deflection at mid-span)	3–5	3–5

Source: DOE announcement No: DE-PS36-06GO96017, Research and development of fuel cell technology for the hydrogen economy, http/www.cere.energy.gov/hydrogenandfuelcells/mypp.

10.5 Bipolar Plate Design

Bipolar plates not only play a key role in the performance of the fuel cell but also contribute significantly to the cost, weight, reliability, and durability of the fuel cell. The primary purpose of the bipolar plate is to distribute the reactant gases hydrogen and oxygen uniformly over the entire active area of the electrode surfaces at all current densities and provide separation and containment of anode and cathode gases. The gas flow-field design also plays a critical role in the effective removal of water, in the heat generated during the electrochemical reactions, in any unreacted gas, and, to some extent, for the removal of any corroded ions from the metallic bipolar plates or any metallic interconnect material.

10.5.1 Major Design Considerations

Bipolar plates include separate gas flow channels engraved, milled, or molded on the surfaces facing the anode and cathode electrodes of a fuel cell. These gas flow channels supply reactant gas as well as remove products like water and heat to and from the gas diffusion layer of the anode and cathode electrodes of the fuel cell. The flow-field as well as the energy and mass transport in the gas channels play a significant role in the distribution of gas species and current density at the electrode–electrolyte interface and have a significant effect on fuel cell performance, particularly to the mass transport loss as discussed in Chapter 6. The mass transfer loss results in a lower fuel cell output voltage when the gas flow field and conditions cannot sustain a high mass transport rate to supply the reactants to the electrode reaction sites. The flow of gas streams through the gas flow channels also has to be effective in taking the product water away from the electrode surfaces, preventing any water flooding the electrode. This is particularly critical on the cathode side of a PEM fuel cell and for operations at higher current density.

The bipolar plate should be thin with a good contact surface with adjacent gas diffusion layers for reduced electrical and thermal resistance. The reactant gases should provide efficient transport processes with a reduced pressure drop, thus resulting in a uniform current distribution over the electrode surfaces, reduced mass transfer losses, and higher limiting current density. The concentration distribution of reactant gases in the gas channels and in the gas diffusion layers is controlled by the mass transport processes in the gas channel and in the gas diffusion layers and leads to the performance degradation owing to mass transfer losses. For a given gas diffusion layer electrode, mass transfer loss is controlled by the design of the gas flow channel, which establishes the flow field in terms of velocity, pressure, and concentration.

The ideal design of a bipolar plate should address the following key elements and requirements: (a) a high-performance gas flow field for the efficient transport of reactant gases to gas diffusion layers of electrodes, removal of heat from the cell, and removal of water, specifically maintain adequate moisture level on both cathode and anode sides of a Nafion membrane; (b) integrated cooling channels with efficient cooling mechanisms for the efficient removal of heat generated during operation, as well as for heating the cell during start-up operation in significantly cooler ambient conditions; (c) materials with higher electrical and higher thermal conductivity and with low thermal resistance; (d) materials that are impermeable to gases, are corrosion resistant, and have good structural strength to withstand shock, vibration, and mechanical loads owing to clamping forces and thermal stress cycling; (e) materials thinner in construction for reduced volume and weight, and lower in cost and ease of manufacturability.

One of the major challenges in the bipolar plate design is to house reduced-size and highly complex gas flow channels with complex patterns for both fuel and oxidant gas flows as well as house cooling/heating channels if required. Additionally, it has to integrate well with the internal supply and return manifolds.

Metallic bipolar plate design with conventional channels and use of metal foams in the flow fields are also considered (Kumar and Reddy, 2004). Results show a superior performance of the fuel cell with the use of metal foam bipolar plates with lower permeability. Bipolar plates with conventional channels are restricted to higher permeability owing to machining limitations.

The use of SS-316 bipolar plates with multiple-parallel straight channels is very common in the fabrication of PEM fuel cell stack. A porous bipolar plate design to transport liquid water from the cell to the coolant stream is also considered (Wheeler et al., 2001). Reviews of bipolar plate design and materials are given by Ajersch et al. (2003) and Kumar and Reddy (2004).

There are two major considerations for the design of the bipolar plate: (1) selection of materials for bipolar plates and (2) gas flow-field design. A detailed description of these two aspects of the bipolar plate design is given in the following section.

10.5.2 Bipolar Plate Materials

As we have mentioned before, a bipolar plate serves multiple purposes in a fuel cell stack. It serves as a current collector, dissipates the heat generated in the fuel cell by conduction through the solid material and by convection through gas channels and cooling channels, and provides the structural strength and support for the thin MEAs in the stack. Selection of a material that satisfies all these requirements is to some extent challenging and requires extensive analysis depending on the applications.

Major requirements for the selected materials for the bipolar plate are high electrical conductivity for current collection, high thermal conductivity for dissipation of heat dissipated, and adequate mechanical strength to provide structural strength for the thin MEAs. The material has to be impervious to the reactant gases, stable, and corrosion resistant to the electrochemical reactions. Additionally, the bipolar plate should be thin with uniform contact surfaces at the electrodes for reduced electrical and thermal contact resistance but thick enough to prevent any permeation of reactant gases across the material. For example, bipolar plates made of graphite tend to be thicker compared to metallic plates because of their low resistance to permeability of gases. For automobile applications, lower weight is also an additional requirement. For example, aluminum bipolar plates offer a potential weight reduction over more common stainless steel plates while still having the desired structural properties that a metal plate can offer.

There are a number of materials that satisfy a majority of these requirements, namely, metals with and without coatings, metal alloys, carbon-based materials such as graphite, and metal and carbon composites.

A brief discussion of the most competent bipolar plate materials that are attractive in terms of meeting all the requirements is given here.

10.5.2.1 Metallic Bipolar Plates

The most common bipolar plate materials are metal or metal alloys because of their high electrical and thermal conductivity, low gas permeability, high structural strength, and lower cost and ease of machinability for thinner plates with integrated gas channels. The major disadvantages of metallic bipolar plates are their higher density and weight, higher contact resistance at the interface with the gas diffusion layer/electrode, and the critical issue of low corrosion resistance.

Some of the most common metallic bipolar plates are made of carbon-steel, stainless steel SS-316, nickel/chromium alloy, titanium, and aluminum using material processing techniques such as CNC milling, casting, etching, and stamping. Among these, stamped metallic plates are the most common and popular fabrication method. The process involves stamping a number of required features such gas flow channels and internal manifolds and ports using a series of stamping stations on this sheet metal, making it very effective in producing a large number of bipolar plates for a fuel cell stack in a short time and cost-effective manner.

One of the major issues with the use of metallic bipolar plate is the possibility of corrosion. Corrosion occurs in a fuel cell owing to high humidity conditions and formation of oxide layers leading to increased interface contact resistance and transport metallic ions toward the electrode/catalyst sites causing degradation of electrochemical kinetics. Corrosion also leads to failure and lower durability of the bipolar plates. Most metal and metal alloys exhibit poor corrosion resistance, and in order to provide corrosion resistance, some types of surface treatment and coatings are employed.

In a metallic plate, the base metals are coated with different coating materials such as noble metals like Au, metal oxides, metal nitrides, metal carbides, carbons, and polymers. An appropriate coating can provide corrosion resistance in a most cost-effective manner.

Coating methods are electrodeposition, electroplating, physical vapor deposition (PVD), chemical vapor deposition, sputtering, spraying, and nitriding.

The *nitriding method* is a surface treatment process that involves diffusion of nitrogen forming a chromium-nitride layer in steel plates for forming an aluminum-nitride layer on aluminum plates. This surface treatment process involves placing the bipolar plates in a nitrogen-rich environment and heating them at a high temperature range to promote reactions and to form a nitride layer on all exposed surfaces. The process is very effective in providing defect-free corrosion-resistant layers on all surfaces including the grooved gas channels.

In the *PVD method,* ion beams are used to form a charged molecular vapor cloud of coating materials such as gold or TiN and then forming the coating layer by settling the charge particles on the base material surface.

The *electroplating method* uses electric currents to deposit a layer on the surface of the bipolar plates immersed in an aqueous metallic ion bath.

Pickling is a surface treatment method of removing scales or corrosion deposit using acid solution.

The advantages of *steel* include lower cost, higher corrosion resistance, higher structural strength, and less difficulty in machining. Because of their higher structural strength, thinner bipolar plates compared to graphite or graphite composites are possible, which would lead to a lighter and smaller fuel cell stack even though its density is four times higher than graphite. Major disadvantages are the higher weight and reduced contact resistance owing to the formation of metal oxide on the plate surface. In order to reduce this effect, steel surfaces are coated or surface treated by etching or sandblasting. Stainless steel stamped alloy foils with chromium-nitride coating exhibit corrosion resistance as compared to machined graphite plates.

Ni-Cr base alloys with Cr-nitride surface coating provide excellent corrosion resistance and low interfacial contact resistance. However, Ni-Cr bipolar plates are too expansive compared to aluminum, carbon-steel, and stainless steel bipolar plates.

Aluminum bipolar plates along with appropriate corrosion-resistant coating not only can offer enough weight reduction but also have the potential to meet the DOE's bipolar plate targets. Aluminum plates can be coated with metal carbides like TiC, metal silicide like $TiSi_2$, and carbon-based material such as graphite, carbon black, and carbon fibers on an appropriate carbide filler. Composite coating made of graphite flakes and titanium carbide offer higher electrical conductivity as well as corrosion resistance.

Since the use of a corrosion-resistant layer may also increase the conduction heat resistant, some developers proposed another approach in an effort to provide both high conductivity and high corrosion resistance of the plate surface. In this approach, a nonconductive corrosion-resistant coating layer is used over a major portion of the metal plate surface and a portion of conductive corrosion-resistant material on small selected spots, known as *conductive vias*, of the metal surface. Some of the most popular conductive vias materials are gold, conductive carbide, and carbon nanotubes.

10.5.2.2 Graphite Bipolar Plate

Another attractive material for bipolar plate construction is the *graphite* due to its high electrical conductivity, higher resistance to corrosion, low contact resistance, and lower density resulting in reduced weight per unit power. Major disadvantages are its difficulty in machining, higher cost of machining, lower structural strength, and higher porosity. The lower structural strength tends to make the plate thicker and heavier. Since graphite has higher gas permeability because of higher porosity, the graphite plates have to be treated or impregnated with resin to reduce gas permeability. Difficulties in machining graphite and forming flow-field channels tend to make the graphite bipolar plate highly expansive.

Flexible or expanded graphite, developed by GrafTech International, involves a compressed structure with expanded graphite planes and impregnation with resins. Bipolar plates made of flexible graphite are used in the fabrication of PEM fuel cell stack (Rajalakshmi et al., 2004). The flow-fields were introduced by a stamping process. An optimization study was carried out for the width and depth of the flow-fields in terms of applied pressure to flexible graphite sheet.

10.5.2.3 Composite Bipolar Plate

Composite bipolar plates are designed to provide high mechanical, electrical, and thermal properties by adding highly conducting fillers such as metals or carbon in an insulating base such as polymers. While *metallic composites* provide higher conductivity and mechanical strength, they also tend to increase corrosion potential to some extent. Some of the recently investigated metallic composites are stainless steel–nylon fibers and copper or aluminum mesh in graphite.

Carbon/polymer or *graphite/polymer composites* can have electrical and thermal conductivities comparable to those of graphite, but it is lighter and less expansive. Carbons or graphites are added in reinforcing polymer to optimize the properties of the composite materials. Two basic types of polymers such as thermoplastics and thermosets have been considered for forming polymer composites. The fillers can be of different sizes and types of carbons or graphite, such as various types of graphite, carbon blacks, carbon fibers, carbon nanotubes (Single Walled NanoTubes [SWCNTs] and Multi-Walled NanoTubes [MWCNTs]), and graphene, a single atomic plane of graphite. For example, a carbon/polymer composite material is composed of carbon powder in polypropylene at the desired volume ratio.

Carbon or graphite/polymer composite bipolar plates are made using simpler processes such as stamping or injection molding and compression molding. Injection-molded bipolar plates are made using a composite mixture in the typical injection molding process. The major advantages of the injection molding process are the low processing time and the low cost of manufacturing. One of the major disadvantages of the injection molding process is the limited use of carbon fillers to ensure lower viscosity to maintain injection flow of composite in the molds. This results in a lower conductivity of the composite. The main advantage of injection molding over compression molding is the shorter cycle time resulting in lower operation cost.

Barbir et al. (1999) developed a single-piece bipolar plate made of a graphite/polymer mixture using a molding process that required very little or no post-machining. Even though the mixture has lower electrical conductivity compared to pure graphite, the resistance of a molded plate sandwiched with backing layers was reduced.

The compression molding process involves compressing the mixture of filler and polymer over a heated mold. The process require some curing time, but is significantly faster than typical machining processes. The compression molding process has the advantage of having higher percentage of carbon fibers and hence higher conductivity than the injection molding process.

10.5.3 Material Selection

Selection of one of these materials is not straightforward because of multiple competing factors and criteria. It requires optimization of the thin bipolar plate with respect to its functionality such as heat dissipation characteristics and structural strength subjected to different loading conditions including thermal stress and mechanical loading owing to clamping forces, vibration, and shocks. Table 10.4 presents the properties of some of the typical materials for bipolar plates.

TABLE 10.4

Thermal Properties of Bipolar Plate Materials

Materials	Density (kg/m³)	Specific Heat (J/kg·K)	Thermal Conductivity (W/m·K)
Aluminum	2702	903	237
Carbon steel	7854	434	60.5
SS-304	7900	477	14.9
SS-316	8238	468	13.4
Titanium	4500	522	21.9
Graphite	2210	709	1950 ∥ to layer 5.7 ⊥ to layer
Graphite fiber epoxy (25% vol.) composite	1400	935	11 ∥ to fibers 0.87 ⊥ to fibers

10.6 Gas Flow-Field

The gas flow channels are designed for effective transport of gas species through the gas diffusion layer to get high current density. The gas flow-field design in a bipolar plate has a significant effect on the variation of gas pressure along the channel, on the mass transport rates of reactant gases and products to/from the electrode surfaces, on the gas species concentration, and hence on the current density at the electrode–electrolyte interface. Improvements can occur in the performance of a fuel cell through optimization of the channel dimensions and shape in the flow-field of bipolar plates.

One of the major losses in PEM fuel cells is the mass transfer loss, which is caused by the lack of reactant gas concentration distribution at the electrode–catalyst reaction surface. In order to reduce this resistance, a high-performance gas flow channel design has to be developed.

The mass transfer loss results in a decrease in fuel cell output voltage when gas flow-field and conditions cannot sustain a high mass transport rate to supply the reactants and remove the products to/from the electrode reaction sites effectively. This is more critical for the operation of the fuel cell at higher current density, which is desirable for achieving higher power density. An effective gas flow-field design obtained through optimization of the channel dimensions, shape, and configuration may result in an improved bipolar plate. The effective contact surface area of the reactant gas or the land area of the electrode surface has a direct effect on the gas concentration distributions in the electrodes and hence on overall electrochemical reaction.

The contact surface area of the reactant gas on the bipolar plates is increased by decreasing spacing between the channels and increasing the number of channels, which contributes to the local distribution of gas, overall reaction of the gas, and the local current density.

Removal of water from the electrode surface in the form of water vapor or liquid water droplets may lead to a complex multiphase flow depending on the size and type of gas flow channels used. The multiphase flow is generally characterized by a number of different flow regimes such as bubbly flow, slug flow, plug flow, and annular flow.

Recent research effort includes development of bipolar plates with optimized gas flow channels in terms of flow configurations and channel sizes. The main focus was to improve gas flow, heat, and mass transport characteristics, and to bring down the size of the bipolar plate while reducing the channel size, in turn reducing the overall weight of the fuel cell. The flow channels can be machined or molded out of solid plates and function to distribute reactant gas as well as to circulate coolant to dissipate heat from the fuel cell. Design optimization should involve selecting the appropriate thickness of the bipolar plate that houses channels for reactant gas and coolants as well as provides enough structural strength. Considerations should be made for the suitability of different materials having different density, thermophysical property, structural strength, and feasibility for manufacturing in an efficient and cost-effective manner.

10.6.1 Gas Flow Channel Design

Considerable attention has been given to the design of gas flow channel. Figure 10.10 shows reactant gas supply channels adjacent to the gas diffusion layer of electrodes. While a single gas supply channel adjacent to the gas diffusion layer of the electrode, shown in Figure 10.10a, can provide maximum contact area with the gas diffusion layer and seems preferable, it has a number of deficiencies in terms of poor gas flow distribution, lower average velocity, and hence low heat and mass transport coefficients. Since this design provides very low *wall contact area* or *land area* with the electrodes, it is a poor conductor of electrons as well as a poor heat conductor for dissipation of heat generated in the cell.

In order to improve the gas flow distributions and achieve higher heat and mass transport coefficients, a number of smaller gas flow channels separated by the solid metal walls of the bipolar plate, named as the *land area* of the plate, are generally used as shown in Figure 10.10b. While such a design provides additional solid land area for efficient electron and heat transport, it also creates a region adjacent to the land area with poor gas concentration distributions that results in non-reacting regions at the electrode–membrane interface as shown in Figure 10.10c.

As gas enters the gas diffusion layer through the gas contact area, it diffuses toward the reaction surfaces as well as into the land area to some extent. The overall performance of the fuel cell is directly proportional to the area of the gases in contact with the bipolar plate.

A ratio of reactant *gas contact area* to the *land area* has to be selected based on a balance of effective heat and mass transport rate through the gas contact

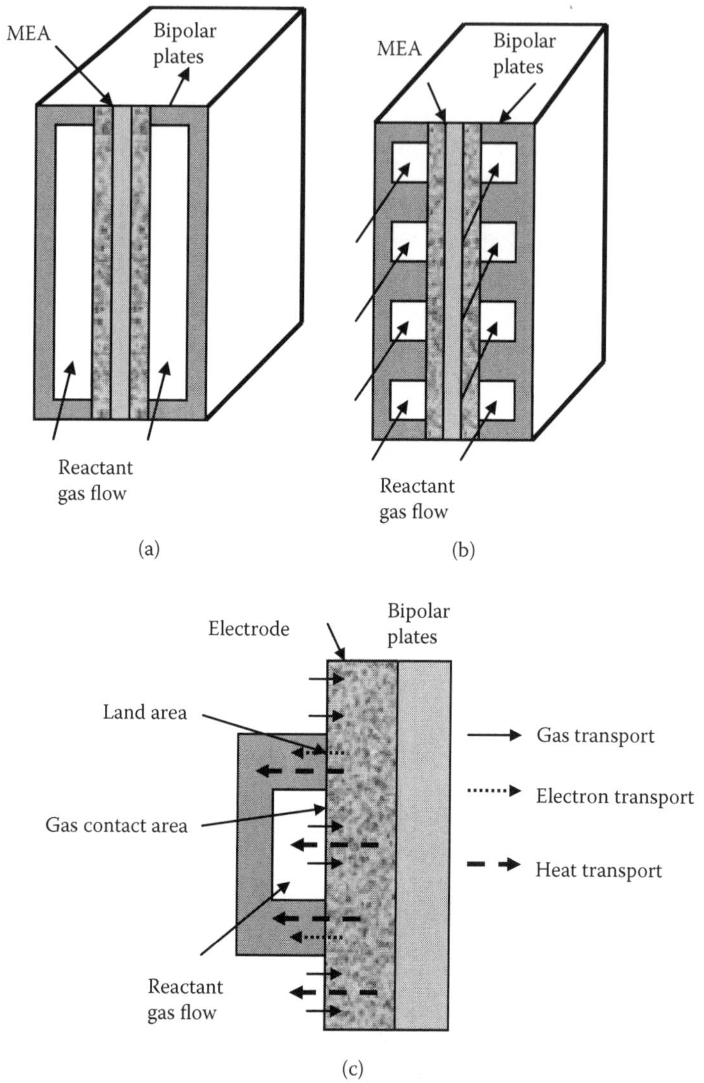

FIGURE 10.10
MEA and bipolar plate with integrated gas flow channels. (a) Open single gas flow channels. (b) Multiple separated gas flow channels. (c) Gas channel with direct contact with GDL and land area.

area and effective electron and heat dissipation by conduction through the land area. The ratio of contact area to land area is defined by the geometrical parameters such as the height, h, and width, w, of the channel and the height of the land area, l, as shown in Figure 10.11a. The channel geometry can be of any shape such as square or rectangle of different aspect ratios or trapezoid as shown in Figure 10.11b.

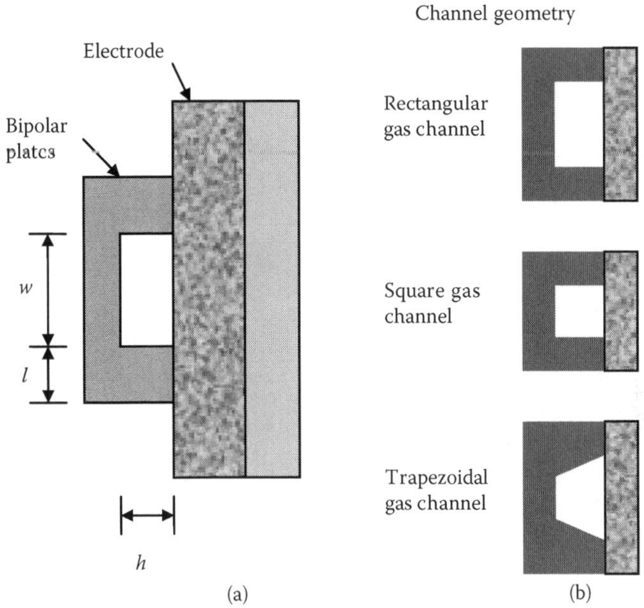

FIGURE 10.11
Bipolar plates with gas flow channels in contact with anode and cathode electrodes. (a) Geometrical parameters for contact and land areas. (b) Different channel geometries.

10.6.2 Flow-Field Channel Layout Configurations

The overall contact surface area of the flow-field plate varies with the channel configuration, number, and size of the channel. The basic gas flow-field design configurations include the following: (1) straight parallel channels, (2) serpentine channels, (3) multiple parallel serpentine channels, (4) pin design, and (5) interdigitated design.

The three-dimensional flow configuration and geometry of straight parallel channels and serpentine channels with different channel sizes are depicted in Figure 10.12. The current densities as well as the activation losses depend on the primary contact surface area of the gas channels adjacent to the gas diffusion layer.

10.6.2.1 Straight Parallel Channels

A straight parallel design involves a number of straight parallel flow channels that run from the inlet port of the gas feeder to the outlet port as shown in Figure 10.13.

Reactant gas enters through the inlet port of the gas feeder, takes multiple parallel flow paths through the channels, and exits through the outlet port. One major advantage of this design is the lower pressure drop because of the parallel nature of the flow. However, the gas distribution through the

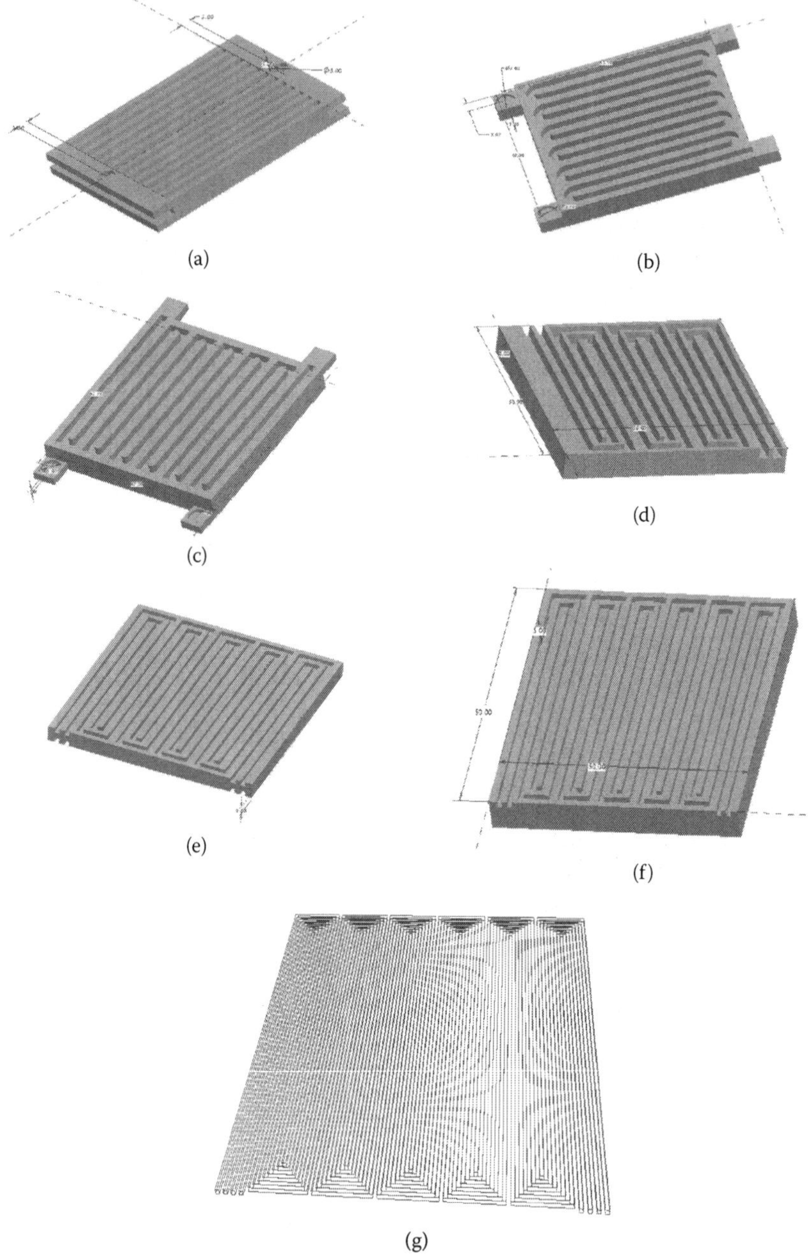

FIGURE 10.12
Different flow channel design configurations. (a) Straight parallel channels. (b) Single serpentine curvilinear bends. (c) Single serpentine channel with square bends. (d) Dual 2-mm serpentine parallel channels. (e) Dual 1.2-mm serpentine parallel channels. (f) Dual 1-mm serpentine parallel channels. (g) Four 0.5-mm serpentine parallel channels.

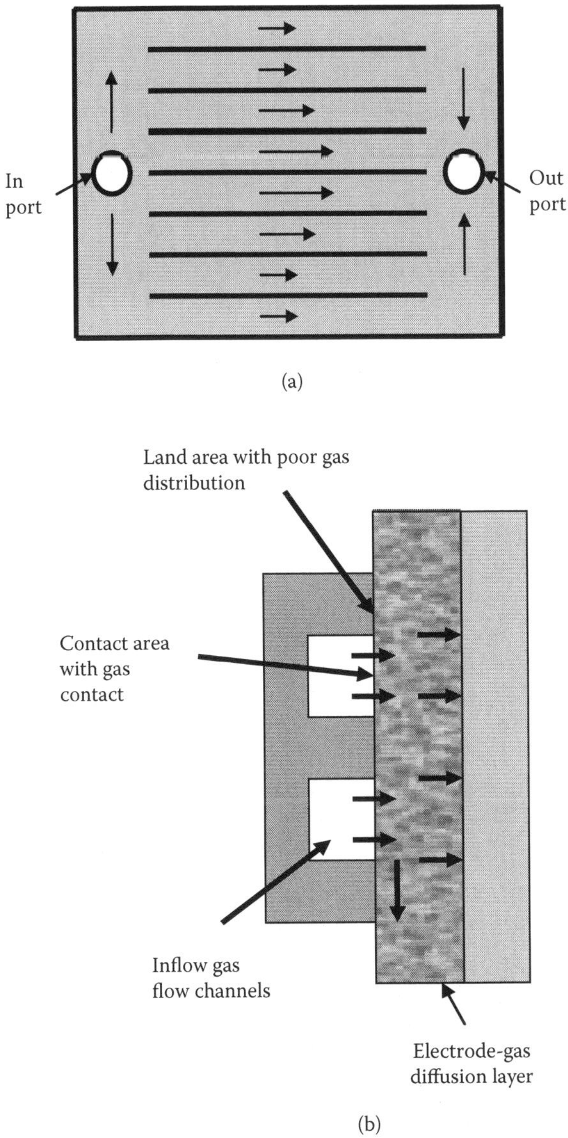

FIGURE 10.13
Straight parallel channel gas flow-field design. (a) Straight parallel flow channels. (b) Gas distribution primarily by diffusion in the gas diffusion layer.

electrode gas diffusion layer is by diffusion and may result in a poor gas concentration distribution in the electrode, particularly under the land area of the plate as shown in Figure 10.12b. Additionally, the design also tends to create nonuniform gas flow distributions through the channels as the flow channels that are located further away from the inlet port involve longer flow length. This effect could be quite significant in a larger-scale fuel cell. This causes nonuniform gas concentration distribution, water accumulation in some channels, and increased mass transfer losses. The problem, to some extent, can be reduced while using a higher stoichiometric ratio for the reactant gas flow.

10.6.2.2 Serpentine Flow Channel Design

A serpentine channel configuration involves a single flow channel that is laid out over the entire plate in a serpentine manner as shown in Figure 10.14. In serpentine channels, the effective consumption of gases for the reaction is increased as compared to the straight through channels. The overall contact surface area on a bipolar plate varies as the channel configuration, number, and size are changed. The channel bends can be either curvilinear or square.

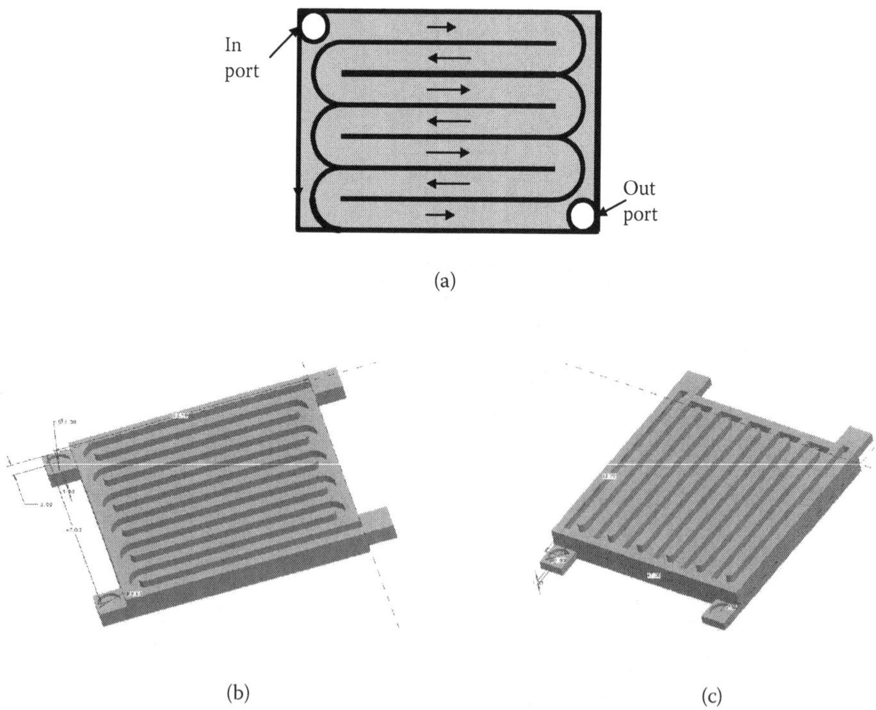

FIGURE 10.14
Single serpentine channels. (a) Serpentine flow channels. (b) Single channel curvilinear bends. (c) Single channel square bends.

10.6.2.3 Multiple Parallel Serpentine Channels with Square Bends

Since the serpentine channel with square bends provides a more effective contact area, multiple parallel serpentine channels are considered with square bends in order to take advantage of increased contact area as well as decreased pressure drop because of the parallel nature of the flow as shown in Figure 10.12d.

By decreasing the channel size from 2 mm to 1 mm, the number of dual channels and the ratio of gas contact area to land area can be progressively increased considerably as shown in Figure 10.12d through 10.12f. A high-performance bipolar plate with four 0.5 mm serpentine parallel channels is also considered (Boddu et al., 2009) as shown in Figure 10.12g. With a combination of an increase in number parallel channels and a decrease in channel size, the effective gas contact area can be optimized in terms of heat and mass transport and pressure drop in the gas flow channels and performance of the fuel cell.

10.6.2.4 Pin-Array Flow-Field

In this flow-field design, the flow network is formed by causing the fluid to move around a number of pins arranged in a staggered or in-line manner as shown in Figure 10.15. The pin can be of cubical or circular or any other shape with different spacing between them to achieve desired local velocities. Such a design creates a series-parallel flow network for the gas flow and hence results in a relatively lower pressure drop compared to the straight series flow design. However, one of the drawbacks of this design is that the flow tends to follow a least resistant path and create a slower-moving and even stagnant flow region. This may lead to uneven gas concentration distribution over the contact area and inadequate removal of heat and water from all parts of the contact surface.

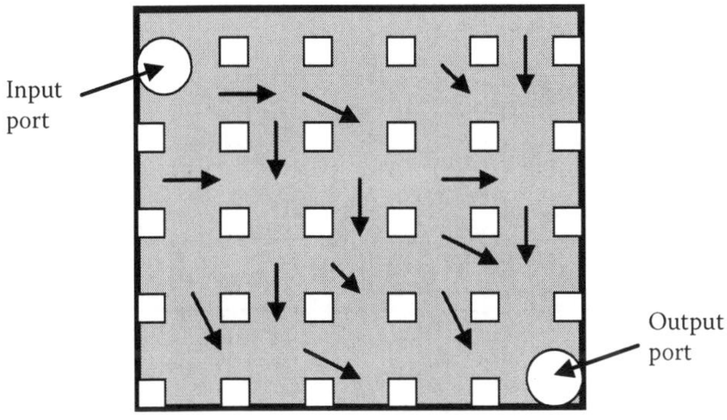

FIGURE 10.15
Pin-array flow-fields.

10.6.2.5 Interdigitated Flow-Field

The interdigitated flow-field design consists of two sets of dead-ended gas flow channels as shown in Figure 10.16a. The first set of dead-ended inflow channels carries the gas stream from the inlet ports and transfers to the electrode gas diffusion layer. The gas stream is forced by advection through the porous gas diffusion layer toward the electrode–electrolyte interface and toward the second set of dead-ended outflow channels and moves toward the gas stream outlet port.

As demonstrated in Figure 10.16b, better gas distribution can be achieved over the entire electrode–electrolyte reaction sites, including in the region adjacent to the land area of the gas flow-field plate. This channel design forces gas stream to transmit through the gas diffusion layer and becomes very effective in removing water from the electrode layer. Major disadvantages of this design include the excessive pressure drop and the fact that applications can be limited by any restriction on the parasitic power requirement.

10.6.3 Simulation Analysis of Flow-Field

The computational fluid dynamic (CFD) analysis of bipolar plate flow-field designs is often performed to evaluate the effectiveness of the flow-field design in terms of developing heat and mass transfer coefficient and

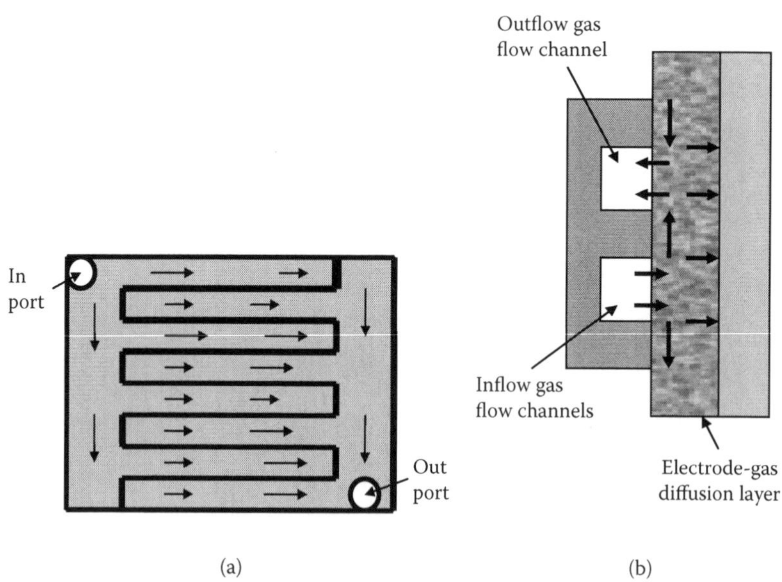

FIGURE 10.16
Interdigitated gas flow-field design. (a) Interdigitated gas flow channel configuration. (b) Gas distribution by convection and diffusion gas diffusion layer.

pressure drop. Further, the CFD model for the gas flow channel is coupled with the electrochemical and heat and mass transport simulation model for the MEA to analyze the performance of the fuel cell (Dutta et al., 2000, 2001; Nguyen et al., 2004; Shimpalee and Dutta, 2000).

The computer simulation model for studying the hydro-dynamically developing flow-field and developing heat and mass transport phenomena in the gas flow channel is given based on the incompressible Navier–Stokes equation for fluid flow and heat and mass transport equations as described below.

10.6.3.1 Gas Channel

10.6.3.1.1 Fluid Flow Model

Gas flow in anode and cathode gas channels is assumed to be incompressible fluid flow with constant fluid viscosity (μ) and is governed by the Navier–Stokes equation given as follows:

Mass Continuity:

$$\nabla \cdot \vec{V} \tag{10.1a}$$

or

$$\frac{\partial u}{\partial x} + \frac{\partial v}{\partial y} + \frac{\partial w}{\partial z} = 0 \tag{10.1b}$$

Momentum:

$$\rho\left(\vec{V} \cdot \nabla\right)\vec{V} = \rho\vec{g} - \nabla P + (\nabla \cdot (\mu \nabla))\vec{V} \tag{10.2a}$$

and in scalar form as

$$x\text{-component: } \rho\left(u\frac{\partial u}{\partial x} + v\frac{\partial u}{\partial y} + w\frac{\partial u}{\partial z}\right)$$

$$= -\frac{\partial P}{\partial x} + \frac{\partial}{\partial x}\left(\mu\frac{\partial u}{\partial x}\right) + \frac{\partial}{\partial y}\left(\mu\frac{\partial u}{\partial y}\right) + \frac{\partial}{\partial z}\left(\mu\frac{\partial u}{\partial z}\right) \tag{10.2b}$$

$$y\text{-component: } \rho\left(u\frac{\partial v}{\partial x} + v\frac{\partial v}{\partial y} + w\frac{\partial v}{\partial z}\right)$$

$$= -\frac{\partial P}{\partial y} + \frac{\partial}{\partial x}\left(\mu\frac{\partial v}{\partial x}\right) + \frac{\partial}{\partial y}\left(\mu\frac{\partial v}{\partial x}\right) + \frac{\partial}{\partial z}\left(\mu\frac{\partial v}{\partial z}\right) \tag{10.2c}$$

$$z\text{-component: } \rho\left(u\frac{\partial w}{\partial x}+v\frac{\partial w}{\partial y}+w\frac{\partial w}{\partial z}\right)$$

$$=-\frac{\partial P}{\partial z}+\frac{\partial}{\partial x}\left(\mu\frac{\partial w}{\partial x}\right)+\frac{\partial}{\partial y}\left(\mu\frac{\partial w}{\partial y}\right)+\frac{\partial}{\partial z}\left(\mu\frac{\partial w}{\partial z}\right) \quad (10.2d)$$

10.6.3.1.2 Mass Transport

Mass concentration distribution in gas channel is given considering diffusion and convection mass transfers as

$$\rho\vec{V}\cdot(\nabla C_i)=\nabla\cdot(D_{\text{eff}}\nabla C_i) \quad (10.3a)$$

or

$$\rho\left(u\frac{\partial C_i}{\partial x}+v\frac{\partial C_i}{\partial y}+w\frac{\partial C}{\partial z}\right)=\frac{\partial}{\partial x}\left(D_{ij}\frac{\partial C_i}{\partial x}\right)+\frac{\partial}{\partial x}\left(D_{ij}\frac{\partial C_i}{\partial x}\right)+\frac{\partial}{\partial x}\left(D_{ij}\frac{\partial C_i}{\partial x}\right), \quad (10.3b)$$

where i stands for the species concentration in the anode and cathode gas channels.

10.6.3.1.3 Heat Transport Equation

The thermal heat equation for the gas flow stream is given considering conduction and convection, and expressed as

$$\rho C_p\vec{V}\cdot(\nabla T_i)=\nabla\cdot(k\nabla T) \quad (10.4a)$$

or

$$\rho C_p\left(u\frac{\partial T_i}{\partial x}+v\frac{\partial T_i}{\partial y}+w\frac{\partial T_i}{\partial z}\right)=\frac{\partial}{\partial x}\left(k_i\frac{\partial T_i}{\partial T_x}\right)+\frac{\partial}{\partial y}\left(k_i\frac{\partial T_i}{\partial y}\right)+\frac{\partial}{\partial z}\left(k_i\frac{\partial T_i}{\partial z}\right). \quad (10.4b)$$

10.6.3.1.4 Inlet and Boundary Conditions

Uniform inlet velocity or constant pressure is often used at the inlet along with no-slip velocity at all walls. At the outlet, mass flow rate is specified. The mass flow outlet adjusts the exit pressure such that a target mass flow rate (i.e., mass flow at the inlet) is obtained at convergence. The top wall of the channel is maintained at a constant surface temperature or surface gas concentration and the rest of the walls are subjected to adiabatic condition. Such a boundary condition is used just to analyze the gas flow channel in comparing different channel designs. Typical results for flow and temperature distributions in some of these flow channels are presented.

10.6.3.2 Flow in Parallel Straight Channels

The mathematical model for studying the hydro-dynamically developing flow-field and developing heat and mass transport phenomena is solved for different bipolar plate designs (Boddu and Majumdar, 2008). Figure 10.17 shows a three-dimensional model for simulation and a typical computational mesh size distribution in the feeder section of the channels.

Typical pressure and velocity distribution of gas flow through the parallel straight gas flow channels are shown in Figure 10.18. Results show a nonuniform distribution of gas flows in the channels with most dominant

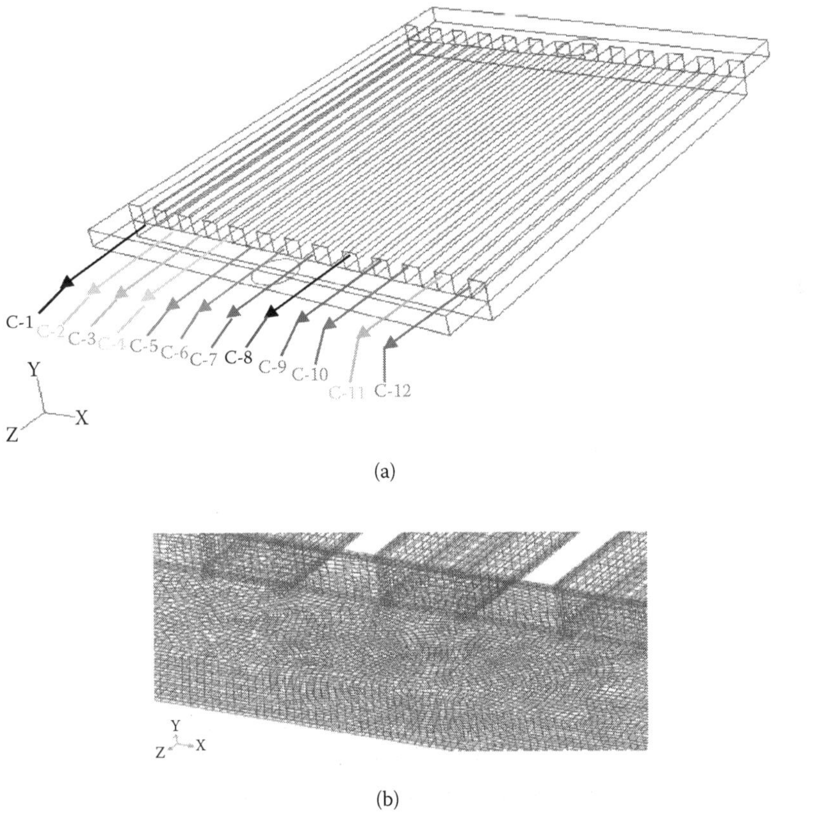

FIGURE 10.17
Computational model and mesh straight parallel channels. (a) Computational model for the bipolar plate with straight parallel channels. (b) Computational mesh for straight parallel channels.

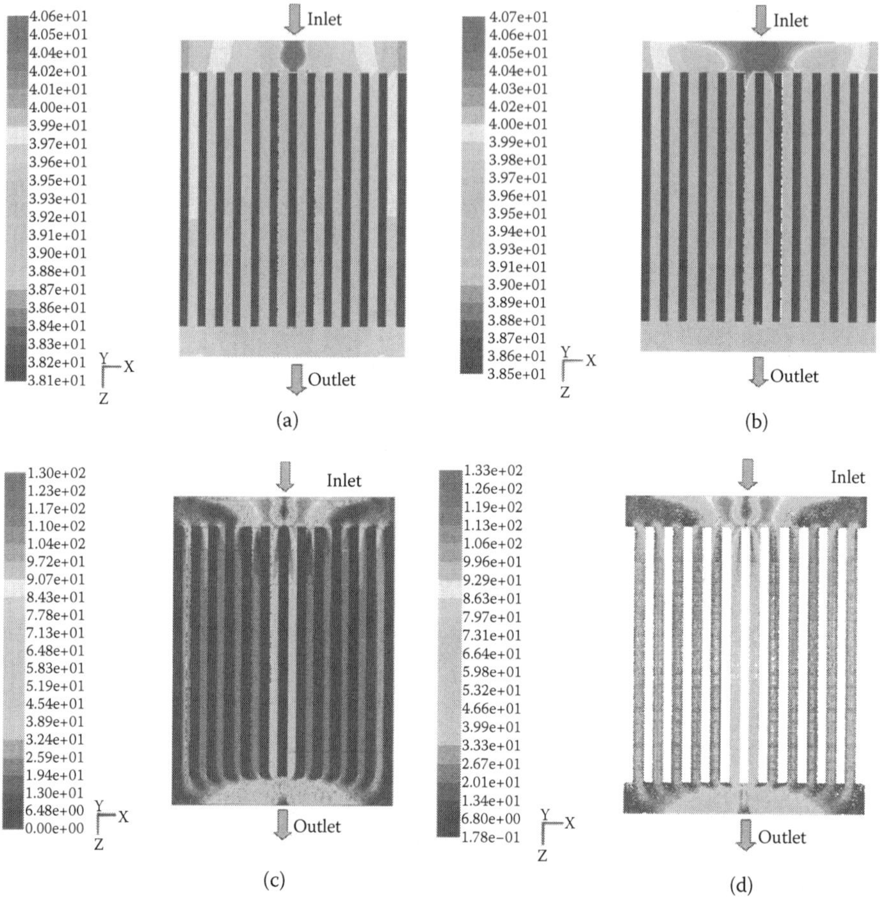

FIGURE 10.18
Pressure and velocity distribution in the parallel straight channels. (a) Static pressure distribution. (b) Total pressure distribution. (c) Distribution of velocity magnitude. (d) Velocity vectors in the flow channels.

flow in the center channels (6, 7) followed by the outermost channels (1, 12). Pressure and velocity vector plots show considerable turbulence mixing and flow separation in the feeder near the flow channel inlets. Also, there is a strong impingement of the flow near the inlet of the channel, especially near the center channels. As the flow approaches the outer channels, there is conversion of dynamic head into static pressure and this contributed toward an increased flow in the outermost channels. In general, results show non uniform distribution of mass flow rates in the channels. Similar results are also obtained for all other cases.

10.6.3.3 Single Serpentine Channel

Contour plots for pressure and velocity for single serpentine channels are shown in Figure 10.19. A closer look at the pressure drop data shows developing flow phenomena in each channel. However, when the total pressure drop of each channel is considered, a periodic, fully developed pressure drop is noticed beyond the third channel. It can be noticed from the contour plots that the serpentine channels give rise to secondary flows near the inner side of the bends, and this causes additional losses (Boddu et al., 2009). However, flow displays a periodically, fully developed flow by the second and third turns.

In order to get a clearer picture of this phenomenon and the subsequent effect on transport field, velocity and temperature profiles are analyzed at different flow sections along the length of the channel. Contour plots for velocity and temperature across the channel presented at different sections along the flow channel are presented in Figure 10.20. The contour plots demonstrate that the serpentine channel gives rise to the presence of secondary flows with flow separation near the inner side of each bend. For instance,

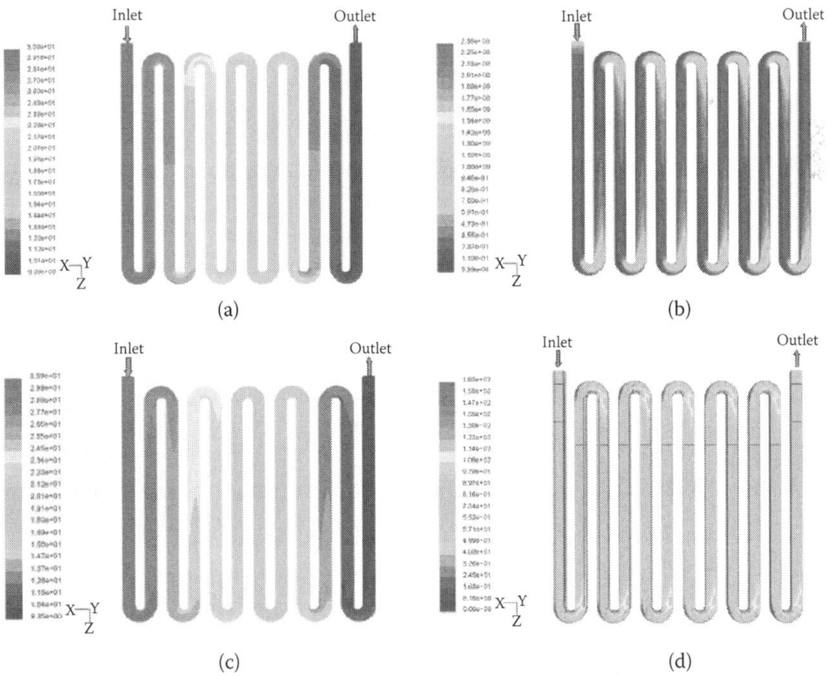

FIGURE 10.19
Contour plots for pressure and velocity in a single serpentine channel. (a) Static pressure. (b) Dynamic pressure. (c) Total pressure. (d) Velocity magnitude.

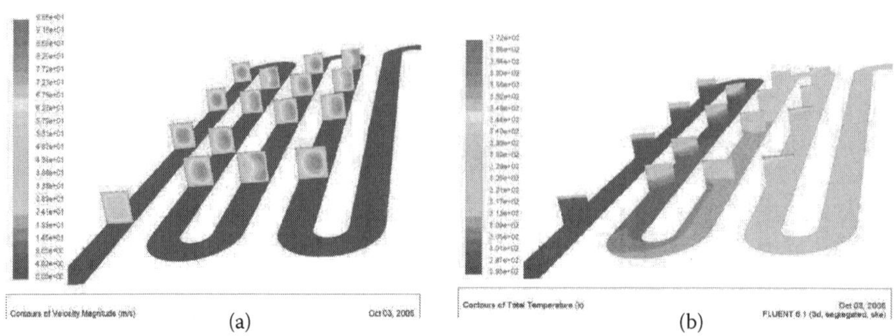

FIGURE 10.20
Typical velocity contours at different sections of the developing flow-fields.

flow is approaching fully developed flow before entering the first turn. As the flow passes along the first bend, it undergoes flow separation, giving rise to a secondary recirculation flow near the inner wall. At this point, the velocity contours clearly show non-symmetric velocity distribution. As the flow approaches the second turn, it regains the symmetrical shape, approaching a fully developed flow pattern before entering the send bend. This is continued in the subsequent channels and bends.

Results for temperature contours are presented in Figure 10.20b. Results show a symmetric thermal penetration layer at the top surface before the first bend, as expected, and considerable flow separation and subsequent recovery at the entrance of the next channel.

In a serpentine channel design, the reactant gas travels over the entire contact surface through a single channel. This channel design avoids any stagnation flow but involves excessive pressure drop owing to longer series flow length and experiences a larger drop in gas concentration from the inlet section to the outlet section. Additionally, the water removed from the electrode surface in the form of small water droplets tends to coalesce and form larger water droplets while traveling down the single channel. This demands an increased pressure level to flush out the water droplets through the channel. Additionally, such a complex two-phase gas–water flow experiences increased pressure drop, leads to inadequate water removal, and results in insufficient gas concentration distribution in the electrode active layers.

10.6.3.4 Single Serpentine Channel with Square Bends

For a single serpentine channel, the effective contact surface area of the gas channels can be increased by replacing circular bends with square bends. A schematic view of the single serpentine channel with square bends is shown in Figure 10.21. The operating conditions are the same as those of the serpentine channels with curvilinear bends.

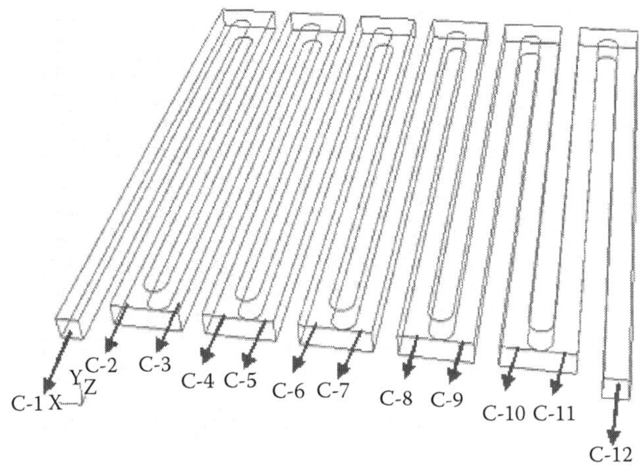

FIGURE 10.21
Serpentine channel designs with square bends.

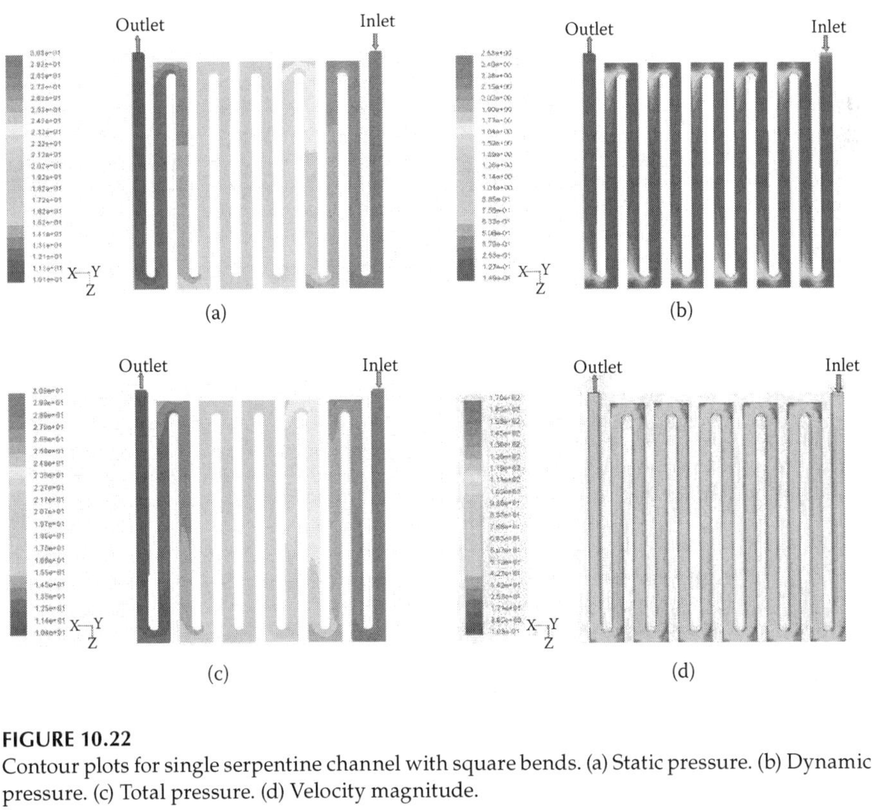

FIGURE 10.22
Contour plots for single serpentine channel with square bends. (a) Static pressure. (b) Dynamic pressure. (c) Total pressure. (d) Velocity magnitude.

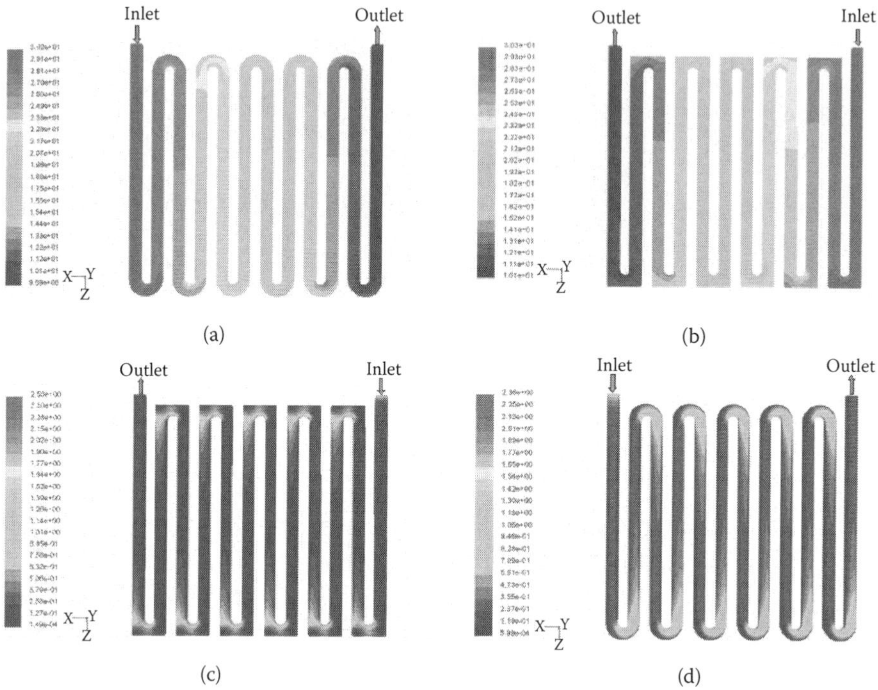

FIGURE 10.23
Static and dynamic pressure distributions a single serpentine channel. (a) Static pressure for curvilinear bends. (b) Static pressure for square bends. (c) Dynamic pressure for square bends. (d) Dynamic pressure for curvilinear bends.

It can be noticed from the contour plots in Figure 10.22 that serpentine channels with square bends also display strong secondary flows near the inner side of each bend as well as periodic, fully developed flows.

For comparison of serpentine curvilinear and serpentine square bends, contour plots of static pressure and dynamic pressure distributions are shown in Figure 10.23 for a single serpentine channel.

The square bends not only provide more contact but also exhibited consistently lower pressure drops compared to curvilinear bends.

10.6.3.5 Multiple Parallel Serpentine Channels with Square Bends

The main idea for considering the multiple parallel serpentine channels is to increase contact surface area with the electrode surface, but with reduced pressure drops.

A more advanced serpentine design involves multiple parallel serpentine channels to carry the gas over the surface. Such a design not only reduces the pressure drop in the channels owing to the parallel nature of the flow but also creates more uniform gas concentration distribution over the contact surface and over the active area of the electrodes. The main idea for considering the multiple parallel serpentine channels is to increase the contact surface area with the electrode surface, but with reduced pressure drops. It is expected that a decrease in cross-sectional area would lead to a more uniform distribution gas concentration at electrode surfaces and increased pressure drop. The smallest size will only be limited by the decrease in fuel cell performance and by limitations in the machining and fabrication processes.

A flow-field design with two parallel serpentine channels with square bends is shown in Figure 10.24a. Contour plots for static pressure, dynamic pressure, and total pressure and velocity distributions are presented in Figure 10.24b through 10.24e.

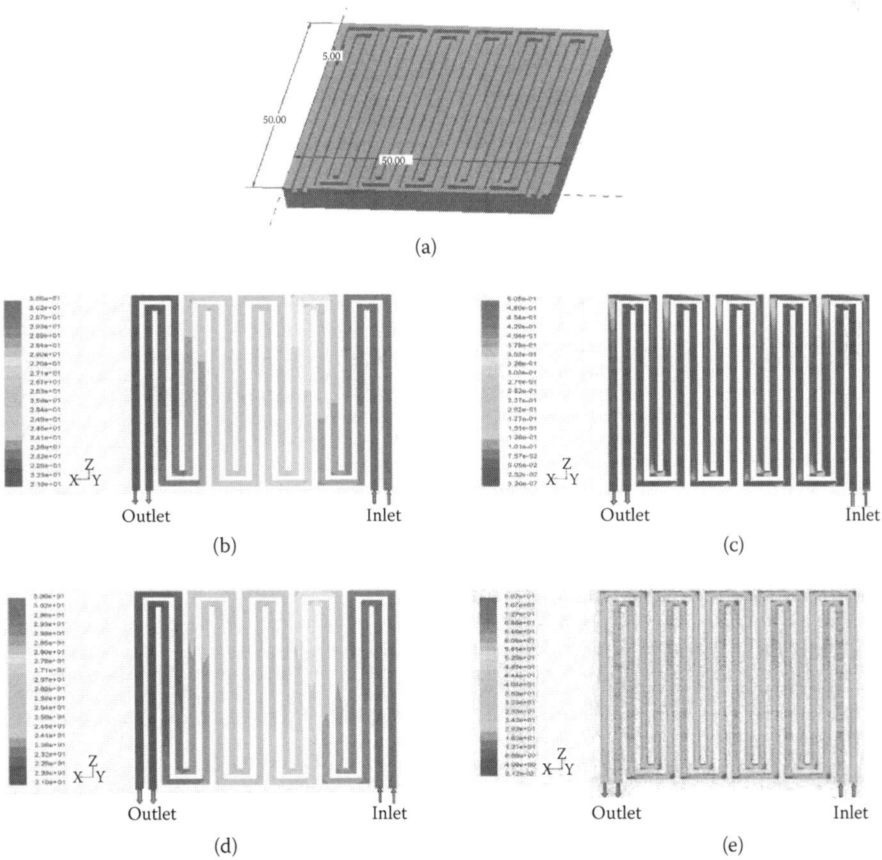

FIGURE 10.24
Multiple serpentine gas channels flow-field design. (a) Dual serpentine channel geometry. (b) Static pressure distribution. (c) Dynamic pressure distribution. (d) Total pressure. (e) Velocity magnitude. (From Rajesh et al., 2009.)

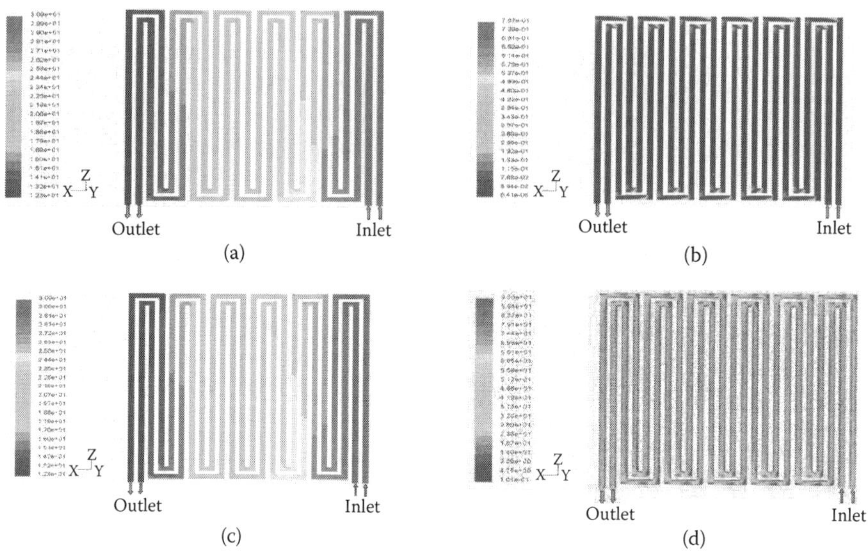

FIGURE 10.25
Contour plots for dual serpentine channel with square bends. (a) Static pressure. (b) Dynamic pressure. (c) Total pressure. (d) Velocity magnitude.

It can be noticed from the contour plots that the dual serpentine channels give rise to similar periodically, fully developed flow in the second and third turns. However, the presence of flow separations and secondary flows in each bend is significantly reduced compared to single square serpentine channels.

In order to increase the contact surface area further, the size of the dual square serpentine channel is reduced to 1 mm. Contour plots for pressure and velocity are presented in Figure 10.25. Results show further reduction in flow separation and recirculation decrease in the square channels.

The pressure drop is reduced significantly as we consider a dual square serpentine channel. The percent reduction is about 67% for a dual channel compared to the single channel with a channel size of 2 mm. When the channel size is reduced from 2 mm to 1 mm to increase contact surface area, the pressure drop is increased by 31.22%. However, the pressure drop is still less than that of the single curvilinear and square serpentine channels.

Increased contact surface area and hence a more uniform gas concentration distribution can be achieved with an increased number of parallel channels and with a decrease in channel size while keeping the pressure drop on the same order as that of a single serpentine channel.

The bipolar plates with integrated cooling channels need to be characterized on the basis of temperature distribution and heat transfer rates, induced stresses during thermal cycles, and mechanical loading.

PROBLEMS

1. Consider an 80 kW fuel cell stack operating at a temperature of 90°C with an operating voltage and current density of 0.65 V and $0.4\frac{W}{cm^2}$. Estimate the following: (a) mass and volume flow rates of hydrogen and oxygen and flow rate if oxygen is supplied as air, (b) water production rate, (c) heat generation rate, and (d) number of unit fuel cell for a unit cell size of 20 cm × 20 cm.
2. Consider a 10 kWe PEM fuel cell stack with an operating power efficiency of 60%. Determine (i) the heat generation rate and (ii) the mass flow rate of water as coolant assuming 50% of the heat is picked up by the coolant fluid and assuming a coolant temperature rise of $\Delta T_c = 15°C$.

References

Ajersch, M. J., M. W. Fowler, K. Karan and B. A. Peppley. PEM Fuel cell bipolar plate reliability and material selection. Proceedings of the Fuel Cell Science, Engineering and Technology, FUELCELL2003-1727, Rochester, New York, 2003.

Barbir, F., J. Braun and J. Neutzler. Properties of molded graphite bi-polar plates for PEM fuel cell stacks. *Journal of New Materials for Electrochemical Reactions* 2: 197–200, 1999.

Boddu, R. and P. Majumdar. Computational flow analysis of bi-polar plate for fuel cells. *ASME Journal of Fuel Cell Science and Technology* 8(4): 2008.

Boddu, R., U. K. Marupakula, B. Summers and P. Majumdar. Development of bipolar plates with different channel configurations for fuel cells. *Journal of Power Sources* 189: 1083–1092, 2009.

DOE announcement No: DE-PS36-06GO96017. Research and development of fuel cell technology for the hydrogen economy. Available at http://www.cere.energy.gov/hydrogenandfuelcells/mypp.

Dutta, S., S. Shimpalee and J. W. Van Zee. Three-dimensional numerical simulation of straight channel PEM fuel cells. *Journal of Applied Electrochemistry* 30: 135–146, 2000.

Dutta, S., S. Shimpalee and J. W. Van Zee. Numerical prediction of mass-exchange between cathode and anode channels in a PEM fuel cell. *International Journal of Heat and Mass Transfer* 44: 2029–2042, 2001.

Kumar, A. and R. G. Reddy. PEM fuel cell bipolar plate-material selection, design and integration. *Fundamentals of Advanced Materials for Energy Conversion II*, TMS, 41–51, 2002.

Kumar, A. and R. G. Reddy. Recent developments in materials, design and concepts for bipolar/end plates in PEM fuel cells. *Advanced Materials for Energy Conversion II*, TMS, 317–324, 2004.

Nguyen, P. T., T. Berning and N. Djilali. Computational model of a PEM fuel cell with serpentine gas flow channels. *Journal of Power Sources* 130: 149–157, 2004.

Oseen-Senda, K. M., J. Pauchet, M. Feidt and O. Lottin. An evaluation of cooling systems for PEMFC in transport applications. Proceedings of the Fuel Cell Science, Engineering and Technology, FUELCELL2003-1763, 2003.

Rajalakshmi, N., V. Vijay, S. Pandian and S. Dhathathreyan. PEM fuel cell stack development—Grafoil bi-polar materials—A feasibility study. Proceedings of the Fuel Cell Science, Engineering and Technology—2004 Conference, FUELCELL2004-2505, Rochester, pp. 449–452, 2004.

Shimpalee, S. and S. Dutta. Numerical prediction of temperature distribution in PEM fuel cells. *Numerical Heat Transfer, Part A* 38: 111–128, 2000.

Wheeler, D. J., J. S. Yi, R. Fredley, D. Yang, T. Patterson and L. VanDine. Advancements in fuel cell stack technology at international fuel cells. *Journal of New Materials for Electrochemical Systems* 4: 233–238, 2001.

Further Reading

Berning, T., D. M. Lu and N. Djilali. Three-dimensional computational analysis of transport phenomena in a PEM fuel cell. *Journal of Power Sources* 106: 284–294, 2002.

Borup, R. L. and N. E. Vanderborgh. Design and testing criteria for bipolar plate materials for PEM fuel cell application. *Mater Res Soc Proc* 393: 151, 1995.

Cooper, J. S. Design analysis of PEMFC bipolar plates considering stack manufacturing and environment impact. *Journal of Power Sources* 129: 152–169, 2004.

Cunningham, B. and D. G. Baird. Development of economical bipolar plates for fuel cells. *Journal of Materials Chemistry* 16: 4385–4388, 2006.

Curtis, M. and L. Xianguo. Performance modeling of a proton exchange membrane fuel cell. Proceedings of the 1998 ASME Energy Source Technology Conference, 1998.

De Bruijn, F. A., V. A. T. Dam and G. J. M. Janssen. Review: Durability and degradation issues of PEM fuel cell components. *Fuel Cell* 8(1): 3–22, 2008.

Dodge, C. E. Tubular fuel cell with structural current collectors. U.S. Patent 5,458,989, 1995.

Fuel Cell Technological Program Multi-Year Research, Development and Demonstration Plan. U.S. Department of Energy, Fuel Cell Technologies Program, Washington, DC, 2007.

Hamilton, P. J. and B. G. Pollet. Polymer electrolyte membrane fuel cell (PEMFC) flow field plate: Design, materials and characterization. *Fuel Cells* 10(4): 489–509, 2010.

Kalaga, S. and P. Majumdar. Developing flow in micro-channels using continuum model with slip boundary conditions. Proceedings of the 18th National and 7th IHMT-ASME Heat and Mass Transfer Conference, HMT-2006-C333, pp. 2399–2406, 2006.

Koncar, G. J. and L. G. Marianowski. Proton exchange membrane fuel cell separator plate. U.S. Patent 5,942,347, 1999.

Li, X. and I. Sabir. Review of bipolar plates in PEM fuel cells flow field designs. *International Journal of Hydrogen Energy* 359–371, 2005.

Mehta, V. and J. S. Cooper. Review and analysis of PEM fuel Cell design and manufacturing. *Journal of Power Sources* 114: 32–53, 2003.

Mehta, V. and J. S. Cooper. *Journal of Power Sources* 129: 152, 2004.

Peng, L., X. Lai, J. Ni and Z. Lin. Flow channel shape design of stamped bipolar plates for PEM fuel cell by micro-forming simulation. Proceedings of the 4th ASME International Conference on Fuel Cell Science, Engineering and Technology, Irvine, 2006.

Tawfik, H., Y. Hung and D. Mahajan. Metallic bipolar plates for PEM fuel cells—A review. *Journal of Power Sources* 163: 755–767, 2007.

Watkins, D. S., K. W. Diycks and D. G. Epp. Novel fuel cell fluid flow field plate. U.S. Patent 4,988,583, 1991.

Wu, J., X. Z. Yuan, J. J. Martin, H. Wang, J. Zhang, J. Shen, S. Wu and W. Merida. A review of PEM fuel cell durability: Degradation mechanism and mitigation strategies. *Journal of Power Sources* 184: 104–119, 2008.

Xue, D. and Z. Dong. Optimal fuel cell system design considering functional performance and production costs. *Journal of Power Sources* 76: 69–80, 1998.

11

Simulation Model for Analysis and Design of Fuel Cells

In this chapter, we will present simulation models for the analysis and design of a fuel cell on the basis of the basic principles and theory that we have described in the last several chapters on the thermodynamics of a fuel cell, electrochemical kinetics, and charge and species transport through the gas supply channels, electrode/gas diffusion layer, and electrolyte. Simulation models serve as an excellent tool to analyze and obtain in-depth understanding of the operation of a fuel cell. A simulation model can be used to develop polarization characteristics such as voltage–current curve and power–current curve on the basis of input parameters for material properties, transport properties, empirical constants for electrode kinetics, and dimensions. A simulation model can be classified into *zero-order model, one-dimensional model,* and *multidimensional and computational fluid dynamics (CFD) model* depending on the number of space variables used in describing the operation of the fuel cell. Models can also be classified as an isothermal model with no temperature dependency and as a non-isothermal coupled heat and mass transport multi-physics model. A CFD-based model takes into account of the velocity field in the gas channels on the basis of *Navier–Stokes equations* and velocity field within the gas diffusion layers on the basis of *Darcy* or *Brinkman equations.* While a zero-order model is the simplest simulation model that does not take into account any variations in gas species and charge concentration across the channel and cell, a multi-physics three-dimensional model is the most comprehensive model that takes into account the gas species concentration, temperature variations with the gas channels and the cell, and variation of electrical potential and current density within the electrode–electrolyte layers. In the following sections, we present descriptions of these models followed by general algorithms for computations and fuel cell design for different applications.

11.1 Zero-Order Fuel Cell Analysis Model

In a zero-order model, transport equations for gas concentration and temperature are not considered and performance characteristic in terms of voltage–current polarization is derived on the basis of the reversible open circuit voltage and different components of fuel cell overpotenials and is given in the form of an algebraic equation as

$$V = E_0 - \eta_{act} - \eta_{ohm} - \eta_{conc}, \tag{11.1}$$

where

E_0 = open circuit voltage given by the *Nernst Equation* (Equation 4.49) as

$$E_0 = E_{rev}(T) - \frac{\bar{R}T}{n_e F} \ln \frac{c_M^{n_M} c_N^{n_N}}{c_A^{n_A} c_B^{n_B}} \tag{11.2a}$$

and for hydrogen–oxygen fuel cell as

$$E_0 = E_{rev}(T) - \frac{\bar{R}T}{2F} \ln \frac{P_{H_2O}}{P_{H_2} P_{O_2}^{1/2}} \tag{11.2b}$$

$E_{rev}(T)$ = reversible voltage at a temperature, T
η_{act} = activation losses for electrode kinetics
$= \eta_{act,a} + \eta_{act,c}$
$\eta_{act,a}$ = activation loss owing to anode electrode kinetics
$\eta_{act,c}$ = activation loss owing to cathode electrode kinetics
η_{ohm} = ohmic loss owing to charge transport
η_{conc} = concentration loss owing to mass transport and variation of gas concentration

11.1.1 Activation Loss: η_{act}

Activation loss is given by the *Butler–Volmer equation* (Equation 7.19) or by the *Tafel equation* with empirical constants obtained on the basis of the fit of experimental data. For anode and cathode electrodes, the Butler–Volmer equation (Equation 7.19b) is written as

$$\textit{Anode:}\ j_a = j_{o,a}\left(e^{\alpha_a n_{e,a} F \eta_{act,a}/(RT)} - e^{-(1-\alpha_a) n_{e,a} F \eta_{act,a}/(RT)}\right) \tag{11.3a}$$

$$\textit{Cathode:}\ j_c = j_{o,c}\left(e^{\alpha_c n_{e,c} F \eta_{act,c}/(RT)} - e^{-(1-\alpha_c) n_{e,c} F \eta_{act,c}/(RT)}\right), \tag{11.3b}$$

where the first term represents the forward activation rate and the second term represents the reverse activation rate. α_a and α_c are the *transfer coefficients* for the anode and cathode reactions and represent the order of magnitude of the forward and reverse activation rates for an electrode. Typical values of the transfer coefficient lie in the range of 0–1. For symmetric reactions, the value is 0.5. $j_{o,a}$ and $j_{o,c}$ represent the *exchange current densities* of the anode and cathode electrodes, representing the current that flows equally in the forward and reverse directions at equilibrium at standard concentration. As discussed in a previous chapter, exchange current density, j_o, and transfer function α are the kinetic parameters of the electrochemical reactions and obtained experimentally.

A simplified form of the Butler–Volmer equations for cases with very small activation loss and for very large activation loss are derived and given as follows.

11.1.2 Simplified Butler–Volmer Equation: Very Small η_{act}

The exponential functions in the Butler–Volmer equation can expanded in series and simplified by neglecting higher-order terms for a small value of η_{act} as

$$j = j_{o,i}[\alpha_i n_{e,i} F \eta_{act,i}/RT - (1-\alpha_i) n_{e,i} F \eta_{act,i}/RT]. \tag{11.4}$$

Simplifying further,

$$j = j_{o,i} \frac{n_{e,i} F \eta_{act}}{RT}. \tag{11.5}$$

Rewriting, the activation overpotential is expressed as

$$\eta_{act,i} = \frac{RT}{n_{e,i} F}\left(\frac{j}{j_{o,i}}\right). \tag{11.6}$$

11.1.3 Simplified Butler–Volmer Equation: Very Large η_{act}

For a very large value of η_{act}, the second term in the Butler–Volmer equation for reverse rate can be neglected as compared to the first term for the forward rate and Equation 11.3 can be approximated as

$$j = j_{o,i} e^{\alpha_i n_{e,i} F \eta_{act,i}/(RT)}. \tag{11.7}$$

Rewriting the equation, the activation overpotential is expressed as

$$\eta_{act,i} = \frac{RT}{\alpha_i n_{e,i} F} \ln \frac{j}{j_{o,i}}. \tag{11.8}$$

A simplified Butler–Volmer equation (Equation 11.3) can also be written in the form of the *Tafel equation* as

$$\eta_{\text{act},i} = -\frac{RT}{\alpha_i n_{\text{e},i} F} \ln i_{\text{o},i} + \frac{RT}{\alpha_i n_{\text{e},i} F} \ln i \tag{11.9}$$

or

$$\eta_{\text{act},i} = a_i + b_i \ln i, \tag{11.10}$$

where

$$a_i = -\frac{RT}{\alpha_i n_{\text{e},i} F} \ln i_{\text{o},i} \quad \text{and} \quad b_i = \frac{RT}{\alpha_i n_{\text{e},i} F}. \tag{11.11}$$

While the constants a_i and b_i in the *Tafel* equation (Equation 11.10) can be estimated from the reaction kinetic parameters, these are normally given directly by the linear fit of the *Tafel plot* for current–polarization or η–ln j measurement for the electrochemical reaction at higher overpotential values. Also, with the known values of the j-axis intercept and the Tafel slope of the linear part of the Tafel plot from the linear fit equation, the *exchange current density*, j_0, and *transfer coefficient*, α, can be computed.

The net activation loss of a fuel cell including both anode and cathode reactions can be written on the basis of the *Tafel equation* as

$$\eta_{\text{act},i} = (a_{i,\text{a}} + b_{i,\text{a}} \ln i) + (a_{i,\text{c}} + b_{i,\text{c}} \ln i), \tag{11.12}$$

where

$a_{i,\text{a}}$, $b_{i,\text{a}}$ = reaction kinetic parameters for anode reaction
$a_{i,\text{c}}$, $b_{i,\text{c}}$ = reaction kinetic parameters for cathode reaction

11.1.4 Simplified Butler–Volmer Equation with Identical Charge Transfer Coefficient

For the equal charge transfer coefficient on the anode and cathode electrode, that is, $\alpha_\text{a} = \alpha_\text{c} = \alpha$, the general Butler–Volmer equation (Equation 11.3) can be written as

$$j = j_\text{o}\left(e^{\alpha\, n_\text{e} F \eta_{\text{act}}/(RT)} - e^{-\alpha\, n_\text{e} F \eta_{\text{act}}/(RT)}\right) \tag{11.13}$$

or

$$j = 2 j_\text{o} \sinh\left(\frac{\alpha n_\text{e} F}{RT} \eta_{\text{act}}\right). \tag{11.14}$$

The activation loss can be expressed as

$$\eta_{act} = \frac{RT}{\alpha n_e F}\sinh^{-1}\left(\frac{j}{2j_o}\right). \tag{11.15}$$

11.1.5 Ohmic Loss: η_{ohm}

As we have discussed in Chapter 7, ohmic voltage loss in a fuel cell takes place owing to the resistance of the material to charge transport. This includes electron transport from the anode side to the cathode side of the cell through the electrodes and interconnects, and ion transport through the electrolyte. The net ohmic voltage loss is expressed as

$$\eta_{ohm} = IR = I(R_{elec} + R_{ionic}), \tag{11.16}$$

where

I = cell current in amperes (A)

R_{elec} = electronic resistance = $\dfrac{L}{\sigma_{elec}A}(\Omega)$

R_{ionic} = ionic resistance of electrolyte = $\dfrac{L}{\sigma_{ionic}A}(\Omega)$

σ_{elec} = electronic conductivity (Ω^{-1} cm^{-1})

σ_{ionic} = ionic conductivity (Ω^{-1} cm^{-1})

L = thickness of the respective electronic conductor or ionic conductor, that is, the electrolyte

A = cross-sectional area of the conductor or the electrolyte membrane surface area for the ohmic loss owing to the charge transport through the electrolyte membrane and electrodes. Cross-sectional area for the interconnects for ohmic loss owing to electron transfer varies with the design.

Ohmic resistance is often written in terms of current density and area-specific resistance (ASR) as

$$\eta_{ohm} = j(\text{ASR}_{ohmic}), \tag{11.17}$$

where

j = current density (A/cm^2)

ASR_{ohmic} = area-specific resistance = $A_{cell}R_{ohm}$

The principal thing that should be remembered here is that the ohmic resistance or voltage loss increases with the thickness of the electrolyte and the thickness or the length of the electronic conductor and decreases with higher ionic and electronic conductivities. Since the ionic conductivity is order of magnitudes lower than the electronic conductivity, reducing the ohmic loss entails reducing the thickness of electrolyte membrane or developing electrolyte materials with higher ionic conductivity.

11.1.6 Concentration Loss: η_{conc}

The concentration voltage loss takes place because of the variation of reactant and product concentrations at the reaction sites. As discussed in Chapter 6, the variation of reactant concentrations and depletion of reactants cause voltage loss in two ways: first, there is decrease in voltage loss because of the reaction kinetics, and second, the voltage loss owing to concentration effects given in the Nernst equation.

The concentration voltage loss owing to the reaction kinetics is derived from the Butler–Volmer equation at higher current densities as

$$\eta_{\mathrm{conc,k}} = \frac{RT}{\alpha n_e F} \ln \frac{j_{\mathrm{L}}}{j_{\mathrm{L}} - j}. \tag{11.18}$$

The concentration voltage loss owing to the concentration effects in the Nernst equation is

$$\eta_{\mathrm{conc,N}} = \frac{RT}{n_e F} \ln \frac{i_{\mathrm{L}}}{j_{\mathrm{L}} - j}. \tag{11.19}$$

Combining Equations 11.18 and 11.19, we have the total concentration voltage loss as

$$\eta_{\mathrm{conc}} = \eta_{\mathrm{cocn,k}} + \eta_{\mathrm{conc,N}} \tag{11.20}$$

or

$$\eta_{\mathrm{conc}} = \left((1 + \frac{1}{\alpha} \right) \left(\frac{RT}{n_e F} \right) \ln \frac{j_{\mathrm{L}}}{j_{\mathrm{L}} - j}. \tag{11.21}$$

In Equations 11.18 through 11.21, j_{L} is the limiting current density given by Equation 6.81a as

$$j_{\mathrm{L}} = n_{\mathrm{e}} F \frac{C_{i,\mathrm{ch}}}{\left(\frac{1}{h_{\mathrm{m}}} + \frac{L_{\mathrm{el}}}{D_{ij}^{\mathrm{eff}}} \right)}, \tag{11.22}$$

where

$C_{i,\mathrm{ch}}$ = reactant concentration in the gas channel
D_{ij}^{eff} = diffusion coefficient in cathode electrode
h_{m} = convective mass transfer coefficient
L_{el} = electrode thickness

In situations where the diffusion resistance in the electrode is significantly higher than the mass transfer resistance in the gas channel, the *limiting current density* is approximated as

$$j_{\mathrm{L}} = n_{\mathrm{e}} F D_{ij}^{\mathrm{eff}} \left(\frac{C_{i,\mathrm{ch}}}{L_{\mathrm{el}}} \right). \tag{11.23}$$

For dominant mass concentration loss in the cathode electrode,

$$j_{\mathrm{L}} = n_{\mathrm{e}} F D_{\mathrm{O}_2}^{\mathrm{eff}} \left[\frac{C_{\mathrm{O}_2,\mathrm{ch}}}{L_{\mathrm{c}}} \right], \tag{11.24a}$$

where

$C_{\mathrm{O}_2,\mathrm{ch}}$ = oxygen concentration in the cathode gas channel
$D_{\mathrm{O}_2}^{\mathrm{eff}}$ = oxygen diffusion coefficient in cathode electrode
L_{c} = thickness of cathode electrode

Substituting $C_{\mathrm{O}_{2,\mathrm{ch}}} = \rho_{\mathrm{c}} x_{\mathrm{O}_{2,\mathrm{ch}}}$, the limiting current density is also given as

$$j_{\mathrm{L}} = n_{\mathrm{e}} F D_{\mathrm{O}_2}^{\mathrm{eff}} \left[\frac{\rho_{\mathrm{c}} x_{\mathrm{O}_{2,\mathrm{ch}}}}{L_{\mathrm{c}}} \right]. \tag{11.24b}$$

Finally, we can combine all components of voltage loss and express the fuel cell operating voltage as a function of current density as

Very small η_{act}:

$$V = E - \frac{RT}{n_{\mathrm{e},i} F} \left(\frac{j}{j_{\mathrm{o},i}} \right) - j(\mathrm{ASR}_{\mathrm{ohmic}}) - \left(1 + \frac{1}{\alpha} \right) \left(\frac{RT}{n_{\mathrm{e}} F} \right) \ln \frac{j_{\mathrm{L}}}{j_{\mathrm{L}} - j} \tag{11.25a}$$

Very large η_{act}:

$$V = E - \frac{RT}{\alpha_i n_{\mathrm{e},i} F} \ln \frac{j}{j_{\mathrm{o}}} - j(\mathrm{ASR}_{\mathrm{ohmic}}) - \left(1 + \frac{1}{\alpha} \right) \left(\frac{RT}{n_{\mathrm{e}} F} \right) \ln \frac{j_{\mathrm{L}}}{j_{\mathrm{L}} - j} \tag{11.25b}$$

and in terms of Tafel parameters for activation loss as

$$V = E - (a_{i,a} + b_{i,a} \ln i) - (a_{i,c} + b_{i,c} \ln j) - j(\text{ASR}_{\text{ohmic}}) \\ - \left(1 + \frac{1}{\alpha}\right)\left(\frac{RT}{\alpha n F}\right) \ln \frac{j_L}{j_L - j}. \tag{11.25c}$$

Example 11.1

Consider a SOFC fuel cell with an operating temperature of 800°C and open circuit voltage of $E_0 = 1.15$ V. Compute the operating cell voltage at an operating current density of 0.94 A/cm^2 and based on the following cell data: Exchange current density, j_o = 011 A/cm^2, activity coefficient, $\alpha = 0.5$, area specific resistance, $\text{ASR}_{\text{ohmic}} = 0.045\ \Omega\ \text{cm}^2$ and limiting current density, $j_L = 1.88$ A/cm^2.

Solution

Activation loss based on Equation 11.15 for the cathode electrode

$$\eta_{\text{act}} = \frac{8.314 \times 1073.2}{0.5 \times 4 \times 96,485} \sinh^{-1}\left(\frac{0.94}{2 \times 0.11}\right)$$

$$\eta_{\text{act}} = 0.0462 \sinh^{-1}(4.2727) \\ = 0.0462 \times 2.15881$$

$$\eta_{\text{act}} = 0.0998 \text{ V}$$

Ohmic loss based on Equation 11.17

$$\eta_{\text{ohm}} = j(\text{ASR}_{\text{ohmic}})$$

$$\eta_{\text{ohm}} = 0.94 \times 0.045$$

$$\eta_{\text{ohm}} = 0.0423 \text{ V}$$

Concentration loss based on Equation 11.21

$$\eta_{\text{conc}} = \left(1 + \frac{1}{0.5}\right)\left(\frac{8.314 \text{ J/mol}\cdot\text{K} \times 1073.2 \text{ K}}{4 \times 96,485}\right) \ln\left(\frac{1.88}{1.88 - 0.94}\right)$$

$$\eta_{\text{conc}} = (3.0)(0.023119)\ln(2)$$

$$\eta_{\text{conc}} = 0.048 \text{ V}$$

The operating voltage is given as

$$E = E_0 - \eta_{\text{act}} - \eta_{\text{ohm}} - \eta_{\text{conc}}$$

$$E = 1.15 - 0.0998 - 0.0423 - 0.048 \\ = 1.15 - 0.1901$$

$$E = 0.9599 \text{ V}$$

11.2 One-Dimensional Fuel Cell Analysis Model

In the one-dimensional fuel cell model, reactant gas concentration variations across the cell from the gas channel to the electrode–electrolyte interface, shown in Figure 11.1, are taken into account. A one-dimensional model provides basic qualitative understanding of the internal mechanics of the fuel cell operation, and it provides a base for developing a multi-dimensional model. In such a one-dimensional model, variations are only considered across the cell in the *z*-direction as shown in Figure 11.1. Hydrogen and oxygen gas concentrations vary from the bulk gas concentration in the gas flow channel owing to a number of mass transfer resistances: (1) convective mass transfer resistance between the flowing gas stream and the porous electrode surface, and (2) convective and diffusion mass transfer resistance within the porous electrodes.

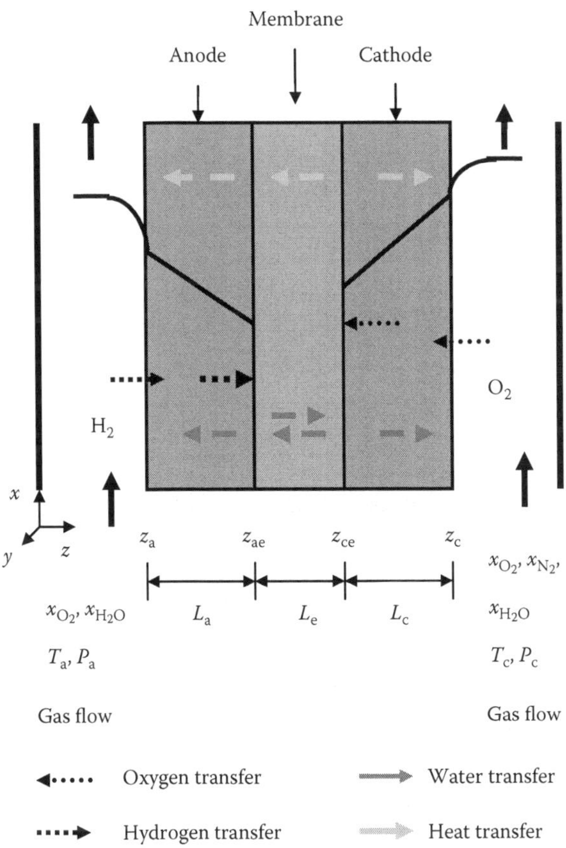

FIGURE 11.1
Tri-layer fuel cell with adjacent reactant gas flow channels.

A simplified one-dimensional model is presented here on the basis of the following assumptions:

1. Humidified hydrogen enters the anode gas channel while oxygen in the form of air enters the cathode gas channel.
2. The catalyst layer and the reaction region are very thin and reactant gas concentration within this layer can be neglected. Electrochemical reaction is assumed to take place at the electrode–electrolyte interface as a surface reaction.
3. The reactant gas stream is assumed as an ideal gas mixture.
4. Diffusion mass transfer controls gas transport within the electrode with negligible advection mass transfer.
5. Ohmic loss is primarily due to the ionic transport through electrolyte membrane. Ohmic losses owing to the electron transfer through the electrode and interconnect are neglected.
6. Activation loss is primarily caused by the oxygen reduction reaction at the cathode. Hydrogen oxidation reaction at the anode is several orders of magnitude higher than the oxygen reduction reaction at the cathode, and the anodic activation loss can be neglected.
7. Under steady-state operation, the net mass flux of the hydrogen transfer rate is identical to the transfer rate from the gas channel to the electrode surface and the transfer rate through the electrode layer, and equal to the gas consumption rate at the electrode–electrolyte interface because of the electrochemical reaction.
8. Convective mass transfer in the gas channels is defined in terms of constant convective transfer coefficient (h_m) as defined in Chapter 6.
9. Water is assumed to remain in vapor form and a single-phase model is considered for water transport.

Based on these assumptions, the governing equations for the one-dimensional fuel cell model are given below.

11.2.1 Anode Gas Channel

Hydrogen mass transfer rate from the gas channel to the anode electrode surface is given in terms of the convective mass transfer coefficient as

$$J_{H_2} = h_{m,a}\left(x_{H_2,ah} - x_{H_2,a}\right), \tag{11.26}$$

where

$x_{H_2,ah}$ = hydrogen concentration in the anode gas channel

$x_{H_2,a}$ = hydrogen concentration at the anode electrode surface adjacent to the gas channel at $Z = Z_a$
$h_{m,a}$ = convective mass transfer coefficient in the anode gas channel

11.2.2 Anode Electrode

The mass flux through the anode electrode is given based on Fick's law of diffusion as

$$J_{H_2} = \rho_a D_{ij}^{eff} \frac{\left(x_{H_2,a} - x_{H_2,ae}\right)}{L_a}, \tag{11.27}$$

where
D_{ij}^{eff} = effective diffusion coefficient of hydrogen in porous anode electrode
L_a = thickness of anode electrode layer
$\rho_a = \dfrac{P_a}{RT_a}$ = density of hydrogen gas stream

Hydrogen gas consumption at the anode electrode–electrolyte

$$J_{H_2} = \frac{j}{2F}. \tag{11.28}$$

Combining Equations 11.26 through 11.28 and eliminating $x_{H_2,ae}$, we get

$$J_{H_2} = \frac{j}{2F} = \frac{x_{H_2,ch} - x_{H_2,ae}}{\dfrac{L_a}{\rho_a D_{ij}^{eff}} + \dfrac{1}{h_{m,a}}} \tag{11.29}$$

Solving for the hydrogen concentration at the anode–electrolyte interface, $x_{H_2,ae}$

$$x_{H_2,ae} = x_{H_2,ch} - \frac{j}{2F}\left(\frac{L_a}{\rho_a D_{ij}^{eff}} + \frac{1}{h_{m,a}}\right). \tag{11.30}$$

For negligible mass transfer resistance in the channel compared to the mass transfer resistance in the electrode, the profile for the gas concentration variation within the cell differs significantly as demonstrated in Figure 11.2.

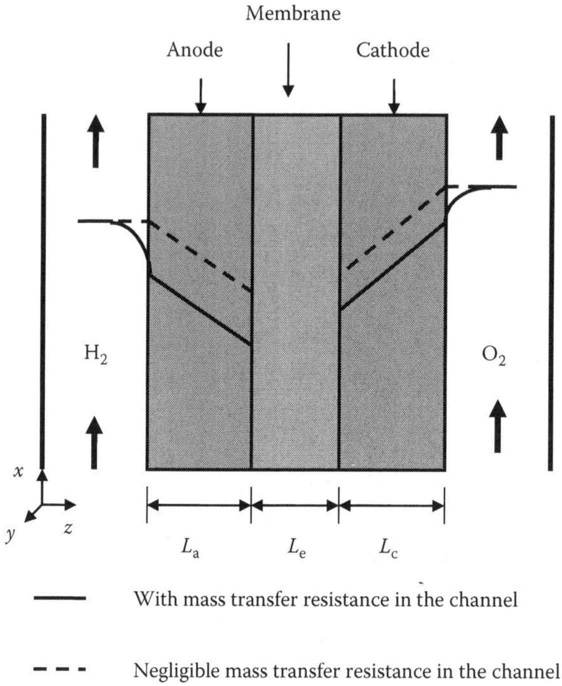

FIGURE 11.2
Gas concentration variations across the fuel cell and gas channels.

For negligible mass transfer resistance, $C_{H_2,ch} \approx C_{H_2,a}$, as demonstrated in Figure 11.2, Equation 11.30 reduces to

$$x_{H_2,ae} = x_{H_2,a} - \frac{j}{2F}\frac{L_a}{\rho_a D_{ij}^{eff}}. \tag{11.31}$$

11.2.3 Cathode Gas Channel

Oxygen mass transfer rate from the gas channel to the cathode electrode surface is given in terms of convective mass transfer coefficient as

$$J_{O_2} = h_{m,c}\left(x_{O_2,ch} - x_{O_2,c}\right), \tag{11.32}$$

where

$x_{O_2,ch}$ = oxygen concentration in the cathode gas channel
$x_{O_2,c}$ = oxygen concentration at the cathode electrode surface adjacent to the gas channel at $Z = Z_c$
$h_{m,c}$ = convective mass transfer coefficient in cathode gas channel

11.2.4 Cathode Electrode

The variation of oxygen concentration within the cathode electrode is given based on Fick's law of diffusion as

$$J_{O_2} = \rho_c D_{O_2}^{\text{eff}} \frac{\left(x_{O_2,c} - x_{O_2,ce}\right)}{L_c}, \tag{11.33}$$

where

$D_{O_2}^{\text{eff}}$ = effective diffusion coefficient of oxygen in porous anode electrode
L_c = thickness of cathode electrode layer
$\rho_c = \dfrac{P_c}{RT_c}$ = density of oxygen gas stream

Oxygen gas consumption at the cathode electrode–electrolyte

$$J_{O_2} = \frac{j}{4F} \tag{11.34}$$

Solving Equations 11.32 through 11.34 and eliminating $x_{O_2,c}$,

$$J_{O_2} = \frac{j}{4F} = \frac{x_{O_2,ch} - x_{O_2,ce}}{\dfrac{L_c}{\rho_a D_{O_2}^{\text{eff}}} + \dfrac{1}{h_{m,c}}}. \tag{11.35}$$

Oxygen gas concentration at the interface of the cathode electrode and electrolyte is expressed as

$$x_{O_2,ce} = x_{O_2,ch} - \frac{j}{4F}\left(\frac{L_c}{\rho_c D_{O_2}^{\text{eff}}} + \frac{1}{h_{m,c}}\right). \tag{11.36}$$

As demonstrated for the anode side for the case where diffusion resistance within the electrode is significantly greater than convective channel resistance, $C_{O_2,ch} \approx C_{O_2,c}$, and Equation 11.36 reduces to

$$x_{O_2,ce} = x_{O_2,ch} - \frac{j}{4F}\frac{L_c}{\rho_c D_{O_2}^{\text{eff}}}. \tag{11.37}$$

11.3 One-Dimensional Water Transport Model

During the operation of the fuel cell, water is generated within the cell as discussed in Chapter 4. Water transport through the cell is quite complex, particularly for the case of PEM fuel cell as discussed in Chapters 7 and 10. Water is generated at the electrolyte–cathode interface for the case of a PEM fuel cell and at the anode–electrolyte interface for the case of SOFC. Water is transported through the cell primarily by diffusion through the two electrode layers of the cell and by convection to moving gas streams through the gas channels. While water transport through the porous solid oxide membrane is by diffusion, the mass transport though the *Nafion* polymer membrane is more complex as discussed in Chapter 10. Figure 11.3a and b show water transport and a typical water distribution through the SOFC and PEMFC, respectively.

One-dimensional differential water transport model for the fuel cell is given in the following subsections.

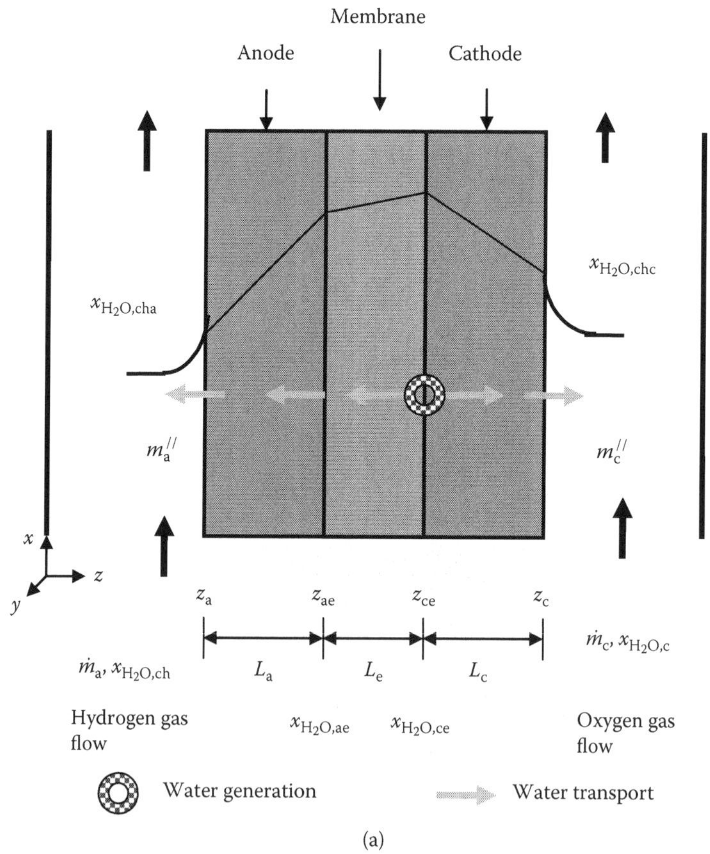

FIGURE 11.3
(a) One-dimensional water transport and distribution in a one-dimensional PEM fuel cell model. (b) One-dimensional water transport and distribution in a one-dimensional SOFC model.

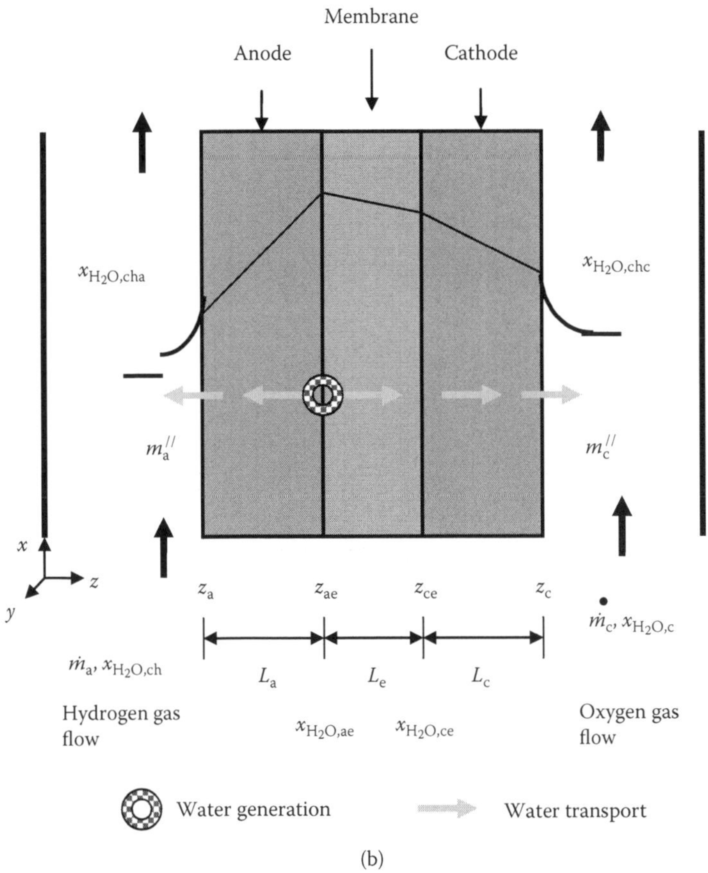

FIGURE 11.3 (Continued)
(a) One-dimensional water transport and distribution in a one-dimensional PEM fuel cell model. (b) One-dimensional water transport and distribution in a one-dimensional SOFC model.

11.3.1 Anode Gas Channel

Water transfer from anode surface to anode gas stream by convective mass transfer:

$$J_{H_2O,a} = h_{m,a}A(x_{H_2O,cha} - x_{H_2O,a}), \tag{11.38}$$

where
$h_{m,a}$ = convective water mass transfer coefficient in the anode channel
$x_{H_2O,cha}$ = water concentration in the anode gas channel
$x_{H_2O,a}$ = water concentration at the anode surface

11.3.2 Anode Electrode

Water transport by diffusion mass transport

$$\frac{d^2 x_{H_2O,a}}{dz^2} = 0. \tag{11.39a}$$

Integration of this equation and substituting boundary conditions lead to the expression for water mass flux based on Fick's law of diffusion as

$$J_{H_2O,a} = \rho_a D^{eff}_{H_2O,H_2} \frac{\left(x_{H_2O,a} - x_{H_2O,ae}\right)}{L_a}, \tag{11.39b}$$

where

$x_{H_2O,a}$ = water concentration at the anode surface

$x_{H_2O,ae}$ = water concentration at the anode–electrolyte interface

This equation (Equation 11.39b) for water mass flux through the anode can also be expressed in a form analogous to an electrical circuit as

$$J_{H_2O,a} = \rho_a \frac{\left(x_{H_2O,a} - x_{O_2,ae}\right)}{R_{H_2O,a}}, \tag{11.40a}$$

where

$$R_{H_2O,a} = \frac{L_a}{D^{eff}_{H_2O,H_2}}. \tag{11.40b}$$

Considering that water transport through the anode side is a fraction of the water generation at the electrode–electrolyte interface, the water mass flux toward the anode side is given as

$$J_{H_2O,a} = \beta_o J_{H_2O,gen} = \beta_o \frac{i}{2F} \tag{11.41}$$

where

β_o = fraction of water generation that transports through the membrane, anode electrode and transferred to or from the anode gas stream.

Substituting Equation 11.41 into Equation 11.39b, we get the expression for water concentration at the anode–electrolyte interface as

$$x_{H_2O,ae} = x_{H_2O,a} - \beta_o \frac{j}{2F}\left(\frac{L_a}{\rho_a D^{eff}_{H_2O,H_2}}\right) \tag{11.42a}$$

and in terms of water concentration in the channels as

$$x_{H_2O,ae} = x_{H_2,cha} - \beta_o \frac{j}{2F}\left(\frac{L_a}{\rho_a D^{eff}_{H_2O,H_2}} + \frac{1}{h_{m,a}}\right). \tag{11.42b}$$

11.3.3 Cathode Gas Channel

Water transports to or from the cathode surface to the cathode gas stream by convective mass transfer and is given as

$$J_{H_2O,c} = h_{m,c}A(x_{H_2O,chc} - x_{H_2O,c}). \tag{11.43}$$

11.3.4 Cathode Electrode

Water transport by diffusion mass transport

$$\frac{d^2 x_{H_2O,c}}{dz^2} = 0. \tag{11.44a}$$

Integration of this equation and substituting boundary conditions lead to

$$J_{H_2O,c} = \rho_a D^{eff}_{H_2O,O_2} \frac{\left(x_{H_2O,c} - x_{H_2O,ce}\right)}{L_c} \tag{11.44b}$$

where

$x_{H_2O,c}$ = water concentration at the cathode surface
$x_{H_2O,ce}$ = water concentration at the cathode–electrolyte interface

Writing this expression in the form of an electrical circuit as

$$J_{H_2O,c} = \rho_c \frac{\left(x_{H_2O,c} - x_{O_2,ce}\right)}{R_{H_2O,c}}, \tag{11.44c}$$

where

$$R_{H_2O,c} = \frac{L_c}{D^{eff}_{H_2O,O_2}}. \tag{11.44d}$$

Water transport through the cathode side is a fraction of the water generation at the electrode–electrolyte interface and given as

$$J_{H_2O,c} = (1-\beta_o)J_{H_2O,gen} = (1-\beta_o)\frac{j}{2F}. \tag{11.45}$$

Substituting Equation 11.45 into Equation 11.44b, we get the expression for water concentration at the cathode–electrolyte interface as

$$x_{H_2O,ce} = x_{H_2O,c} + (1-\beta_o)\frac{j}{2F}\left(\frac{L_c}{\rho_c D^{eff}_{H_2O,O_2}}\right) \tag{11.46a}$$

and in terms of water concentration in the channels as

$$x_{H_2O,ce} = x_{H_2O,chc} + (1-\beta_o)\frac{j}{2F}\left(\frac{L_c}{\rho_c D^{eff}_{H_2O,O_2}} + \frac{1}{h_{m,c}}\right). \tag{11.46b}$$

11.3.5 Electrolyte Membrane

11.3.5.1 SOFC Electrolyte Membrane

Water transport in the SOFC electrolyte membrane is primarily by diffusion mass transport

$$\frac{d^2 x_{H_2O,e}}{dz^2} = 0. \tag{11.47a}$$

Integration of this equation and substituting boundary conditions lead to

$$J_{H_2O,e} = \rho_e D^{eff}_{H_2O}\frac{\left(x_{H_2O,ae} - x_{H_2O,ce}\right)}{L_e}, \tag{11.47b}$$

where

$x_{H_2O,ae}$ = water concentration at the anode–electrolyte interface
$x_{H_2O,ce}$ = water concentration at the cathode–electrolyte interface

Writing this expression in the form of an electrical circuit as

$$J_{H_2O,e} = \frac{\left(x_{H_2O,ae} - x_{H_2O,ce}\right)}{R_{H_2O,e}} \tag{11.48a}$$

where

$$R_{H_2O,e} = \frac{L_e}{D^{eff}_{H_2O}}. \tag{11.48b}$$

Noting that water transport through the membrane is identical to that through the cathode membrane, we can write

$$J_{H_2O,e} = J_{H_2O,c} = (1-\beta_o)J_{H_2O,gen} = (1-\beta_o)\frac{j}{2F}, \quad (11.49)$$

where
β_o = fraction of water generation that transport to or from the anode side.

11.3.5.2 PEM Electrolyte Membrane

Water is produced at the cathode–membrane interface in a PEM fuel cell, and as we have discussed in Chapter 7, water transport through the *Nafion* polymer membrane is more complex. A fraction of the water produced transports through the cathode electrode by diffusion and transfers to the oxygen gas stream in the cathode channel by convection. Direction and amount of water transport through the *Nafion* polymer electrolyte membrane is based on the net balance of electro-osmotic drag, convection owing to pressure gradient and back diffusion owing to water concentration gradient. The net water mass flux through the Nafion membrane based on electro-osmotic drag and back diffusion is given based on Equation 7.77 and written in a one-dimensional form as

$$J_{H_2O,m} = 2n_{drag}\frac{j}{2F} - \frac{D_{H_2O}(\lambda_m)\rho_{dry}}{M_m}\frac{d\lambda_m}{dz}. \quad (11.50a)$$

In Equation 11.50a, the water flux component based on electro-osmotic drag is computed on the basis of the fact that the proton flux is two times the hydrogen consumption rate of $\frac{j}{2F}$. Substituting Equation 9.24 for electro-osmotic drag coefficient for a Nafion membrane, we get

$$J_{H_2O,m} = 2n_{drag}^{sat}\frac{j}{2F}\frac{\lambda_m}{22} - \frac{D_{H_2O}(\lambda_m)\rho_{dry}}{M_m}\frac{d\lambda_m}{dz}. \quad (11.50b)$$

Rearranging,

$$\frac{D_{H_2O}(\lambda_m)\rho_{dry}}{M_m}\frac{d\lambda_m}{dz} = 2n_{drag}^{sat}\frac{j}{2F}\frac{\lambda_m}{22} - J_{H_2O,m}. \quad (11.50c)$$

We noticed that the net mass flux through the membrane is the same as the net mass flux through the anode side of the cell under steady-state operation.

As discussed before, depending on the net balance of the water flux, the net water flux through the membrane and through the anode electrode may transport toward the anode–electrolyte interface and add up to the water generation or it may take away a fraction of the water generated at the interface and transport toward the anode and anode gas channel. Let us express the net water flux toward the membrane and anode side as follows:

$$J_{H_2O,m} = J_{H_2O,a} = \beta_o J_{H_2O,gen} = \beta_o \left(\frac{j}{2F} \right), \tag{11.51}$$

where

β_o = fraction of water generation that transport to or from the anode side.

Substituting Equation 11.51 into Equation 11.50c,

$$\frac{d\lambda_m}{dz} - \left(\frac{j}{2F} \frac{M_m}{\rho_{dry} D_{H_2O}(\lambda_m)} \frac{n_{drag}^{sat}}{11} \right) \lambda_m = -\left(\frac{j}{2F} \frac{M_m}{\rho_{dry} D_{H_2O}(\lambda_m)} \right) \beta_0. \tag{11.52}$$

Equation 11.52 represents a first-order nonhomogeneous ordinary differential equation for water content in *Nafion* subject to the following two boundary conditions based on known water content at the anode–membrane and cathode–membrane interfaces:

$$\text{Boundary condition 1: at } z = z_{am}, \lambda_m = \lambda_{am} \tag{11.53a}$$

and

$$\text{Boundary condition 2: at } z = z_{cm}, \lambda_m = \lambda_{cm}, \tag{11.53b}$$

where λ_{am} and λ_{cm} are given based on water activity ratio and based on Equation 9.13 as

$$\lambda_m = 0.043 + 17.81a - 39.85a^2 + 36.0a^3, 0 < a \le 1 \ (\textit{unsaturated}) \tag{9.13a}$$

and a linear relation is suggested for saturated range as

$$\lambda_m = 14 + 1.4(a - 1), 1 \le a \le 3 \ (\textit{saturated}). \tag{9.13b}$$

Solution to the first-order ordinary differential equation (Equation 11.52) is obtained based on a homogeneous and particular solution.

Homogeneous equation and solution:

$$\frac{d\lambda_{m,h}}{dz} - \left(\frac{j}{2F} \frac{M_m}{\rho_{dry} D_{H_2O}(\lambda_m)} \frac{n_{drag}^{sat}}{11} \right) \lambda_{m,h} = 0 \tag{11.54a}$$

or

$$\frac{d\lambda_{m,h}}{dz} - C_1 \lambda_{m,h} = 0, \tag{11.54b}$$

where

$$C_1 = \frac{j}{2F}\frac{M_m}{\rho_{dry} D_{H_2O}(\lambda_m)}\frac{n_{drag}^{sat}}{11}. \tag{11.54c}$$

Integrating,

$$\lambda_{m,h} = C_2 \exp(C_1 z), \tag{11.55}$$

where C_2 is the constant of integration.

The particular solution is given as

$$\lambda_{m,p} = \frac{11\beta_o}{n_{drag}^{sat}}. \tag{11.56}$$

Combining a homogeneous and particular solution, the solution of water distribution in the polymer membrane is given as

$$\lambda_m = C_2 \exp\left(\frac{j}{2F}\frac{M_m}{\rho_{dry} D_{H_2O}}\frac{n_{drag}^{sat}}{11} z\right) + \frac{11\beta_o}{n_{drag}^{sat}}. \tag{11.57}$$

Equation 11.57 represents the variation of water distribution in the Nafion membrane involving two unknowns β_o and C_2, which are determined using boundary conditions given by Equation 11.53.

Example 11.2

Determine water content in the Nafion membrane at the anode–membrane and cathode–membrane interfaces for PEM fuel cell operation at 80°C and operating pressure of 2 atm on both anode and cathode sides. Assume water mole fractions as 0.1 and 0.25 at the anode–membrane and cathode–membrane interfaces, respectively.

Solution

At an operating temperature of 80°C, the corresponding water saturation pressure is $P_{sat} = 47.39$ kPa. Based on this water vapor pressure, the water activity ratios at the anode–membrane and cathode–membrane interfaces are computed as

$$a_{am} = x_{H_2O,a}\frac{P_a}{P_{sat}} = 0.1 \times \frac{2 \times 101.3}{47.39} = 0.4275.$$

And assuming that water is in liquid form at the cathode–membrane interface,

$$a_{cm} = x_{H_2O,c} \frac{P_c}{P_{sat}} = 0.25 \times \frac{2 \times 101.3}{47.39} = 1.068.$$

Water content at the interface is computed using Equation 9.14a for the anode–membrane interface as

$$\begin{aligned}\lambda_{am} &= 0.043 + 17.81\, a_{am} - 39.85\, a_{am}^2 + 36.0\, a_{am}^3 \\ &= 0.043 + 17.81 \times 0.4275 - 39.85 \times 0.4275^2 + 36.0 \times 0.4275^3 \\ &= 0.043 + 7.613775 - 7.28283 + 2.81261 \\ \lambda_{am} &= 3.1866\end{aligned}$$

and using Equation 9.14b for the cathode interface as

$$\lambda_{cm} = 14 + 1.4(1.068 - 1) = 14.095.$$

11.4 One-Dimensional Electrochemical Model

The electrochemical model for the voltage–current polarization is as discussed for the zero-order model and given by

$$V = E_o - \eta_{act} - \eta_{ohm} - \eta_{conc}. \tag{11.1}$$

11.4.1 Activation Loss: η_{act}

Following the procedure outlined for the zero-order model, the activation loss is given by the simplified *Butler–Volker equation* (Equation 11.39) as

$$\textit{Anode: } j = j_{o,a}\left(\frac{C_{H_2}}{C_{H_2}^0} e^{\alpha_a n_{e,a} F \eta_{act,a}/(RT)}\right) \tag{11.39a}$$

$$\textit{Cathode: } j = j_{o,c}\left(\frac{C_{O_2}}{C_{O_2}^0} e^{\alpha_c n_{e,c} F \eta_{act,c}/(RT)}\right) \tag{11.39b}$$

Note that Butler–Volmer equations (Equations 11.39a and 11.39b) include the ratio of reactant concentration at the electrode reaction surface, C_R, to the bulk concentration at the channel, C_R^0. Solving for the activation overpotential:

$$\textit{Anode: } \eta_{act,a} = \frac{RT}{\alpha_a n_{e,a} F} \ln\left(\frac{j}{j_{o,a}} \frac{C_{H_2}^0}{C_{H_2}}\right) \tag{11.40a}$$

$$Cathode:\ \eta_{act,c} = \frac{RT}{\alpha_c n_{e,c} F} \ln\left(\frac{j}{j_{o,c}} \frac{C_{O_2}^0}{C_{O_2}} \right) \tag{11.40b}$$

The mass concentration ratio of reactants, $\frac{C_R^0}{C_R}$, can be expressed in terms of mole concentration ratio using ideal gas equation of state $P = \rho RT$ or $P = CRT$. Based on this equation, the mass concentration ratio can be expressed as

$$\frac{C_{H_2}^0}{C_{H_2}} = \frac{P_{a,H_2}^0}{P_{a,H_2}} \quad \text{and} \quad \frac{C_{O_2}^0}{C_{O_2}} = \frac{P_{c,O_2}^0}{P_{c,O_2}}.$$

Writing $P_{a,H_2} = x_{H_2} P_a$ and $P_{c,O_2} = x_{O_2} P_c$, Equations 11.40a and 11.40b for activation potentials can be expressed as

$$Anode:\ \eta_{act,a} = \frac{RT}{\alpha_a n_{e,a} F} \ln\left(\frac{j}{j_{o,a}} \frac{P_a^0}{x_{H_2} P_a} \right) \tag{11.41a}$$

$$Cathode:\ \eta_{act,c} = \frac{RT}{\alpha_c n_{e,c} F} \ln\left(\frac{j}{j_{o,c}} \frac{P_c^0}{x_{O_2} P_c} \right) \tag{11.41b}$$

Substituting the reactant concentrations at the electrode–electrolyte interfaces $x_{H_2} = x_{H_2,ae}$ and $x_{O_2} = x_{O_2,ae}$ based on Equations 11.31 and 11.37 respectively, we get the expression for the activation overpotential as

$$Anode:\ \eta_{act,a} = \frac{RT}{\alpha_a n_{e,a} F} \ln\left(\frac{j}{j_{o,a}} \frac{P_a^0}{P_a \left[x_{H_2,ch} - \dfrac{i}{2F} \dfrac{L_a}{\rho_a D_{H_2}^{eff}} \right]} \right) \tag{11.42a}$$

and

$$Cathode:\ \eta_{act,c} = \frac{RT}{\alpha_c n_{e,c} F} \ln\left(\frac{j}{j_{o,c}} \frac{P_c^0}{P_c \left[x_{O_2,ch} - \dfrac{j}{4F} \dfrac{L_c}{\rho_c D_{O_2}^{eff}} \right]} \right) \tag{11.42b}$$

11.4.2 Ohmic Loss: η_{ohm}

Ohmic loss is given in terms of current density and ASR from Equation 11.17 as

$$\eta_{ohm} = j(\text{ASR}_{ohmic}), \tag{11.17}$$

where
j = current density (A/cm^2)
ASR_{ohmic} = area-specific resistance = $A_{cell}R_{ohm}$

11.4.3 Ohmic Loss η_{ohm} in Polymer Membrane

Assuming that the ohmic loss is primarily given by the ionic conductivity of the electrolyte membrane, the ohmic loss is given by Equation 11.17

$$\eta_{ohm} = j\text{ASR} = j\frac{L_e}{\sigma_i}$$

For the polymer membrane, the ionic conductivity is given as

$$\sigma(T,\lambda) = \sigma_{303\text{K}}(\lambda)\exp\left[1268\left(\frac{1}{303} - \frac{1}{T}\right)\right],$$

where

$$\sigma_{303\text{K}}(\lambda) = 0.005139\lambda - 0.00326 \text{ for } \lambda > 1.$$

Since ionic conductivity varies with water content (λ), which also varies locally throughout the thickness of the membrane, we can employ two approaches in determining the average conductivity of the membrane:

Approach 1

Average conductivity based on the average water content in the membrane as

$$\sigma(T,\bar{\lambda}) = \sigma_{303\text{K}}(\bar{\lambda})\exp\left[1268\left(\frac{1}{303} - \frac{1}{T}\right)\right],$$

where the average water content in the membrane is given as

$$\bar{\lambda} = \frac{1}{L_e} \int_0^{L_e} \lambda \, dx.$$

Approach 2

Estimate the membrane ionic resistance based on

$$R_e = \int_0^{L_e} \frac{dz}{\sigma(z)},$$

where $\sigma(z)$ is the ionic conductivity as a function membrane thickness, which can be derived by substituting the relationship of membrane water content as a spatial variation with the thickness of the membrane.

11.4.4 Water Content in Nafion–PEM

The water absorbing capacity of Nafion-117 is given by Equation 9.13 as

$$\lambda = 0.043 + 17.81a - 39.85a^2 + 36.0a^3,\ 0 < a \le 1 \tag{9.13a}$$

$$\lambda = 14 + 1.4(a - 1),\ 1 \le a \le 3. \tag{9.13b}$$

The humidity condition at the gas diffusion catalyst layer interface is given as the water activity or relative humidity (a) and defined as

$$a = \frac{P_{H_2O}}{P_{sat}} \tag{9.14a}$$

or

$$a = \frac{y_{H_2O} P}{P_{sat}}, \tag{9.14b}$$

where y_{H_2O} is the mole fraction of water, P is the total pressure, and P_{sat} is the saturation pressure corresponding to the temperature of water.

11.4.5 Mass Concentration Loss: η_{conc}

Mass concentration loss is given by Equation 11.21

$$\eta_{conc} = \left(1 + \frac{1}{\alpha}\right)\left(\frac{RT}{n_e F}\right) \ln \frac{j_L}{j_L - j}. \tag{11.21}$$

For dominant mass concentration loss in the cathode electrode, the limiting current density is given by Equation 11.24b

$$j_L = n_e F D_{O_2}^{eff} \left[\frac{\rho_c x_{O_2,ch}}{L_c} \right]. \tag{11.24b}$$

Example 11.3: One-Dimensional SOFC Model

Consider the electrolyte as 8% YSZ solid oxide fuel cell with the following known data:

Anode gas stream:

Inlet gas composition: 95% H_2 and 5%H_2O

Inlet pressure, P_a = 1 atm (101.3 Pa)

Cathode gas stream:

Inlet gas composition: Air with 21% O_2 and 79% N_2

Inlet pressure, P_c = 1 atm (101.3 Pa)

Diffusivities:

$$D_{H_2,H_2O}^{eff} = 1 \times 10^{-4}\ m^2/s \text{ and } D_{O_2,N_2}^{eff} = 0.2\ cm^2/s$$

Kinetic parameters:

Cathode exchange transfer coefficient, $\alpha = 0.5$

Cathode exchange current density, $j_o = 0.1\ A/cm^2$

Layer thicknesses:

Anode thickness, $L_a = 400\ \mu m$

Cathode thickness, $L_c = 400\ \mu m$

Electrolyte, $L_e = 50\ \mu m$

Determine the operating voltage of the fuel cell for the current density of 0.5 A/cm² and a temperature of T = 800°C.

Solution

Considering hydrogen gas stream as a mixture of 95% H_2 and 5% H_2O and air as a mixture of 21% oxygen and 79% nitrogen, the overall reaction in the fuel cell is written as

$$0.95\ H_2 + 0.475\ O_2 + (0.475)(3.76)\ N_2 = 0.95\ H_2O + (0.475)(3.76)\ N_2.$$

For this reaction, the mole fractions and the partial pressures are given as

Hydrogen: $x_{H_2} = 0.95$ and $P_{H_2} = 0.95 \times 1 \text{ atm} = 0.95 \text{ atm}$

Oxygen: $x_{O_2} = \dfrac{n_{O_2}}{n_{O_2} + n_{N_2}} = \dfrac{0.475}{0.475 + (0.475)(3.76)} = 0.21$ and

$P_{O_2} = 0.21 \times 1 \text{ atm} = 0.21 \text{ atm} = 0.21 \times 1 \text{ atm} = 0.21 \text{ atm}$

Water: $x_{H_2O} = \dfrac{n_{H_2O}}{n_{H_2O} + n_{N_2}} = \dfrac{0.95}{0.95 + 0.475 \times 3.76} = 0.347$

and $P_{H_2O} = 0.347 \times 1$ atm $= 0.347$ atm.

Open circuit voltage is given as

$$E_0 = E_{rev} - \frac{\bar{R}T}{2F} \ln \frac{P_{H_2O}}{P_{H_2} P_{O_2}^{1/2}}.$$

Reversible voltage is given as

$$E_{rev} = \frac{-\Delta \bar{g}_f^0}{n_e F}.$$

At an operating temperature of $T = 800°C = 1073$ K, $\Delta \bar{g}_f^0 = 188,500$ J/mol from Table 4.3 and reversible voltage is given as

$$E_{rev} = \frac{-\Delta \bar{g}_f^0}{n_e \Phi} = -\frac{(-188,500)}{(2)(96,485)}$$

$$E_{rev}^0 = 0.9768 \text{ V}.$$

The open circuit voltage for this fuel cell is given as

$$E_0 = E_{rev}^0 - \frac{\bar{R}T}{2F} \ln \frac{P_{H_2O}}{P_{H_2} P_{O_2}^{1/2}}$$

$$E_0 = 0.9768 - \frac{8.314 \times 1098}{(2)(96,485)} \ln \frac{0.347}{(0.95)(0.21)^{1/2}}$$

$$E_0 = 0.9768 - 0.03601129 \ \ln(0.797)$$

$$E_0 = 0.9768 - 0.0729040$$

$$E_0 = 0.9039 \text{ V}.$$

Activation overpotential based on cathode electrode is given by Equation 11.41b as

$$\eta_{act,c} = \frac{RT}{\alpha_c n_{e,c} F} \ln \left(\frac{j}{j_{o,c}} \frac{P_c^0}{P_c x_{O_2,ce}} \right).$$

Oxygen concentration at the cathode–electrolyte interface is given based on Equation 11.37 for the case with dominant diffusion mass transfer resistance as

$$x_{O_2,ce} = x_{O_2,ch} - \frac{j}{4F} \frac{L_c}{\rho_c D_{O_2}^{eff}},$$

where

$$\rho_c = \frac{P_c}{\bar{R}T} = \frac{101,325 \text{ Pa}}{8.3145 \text{ kJ/kmol K} \times 1073\text{K}} = 11.357 \text{ kmol/m}^3$$

$$\rho_c = 0.01136 \times 10^{-3} \text{ mol/cm}^3$$

$$\begin{aligned} x_{O_2,ce} &= x_{O_2,ch} - \frac{j}{4F}\frac{L_c}{\rho_c D_{O_2}^{eff}} \\ &= 0.21 - \frac{0.5 \text{ A/cm}^2}{4 \times 96,485 \text{ C/mol}} \frac{0.04 \text{ cm}}{0.01136 \times 10^{-3} \times 0.2 \text{ cm}^2/\text{s}} \\ &= 0.21 - 0.0228 \\ x_{O_2,ce} &= 0.1872. \end{aligned}$$

Activation overpotential based on cathode electrode is computed as

$$\begin{aligned} \eta_{act,c} &= \frac{RT}{\alpha_c n_{e,c} F} \ln\left(\frac{j}{j_{o,c}} \frac{P_c^0}{P_c x_{O_2,ce}}\right) \\ &= \frac{8.3145 \text{ J/mol K} \times 1073 \text{ K}}{0.5 \times 4 \times 96,485 \text{ C/mol}} \ln\left(\frac{0.5 \text{ A/cm}^2}{0.1 \text{ A/cm}^2} \frac{1 \text{ atm}}{1 \text{ atm} \times 0.1872}\right) \end{aligned}$$

$$\eta_{act,c} = \frac{8.3145 \text{ J/mol K} \times 1073 \text{ K}}{0.5 \times 4 \times 96,485 \text{ C/mol}} \ln(26.709)$$

$$\eta_{act,c} = 0.1159 \text{ V}.$$

Ohmic loss:

Assuming that the ohmic loss is primarily given by the ionic conductivity of the electrolyte membrane, the ohmic loss is given by

$$\eta_{ohm} = j \text{ ASR} = j\frac{L_e}{\sigma_i}.$$

Let us compute the ionic conductivity based on Equation 7.48 for 8% YSZ electrolyte:

$$\sigma_i = a\exp\left(-\frac{b}{T}\right), \tag{7.48}$$

where the empirical coefficients are given as $a = 3.34 \times 10^4$ $(\Omega\text{-m})^{-1}$ and $b = 1.03 \times 10^4$ K,

$$\sigma_i = 3.34\times 10^4 \exp\left(-\frac{1.03\times 10^4}{1073}\right)$$

$$\sigma_i = 2.2632\ \Omega m^{-1} = 2.2632\times 10^{-2}\ (\Omega cm^{-1})$$

$$\eta_{ohm} = j\frac{L_e}{\sigma_i}$$

$$\eta_{ohm} = 0.5\ \text{A/cm}^2 \frac{50\times 10^{-4}\ \text{cm}}{2.26321\times 10^{-2}\Omega\text{cm}^{-1}}$$

$$\eta_{ohm} = 0.1105\ \text{V}.$$

Mass concentration loss is given by

$$\eta_{conc} = \left(1+\frac{1}{\alpha}\right)\left(\frac{RT}{n_e F}\right)\ln\frac{j_L}{j_L - j}. \tag{11.21}$$

For dominant mass concentration loss in the cathode electrode, the limiting current density is given by Equation 11.24b

$$j_L = n_e F D_{O_2}^{eff}\left[\frac{\rho_c x_{O_{2,ch}}}{L_c}\right]$$

$$j_L = 4\times 96,485\ \text{C/mol}\times 0.2\ \text{cm}^2/\text{s}\left[\frac{0.08805\times 10^{-3}\ \text{mol/cm}^3\times 0.21}{400\times 10^{-4}\text{cm}}\right]$$

$$j_L = 35.677\ \text{A/cm}^2.$$

Mass concentration loss is computed as

$$\eta_{conc} = \left(1+\frac{1}{0.5}\right)\left(\frac{8.314\times 1073.2}{n_{e,c}F}\right)\ln\frac{j_L}{j_L - j}$$
$$\eta_{conc} = \left(1+\frac{1}{0.5}\right)\left(\frac{8.314\ \text{J/mol K}\times 1073.2\ \text{K}}{4\times 96,485}\right)\ln\left(\frac{35.677}{35.677-0.5}\right)$$
$$\eta_{conc} = (3.0)(0.023119)\ln(1.014)$$
$$\eta_{conc} = 0.000964\ \text{V}.$$

Cell operating voltage is given as

$$E = E_0 - \eta_{act} - \eta_{ohm} - \eta_{conc}$$
$$= 0.9039\ \text{V} - 0.1159 - 0.1105 - 0.000964$$
$$E = 0.6765\ \text{V}$$

Example 11.4: One-Dimensional PEM Fuel Cell Model

Consider the polymer electrolyte membrane, *Nafion-117* (PEM) fuel cell operating at a temperature of 80°C with the following known data:

Anode gas stream:
Inlet gas composition: 95% H_2 and 5% H_2O
Inlet pressure, P_a = 2 atm (202.6 kPa)
Cathode gas stream:
Inlet gas composition: Air with 21% O_2 and 79% N_2
Inlet pressure, P_c = 2 atm (202.6 kPa)
Diffusivities:

$$D^{\text{eff}}_{\text{H}_2,\text{H}_2\text{O}} = 0.15\ \text{cm}^2/\text{s and } D^{\text{eff}}_{\text{O}_2,\text{H}_2\text{O}} = 0.03\ \text{cm}^2/\text{s}$$

$$D^{\text{eff}}_{\text{H}_2\text{O}} = 3.8\times 10^{-6}\ \text{cm}^2/\text{s}$$

Kinetic parameters:
Cathode exchange transfer coefficient, α = 0.5
Cathode exchange current density, j_o = 0.0001 A/cm²
Layer thicknesses:
Anode thickness, L_a = 300 μm
Cathode thickness, L_c = 300 μm
Electrolyte, L_e = 100 μm

Determine the operating voltage of the PEM fuel cell for the current density of 0.5 A/cm² and a temperature of T = 80°C.

Solution

Considering hydrogen gas stream as a mixture of 95% H_2 and 5% H_2O and air as a mixture of 21% oxygen and 79% nitrogen, the overall reaction in the fuel cell is written as

$$0.95\ \text{H}_2 + 0.475\ \text{O}_2 + (0.475)(3.76)\ \text{N}_2 = 095\ \text{H}_2\text{O} + (0.475)(3.76)\ \text{N}_2.$$

For this reaction, the mole fractions and the partial pressures are given as

$$\text{Hydrogen: } x_{\text{H}_2} = 0.95 \text{ and } P_{\text{H}_2} = 0.95\times 2\text{ atm} = 1.9\text{ atm}$$

$$\text{Oxygen: } x_{\text{O}_2} = \frac{n_{\text{O}_2}}{n_{\text{O}_2}+n_{\text{N}_2}} = \frac{0.475}{0.475+(0.475)(3.76)} = 0.21 \text{ and}$$

$$P_{\text{O}_2} = 0.21\times 2\text{ atm} = 0.42\text{ atm}$$

$$\text{Water: } x_{\text{H}_2\text{O}} = \frac{n_{\text{H}_2\text{O}}}{n_{\text{H}_2\text{O}}+n_{\text{N}_2}} = \frac{0.95}{0.95+0.475\times 3.76} = 0.347 \text{ and}$$

$$P_{\text{H}_2\text{O}} = 0.347\times 2\text{ atm} = 0.684\text{ atm}.$$

Open circuit voltage is given as

$$E_0 = E_{rev} - \frac{\bar{R}T}{2F}\ln\frac{P_{H_2O}}{P_{H_2}P_{O_2}^{1/2}}.$$

Reversible voltage is given as

$$E_{rev} = \frac{-\Delta\bar{g}_f^0}{n_e F}.$$

At an operating temperature of $T = 80°C = 353$ K, $\Delta\bar{g}_f^0 = -228{,}350$ J/mol from Table 4.3 and reversible voltage is given as

$$E_{rev} = \frac{-\Delta\bar{g}_f^0}{n_e F} = -\frac{(-228{,}350)}{(2)(96{,}485)}$$

$$E_{rev}^0 = 1.183 \text{ V}.$$

The open circuit voltage for this fuel cell is given as

$$E_0 = E_{rev}^0 - \frac{\bar{R}T}{2F}\ln\frac{P_{H_2O}}{P_{H2}P_{O_2}^{1/2}}$$
$$E_0 = 1.183 - \frac{8.314 \times 353}{(2)(96{,}485)}\ln\frac{0.347}{(0.1)(0.42)^{1/2}}$$
$$E_0 = 1.183 - 0.01521\ \ln(7.57216)$$
$$E_0 = 1.183 - 0.03077$$
$$E_0 = 1.152 \text{ V}.$$

In order to estimate the activation loss, ohmic loss, and mass concentration loss, it is necessary to compute gas and water concentrations at the electrode–membrane interfaces.

Hydrogen gas concentration at the anode–membrane interface is given by Equation 11.31 as

$$x_{H_2,ae} = x_{H_2,a} - \frac{j}{2F}\frac{L_a}{\rho_a D_{H_2}^{eff}}.$$

Hydrogen gas density is computed using ideal gas equation of state as

$$\rho_a = \frac{P_a}{\bar{R}T} = \frac{2 \times 101{,}325 \text{ Pa}}{8.3145 \text{ J/mol K} \times 353 \text{ K}} = 69.046 \text{ mol/m}^3$$

$$\rho_a = 69.046 \times 10^{-6} \text{ mol/cm}^3$$

$$x_{H_2,ae} = x_{H_2,a} - \frac{j}{2F}\frac{L_a}{\rho_a D_{H_2}^{eff}}$$
$$= 0.95 - \frac{0.5\ \text{A/cm}^2}{2\times 96,485\ \text{C/mol}}\frac{0.03\ \text{cm}}{69.046\times 10^{-6}\times 0.15\ \text{cm}^2/\text{s}}$$
$$x_{H_2,ae} = 0.95 - 0.0075$$
$$x_{H_2,ae} = 0.9425.$$

Oxygen gas concentration at the cathode–membrane interface is given by Equation 11.47 as

$$x_{O_2,ce} = x_{O_2,ch} - \frac{j}{4F}\frac{L_c}{\rho_c D_{O_2}^{eff}}.$$

Oxygen gas density is computed using ideal gas equation of state as

$$\rho_c = \frac{P_c}{RT} = \frac{2\times 101,325\ \text{Pa}}{8.3145\ \text{J/mol K}\times 353\ \text{K}} = 69.04556\ \text{mol/m}^3$$

$$\rho_c = 69.04556\times 10^{-6}\text{mol/cm}^3.$$

Substituting, we get the oxygen gas concentration at the cathode–membrane interface as

$$x_{O_2,ce} = x_{O_2,ch} - \frac{j}{4F}\frac{L_c}{\rho_c D_{O_2}^{eff}}$$
$$x_{O_2,ch} = 0.21 - \frac{0.5\ \text{A/cm}^2}{4\times 96,485\ \text{C/mol}}\frac{0.03\ \text{cm}}{69.04556\times 10^{-6}\ \text{mol/cm}^3\times 0.03\ \text{cm}^2/\text{s}}$$
$$= 0.21 - 0.01876$$
$$x_{O_2,ce} = 0.19124.$$

Activation overpotential based on cathode electrode:

$$\eta_{act,c} = \frac{RT}{\alpha_c n_c F}\ln\left(\frac{j}{j_{o,c}}\frac{P_c^0}{P_c x_{O_2,ce}}\right)$$
$$= \frac{8.3145\ \text{J/mol K}\times 1073\ \text{K}}{0.5\times 4\times 96485\ \text{C/mol}}\ln\left(\frac{0.5\ \text{A/cm}^2}{0.0001\ \text{A/cm}^2}\frac{1\ \text{atm}}{2\ \text{atm}\times 0.1912}\right)$$
$$\eta_{act,c} = \frac{8.3145\ \text{J/mol K}\times 353\ \text{K}}{0.5\times 4\times 96,485\ \text{C/mol}}\ln(13,075.31)$$
$$\eta_{act,c} = 0.1441\ \text{V}.$$

Ohmic loss in polymer membrane:

Assuming that the ohmic loss is primarily given by the ionic conductivity of the electrolyte membrane, the ohmic loss is given by

$$\eta_{\text{ohm}} = j\ \text{ASR} = j\frac{L_e}{\sigma_i}$$

For polymer membrane, the ionic conductivity is given by Equation 9.4 as

$$\sigma(T,\lambda) = \sigma_{303\text{K}}(\lambda)\exp\left[1268\left(\frac{1}{303}-\frac{1}{T}\right)\right],$$

where

$$\sigma_{303\text{K}}(\lambda) = 0.005139\ \lambda - 0.00326.$$

Since ionic conductivity varies with water content (λ), which also varies locally throughout the thickness of the membrane, we can employ two approaches in determining the average conductivity of the membrane:

Approach 1

Estimate average conductivity based on the average water content in the membrane as

$$\sigma(T,\bar{\lambda}) = \sigma_{303\text{K}}(\bar{\lambda})\exp\left[1268\left(\frac{1}{303}-\frac{1}{T}\right)\right].$$

The average water content in the membrane is given as

$$\bar{\lambda} = \frac{1}{L_e}\int_0^{L_e}\lambda\, dx,$$

where $\lambda = \lambda(z)$ represents local variation water content within the membrane.

Water content in the Nafion-117 membrane is given by Equation 9.13 as

$$\lambda = 0.043 + 17.81a - 39.85a^2 + 36.0a^3,\ 0 < a \le 1$$

$$\lambda = 14 + 1.4(a-1),\ 1 \le a \le 3.$$

The humidity condition at the gas diffusion catalyst layer interface is given as the water activity or relative humidity (a) and defined as

$$a = \frac{P_{H_2O}}{P_{\text{sat}}} = \frac{x_{H_2O}P}{P_{\text{sat}}}, \tag{9.14}$$

where x_{H_2O} is the mole fraction of water, P is the total pressure, and P_{sat} is the water saturation pressure.

Approach 2

Estimate the membrane ionic resistance based on

$$R_e = \int_0^{L_e} \frac{dz}{\sigma(z)},$$

where $\sigma = \sigma(z)$ represents the local variation of ionic conductivity within the membrane.

Diffusion coefficient of water in Nafion-117 is given by Equation 9.16 as

$$D_{H_2O} = D_\lambda \exp\left[2416\left(\frac{1}{303} - \frac{1}{T}\right)\right] \ (\text{cm}^2/\text{s}),$$

where

$$D_\lambda = (2.563 - 0.33\lambda + 0.0264\lambda^2 - 0.000671\lambda^3) \times 10^{-6} \text{ for } \lambda > 4.$$

Water concentration at the anode–membrane and cathode–membrane interfaces is given as

Anode–membrane interface:

$$x_{H_2O,ae} = x_{H_2O,ch-a} - \beta_o \frac{j}{2F}\left(\frac{L_a}{\rho_a D^{eff}_{H_2O,H_2}}\right)$$

$$x_{H_2O,ae} = x_{H_2O,ch-a} - \beta_o \frac{J}{2F}\frac{L_a RT}{P_a D^{eff}_{H_2O,H_2}}$$

Cathode–membrane interface:

$$x_{H_2O,ce} = x_{H_2O,c} + (1-\beta_o)\frac{J}{2F}\left(\frac{L_c}{\rho_c D^{eff}_{H_2O,O_2}}\right)$$

$$x_{H_2O,ce} = x_{H_2O,ch} + (1-\beta_o)\frac{j}{2F}\frac{L_c RT}{P_c D^{eff}_{H_2O,O_2}}$$

At an operating temperature of 80°C, the corresponding water saturation pressure is $P_{sat} = 47.39$ kPa. Based on this water vapor pressure, the water activity ratios at the anode–membrane and cathode–membrane interfaces are computed as

$$a_{am} = x_{H_2O,ae}\frac{P_a}{P_{sat}} = 0.1 \times \frac{2 \times 101.3}{47.39} = 0.4275.$$

And assuming that water is in liquid form at the cathode–membrane interface,

$$a_{cm} = x_{H_2O,ce}\frac{P_c}{P_{sat}} = 0.25\times\frac{2\times 101.3}{47.39} = 1.068.$$

Water content at the interface is computed using Equation 9.14a for the anode–membrane interface as

$$\begin{aligned}\lambda_{am} &= 0.043 + 17.81a_{am} - 39.85a_{am}^2 + 36.0a_{am}^3\\ &= 0.043 + 17.81\times 0.4275 - 39.85\times 0.4275^2 + 36.0\times 0.4275^3\\ &= 0.043 + 7.613775 - 7.28283 + 2.81261\\ \lambda_{am}(0) &= 3.1866\end{aligned}$$

and using Equation 9.14b as

$$\lambda_{cm}(L_e) = 14 + 1.4(1.068 - 1) = 14.095.$$

Based on the known values of $\lambda_{cm}(0) = 3.1866$ and $\lambda_{cm}(L_e) = 14.095$, the variation of water distribution in the Nafion membrane can be determined from Equation 11.57 by determining the two unknowns β_o and C_2 as follows:

$$\lambda_m = C_2\exp\left(\frac{j}{2F}\frac{M_m}{\rho_{dry}D_{H_2O}}\frac{n_{drag}^{sat}}{11}z\right) + \frac{11\beta_o}{n_{drag}^{sat}} \tag{11.57}$$

$$\lambda_m = C_2\exp\left(\frac{j}{2\times 96{,}485}\frac{1.0\ \text{kg/mol}}{0.00197\ \text{kg/cm}^3 D_{H_2O}}\frac{2.5}{11}z\right) + \frac{11\beta_o}{2.5}$$

$$\lambda_m = 4.4\beta_o + C_2\exp\left(5.978\times 10^{-4}\frac{j}{D_{H_2O}}z\right) \tag{E11.3-1}$$

at $z = 0$, $\lambda_m(0) = C_2 + 4.4\beta_o$

$$C_2 + 4.4\beta_o = 3.1866 \tag{E11.3-2}$$

and at $z = L_e$,

$$\lambda_m(L_e) = 4.4\beta_o + C_2\exp\left(5.978\times 10^{-4}\frac{j}{D_{H_2O}}L_e\right)$$

or

$$4.4\beta_o + C_2\exp\left(5.978\times 10^{-4}\frac{j}{D_{H_2O}}L_e\right) = 14.095. \tag{E11.3-3}$$

For $j = 0.5$ A/cm², $D^{\text{eff}}_{\text{H}_2\text{O}} = 3.8\times10^{-6}\ \text{cm}^2/\text{s}$ and $L_e = 0.01$ cm, Equation E11.3-3 reduces to

$$4.4\beta_o + C_2\exp\left(5.978\times10^{-4}\frac{0.5\times0.01}{3.86\times10^{-6}}\right) = 14.095$$

$$4.4\beta_o + 2.169\ C_2 = 14.095 \tag{E11.3-4}$$

Solving Equations E11.3-2 and E11.3-4, we get

$$\beta_o = -1.3965 \text{ and } C_2 = 9.3314.$$

Variation of water content in the membrane is given by Equation E11.3-1 as

$$\lambda_m = -6.1446 + 9.3314\ \exp(154.8705jz). \tag{E11.3-5}$$

For $j = 0.5$ A/cm², the equation reduces to

$$\lambda_m = -6.1446 + 9.3314\ \exp(77.435z). \tag{E11.3-6}$$

Substituting, the average water content in the membrane is given as

$$\bar{\lambda}_m = \frac{1}{L_e}\int_0^{L_e}\lambda_m\,dx$$

$$\bar{\lambda}_m = \frac{1}{L_e}\int_0^{L_e}(-6.1446 + 9.3314\ e^{77.435z})\ dz$$

$$\bar{\lambda}_m = -6.1446 + \frac{722.577}{L_e}(e^{77.435L_e} - 1).$$

For a membrane thickness of $L_e = 0.01$ cm,

$$\bar{\lambda}_m = -6.1446 + \frac{722.577}{0.01}(e^{77.435\times0.01} - 1)$$

$$\bar{\lambda}_m = 7.944.$$

The membrane ionic conductivity can now be computed on the basis of the average water content based on Equation 9.17:

$$\sigma_i = \sigma(T, \bar{\lambda}) = (0.005139\ \bar{\lambda}_{\mathrm{m}} - 0.00326)\ \exp\left[1268\left(\frac{1}{303} - \frac{1}{353}\right)\right]$$

$$\sigma_i = (0.005139 \times 7.944 - 0.00326)\ \exp\left[1268\left(\frac{1}{303} - \frac{1}{353}\right)\right]$$

$$\sigma_i = 0.06447$$

The ohmic loss is given by

$$\eta_{\mathrm{ohm}} = j\frac{L_{\mathrm{e}}}{\sigma_i} = 5 \times \frac{0.1}{0.06447}$$

$$\eta_{\mathrm{ohm}} = 0.0776\ \mathrm{V}.$$

Mass concentration loss is given by Equation 11.21:

$$\eta_{\mathrm{conc}} = \left(1 + \frac{1}{\alpha}\right)\left(\frac{RT}{nF}\right)\ln\frac{j_{\mathrm{L}}}{j_{\mathrm{L}} - j}. \tag{11.21}$$

For dominant mass concentration loss in the cathode electrode, the limiting current density is given by Equation 11.24b

$$j_{\mathrm{L}} = n_{\mathrm{e}} F D_{\mathrm{O_2}}^{\mathrm{eff}}\left[\frac{\rho_{\mathrm{c}} x_{\mathrm{O_{2,ch}}}}{L_{\mathrm{c}}}\right]$$

$$j_{\mathrm{L}} = 4 \times 96{,}485\ \mathrm{C/mol} \times 0.03\ \mathrm{cm^2/s}\left[\frac{69.04556 \times 10^{-6}\ \mathrm{mol/cm^3} \times 0.21}{300 \times 10^{-4}\,\mathrm{cm}}\right]$$

$$j_{\mathrm{L}} = 5.596\ \mathrm{A/cm^2}$$

Mass concentration loss is computed as

$$\eta_{\mathrm{conc}} = \left(1 + \frac{1}{0.5}\right)\left(\frac{8.314 \times 1073.2}{n_{\mathrm{e}} F}\right)\ln\frac{j_{\mathrm{L}}}{j_{\mathrm{L}} - j}$$

$$\eta_{\mathrm{conc}} = \left(1 + \frac{1}{0.5}\right)\left(\frac{8.314\ \mathrm{J/mol \cdot K} \times 353.2\ \mathrm{K}}{4 \times 96{,}485}\right)\ln\left(\frac{5.596}{5.596 - 0.5}\right)$$

$$\eta_{\mathrm{conc}} = (3.0)(0.0076)\ln(1.098)$$

$$\eta_{\mathrm{conc}} = 0.00213\ \mathrm{V}.$$

Cell operating voltage is given as

$$
\begin{aligned}
E &= E_0 - \eta_{\text{act}} - \eta_{\text{ohm}} - \eta_{\text{conc}} \\
&= 1.152\ \text{V} - 0.1441 - 0.0776 - 0.00213 \\
E &= 0.931\ \text{V}.
\end{aligned}
$$

11.5 One-Dimensional Fuel Cell Thermal Analysis Model

During the operation of the fuel cell, reversible and irreversible components of heat are generated within the cell as discussed in Chapters 4 and 6. While the reversible part of the heat generation is due to entropy change of the electrochemical reaction, the irreversible part of the heat generation is due to the irreversibilities associated with the electrochemical reaction and activation loss at the electrode–electrolyte interfaces and heat generated owing to the transport of ions and electrons through the electrolyte and electrodes, respectively, resulting in an ohmic volume heating of the cell. While heat generation owing to reaction kinetics can often be assumed as surface heat generation, the ohmic heating is primarily a volumetric heat generation. Heat generated is dissipated through the cell primarily by conduction through the three layers of the cell and by convection to moving gas streams through the gas channels. Figure 11.4 shows heat transfer and a typical temperature distribution across the cell.

One-dimensional differential heat model for the fuel cell is given as follows:

Anode gas channel:

Heat transfer from the anode surface to the anode gas stream:

$$q_{\text{a}} = h_{\text{e,a}} A (T_{\text{a,a}} - T_{\text{ch,a}}) \tag{11.58}$$

Anode electrode:

Heat transport through electrode involves heat conduction and volumetric heat generation due to ohmic heating caused by electron transport

$$\frac{d^2 T_{\text{a}}}{dz^2} + \dot{Q}_{\text{a}} = 0, \tag{11.59a}$$

where

$$\dot{Q}_{\text{a}} = I^2 R_{\text{a}} \quad \text{and} \quad R_{\text{a}} = \frac{L_{\text{a}}}{\sigma_{\text{a,e}} A}. \tag{11.59b}$$

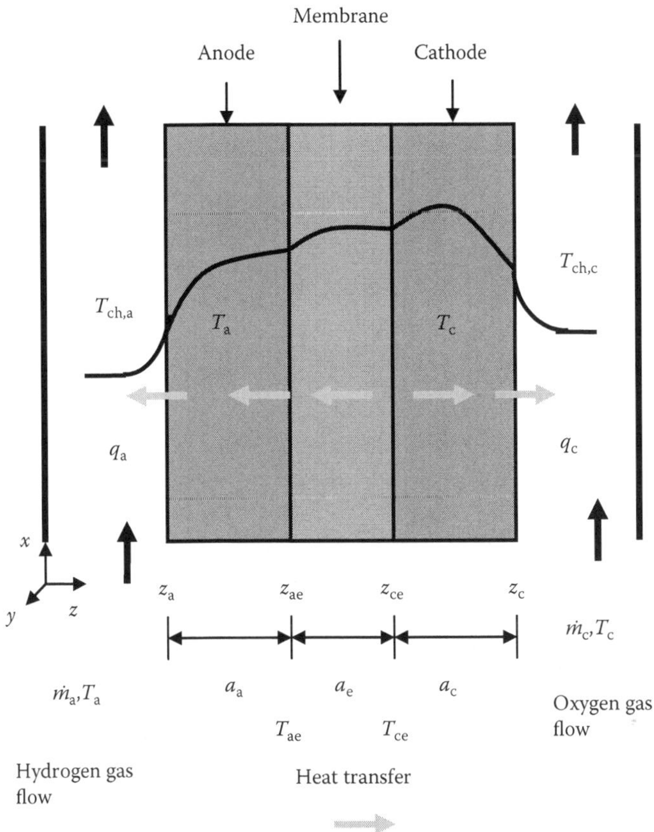

FIGURE 11.4
One-dimensional heat transfer and temperature distribution in a one-dimensional fuel cell model.

Electrolyte membrane:

Heat transport in the electrolyte membrane involves heat conduction and volumetric heat generation due to ohmic heating caused by ion transport

$$\frac{d^2T_e}{dz^2} + \dot{Q}_e = 0, \tag{11.60a}$$

where

$$\dot{Q}_e = I^2 R_e \quad \text{and} \quad R_e = \frac{L_e}{\sigma_i A}. \tag{11.60b}$$

Cathode electrode:

Heat transport in the cathode electrode involves heat conduction and volumetric heat generation due to ohmic heating caused by electron transport

$$\frac{d^2T_c}{dz^2} + \dot{Q}_c = 0, \tag{11.61}$$

where

$$\dot{Q}_c = I^2R_c \quad \text{and} \quad R_c = \frac{L_c}{\sigma_{c,e}A}. \tag{11.62}$$

Cathode gas channel:

Heat transfer from the cathode electrode surface to the cathode gas stream convection heat transfer:

$$q_c = h_{c,c}A(T_{c,c} - T_{ch,c}). \tag{11.63}$$

Boundary conditions:

Boundary conditions at the electrode–electrolyte membrane interfaces are derived on the basis of the energy balance of the sum of heat losses through anode and cathode sides with the surface heat generation owing to electrochemical reactions and activation losses as follows:

$$\text{At } z = z_{ae},\ k_a\frac{dT_a}{dz} - k_e\frac{dT_e}{dz} = \frac{Q_{gen,a}}{A} \tag{11.64a}$$

$$\text{At } z = z_{ce},\ k_e\frac{dT_e}{dz} - k_c\frac{dT_c}{dz} = \frac{Q_{gen,c}}{A}, \tag{11.64b}$$

where $Q_{gen,a}$ and $Q_{gen,c}$ are the surface heat generation owing to electrochemical reactions and activation loss at the anode electrode–electrolyte interface and cathode electrode–electrolyte interface, respectively, and are expressed as follows:

$$Q_{gen,a} = \left[\frac{T(-\Delta S)_a}{n_e F} + \eta_{act,a}\right]i_a \tag{11.65a}$$

and

$$Q_{gen,c} = \left[\frac{T(-\Delta S)_c}{n_e F} + \eta_{act,c}\right]i_c. \tag{11.65b}$$

Also, in deriving the interface boundary conditions, it is assumed that the catalyst layer or the active layer thicknesses are negligible.

Boundary conditions at the electrode and gas channel interfaces can be given by the convective conditions as described below:

$$\text{At } z = z_a,\ k_a \frac{dT_a}{dz} = h_{c,a}(T_{a,a} - T_{ch,a}) \quad (11.66a)$$

$$\text{At } z = z_c,\ -k_c \frac{dT_c}{dz} = h_{c,c}(T_{c,c} - T_{ch,c}), \quad (11.66b)$$

where $h_{c,a}$ and $h_{c,c}$ are the convection heat transfer coefficients in the anode and cathode gas channels, respectively.

11.5.1 A Simplified One-Dimensional Heat Transfer Model

A simplified one-dimensional model can be derived on the basis of neglecting the volumetric ohmic heat generation and considering only the reversible and irreversible surface heat generation caused by the electrochemical reaction and activation loss at the cathode–electrolyte interface. Assuming a surface heat generation, Q_{gen}, at the cathode–electrolyte interface, $z = z_{ce}$, and assuming that q_a amount of heat dissipates through the electrolyte and anode electrode toward anode gas stream and q_c amount of heat dissipates through the cathode electrode toward the cathode gas stream, an expected temperature distribution with in the cell and channel is demonstrated in Figure 11.5.

The figure also shows the equivalent thermal circuit based on conduction and convection resistances, which are given as follows:

$$\text{Convection resistance in anode gas channel, } R_{conv,a} = \frac{1}{h_{c,a}A} \quad (11.67a)$$

$$\text{Anode conduction resistance, } R_a = \frac{L_a}{k_a A} \quad (11.67b)$$

$$\text{Electrolyte membrane conduction resistance, } R_e = \frac{L_e}{k_e A} \quad (11.67c)$$

$$\text{Cathode conduction resistance, } R_c = \frac{L_c}{k_c A} \quad (11.67d)$$

$$\text{Convection resistance in cathode gas channel, } R_{conv,c} = \frac{1}{h_{c,c}A} \quad (11.67e)$$

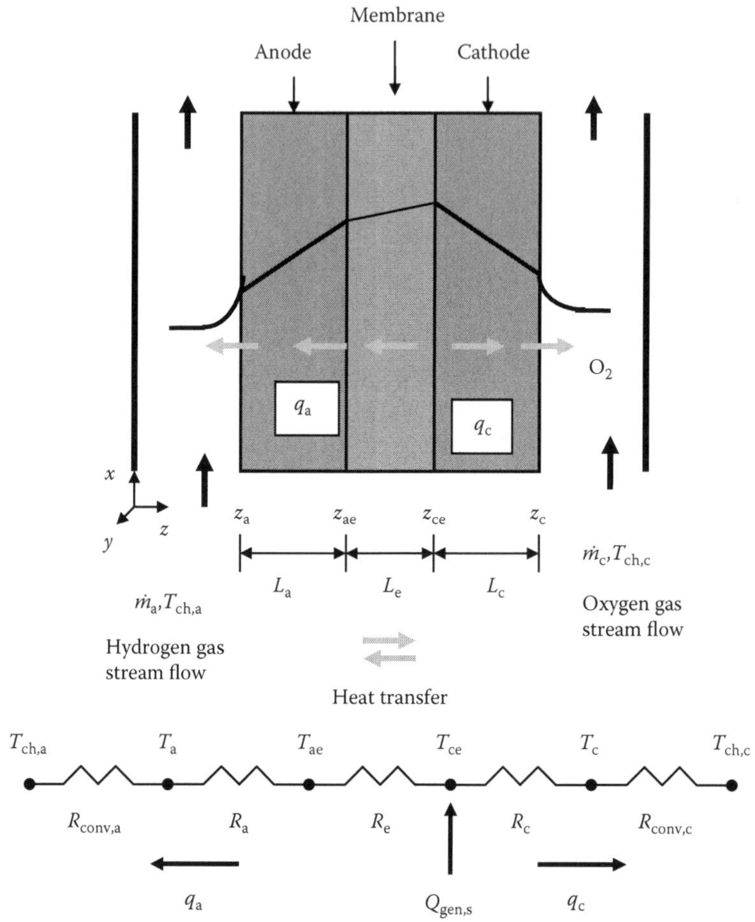

FIGURE 11.5
A simplified one-dimensional heat transfer model and resistance diagram.

Based on the one-dimensional steady-state conduction analysis, we can express the heat transfer rates q_a and q_c as follows:

$$q_a = \frac{T_{ce} - T_{ch,a}}{R_{conv,a} + R_a + R_{ce}} = \frac{T_{ce} - T_{ch,a}}{\dfrac{1}{h_{c,a}A} + \dfrac{L_a}{k_a A} + \dfrac{L_e}{k_e A}} \tag{11.68}$$

and

$$q_c = \frac{T_{ce} - T_{ch,c}}{R_{conv,c} + R_c} = \frac{T_{ce} - T_{ch,c}}{\dfrac{1}{h_{c,a}A} + \dfrac{L_c}{k_a A}}. \tag{11.69}$$

Additionally, we have the overall energy balance equation given by

$$q_a + q_c = Q_{gen}. \tag{11.70}$$

As we can see, the set of three equations (Equations 11.68 through 11.70) involve three unknowns, q_a, q_c, and T_{ce}, and can be determined by solving the three equations simultaneously for known thicknesses and thermal conductivity values of the electrodes and electrolyte membrane, and convective heat transfer coefficients in the gas flow channels.

Substituting Equations 11.68 and 11.69 into Equation 11.70,

$$\frac{T_{ce} - T_{ch,a}}{\frac{1}{h_{c,a}A} + \frac{L_a}{k_a A} + \frac{L_e}{k_e A}} + \frac{T_{ce} - T_{ch,c}}{\frac{1}{h_{c,a}A} + \frac{L_a}{k_a A}} = Q_{gen} \tag{11.71a}$$

or

$$\frac{T_{ce} - T_{ch,a}}{\sum R_a} + \frac{T_{ce} - T_{ch,c}}{\sum R_c} = Q_{gen}, \tag{11.71b}$$

where

$$\sum R_a = \frac{1}{h_{c,a}A} + \frac{L_a}{k_a A} + \frac{L_e}{k_e A} \tag{11.72a}$$

and

$$\sum R_c = \frac{1}{h_{c,a}A} + \frac{L_a}{k_a A} \tag{11.72b}$$

Solving for the cathode–electrolyte interface temperature, T_{ce}, as

$$T_{ce} = \frac{Q_{gen} + \frac{T_{ch,a}}{\sum R_a} + \frac{T_{ch,c}}{\sum R_c}}{\left(\frac{1}{\sum R_a} + \frac{1}{\sum R_c}\right)}. \tag{11.73}$$

With the known interface temperature, T_{ce}, given by Equation 11.73, heat transfer rates are given as

$$q''_a = \frac{q_a}{A} = \frac{T_{ce} - T_{ch,a}}{\sum R_a} \tag{11.74a}$$

and

$$q_c'' = \frac{q_c}{A} = \frac{T_{ce} - T_{ch,c}}{\sum R_c}. \tag{11.74b}$$

Example 11.5

Consider a fuel cell with three layers: *anode electrode, electrolyte membrane,* and *cathode electrode* as shown in Figure 11.6. The tri-layer has a surface area of 10 cm × 10 cm. Consider that the only component of heat generation owing to irreversible electrochemical reaction *at the cathode–membrane interface* provides a heat flux $q''_{gen} = \frac{Q_{gen}}{A} = 3.5 \times 10^5\ \text{W/m}^2$ at the interface area. The flow condition of the hydrogen and air streams is given by stream temperature T_∞ and a convection coefficient, h.

Assume the following conditions: $T_{ch,a} = 300°\text{C}$, $h_a = 500\ \text{W/m}^2 \times \text{K}$ and $T_{ch,c} = 400°\text{C}$, $h_c = 600\ \text{W/m}^2 \times \text{K}$.

The thermal conductivities of the three layers are as follows: anode: $k_a = 90.5\ \text{W/m} \times \text{K}$; electrolyte membrane: $k_e = 1.62\ \text{W/m} \times \text{K}$; cathode, $k_c = 1.95\ \text{W/m} \times \text{K}$.

a. Calculate the heat transferred to hydrogen and oxygen gas streams.
b. What are the temperatures at the anode–membrane and cathode–membrane interfaces?

Solution

Area of the cell is

$$A_{cell} = 0.01 \times 0.01 = 1.0 \times 10^{-2}\ \text{m}^2$$

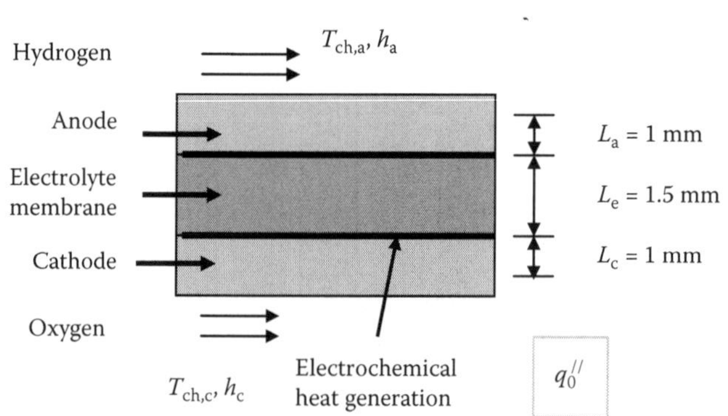

FIGURE 11.6
Three-layer fuel cell thermal model and flow conditions.

Heat generation,

$$Q_{\text{gen}} = q''_{\text{gen}} \times A_{\text{cell}} = 3.5 \times 10^5 \text{ W/m}^2 \times 1.0 \times 10^{-2} \text{ m}^2 = 3.5 \times 10^3 \text{ W}$$

Thermal resistances are computed as follows:

$$R_{\text{conv,a}} = \frac{1}{h_{\text{a}} \times A_{\text{cell}}} = \frac{1}{500 \times 1.0 \times 10^{-2}} = 0.2 \text{ K/W}$$

$$R_{\text{a}} = \frac{L_{\text{a}}}{k_{\text{a}} \times A_{\text{cell}}} = \frac{0.001}{90.5 \times 1.0 \times 10^{-2}} = 0.001105 \text{ K/W}$$

$$R_{\text{e}} = \frac{L_{\text{e}}}{k_{\text{e}} \times A_{\text{cell}}} = \frac{0.0015}{1.62 \times 1.0 \times 10^{-2}} = 0.09259 \text{ K/W}$$

$$R_{\text{c}} = \frac{L_{\text{c}}}{k_{\text{c}} \times A_{\text{cell}}} = \frac{0.001}{1.95 \times 1.0 \times 10^{-2}} = 0.05128 \text{ K/W}$$

$$R_{\text{conv,c}} = \frac{1}{h_{\text{c}} \times A_{\text{cell}}} = \frac{1}{600 \times 1.0 \times 10^{-2}} = 0.16666$$

The sum of the resistances on the anode side is given as

$$\sum R_{\text{a}} = R_{\text{conv,a}} + R_{\text{a}} + R_{\text{e}} = 0.2 + 0.001105 + 0.09259 = 0.2937 \frac{^\circ\text{C}}{\text{W}}.$$

The sum of the resistances on the cathode side is given as

$$\sum R_{\text{c}} = R_{\text{c}} + R_{\text{conv,c}} = 0.05128 + 0.16666 = 0.21794 \frac{^\circ\text{C}}{\text{W}}.$$

Based on the resistance diagram shown in Figure 11.5, the heat transfers in the anode and cathode sides are given as

$$q_{\text{a}} = \frac{T_{\text{ce}} - T_{\text{ch,a}}}{\dfrac{1}{h_{\text{a}} A_{\text{c}}} + \dfrac{L_{\text{a}}}{k_{\text{a}} A_{\text{c}}} + \dfrac{L_{\text{e}}}{k_{\text{e}} A_{\text{c}}}}$$

and

$$q_{\text{c}} = \frac{T_{\text{ce}} - T_{\text{ch,c}}}{\dfrac{1}{h_{\text{c}} A_{\text{c}}} + \dfrac{L_{\text{c}}}{k_{\text{c}} A_{\text{c}}}}.$$

From the over energy balance equation, we get

$$\frac{T_{ce}-T_{ch,a}}{\sum R_a}+\frac{T_{ce}-T_{ch,c}}{\sum R_c}=Q_{gen}$$

$$\frac{T_{ce}-300}{0.2937}+\frac{T_{ce}-400}{0.21784}=3.5\times10^{5}\times1.0\times10^{-2}.$$

Solving for the cathode–electrolyte interface temperatures:

$$T_{ce}=\frac{q_{gen}+\dfrac{T_{ch,a}}{\sum R_a}+\dfrac{T_{ch,c}}{\sum R_c}}{\left(\dfrac{1}{\sum R_a}+\dfrac{1}{\sum R_c}\right)}$$

$$T_{ce}=\frac{3.5\times10^{3}+\dfrac{300}{0.2937}+\dfrac{400}{0.2178}}{\left(\dfrac{1}{0.2937}+\dfrac{1}{0.2178}\right)}$$

$$T_{ce}=795.1687°\text{C}.$$

Heat transfer to the anode and cathode side is computed as follows:

$$q_a=\frac{T_{ce}-T_{ch,a}}{\dfrac{1}{h_aA_c}+\dfrac{L_a}{k_aA_c}+\dfrac{L_e}{k_eA_c}}=\frac{795.1687-300}{0.2937}=1685.9676\text{ W}$$

and

$$q_c=\frac{T_{ce}-T_{ch,c}}{\dfrac{1}{h_cA_c}+\dfrac{L_c}{k_cA_c}}=\frac{795.1687-400}{0.21784}=1814.0318\text{ W}.$$

Temperature at the anode–electrolyte is computed as

$$q_a=\frac{T_{ae}-T_{ch,a}}{\dfrac{1}{h_aA_c}+\dfrac{L_a}{k_aA_c}}$$

$$1685.9676=\frac{T_{ae}-300}{0.2+0.001105}$$

$$T_{ae}=639.056°\text{C}.$$

Temperature at the anode channel surface is given as

$$q_a = \frac{T_a - T_{ch,a}}{R_{conv,a}}$$

$$1685.9606 = \frac{T_a - 300}{0.2}$$

$$T_a = 637.1921°C.$$

Similarly, the temperature at the cathode–channel interface is given as

$$q_c = \frac{T_c - T_{ch,c}}{R_{conv,c}} = \frac{T_c - 400}{0.1666} = 1814.0318 \text{ W}$$

$$T_c = 702.22°C.$$

11.6 Multi-Dimensional Model

Figure 11.7 shows a three-dimensional geometry of a fuel cell model consisting of three-layer MEA and two half-sections of bipolar plates on either side.

In this diagram, *z*-coordinate represents the space variation across the cell from the anode side to the cathode side along the thickness of the cell

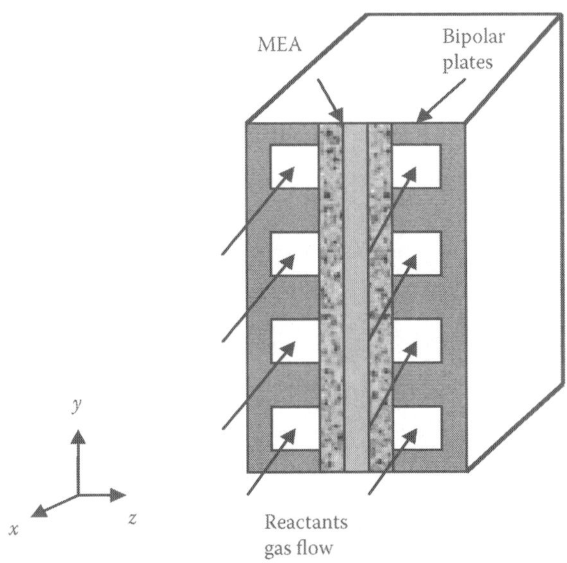

FIGURE 11.7
Three-dimensional geometry of fuel cell.

and bipolar plates. The y-coordinate represents the space variation along the height of the cell and bipolar plates. The x-direction represents the space variation along the length of the cell and along the length of the bipolar plate or along the flow direction of the reactant gas channels.

11.6.1 Two-Dimensional Model

In order to reduce the computational time and expense, simulation analysis is often performed using a two-dimensional model. In the one-dimensional fuel cell model, we are only concerned with the variation along the thickness of the cell or z-direction for reactant species concentration, water mass distribution, and temperature distribution. In a two-dimensional model, variations of these quantities in either y- and z-directions or x- and z-directions are considered as demonstrated in Figure 11.8.

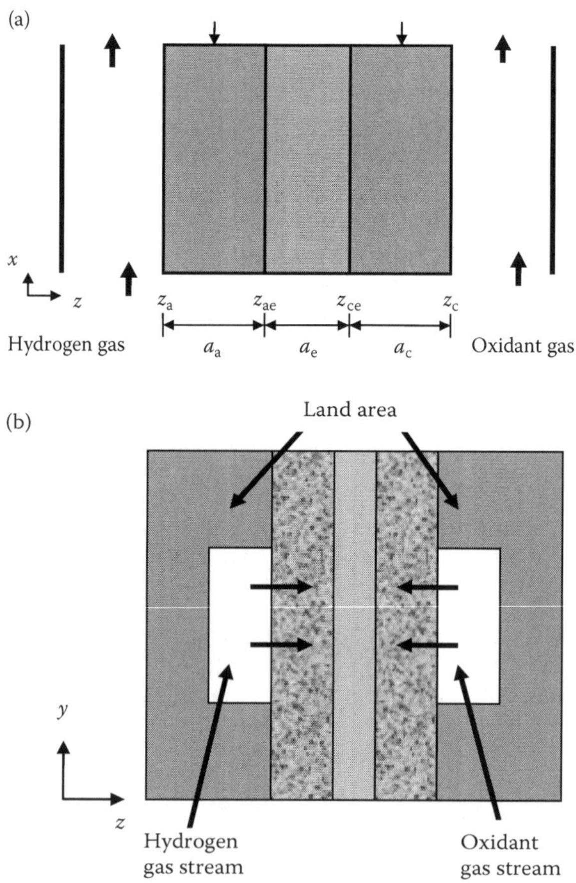

FIGURE 11.8
Two-dimensional model: (a) x–z plane and (b) y–z plane.

In the two-dimensional model shown in Figure 11.8a, variations are only taken into account across the cell and along the length of the bulk gas flow through the gas channels. In such a two-dimensional model, development of the flow, depletion of the reactant gas, and pressure drop along the channel are taken into account. Hence, the variable entrance region convective heat and mass transfer conditions along the channel are used as the variable channel–electrode interface conditions. However, one of the deficiencies of such a model is that the variation of gas concentrations in the diffusion layers adjacent to the land area of the bipolar plate or regions in between two gas channels is not taken into account, whereas in the two-dimensional model in the y–z plane, shown in Figure 11.8b, variation in the regions of gas diffusion layer adjacent to the land area of the bipolar plate is taken into account. Variations in gas and water concentration in such regions have a significant effect on the electro-chemical reactions, local current densities, and performance of the fuel cell. This two-dimensional model includes two-symmetric mid-sections of bipolar plates on both sides and two horizontal symmetric lines.

11.6.2 Three-Dimensional Model

One- and two-dimensional models are simple, computationally less expensive, and can provide qualitative analysis of the fuel cell performance. However, a more accurate quantitative analysis and design of a fuel cell can only be achieved with a three-dimensional model, which includes flow, mass, charge, and heat transports in all three directions including the diffusion and active reaction regions adjacent to the land area of the bipolar plates. A complete three-dimensional simulation analysis model of a fuel cell unit cell composed of two half-bipolar plates with multiple flow channels is desirable taking into account the variation in flow, heat, and mass transfer characteristics from channel to channel (Boddu and Majumdar, 2006). However, a simplified three-dimensional model is often used to reduce computational time by assuming all channels as identical in terms of flow, heat, and mass transfer characteristics. Figure 11.9 shows a symmetric three-dimensional model of the fuel cell that includes symmetric mid-sections of bipolar plates on both sides and two horizontal symmetric lines that runs through the mid-section of the land area between two gas flow channels.

Another important aspect and requirement of a fuel cell model is the two-phase nature of water transport in a PEM fuel cell, especially for operation at a higher current density. Water tends to accumulate in a liquid state at the cathode side owing to higher rate of water generation at the cathode–membrane interface and a higher influx of water transport through the membrane from the anode side to the cathode side. Water may also appear as liquid state at the anode side if the dew point temperature of the anode gas stream is high because of higher humidification level. As we have discussed in earlier sections and chapters, high-performance gas channel designs play

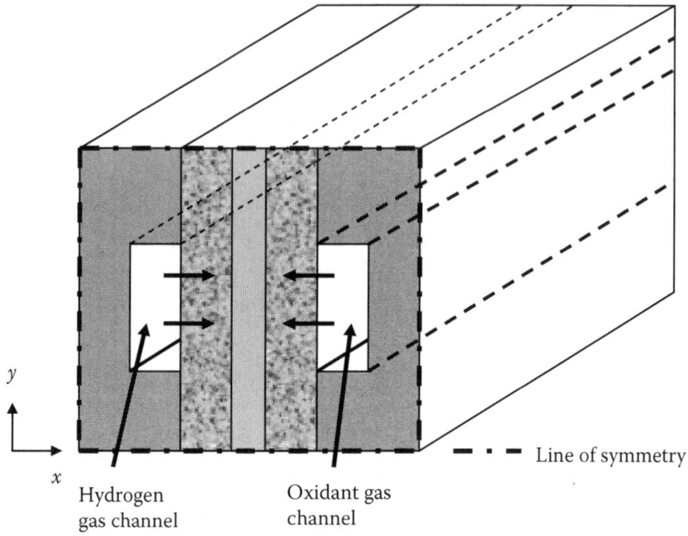

FIGURE 11.9
Three-dimensional model with symmetric lines.

a significant role in removing excess water collected in regions of gas diffusion layer and electrode–membrane interface, particularly from the regions adjacent to land area of the bipolar plate, and in preventing flood and reducing mass concentration loss. In order to analyze such a water management issue within the PEM fuel cell, the simulation requires a two-phase flow and transport model (Hu et al., 2004a, b).

The three-dimensional simulation model for the fuel cell based on steady-state, single-phase and incompressible flow analysis is presented here. The other basic assumptions that have been made in formulating the model are as follows: (1) gas flows in the channels are assumed to be incompressible, (2) water generation takes place only at the anode membrane interface, (3) there is no water generation and water transport in the electrolyte, (4) water exists only in the gas phase in the fuel cell, and (5) humidified hydrogen and air are assumed to be ideal gases.

The governing set of equations for fluid flow, heat, and mass transport through the gas channels, electrode, and electrolyte membrane of unit fuel cell is given as follows:

11.6.2.1 Gas Channel

11.6.2.1.1 Fluid Flow Model

Gas flow in anode and cathode gas channels is assumed to be incompressible fluid flow with constant fluid viscosity (μ) and is governed by the Navier–Stokes equation given as follows:

Mass Continuity:

$$\nabla \cdot \vec{V} \tag{11.75a}$$

or

$$\frac{\partial u}{\partial x}+\frac{\partial v}{\partial y}+\frac{\partial w}{\partial z}=0 \tag{11.75b}$$

Momentum:

$$\rho(\vec{V}\cdot\nabla)\vec{V}=\rho\vec{g}-\nabla P+\mu\nabla^2\vec{V} \tag{11.76a}$$

and in scalar form as

$$x\text{-component: } \rho\left(u\frac{\partial u}{\partial x}+v\frac{\partial u}{\partial y}+w\frac{\partial u}{\partial z}\right)=\rho g_x-\frac{\partial P}{\partial x}+\mu\left(\frac{\partial^2 u}{\partial x^2}+\frac{\partial^2 u}{\partial y^2}+\frac{\partial^2 u}{\partial z^2}\right) \tag{11.76b}$$

$$y\text{-component: } \rho\left(u\frac{\partial v}{\partial x}+v\frac{\partial v}{\partial y}+w\frac{\partial v}{\partial z}\right)=\rho g_y-\frac{\partial P}{\partial y}+\mu\left(\frac{\partial^2 v}{\partial x^2}+\frac{\partial^2 v}{\partial y^2}+\frac{\partial^2 v}{\partial z^2}\right) \tag{11.76c}$$

$$z\text{-component: } \rho\left(u\frac{\partial w}{\partial x}+v\frac{\partial w}{\partial y}+w\frac{\partial w}{\partial z}\right)=\rho g_x-\frac{\partial P}{\partial z}+\mu\left(\frac{\partial^2 w}{\partial x^2}+\frac{\partial^2 w}{\partial y^2}+\frac{\partial^2 w}{\partial z^2}\right) \tag{11.76d}$$

11.6.2.1.2 Mass Transport

Mass concentration distribution in a gas channel is given considering diffusion and convection mass transfers given as

$$\rho\vec{V}\cdot(\nabla C_i)=\nabla\cdot(D_{\text{eff}}\nabla C_i) \tag{11.77a}$$

or

$$\rho\left(u\frac{\partial C_i}{\partial x}+v\frac{\partial C_i}{\partial y}+w\frac{\partial C}{\partial z}\right)=D_{ij}\left(\frac{\partial^2 C_i}{\partial x^2}+\frac{\partial^2 C_i}{\partial y^2}+\frac{\partial^2 C_i}{\partial z^2}\right). \tag{11.77b}$$

11.6.2.1.3 Heat Transport Equation

The thermal heat equation for the gas flow stream is given considering conduction and convection, and expressed as

$$\rho C_p\vec{V}\cdot(\nabla T_i)=\nabla\cdot(k\nabla T) \tag{11.78a}$$

or

$$\rho C_p\left(u\frac{\partial T_i}{\partial x}+v\frac{\partial T_i}{\partial y}+w\frac{\partial T_i}{\partial z}\right)=\frac{\partial}{\partial x}\left(k_i\frac{\partial T_i}{\partial T_x}\right)+\frac{\partial}{\partial y}\left(k_i\frac{\partial T_i}{\partial y}\right)+\frac{\partial}{\partial z}\left(k_i\frac{\partial T_i}{\partial z}\right). \quad (11.78b)$$

11.6.2.2 Flow in Porous Electrodes

The mass continuity and momentum equations for the bulk fluid flow in a channel are significantly altered in a porous media owing to the presence of complex flow geometries. A detailed description of fluid in a gas diffusion layer electrode is given in Chapter 6.

Most general fluid flow equations in porous media are given by the modified Navier–Stokes equation given by Equation 6.22 as

$$\rho(\vec{V}\cdot\nabla)\vec{V}=\rho\vec{g}-\nabla P+\mu_e\nabla^2\vec{V}-\frac{\mu}{\kappa}\vec{V}, \quad (11.79)$$

where μ_e = effective viscosity to be used for the fluid in the porous media, κ = permeability of the porous media, and μ = dynamic viscosity of the fluid.

For negligible inertia force, the flow in porous media is modeled using *Brinkman's equation* as

$$\nabla P=-\frac{\mu}{\kappa}\vec{V}+\mu_e\nabla^2\vec{V} \quad (11.80)$$

and for reduced inertia and viscous force terms, it reduces to Darcy's law given by Equation 6.17

$$\vec{V}=\frac{\kappa}{\mu}\nabla P. \quad (11.81)$$

The mass continuity equation for flow in a porous media is given as

$$\nabla\cdot\vec{V}=S_i, \quad (11.82)$$

where S_i represents the reactant gas consumption rates for hydrogen and oxygen and the mass source term for the water transport. For the GDL with no volume reaction zone, the source term is dropped and included as a boundary condition at the electrode–membrane interfaces.

11.6.2.3 Mass Transport

11.6.2.3.1 Mass Species Transport in Fuel Cells

In order to determine the reactant gas transport rates in the electrode/gas diffusion layers and the consumption rates at electrode–membrane interfaces, it is necessary to determine the gas concentration distributions from mass species Mass species transport based on mass convection and diffusion is given by

$$\rho\vec{V}\cdot\left(\nabla C_i\right)-\nabla\cdot\left(D_{ij}^{\text{eff}}\nabla C_i\right)+S_i. \tag{11.83}$$

S_i represents the reactant gas consumption rates in the active layer for hydrogen and oxygen and the mass source term for the water transport. For the GDL with no volume reaction zone, the source term is dropped and included as a boundary condition at the electrode–membrane interfaces.

In the electrode–electrolyte interface layers, the source term is given by the electrochemical reaction rate. Various consumption rates at the electrolyte–catalyst interface are specified as follows:

$$\text{Oxygen consumption rate: } S_{O_2}=-\frac{M_{O_2}}{nF}j \tag{11.84a}$$

$$\text{Hydrogen consumption rate: } S_{H_2}=-\frac{M_{H_2}}{nF}j \tag{11.84b}$$

$$\text{Water generation: } S_{H_2O}=\frac{M_{H_2O}}{nF}j \tag{11.84c}$$

D is the multi-component diffusion coefficient of the material of gas diffusion layer given by the Bruggeman equation

$$D_{ij}^{\text{eff}}=D_{ij}\varepsilon^{1.5}. \tag{11.85}$$

11.6.2.4 Heat Transport Equation

The thermal heat equation for the gas diffusion layer is given considering conduction, convection, and heat generation, and expressed as

$$\rho C_p\vec{V}\cdot(\nabla T_i)=\nabla\cdot(k\nabla \mathrm{T})+\dot{Q}_i \tag{11.86a}$$

or

$$\rho C_p\left(u\frac{\partial T_i}{\partial x}+v\frac{\partial T_i}{\partial y}+w\frac{\partial T_i}{\partial z}\right)=\frac{\partial}{\partial x}\left(k_i\frac{\partial T_i}{\partial T_x}\right)+\frac{\partial}{\partial y}\left(k_i\frac{\partial T_i}{\partial y}\right)+\frac{\partial}{\partial z}\left(k_i\frac{\partial T_i}{\partial z}\right)+\dot{Q}_i \tag{11.86b}$$

where

i = index for the anode and cathode electrode

$\dot{Q}_i$ = volume heat generation due to ohmic heating of electron transport

11.6.2.5 Electrolyte Membrane

11.6.2.5.1 Water Transport

As we have discussed in Chapter 7, the water transport in the electrolyte is governed primarily by diffusion owing to concentration difference and migration owing to pressure difference. The governing equation is given as

$$\nabla \cdot J_{H_2O} = 0, \tag{11.87}$$

where the net water flux owing to the combined effect of water diffusion and convection is given based on Equation 7.73 as

$$j_{H_2O} = D_{H_2O}\nabla c_{H_2O} - c_{H_2O}\frac{k_{H_2O}}{\mu_{H_2O}}\nabla P. \tag{11.88}$$

For water transport through the *Nafion polymer membrane* in a PEM fuel cell, there is an additional driving force such as electro-osmotic drag, and the net water flux is given as

$$J_{H_2O} = 2n_d\frac{j}{n_e F} - D_{H_2O}\nabla c_{H_2O} - c_{H_2O}\frac{k_{H_2O}}{\mu_{H_2O}}\nabla P \tag{11.89}$$

and in terms of water sorption capacity of the Nafion-117 polymer electrolyte membrane the net flux is given based on Equation 7.77 as

$$J_{H_2O} = 2n_{drag}^{sat}\frac{\lambda_m}{22}\frac{j}{n_e F} - \frac{\rho_{dry}D_{H_2O}(\lambda_m)}{M_m}\nabla\lambda_m - \lambda_m\frac{\rho_{air}}{M_m}\frac{K_{H_2O}}{\mu_{H_2O}}\nabla P. \tag{11.90}$$

11.6.2.5.2 Heat Transport Equation

The thermal heat equation for the electrolyte membrane is given by considering conduction, convection, and heat generation, and expressed as

$$\nabla \cdot (k_i \nabla T) + \dot{Q}_i = 0 \tag{11.91a}$$

or

$$\frac{\partial}{\partial x}\left(k_i\frac{\partial T_i}{\partial T_x}\right) + \frac{\partial}{\partial y}\left(k_i\frac{\partial T_i}{\partial y}\right) + \frac{\partial}{\partial z}\left(k_i\frac{\partial T_i}{\partial z}\right) + \dot{Q}_i \tag{11.91b}$$

where

i = index for the anode and cathode electrode

$\dot{Q}_i$ = volume heat generation owing to ohmic heating of ion transport

11.6.2.6 Boundary Conditions

11.6.2.6.1 *Flow Boundary Conditions*

Non-slip velocity conditions, that is, $\vec{V} = 0$ on all solid impermeable walls. A continuity in velocity and shear stress conditions are used at all permeable walls such as the gas channel and porous gas diffusion layer of the electrode.

11.6.2.6.2 *Concentration Boundary Conditions*

Symmetric surfaces:

A zero net flux is used at all symmetric surfaces. This condition is given as

$$n \cdot J_i = 0, \quad J_i = -D_i \nabla C_i + C_i \mathbf{u}. \tag{11.92}$$

Interfaces:

At the channel and gas diffusion layer interfaces, a continuity condition is applied as

$$n \cdot (J_1 - J_2) = 0, \quad J_i = -D_i \nabla C_i + C_i \mathbf{u}. \tag{11.93}$$

Mass flux discontinuity condition is used at the electrode–membrane interface

$$-n \cdot (J_1 - J_2) = S, \tag{11.94}$$

where S is a sink term for gas consumption rates at membrane and electrode interfaces given by Equations 11.84a and 11.84b for cathode and anode, respectively. S is a source term for water given by Equation 11.84c at the electrode–membrane interface.

11.6.2.6.3 *Thermal Boundary Conditions*

Symmetric surfaces:

An adiabatic condition is used at all symmetric surfaces. This condition is given as

$$n \cdot q = 0; \quad q = -k \nabla T + \rho C_p T \mathbf{u}. \tag{11.95}$$

Interfaces:

Continuity in heat flux is used at all interfaces except at the cathode and membrane interface. For example, at the channel and gas diffusion layer interface, at the channel and bipolar plate interface, and at the anode–membrane interface, the continuity condition is given as

$$n \cdot (q_1 - q_2) = 0. \tag{11.96}$$

At the electrode–membrane interface, a heat flux discontinuity condition is given as

$$n \cdot (q_1 - q_2) = Q_{gen}, \tag{11.97}$$

where

$$q_i = -k_i \nabla T_i + \rho_i C_{pi} T_i \mathbf{u}_i. \tag{11.98}$$

Channel Inlet conditions:

Uniform temperature, $T_i = T_o$, and concentration, $C_i = C_o$, are used at the channel inlets.

Inlet condition:

A uniform gas velocity condition is used at the inlet to the gas flow channel, given as

$$u = U_{in}. \tag{11.99}$$

The inlet velocities for reactant gases are calculated based on the total consumption rate given in terms of the total current density and stoichiometric ratio.

$$u = U_{in} = \frac{\xi I}{nF} \frac{A_{mea}}{A_{ch}} \frac{RT}{x_i P}, \tag{11.100}$$

where ξ = stoichiometric ratio, A_{mea} = area of membrane electrode interface, A_{ch} = area of gas channel, and x_i = mole fraction of reactant.

Heat generation:

Heat generation at the electrode–electrolyte interface owing to electrochemical reaction can be expressed in terms of dominant cathode reaction and negligible anode reaction from Equation 4.94d as

$$Q_{gen} = \left[\frac{T(-\Delta S)}{n_e F} + \eta_{act,c} \right] i_c. \tag{11.101}$$

Local current density distribution at the reaction zone is given by the *Butler–Volmer equation* as

Cathode current:

$$j_c = j_{o.c}^{ref} \left(\frac{C_{O_2}}{C_{O_{2,ref}}} \right) \left[\exp\left(\frac{\alpha_a F}{RT} \eta_{act,c} \right) - \exp\left(-\frac{\alpha_c F}{RT} \eta_{act,c} \right) \right]$$

Anode current:

$$j_{\mathrm{a}} = j_{\mathrm{o.a}}^{\mathrm{ref}}\left(\frac{C_{\mathrm{H_2}}}{C_{\mathrm{H_2,ref}}}\right)\left[\exp\left(\frac{\alpha_{\mathrm{a}}F}{RT}\eta_{\mathrm{act,a}}\right)-\exp\left(-\frac{\alpha_{\mathrm{c}}F}{RT}\eta_{\mathrm{act,a}}\right)\right]$$

PROBLEMS

1. Consider a SOFC fuel cell with an operating temperature of 900°C and an open circuit voltage of $E_0 = 0.95$ V. Compute the operating cell voltage at an operating current density of 0.5 A/cm^2 and based on the following cell data: exchange current density, $j_o = 0.15$ A/cm^2; activity coefficient, $\alpha = 0.5$; area specific resistance, $ASR_{\mathrm{ohmic}} = 0.06$ $\Omega \cdot$ cm^2; and limiting current density, $j_L = 2.0$ A/cm^2.
2. Consider the electrolyte as 8% YSZ solid oxide fuel cell with the following known data:

   Anode gas stream: Inlet gas composition: 98% H_2 and 2% H_2O; inlet pressure: $P_a = 2$ atm

   Cathode gas stream: Inlet gas composition: Air with 21% O_2 and 79% N_2; inlet pressure: $P_c = 2$ atm

   Kinetic parameters: Cathode exchange transfer coefficient, $\alpha = 0.5$; cathode exchange current density, $j_o = 0.2$ A/cm^2

   Layer thicknesses: Anode thickness, $L_a = 500$ μm; cathode thickness, $L_c = 200$ μm; and electrolyte, $L_e = 100$ μm

   Determine the operating voltage of the fuel cell for the current density of 0.4 A/cm^2 and a temperature of $T = 900$°C.
3. Consider the polymer electrolyte membrane *Nafion-117* (PEM) fuel cell operating at a temperature of 70°C with the following known data:

   Anode gas stream: Inlet gas composition: 90% H_2 and 10% H_2O, inlet pressure, $P_a = 2.5$ atm

   Cathode gas stream: Inlet gas composition: Air with 21% O_2 and 79% N_2, inlet pressure, $P_c = 3$ atm

   Kinetic parameters: Cathode exchange transfer coefficient, $\alpha = 0.5$; cathode exchange current density, $j_o = 0.00015$ A/cm^2

   Layer thicknesses: Anode thickness, $L_a = 300$ μm; cathode thickness, $L_c = 200$ μm; electrolyte, $L_e = 150$ μm

   Determine the operating voltage of the PEM fuel cell for the current density of 0.45 A/cm^2 and a temperature of $T = 70$°C.
4. Consider a fuel cell with three layers: *anode electrode, electrolyte membrane*, and *cathode electrode*. The tri-layer has a surface area

of 15 cm × 15 cm. The heat generation owing to the electrochemical reaction *at the cathode–membrane interface* provides a heat flux $q''_{gen} = \frac{Q_{gen}}{A} = 5.0 \times 10^5\ \text{W/m}^2$ at the interface area. The flow condition of the hydrogen and air streams is given by stream temperature T_∞ and a convection coefficient, h. Assume the following conditions: $T_{ch,a} = 70°C$, $h_a = 1000\ \text{W/m}^2 \times \text{K}$ and $T_{ch,c} = 65°C$, $h_c = 1100\ \text{W/m}^2 \times \text{K}$. Thermal conductivities of the three layers are as follows: Anode: $k_a = 60.0\ \text{W/m} \times \text{K}$; electrolyte membrane: $k_e = 1.80\ \text{W/m} \times \text{K}$; cathode, $k_c = 55\ \text{W/m} \times \text{K}$.

a. Calculate the heat transferred to hydrogen and oxygen gas stream.
b. What are the temperatures at the anode–membrane and cathode–membrane interfaces?

References

Boddu, R. and P. Majumdar. Gas flow analysis of bi-polar plates designs for fuel cells. *Proceedings of the 4th International Conference on Fuel Cell Science, Engineering and Technology*, Irvine, California, June 19–21, 2006.

Hu, M., A. Gu, M. Wang, X. Zhu and L. Yu. Three dimensional, two phase flow mathematical model for PEM fuel cell: Part I, Model development. *Energy Conservation and Management* 45: 1861–1882, 2004a.

Hu, M., X. Zhu, M. Wang, A. Gu and L. Yu. Three dimensional, two phase flow mathematical model for PEM fuel cell: Part II, Analysis and discussion of the internal transport mechanisms. *Energy Conversion and Management* 45: 1883–1916, 2004b.

Further Reading

Bernardi, D. M. and M. W. Verbrugge. Mathematical model of a gas diffusion electrode bonded to a polymer electrolyte. *AIChE Journal* 37(8): 1151–1163, 1991.

Chen, K. S. and M. A. Hickner. Modeling PEM fuel cell performance using the finite-element method and a fully-coupled implicit solution scheme via Newton's technique. *Proceedings of the 4th International Conference on Fuel Cell Science, Engineering and Technology*, Irvine, California, July 19–21, 2006.

Dannenberg, K., P. Ekdunge and G. Lindbergh. Mathematical model of the PEMFC. *Journal of Applied Electrochemistry* 30: 1377–1387, 2000.

Dutta, S., S. Shampalee and J. W. Van Zee. Three-dimensional numerical simulation of straight channel PEM fuel cells. *Journal of Applied Electrochemistry* 30: 135–146, 2000.

Dutta, S., S. Shampalee and J. W. Van Zee. Numerical prediction of mass-exchange between cathode and anode channels in a PEM fuel cells. *International Journal of Heat and Mass Transfer* 44: 2029–2042, 2001.

Fuller, T. F. and J. Newman. Water and thermal management in solid-polymer-electrolyte fuel cells. *Journal of the Electrochemical Society* 140(5): 1218–1225, 1993.

Konnepart, P. K. and P. Majumdar. Heat and mass transfer analysis of polymer electrolyte membrane fuel cell with bipolar plates. *Proceedings of ASME 2009 Heat Transfer Summer Conference*, HT2009-88630, 2009.

Majumdar, P., U. Marupakula and P. Vohra. *Proceedings of the ICCES '10: Simulation and Design of Bi-polar Plates Integrated PEM Fuel Cell, International Conference on Computational & Experimental Engineering and Sciences*, pp. 422–465, 2010.

Nguyen, P. T., T. Berning and N. Djilali. Computational model of a PEM fuel cell with serpentine gas flow channels. *Journal of Power Sources* 130: 149–157, 2004.

O'Hayre, R., S.-W. Cha, W. Cotella and F. B. Prinz. *Fuel Cell Fundamentals*. John Wiley & Sons, Inc., New Jersey, 2006.

Shimpalee, S. and S. Dutta. Numerical prediction of temperature distribution in PEM fuel cell. *Numerical Heat Transfer, Part A* 38: 111–128, 2000.

Springer, T. E., T. A. Zawodzinski and S. Gottesfeld. Polymer electrolyte fuel cell model. *Journal of the Electrochemical Society* 138(8): 2334–2342, 1991.

Verbrugge, M. W. and R. F. Hill. Transport phenomena in perfluorosulfonic acid membranes during the passage of current. *Journal of the Electrochemical Society* 137(4): 1131–1138, 1990.

Yi, J. S. and T. V. Nguyen. An along-the-channel model for proton exchange membrane fuel cells. *Journal of the Electrochemical Society* 145(4): 1149–1159, 1998.

You, L. and H. Liu. A two-phase flow and transport model for the cathode of PEM fuel cells. *International Journal of Heat and Mass Transfer* 45: 2277–2287, 2002.

12

Dynamic Simulation and Fuel Cell Control System

The dynamic behavior of a fuel cell can be obtained from experimental measurements or from sound physical models. Typically, the measured behavior of the fuel cell only applies to the particular cell being measured. However, the physical models can apply to various fuel cell systems and the physical models can be tuned to a particular fuel cell to predict its behavior under different operating conditions. By developing a physically based dynamic model, the transient behavior of a fuel cell can be accurately predicted and the design envelopes can be optimized. The dynamic model is especially beneficial for control testing in the development stage of the fuel cell. Hence, there is great interest in developing dynamic models. In this chapter, methods employed in the dynamic simulation of fuel cell systems are described. The MATLAB® simulation models are discussed for PEMFC and MCFC fuel cell–based hybrid power systems. Control strategies including advanced neural network and fuzzy logic application on control systems are presented.

12.1 Dynamic Simulation Model for Fuel Cell Systems

Fuel cell systems are heterogeneous systems consisting of various chemical, thermal, mechanical, and electrical systems. The fuel cell system involves a large number of these components besides the cell itself depending on its application. Some of the major components that are typical in a fuel cell power system are the fuel cell stack, fuel processing module, power conditioning unit, and of course many auxiliary components such as pumps, pipes, filters, valves, and sensors.

The fuel cell needs streams of fuel and oxidant such as hydrogen or hydrogen-rich gas and oxygen to electrochemically convert into electrical power. There is also a product of a chemical reaction such as water if hydrogen and oxygen are the reactants, and carbon dioxide if the fuel contains carbon molecules. The oxygen for the fuel cell is normally supplied using air, which is readily available. Depending on the type of fuel cell, the air may have to be humidified, pressurized, and brought close to cell temperature. The hydrogen is not available in free form with current energy distribution infrastructures and therefore many fuel cell systems include a *fuel processing system*. The fuel processing system converts available fossil fuels such as natural gas or gasoline into hydrogen. The complexity of the fuel processing system can easily contribute to more than half of the components in the full system.

Because of the exothermic nature of fuel cells, heat exhaust from the fuel cell power system can also be used as an energy source to turn a turbine or provide heat. These systems, also called *combined heat and power systems*, can achieve higher efficiencies than the fuel cell by itself. In the case of the high-temperature stationary fuel cell power systems that make use of solid oxide or molten carbonate fuel cells, the heat can be collected along with fuel exhaust and unused fuel from the fuel cell and used in a combustion process or turbine to create what is called a *hybrid fuel cell system*. Thus, fuel cell power systems usually also include a *thermal management system* to manage the heat generated by the fuel cell stack assembly and the fuel processing system. To convert the direct current from the fuel cell stack into alternating current for supply to the power system or the grid, the system also needs to include a *power conditioning system*.

In order to understand or predict the behavior of the fuel cell system, system level modeling of the fuel cell system is required. Given the heterogeneous nature of the fuel cell system and interactions between mechanical, chemical, thermal, and electrical components and subsystems, it is a challenge to study the dynamic behavior of the fuel cell system (Machowski et al., 1997; Thomas, 2007). There is also good control structure associated with the fuel cell system to establish the required operation characteristics with strongly coupled systems. The modeling challenges consist of both numerical and model handling issues. In this chapter, several aspects of dynamic simulation and its application to fuel power systems are presented. Dynamic simulation can be predicting the behavior, performance, and limitations of a fuel cell power system.

12.1.1 System Dynamics

System dynamics is a methodology used for understanding the dynamic behavior of complex systems. This methodology was originally developed during the mid-1950s by Professor Jay Forrester of the Massachusetts Institute of Technology to help corporate managers improve their understanding of industrial processes (Forrester, 1961). The system dynamics principle lies in recognition that the structure of any system is often just as important in determining its behavior as the individual components themselves. It takes into account internal feedback loops that connect components of the system and time delays that affect the behavior of the entire system.

A system dynamics diagram is made of four kinds of elements: stocks, variables, flows and links. A *stock* is a collection of stuff, an aggregate. For example, a stock can represent a population of sheep, the water in a lake, or the number of widgets in a factory. A *flow* brings things into, or out of a stock. Flows look like pipes with a faucet because the faucet controls how much stuff passes through the pipe. A *variable* is a value used in the diagram. It can be an equation that depends on other variables, or it can be a constant. A *link* makes a value from one part of the diagram available to another. A link transmits a number from a variable or a stock into a stock or a flow. The common tools used in the system dynamics model are *causal loop diagrams* and *stock and flow diagrams*.

The *causal loop diagram* explores the structural interrelationships between the parts of a system. It captures graphically how each factor or variable in a system influences the others. A variable is defined as anything that can increase or decrease over time in the system. Causal loop diagrams represent the system in the form of feedback loops. There are two generic types of feedback loops: reinforcing and balancing. Reinforcing feedback loops compound change with even more change. They result in either a vicious or virtuous cycle and are often referred to as snowball or bandwagon effects. They are responsible for the exponential growth or decline of various metrics. Common reinforcing loops include compound interest and population growth. Balancing loops on the other hand maintain balance in a system and negate a metric's change in one direction by pushing the system to drive the metric in the opposite direction. The laws of supply and demand are both balancing loops.

Stock and flow diagrams are used for exploring system interrelationships. The tool underscores the difference between "stock" variables and their rates of change, or "flow." This type of diagramming is the basis for computer simulation model feedback, accumulation of flows into stocks, and time delays. These elements help describe how even seemingly simple systems display baffling nonlinearity.

The modeling system dynamics is a computer simulation that quantifies the relationships of the causal loop diagrams. Because of the complex nature of most systems, it is difficult to visually sort through and analyze the behavior of the causal loop diagrams. Using system dynamics software, these relationships can be represented in the form of stock and flow diagrams and simulated. *Dynamic simulation* is the use of a computer program to model the time-varying behavior of a system (Sterman, 2001). The systems are typically described by ordinary differential equations or partial differential equations. As mathematical models incorporate real-world constraints, equations become nonlinear. This requires numerical methods to solve the equations.

12.1.2 Block and Information Flow Diagram

Block diagram is a convenient graphical representation of input–output behavior of a system, where the signal into the block represents the input and the signal out of the block represents the output. The block or rectangles used represent a unit operation. The blocks are connected by straight lines that represent the process flow streams that flow between the units. These process flow streams may be mixtures of liquids, gases, and solids or information flowing in pipes or ducts in a particular direction. The flow of information (the signal) is from the input to the output. The primary use of the block diagram is to portray the interrelationship of distinct parts of the system. A block diagram consists of two basic functional units that represent system operations. The individual block symbols portray the dynamic relations between the input and output signals. The second type of unit, called a summing point, is represented by a circle with arrows feeding into it. The operation that results is a linear combination of incoming signals to generate the output signal. The sign appearing alongside each input to the summing point indicates the sign of that signal as it appears in the output. Figure 12.1 shows a graphical representation of a block. There are several simple rules that should be followed to develop easy to understand and unambiguous block flow diagrams as listed below.

1. Unit operations such as mixers, separators, reactors, fuel cell, distillation columns, and heat exchangers are usually denoted by a simple block or rectangle.
2. Groups of unit operations may be noted by a single block or rectangle.
3. Process or information flow streams flowing into and out of the blocks are represented by neatly drawn straight lines. These lines should either be horizontal or vertical.
4. The direction of flow of each of the process flow streams must be clearly indicated by arrows.
5. Flow streams should be numbered sequentially in a logical order.
6. Unit operations (i.e., blocks) should be labeled.
7. Where possible, the diagram should be arranged so that the process material flows from left to right, with upstream units on the left and downstream units on the right.

In Figure 12.2, a small simplified block flow diagram is shown.

Similarly, an *information flow diagram (IFD)* is an illustration of information flow throughout an organization or system. An IFD shows the relationship between external and internal information flows between organizations. It also shows the relationship between the internal departments and subsystems. The IFD is generally laid out in diagrammatic form usually using "blobs" to explain in more detail the system and subsystems to elemental parts. Following on from this, one can add in lines to show how the information travels from one system to another.

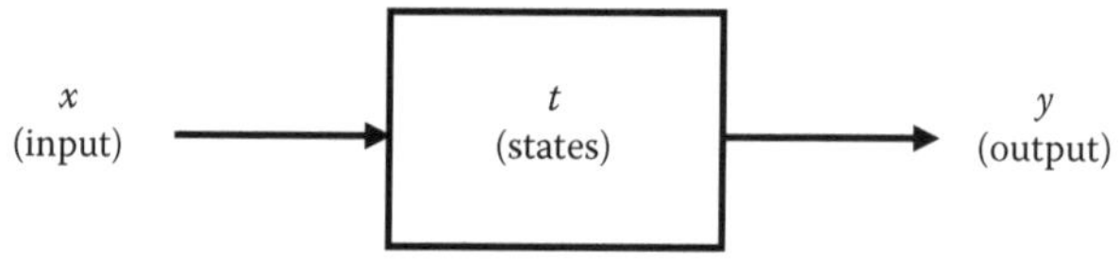

FIGURE 12.1
Block diagram.

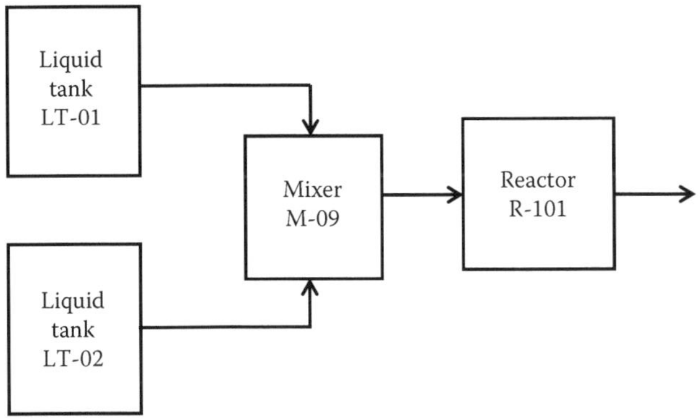

FIGURE 12.2
Example of block flow diagram.

Typically, the current values of some system, and hence model, outputs are functions of the previous values of temporal variables. Such variables are called states. Computing a model's outputs from a block diagram hence entails saving the value of states at the current time step for use in computing the outputs at a subsequent time step. This task is performed during simulation for models that define states. The total number of a model's states is the sum of all the states defined by all its blocks. Determining the number of states in a diagram requires parsing the diagram to determine the types of blocks that it contains and then aggregating the number of states defined by each instance of a block type that defines states.

In modeling a system, some parameters are first defined and equations governing system behavior are obtained. A block diagram is constructed, and the transfer function for the whole system is determined. If a system has two or more input variables and two or more output variables, simultaneous equations for the output variables can be written. In general, when the number of inputs and outputs is large, the simultaneous equations are written in matrix form. Block diagrams can be used to portray nonlinear as well as linear systems.

The block diagram is typically used for a higher-level, less detailed description aimed more at understanding the overall concepts and less at understanding the details of implementation. Because block diagrams are a visual language for describing actions in a complex system, it is possible to formalize them into a specialized programmable logic controller programming language. A function block diagram is one of the five programming languages. As a bona fide computer programming language, it is highly formalized with strict rules for how diagrams are to be built. Directed lines are used to connect input variables to function inputs, function outputs to output variables, and function outputs to inputs of other functions. These blocks portray mathematical or logical operations that occur in time sequence. They do not represent the physical entities, such as processors or relays that perform those operations. Each block is therefore a black box. The rules require the logical sequence to go from left to right and top to bottom.

For various engineering processes, a process flow diagram (PFD) is used to indicate the general flow of plant process streams and equipment (Thomas, 2007). The PFD indicates the relationship between *major* equipment of a plant facility and does not show details such as minor equipment items, piping materials of construction, and piping sizes. Another commonly used term for a PFD is a *process flow sheet*. Typically, PFDs of a single unit process will include process piping, fluid composition, flow direction, major bypass and recirculation lines, major equipment symbols, names and identification numbers, control loops, interconnections, system conditions such as range and nominal values of flow, temperature and pressure, and ratings.

12.1.3 Solution Methodology for Dynamic Simulation

In the simulation, the first step is to develop mathematical modeling. The modeling, based on first principles, is done by applying a standard input/output approach for time-dependent systems with one or multiple inputs $x(t)$ and one or multiple outputs $y(t)$. The mathematical descriptions of components and hardware are formulated in the form of ordinary differential equations with the time t as the independent variable. The system description is represented mathematically by a system of coupled, nonlinear, first-order differential (or integral) equations:

$$\frac{dx(t)}{dt} = f(x, p), \tag{12.1}$$

where x is a vector of levels (stocks or state variables), p is a set of parameters, and f is a nonlinear vector-valued function. Example of single input and output model is shown in Figure 12.3.

In general, most of the dynamic models consist of a coupled set of partial differential or integral equations that are derived from the balance equations of charge, mass, momentum, and energy. Such modeling, however, can be split into two main categories, such as finite element modeling– or computational fluid dynamics (CFD) modeling–based control volume analysis. Each of these in fact can also be split into higher-order and lower-order modeling regimes as well as one, two, and three dimensions. While CFD-based

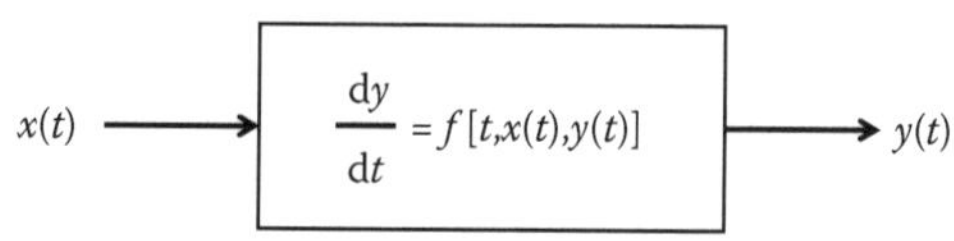

FIGURE 12.3
Time invariant input–output system for one input $x(t)$ and one output $y(t)$ and a first-order differential equation relating output and input function.

modeling is essential to understanding the flow paths, for example in an anode or cathode, and can be helpful in optimizing an individual cell of a fuel cell, it is very computationally expensive and in most cases cannot be used to predict the behavior of an entire fuel cell power system or even a multi-cell fuel cell stack. Therefore, block methods are used to study entire system dynamics.

The equations used to model the different elements are collected together to form a set of differential equations

$$x = f(x, y) \tag{12.2}$$

and a set of algebraic equations

$$0 = g(x, y). \tag{12.3}$$

The differential equations describe the system dynamics and are primarily contributed by the generating units and the dynamic loads while the algebraic equations describe the network, the static loads, and the generator algebraic equations. The solution of these two equations defines the electromechanical state of the power system at any instant in time. A disturbance in the network usually requires a change to both the network configuration and the boundary conditions. These are modeled by changing the coefficients in the functions appearing on the right-hand side of Equations 12.2 and 12.3. The dynamic power system simulation computer program must then solve the differential and algebraic equations over a period of time for a given sequence of network disturbances. Equations 12.2 and 12.3 can be solved using either a partitioned solution or a simultaneous solution. In the partitioned solution, the simulated time is partitioned into discrete intervals of length $\mathrm{d}t$ and stepping the system through time one $\mathrm{d}t$ at a time. Each state variable is computed from its previous value and its net rate of change $x'(t)$ as

$$x(t) = x(t - \mathrm{d}t) + \mathrm{d}t \times x'(t - \mathrm{d}t). \tag{12.4}$$

In the *partitioned solution*, the differential equations are solved using a standard explicit numerical integration method with the algebraic equations (Equation 12.3) being solved separately at each time step. The *simultaneous solution* uses implicit integration methods to convert the differential equations (Equation 12.2) into a set of algebraic equations that are then combined with the algebraic network equations (Equation 12.3) to be solved as one set of simultaneous algebraic equations. The effectiveness of these two solutions depends on both the generator model used and the method of numerical integration. In order to select the most appropriate integration method to use, it is necessary to understand the time scale of the dynamics included in the model of the generating unit. The solution of any set of linear differential equations is in the form of a linear combination of exponential functions each of which describes the individual system modes. These modes are themselves defined by the system eigenvalues that are linked to the time scale of the different dynamics in the model. When the eigenvalues have a range of values that are widely distributed on the complex plane, the solution will consist of the sum of fast-changing dynamics, corresponding to large eigenvalues, and slow-changing dynamics, corresponding to small eigenvalues. In this instance, the system of differential equations is referred to as a *stiff system*. A nonlinear system is referred to as stiff if its linear approximation is stiff.

In the block diagram, each block represents multiple equations. These equations are represented as block methods. These block methods are evaluated (executed) during the execution of a block diagram. The evaluation of these block methods is performed within a simulation loop, where each cycle through the simulation loop represents the evaluation of the block diagram at a given point in time. There are several tools to handle the time-dependent variable equation sets in each block such as fixed step solvers and variable step solvers.

In equation-based simulators, the mathematical equations that describe the physical process are entered into an equation solver that then uses appropriate techniques to solve them. In modular-based process simulators, the mathematical equations that describe the physical process are coded into modules that the user "flow sheets" together. Modular-based process simulators are preferred over equation-based simulators because it is easier for the user to "map" the real world into the virtual one, and programming and debugging of the modules are easier than analyzing sets of equations (Popovic and Bhatkar, 1997). However, equation-based simulators have proved highly successful in the field of optimum process control. Equation-based models handle instrument error and incorrect or errors from modeling simplifications better than modular-based simulators. Modular-based simulators invariably have a data reconciliation step, where the model is run against values obtained from the instrumentation system and then a least-squares fit is performed to fit the model to the process.

12.2 Simulation of the Fuel Cell–Powered Vehicle

12.2.1 Fuel Cell Vehicle Simulation

For application of the fuel cell as the power source for vehicles, the modeling tool needs to address the transient dynamic interaction between the electric drive train and the fuel cell system. Thus, the mathematical model that simulates the fuel cell behavior should, at the same time, simulate the vehicle power train working conditions. This allows designing in an integrated manner the whole system by simulating it in different working conditions to stress and optimize the choice of components.

For vehicle application, the PEMFC and SOFC technologies appear to be best suited and are chosen according to the requirement on characteristics of starting and response times. The drive application is the most stringent requirement. The PEMFC technology fueled with hydrogen offers starting and response times quasi-instantaneously and compatible with the drive application. There are, however, still problems to be solved such as the high cost of the PEMFC stack and the system thermal management. The SOFC technology offers starting and response times hardly compatible with drive application. Maintaining the SOFC power plant at its working temperature as well as operating without excessive demand variations can be considered in order to preserve the stack integrity. Its advantage is its simplified reforming characteristic, which makes this technology an interesting option for application as an auxiliary power source.

The main components of the fuel cell vehicle include Power Electronic, Motor Controller and Motor Control Algorithm, Transmission, and fuel cell.

The fuel cell vehicle model is generally described first with the uppermost level, the vehicle. At this level, the driver interaction and drive cycle are modeled. The uppermost level of the fuel cell vehicle model consists of the following main blocks: "Specified Drive Cycle," "Driver," and "Vehicle" (Figure 12.4).

Each of these high level systems consists of several components and each individual component may consist of several subcomponents. In Figure 12.5, the content of the Vehicle block is shown. The FC Power Source block contains the FC system and stack current feed and voltage output. The fuel cell system generally refers to the combined system of fuel cell stack, water and thermal management, air supply (including compressor and expander), and fuel processor (including reformer, cleanup stages, and the burner for the anode off-gas).

In the model, each of the component interfaces transfers physical properties from one component to the other components. In addition to component models, the control strategies determine the nature of the interaction of the individual components. The modularity of design allows one to test individual component models off-line and replace it with a different model for the same

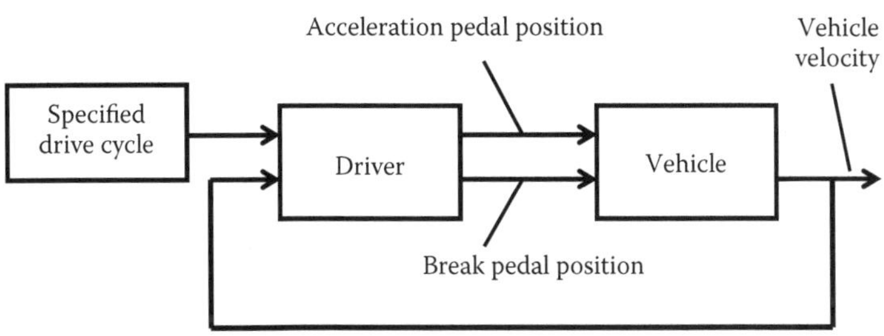

FIGURE 12.4
Uppermost level of the fuel cell vehicle model.

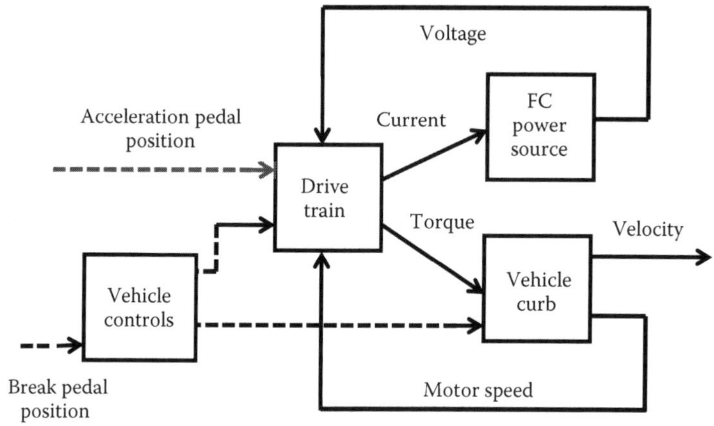

FIGURE 12.5
Content of the block vehicle.

component if the interfaces of the replacement model match the interfaces of the model replaced. The modeling effort needs to (i) estimate fuel consumption, energy flows and losses, and the vehicle dynamics such as acceleration time and speed, and (ii) set up control strategies for safe operation and estimate theoretical limits. This latter part enables application of existing hardware.

There are various dynamic vehicle system and vehicle component models for hybrid electrical vehicles (Cole, 1993; Cuddy and Wipke, 1996; Hauer, 2001; Hauer and Moore, 2003; Murrell, 1995; Rousseau and Larsen, 2000; Wipke et al., 1999). Two state-of-the art simulations packages, PSAT and ADVISOR, are widely used for fuel cell vehicle modeling as they are flexible, are reusable, and have been benchmarked with a variety of databases. The dynamic power train vehicle model PSAT (Powertrain Systems Analysis Toolkit), was developed under the direction and with the contribution of Ford, General Motors, and DaimlerChrysler for the Partnership for a New Generation of Vehicle (PNGV). *PSAT* allows dynamic analysis of vehicle performance and efficiency to support detailed design, hardware development, and validation. A driver model attempts to follow a vehicle driving cycle, sending a power demand to the vehicle controller, which, in turn, sends a demand to the propulsion components (commonly referred to as "forward-facing" simulation). Dynamic component models react to the demand (using transient equation-based and physics-based models) and feedback their status to the controller, and the process iterates on a subsecond basis to achieve the desired result (similar to the operation of a real vehicle). The National Renewable Energy Laboratory has developed ADVISOR as an Advanced Vehicle Simulator, which is a very useful computer simulation tool for the analysis of energy use and emissions in both conventional and advanced vehicles.

By incorporating various vehicle performance and control information into a modular environment within MATLAB and Simulink®, ADVISOR allows the user to interchange a variety of components, vehicle configurations, and control strategies. Other unique and valuable features of ADVISOR include the ability to quickly perform parametric and sensitivity studies of vehicle parameters on overall performance and economy. The vehicle models can be integrated with fuel cell models to incorporate the fuel cell as a power source in the vehicle. The ADVISOR model can be used to simulate energy storage, energy generation, and energy flow within a vehicle that are required to propel it along at a particular speed versus time trace.

12.2.2 Simulation Model for PEMFC System

There are several fuel cell system dynamic models in the literature that are specifically developed for vehicle propulsion. The simplest dynamic model of a fuel cell stack is the representation of the stack with an equivalent circuit whose operating parameters are based on the polarization curve obtained from the manufacturer data sheet at nominal conditions of temperature and pressure. An equivalent single-cell model with semiempirical equation for the fuel cell polarization curve can also be used for this purpose.

The simplest empirical model for single fuel cell voltage as given in Equation 5.159 is

$$E = E_{OCV} - A'\ln j - jr - C\ln nj, \tag{12.5}$$

where A', r, B, C, n, and j_L are empirical parameters obtained from curve-fits of experimental data. The fuel cell open circuit voltage or thermodynamic potential E_{OCV} can be obtained from experiments or theoretically calculated.

The fuel cell performance curve can also be developed from first principles and mechanistic modeling combined with empirical relations. Since there are a number of variables in a given design of a fuel cell, a semiempirical approach seems to be more appropriate. Springer et al. (1993, 1996), Amphlett et al. (1995a, b, 1996) and Mann et al. (2000) have carried out such semiempirical models for PEMFC.

These models were developed for single PEM fuel cell taking into account interfacial kinetics at the Pt/ionomer interface, gas-transport and ionic-conductivity limitations in the catalyst layer, and gas-transport limitations in the cathode backing. These models have been improved for dynamic simulation of fuel cell stack with additional feature on anode fuel flow and cathode flow, and membrane hydration air supply models (Pukrushpan, 2003).

From Equation 5.154, the single fuel cell voltage E is given as

$$E = E_{OCV} - \eta, \tag{12.6}$$

where the overpotential η is given from Equation 5.149

$$\eta = \eta_{act,a} + \eta_{act,c} + \eta_{ohm,} + \eta_{mt,a} + \eta_{mt,c} + \eta_{other}. \tag{12.7}$$

The open circuit voltage can be given by the Nernst equation. The Nernst equation for a hydrogen–oxygen PEM fuel cell, using the thermodynamic values of standard state entropy change, is given by (Amphlett et al., 1995b)

$$E_{\mathrm{OCV}} = 1.229 - (8.5\times10^{-4})(T-298.15) + (4.308\times10^{-5})T\left[\ln p_{\mathrm{H_2}} + \frac{1}{2}\ln p_{\mathrm{O_2}}\right], \quad (12.8)$$

where fuel cell temperature T is expressed in Kelvin and the partial pressure of hydrogen at the anode catalyst–gas interface $p_{\mathrm{H_2}}$ and the partial pressure of oxygen at the cathode catalyst–gas interface $p_{\mathrm{O_2}}$ are expressed in atmospheres.

The sum of the anode and cathode activation overvoltage is calculated as

$$\eta_{\mathrm{act}} = \eta_{\mathrm{act,a}} + \eta_{\mathrm{act,c}} = \xi_1 + \xi_2 T + \xi_3 T\ln(c_{\mathrm{O_2}}) + \xi_4 T\ln(j), \quad (12.9)$$

where ξ values represent parametric coefficients for each cell model given as

$$\xi_1 = \frac{\Delta G_{\mathrm{c}}}{2F} - \frac{\Delta G_{\mathrm{a}}}{\alpha_{\mathrm{c}} nF} \quad (12.10)$$

$$\xi_2 = \frac{R}{\alpha_{\mathrm{c}} nF}\ln\left[nFAk_{\mathrm{c}}^{\mathrm{o}}(c_{\mathrm{H^+}})^{(1-\alpha_{\mathrm{c}})}(c_{\mathrm{H_2O}})^{\alpha_{\mathrm{c}}}\right] + \frac{R}{2F}\ln\left[4FAk_{\mathrm{a}}^{\mathrm{o}}c_{\mathrm{H_2}}\right] \quad (12.11)$$

$$\xi_3 = \frac{R(1-\alpha_{\mathrm{c}})}{\alpha_{\mathrm{c}} nF} \quad (12.12)$$

$$\xi_4 = -\left(\frac{R}{\alpha_{\mathrm{c}} nF} + \frac{R}{2F}\right) \quad (12.13)$$

Here, $c_{\mathrm{O_2}}$ is the concentration (mol/cm^3) of dissolved oxygen at the gas/liquid interface at the cathode side and $c_{\mathrm{H_2}}$ is the concentration of dissolved hydrogen at the gas/liquid interface at the anode side. These concentrations are given in terms of the partial pressures of oxygen $P_{\mathrm{O_2}}$ and partial pressure of hydrogen $P_{\mathrm{H_2}}$ at the cathode and anode interfaces, respectively, as

$$c_{\mathrm{O_2}} = \frac{P_{\mathrm{O_2}}}{5.08\times10^{6} e^{\frac{-498}{T}}} \quad (12.14)$$

$$c_{\mathrm{H_2}} = \frac{P_{\mathrm{H_2}}}{1.09\times10^{6} e^{\frac{77}{T}}}. \quad (12.15)$$

The parameters in Equations 12.10 through 12.14 are defined as follows. ΔG_c and ΔG_a are the standard-state free energy of activation for chemisorptions (J/mol) at the cathode and anode, respectively, and A is the active cell area (cm^2). k_c^o and k_a^o are the intrinsic rate constants (cm/s) for the cathode and anode reaction, respectively. c_{H^+} and c_{H_2O} are respectively the concentration of hydrogen ion and water at the membrane gas interface on the cathode side of the cell. α_c is called the cathodic charge transfer coefficient or chemical activity parameter. R is the gas constant. The ohmic overvoltage can be represented using Ohm's law as discussed in Chapter 7 as

$$\eta_{ohm} = j(R_{el} + R_{pr}), \tag{12.16}$$

where R_{el} is the resistance of the membrane to electron flow and R_{pr} is the resistance to proton flow. The parameter R_{pr} is usually considered constant. The parameter R_{el} is calculated as

$$R_{el} = \frac{\rho_M L_e}{A}, \tag{12.17}$$

where L_e is the thickness of the electrolyte membrane (cm) and ρ_M is the specific resistivity of the membrane for the electron flow (Ω·cm). The specific resistivity for the membranes of the type *Nafion-117* can be expressed as

$$\rho_M = \frac{181.6\left[1 + 0.03\left(\frac{i}{A}\right) + 0.062\left(\frac{T}{303}\right)^2\left(\frac{i}{A}\right)^{2.5}\right]}{\left[\lambda - 0.634 - 3\left(\frac{i}{A}\right)\right]\exp\left(4.18\left\{\frac{T-303}{T}\right\}\right)}. \tag{12.18}$$

For zero current, the specific resistivity is 181.6/(λ – 0.634). The exponential term in the denominator is the temperature correction factor if the cell is not at 30°C.

The parameter λ is an adjustable parameter with a value ranging from 14 under the ideal condition of 100% humidity to 23 under oversaturated conditions. It is influenced by the membrane preparation procedure and is a function of the relative humidity and stoichiometric ratio of the anode feed gas. It will likely also be a function of the age (time in service) of the membrane.

The mass transport affects the concentrations of hydrogen and oxygen, which causes a decrease of the partial pressures of these gases. The decrease in pressures of hydrogen and oxygen depends on the load current and other physical characteristics of the system. Larger electrical current will have larger reduction in the pressures of oxygen and hydrogen. The maximum current density j_{max} can be defined as the current under which the fuel is used at a rate of maximum supply speed. The current density is limited to

TABLE 12.1
Fuel Cell Parameters

Parameter	Value	Parameter	Value
A	50.6 cm^2	ξ_1	–0.9477
L_e	0.0178 cm	ξ_2	0.0033
T	343 K	ξ_3	7.5×10^{-5}
P_{O_2}	1 atm	ξ_4	-1.915×10^{-4}
P_{H_2}	1 atm	ψ	23
R_{pr}	0.0003 Ω	j_{max}	1.5 A
B	0.016 V	C	3 F

this value because the fuel cannot be supplied at a larger rate. The mass transfer overvoltage is given as

$$\eta_{mt} = -B\ln\left(1-\frac{j}{j_{max}}\right), \tag{12.19}$$

where parametric coefficient B depends on the cell and its operation state and j is the actual current density of the cell (A/cm^2). A list of typical parameters used for a simulation are presented in Table 12.1.

The polarization curve represents the fuel cell output voltage as a function of the current density in steady state. Electrical power, in watts, supplied by the fuel cell to the load, can be determined using the following equation:

$$P_{FC} = V_{FC} i_{FC}. \tag{12.20}$$

The fuel cell efficiency can be determined from

$$\mu = \mu_f \frac{V_{FC}}{1.48}, \tag{12.21}$$

where μ_f is the fuel utilization coefficient, generally in range of 95%, and 1.48 V represents the maximum voltage that can be obtained using the higher heating value for the hydrogen enthalpy.

12.2.3 Dynamic Simulation Model of the PEMFC Cell

An electrical circuit can be used to model the fuel cell dynamic behavior as represented in Figure 12.6. In the equivalent circuit in Figure 12.6, there is a first-order delay in the activation and the concentration voltage components represented by the resistances R_{act} and R_{conc}, respectively. This delay is caused by the charge double layer effect. The charge double layer on the interface electrode/electrolyte (or close to the interface) acts as storage of electrical charges and energy and, in this way, it behaves as an electrical capacitor (Correa et al., 2005). If the voltage changes, there will be some time for

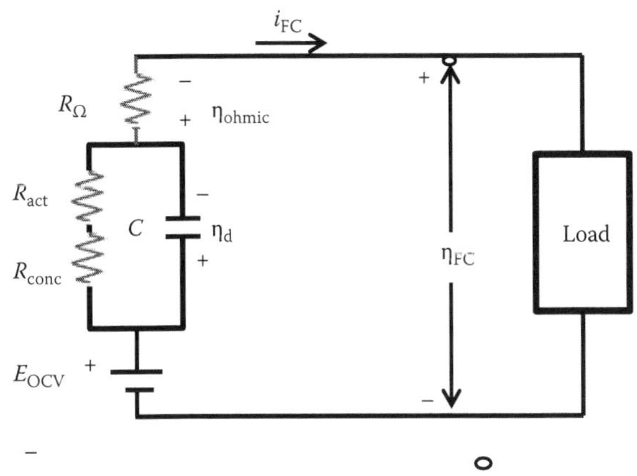

FIGURE 12.6
Fuel cell dynamic model—electrical equivalent circuit.

the charge to decrease if the voltage increases or to increase if the voltage decreases. The activation and concentration potentials are affected by this delay and ohmic overpotential is not affected, since it is linearly related to the cell current through Ohm's law. Thus, a change in the current causes an immediate change in the ohmic voltage drop. In this way, it can be considered that a first-order delay exists in the activation and concentration voltages. The time constant τ associated with this delay is given as

$$\tau = C(R_{act} + R_{conc}) = C\frac{(\eta_{act} + \eta_{conc})}{j}, \tag{12.22}$$

where C represents the equivalent capacitance (F) of the system. In Figure 12.6, the capacitor is positioned in parallel with the activation and concentration voltages (represented by their equivalent resistor) to take into account the dynamic effect of these voltage drops.

This resulting loop is then connected in a series with the Nernst potential (thermodynamic potential) and with the ohmic voltage drop (represented by its equivalent resistance). The dynamic equations of the model presented in Figure 12.6 are given by

$$\begin{aligned} \frac{d\eta_d}{dt} &= \frac{1}{C}j - \frac{1}{\tau}\eta_d \\ \eta_{FC} &= E_{OCV} - \eta_{ohmic} - \eta_d, \end{aligned} \tag{12.23}$$

where η_d is the dynamic voltage of equivalent electrical capacitance. All the equivalent resistors R_{act}, R_{conc}, R_Ω, are defined using nonlinear equations and, as a result, the dynamic model (Equation 12.19) is nonlinear too.

12.3 Dynamic Simulation of Integrated Fuel Cell Systems

The integrated fuel cell systems consist of a fuel cell and other power systems or fuel processing units to support the fuel cell. For example, a regenerative solar PEMFC system serves as an energy storage and conversion device using hydrogen as the energy medium and a solar light source as the primary energy. High-temperature fuel cells (MCFC and SOFC) operate at temperatures exceeding 950 K and produce a large amount of heat that can be either given off as waste heat or harnessed for use in another system. For this reason, high-temperature fuel cells are more apt to be used in a hybrid or combined cycle design. Here, two such systems are discussed, the PEM regenerative FC (RFC) system and the MCFC-GT hybrid system. Dynamic modeling of such systems requires models for individual components and detailed control systems.

There are many different levels and approaches of modeling fuel cells. More detailed single-cell models capture the micro-scale phenomena; however, these typically use a CFD approach, which is computationally expensive. This makes it difficult to study long time dynamics and is limited to modeling specific components alone. A lumped parameter thermodynamic approach is therefore more applicable for studying different hybrid system configurations as well as part load dynamics. The benefit of using a lumped approach is its simplicity and fast calculation time while still capturing the overall dynamic response of a system. Also, lumped models are easily adapted to different geometries and can be easily fitted to experimental data by changing geometry-related parameters. The main disadvantage of lumped models is that their results represent mean values of parameters, and therefore more detailed models need to be considered to investigate local effects that may be undesirable. Because the integrated systems incorporate many components such as the fuel cell, oxidizer, turbine, and compressor, a lumped model approach with control volume analysis is more suitable.

12.3.1 Regenerative PEM Fuel Cell System

A proton exchange membrane (PEM) solar regenerative fuel cell (SRFC) is an energy storage and conversion device using an energy medium such as hydrogen. An S system is capable of storing energy for later usage. The PEM H_2/O_2–based fuel cell system is the highest storage capacity and lowest weight storage system for extraterrestrial applications. For the regenerative system design, the electrolyzer produces hydrogen and oxygen from the electrolysis of water using renewable energy sources such as wind energy or solar energy. The present system is a solar-powered RFC system composed of a PEMFC stack, a PEM electrolyzer, solar panels, a power control bus, multi-stage compressors, storage tanks, and a heat exchanger.

Figure 12.7 shows the simplified schematic layout of the SRFC system. The key subsystems of the SRFC system are as follows: photovoltaic system, power

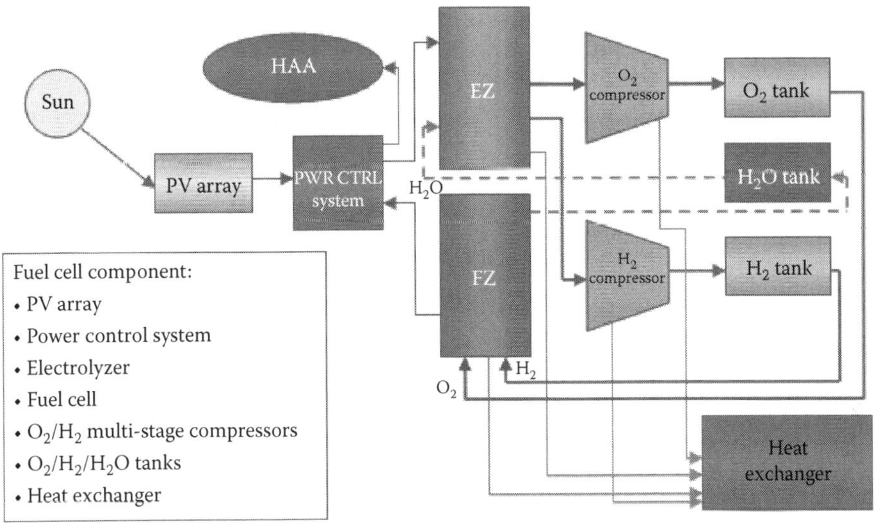

FIGURE 12.7
Schematic layout of the final SRFC model.

control system, electrolysis subsystem, fuel cell subsystem, compressor-tank subsystem, and heat exchanger subsystem. Here, the dynamic models for the photovoltaic system and PEMFC are discussed in detail along with implementation of models in MATLAB/Simulink (Klee, 2002; Kota, 2006; Spiegel, 2008).

12.3.2 Photovoltaic System

12.3.2.1 Solar Cell

The solar cell is the basic unit of the photovoltaic generator. The solar cell is the device that transforms the sun's rays or photons directly into electricity. There are various models of solar cells made with different technologies available in the market today. These models have varying electrical and physical characteristics depending on the manufacturer. The photocurrent generated by a solar cell under illumination at short circuit is dependent on the incident light. To relate the photocurrent density, J_{sc}, to the incident spectrum, we need the cell's quantum efficiency (*QE*). *QE* (*E*) is the probability that an incident photon of energy *E* will deliver one electron to the external circuit. Then,

$$J_{sc} = q\int b_s(E)QE(E)\mathrm{d}E, \tag{12.24}$$

where $b_s(E)$ is the incident spectral photon flux density, the number of photons in the range E to $E + \mathrm{d}E$, which are incident on unit area in unit time, and q is the electronic charge. *QE* depends upon the absorption coefficient of the solar cell material, the efficiency of charge separation, and the efficiency of charge collection in the device but does not depend on the incident spectrum. It is therefore a key quantity in describing solar cell performance under different conditions. *QE* and spectrum can be given as functions to photon energy or wavelength, λ. The relationship between *E* and λ is defined by

$$E = \frac{hc}{\lambda}. \tag{12.25}$$

When a load is present, a potential difference develops between the terminals of the cell. This potential difference generates a current, which acts in the opposite direction to the photocurrent, and the net current is reduced from its short circuit value. This reverse current is usually called *dark current* in analogy with the current I_{dark} (V), which flows across the device under an applied voltage, or *bias*, V, in the dark. Most solar cells behave like a diode in the dark, admitting a much larger current under forward bias ($V > 0$) than under reverse bias ($V < 0$). This rectifying behavior is a feature of photovoltaic devices, since an asymmetric junction is needed to achieve charge separation. For an ideal diode, the *dark current density* $J_{\text{dark}}(V)$ varies like

$$J_{\text{dark}}(V) = J_{\text{o}}\left[\exp\left(\frac{qv}{kT}\right) - 1\right]. \tag{12.26}$$

The overall current voltage response of the cell, its *current–voltage characteristic*, can be approximated as the sum of the short circuit photocurrent and the dark current. The sign convention for photocurrent and voltage in photovoltaic is such that the photocurrent is positive. This is the opposite of the usual convention for electronic devices. With this sign convention, the net current density in the cell is given by

$$J(V) = J_{\text{sc}} - J_{\text{dark}}(V). \tag{12.27}$$

This becomes, for an ideal diode,

$$J = J_{\text{sc}} - J_{\text{o}}\left(\exp\left(\frac{qv}{kT}\right) - 1\right). \tag{12.28}$$

Figure 12.8a shows the equivalent circuit of an ideal solar cell, and in Figure 12.8b, the voltage–current circuit is shown. The current–voltage product is positive, and the cell generates power, when the voltage is between 0 and V_{oc}.

The open circuit voltage V_{oc} is when the contacts are isolated, and this potential difference has its maximum value. This is equivalent to the condition

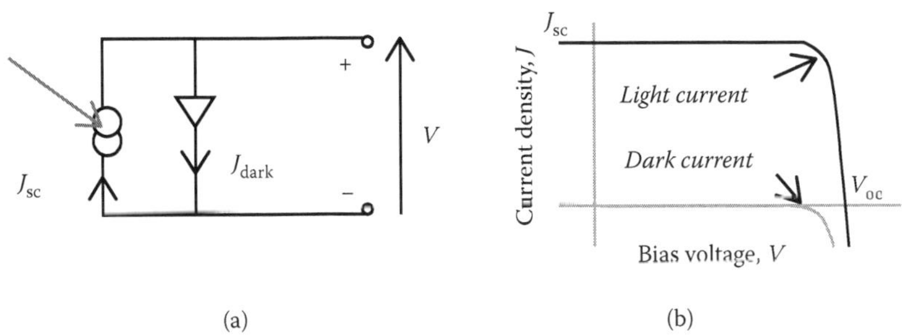

FIGURE 12.8
(a) Equivalent circuit of ideal solar cell. (b) Voltage–current characteristics of an ideal solar cell.

when the dark current and short circuit photocurrent exactly cancel out. For an ideal diode, from the above equation, we get

$$V_{oc} = \frac{kT}{q}\ln\left(\frac{J_{sc}}{J_o}+1\right). \tag{12.29}$$

The above equation shows that V_{oc} increases logarithmically with light intensity. Note that the voltage is defined so that the photovoltage occurs in the forward bias, where $V > 0$.

The operating regime of the solar cell is the range of bias, from 0 to V_{oc}, in which the cell delivers power. The cell power is given by

$$P = (J \times A_{cell} \times V), \tag{12.30}$$

where A_{cell} is the area of the cell.

The cell power, P, reaches a maximum at the cell's operating point or maximum power point. This occurs at some voltage V_m with a corresponding current density J_m. The optimum load thus has sheet resistance given by V_m/J_m. The *fill factor* is defined as the ratio

$$FF = \frac{J_m V_m}{J_{sc} V_{oc}}. \tag{12.31}$$

The efficiency of the cell is the power density delivered at operating point as a fraction of the incident light power density, P_s

$$\eta = \frac{J_m V_m}{P_s}. \tag{12.32}$$

These four quantities, J_{sc}, V_{oc}, FF, and η, are the key performance characteristics of a solar cell. The standard test conditions for a solar cell are as follows: air mass 1.5 spectrum, an incident power density of 1000 W/m^2, and a cell temperature of 25°C.

12.3.2.2 Simulink Model of PV System

Cells are normally grouped into "modules," which are encapsulated with various materials to protect the cells and the electrical connectors from the environment. The manufacturers supply PV cells in modules, consisting of N_{PM} parallel branches, each with N_{SM} solar cells in series, as shown in Figure 12.9.

In order to have a clear specification of which element (cell or module) the parameters in the mathematical model refer to, the following notation is used from now on: the parameters with superscript "M" refer to the PV module, while the parameters with superscript "C" refer to the solar cell. Thus, the applied voltage at the module's terminals is denoted by V^M, while the total generated current by the module is denoted by I^M. A model for the PV module is obtained by replacing each cell in Figure 12.9 by the equivalent diagram from Figure 12.8a. In the following, the mathematical model of a PV module, suggested by Lorenzo (1994), is given. The advantage of this model is that it can be established applying only standard manufacturer supplied data for the modules and cells.

The photovoltaic array is nothing but modules connected in series and parallel. The PV array for the SRFCs consists of several modules, which in turn consist of solar cells. The solar cells used in designing the present PV array are *crystalline Si* solar cells. The appropriate cell details are given in the Table 12.2.

But, during the day, the solar irradiation from the sun is not at a constant wattage of 1000 W/m^2. It varies from a minimum to a maximum. The solar irradiation or the flux for any given day is like a positive sine curve. For the Simulink model, the flux is the input and the output is the power produced by an array. Just changing the number of cells or modules that are connected in series or parallel can alter this power.

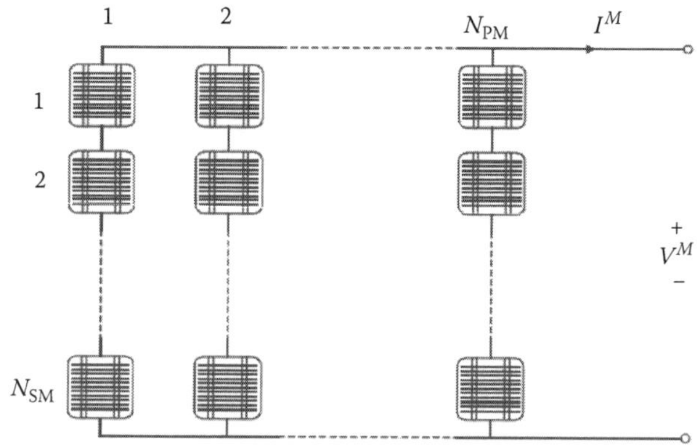

FIGURE 12.9
The PV module consists of N_{PM} parallel branches, each of N_{SM} solar cells in series. (From Hansen, A. D. et al., Models for stand-alone PV system. Riso National Laboratory, Report Riso-R-1219(EN)/SEC-R-12, Roskilde, Denmark, 2000.)

TABLE 12.2

Performance of Some Types of PV Cell

Cell Type	Area (cm²)	V_{oc} (V)	J_{sc} (mA/cm²)	FF	Efficiency (%)
Crystalline Si	4.0	0.706	42.2	82.8	24.7
Crystalline GaAs	3.9	1.022	28.2	87.1	25.1
Poly-Si	1.1	0.654	38.1	79.5	19.8
a-Si	1.0	0.887	19.4	74.1	12.7
$CuInGaSe_2$	1.0	0.669	35.7	77.0	18.4
CdTe	1.1	0.848	25.9	74.5	16.4

Source: Green, M. A. et al., *Progress in Photovoltaics: Research and Applications* 9: 49–56, 2001.

The operating cell temperature as a function of solar flux is given as

$$T^C = T_a + C_2 \times G_a \text{ and } C_2 = \frac{T^C_{ref} - T_{a,ref}}{G_{a,ref}}. \tag{12.33}$$

If T^C_{ref}—reference cell temperature—is not known, it is reasonable to approximate C_2 = 0.03°C·m²/W. The operating cell open circuit voltage depends exclusively on the operating cell temperature

$$V^C_{OC} = V^C_{OC,0} + C_3(Tc - T^C_0), \tag{12.34}$$

where C_3 = –2.3 mV/C. The operating short circuit current

$$I^C_{SC} = C_1 * G_a \text{ and } C_1 = \frac{I^C_{sc,0}}{G_{a,0}}. \tag{12.35}$$

In Figure 12.10, the Simulink model of the photovoltaic array is shown.

The simulated solar flux for 24 h time is shown in Figure 12.11a and the corresponding short circuit current is shown in Figure 12.11b. Using the solar cell efficiency shown in Figure 12.12a, the cell-generated voltage is shown in Figure 12.12b.

12.3.2.3 Fuel Cell Subsystem

In order to simplify the process of modeling, the properties of the three control volumes, or sections, of the fuel cell—anode, cathode, and membrane—will be considered to be uniform. In other words, the model will not describe what happens at an exact point along the membrane but will attempt to analyze the overall effects and the average values. In addition, several

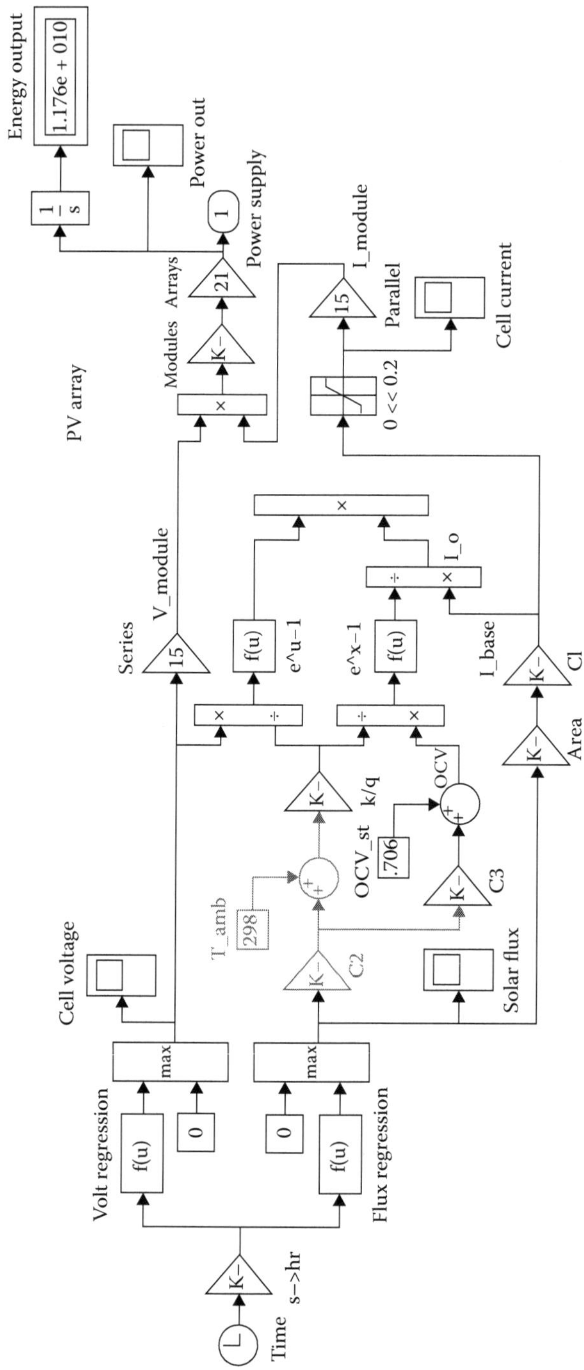

FIGURE 12.10
Photovoltaic array subsystem.

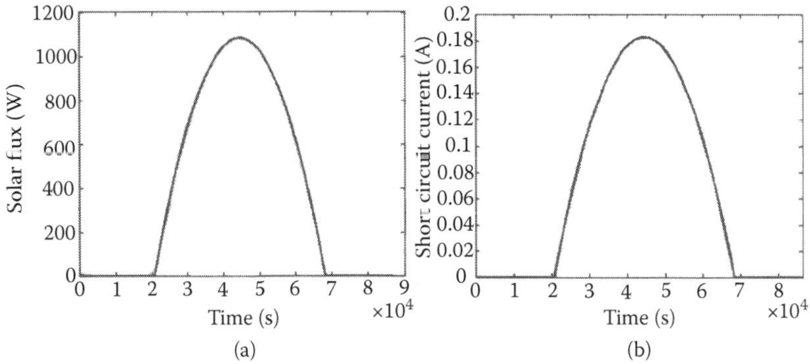

FIGURE 12.11
(a) Characteristic curve for the solar flux when simulated for 24 h. (b) Short circuit current curve.

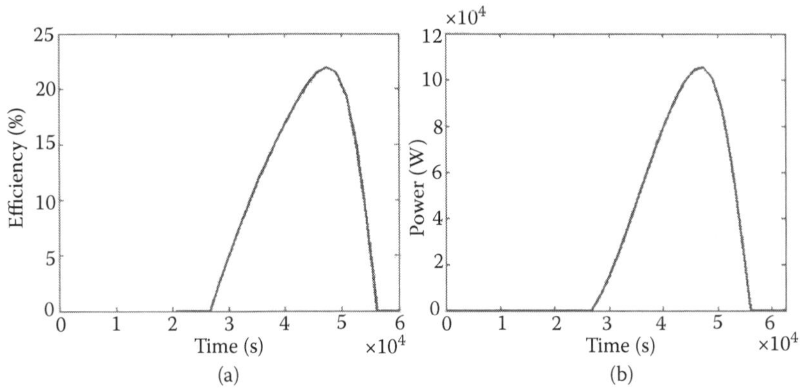

FIGURE 12.12
(a) Efficiency of the solar cell during the 24 h. (b) Power produced (watts) by solar array.

assumptions can be made that will greatly simplify the calculations without degrading the accuracy of the model:

- All gases are assumed to be ideal.
- Kinetic and potential energy of the gases are insignificant and will be ignored.
- Heat transfer by radiation is much smaller than by convection and can be neglected.
- Water is assumed to remain in a gaseous state to maintain simplicity.

The fundamental governing equations for a control volume are the continuity equation, based on the conservation of mass, the energy equation, and the momentum equation. Because this study considers the overall effects, the momentum equation will be omitted. The continuity and energy equations are shown below in integral form, respectively,

$$\frac{\partial}{\partial t}\iiint\limits_{CV}\rho \mathrm{d}V+\oiint\limits_{CS}\rho(\vec{V}\cdot\vec{n})\mathrm{d}A=0 \tag{12.36}$$

$$\frac{\partial}{\partial t}\iiint\limits_{CV}\rho e_{\mathrm{t}}\,\mathrm{d}V+\oiint\limits_{CS}\rho h(\vec{V}\cdot\vec{n})\mathrm{d}A=\frac{\mathrm{d}Q}{dt}+\frac{\mathrm{d}W}{dt}. \tag{12.37}$$

These two basic governing equations can be applied to the three control volumes of the fuel cell, but they are very abstract and generic. A more detailed discussion of what occurs in each of the control volumes continues below.

The mass flow in the anode and cathode consists of mass entering at the inlet, mass exiting at the outlet, and mass crossing the membrane. The inlet and outlet mass flow can be simplified using the nozzle flow rate equation as

$$\iint\limits_{\mathrm{inlet}}\rho h(\vec{V}\cdot\vec{n})\mathrm{d}A=k_{\mathrm{flow}}(\Delta p), \tag{12.38}$$

where k_{flow} is a flow coefficient and Δp is the pressure difference across the inlet or outlet.

The consumption of hydrogen and oxygen is directly related to the electrical current, as shown in the equation

$$\iint\limits_{\mathrm{elect}}\rho(\vec{V}\cdot\vec{n})\mathrm{d}A=N\frac{i}{n_{\mathrm{e}}F}M, \tag{12.39}$$

where n_{e} is the number of electrons transferred for each mole of the reactant and N is stoichiometric factor.

The energy balance in the anode and cathode channels consists of the energies of the masses entering and leaving the control volume and the heat transfer between the anode or cathode and the fuel cell body. These are given below:

$$\frac{\partial}{\partial t}\iiint\limits_{CV}\rho e_{\mathrm{t}}\,\mathrm{d}v=\sum_{i}\left(\frac{\mathrm{d}m_i}{dt}C_{\mathrm{v}_i}T_i\right) \tag{12.40}$$

$$\frac{\mathrm{d}Q}{dt}=h_{\mathrm{conv}}A(T_{\mathrm{an}}-T_{\mathrm{H_2}}). \tag{12.41}$$

When determining heat transfer in the cathode, the subscripts "an" and "H_2" would be replaced by "ca" and "O_2," respectively. The current model calculates separate temperatures for the anode channel, cathode channel, and cell body.

The mass of the cell body does not change; thus, the general continuity equation (Equation 12.36) can be ignored. The energy equation relates the heat transferred by convection to the anode, cathode, and environment; the heat generated during the chemical reaction; and the electrical power generated. This can be characterized as

$$\frac{dQ}{dt}+\frac{dW}{dt}=h_{an}A(T_{an}-T_{body})+h_{ca}A(T_{ca}-T_{body})+h_{amb}A(T_{amb}-T_{body})+\Delta H_{R,T}-Vi \tag{12.42}$$

where h represents the corresponding convective heat transfer coefficient and $\Delta H_{R,T}$ is the lower heating value of hydrogen gas.

In order to determine the cell voltage, three voltage values must be calculated: the open circuit voltage, the activation loss, and the ohmic loss. The open circuit voltage is the ideal voltage dictated by thermochemistry. It can be found by the Nernst equation (Xue et al., 2004)

$$\begin{aligned} V_{Nernst} &= 1.229-(8.5\times10^{-4})(T_{body}-298.15) \\ &\quad +(4.308\times10^{-5})T_{body}[\ln(P_{H_2})+0.5\ln(P_{O_2})], \end{aligned} \tag{12.43}$$

where the partial pressures within the anode and cathode channels are used for P_{H2} and P_{O2}. The activation loss represents the voltage lost in generating the chemical reaction, and the ohmic loss measures the electrical resistances within the fuel cell. There are many methods used to calculate these two values, most of which involve at least a few empirical constants. The current model determines activation as a function of temperature, current, and reactant pressure, and ohmic losses as a function of temperature and current. These equations neglect the effects of membrane water content, which is accounted for in the seven empirical constants.

12.3.2.4 Simulink Model and Results

The MATLAB-Simulink model for the fuel cell system is broken down into four subsystems—the anode channel, the cathode channel, stack temperature, and stack voltage. Figure 12.13 shows the Simulink model of fuel cell breakdown. In Figures 12.14 and 12.15, the Simulink models for the anode subsystem and stack voltage, respectively, are shown. The anode and cathode subsystem models are similar. The model can be set to model many different dynamic parameters of the fuel cell, including anode pressure, reactant flow, temperature, and voltage.

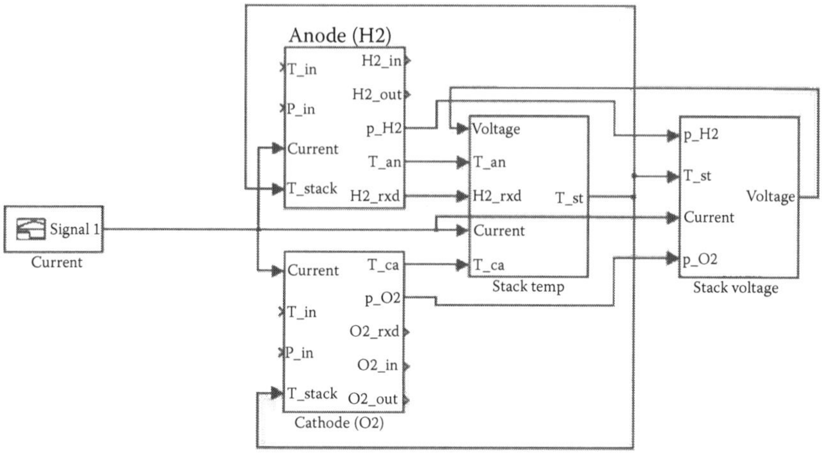

FIGURE 12.13
Fuel cell subsystem breakdown.

Figures 5.2 and 5.3 demonstrate the results of a test of length 5000 s. Simulink results for the cell voltage current characteristics and the corresponding stack temperature are shown in Figure 12.16 and 12.17, respectively.

The integrated closed-loop SRFC system Simulink model is shown in Figure 12.18. The closed-loop Simulink model was run for one complete day,

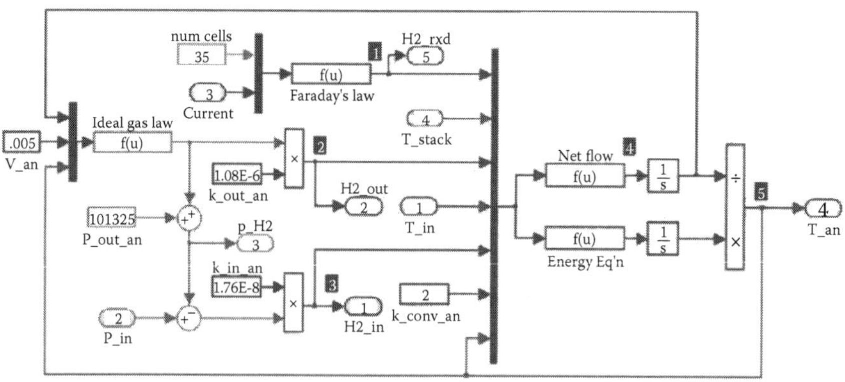

1. The rate of the chemical reaction in mol/s.
2. The molar outflow of hydrogen gas.
3. The molar inflow of hydrogen gas.
4. The net molar flow of hydrogen gas.
5. The temperature of the anode.

FIGURE 12.14
Anode subsystem for fuel cell.

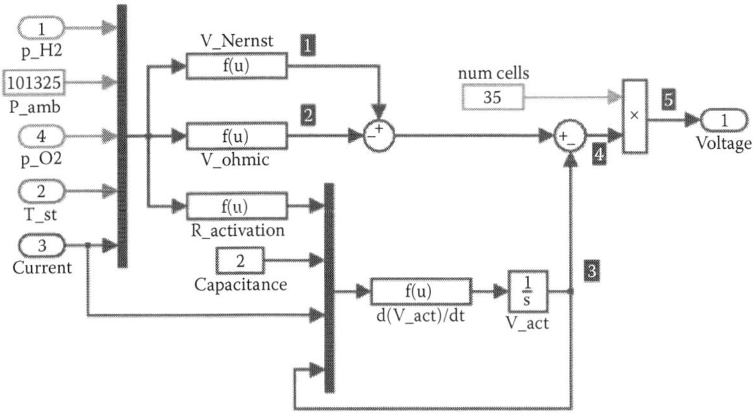

1. The rate of the chemical reaction in mol/s.
2. The molar outflow of hydrogen gas.
3. The molar inflow of hydrogen gas.
4. The net molar flow of hydrogen gas.
5. The total voltage output of the fuel cell.

FIGURE 12.15
Stack voltage subsystem for fuel cell.

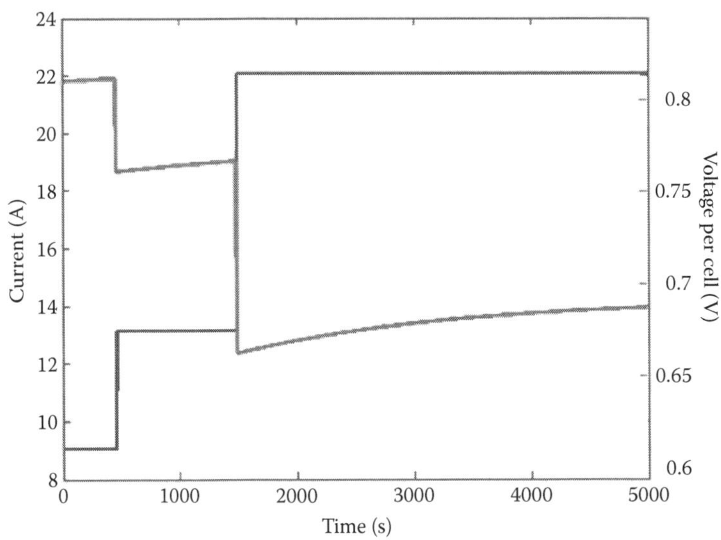

FIGURE 12.16
Current load on fuel cell and corresponding cell voltage.

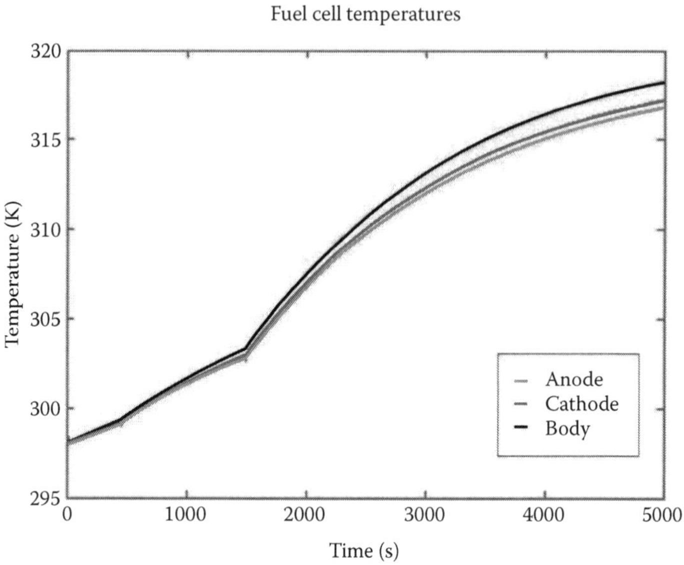

FIGURE 12.17
Anode, cathode, and stack temperatures.

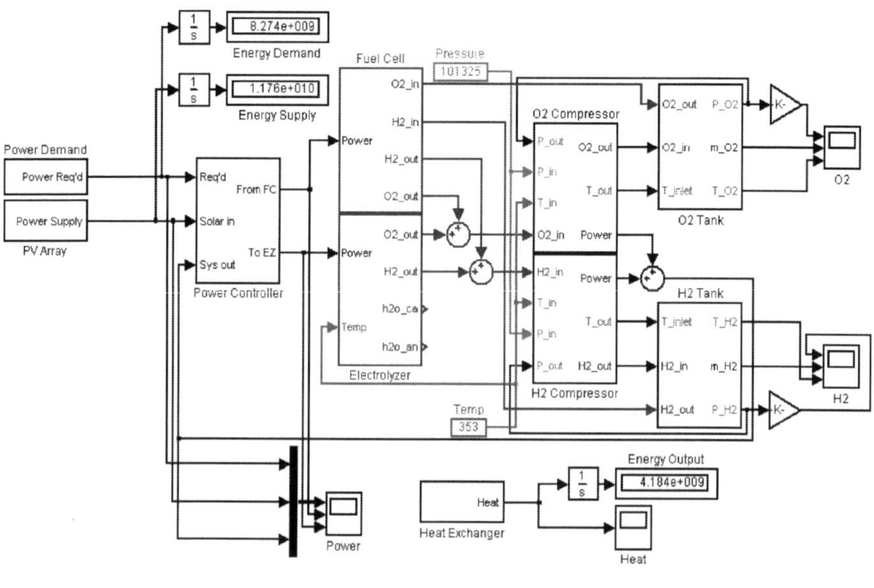

FIGURE 12.18
Closed-loop SRFC system after interfacing.

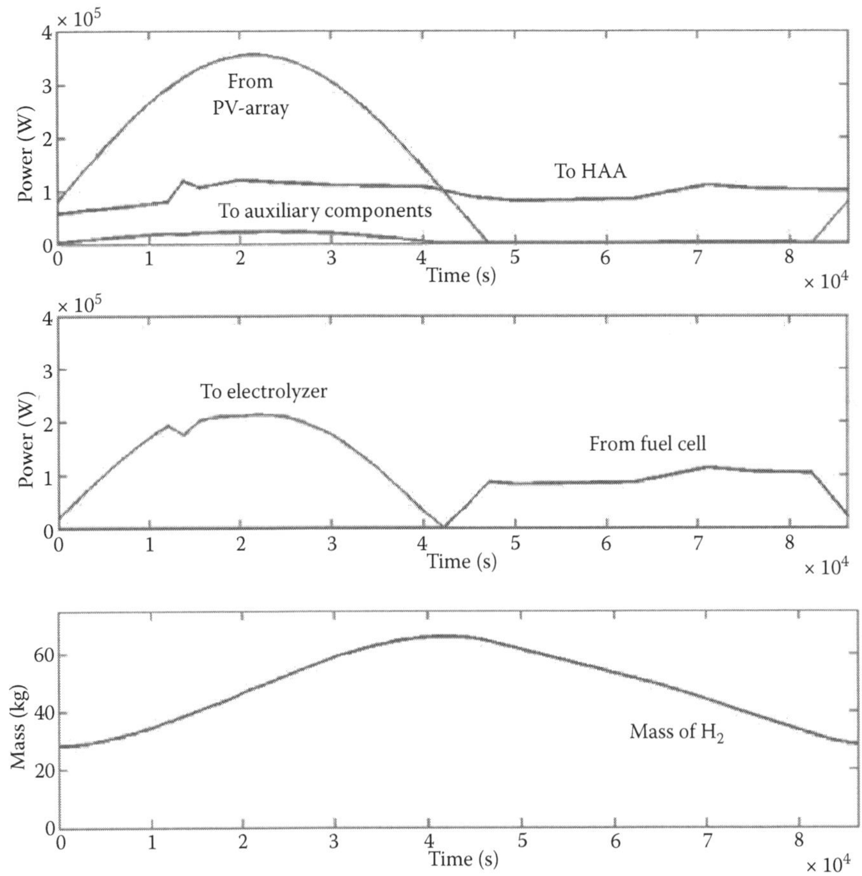

FIGURE 12.19
Power distribution in the SRFC system.

that is, 24 h, and the results are shown in Figure 12.19. This figure shows the electrical power received by the PV array, the power generated by the fuel cell, the power required for the load (HAA) and the auxiliary components, and the hydrogen inventory in the storage tank. During daytime, the power from the PV arrays is used to power the electrolyzer, load, and the auxiliary system. During nighttime, the fuel cell supplies the needed power to load and the auxiliary systems using the hydrogen produced and stored during daytime from the electrolyzer.

12.3.3 Molten Carbonate Fuel Cell System Model

High-temperature fuel cells (MCFC and SOFC) operate at temperatures exceeding 950 K and produce an exuberant amount of heat that can be either given off as waste heat or harnessed for use in another system. For this reason, high-temperature fuel cells are more apt to be used in a hybrid or combined cycle design. This does not limit hybrid systems to only use high-temperature fuel cells, however. Hybrid systems have proven to generate high efficiencies with lower levels or pollutant emissions than current generating technologies. This allows for greater fuel energy savings than non-hybrid systems because of the higher fuel utilization. Fuel cell hybrid systems are expected to help alleviate future potential electricity generation shortages, strain on the transmission grid, dependence on foreign oil, and growing environmental concerns.

MCFC combined with a turbine technology has been considered since the mid-1990s. Figure 12.20 shows the process flow sheet, where natural gas is fed to an external reformer and then to the anode. The anode exhaust is then oxidized in a catalytic burner that is mixed with air from the compressor and fed to the cathode side of the fuel cell. The now heated cathode exhaust is used to turn a turbine to produce power in a bottoming cycle. Such configuration has been found to have a steady-state electrical efficiency of 54.8%.

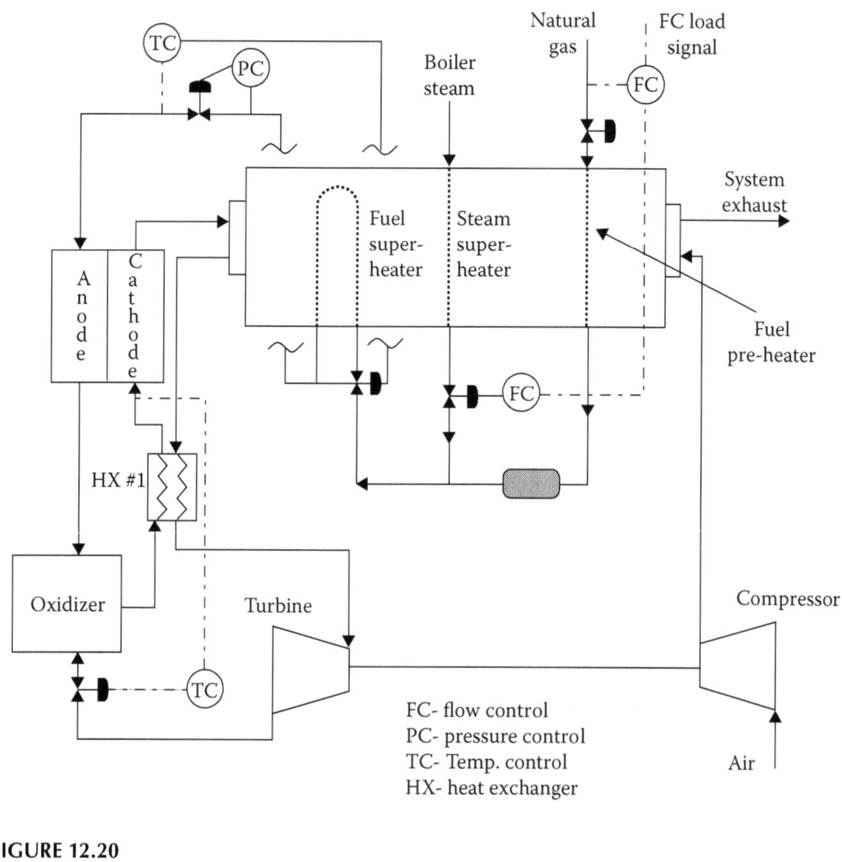

FIGURE 12.20
PFD of MCFC–GT hybrid system.

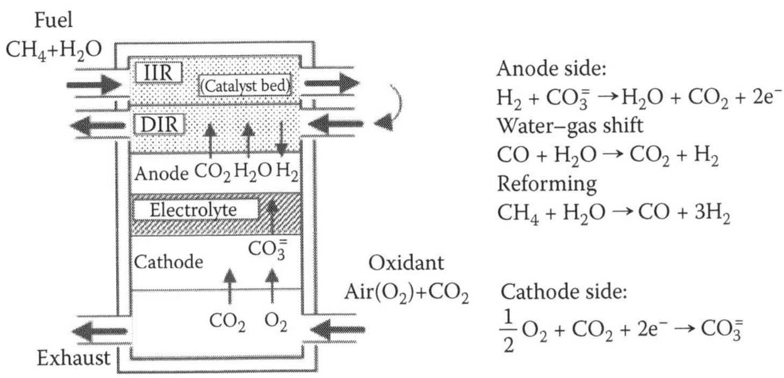

FIGURE 12.21
IIR/DIR–MCFC model and reaction.

12.3.3.1 Geometry

Here, an MCFC system with natural gas partially reformed in an internal reformer and partially in the anode compartment as shown in Figure 12.21 is considered (Wolf, 2007). This is a combination of indirect internal reforming (IIR) and direct internal reforming (DIR), which leads to better thermal management. In the IIR step, a reforming unit is placed between every 10 cells in the stack and converts about 50% of the natural gas to hydrogen-rich gas prior to entering the anode section. The rest of the reforming is left for the anode, which contains reforming catalysts. In the model, the fuel cell geometry is split into two compartments. One compartment is used to model both the anode sides IIR reforming and DIR reforming volume, while the other compartment represents the cathode volume. The model is a lumped parameter, zero-dimensional model with each volume representing a continuously stirred tank reactor (CSTR).

12.3.3.2 Mass Balance

Gas component balances are derived using the CSTR. Assuming ideal gas for a CSTR, a mole balance equation is given by

$$\frac{\mathrm{d}N}{\mathrm{d}t} = N^{\mathrm{in}} - N^{\mathrm{out}} + R, \tag{12.44}$$

where N is the total change of moles in the reactor and R is the total rate of production of species in the reactor. For a single species n_i, this can be written as

$$\frac{\mathrm{d}n_i}{\mathrm{d}t} = n_i^{\mathrm{in}} - n_i^{\mathrm{out}} + R_i. \tag{12.45}$$

Under the ideal gas assumption, one can write the total concentration of species in a volume as

$$\tilde{C}_{\mathrm{t}} = \frac{P}{RT} = \frac{N}{V}. \tag{12.46}$$

From the ideas gas law, $PV = nRT$, and $V\tilde{C}_{\mathrm{t}} = N$, where N is the total number of species in the reactor. Applying this to Equation 12.45, the change in concentration of each species in a constant volume reactor is then

$$V\frac{\mathrm{d}C_i}{\mathrm{d}t} = \frac{\mathrm{d}n_i}{RT} = n_i^{\mathrm{in}} - n_i^{\mathrm{out}} + R_i. \tag{12.47}$$

The total rate of production of each species i is dependent on all chemical reactions occurring in the volume. Assuming that there are a total of μ independent reactions, then the total rate of production of each species i is generally

$$R_i = \sum_{j=1}^{\mu} \upsilon_{ij} r_j, \tag{12.48}$$

where μ is the total number of chemical reactions, r_j is rate of reaction j, and υ_{ij} are the stoichiometric coefficients of species i in reaction j. For example, in the reforming reaction of $CH_4 + H_2O \rightarrow CO + 3H_2$, H_2 has a coefficient of 3, and r_j is the molar extent of the reaction or reaction rate of reaction j. The total flow rate of a species is simply the species mole fraction multiplied by the total flow rate of all species ($N_i = x_i\, N_{\mathrm{tot}}$). The concentration of an individual species can be written as $C_i = x_i C_{\mathrm{tot}}$. Using the lumped parameter assumption, the total flow rate out of the reactor can be written as the total flow in plus the sum of the rate of production over all species and reactions in the volume.

$$N_{\mathrm{tot}}^{\mathrm{out}} = N_{\mathrm{tot}}^{\mathrm{in}} + \sum_{i=1}^{\xi} R_i \tag{12.49}$$

Substituting these relations into Equation 12.27 gives us a component material balance

$$V\tilde{C}_{\mathrm{t}}\frac{\mathrm{d}x_i}{\mathrm{d}t} = N^{\mathrm{in}}(x_i^{\mathrm{in}} - x_i^{\mathrm{s}}) - x_i^{\mathrm{s}}\sum_{i=1}^{\xi} R_i + R_i, \quad i = 1,\ldots.\xi. \tag{12.50}$$

12.3.3.3 Reaction Rates

In order to fulfill Equation 12.50, we must define the reaction rates of the system volumes. The anode and cathode reactions are electrochemical and therefore proportional to load current through Faraday's law. This allows them to be rather easily determined.

The reforming reaction and water–gas shift reaction, on the other hand, are dependent on many parameters such as pressure, temperature, and gas composition at the anode compartment. If these were defined separately, it could add much complexity to the model. It is therefore important to simplify the reaction kinetics as much as possible. The water–gas shift reaction is assumed to be in equilibrium owing to the fact that it has a very fast reaction rate compared to that of the reforming reaction. This imposes a simple constraint on the gas species that appear on the anode side described below.

$$K_{\mathrm{wgs}} = \frac{x^{\mathrm{s}}_{\mathrm{CO_2}} x^{\mathrm{s}}_{\mathrm{H_2}}}{x^{\mathrm{s}}_{\mathrm{CO}}\ x^{\mathrm{s}}_{\mathrm{H_2O}}} = e^{(E_1 + E_2/T^{\mathrm{S}})} \tag{12.51}$$

Here, K_{wgs} is the water–gas shift reaction's equilibrium constant dependent on Gibbs free energy and stack temperature, where E_1 and E_2 are constants related to Gibbs free energy change. The assumption that this reaction is in equilibrium allows us to remove its reaction rate from the equation set.

At the anode side, there are three reactions occurring simultaneously. In the following, these will be defined as reactions 1, 2, and 3, respectively. The rate of reaction for the water–gas shift reaction can then be separated from the term for total rate of production at the anode as follows:

$$R_{ai} = \sum_{j=1}^{3} v_{ij} r_j = \sum_{\substack{i=1 \\ j \neq 2}}^{\xi} v_{ij} r_j + v_{i2} r_2 \equiv \overline{R}_{ai} + v_{i2} r_2. \tag{12.52}$$

Now, r_2 can be separated and eliminated from the equation set. First, the set of all gas species that can be present in either the anode or cathode is defined:

$$\mathbf{S} \equiv [\mathrm{H_2\ CH_4\ CO\ CO_2\ H_2O\ N_2\ O_2}], \tag{12.53}$$

where the ordering is equivalent to the index i. On the anode side, the component balance for CO is chosen to express r_2 by substituting Equation 12.52 into Equation 12.50; then, the resulting equation is substituted into the remaining component balances of the anode side to eliminate r_2:

$$\frac{P_a^s V_a}{RT^s}\frac{dx_{ai}^s}{dt} = N_{ta}^{in}(x_{ai}^{in} - x_{ai}^s) - \sum_{i=1}^{\xi} \bar{R}_{ai} - \bar{R}_{ai} + \frac{(v_{i2} - x_{ai}^s \bar{v}_2)}{(v_{32} - x_{CO}^s \bar{v}_2)}$$
$$\times \left[\frac{P_a^s V_a}{RT^s}\frac{dx_{CO}^s}{dt} - N_{ta}^{in}(x_{CO}^{in} - x_{CO}^s) - x_{CO}^s \sum_{i=1}^{\xi} \bar{R}_{ai} - \bar{R}_{a3} \right], \tag{12.54}$$

where

$$\bar{v}_2 = \sum_{i=1}^{\xi} v_{i2}. \tag{12.55}$$

The stoichiometric coefficients of the water gas shift reaction are

$$v_2 = \{1 \quad 0 \quad -1 \quad 1 \quad -1 \quad 0 \quad 0\}, \tag{12.56}$$

where negative coefficients represent reactants and positive coefficients represent products. This summation of Equation 12.55 is then $\bar{v}_2 = 0$, which implies that when $i = 2$, 6, or 7, the bracketed term in Equation 12.54 is multiplied by zero. Thus, a coupled set of implicit differential equations for $x_{H_2}^s$, x_{CO}^s, $x_{CO_2}^s$, $x_{H_2O}^s$, and T^s exists. In order to create an explicit set, the water–gas shift equilibrium constraint is differentiated with respect to time, resulting in

$$\frac{dx_{CO}^s}{dt} = \frac{x_{CO}^s}{x_{H_2}^s}\frac{dx_{H_2}^s}{dt} + \frac{x_{CO}^s}{x_{CO_2}^s}\frac{dx_{CO_2}^s}{dt} - \frac{x_{CO}^s}{x_{H_2O}^s}\frac{dx_{H_2O}^s}{dt} + \frac{x_{CO}^s E_2}{(T^s)^2}\frac{dT^s}{dt}. \tag{12.57}$$

The last step to forming an explicit step is to substitute $x_{H_2}^s$, $x_{CO_2}^s$, $x_{H_2O}^s$, and T^s into Equation 12.57, and solving for the only derivative left, x_{CO}^s. It is important to note that this derivation is only needed for the anode side, since the cathode side only has one electrochemical reaction rate. The last requirement of the anode side balance is the definition of the reaction rate for the reforming reaction and electrochemical reaction.

Assuming equilibrium allows us to lump the water–gas shift reaction and reforming reaction into one equivalent expression,

$$CH_4 + 2H_2O \rightarrow CO_2 + 4H_2. \tag{12.58}$$

This reaction has been studied extensively, and in 1975, Rostrup-Nielsen (2005) proposed a simple expression to describe its reaction rate below.

$$r_{3'} = kP_{CH_4}[(1-Q)/K_{REF}] \quad (mol/s) \tag{12.59}$$

In this expression, k is the reforming rate constant (mol/s), kP_{CH_4} is the methane partial pressure, Q is the mass action expression, and K_{REF} is the reforming reaction rate equilibrium constant. The reaction rates of both the electrochemical reactions can be described through Faraday's law as

$$r = \frac{-A_e c_{cell} n_{stack} i}{2F}, \tag{12.60}$$

where A_e is the fuel cell active area (cm^2), n_{cell} is the number of cells in the stack, n_{stack} is the number of stack, i is current density (A/cm^2), and F is the Faraday constant (J/mol V).

12.3.3.4 Energy Balance

The fuel cell stack hardware has a large mass-specific heat product and is many more magnitudes larger than the mass-specific heat product of the gases inside the fuel cell volumes at any given time. Because of this, along with the assumptions of ideal gas mixtures, and a cell stack temperature with no gradient between the anode and cathode, a single energy balance is written accounting for energy storage only within the fuel cell stack hardware.

$$\begin{aligned} M^s C_p^s \frac{dT^s}{dt} &= N_a^{in}\left[\sum_{i=1}^{\xi} x_{ai}^{in}(\bar{h}_{ai}^{in} - \bar{h}_{ai}^{s})\right] - \sum_{i=1}^{\xi} \bar{h}_i^s R_{ai} \\ &\quad - N_c^{in}\left[\sum_{i=1}^{\xi} x_{ci}^{in}(\bar{h}_{ci}^{in} - \bar{h}_{ci}^{s})\right] - \sum_{i=1}^{\xi} \bar{h}_i^s R_{ci} - Q_l - P_{dc}, \end{aligned} \tag{12.61}$$

where

$M^s C_p^s$ = stack solid mass-specific heat product (J/K)
T^s = stack solid average temperature (K)
$N_a^{in}(N_c^{in})$ = anode (cathode) total inlet molar flow (mol/s)
$x_{ai}^{in}(x_{ci}^{in})$ = anode (cathode) inlet mole fractions
$\bar{h}_{ai}^{in}(\bar{h}_{ci}^{in})$ = anode (cathode) inlet partial molar enthalpies (J/mol)
$\bar{h}_i^s$ = partial molar enthalpies at stack temperature (J/mol)
ξ = total gas components in anode or cathode
Q_l = convective heat loss (W)
P_{dc} = stack dc power (W)
$R_{ai}(R_{ci})$ = anode (cathode) total rate of production of species i (mol/s)

The first and third terms on the right-hand side of Equation 12.61 are the energy changes as a result of flow into and out of the anode and cathode based on inlet molar flow rate. These terms do not account for changes owing to reactions and therefore the second and fourth terms correspond to the heat produced or lost by reaction in both the anode and cathode. It is assumed that the anode and cathode exhaust exit at the stack temperature.

The ideal gas assumption allows the partial molar enthalpies to be written as a function of specific heats at a constant pressure. The partial molar enthalpies are computed using

$$\bar{h}_i = \bar{h}_i^{\text{ref}} + \int_{T_{\text{ref}}}^{T} c_{\text{p},i}(u)\,\mathrm{d}u, \tag{12.62}$$

where the specific heats are calculated using

$$c_{\text{p},i} = a_i + b_i T + c_i T^2 + \text{d}_i T^3. \tag{12.63}$$

The coefficients of specific heats are found in standard reference tables. Each of the chemical species involved may undergo a chemical reaction. In chemical reacting systems, it is important to account for the energy change owing to chemical reaction. This energy change is taken into account by adding heat of formation for each gas species to the reference enthalpy at standard reference temperature. The total rates of production R_i is given by Equation 12.48.

12.3.3.5 Performance

Quantifying performance characteristics is an important step in developing a simulation model for a fuel cell. Typically, the first step is to find the ideal performance and then subtract the losses from the ideal performance to describe the actual behavior of the fuel cell. The ideal performance of a fuel cell is described by the Nernst equation:

$$V_{\text{FC}} = E_{\text{o}} + \frac{RT}{2F}\ln\frac{P_{\text{H}_{2,\text{a}}} P_{\text{O}_{2,\text{c}}}^{1/2}}{P_{\text{H}_2\text{O}_{,\text{a}}} P_{\text{CO}_{2,\text{a}}}}, \tag{12.64}$$

where E_{o} is ideal standard potential [V] and P_i is partial pressure ($P_{i,\text{a}} = x_{\text{a}i}P_{\text{a}}/P_{\text{atm}}$, $P_{i,\text{c}} = x_{\text{c}i}P_{\text{c}}/P_{\text{atm}}$). The process considered is for the overall fuel cell reaction. The ideal standard potential is a function of Gibbs free energy and can be written algebraically as a function of temperature. Here, the following relation was used to determine the ideal standard potential for reaction.

$$E_{\text{o}} = 1.2723 - 2.7645 \times 10^{-4}\, T \tag{12.65}$$

From this, the cell voltage can be described as

$$V_{cell} = V_0 - i\,(R_{ohm} + Z_{anode} + Z_{cathode}), \tag{12.66}$$

where i is cell current density and R_{ohm}, Z_{anode}, and $Z_{cathode}$ are cell ohmic impedance or internal resistance, anode and cathode polarization and activation resistance, respectively.

Because of the fact that the internal resistance of a fuel cell is strongly dependent on the cell materials, contact resistance, and temperature, one must rely strongly on experimental data and literature to model fuel cell voltage. The polarization resistances are not well established physically; thus, correlations are generally used to describe the losses involved. The following relations listed are from experimental data and literature:

$$R_{ohm} = 0.5\times10^{-4}\exp\left[3016\left(\frac{1}{T}-\frac{1}{923}\right)\right] \text{ (Koh et al., 2002)} \tag{12.67}$$

$$Z_{anode} = 2.27\times10^{-9}\exp\left(\frac{6435}{T}\right)P_{H_2}^{-0.42}P_{CO_2}^{-0.17}P_{H_2O}^{-1.0} \text{ (Yuh and Selman, 1991)} \tag{12.68}$$

$$Z_{cathode} = 7.505\times10^{-10}\exp\left(\frac{9298}{T}\right)P_{O_2}^{-0.43}P_{CO_2}^{-0.09} \text{ (Yuh and Selman, 1991)} \tag{12.69}$$

The anode side consists of a reforming unit and the anode; however, in this simulation, the two are lumped together in one volume. Since the reforming unit is not directly in contact with the electrolyte as seen in Figure 12.8, it must be separated from the performance calculations. This is done by some assumptions to account for the gas composition within the reforming unit. For performance calculations, the following assumptions are used:

$$T = \frac{(T^s + T_c^{in})}{2}, \quad P_{i,c} = \frac{P_c^s(x_{ci}^s + x_{ci}^{in})}{2}, \quad P_{i,a} = \frac{P_a^s(x_{ai}^s + x_{ai}^{ru})}{2}. \tag{12.70}$$

In the equations above, T is the arithmetic average of cathode inlet and exit temperatures. Since the cathode does not have a reforming unit, the partial pressures are treated as arithmetic averages of inlet and exit gas partial pressures. In the anode, however, x_{ai}^{ru} denotes the reforming unit exit mole fraction, which is essentially the entrance to the anode. These are solved for by assuming that the water gas shift and reforming reactions are at equilibrium, corresponding to a pre-equilibrium temperature and pressure. The assumptions made for pre-equilibrium temperature and pressure are

$$T_{pre} = T_a^{in}, \quad P_{pre} = P_a^s. \tag{12.71}$$

The calculation of x_{ai}^{ru} becomes an equilibrium problem, solved for each time step by a Newton–Raphson iteration. Using these correlations in Equation 12.66 yields a mathematical description of the voltage for simulation.

12.3.4 MATLAB/Simulink Simulation of MCFC

The model described in the above section for the MCFC with IIR/DIR is implemented in MATLAB/Simulink (Figure 12.22). Simulink is a modeling environment that is capable of solving mixed systems of integral, partial differential, and algebraic equations. It can be used for both steady-state and dynamic calculations. For dynamic simulation, inputs or external actions are used to put a disturbance on the system. The overall system model is broken down into interconnected subsystems. Each component is a subsystem in itself and may also carry subsystems within it. The fuel cell model consists of 10 differential state equations and over 150 algebraic calculations.

12.3.4.1 Steady-State Analysis

For steady-state analysis, the inputs of the system are held constant. No external controls accompany the system; however, the steam-to-carbon ratio is set equal to 2 at the anode inlet. Also, 75% fuel utilization was assumed and also held constant to determine the flow rate into the anode. The steady-state results correspond to a constant full load of 2 MW DC power, which corresponds to a 160 mA/cm^2 load current density. The model represents 16,125 kW stacks connected four in series and four in parallel. Table 12.3

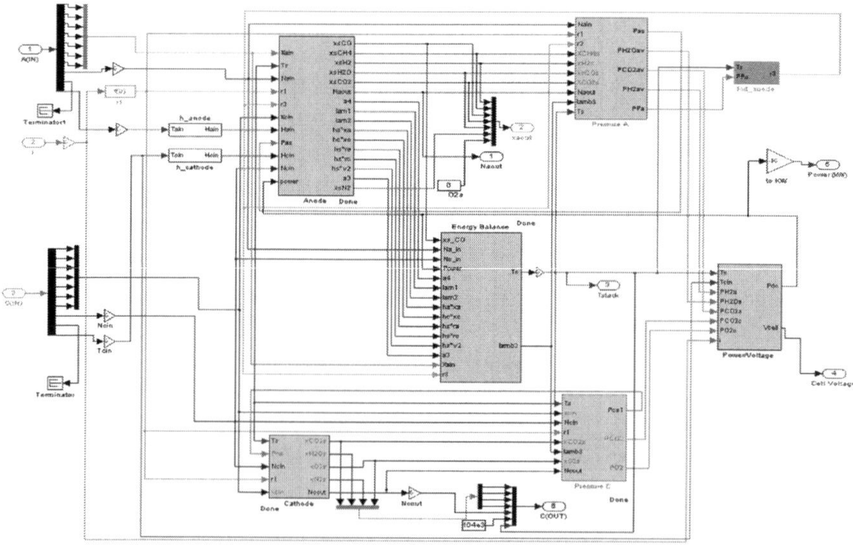

FIGURE 12.22
Fuel cell simulation model in Simulink.

TABLE 12.3
Steady-State Results from the MCFC Simulation Model

	Current = 160 mA/cm²; Power = 2139 kW			
Mole Fraction	**Anode Inlet**	**Anode Outlet**	**Cathode Inlet**	**Cathode Outlet**
H_2	0.1168	0.074	0	0
CH_4	0.2798	0.0011	0	0
CO	0.0005	0.0459	0	0
CO_2	0.0346	0.4537	0.1553	0.0476
H_2O	0.5662	0.4244	0.1553	0.188
N_2	0	0	0.5599	0.6778
O_2	0	0	0.1294	0.0865
Temperature (K)	849	949	838	949
Flow rate (mol/s)	17.61	141	43.7	116.4

shows the steady-state model results where the anode and cathode inlet mole fractions, flow rate, and temperature were held constant. Inlet and outlet mole fractions are listed corresponding to each species. Results of the simulation show that 99% of the methane is reformed in the anode. Also, the electrochemical reactions raised the stack temperature 100 K above inlet temperatures. The fuel cell efficiency is calculated as 46.3%.

12.3.4.2 *Transient Simulation*

The transient response of the fuel cell to a load trip is simulated, where the full load current of 160 mA/cm² is instantaneously removed during steady-state operation. In the transient, the anode and cathode inlet properties are held constant. Figure 12.23 shows the effect of the load change on the gas

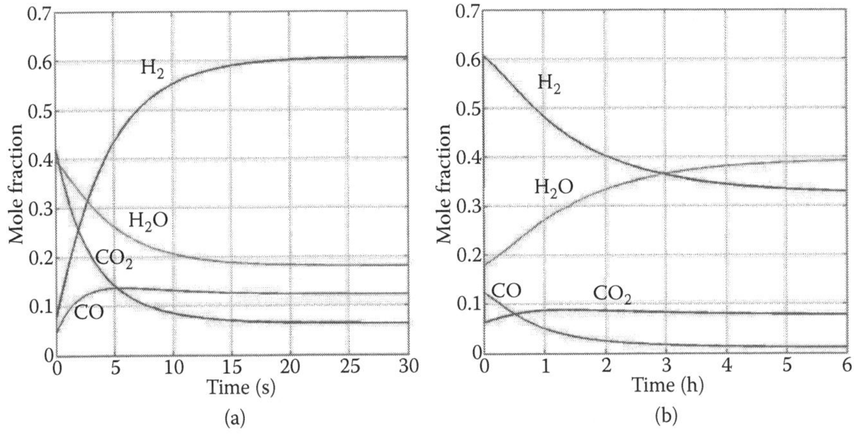

FIGURE 12.23
Response of the anode section for 100% load trip, (a) first 30 s and (b) for a 6 h period.

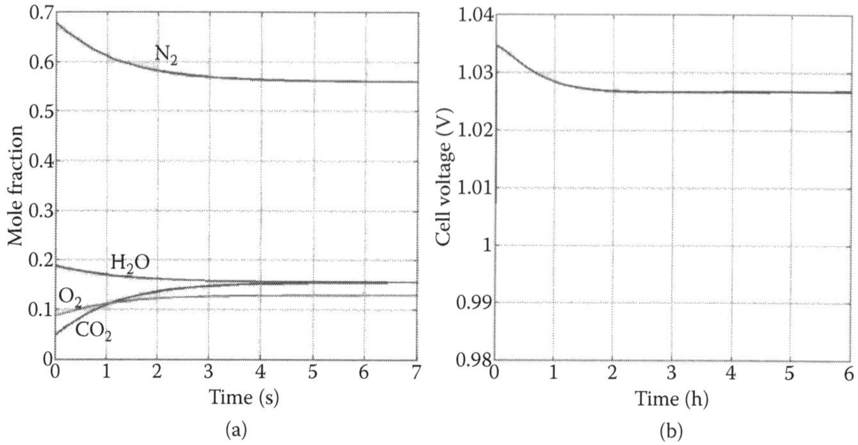

FIGURE 12.24
(a) Response stack temperature and (b) cell voltage for 100% load trip.

composition when the electrochemical reaction rate at the anode decreases to zero. The fuel cell consumes less hydrogen while the reforming reaction, driven by temperature, keeps producing hydrogen at nearly the same rate. Figure 5.2 is shown to analyze the long-term effects of the relatively slow response of the reforming reaction rate in the anode compartment.

This slow reforming rate can be explained by studying the stack temperature as seen in Figure 12.24a. This slow response is due to the ability of the stack to store massive amounts of energy because of its very large mass-specific heat product. This is an excellent quality of the fuel cell, because during a transient such as this, heat can still be generated to maintain partial turbine side power. Also, the combustible fuel not consumed in the anode can be transformed to heat by the catalytic combustor. The composition in the cathode responds very quickly because only electrochemical reactions take place; therefore, no slow response occurs. Because of the dependence of the cell voltage on both anode and cathode compositions as well as stack temperature, there exists both fast and slow time behavior as seen in Figure 12.24b.

12.4 Control System

12.4.1 Fuel Cell System Control

A fuel cell system may be well understood, but it requires a control strategy for predictable operation. Fuel cell systems are complex and must have some sort of regulation to reliably achieve desired power levels. Without regulation, fuel cell power output will drift over time, even if reactant flow rates are held steady. Fuel cells for transportation definitely require controls since they will experience widely varying changes in power demand just like internal combustion engines in present vehicles. A fuel cell stack has to be operated properly to get good power efficiency, reliability, and smooth operation; hence, a control system is required during steady-state operation and current level changes. Controls can be used to regulate the ideal amounts of reactants, cell temperature, inlet humidity, and output power, for example. Many different control configurations can be used to accomplish these goals.

In a PEM fuel cell, three key control systems are required for air and fuel delivery, water management, and heat management. Air delivery control is the largest factor in maximizing fuel cell efficiency. Thus, there are three key parameters of fuel cell stack, the stack current *I*, stack voltage *E*, and the oxygen stoichiometric ratio λ_{O2}. One of the main disadvantages of the fuel cells stack system is oxygen starvation. Since current is instantaneously drawn from the load source connected to the fuel cell, the control system is required to maintain optimal stack temperature, membrane hydration, and partial pressure of the reactants across the membrane to avoid degradation of the cell voltage that can reduce efficiency. Various control systems have been developed for the optimal and efficient operation of the fuel cell stacks. For example, when the current is drawn from the fuel cell, different methods are employed to control the breathing of the fuel cell stack in order to prevent the problem of oxygen starvation.

If fuel cell stacks are coupled with other systems such as a distribution generation system, the resulting system complexity increases and the transient behaviors of the fuel cell systems become highly nonlinear. Numbers of models have been advanced and control algorithms have been developed to control operation of the fuel cell system in the desired fashion. For example, Pukrushpan et al. (2004) have implemented a nonlinear fuel cell dynamic model for a control study of the fuel cell, where the model captures the transient phenomena that include the flow characteristics and inertia dynamics of the compressor, the manifold filling dynamics, and, consequently, the reactant partial pressures. This model included an observer-based feedback and feed-forward controller that manages the trade-off between reduction of parasitic losses and fast fuel cell net power response during rapid current (load) demands. An air flow controller to protect the fuel cell stack from oxygen starvation during step changes of current demand is designed by assigning an integrator to the compressor flow. Linear operability techniques were employed to demonstrate improvements in transient oxygen regulation when the stack voltage is included as a measurement for the feedback controller. The limitation of this method arises when the fuel cell stack system architecture dictates that all auxiliary equipment is powered directly from the fuel cell with no secondary power sources. Recently, methods for controlling a nonlinear under-actuated system using augmented sliding mode control have been investigated. This control approach involves introducing a transformation matrix mapping the system's input influence matrix to a transformed system that is square and thus invertible; this approach is shown to control selectable states with proper choice of the transformation matrix yielding good control performance.

Neural networks and fuzzy logic have also been applied in the development of the control system of the fuel cell.

Fuzzy logic controller techniques have been found to be a good replacement for conventional control techniques, owing to their low computational burden and ease of implementation using microcomputers. The fuzzy logic–based controller overcomes system ambiguities and parameter variations by modeling the control objective based on a human operator experience, common sense, observation, and understanding how the system responses, thereby eliminating the need for an explicit mathematical model for the system dynamics. In the following sections, these various control techniques are discussed.

12.4.2 Control Techniques

12.4.2.1 Control Problem Formulation

In the control technique, the first one has to define the objective or objectives for the control. Once the objective is identified, the key parameters affecting the objective are then identified. These parameters are studied for how they affect the objective, and based on that, the design of the control system is devised. Let us consider the control problem of maintaining the oxygen stoichiometric ratio desired $\lambda_{O2,des} = 2$ with compressor motor input voltage $V_{cp,m}$ and achieving a desired fuel cell net power of $P_{net,des}$. This control strategy for air delivery is discussed following the work of Pukrushpan (2003). The required stack current I for this desired net power can be expressed as a function of net power P_{net} as $I = f(P_{net})$. The current is considered as an external input or disturbance to the system. The control problem is formulated as shown in Figure 12.25. The state space formulation, system states, system measurements required, performance variables to be optimized, control input, and system disturbance are given through the following set of equations.

$$\text{State space: } \dot{\underline{x}} = f(\underline{x}, \underline{u}, \underline{w}) \tag{12.72}$$

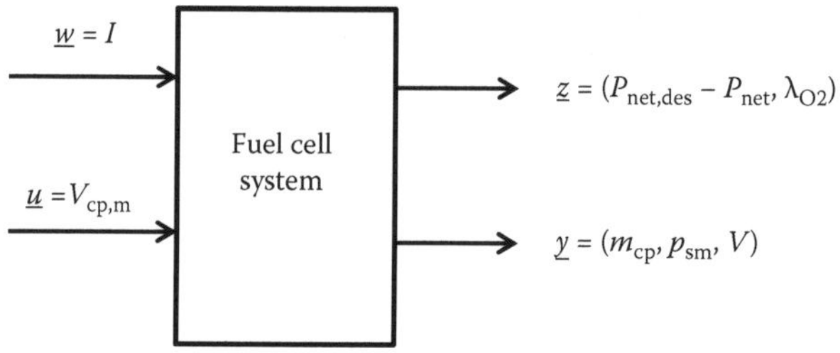

FIGURE 12.25
Compressor voltage control problem formulation to achieve desired net power.

$$\text{System states: } \underline{x} = [m_{O_2}, m_{H_2}, m_{N_2}, m_{cp}, p_{sm}, m_{sm}, m_{w,an}, m_{w,ca}, p_{rm}]^T \quad (12.73)$$

$$\text{System measurements: } \underline{y} = [m_{cp}, p_{sm}, V]^T \quad (12.74)$$

$$\text{Performance variables to be optimized: } \underline{z} = [P_{net,des} - P_{net}, \lambda_{O_2,des} - \lambda_{O_2}]^T \quad (12.75)$$

$$\text{Control input: } \underline{u} = V_{cp,m} \quad (12.76)$$

$$\text{System disturbance: } \underline{w} = I \quad (12.77)$$

Here in Equation 12.73, the state variables governing the problem are mass flow rates of oxygen, m_{O_2}, hydrogen, m_{H_2}, and nitrogen, m_{N_2}; air flow rate through compressor, m_{cp}; supply manifold pressure, p_{sm}; mass flow rate in supply manifold, m_{sm}; water mass flow rate in anode, $m_{w,an}$; water flow rate in cathode, $m_{w,ca}$; and pressure in return manifold, p_{rm}. The control input is compressor motor voltage, $V_{cp,m}$. The objectives are to achieve the following: $z_1 = P_{net,des} - P_{net} = 0$ and $z_2 = \lambda_{O2} = 2$.

12.4.2.2 Control Configuration

The three basic modes of control schemes used by Pukrushpan et al. (2004) are illustrated in Figure 12.26. These were techniques employed to regulate fuel cell inlet air with simple static feed-forward, dynamic feed-forward, dynamic feed-forward with proportional integral (PI) feedback, and state feedback control methods (Pukrushpan et al., 2004; Rgab et al., 2010). In the static feed-forward architecture, a static function that correlates the steady-state value between the control input, the compressor motor voltage, $V_{cp,m}$, and the disturbance in fuel cell stack current, I, is implemented with a functional relation area look-up table. A mapping to relate $\underline{w}$ (fuel cell current I) and $\underline{u}$ (compressor pump motor voltage $V_{cp,m}$) is required where for a set of values the required z is achieved at steady state. For such mapping, a look-up table or data on compressor voltage for each current value are required as illustrated in Figure 12.27.

Here, the desired stack current is used to determine the desired air flow for the reaction and the compressor speed is adjusted accordingly via the compressor motor voltage, $m_{cp} = f_{cp}(I(t))$ without effort to meet desired values of performance variables z_1 and z_2.

Dynamic feed-forward attempts to manage the z_2 goal by adding a filter to cancel disturbances to the desired $z_2 = \lambda_{O2}$ resulting from changes in stack current. Both of these methods are open loop feed-forward controllers; hence, they are not very robust with respect to external disturbances, system parameter variations, or modeling errors. They are not guaranteed to eliminate any steady-state error of the $\underline{y}$ terms from desired values. For this feedback, controls are used where measurable values $\underline{y}$ are fed back to the

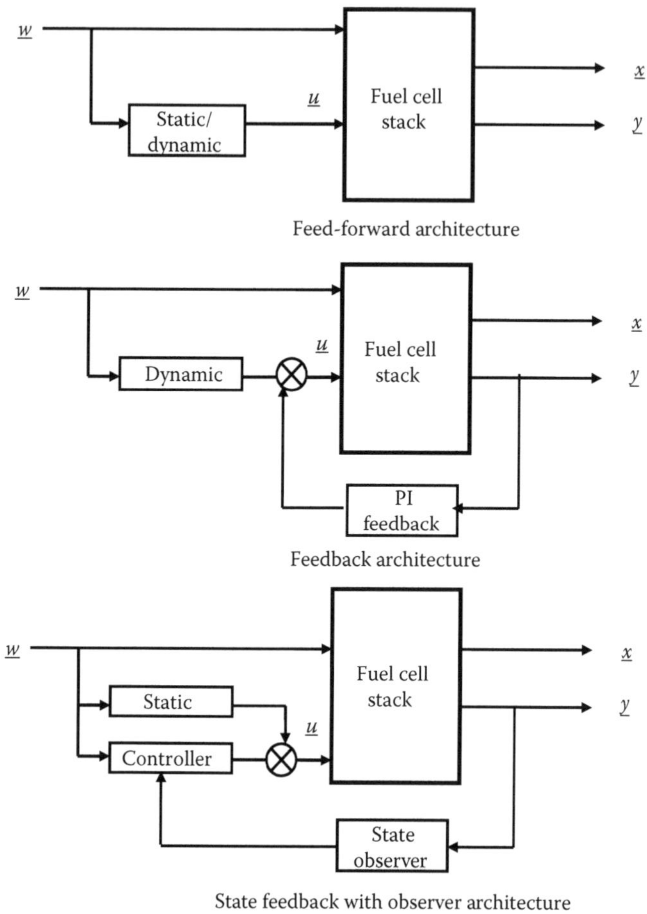

FIGURE 12.26
Control configurations for fuel cell. (With kind permission from Springer Science+Business Media: *Control of Fuel Cell Power Systems: Principles, Modeling, Analysis, and Feedback Design*, 2004, Pukrushpan, J. T., A. G. Stefanopoulou, and H. Peng.)

controller. In the feedback control design, linearization is done for inputs and outputs. Three basic modes of feedback control are proportional (P), integral (I), and derivative (D) control. Proportional–integral (PI) and proportional–integral–derivative (PID) controls are commonly used in practice. PID control is introduced in the next section.

The added PI feedback controller to the dynamic feed-forward control gives better results. The controller input u for this is given as

$$u = K_{\text{dff}}\left(I(t) - I^{\text{nom}}\right) + k_{\text{P}}\left(f_{\text{cp}}(I(t)) - m_{\text{cp}}(t)\right) + k_{\text{I}}\int_0^1 \left(f_{\text{cp}}(I(\tau)) - m_{\text{cp}}(\tau)\right)\text{d}\tau. \quad (12.78)$$

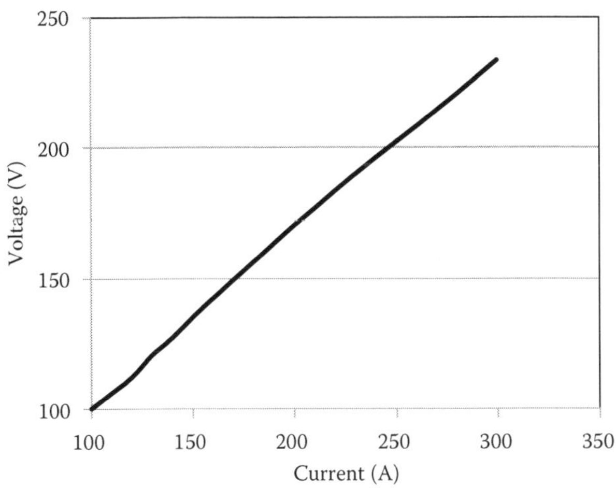

FIGURE 12.27
Compressor voltage ($V_{cp,m}$) and current (I) relation.

The first term is an abstract representation of the dynamic feed-forward controller while the remaining terms are for PI control. The air mass flow rate is the output flow rate from the compressor. This controller provides air stoichiometry recovery that is twice as fast as the static feed-forward controller when I is changed in step increments. However, this controller suffers from a low bandwidth so it will not be able to reject rapid changes in I, the system disturbances.

The state feedback controller with integration that includes the static feed-forward for calculating compressor air flow rate state error minimizes error between desired and actual air flow rate through the compressor.

$$\dot{x}_I = \dot{m}_{\text{air,des}} - \dot{m}_{\text{air}} \tag{12.79}$$

The state feedback controller is designed using Linear Quadratic Regulator (LQR) optimal design to minimize the performance variables in $\underline{z}$ without using excessive control input. There are two controller options. The first one minimizes the λ_{O2} deviation from desired while the second minimizes deviations of both λ_{O2} and net power. The goal for each is formulated with a suitable cost function. The cost function for the first is given.

$$J = \int_0^\infty \delta z_2^T Q_z \, \delta z_2 + \delta u^T R_u \, \delta u \, \mathrm{d}t \tag{12.80}$$

The δ in front of a variable indicates a linearized quantity. Q_z and R_u are weighting matrices used to set the relative importance of minimization. If a value of Q_z is greater than a value of R_u, then minimizing the performance variable will be more important than reducing the control effort required and vice versa. This can be used to adjust the importance of control parameters and their resulting gains. The Algebraic Riccati Equation is used to solve for gains that satisfy a cost function. The result is a state feedback controller.

$$\delta \underline{u} = -\underline{k}_p \delta \underline{x} - k_I x_I \tag{12.81}$$

The state feedback controllers assume that all linearized states are available to them at all times. Hence, a state observer needs to be designed. A state observer estimates unknown values in $\underline{x}$, those not directly measured with a sensor, so that accurate solutions to the control state equations can be found. The states directly measured are given by $\underline{y}$ in Equation 12.74. A similar method to LQR is used to determine the observer design. The method used is called Linear Quadratic Estimator or Kalman estimator design. However, a reduced order estimator is designed because one of the eight linearized states, $m_{w,ca}$, is unobservable. The amount of water in the anode is determined with an input estimator. The use of three measurement states in $\underline{y}$ allows the remaining states to be estimated within half a second. The observer plus state feedback controller that minimizes only λ_{O2} deviation has a response marginally better than that of dynamic feed-forward plus PI.

12.4.3 PID, Fuzzy Logic, and Neural Networks–Based Control Systems

As discussed above in combination of a feed-forward and a feedback control configuration, the feedback controllers can be designed with PID as well. Recently, in the feedback control design, fuzzy logic and neural network methods have been applied with superior control capabilities (Chen and Pham, 2001; Hagan et al., 1999). The fuzzy logic and neural network methods can also be used in feed-forward architecture. In Figure 12.28, the options for the feed-forward and feedback controllers using a look-up table, PID, and fuzzy logic or neural network are shown. Here, first we consider a case of PID feedback controller and feed-forward controller with options: look-up table, fuzzy logic, or neural network control. The feed-forward controller can be implemented with a look-up table or graph such as the one shown in Figure 12.28 showing fuzzy logic or neural network control.

12.4.3.1 The PID Controller

The PID controller algorithm is described in time domain by

$$u(t) = K\left(e(t) + \frac{1}{T_i}\int_0^1 e(\tau)\,d\tau + T_d \frac{de(t)}{dt} \right), \tag{12.82}$$

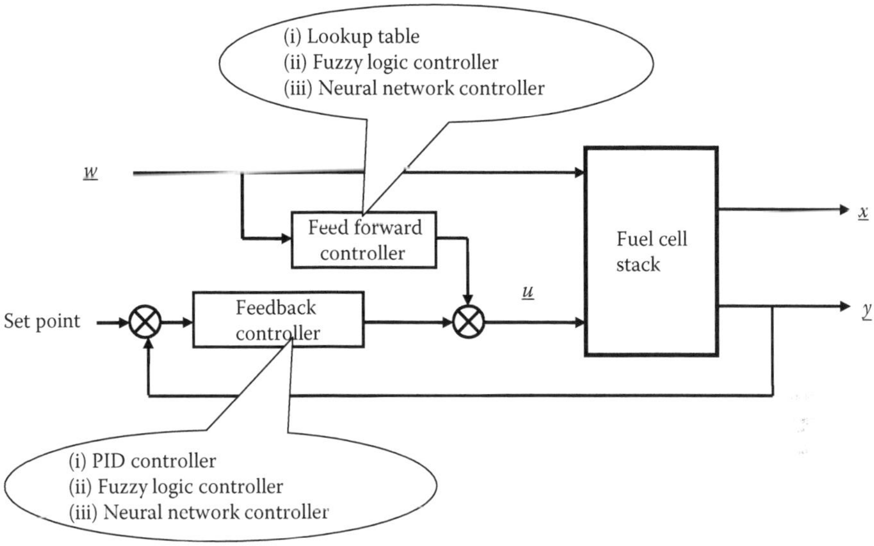

FIGURE 12.28
Fuel cell control options with feed-forward and feedback configurations using a look-up table, PID, fuzzy logic, and neural network controllers.

where u is the control signal, e is the control error ($e = y_{sp} - y$), y is the measured process variable, and y_{sp} is the reference variable or the desired set point. The control signal is thus a sum of three terms: the P-term (which is proportional to the error), the I-term (which is proportional to the integral of the error), and the D-term (which is proportional to the derivative of the error). The controller parameters are proportional gain K, integral time T_i, and derivative time T_d. The integral, proportional, and derivative part can be interpreted as control actions based on the past, the present, and the future. The PID algorithm given by Equation 12.82 can be represented by the Laplace transfer function

$$Wc(s) = K_p + \frac{K_i}{s} + K_d s. \tag{12.83}$$

The parameters are related to the parameters of the standard form through

$$K_p = K \quad K_i = \frac{K}{T_i} \quad K_d = KT_d. \tag{12.84}$$

Here, K_p, K_i, and K_d are proportional, integral, and differential gains, respectively. The PID is fine-tuned using some preconceived "ideal" response profile for the closed-loop system.

12.4.3.2 Fuzzy Logic Control

The fuzzy logic controller is used to overcome inherent disadvantages such as uncontrollable large overshoot and large current ripple. The fuzzy control scheme does not need an accurate mathematical plant model. Therefore, it is applicable to a process where the plant model is unknown or ill defined. The fuzzy control is also nonlinear and adaptive in nature and offers robust performance under parameter variations and load disturbances. The fuzzy logic controller structure as shown in Figure 12.29 has four blocks. Crisp input information from the device is converted into fuzzy values for each input fuzzy set with the fuzzification block. The universe of discourse of the input variables determines the required scaling for correct per-unit operation. The scaling is very important because the fuzzy system can be retrofitted with other devices or ranges of operation by just changing the scaling of the input and output. The decision-making logic determines how the fuzzy logic operations are performed and, together with the knowledge base, determines the outputs of each fuzzy IF–THEN rules. Those are combined and converted to crispy values with the defuzzification block. The output crisp value can be calculated by the center of gravity or the weighted average.

In order to process the input to obtain the output reasoning, there are six steps involved in the creation of a rule based fuzzy system:

1. Identify the inputs and their ranges and name them.
2. Identify the outputs and their ranges and name them.
3. Create the degree of fuzzy membership function for each input and output.
4. Construct the rule base that the system will operate under.
5. Decide how the action will be executed by assigning strengths to the rules.
6. Combine the rules and defuzzify the output.

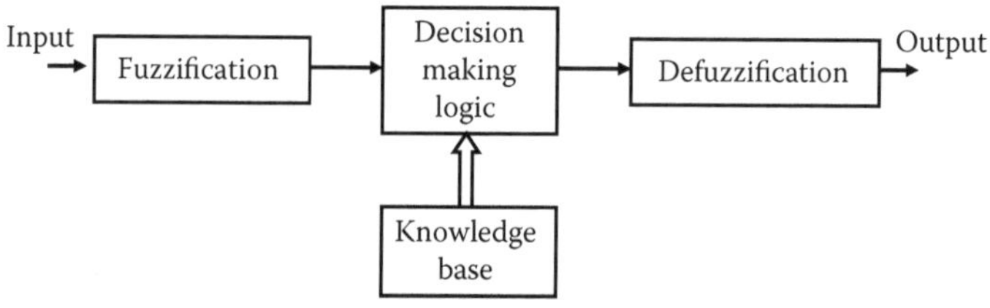

FIGURE 12.29
Fuzzy logic controller block diagram.

12.4.3.3 Input and Output Variables

The user decides what information he or she will use as inputs to the decision-making process. And then the output is chosen based on what parameters or variable is to be controlled. As in the case of the above example of control of oxygen ratio through the control of the compressor voltage, the fuzzy controller inputs are the error $e(k)$, and change of error $ce(k)$. The output from the controller is used to control the compressor voltage $u(k) = V_{cp,m}$. The error is the difference between fuel cell output $y = \lambda_{O2}$ and set point value, for example, $\lambda_{O2} = 2$. The two inputs and single output of the controller are given as

$$\left.\begin{aligned} e(k) &= \lambda_{O2} - 2 \\ ce(k) &= e(k) - e(k-1) \\ u(k) &= u(k-1) + \rho \Delta u(k). \end{aligned}\right\} \tag{12.85}$$

Here, $\Delta u(k)$ is the inferred change of output and ρ is the gain factor of the controller. Single or multiple inputs can be used and single or multiple outputs can be chosen depending on the control problem.

12.4.3.4 Membership Functions

These are simple mathematical tools for indicating flexible memberships to a set. The universe of discourse is the space where the fuzzy variables are defined. The membership function gives the grade, or degree, of membership within the set, of any element of the universe of discourse. The membership function maps the elements of the universe onto numerical values in the interval [0, 1]. They have a peak or plateau with membership grade 1, over which the members of the universe are completely in the set. The membership function is increasing toward the peak and decreasing away from it. A typical case is the triangular fuzzy membership function described by

$$\mu(x) = \begin{cases} \dfrac{x-a}{b-a} & \text{if } a \le x \le b \\ \dfrac{x-c}{b-c} & \text{if } b \le x \le c \\ 0 & \text{otherwise} \end{cases} \tag{12.86}$$

where x represents the input. The membership function for the input and output fuzzy logic controller is generally divided into 3, 5, 7, 9, or more membership levels. Higher membership gives better results for transient where overshoots and undershoots are minimized.

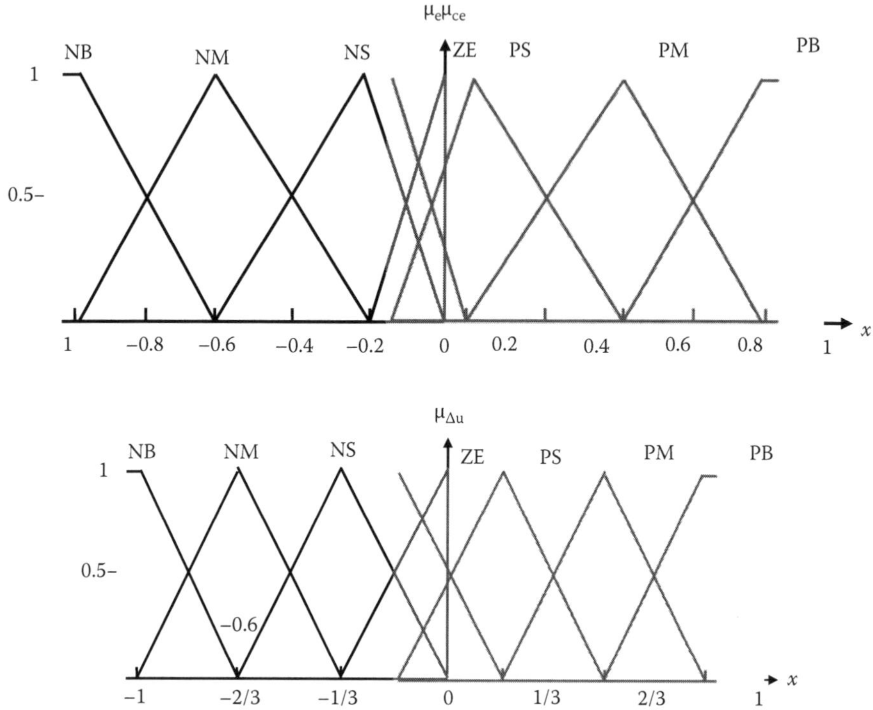

FIGURE 12.30
Membership function for $e(k)$, $ce(k)$, $\Delta u(k)$.

As shown in Figure 12.30 with seven fuzzy subsets, the collections of the reference fuzzy sets for the error, the change of error, and the control input can be same, but their scale factors can have little difference. As shown in Figure 12.30, seven fuzzy subsets given in linguistic terms, positive big (PB), positive medium (PM), positive small (PS), zero (ZE), negative small (NS), negative medium (NM), and negative big (NB), are selected for the input and output variables $e(k)$, $ce(k)$, $\Delta u(k)$.

12.4.3.5 Design of Fuzzy Control Rules

Fuzzy control rules are obtained from the behavior analysis of the PEMFC system. A set of control rules are logics such as *IF (condition) Then (control actions)*, where "condition" defines the state of the process, for which the control adjustment specified in the control action should be executed. These rules are derived from the knowledge of experts with substantial experience in the system. Say, for example, when the output is far from the set point ($e(k)$ is PB or NB), the corrective action must be strong; this means that u should be NB (or PB), in order to prevent the continuous increase (or decrease) of integral term that would cause overshoots. In this case, the change of error plays little part. The basic control rules are as follows:

If $e(k)$ is PB, then Δu is PB
If $e(k)$ is NB, then Δu is NB

When the output is close to the set point, the change of error must be properly taken into account in order to ensure stability and speed of response. The goal of the fuzzy controller is to achieve a satisfactory dynamic performance with small sensitivity to parameter variations. The control rules are as follows:

If both $e(k)$ and $ce(k)$ are ZE, then Δu *is* ZE
If both $e(k)$ and $ce(k)$ are negative, Δu is negative
If both $e(k)$ and $ce(k)$ are positive, Δu *is* positive

According to these criteria, the rule sets are shown in Table 12.4, where control action is proportional to both $e(k)$ and $ce(k)$. For example, for some constants $a > 0$ and $b > 0$, we have control output as

$$\rho\Delta u(k) = a\, e(k-1) + b\, ce(k-1). \tag{12.87}$$

12.4.3.6 Inference

The inference method used is basic and simple; it is developed from the minimum operation function rule as a fuzzy implementing function. The commonly used fuzzy inference methods are Max–Min fuzzy inference reasoning, Max–Product inference reasoning, and Sum–Product fuzzy reasoning. For example, if membership functions of e and ce are given by μ_{ci} and

TABLE 12.4

Linguistic Control Rule Table for Δu

	ce(k)						
e(k)	**NB**	**NM**	**NS**	**ZE**	**PS**	**PM**	**PB**
NB	NB	NB	NB	NB	NM	ZE	PS
NM	NB	NB	NB	NM	ZE	PS	PM
NS	NB	NB	NM	ZE	PS	PM	PB
ZE	NB	NM	ZE	PS	PM	PB	PB
PS	NM	ZE	NS	PM	PB	PB	PB
PM	ZE	PS	PM	PB	PB	PB	PB
PB	PS	PM	PB	PB	PB	PB	PB

μ_{cei}, respectively, and that for Δu are $\mu_{\Delta ui}$ and μ_{ci}, then the Min–Max method is given for rule $i = 1 \ldots n$ (here, $n = 49$ for seven sets of membership levels)

$$\mu_{Ri}(e, ce) = \min[\mu_{ei}(e), \mu_{cei}(ce)] \tag{12.88}$$

$$\mu_{ci}(\Delta u) = \max[\mu_{Rii}(e, ce), \mu_{\Delta ui}(\Delta u)] \tag{12.89}$$

12.4.3.7 Defuzzification

After fuzzy reasoning, we have a linguistic output variable that needs to be translated into a crisp value. The objective is to derive a single crisp numeric value that best represents the inferred fuzzy values of the linguistic output variable. Defuzzification is such an inverse transformation that maps the output from the fuzzy domain back into the crisp domain. Some defuzzification methods tend to produce an integral output considering all the elements of the resulting fuzzy set with the corresponding weights. Other methods take into account just the elements corresponding to the maximum points of the resulting membership functions. The following defuzzification methods are of practical importance:

Center-of-Gravity (C-o-G): The C-o-G method (centroid defuzzification) is often referred to as the Center-of-Area method because it computes the centroid of the composite area representing the output fuzzy term.

Center-of-Maximum (C-o-M): In the C-o-M method, only the peaks of the membership functions are used. The defuzzified crisp compromise value is determined by finding the place where the weights are balanced. Thus, the areas of the membership functions play no role and only the maxima (singleton memberships) are used. The crisp output is computed as a weighted mean of the term membership maxima, weighted by the inference results.

Mean-of-Maximum (M-o-M): The M-o-M is used only in some cases where the C-o-M approach does not work. This occurs whenever the maxima of the membership functions are not unique and the question is as to which one of the equal choices one should take.

For the above case of output membership function, the centroid defuzzification method is given by the expression

$$\Delta u = \frac{\sum_{i}^{n} \mu_{ci}(\Delta u)\Delta u_i}{\sum_{i}^{n} \mu_{ci}(\Delta u)}. \tag{12.90}$$

12.4.3.8 Neural Networks

The neural network predictive controller is one of the promising strategies for complex fuel cell systems. The neural network predictive controller strategy includes the specification of the reference model with the desired dynamic, on-line parameter estimation and calculation of control signals. The first step in model predictive control is to determine the neural network plant model (system identification). In this stage, the prediction error between the plant output and the neural network output is used as the neural network training signal. The process is represented by Figure 12.31. A neural network is an organization of sequential layers, with each "hidden layer" between the inputs and output layers containing neurons. Inputs to a neuron undergo a weighted summation before an activation function is applied to determine the neuron output. Eventually, the output layer is reached and the outputs from the last hidden layer are summed with weights. A final activation function is then applied. The summation weights are found through training, which seeks to minimize the difference between desired and actual network output. The neural network plant model uses previous inputs and previous plant outputs to predict future values of the plant output.

The multilayer neural network is made up of simple components. A single-layer network of neurons having numbers of neutron S, with multiple inputs R, is shown in Figure 12.32. Each scalar input p_i (i = 1,... R) is multiplied by the scalar *weight* w_i to form w_ip_i, which is sent to the summer. The other input, 1, is multiplied by a *bias* b_j (j = 1,... S) and is then passed to the summer. The summer output, often referred to as the *net input*, goes into a *transfer function*, which produces the scalar neuron output a_{ij}, or in matrix form:

$$\mathbf{a} = \mathbf{f}(\mathbf{wp} + \mathbf{b}). \tag{12.91}$$

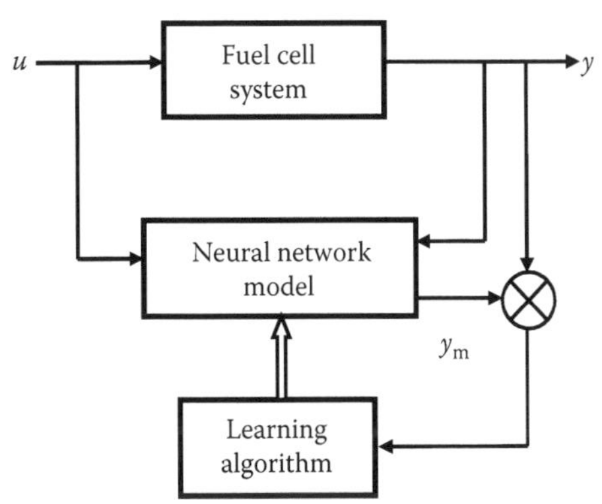

FIGURE 12.31
Process of neural network.

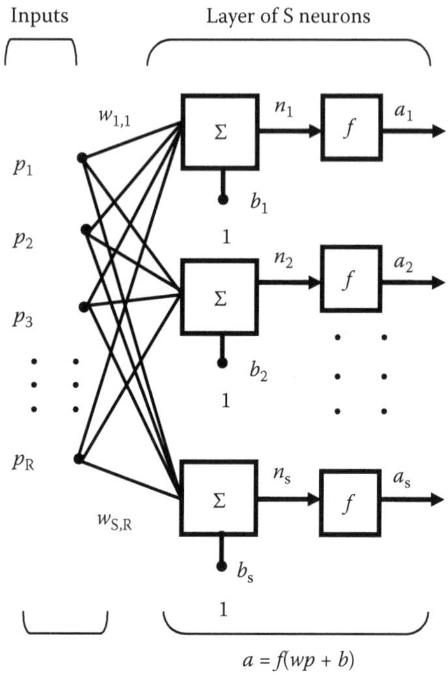

FIGURE 12.32
Layers of S neurons.

Now, for multilayer networks, the output of one layer becomes the input to the following. The equations that describe this operation are

$$\mathbf{A}^{m+1} = \mathbf{f}^{m+1}\left(\mathbf{w}^{m+1}\,\mathbf{a}^{m} + \mathbf{b}^{m+1}\right) \quad \text{for } m = 1, \ldots M-1 \tag{12.92}$$

where M is the number of layers in the network. The neurons in the first layer receive external inputs:

$$\mathbf{a}^0 = \mathbf{p}, \tag{12.93}$$

which provides the starting point for Equation 12.84. The outputs of the neurons in the last layer are considered the network outputs:

$$\mathbf{a} = \mathbf{a}^M. \tag{12.94}$$

The feed-forward network can be trained offline in batch mode, using data or a look-up table with any of the training algorithms in Back Propagation. The back propagation algorithm for multilayer networks is a gradient descent optimization procedure in which minimization of a mean square error performance index is done. The algorithm is provided with a set of examples of proper network behavior. For example, the data set can be the {*input* = *current*, *target* = *compressor voltage*} as in the case of oxygen ratio control problem discussed previously.

$$\{\mathbf{p}_1, \mathbf{t}_1\}, \{\mathbf{p}_2, \mathbf{t}_2\}, \ldots\ldots\ldots\ldots\ldots\ldots \{\mathbf{p}_Q, \mathbf{t}_Q\} \tag{12.95}$$

where $\mathbf{p}_Q$ is an input to the network and $\mathbf{t}_Q$ is the corresponding target output. As each input is applied to the network, the network output is compared to the target. The algorithm should adjust the network parameters in order to minimize the sum squared error indicated by performance index:

$$F(x) = \sum_{q=1}^{Q} e_q^2 = \sum_{q=1}^{Q} (t_q - a_q)^2, \tag{12.96}$$

where x is a vector containing all network weights and biases.

For training, there are several back propagation algorithms available. For example, the Levenberg–Marquardt algorithm uses a nonlinear least-squares algorithm to the batch training of the network and is efficient to obtain lower mean square errors and faster convergence. The receding horizon technique is one good method employed in the neural network predictive control method. The neural network model predicts the fuel cell system response over a specified time horizon. The predictions are used by a numerical optimization program to determine the control signal that minimizes the following performance criterion over the specified horizon.

$$J = \sum_{j=N_1}^{N_2} (y_\mathrm{r}(k+j) - y_\mathrm{m}(k+j))^2 + \rho \sum_{j=N_1}^{N_u} (u'(k+j-1) - u'(k+j-2))^2, \tag{12.97}$$

where, N_1, N_2, and N_u define the horizons over which the tracking error and the control increments are evaluated. The u' variable is the tentative control signal, y_r is the desired response, and y_m is the network model response. The ρ value determines the contribution that the sum of the squares of the control increments has on the performance index. The block diagram shown in Figure 12.33 illustrates the model predictive control process. The controller consists of the neural network plant model and the optimization block. The optimization block determines the values that minimize, and then the optimal is input to the plant.

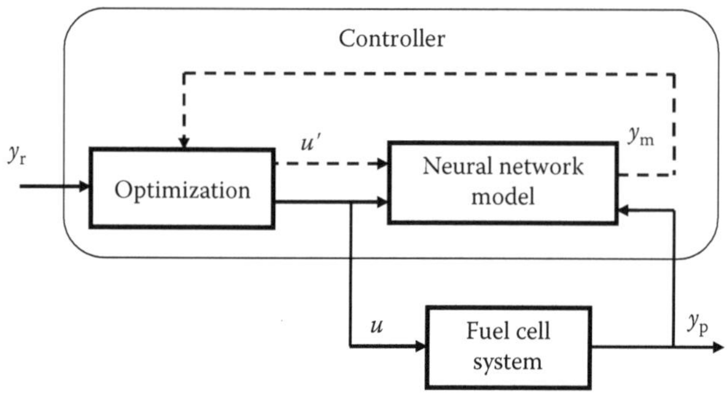

FIGURE 12.33
Neural network predictive control configuration.

References

Amphlett, J. C., R. M. Baumert, R. F. Mann, B. A. Peppley, P. R. Roberge and T. J. Harris. Performance modeling of the Ballard Mark IV solid polymer electrolyte fuel cell I. Mechanistic model development. *Journal of the Electrochemical Society* 142(1): 1–8, 1995a.

Amphlett, J. C., R. M. Baumert, R. F. Mann, B. A. Peppley, P. R. Roberge and T. J. Harris. Performance modeling of the Ballard Mark IV solid polymer electrolyte fuel cell II. Empirical model development. *Journal of the Electrochemical Society* 142(1): 9–15, 1995b.

Amphlett, J. C., R. F. Mann, B. A. Peppley, P. R. Roberge and A. Rodrigues. A model predicting transient responses of proton exchange membrane fuel cells. *Journal of Power Sources* 61: 183–188, 1996.

Chen, G. and T. T. Pham. *Introduction to Fuzzy Sets, Fuzzy Logic, and Fuzzy Control Systems*. CRC Press, Boca Raton, FL, 2001.

Cole, G. H. *SIMPLEV: A Simple Electric Vehicle Simulation Program, Version 2.0*. EG&G Idaho, Inc., Idaho Falls, ID, 1993.

Correa, J. M., F. A. Farret, V. A. Popov and M. G. Simoes. Sensitivity analysis of the modeling parameters used in simulation of proton exchange membrane fuel cells. *IEEE Transactions on Energy Conversion* 20(1): 211–218, 2005.

Cuddy, M. and K. Wipke. ADVISOR, Advanced vehicle simulator, New opportunities for screening advanced components. Proceedings of the Annual Automotive Technology Development Customer's Coordination Meeting, Dearborn, MI, October 28–November 1, 1996.

Forrester, J. W. *Industrial Dynamics*. The MIT Press, Cambridge, MA. Reprinted by Pegasus Communications, Waltham, MA, 1961.

Green, M. A., K. Emer, D. L. King and S. Igari. Solar cell efficiency tables (version 17). *Progress in Photovoltaics: Research and Applications* 9: 49–56, 2001.

Hagan, M. T., O. De Jesus and R. Schultz. Training recurrent networks for filtering and control, Chapter 12. In: *Recurrent Neural Networks: Design and Applications*, Editors L. Medsker and L. C. Jain. CRC Press, Boca Raton, FL, pp. 311–340, 1999.

Hansen, A. D., P. Sorensen, L.H. Hansen and H. Bindner. Models for stand-alone PV system. Riso National Laboratory, Report Riso-R-1219(EN)/SEC-R-12, Roskilde, Denmark, 2000.

Hauer, K. H. Analysis tool for fuel cell vehicle hardware and software (controls) with an application to fuel economy comparisons of alternative system designs. Ph.D. Dissertation, University of California Davis, Institute of Transportation Studies, Davis, CA, 2001.

Hauer, K. H. and R. M. Moore. Fuel cell vehicle simulation—Part 1: Benchmarking available fuel cell vehicle simulation tools. *Fuel Cells* 3(3): 84–94, 2003.

Klee, H. *Simulation of Dynamic Systems with MATLAB® and Simulink®*. CRC Press Taylor & Francis Group, Boca Raton, FL, 2002.

Koh, J. H., H. K. Seo, Y. S. Yoo and H. C. Lim. Considerations of numerical simulation parameters and heat transfer models for a molten carbonate fuel cell stack. *Chemical Engineering Journal* 87: 367–379, 2002.

Kota, R. Simulation of solar regenerative fuel cell system for a high altitude airship. MS Thesis, Purdue University, West Lafayette, IN, 2006.

Lorenzo, E. *Solar Electricity: Engineering of Photovoltaic Systems*, Institute of Solar Energy, PROGENSA, Polytechnic University of Madrid, Spain, 1994.

Machowski, J., J. W. Bialek and J. R. Bumby. *Power System Dynamics and Stability*. John Wiley & Sons, Chichester, West Sussex, UK, 1997.

Mann, R. F., J. C. Amphlett, M. A. I. Hooper, H. M. Jensen, B. A. Peppley and P. R. Roberge. Development and application of a generalized steady-state electrochemical model for a PEM fuel cell. *Journal of Power Sources* 86: 173–180, 2000.

Murrell, J. D. Vehicle powertrain modeling. Letter Rep. under Consultant Agreement CCD-4-1403-01 to NREL, 1995.

Popovic, D. and V. P. Bhatkar. *Distributed Computer Control for Industrial Automation*. CRC Press, New York, 1997.

Pukrushpan, J. T. Modeling and control of fuel cell systems and fuel processors. Ph.D. Thesis, University of Michigan, Ann Arbor, MI, 2003.

Pukrushpan, J. T., A. G. Stefanopoulou and H. Peng. *Control of Fuel Cell Power Systems: Principles, Modeling, Analysis, and Feedback Design*. Springer-Verlag, London, England, 2004.

Rgab, O., D. L. Yu and J. B. Gomm. Polymer electrolyte membrane fuel cell control with feed-forward and feedback strategy. *International Journal of Engineering, Science and Technology* 2(10): 56–66, 2010.

Rostrup-Nielsen, J. R. Manufacture of hydrogen. *Catalysis Today* 106: 293–296, 2005.

Rousseau, A. and R. Larsen. Simulation and validation of hybrid electric vehicles using PSAT. GPC Conference, Detroit, MI, June 6–8, 2000.

Spiegel, C. *PEM Fuel Cell Modeling and Simulation Using MATLAB*. Academic Press is an imprint of Elsevier, San Diego, CA, 2008.

Springer, T. E., M. S. Wilson and S. Gottesfeld. Modeling and experimental diagnostics in polymer electrolyte fuel cells. *Journal of Electrochemical Society* 140(12): 3513–3526, 1993.

Springer, T. E., T. A. Zawodzinski, M. S. Wilson and S. Gottesfeld. Characterization of polymer electrolyte fuel cells using AC impedance spectroscopy. *Journal of Electrochemical Society* 143(2): 587–599, 1996.

Sterman, J. D. System dynamics modeling: Tools for learning in a complex world. *California Management Review* 43(4): 8–25, 2001.

Thomas, C. E. *Process Technology Equipment and Systems*, 2nd Edition. Delmar Cengage Learning, Clifton Park, NY, 2007. ISBN 1-4180-3067-8.

Wipke, K. B., M. R. Cuddy and S. D. Burch. ADVISOR 2.1: A user-friendly advanced powertrain simulation using a combined backward/forward approach. *IEEE Transactions on Vehicular Technology* 48(6): 1751–1761, 1999.

Wolf, B. J. Simulation of hybrid power system using molten carbonate fuel cell for distributive energy generation. MS Thesis, Purdue University, West Lafayette, IN, 2007.

Xue, X., J. Tang, A. Smirnova, R. England and N. Sammes. System level lumped-parameter dynamic modeling of PEM fuel cell. *Journal of Power Sources* 133(2): 188–204, 2004.

Yuh, C.-Y. and J.-R. Selman. The polarization of molten carbonate fuel cell electrodes 1. Analysis of steady-state polarization data. *Journal of the Electrochemical Society* 138: 3642–3648, 1991.

13

Fuel Cell Power Generation Systems

The fuel cell as a power system contains multiple components besides the fuel cell unit. These include a fuel supply system that may contain a fuel processor and fuel conditioners, the reactant flow and conditioning system, waste effluent handling system, electrical power conditioning and delivery system, and thermal management system. In this chapter, these basic components of a fuel power system are discussed. The key fuel for the fuel cell, hydrogen, as an energy carrier is presented, where its generation, storage, delivery, and application are given in detail. For the fuel cell technology to mature, each of these components has to be economically viable and safe to use.

13.1 Fuel Cell Subsystems

As a power source either for stationary application or for mobile application, the fuel cell stack is one of the many components in the fuel cell power system. The fuel cell power system requires the integration of many other components beyond the fuel cell stack itself, such as a fuel processor unit; auxiliary systems such as humidifiers, pumps, and blowers; a fuel power conditioning unit; and heat management systems. Depending on the type of fuel cell, only certain processed fuel is used. For example, a proton exchange membrane (PEM) fuel cell requires pure hydrogen whereas SOFC and MCFC can utilize a mixture of gases including hydrogen and carbon monoxide or directly methane. Even in the case of methane as a fuel, a processor is required to scrub sulfur and other impurities. The fuel cell stack will produce only DC power, and hence the cell power has to be conditioned to tie into the AC power grid or as a power supply to the AC-driven motors or equipment. The heat management system could be part of a cogeneration or bottoming cycle to utilize the rejected heat. A schematic of these basic systems and their interconnections is shown in Figure 13.1.

In the following sections, system components such as fuel processors, heat management, and utilization; the power conditioner units; and auxiliary equipment are introduced.

13.1.1 Fuel Processing

In fuel processing, commercially available gas, liquid, or solid fuels are converted to a fuel gas reformate suitable for the fuel cell anode reaction. Fuel

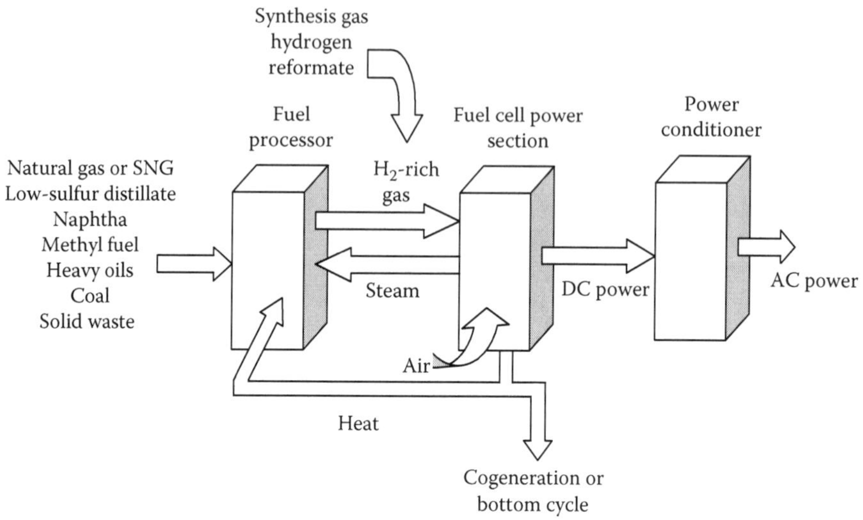

FIGURE 13.1
The key components a fuel cell power system.

processing may involve the following one or more steps: the cleaning and removal of unwanted and often harmful species in the fuel, the conversion of the fuel to the fuel gas reformate, and downstream processing to alter the fuel gas reformate according to specific fuel cell requirements. In the fuel cleaning step, sulfur, halides, and ammonia are removed to prevent fuel processor and fuel cell catalyst degradation. The fuel conversion involves converting a primary fuel, typically a hydrocarbon to a hydrogen-rich gas reformate. The carbon monoxide (CO) and water (H_2O) in the fuel gas reformate are then converted to hydrogen (H_2) and carbon dioxide (CO_2) via the water–gas shift reaction, or selective oxidation to reduce CO to a few parts per million, or removal of water by condensing to increase the H_2 concentration. The various fuel conversion methods are described in the following sections.

Fuel processors are being developed to allow a wide range of commercial fuels suitable for stationary, vehicle, and military applications. Technology from large chemical installations has been successfully transferred to small, compact fuel cells to convert pipeline natural gas, the fuel of choice for small stationary power generators. The fuel processor unit design feature may include high thermal efficiency, high hydrogen yield, or, for some fuel cells, hydrogen plus carbon monoxide yield, multi-cycling, compactness, low weight, and quick starting capability, depending on the application. Often fuel processors make use of the chemical and heat energy left in the fuel cell effluent to provide heat for fuel processing, thus enhancing system efficiency. The catalysts used in fuel conversions and reformate alterations are normally susceptible to deactivation by impurities; hence, the fuel cleaning process takes place upstream or within the fuel conversion process. The fuel conversion and reformate gas alteration processes are often placed either external to the fuel cell referred to as external reforming fuel cell or within the fuel cell anode compartment referred to as an internal reforming fuel cell.

13.1.2 Fuel Cell Auxiliary

The fuel cell stack is integrated with a reactant supply and exhaust system, a coolant inlet and outlet system, associated piping, and power interconnector and feed lines. Depending on the type of fuel cell, the reactant exhaust systems are rerouted to a reactant feed line via a combustor to make use of the unspent fuel and feed carbon dioxide as in the case of MCFC. The high-temperature fuel cells will generally have an external combustor associated with a reactant exhaust line. The heat from this combustor is often used to heat inlet air supply to match the fuel cell operating temperature. In addition, the energy of exhaust gases from a fuel cell can sometimes be harnessed using a turbine, making use of what would otherwise go to waste. In PEM fuel cells, the air supply and the hydrogen supply have to be humidified to maintain the hydration in the electrolyte. The humidifiers can be passive systems or a mechanical or electrical device to humidify the inlet.

Among various mechanical devices, ejectors, compressor, blowers, fans, and pumps are commonly used in fuel cell power systems. An ejector, a very simple type of pump, is often used to circulate hydrogen gas if it comes from a high-pressure store, or for recycling anode gases. Fans and blowers are used for cooling and for cathode gas supply in small fuel cells. Membrane or diaphragm pumps are used to pump reactant air and hydrogen through small (200 W) to medium (3 kW) PEM fuel cells.

The auxiliary units thus contain feed and exhaust piping for reactant, coolant piping, pumps to flow coolant or liquid fuels, and fans, compressors, and blowers for the gaseous reactant supply or exhaust. If air is used as feed gas to the cathode side of the fuel cell, it needs to be filtered for particulates before it is sent to the blower or compressor.

13.1.3 Power Electronics and Power Conditioning

The power electronics and power conditioning system is one of the key subsystems of the fuel cell power system that is required to convert DC electrical power generated by a fuel cell into usable AC power for stationary loads, automotive applications, and interfaces with electric utilities. Depending on the application of the system, the power electronics and power conditioning architecture may involve sets of power controls as well as conditioning and processing electronic units (Kordesch and Simader, 1966).

The DC voltage generated by a fuel cell stack is low in magnitude and varies widely in range, typically less than 50 V for a 5 to 10 kW system and greater than 350 V for a 300 kW system. Hence, a step up DC–DC converter is essential to generate a regulated higher voltage DC to higher than 350 V required for 120/240 V AC output. The DC–DC converter is responsible for drawing power from the fuel cell and therefore should be designed to match fuel cell ripple current specifications. Further, the DC–DC converter should not introduce any negative current into the fuel cell. A DC–AC inverter is essential to provide the DC useful AC power at 60 Hz or 50 Hz frequency. An output filter connected to the inverter filters the switching frequency harmonics and generates a high-quality sinusoidal AC waveform suitable for the load (Figure 13.2).

The power conversion unit must be capable of operating in the required range and, in particular, be able to deliver rated power while regulating output voltage. Output from the power conversion unit is expected to be high-quality power with less than 5% total harmonic distortion. For domestic loads, a 5:1 or better peak to average power capability for tripping breakers and starting motors is desired. This puts an additional constraint on the design of the power conditioning unit for stand-alone loads. Table 13.1 shows a typical specification for a stand-alone fuel cell power conditioning unit for US domestic loads.

In a dedicated power supply unit for a motor, the fuel cell output DC (e.g., 29 V to 39 V) is converted to a regulated DC output (e.g., 50 V) by means of a simple DC–DC boost converter. The output of the DC–DC converter is processed via a pulse width modulation DC–AC inverter to generate a low-voltage sinusoidal AC of power 35 V AC (rms); a line frequency isolation transformer with a turn ratio of 1:3.5 is then employed to generate 120 V/240 V

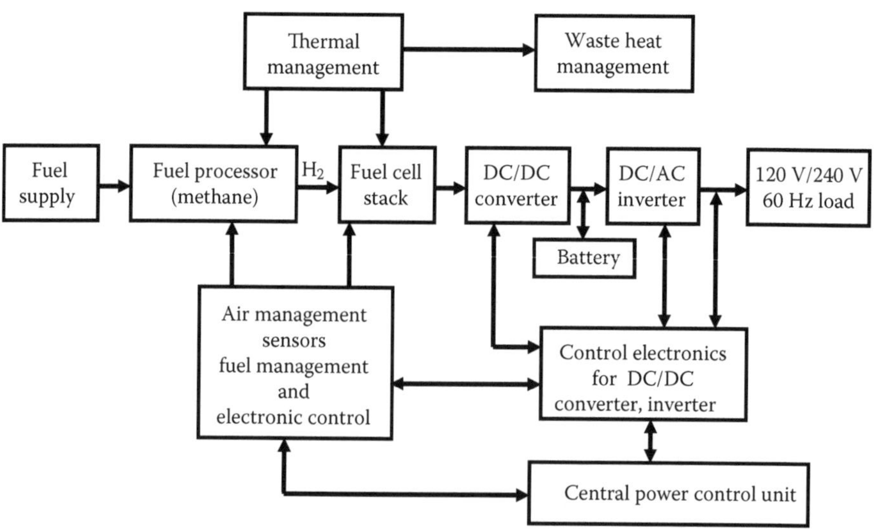

FIGURE 13.2
Block diagram of fuel cell power electronics and power conditioning system.

TABLE 13.1

Specifications of a Typical Fuel Cell Power Conditioning Unit for Stand-Alone US Domestic Loads

Output power	10 kW continuous
Output phase(s)	Split single phase, each output rated for 0 to 5000 VA, not to exceed 10,000 VA total
Output voltage	120 V, 240 V sinusoidal AC. Output voltage tolerance is no wider than ±6% over the full allowed line voltage and temperature ranges, from no load to full load. Frequency, 60 ± 0.1 Hz
Output frequency	60 Hz with enough precision to run AC clock accuracy
Fuel cell current ripple (fuel cell dependent)	120 Hz ripple <15% from 10% to 100% load 60 Hz ripple <10% from 10% to 100% load 10 kHz and above <60% from 10% to 100% load
Output total harmonic distortion	<5%
Protection	Overcurrent, overvoltage, short circuit, overtemperature, and undervoltage. No damage caused by output short circuit. The inverter must shut down if the input voltage dips below the minimum input of 42 V. Inverter should not self-reset after a load-side fault
Acoustic noise	No louder than a conventional domestic refrigerator. Less than 50 dBA sound level measured 1.5 m from the unit
Environment	Suitable for indoor installation in domestic applications, 10°C to 40°C possible ambient range
Electromagnetic interference	Per FCC 18 Class A—industrial
Efficiency	Greater than 90% for 5 kW resistive load
Safety	The system is intended for safe, routine use in a home or small business by non-technical customers
Life	The system should function for at least 10 years with routine maintenance when subjected to normal use in a 20°C to 30°C ambient environment

Source: Adapted from *Fuel Cell Handbook* DOE, 2007.

AC output. A 42 to 48 V battery is connected to the output terminals of the DC–DC converter to provide additional power at the output terminals for motor start-ups and so on. During steady state, the DC–DC converter regulates its output to 50 V and the battery operates in a float mode. The fuel cell and the DC–DC converter are rated for steady-state power (say 10 kW), while the DC–AC inverter section is rated to supply the motor-starting VA. The DC–DC boost converter is operated in current mode control. During a motor start-up operation, the current mode control goes into saturation and limits the maximum current supplied from the cell. During this time, the additional energy from the battery is utilized. During steady-state operation, the fuel cell energy is used to charge the battery when the output load is low. Currently, fuel cells supply only average power from the fuel cell. Thus, peak power must be supplied from some other energy source such as a battery or supercapacitor. The power conditioning unit must therefore provide means for interfacing a battery and also ensure its charge maintenance.

13.1.4 Thermal and Water Management

Water is generated at the cathode reaction and must be removed from the fuel cell to prevent blockage of reaction sites. In the case of PEM fuel cell, proper humidification of the membrane is necessary to improve proton transfer (and efficiency), so commonly exhaust gas water is recycled to inlet air and hydrogen streams to carry water to the membrane. Thus, water management is very important in PEM fuel cells. The water management system in PEM fuel cells will contain a humidification system for both air and hydrogen streams.

Even though the fuel cell reaction is more efficient than the combustion engine, heat is generated in the fuel cell as discussed in previous chapters. There are several heat flows to consider in fuel cell systems. The thermal energy generated in the fuel cell stack is removed from the fuel cell (in most designs) with the use of a coolant fluid that flows through channels in cooling plates sandwiched between some of the cells or through integrated channels in bipolar plates as discussed in Chapter 10. This coolant picks up the thermal energy at the fuel cell and exhausts it through either a heat exchanger or a radiator. The air entering the system is preheated in order to retain more water from any humidification. Also, the humidification water, if any, is preheated as well. There are reformers that require temperatures of at least 300°C to operate and produce net heat. Heat must be supplied to the metal hydride system, if one exists, in order to desorb the hydrogen. Thus, this subsystem requires pumps, a heat exchanger unit, piping, and control valves.

13.1.5 System Efficiency

The efficiency of the total fuel cell system can be calculated by accounting for the generation efficiency of the fuel cell stack, the efficiency of power conversion devices, and the accessory load of the related subsystems. In order to obtain the best overall fuel cell power system efficiency and better economics, a system optimization is required. The optimization involves minimizing the cost of electricity or heat and electric products as in a cogeneration system; all the component processes of the system should be integrated into an efficient plant with low capital cost. Often, these objectives are conflicting, so compromises, or design decisions, have to be made. In addition, project-specific objectives, such as desired fuel, environmental emission levels, potential uses of rejected heat, desired output levels, volume (volume/kW) or weight criteria (weight/kW), tolerance for risk, and life cycle all determine the design of the fuel cell power system. Thus, the process of optimization of a fuel cell power system has to account for possible design options and trade-offs that affect unit capital cost, operating cost, efficiency, parasitic power consumption, complexity, reliability, availability, fuel cell life, and operational flexibility.

As shown in Figure 13.3, the fuel cell itself has many trade-off options. Because fuel cell efficiency is directly proportional to cell voltage, this operating condition has a detrimental impact on overall system efficiency. A fuel cell stack can be rated at any cell voltage, which means that while a higher selected nominal cell voltage would result in a larger stack size, it would be a more efficient system overall. As the operating point moves up in voltage by moving to a lower current density, the system becomes more efficient but requires a greater fuel cell area to produce the same amount of power. Thus, by moving to a higher voltage current density line, the system will experience lower operating costs at the expense of higher capital costs. Many system parameters can be varied simultaneously to achieve the desired operating point. Some of the significant fuel cell parameters that can be varied are pressure, temperature, fuel composition and utilization, and oxidant composition and utilization.

The increase in system pressure improves process performance and hence better fuel cell output. The pressurization is a trade-off between the improved performance and/or reduced cell area and the reduced piping volume, insulation, and heat loss compared to the increased parasitic load, capital cost of the compressor, and pressure-rated equipment and design of the cell to withstand higher pressure.

Similarly, the fuel cell performance at operating current densities increases with increasing temperature owing to reduced mass transfer polarizations and ohmic losses. The increased temperature also yields higher-quality rejected heat. However, temperatures at which the various fuel cells can

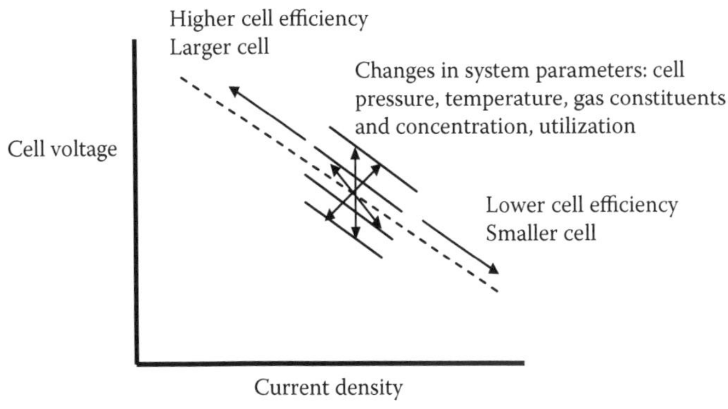

FIGURE 13.3
Fuel cell system power optimization options.

operate are, however, limited by material constraints where an increase in temperature leads to higher corrosion with cell materials.

Both fuel and oxidant high utilizations are considered to be desirable because they minimize the required fuel and oxidant flow, for a minimum fuel cost and compressor/blower load and size. However, utilizations that are pushed too high result in significant voltage drops. Optimum utilization requires an engineering trade-off that considers the specifics of each case.

Depending upon the size of the system, the temperature of the available heat, and the requirements of the particular site, heat produced from the fuel cell can be rejected, used to produce steam or hot water, or converted to electricity via a gas turbine or steam bottoming cycle.

There are several methods of improvement that would significantly increase the overall fuel cell system efficiency. One is implementing an air compressor with load-following control. At high loads, the air compressor must supply maximum flow, consuming large amounts of power. However, if the stack is under a light load, it is ideal for the air compressor to provide a minimum flow to boost system efficiency. Some approaches to air compression include operation at ambient or near-ambient conditions, eliminating the need for an expensive compression system with high parasitic loads.

An issue related to the laws of thermodynamics and the balance of water in the system arises where at low system pressure not enough water can be collected through simple means to supply a pre-stack standard humidification system. The use of membranes that are self-humidifying or by using a stack design that does the humidification internally will help this problem. A thermal management subsystem presents one of the greatest technical challenges for fuel cell systems. Low stack-to-ambient temperature differentials, especially in hot climates, and the fact that nearly 100% of waste heat exits through the cooling fluid require a greater heat exchanger or radiator area.

13.1.6 System Integration

Successful fuel cell integration depends on the successful operation of several important subsystems. While the fuel cell stack is certainly the heart of the system, the other subsystems such as air flow systems, fuel supply system, heat management system, water management system, and power conversion/conditioning system, along with control systems, should be compatible and work harmoniously. The design and construction of these subsystems are based on the required operating conditions of the fuel cell power system. The operating conditions are generally prescribed in terms of fuel cell operating pressure, air injection pressure and temperature with required relative humidity, fuel type and its contents, fuel injection pressure and temperature, and its humidity if any, as well as coolant water flow rates with inlet and outlet temperatures. Additional conditions may be there depending on the type of the subsystems such as cogeneration, or internal or external fuel reformed system, and fuel exhaust conditions among others.

The materials used in the subsystems should be compatible with reactant and exhaust systems. For example, in PEM fuel cells, the fuel cell will become contaminated if any of the components in the air, water, or hydrogen systems are made of anything other than an inert substance such as stainless steel, titanium, certain grades of rubber, or Teflon. This specification dictates that many of the components have to be custom built or ordered from laboratory supply companies.

13.2 Fuels and Fuel Processing

Though fuel cells are being developed to directly use commercially available gas and liquid fuels, the chemically preferred reaction of present fuel cells is via hydrogen-rich gas. The fuel processing is defined as the conversion of the raw primary fuel supplied to a fuel cell system into the fuel required by the stack. There are a number of well-established fuel processing methods that are commercially available to produce hydrogen or hydrogen-rich gas. Natural gas fuel reforming for fuel cells is essentially mature. Recent fuel processor research and development has focused on fuels for transportation and military applications. In this section, the types of fuel available and the fuel processing methods are described.

13.2.1 Basic Fuels and Processes

Current basic fuels can be broadly categorized as renewable fuel sources, such as biomass and ethanol, and nonrenewable fuel sources, such as natural gas. The potential candidate fuels for H_2 manufacture are natural gas, gasoline, diesel, methanol, ethanol, ammonia, and biomass. The choice of the actual fuel feed for the fuel cell depends on its application. Table 13.2 lists the type of fuel and their specification for different fuel cells. There is a strong relation between viable applications of the fuel cell and the infrastructure of available fuels. General application areas of present interest to the fuel cell community are multi-kWe residential, commercial, and light industrial stationary power; transportation prime and auxiliary power; and military uses. The initial terrestrial application was to increase power generation efficiency and to improve the environment by lowering fossil-fueled power generation exhaust emission. Although coal-derived gas was recognized as a viable fuel, early fuel cell development was based on conveniently accessible pipeline gas prior to turning attention to coal-derived gas.

Hydrogen is preferable for a closed environment such as space vehicle application. For strategic applications such as for onboard electric power in space vehicles, hydrogen has been considered as the best fuel because it does not require a fuel processor other than storage and pressurization and it is relatively

TABLE 13.2
Types of Fuel and Their Specification Required by Fuel Cells (Haji et al. 2006)

Fuel Cell Type	Operating Temperature (°C)	H_2	CO	CH_4	CO_2 and H_2O	Sulfur (as H_2S and COS)
PEMFC	60–100	Fuel	Poison (>10 ppm)	Diluent	Diluent	Poison (>0.1 ppm)
DMFC[a]	25–60	Methanol + H_2O	Poison (>10 ppm)	Diluent	Diluent	Poison (>0.1 ppm)
AFC	90–100	Fuel	Poison	Diluent	Poison[b]	Unknown
PAFC	175–200	Fuel	Poison (>0.5%)	Diluent	Diluent	Poison (>50 ppm)
MCFC	~650	Fuel	Fuel[c]	Diluent[c,d]	Diluent	Poison (>0.5 ppm)
SOFC	600–1000	Fuel	Fuel[c]	Diluent[c,d]	Diluent	Poison (>1 ppm)

Source: With kind permission from Springer-Verlag: *Fuel Cell Technology*, Fuels and Fuel Processing, Chapter 5, 2006, 165–211, Nigel Sammes (Ed.).
[a] Methanol with water is fuel for DMFC.
[b] CO_2 is a poison for AFC.
[c] CO reacts with H_2O to produce H_2 and CO_2 by water–gas shift reaction and CH_4 reacts with H_2O to form H_2 and CO faster than reacting as a fuel at the electrode.
[d] A fuel in the internal reforming MCFC and SOFC.

lightweight when stored under pressure. Hydrogen was also the best fuel for the early-developed alkaline fuel cell. There are sources of hydrogen-rich gases, such as an off-gas at a chemical plant, that require only fuel cleaning.

The fuel choice for small, stationary power plants is pipeline gas because of its availability for multiple commercial, light-industrial, and residential applications. Some users request that the fuel processor converts at least one additional fuel, that is, a light distillate. For light vehicles, the fuel choices include the use of onboard hydrogen and liquid fuels such as gasoline methanol with onboard reformation. For heavy vehicles, fuel could be sulfur-free diesel or gasoline if such a fuel cell system could compete. For megawatt-size power stations and fuel cells that are scaled larger in size, one possibility is the use of coal as a basic fuel with cleaning and capture of impurities and emissions. There are also other fuels such as light distillates, ethanol, anaerobic digester gas, biomass, and refuse-derived fuel.

The process of converting liquid or gaseous light hydrocarbon fuels to hydrogen and carbon monoxide is generally referred to as *"reforming."* There are a number of methods to reform fuel. The three most commercially developed and popular methods are (1) *steam reforming*, (2) *partial-oxidation reforming*, and (3) *autothermal reforming*. Steam reforming (SR) provides the highest concentration of hydrogen and has approximately 80% conversion efficiency. Partial oxidation (POX) is a fast process requiring relatively small reactor size and hence has fast startup time. Noncatalytic POX operates at temperatures of approximately 1400°C, but adding a catalyst (catalytic POX or CPOX) can reduce this temperature to as low as 870°C. Combining steam reforming closely with CPOX is termed autothermal reforming (ATR). The type of fuel processing method employed depends on the following features: (1) choice of commercially available fuels suitable for specific applications, (2) fuel flexibility, (3) catalyst tolerance, (4) fuel cell size, and (5) vaporization of heavy hydrocarbons. Heavy hydrocarbons, such as diesel, require vaporization temperatures much in excess of 350 to 400°C, at which temperature some of the heavier fuels pyrolyze.

A general equation given below describes fuel conversion from steam reformation to the POX.

$$C_nH_mO_p + x(O_2 + 3.76N_2) + (2n - 2x - p)H_2O = nCO_2 + (2n - 2x - p + m/2)H_2 + 3.76xN_2, \tag{13.1}$$

where x is the molar ratio of oxygen to fuel. This ratio is very important because it determines the following: (i) The minimum amount of water that is required to completely convert the carbon in the fuel to carbon dioxide $(2n - 2x - p)$. Excess water is used in practice to ensure the conversion, resulting in water in the reformate on the right side of the equation. Typically, 1 or 2 mol of water for every mole of oxygen is used. (ii) The maximum hydrogen yield $(2n - 2x - p + m/2)$. (iii) The maximum concentration of hydrogen in the reformate $\{[2n - 2x - p + m/2]/[n + (2n - 2x - p + m/2) + 3.76x]\}$, all times 100%. (iv) The heat of reaction $\{\Delta H_r = n(\Delta H_{f,CO_2}) - (2n - 2x - p)\,\Delta H_{f,H_2O} - \Delta H_{f,fuel}\}$.

From Equation 13.1, the following are observed. Decreasing the oxygen-to-fuel ratio, x, results in increasing demand for water (water-to-fuel ratio), with commensurate increases in the yield and concentration of hydrogen in the reformate gas. When $x = 0$, the equation reduces to the strongly endothermic steam reforming reaction. The reaction becomes less endothermic with increasing oxygen. It becomes thermoneutral at $x = x_0$ (0.44 for methane). Above this point, the reaction becomes increasingly exothermic. At $x = 1$ with methane, the pure POX reaction, the feed contains sufficient oxygen to convert all of the carbon in the fuel to CO_2. No water needs to be added. The equation is a mix of the steam reforming reaction and the POX reaction at values of x between 0 and n. Beyond $x = [n - (p/2)] = n$ (when $p = 0$), where water is a product, the heat of reaction is determined by the phase (gas or liquid) of the product water. At still higher values, the excess oxygen oxidizes the hydrogen to produce water. Finally, at stoichiometric combustion, all carbon and hydrogen are converted to carbon dioxide and water. Here, $x = X_c = [n - (p/2) + (m/4)]$. The value of x reduces to 2 with CH_4 as the fuel.

When the function of a fuel processor is to convert a fuel to hydrogen, the fuel conversion efficiency is defined as

$$\text{Efficiency} = (\text{Lower heating value of anode fuel produced})/(\text{Lower heating value of fuel used}). \tag{13.2}$$

TABLE 13.3
Thermoneutral Oxygen-to-Fuel Molar Ratios (x_o) and Maximum Theoretical Efficiencies at x_o for Various Fuels

$C_nH_mO_p$	n	m	p	$\Delta H_{f,fuel}$ (kcal/gmol)	$m/2n$	x_o $\Delta H_r = 0$	Efficiency (%)
Methanol, CH_3OH	1	4	1	–57.1	2	0.230	96.3
Methane, CH_4	1	4	0	–17.9	2	0.443	93.9
Iso-octane, C_8H_{18}	8	18	0	–62.0	1.125	2.947	91.2
Gasoline, $C_{7.3}H_{14.8}O_{0.1}$	7.3	14.8	0.1	–53.0	1.014	2.613	90.8

The fuel conversion efficiency for methane conversion to hydrogen is 93.9% at the thermoneutral point, $x = 0.44$ (an ATR reaction), and 91.7% at $x = 0$ (the SR reaction). The difference between the two efficiency values is exactly equivalent to the loss represented by the latent heat of vaporization of the H_2O that escapes with the combustion products in the SR burner exhaust. The concentration of hydrogen is 53.9% at $x = 0.44$ (ATR) and 80% at $x = 0$ (SR). Table 13.3 presents efficiencies at the thermoneutral point for various hydrocarbon fuels.

Besides their basic fuel reforming function, fuel processors require the removal of impurities that degrade the fuel processor or fuel cell performance. Sulfur is the major contaminant encountered. Carbon monoxide reduction for low-temperature fuel cells and avoidance of carbon deposition are also addressed. A typical processing chain for a low-temperature fuel cell will have a hydrodesulfurizer, a halogen guard, a zinc oxide sulfur absorber, a catalytic reformer, a high-temperature shift converter, a second halogen guard, and a low-temperature shift converter. The activity of catalysts for steam reforming and ATR can be affected by sulfur poisoning and coke formation; this commonly occurs with most fuels used in fuel cells of present interest. Other fuel constituents can also prove detrimental to various fuel cells. Examples of these are halides, hydrogen chloride, and ammonia. Hence, in each reformation process, these impurities are removed.

13.2.2 Desulfurization

Hydrocarbon fuels contain significant amounts of sulfur in the form of compounds such as thiophenes, benzothiophenes, and dibenzothiophenes. For natural gas, sulfur is added in the form of the odorant for safety purposes. Sulfur levels in diesel and gasoline are around 500 and 300 parts per million by weight (ppmw), respectively. As shown in Table 13.2, the presence of such compounds at such concentrations has a detrimental effect on the performance of fuel cells and on catalysts used in different sections of the fuel processor. Thus, a desulfurization unit is required prior to the other fuel processing units.

Two approaches for desulfurization, namely, hydrodesulfurization (HDS) and sulfur removal by adsorption, are used with the former being the most mature technology. HDS is a catalytic hydrogenation process that removes sulfur from fossil fuels. In this process, the organosulfur compound such as ethanethiol (C_2H_5SH), a sulfur compound present in some petroleum products, is converted to H_2S and sulfur-free organic compounds ethane (C_2H_6) by reaction with H_2 in the presence of a catalyst as shown below:

$$C_2H_5SH + H_2 \xrightarrow{\text{Catalyst}} C_2H_6 + H_2S. \tag{13.3}$$

HDS reactors are commonly operated at moderate temperatures (300°C–360°C) and at H_2 pressures of 3.0–5.0 MPa, usually with $CoMo/Al_2O_3$ or $NiMo/Al_2O_3$ catalysts. While in refineries, the resulting H_2S from the HDS reaction is eventually converted to elemental sulfur, in fuel cell applications, H_2S is eliminated by absorbing onto a bed of zinc oxide to form zinc sulfide. In the sulfur removal process by adsorption, the fossil fuel is brought into contact with a solid adsorbent that selectively adsorbs sulfur-containing compounds over paraffinic, olefinic, and aromatic hydrocarbons. Sulfur removal by adsorption was incorporated in the fuel processing train in fuel cell applications. Desulfurization by adsorption faces two major challenges. The first is to develop easy regenerable adsorbents with a high adsorption capacity for sulfur compounds. The second challenge is to find adsorbents that selectively adsorb the sulfur compounds, which are mainly aromatic sulfur compounds that have not been removed in the HDS process in refinery, over the other aromatic and olefinic compounds present in the hydrocarbon fuel. Examples of the adsorbents are zeolites, activated carbon, alumina, zirconia, and silica gel. One of the highest capacity adsorbents in the literature is the combination of Cu(I)-Y zeolite with activated carbon prepared with total sulfur uptake (saturation loading) of 18.9 mg S/g sorbent for commercial diesel fuel (430 ppmw S) fed to the column at room temperature.

13.2.3 Steam Reforming

Steam reforming is one of the principal processes for the production of H_2 and syngas (synthetic gas = mixture of H_2 and CO) using light hydrocarbons. Although POX and coal gasification also play a vital role in hydrogen production, steam reforming is cheaper and more energy efficient than the former two methods.

The reactions for steam reforming of a methane and generic hydrocarbon into a mixture of hydrogen and carbon monoxide (syngas) are as follows:

$$CH_4 + H_2O \rightarrow CO + 3H_2 \ [\Delta H = 206 \text{ kJ mol}^{-1}] \tag{13.4}$$

$$C_nH_m + nH_2O \rightarrow nCO + (n + m/2)\, H_2 \tag{13.5}$$

This reaction is endothermic, which leads to the need for heat transfer. Reactor designs are typically limited by heat transfer rather than the reaction kinetics. Consequently, these reactors are designed to promote heat exchange and tend to be heavy and large. The fuel is heated and vaporized and then injected with superheated steam into the reaction vessel. The steam-to-carbon molar ratio is usually in the neighborhood of 2.5:1 but lower ratios are sought to improve cycle efficiency. Excess steam is used to force the reaction to completion as well as to inhibit soot formation. Like most light hydrocarbons, heavier fuels can be reformed through high-temperature reaction with steam. Steam reforming of most hydrocarbons occurs only over appropriate catalysts and at elevated temperatures 760°C to 980°C (1400°F to 1800°F) where unwanted reforming may occur without a catalyst. A typical steam reformed natural gas reformate is presented in Table 13.4.

Catalysts for SR are usually group VIII metals whereby Ni appears to be the most active of this group. Hydrocarbons are usually contaminated with small amounts of substances that can poison the reformer catalysts. The catalytic activity depends on metal surface area. For nickel, the crystals sinter quickly above the so-called Tamman temperature (590°C), approaching a maximum size related to the pore diameter of the support. Most commercial applications of steam reforming use a catalyst to enhance reaction rates at decreased temperatures. Lower temperatures favor high CO and hydrogen concentration. The reforming catalyst also promotes the water–gas shift reaction. The exothermic water–gas shift reaction occurs in the steam reformer reactor. The combined reaction, steam reforming and water gas shift, is endothermic. As such, an indirect high-temperature heat source is needed to operate the reactor. This heat source usually takes the form of an adjacent, high-temperature furnace that combusts a small portion of the fuel or the fuel effluent from the fuel cell.

Steam reforming of higher hydrocarbons can be used to produce methane suitable for use in high-temperature internal reforming fuel cells. Steam prereforming of hydrocarbons, as a process step in the manufacture of hydrogen, ammonia, methanol, carbon monoxide, and syngas, is an established technology. All higher hydrocarbons are converted over a nickel-based catalyst into a gas mixture containing hydrogen, methane, and carbon oxides. Establishment

TABLE 13.4

Typical Steam Reformed Natural Gas Reformate

Mole Percent	Reformer Effluent	Shifted Reformate
H_2	46.3	52.9
Co	7.1	0.5
CO_2	6.4	13.1
CH_4	2.4	2.4
N_2	0.8	0.8
H_2O	37.0	30.4
Total	100.0	100.0

of methanation and shift reaction equilibria at the process conditions determines the composition of the pre-reformed gas. By proper design of fuel processing systems, a wide variety of fuels may be converted to a suitable reformate.

Sulfur (S) is the most severe poison as it chemisorbs readily on any metal surface. A desulfurization step is carried out before the fuel is led into the reformer to avoid this poisoning. This is usually accomplished with a zinc oxide sulfur polisher and the possible use of a hydrodesulfurizer, if required.

13.2.4 Partial Oxidation Reforming

Methane and other hydrocarbons can be converted to hydrogen via POX, which may be catalyzed or unanalyzed or a combination of both. POX is a reforming process in which the fuel is partially combusted in an exothermic reaction that provides heat for other reactions in the reforming system usually in combination with steam reforming. Figure 13.4 shows schematics

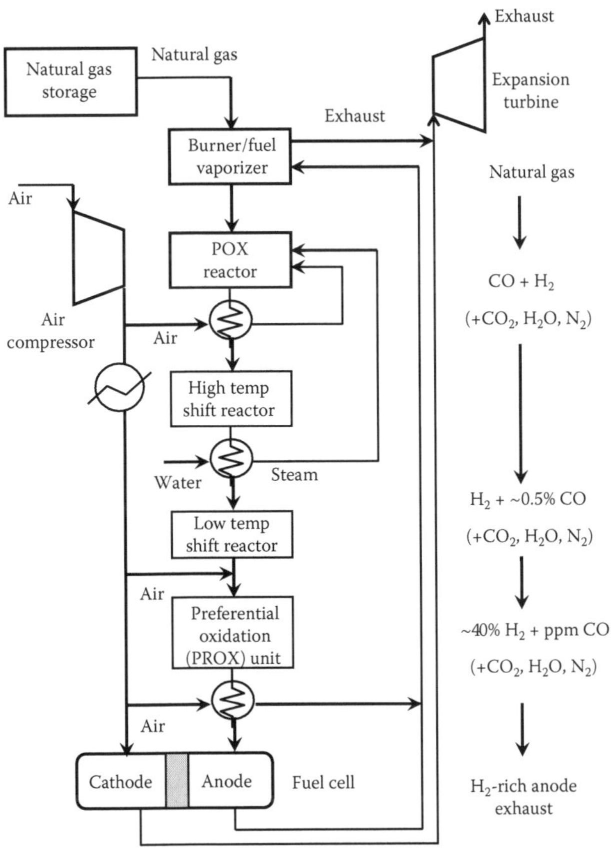

FIGURE 13.4
POX for the formation of hydrogen from natural gas.

of the natural gas–fed POX process. The oxygen fed to the system is substoichiometric such that both CO and CO_2 are formed in the system. A water–gas shift reaction can transform the CO from the first reaction further into CO_2. Reactions are

$$CH_4 + 1/2\ O_2 \rightarrow CO + 2H_2\ [\_H = -247\ \text{kJ mol}^{-1}] \tag{13.6}$$

$$CH_4 + O_2 \rightarrow CO_2 + 2H_2 \tag{13.7}$$

POX is highly exothermic and raises the reactants to a high temperature. The resulting reaction products, still in a reduced state, are then quenched through the introduction of superheated steam. The addition of steam promotes the combined water–gas shift and steam reforming reactions, which further cools the gas. In most cases, and with sufficient preheating of the reactants, the overall reaction is exothermic and self-sustaining. For some applications, however, particularly small-scale configurations, a catalyst can be used to increase reaction rates at lower reaction temperatures.

POX reactor temperatures vary widely. Noncatalytic processes for gasoline reforming require temperatures in excess of 1000°C. These temperatures require the use of special materials and significant preheating and integration of process streams. The use of a catalyst can substantially reduce the operating temperature, allowing the use of more common construction materials such as steel. Lower temperature conversion leads to less carbon monoxide (an important consideration for low-temperature fuel cells), such that the shift reactor can be smaller. Lower temperature conversion will also increase system efficiency. For some heavy hydrocarbon fuels, typical values range from as low as 870°C for catalytic POX upward to 1400°C for noncatalytic POX. For sulfur-bearing diesel fuel, a catalytic POX reactor will usually operate at approximately 925°C. This relatively elevated temperature is needed to overcome catalyst degradation owing to the presence of sulfur. Noncatalytic POX reactors operate at around 1175°C on diesel fuel.

Methane POX occurs over heterogeneous selective catalysts whereby Ni/Al_2O_3 has a high selectivity for CO_2/H_2 at reforming temperatures of 850°C. $Ni/La/Al_2O_3$ was found to have a lower activity and selectivity. At lower temperatures, different catalysts have been tested including $Co/La/Al_2O_3$, which is active at 750°C, and the $Fe/La/Al_2O_3$ catalyst at 600°C.

POX should be reacted so that the overall reaction is exothermic, but at a low oxygen-to-fuel ratio to favor higher hydrogen yields. A POX reformer can also be used to convert gaseous fuels but does not produce as much hydrogen as the steam reformers. For example, a methane-fed POX reformer would produce only approximately 75% of the hydrogen (after shifting) that was produced by an SR. Therefore, POX reformers are typically used only on liquid fuels that are not well suited for steam reformers. POX reformers rank second after steam reformers with respect to their hydrogen yield.

13.2.5 Autothermal Reforming

The coupling of steam reformation with POX is termed autothermal reforming (ATR). ATR is carried out in the presence of a catalyst that controls the reaction pathways and thereby determines the relative extents of the POX and SR reactions. The SR reaction absorbs part of the heat generated by the POX reaction, limiting the maximum temperature in the reactor. The net result can be a slightly exothermic process. ATR provides a fuel processor compromise that operates at a lower O/C and lower temperature than the POX; is smaller, quicker starting, and quicker responding than the SR; and results in high H_2. A catalytic POX reaction must be used to reduce the temperature to a value compatible with the SR temperature.

The ATR conversion of methane to H_2 occurs at 850°C, where a POX process is combined to a catalytic steam reforming process. An improved performance is seen for single-bed configurations as opposed to two-bed systems owing to the better heat-exchange possibilities in a single-bed reactor. A 60% ± 65% methane conversion could be attained with a selectivity of 80% toward hydrogen production. Systems employing the autothermic conversion of hydrocarbons are being used internally for MCFCs and SOFCs and can be used externally for other fuel cell systems.

13.2.6 Water Shift Reaction

After the steam reformation, the product or the POX process, the reformate conversion step, involves conversion of the carbon monoxide to hydrogen via the water–gas shift to carbon dioxide:

$$CO + H_2O \rightarrow H_2 + CO_2 \; [\Delta H = -41 \text{ kJ mol}^{-1}] \qquad (13.8)$$

However, the thermodynamics of the reaction are such that higher temperatures favor the production of carbon monoxide and hence reverse reaction can occur. To avoid this, the product gas from the steam reformer is cooled and is passed through a reactor-containing catalyst, which promotes the shift reaction. This has the effect of converting carbon monoxide into carbon dioxide. Depending on the reformate composition, more than one shift reactor may be needed to reduce the carbon monoxide level to an acceptable level such as for the PAFC fuel cell. The iron–chromium catalyst is found to be effective for promoting the shift reactor at relatively high temperatures (400°C–500°C), and this may be followed by further cooling of the gas before passing to a second, low-temperature reactor (200°C–250°C) containing a copper catalyst. At this temperature, the proportion of carbon monoxide present will typically be approximately 0.25% to 0.5%, and so these two stages of shift conversion are sufficient to decrease the carbon monoxide content to meet the needs of the PAFC. However, this level is equivalent to 2500 to 5000 ppm, which exceeds the limit for PEM fuel cells by a factor of about 100. It is similar to the CO content in the product from a methanol reformer.

13.2.7 Coal Gasification

Coal gasification is an endothermic process requiring high temperature for the reaction to proceed. In this process, solid coal reacts with oxygen and steam to produce a syngas mixture (CO/H_2) according to the following reaction:

$$C + H_2O \rightarrow CO + H_2. \tag{13.9}$$

The reaction process can be very complex, however, as coal is provided in different forms and compositions. The process, therefore, has to be adjusted for each type of coal. The heat required for gasification is supplied by the POX of coal. Overall, the gasification reactions are exothermic, so waste heat boilers are often used at the gasifier effluent. The temperature and, therefore, composition of the product gas depend upon the amount of oxidant and steam, as well as the design of the reactor. Gasifiers typically produce contaminants that must be removed before entering the fuel cell anode. These contaminants include H_2S, COS, NH_3, HCN, particulates, tars, oils, and phenols. The contaminant levels depend on both the fuel composition and the gasifier employed. There are two families of cleanup that remove the sulfur impurities: hot and cold gas cleanup systems. Cold gas cleanup technology is commercial and has been proven over many years. Hot gas cleanup technology is still developmental and would likely need to be joined with low-temperature cleanup systems to remove the non-sulfur impurities in a fuel cell system. For example, tars, oils, phenols, and ammonia could all be removed in a low-temperature water quench followed by gas reheat. A typical cold gas cleanup process following an entrained gasifier would include the following subprocesses: heat exchange (steam generation and regenerative heat exchange), particulate removal (cyclones and particulate scrubbers), COS hydrolysis reactor, ammonia scrubber, acid gas (H_2S) scrubbers (Sulfinol, SELEXOL), sulfur recovery (Claus and SCOT processes), and sulfur polishers (zinc oxide beds). All of these cleanup systems increase process complexity and cost, while decreasing efficiency and reliability (Wilhelm et al., 2001). In addition, many of these systems have specific temperature requirements that necessitate the addition of heat exchangers or direct contact coolers. For example, a COS hydrolysis reactor operates at approximately 180°C (350°F), the ammonia and acid scrubbers operate in the vicinity of 40°C (100°F), while the zinc oxide polisher operates at approximately 370°C (700°F).

The coal gasification systems can be classified as one of three basic types: (1) moving bed, (2) fluidized bed, and (3) entrained bed. All three of these types use steam and either air or oxygen to partially oxidize coal into a gas product. The moving-bed gasifiers produce low-temperature (425°C to 650°C), gas-containing devolatilization products, such as methane and ethane, and hydrocarbons including naphtha, tars, oils, and phenol. Entrained-bed gasifiers produce a gas product at high temperature (>1260°C), composed almost entirely of hydrogen, carbon monoxide, and carbon dioxide. The

TABLE 13.5

Typical Coal Gas Composition for Selected Oxygen-Blown Gasifiers

Gasifier Type	Moving Bed	Fluidized Bed	Entrained Bed			
Manufacturer	Lurgi	Winkler	Destec	Koppers-Totzek	Texaco	Shell
Cola	Illinois No. 6	Texas Lignite	Appalachian Bit.	Illinois No. 6	Illinois No. 6	Illinois No. 6
Mole percent						
Ar	Trace	0.7	0.8	0.9	0.9	1.1
CH_4	3.3	4.6	0.6	—	0.1	—
C_2H_4	0.1	—	—	—	—	—
C_2H_6	0.2	—	—	—	—	—
CO	5.8	33.1	45.2	43.8	39.6	63.1
CO_2	11.8	15.5	8.0	4.6	10.8	1.5
COS	Trace	—	—	0.1	—	0.1
H_2	16.1	28.3	33.9	21.1	30.3	26.7
H_2O	61.8	16.8	9.8	27.5	16.5	2.0
H_2S	0.5	0.2	0.9	1.1	1.0	1.3
N_2	0.1	0.6	0.6	0.9	0.7	4.1
NH_2 + HCN	0.3	0.1	0.2	—	—	—
Total	100.0	100.0	100.0	100.0	100.0	100.0

fluidized-bed gasifier product gas falls between these two other reactor types in composition and temperature (925°C to 1040°C). Typical compositions for several oxygen-blown coal gasification products are shown in Table 13.5.

13.2.8 Carbon Monoxide Removal

As seen above, reforming hydrocarbons into hydrogen always produces small amounts of CO in the gas stream. Though this CO is not a problem for high-temperature fuel cells that can use CO as a fuel for the anode reaction, for low-temperature fuel cells, carbon monoxide must be eliminated (Narusawa et al., 2003). The gas cleanup can be done in several ways. The first step is a water shift conversion of CO into CO_2, thereby producing more hydrogen. The CO content is reduced to the equilibrium value at a specific temperature (in the range of 1% CO; it decreases with decreasing temperature). The next step of CO removal has to be very efficient since it needs to lower the CO content to values in the parts-per-million range so that fuel can be fed into the fuel cell. The three most common techniques are selective oxidation, methanation, and the use of a selective membrane.

Selective removal of CO is done over some catalysts that oxidize mainly CO and not hydrogen. Catalysts used for this purpose are usually alumina (Al_2O_3)-supported structures. Ru and Rh supported on alumina are among the most active catalysts. Near to complete CO conversion can be achieved at temperatures as low as 100°C. Cu particles dispersed on alumina are cheaper, easier to make, and, combined with ZnO, also have a high catalytic activity for CO oxidation. The activity depends on the particle size of Cu with smaller particles being more active.

In the methanation process, CO is reacted with hydrogen to form methane and water. The amount of hydrogen needed to carry out this reaction is, however, three times the amount of CO removed. The reaction for this process is

$$CO + 3H_2 \rightarrow CH_4 + H_2O\ (\Delta H_0 = -206.2\ \text{kJ mol}^{-1}). \tag{13.10}$$

The CO_2 present in the gas stream can react with the produced water in a reverse water shift reaction (Equation 13.4), thereby producing the unwanted CO. Methanation can work for removal of CO but severe restrictions are necessary to make this process viable in a gas cleanup system.

Palladium alloy membranes are very effective in the removal of CO from the fuel stream. The gas is fed through the membrane through which hydrogen permeates selectively. This technique requires a rather large pressure difference and a relatively high temperature that reduces the efficiency of the system. A combined system with different catalytic stages can prove to be the best answer to the problem of gas cleanup. The second cleaning step should take the CO level down to around 5 ± 100 ppm, which can then be fed to the fuel cell without significant losses.

13.3 Hydrogen as Energy Carrier

Hydrogen is the simplest and lightest of all chemical elements on Earth, usually composed of one proton and one electron and forms a diatomic volatile gas, with a boiling point at 20.4 K. It is not a primary source of energy as it occurs only in nature in combination with other elements, primarily with oxygen in water and with carbon, nitrogen, and oxygen in living materials and fossil fuels. However, when split from these other elements to form molecular hydrogen, a process requiring another source of energy, it becomes an environmentally attractive fuel. It can be burned or combined with oxygen in a fuel cell without generating CO_2, producing only water. Like electricity, it is a very clean energy at the point of use. Hydrogen, together with fuel cells, which are very efficient energy conversion devices, is attracting the attention of public authorities and the private industry.

Hydrogen is not a source but a storage medium and an energy carrier. Like any other carrier, it must be produced from a primary energy source, fossil fuel, geothermal, nuclear, and all solar-based energy sources such as photovoltaic, wind, hydro, and biomass. Figure 13.5 shows possible routes for hydrogen generation.

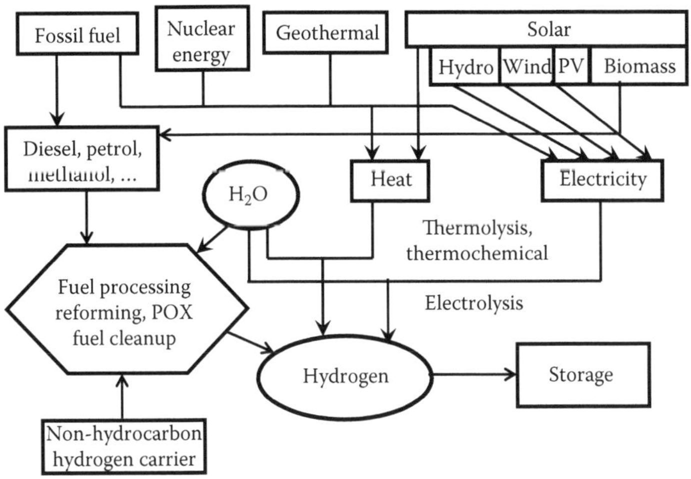

FIGURE 13.5
Possible hydrogen production routes.

Currently more than 80% of the world energy supply comes from fossil fuels, resulting in strong ecological and environmental impacts. There are key drivers to opt for hydrogen as a promising energy carrier. The first one is the diversification of the energy sources and the reduction of dependency on fossil fuels. Since hydrogen can be produced from any primary source, it could improve the reliability of the energy supply and stabilize the energy market. It would make the utilization of energy more independent from the production and allow the economic exploitation of energy sources remote from the consumers. The second reason is the reduction of the environmental impact of the energy system. Indeed, most of the anthropogenic impacts on the environment come from the combustion of fossil fuels in the industrial, domestic, and transport sectors. Using an energy carrier that is carbon free—which hydrogen is—would reduce most of the related environmental problems. The third reason is the control of acceptable costs and the hope of stable prices over time. The supply of energy at reasonable and stable prices is not ensured at all by the producers of crude oil or natural gas. Hydrogen facilitates the diversification of the sources and contributes to the reliability and security of the energy supplies.

13.3.1 Hydrogen Generation Methods

13.3.1.1 Fossil Fuels and Biomass

Various hydrogen generation technologies have been developed over decades. In the previous sections, hydrogen production from fossil fuel has been described, including steam reforming of natural gas, which is currently the least expensive method and is responsible for more than 90% of hydrogen production worldwide, the POX process natural gas or other liquid or gaseous hydrocarbons, the ATR process, and coal gasification. Biomass and biomass-derived fuels such as resources such as wood, agricultural residues, and consumer waste are renewable energy sources that can also be used to produce sustainable hydrogen. Current technologies for converting biomass into molecular hydrogen include gasification or pyrolysis of biomass coupled to subsequent steam reforming. The main conversion processes are indirect-heat gasification, oxygen-blown gasification, and anaerobic fermentation.

13.3.1.2 Electrolysis

As shown in Figure 13.5, any primary energy source (nuclear, solar, geothermal, fossil) converted into electricity can be used to produce hydrogen by the electrolysis of water, where water molecules are split into hydrogen and oxygen. The basic theory and the reactions taking place at the electrodes of the electrolyzer are the same as that for fuel cells except that the reactions happen in an opposite way. Different electrolytes can be used, just as for fuel cells. At low temperatures (<100°C), alkaline liquids and solid PEMs are used as electrolytes with liquid water as feed. At high temperatures, the feed has to be steam and typically solid oxide electrolysis cell (SOEC) (SOFC used to operate in reverse reaction) is used for electrolysis.

A low-temperature PEM-based electrolyzer is shown in Figure 13.6. The basic structure of the electrodes and the electrolyte is the same as that for the PEMFC. The electrode structure of the electrolyzer is different from that of

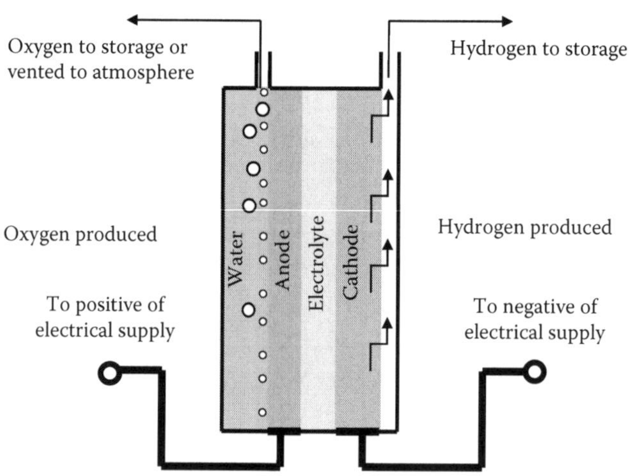

FIGURE 13.6
A PEM-based electrolyzer.

the fuel cell such that the product gases can be expelled quickly as possible. In a PEM electrolyzer, the positive electrode is properly called the anode where oxygen is generated and the negative electrode is called the cathode where hydrogen is generated. At the positive electrode, the water is oxidized (electrons removed) and oxygen is made

$$2H_2O \rightarrow O_2 + 4H^+ + 4e^- \quad (13.11)$$

At the negative electrode, protons are removed from the electrolyte, electrons are provided by the external electrical supply, and hydrogen is formed via the reaction

$$4H^+ + 4e^- \rightarrow 2H_2. \quad (13.12)$$

Unlike PEMFC operation, the water management and cooling problem do not arise in a PEM-based electrolyzer as water supplied to the negative electrode can be circulated for cooling and the positive electrode is flooded with water. The hydrogen generated from the electrolyzer is of high purity though it may have presence of some water vapor.

The efficiency of an electrolyzer is given as

$$\eta = 1.48/V_c, \quad (13.13)$$

where V_c is the applied voltage. There are also kinetic, ohmic, and mass transport losses in electrolyzers exactly the same as for fuel cells. In reality, values of V_c are around 1.6 to 2.0 V, depending on the current density, and cells normally can be operated at current densities of 1.0 A cm^{-2}. Typical operating efficiencies with the low-temperature electrolyzer, including the energy needed to compress the product gas, are around 60% to 70%. However, if the conversion of the primary energy to electrical energy is accounted, then the overall efficiency of the hydrogen generation with low-temperature electrolysis is maximum at around 22%.

A schematic of the high-temperature electrolyzer using SOEC is shown in Figure 13.7. For high-temperature electrolysis (HTE), water is fed as a steam at the cathode electrode. And the oxygen or air is supplied to the anode electrode. Steam is dissociated at the cathode. Simultaneously, oxygen ions migrate through the electrolyte material and oxygen molecules form on the anode surface by releasing electrons.

$$\text{Cathode reaction: } 2H_2O + 4e^- \rightarrow H_2 + 2O^- \quad (13.14)$$

$$\text{Anode reaction: } 2O^- \rightarrow O_2 + 4e^- \quad (13.15)$$

The HTE has high thermal efficiency compared to the low-temperature electrolysis. The overall efficiencies have estimated to increase from 33% to a

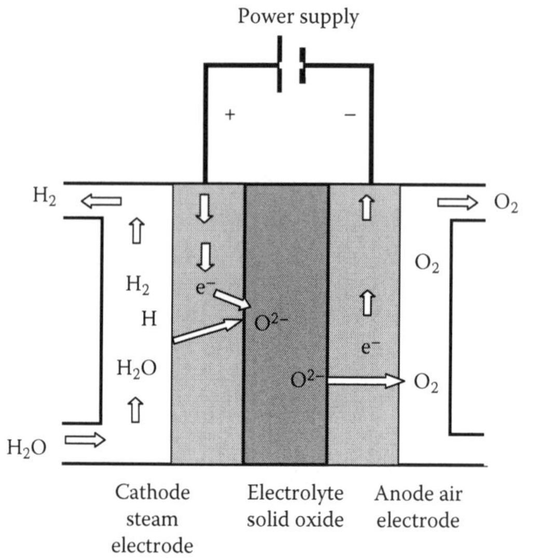

FIGURE 13.7
Solid oxide electrolyte-based high-temperature electrolysis.

maximum of 59% at 1000°C, which is over two times higher than that of the conventional low-temperature water electrolysis (Liu et al., 2008). There are several challenges with HTE. Because of the high operation temperature of 700°C or 800°C to 1000°C, the SOEC components—electrolyte, electrodes, and interconnects—have to withstand chemical stability. The electrolyte should have high ionic and low electronic conductivity, whereas electrodes should have good electronic conduction and coefficient of thermal expansion close to the electrolyte.

Though making electrolytic hydrogen uses considerably more energy than the hydrocarbon processes, electrolysis is of interest for several reasons. As it is compact, the electrolyzer can be located at an existing fueling station and is cost-effective for local hydrogen production. The renewable source (solar, wind, and hydropower) can be used to produce electricity, which in turn can be used to generate hydrogen using electrolysis. Thus, the electrolysis provides a method to store the renewable energy. This also helps in developing new architecture for distributed energy generation.

13.3.1.3 Thermochemical Water Splitting

Various thermochemical water-splitting processes have been proposed to address hydrogen production on a large scale. Most of these cycles utilize a high-temperature heat source for rapid chemical kinetics and high conversion efficiencies. In the thermochemical processes, a set of coupled, thermally-driven chemical reactions decompose water into H_2 and O with various intermediate regents. All reagents returned within the cycle and recycled. A typical cycle with two reagents (A and B) having four reactions (sections) that are driven at temperatures T_1, T_2, T_3, and T_4, respectively, is shown in Equations 13.16 through 13.19

$$2AB + 2H_2O \rightarrow 2AH + 2BOH + \text{heat at } T_1 \quad (13.16)$$

$$2BOH \rightarrow 2B + \frac{1}{2}O_2 + H_2O - \text{heat at } T_2 \quad (13.17)$$

$$2AH \rightarrow 2A + H_2 - \text{heat at } T_3 \quad (13.18)$$

$$2A + 2B \rightarrow 2AB + \text{heat at } T_4 \quad (13.19)$$

These thermochemical water-splitting processes have been explored extensively in the 1970s and some 25 key cycles have been identified (Besenbruch, 1982; Brown et al., 2003). The sulfur–iodine (SI) cycle is currently the leading candidate. The SI cycle consists of three chemical processes expressed as sections:

$$\text{Section 1 (Bunsen reaction, 120°C): } I_2 + SO_2 + 2H_2O \rightarrow 2HI + H_2SO_4 \quad (13.20)$$

Section 2 (sulfuric acid decomposition, 830°C–900°C):

$$H_2SO_4 \rightarrow H_2O + SO_2 + \frac{1}{2}O_2 \quad (13.21)$$

$$\text{Section 3 (hydrogen iodide decomposition, 300°C–450°C): } 2HI \rightarrow H_2 + I_2 \quad (13.22)$$

Temperature on the order of 800°C–1000°C is required for the SI cycle. In the SI cycle, all process fluids are recycled and no greenhouse gases are emitted. Also, the SI cycle has been fully flow sheeted and operated at the bench scale in the United States and Japan. This cycle has the highest efficiency (~52%) of any process that has been simulated through chemical process codes. Nuclear heat energy has been considered as a promising primary energy source for large-scale hydrogen production without carbon dioxide emission with thermochemical process. The high-temperature gas-cooled reactors such as the pebble bed modular reactor, capable of providing heat energy up to 1000°C and that have a high level of inherent safety, are suitable for the large-scale hydrogen production using thermochemical water-splitting processes.

13.3.1.4 Photocatalysis

Photocatalysis is the cleanest way to produce hydrogen by direct splitting of water into hydrogen and oxygen with sunlight. In this process, water splitting occurs on heterogeneous photocatalysts with semiconductor properties (Kudo, 2007; Maeda and Domen, 2007). Semiconductors, as shown in Figure 13.8a, have

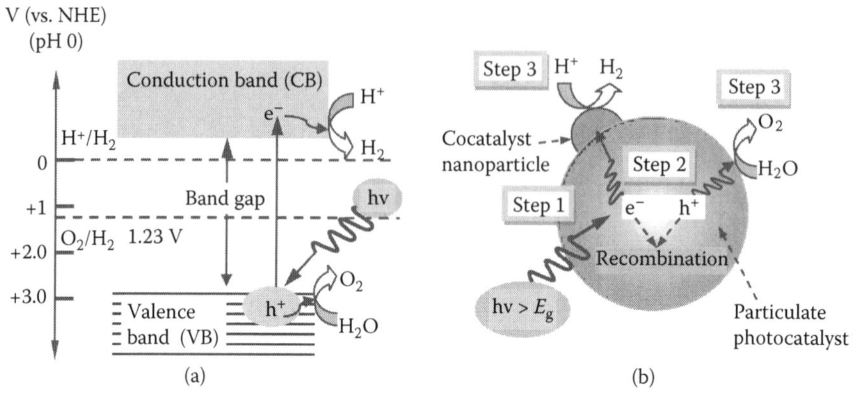

FIGURE 13.8
(a) Principle of water splitting on semiconductor photocatalysts. (b) Processes for photocatalytic reaction in a powdered system.

the band structure in which the conduction band (CB) is separated from the valence band (VB) by a suitable band gap. When light (of which the energy is larger than the band gap) is irradiated, electrons (e^-) and holes (h^+) are generated in the CBs and VBs, respectively. The photogenerated electrons and holes cause redox reactions similarly to electrolysis. Water molecules are reduced by the electrons to form hydrogen and are oxidized by the holes to form oxygen for overall water splitting. Important points in the semiconductor photocatalyst materials are the width of the band gap and energy levels of the CBs and VBs. The bottom level of the CB has to be more negative than the redox potential of H^+/H_2 (0 V vs. normal hydrogen electrode), while the top level of the VB has to be more positive than the redox potential of O_2/H_2O (1.23 V). Therefore, the band gap should be wider than 1.23 eV and sufficiently small to make efficient use of solar spectrum (~<3 eV).

There are three steps in the photocatalysis process. *Step 1:* Photon with energy above 1.23 eV (λ < ~1000 nm) is absorbed; *Step 2:* photoexcited electrons and holes separate and migrate to the surface as shown in Figure 13.8b; and *Step 3:* the adsorbed species (water) is reduced and oxidized by the electrons and holes.

The process of photogenerated electrons and holes has to be separated and migration to the surface is strongly affected by bulk properties such as crystallinity. The better the crystallinity is, the fewer the number of defects that often work as recombination sites between photogenerated electrons and holes, resulting in increases in lifetime and mobility. On the other hand, the number and quality of the active sites for redox reactions by photogenerated electron and hole pairs are kinetically affected by surface properties. Even if the photogenerated electrons and holes possess thermodynamically sufficient potentials for water splitting, they will have to recombine with each other if the active sites for the redox reactions do not exist on the surface. Therefore, a cocatalyst such as Pt, NiO, and RuO_2 is often loaded on the surface in order to introduce the active site. Thus, suitable bulk and surface properties and energy structure are required for photocatalysts.

13.3.1.5 Biohydrogen

Several microbial-driven biochemical reactions, mainly in anaerobic fermentation processes, generate hydrogen as by-product (Das et al., 2008; Mathews and Wang, 2009). In addition, certain microorganisms can produce enzymes that can produce H_2 from water if an outside energy source, like sunlight, is provided to them. Specific processes in which microorganisms can produce H_2 can be classified into four main groups: (i) biophotolysis: direct biophotolysis and indirect biophotolysis, (ii) photofermentation, (iii) dark fermentation, and (iv) hybrid systems: dark fermentative and photofermentative bioreaction, and bioelectrochemically assisted microbial bioreactions. Table 13.6 gives the summary of the general reactions and microorganisms used in these biological hydrogen production processes.

The metabolic process for hydrogen production is dependent on the reduction of the metabolite ferredoxin

$$4H^+ + \text{ferredoxin(ox)} \rightarrow \text{ferredoxin(red)} + 2H_2 \tag{13.23}$$

TABLE 13.6

Different Biological Hydrogen Generation Reactions and Associated Microorganisms

Process	General Reactions and Classifications of Microorganisms Used
i. Biophotolysis of water	
– Direct biophotolysis	$2H_2O + \text{light} \rightarrow 2H_2 + O_2$
– Indirect biophotolysis	Green microalgae (*Chlamydomonas reinhardtii*)
	$6H_2O + 6CO_2 + \text{light} \rightarrow C_6H_{12}O_6 + 6O_2$
	$C_6H_{12}O_6 + 2H_2O \rightarrow 4H_2 + 2CH_3COOH + 2CO_2$
	$2CH_3COOH + 4H_2O + \text{light} \rightarrow 8H_2 + 4CO_2$
	Overall reaction $12H_2O + \text{light} \rightarrow 12H_2 + 6CO_2$
	Microalgae, blue-green algae (cyanobacteria)
ii. Photofermentation	$CH_3COOH + 2H_2O + \text{light} \rightarrow 4H_2 + 2CO_2$
	Purple bacteria, microalgae
iii. Dark fermentation	$C_6H_{12}O_6 + 6H_2O \rightarrow 12H_2 + 6CO_2$
	Fermentative bacteria
iv. Hybrid systems	Stage I: $C_6H_{12}O_6 + 2H_2O \rightarrow 4H_2 + 2CH_3COOH + 2CO_2$
– Combined dark fermentation and photofermentation	Stage II: $CH_3COOH + 2H_2O + \text{light} \rightarrow 4H_2 + 2CO_2$
	Fermentative bacteria followed by anoxygenic phototrophic bacteria
– Bioelectrochemically assisted microbial reactor	$C_6H_{12}O_6 + 2H_2O \rightarrow 4H_2 + 2CO_2 + 2CH_3COOH$
	Anode: $CH_3COOH + 2H_2O \rightarrow 2CO_2 + 8e^- + 8H^+$
	Cathode: $8H^+ + 8e^- \rightarrow 4H_2$
	Bioelectrochemically assisted microbial reactor (BEAMR), reverse process of microbial fuel cell

For this process to run, ferredoxin has to be recycled through oxidation. The recycling process is dependent on the transfer of electrons from nicotinamide adenine dinucleotide (NADH) to ferredoxin.

$$2 \text{ ferredoxin(red)} + 2 \text{ NADH} \rightarrow 2 \text{ ferredoxin(ox)} + H_2 \quad (13.24)$$

The enzymes that catalyze this recycling process are referred to as hydrogen-forming enzymes. There are three main classes of hydrogen-forming enzymes: [FeFe]-hydrogenase, [NiFe]-hydrogenase, and nitrogenase. The hydrogen-forming enzymes are inactivated by molecular oxygen and must be separated from oxygen, to produce hydrogen. In the direct biophotolysis of H_2 production, two photosynthetic systems are responsible for photosynthesis process: (i) photo system I, which produces a reductant for CO_2, and (ii) photo system II which splits water to evolve O_2 as shown in Figure 13.9a. The two photons obtained from the splitting of water can either reduce CO_2 by photo system I or form H_2 in the presence of hydrogenase. In microorganisms, electrons are generated when photo system II absorbs light energy, which is then transferred to ferredoxin. A reversible hydrogenase accepts electrons directly from the reduced ferredoxin (Fd) to generate H_2 in the presence of hydrogenase.

Since hydrogenase is sensitive to oxygen, it is necessary to maintain the oxygen content at a low level (under 0.1%) so that the hydrogen production can be sustained. This condition can be obtained by the use of green algae, *Chlamydomonas reinhardtii*, which can deplete oxygen during the oxidative respiration.

In indirect biophotolysis, the problem of sensitivity of the H_2 evolving process to O_2 is usually circumvented by separating O_2 and H_2 (Figure 13.9b).

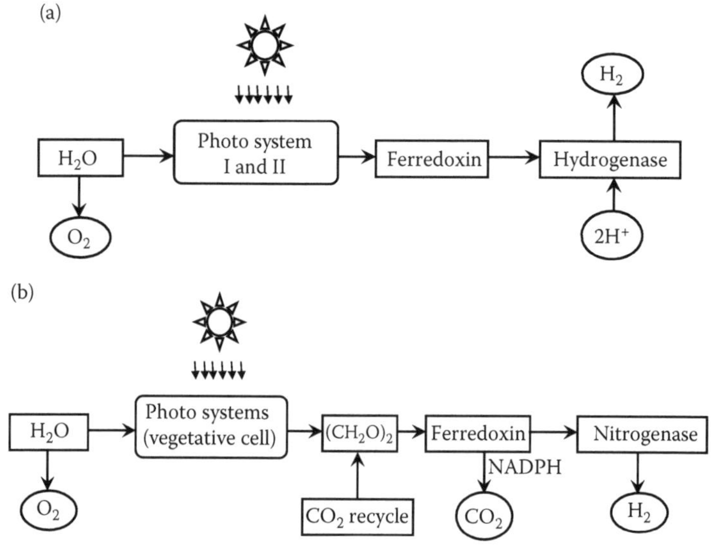

FIGURE 13.9
Biophotolysis of water through (a) direct biophotolysis and (b) indirect biophotolysis.

In this process, CO_2 is intermittently fixed and released, serving as the electron carrier between the O_2 producing (water splitting) reaction and the O_2 sensitive hydrogenase reactions. In such concepts, the algae undergo a cycle of CO_2 fixation into storage carbohydrates (starch, glycogen) followed by its conversion to H_2 by dark anaerobic fermentation processes. Many types of green algae and cyanobacteria, besides having the ability to fix CO_2 via photosynthesis, also have the ability to fix nitrogen from the atmosphere and produce enzymes that can catalyze the second H_2 generating step. Since these nitrogen-fixing enzymes, nitrogenase, are localized within the heterocyst, they provide an O_2-free environment to carry out the H_2 evolution reactions.

13.3.1.6 By-Product of Chemical Production Processes

Various chemical processes do produce hydrogen as a by-product. Hydrogen is a by-product of sodium or potassium chloride electrolysis that produces chlorine and caustic soda or potash:

$$NaCl + H_2O + \text{electricity} \rightarrow \frac{1}{2}CI_2 + NaOH + \frac{1}{2}H_2 \quad (13.25)$$

$$KCl + H_2O + \text{electricity} \rightarrow \frac{1}{2}CI_2 + KOH + \frac{1}{2}H_2 \quad (13.26)$$

Chlorine is one of the most common chemicals in the world. It is produced in huge quantities. Hydrogen is a by-product of catalytic reforming of petrol, made to improve the octane number. It is also a by-product of steam cracking of hydrocarbons for the production of ethylene. This hydrogen is used in other refinery operations. Smaller volumes of hydrogen are by-products in coke ovens and usually are kept for internal use. All by-product hydrogen requires some purification steps before use.

13.3.2 Hydrogen Storage

Energy storage in general enables the provision of additional energy during peak demand when the energy consumption exceeds the production capacity and the storage of energy in the reverse situation. Also, storage capability is required for all transportation and portable applications. Storage of hydrogen has two main methods: physical storage and chemical storage. In the physical storage, hydrogen can be stored as pressurized gas in tanks, as a cryogenic liquid, or by adsorption on high porosity materials or stored in glass microspheres. In chemical storage, it can be stored as metal hydrides and complex hydrides (Niedzwiecki, 2001; Schlapbach and Zuttel, 2001; Teitel, 1981).

Hydrogen being a low-density gas (0.08988 g/L at 1 atm) with a low boiling point (–252.87°C), there are gravimetric and volumetric limitations on storage through compression and liquefaction by the weight of the storage canisters and the insulated tanks. In Figures 13.10 and 13.11, the energy

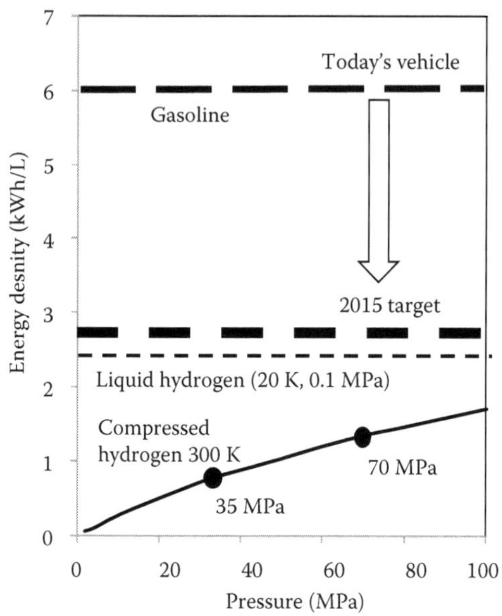

FIGURE 13.10
Energy density as a function of pressure for compressed hydrogen. (Data from http://www.hydrogen.energy.gov/pdfs/review06/2_storage_satyapal.pdf accessed on August 9, 2012.)

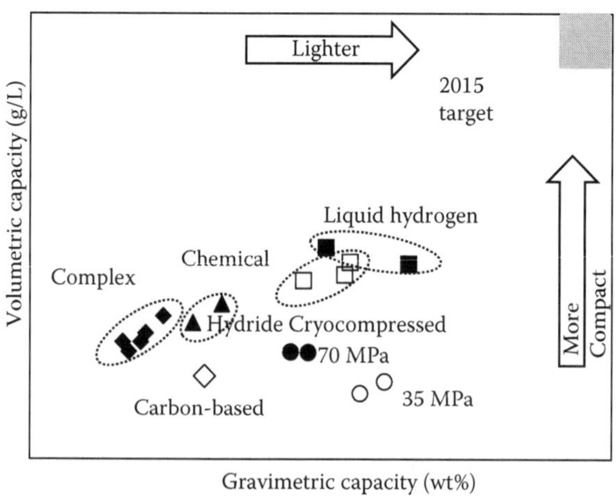

FIGURE 13.11
Volumetric capacity of various hydrogen storage chemicals and gravimetric capacities with current hydrogen storage methods. (Data from http://www.hydrogen.energy.gov/pdfs/review06/2_storage_satyapal.pdf accessed on August 9, 2012.)

density and volumetric and gravimetric capacities with these storage technologies, respectively, are shown. According to the US Department of Energy (DOE), hydrogen storage in one of the three barriers must be overcome to enable industry commercialization of hydrogen fuel cell vehicles (Satyapal et al., 2007). For the transportation application, the vehicle packaging, cost, and performance requirements for onboard hydrogen storage systems require a vehicle driving range of greater than 300 miles (500 km). The DOE onboard hydrogen storage system performance targets were developed through FreedomCAR, a partnership between DOE and the US Council for Automotive Research, and are shown in Figures 13.10 and 13.11, in terms of energy density versus pressure and volumetric capacity versus gravimetric capacity, respectively.

13.3.2.1 Physical Storage

13.3.2.1.1 Compressed Hydrogen

Compared to liquid fuels like gasoline which are put in similar tanks, hydrogen has a low energy-to-volume density. Because of its low density, gaseous hydrogen should be stored at higher pressures to improve density. This higher pressure requires material and special design to ensure tank integrity. Advances in compression technologies are also required to improve efficiencies and reduce the cost of producing high-pressure hydrogen. Today, carbon fiber–reinforced composite tanks of 35 MPa (5000 psi) and 70 MPa (10,000 psi) have been developed by Quantum Technologies and others and have been certified worldwide for storing compressed hydrogen. The inner liner of such tank is a high-molecular-weight polymer that serves as a hydrogen gas permeation barrier. A carbon fiber–epoxy resin composite shell is placed over the liner and constitutes the gas pressure load-bearing component of the tank. An outer shell is placed on the tank for impact and damage resistance. The pressure regulator for the 70 MPa (10,000 psi) tank is located in the interior of the tank. There is also an in-tank gas temperature sensor to monitor the tank temperature during the gas-filling process when tank heating occurs.

The key issues with compressed hydrogen gas tanks are high pressure, weight, volume, conformability, and cost. By increasing the amount and pressure of hydrogen, a greater driving range can be achieved. This requires stronger walled tank at the expense of cost and valuable space within the vehicle. The cost of high-pressure compressed gas tanks is essentially dictated by the cost of the material (carbon fiber) that must be used for lightweight structural reinforcement. The low-cost materials must be capable of meeting tank thickness constraints in order to help meet volumetric capacity targets. Thus, there is trade-off between the cost and weight and volume of the tank. Refueling times, compression energy penalties, and heat-management requirements during compression also need to be considered as the mass and pressure of onboard hydrogen are increased.

To improve gravimetric and volumetric storage capacities of compressed gas tanks, two approaches are being pursued. The first approach uses cryo-compressed tanks where, at fixed pressure and volume, gas tank volumetric capacity increases as the tank temperature decreases. Typically by cooling a tank from room temperature to liquid nitrogen temperature (77°K), its volumetric capacity will increase by a factor of four. The second approach involves the use of conformable tanks whose walls utilize structural support and maximum advantage of available vehicle space. Internal cellular-type load-bearing structures may also be employed for greater degrees of conformability.

13.3.2.1.2 Liquid Hydrogen

The storage energy density of hydrogen can be improved by cooling hydrogen into liquid form at a temperature of 20 K (–253°C). Liquid hydrogen tanks can store more hydrogen in a given volume than compressed gas tanks. The volumetric capacity of liquid hydrogen is 0.070 kg/L, compared to 0.030 kg/L for 70 MPa (10,000 psi) gas tanks. This enables huge amounts of hydrogen to be shipped by a tanker, truck, and rail. Industrial hydrogen liquefaction uses a variety of processes with helium, hydrogen, or gas mixtures as coolant. The hydrogen feed to the process is first cooled by liquid nitrogen and then further cooled in multi-stage heat exchangers, where the cooling power is provided by turbo expanders. Liquefaction is finally accomplished by throttling in a Joule–Thomson valve.

There are issues with liquid hydrogen tanks such as hydrogen boil-off, the energy required for hydrogen liquefaction, volume, weight, and tank cost. The liquefaction process requires a large fraction of the hydrogen energy, requiring approximately 3.8 kWh/kg including the ortho–para conversion of hydrogen during liquefaction. The practical energy requirements are considerably higher, by a factor of about six for small liquefiers and three for large ones. Typically at least 35% of the hydrogen energy is required for the liquefaction process. In order to minimize or eliminate boil-off, special materials for the tank and insulation are required. The addition of insulation reduces system gravimetric and volumetric capacity.

For small amounts of liquid hydrogen (105 to 106 L), Dewar flasks are used to store liquid hydrogen. The flask is then immersed in a second flask containing liquid nitrogen to prevent heat absorption by thermal radiation. Additional heat loss is prevented by using thermal radiation shields positioned within the multilayer insulation. In some tank designs using metal tanks, a vacuum is maintained between two tank layers to minimize conduction and convection heat losses. The largest liquid hydrogen storage vessels at NASA's Kennedy Space Center are two 3–4 million liter spherical-shaped containers. The material for construction of the inner shell (20 m diameter) is austenitic steel and the outer shell is carbon with a diameter of 23 m. The operating pressure of the tank for the delivery of hydrogen is 0.6 MPa and the boil-off rate is approximately 0.02% per day when the tank is full. The capital cost for liquefaction decreases with scale-up but reaches a steady value at approximately 225,000 kg/day. Apart from the use in the NASA program, liquid hydrogen is used on a large scale by the petroleum refining and ammonia production industries.

13.3.2.1.3 Adsorption on High-Porosity Materials

A new class of storage materials is nanostructured materials, such as carbon nanotubes, boron nitride nanotubes, polyaniline nanofibers, and metal organic framework (MOF) systems. These store hydrogen by either physisorption or chemisorption processes. Because of their large surface area and the porous nature of these materials, the amount of physisorbed hydrogen is quite large. Carbon nanotubes are graphite sheets rolled up in cylindrical shape either as single-walled nanotubes with single graphite layer or multi-walled nanotubes with concentric graphite layers. The amount of hydrogen adsorbed by these systems depends on the pore size and the curvature of the tubes. Several research programs aiming at understanding the potential of carbon nanotubes for hydrogen storage are still in progress. MOF materials have also been studied for their hydrogen adsorption properties. These frameworks are porous crystalline materials with a large surface area. They consist of organic linkers and inorganic metal oxide units such as $[ZnO4]_{61}$. Hydrogen molecules are stored in a MOF by adsorbing to its surface. A MOF-filled gas cylinder can store more gas because of adsorption that takes place on the surface of MOFs compared to an empty gas cylinder. The MOFs have fully reversible uptake-and-release behavior.

13.3.2.1.4 Glass Microspheres

High-pressure hydrogen can be stored in glass microspheres. The diameter and the thickness of the glass microspheres are approximately 1 mm and several dozens of micrometers, respectively. The stored hydrogen is released by heating or breaking the glass. Research about this futuristic method is in progress in France at the CEA.

13.3.2.2 Chemical Storage

13.3.2.2.1 Metal Hydrides

Metal hydrides are specific combinations of metallic alloys that adsorb hydrogen under moderate pressure and temperature, creating hydrides. A metal hydride tank contains a granular metal, which adsorbs hydrogen and releases it with the application of heat. Conventional high-capacity metal hydrides require high temperatures (300°C–350°C) to liberate hydrogen. Usually, absorbed hydrogen represents only 1% or 2% of the total weight of the tank. Some metal hydrides reach a storage capacity of 5% to 7%, but in this case, high temperatures are needed to recover the hydrogen. The volume of gas absorbed is relatively low, but the advantage of hydrides is their ability to safely deliver very pure hydrogen at constant pressure.

13.3.2.2.2 Chemical Hydride

Chemical hydrides such as sodium borohydride slurries or solutions can be used as a hydrogen carrier or storage medium. The hydrogen in the hydride is released through a reaction with water. Chemical hydride systems are irreversible and require thermal management and regeneration of the carrier to recharge the hydrogen content. An essential feature of the process is recovery and reuse of spent hydride at a centralized processing plant.

13.3.2.2.3 Other Chemicals

As previously mentioned, hydrogen is the most abundant element in the universe, but it only occurs bounded to other elements in many chemical compounds. Some of these compounds can be produced by chemical reaction and used as hydrogen storage systems. When the reverse reaction occurs, hydrogen is released. The reactions involved are linked to the compound. Ammonia cracking, methanol cracking, and dehydrogenation of cyclohexane into benzene are some examples. With chemical storage, the need for a storage unit for the hydrogen produced is eliminated when hydrogen is produced locally.

13.3.3 Transportation and Distribution

The development of a new infrastructure for transport and distribution of hydrogen is one of the key issues in the advancement of fuel cell technology. Depending on the source from which hydrogen is produced (i.e., from fossil or non-fossil sources) and the form in which it is delivered (gaseous, liquid, or in compound form) as well as dispensed (as a vehicular fuel or for stationary applications, such as electricity generation), various kinds of infrastructures are required.

For portable applications, small-scale hydrogen generation devices or replaceable liquid fuel cartridges are the most likely sources of fuel. For large transportation applications, a hydrogen-specific delivery infrastructure is required. Even if the fuel cell itself is more efficient than the gasoline engine it replaces, the delivery infrastructure must be efficient and economic, or the net well-to-wheel efficiency of the fuel cell system may actually be worse than the gasoline engine it is to replace. Thus, trade-offs in economies of scale between process and distribution technologies, and such issues as operating cost, safety, and materials, dictate the type of infrastructure. Other special concerns in the transportation and delivery are safety and the environment. Hydrogen being an extremely small molecule with very high diffusivity, it is difficult to contain without leakage.

A compressed hydrogen gas delivery truck can carry around 530 kg of hydrogen in the gas phase (or about 60–70 hydrogen car fuel tanks), or 3370 kg (about 425 tanks) in the liquid phase. Hence, the vast majority of pure hydrogen is presently shipped as cryogenic liquid, and a small fraction is shipped as compressed gas over short distances. In small quantities such as delivery to small laboratory volumes, hydrogen is delivered in gas bottles and short travel distances. For delivery from future pipeline or generation stations, some hydrogen transport by truck will be required.

For long distances, hydrogen delivery by pipeline can be the most efficient means for transporting the fuel, similar to the way petroleum or natural gas is currently shipped. Pipeline delivery of hydrogen can be done by three methods: (i) gas-phase hydrogen pipelines, (ii) cryogenic liquid hydrogen pipelines, and (iii) hydrogen carrier pipelines. Gas-phase hydrogen pipelines up to tens of kilometers have existed for over several decades of operation at various gasoline refinery locations all over the world. The design principles and experience for optimized design of natural gas pipelines can be used in developing gas-phase hydrogen pipelines.

The transport of cryogenic liquid hydrogen has the advantage over that of the compressed hydrogen gas as it requires less pumping power, but has the disadvantage of greater initial infrastructure cost to provide an insulated pipeline to prevent boil-off and the use of expensive cryogenic pump system equipment. For special applications such as fueling of liquid rocket engines for the space shuttle, short-distance (subkilometer-long) liquid hydrogen pipelines have been utilized. An energy transport concept called SuperGrid has recently been proposed for long distance electric power transmission with liquid hydrogen distribution that has minimum transmission losses. In this concept, the cryogenic hydrogen would flow around a superconducting metal to deliver large amounts of current with high efficiency. Another option for hydrogen transport is to ship it by hydrogen carrier species in the form of gaseous chemicals such as ammonia, natural gas, or a liquid fuel such as methanol or as metal hydrides if storage densities are high enough.

13.3.4 Hydrogen Safety

Hydrogen gas poses unique challenges because of its ease of leaking, low-energy ignition, wide range of combustible fuel–air mixtures, buoyancy, and its ability to embrittle metals, which must be accounted for to ensure safe operation (Alcock, 2001; Butler et al., 2009; Gregory, 1997; Lewis and von Elbe, 1987). Liquid hydrogen poses additional challenges because of its increased density and the extremely low temperatures needed to keep it in liquid form. Hydrogen, in both the liquid and gaseous states, is particularly subject to leakage because of its low viscosity and low molecular weight. Because of its low viscosity alone, the leakage rate of liquid hydrogen is roughly 100 times that of JP-4 fuel and 10 times that of liquid nitrogen. Hydrogen gas is colorless and odorless; hence, gas detectors are needed to check leakages.

Hydrogen–air or oxygen mixtures have ignition energy (0.02 mJ at 1 atm) that can ignite with very low energy input, one-tenth that required igniting a gasoline–air mixture. For instance, an invisible spark or a static spark from a person can cause ignition. The flammability limits based on the volume percent of hydrogen in air at 1 atm (101 kPa) are 4.0 and 75.0. The flammability limits based on the volume percent of hydrogen in oxygen at 14.7 psia (1 atm, 101 kPa) are 4.0 and 94.0. Hydrogen leaks can support combustion at very low flow rates, as low as 4 μg/s.

Besides the flammability concern, hydrogen simply vented to the atmosphere also poses a potential problem. Hydrogen collects under roofs and overhangs, where it forms an explosion hazard; any building that contains a potential source of hydrogen should have good ventilation, should have strong ignition suppression systems for all electric devices, and should preferably be designed to have a roof that can be safely blown away from the rest of the structure in an explosion. It also enters pipes and can follow them to their destinations. Hydrogen pipes should be located above other pipes to prevent this occurrence. Hydrogen sensors allow for rapid detection of hydrogen leaks to ensure that the hydrogen can be vented and the source of the leak tracked down. As in natural gas, an odorant can be added to hydrogen sources to enable leaks to be detected by smell. Hydrogen burns with a pale blue flame that is nearly invisible in daylight, so it is almost impossible to detect by the human senses, but it shows up readily on UV/IR flame detectors.

In some ways, hydrogen is clearly safer than gasoline. For example, it is very buoyant and disperses quickly from a leak. This contrasts with gasoline, which puddles rather than disperses, and where fumes can build up and persist even outside. Because hydrogen rises and disperses so rapidly, it is unlikely to cause asphyxiation. Hydrogen is nontoxic and nonpoisonous. It will not contaminate groundwater. It is a gas under normal atmospheric conditions, and a release of hydrogen does not contribute to atmospheric or water pollution. Hydrogen can be used as safely as other common fuels we use today when guidelines are observed and users understand its behavior.

13.4 Summary

A fuel cell power system is a hybrid system involving electrochemical processes and electricity generation and delivery. The power system can be very complex or simple depending on the method of fuel supply to the fuel cell and type of fuel cell. A simple small portable DMFC can have fuel in a cartridge fed by surface tension and diffusion to the fuel cell anode and can return the spent fuel to another compartment of the cartridge. The cathode side can be open to atmosphere with a small fan to blow air to feed oxygen. The power outlet can be two terminals from the electrodes of the cell. In another case of large stationary CHP with MCFC, the system may contain a fuel processor and a supply system with natural gas supply, desulfurization unit, heat exchanger, and a steam generator. The exhaust handling from anode will have a combustor, heat exchangers, and piping to feed to the inlet of the cathode/air side. The outlet of cell can have another power conversion unit with gas turbines or heat exchangers to extract heat from the final exhaust from the fuel cell system. These various system performances have to be optimized to produce high-efficiency power from the fuel cell system.

As the hydrogen is not freely available in nature, it must be produced from some form of primary energy. Currently, more than 90% of the hydrogen is produced from fossil fuel, specifically from methane reformation. A large portion of this hydrogen is used in oil refineries and fertilizer manufacturing industries. To replace the existing fossil fuel–based economy with hydrogen-based technology dubbed as hydrogen economy, early efforts are underway. Key components of this technology are efficient methods of generation, storage, transport, and delivery of hydrogen. Currently, there is no large-scale infrastructure for hydrogen economy to affect major industries such as transportation, manufacturing, electrical power, and home utilities. The success or failure of developing these infrastructures will depend on the economy of generation, storage, transport, and delivery of hydrogen.

For carbon-free hydrogen generation, renewable energy seems most ideal. Though electrolysis is well known to produce clean hydrogen, it is inefficient at low temperatures besides needing inexpensive electrical power. New technologies of hydrogen generation are being considered, which include thermochemical processes, high-temperature electrolysis, direct photocatalysis, and biological-based hydrogen generation processes. Hydrogen storage is another key area where numerous methods are being explored; these include compressed gas, cryogenic liquid, metal hydrides, chemical hydrides, and liquid hydrogen carrier fuels. Though none of these technologies have met the goals for the transportation sector, technology developments are progressing such that they are also economically competitive. There are other issues that also require attention for hydrogen economy to mature such as safety and code and standards that should be in place.

References

Alcock, J. L. Compilation of existing safety data on hydrogen and comparative fuels, Report EIHP2—Contract N° ENK6-CT2000-00442, WP N°:5, 2001.

Besenbruch, G. E. General atomic sulfur-iodine thermochemical water-splitting process. *American Chemical Society, Division of Petroleum Chemistry* 271: 48–53, 1982.

Brown, L. C., G. E. Besenbruch, R. D. Lentsch, K. R. Schultz, J. F. Funk, P. S. Pickard, A. C. Marshall and S. K. Showalter. High efficiency generation of hydrogen fuels using nuclear power—Final technical report for the Period August 1, 1999 through September 30, 2002. Prepared under the Nuclear Energy Research Initiative (NERI) Program grant no. DE-FG03-99SF21888 for the US Department of Energy, 2003.

Butler, M. S., C. W. Moran, P. B. Sunderland and R. L. Axelbaum. Limits for hydrogen leaks that can support stable flames. *International Journal of Hydrogen Energy* 34: 5174–5182, 2009.

Das, D., N. Khanna and T. J. Nejat Veziroğlu. Recent developments in biological hydrogen production processes. *Chemical Industry & Chemical Engineering Quarterly* 14: 57–67, 2008.

Gregory, F. D. Safety standard for hydrogen and hydrogen systems, February 12, 1997. Available at http://www.hq.nasa.gov/office/codeq/doctree/canceled/871916.pdf (accessed March 21, 2013).

Haji, S., K. A. Malinger, S. L. Suib and C. Erkey. Fuels and fuel processing, Chapter 5. In: *Fuel Cell Technology*, Editor N. Sammes. Springer-Verlag, London, pp. 165–211, 2006.

Kordesch, K. and G. Simader. *Fuel Cells and Their Applications*. VCH, Weinheim, Germany, 1966.

Kudo, A. Photocatalysis and solar hydrogen production. *Pure and Applied Chemistry* 79: 1917–1927, 2007.

Lewis, B. and G. von Elbe. *Combustion, Flames and Explosions of Gases*, 3rd Edition. Academic Press, Inc., Orlando, FL, p. 532, 1987. ISBN 0-12-446751-2.

Liu, M., B. Yu, J. Xu and J. Chen. Thermodynamic analysis of the efficiency of high-temperature steam electrolysis system for hydrogen production. *Journal of Power Sources* 177: 493–499, 2008.

Maeda, K. and K. Domen. New non-oxide photocatalysts designed for overall water splitting under visible light. *Journal of Physical Chemistry C* 111: 7851–7861, 2007.

Mathews, J. and G. Wang. Metabolic pathway engineering for enhanced biohydrogen production. *International Journal of Hydrogen Energy* 34: 7404–7416, 2009.

Narusawa, K., M. Hayashida, Y. Kamiya, H. Roppongi, D. Kurashima and K. Wakabayashi. Deterioration in fuel cell performance resulting from hydrogen fuel containing impurities: Poisoning effects by CO, CH4, HCHO, and HCOOH. *JSAE Review* 24: 41–46, 2003.

Niedzwiecki, A. "Storage", Quantum Technologies, Proc. Hydrogen Vision Meeting, US DOE, Washington, November 15–16, 2001. Available at http://www.eere.energy.gov/hydrogenandfuelcells/pdfs/hv_report_12-17.pdf (accessed March 21, 2013).

Satyapal, S., J. Petrovic, C. Read, G. Thomas and G. Ordaz. The US Department of Energy's National Hydrogen Storage Project: Progress towards meeting hydrogen-powered vehicle requirements. *Catalysis Today* 120: 246–256, 2007.

Schlapbach, L. and A. Zuttel. Hydrogen-storage materials for mobile applications. *Nature* 414: 353–357, 2001.

Teitel, R. Hydrogen storage in glass microspheres. Rept. BNL 51439, Brookhaven National Laboratories, Upton, New York, 1981.

Wilhelm, D. J., D. R. Simbeck, A. D. Karp and R. L. Dickenson. Syngas production for gas-to liquids applications: Technologies, issues and outlook. *Fuel Processing Technology* 71: 139–148, 2001.

14

Fuel Cell Application, Codes and Standards, and Environmental Effects

There are a host of fuel cell applications from portable devices to transportation to large-scale stationary power systems. The fuel cells are being used as

- Stationary power—for utilities, factories, emergency power for hospitals, communications facilities, banks, and computer installations
- Propulsion/transportation—for cars and buses but also in boats, trains, planes, scooters, and bicycles, as well as highway road signs
- Diverse military applications—for dismounted soldiers and auxiliary power for vehicles
- Domestic power supplies—for individual residences
- Portable devices—mobile phones, laptop computers, and other personal electronic devices
- Portable power—for building sites, camping, and vending machines
- Landfills and wastewater treatment plants (which are using fuel cells to convert the methane gas they produce into electricity)

In order for the fuel cell to be used as consumer products as well as industrial devices, and since they are produced and used by and around workers and the public, they are bound by national and international regulations relating to their safety, use, transportation, and storage. Since fuel cells are a developing technology and are used in various sectors and in various sizes and forms, standards and codes are being developed.

The technology maturity comes with acceptable safety and economics. Moreover, the environmental impact of the implementation of fuel cell technology in various sectors determines the sustainability and overall success of the fuel cell technology and application. In this chapter, the various applications of the fuel cell are discussed. The codes and standards required for the use of fuel cells in a number of applications are reviewed. The chapter concludes with overall environmental considerations of the fuel cell technology.

14.1 Fuel Cell Applications

14.1.1 Stationary Power

Fuel cells for stationary applications fall in the category of distributed generation that is *defined as* small-scale electric generation located close to the load being served. In the case of distributed generation, the potential exists to provide generation at the building where the power is consumed and feed excess power back into the power grid as well as take power from it. Another key advantage of distributed generation is that in addition to the higher efficiency electricity, the heat produced by the fuel cell can be utilized by the site. This is referred to as combined heating and power or CHP.

Fuel cell application for stationary power can be separated into two broad categories, simple cycle and hybrid fuel cell systems. Simple cycle fuel cell systems that directly supply electrical power have been developed for all fuel cell types. They consist of the typical fuel processing, stack, and electric conditioning systems that produce compatible alternating current electricity. Hybrid fuel cell systems combine fuel cell simple cycle technology with conventional gas turbine technology. Hybrid fuel cells can be configured in a number of schemes. The most common hybrid fuel cell configuration is the replacement of the gas turbine combustor with the fuel cell stack. In this configuration, the stack operates at elevated pressures and exhausts hot gases for expansion through the turbine. Hybrid fuel cells simultaneously generate electricity from the gas turbine generator and the fuel cell stack. Hybrid fuel cell gas turbine power plants have been demonstrated at efficiency levels of 60% (~300 kW).

Both high-temperature and low-temperature fuel cells can be utilized for stationary power generation. Compared to high-temperature fuel cells, the low-temperature fuel cells usually have a faster start-up time. High-temperature systems such as the solid oxide fuel cell (SOFC) and MCFC generate high-grade heat that can directly be used in a thermal cycle or indirectly used by incorporating the fuel cell system into a combined cycle. In addition, SOFCs and MCFCs have the advantage that they can operate directly on available fuels such as natural gas without the need for external reforming. For a small distributed power system, for example single-home or multiple-home power generation, a PEM, SOFC, or PAFC combined with a thermal cycle could be used to provide electrical and heat needs.

The PAFC, having a relatively higher temperature than PEM, has a lower start-up time than high-temperature systems, which makes it more attractive for small-power generation. The PAFC can produce enough steam to operate a steam reforming system whereas the PEM system, because of its lower operation temperature, is not able to supply the necessary heat. Power plants in the range above 250 kW can be operated by high-temperature fuel cell systems. The high-grade heat obtained from these systems can be exchanged at a broad temperature range with the possibility of direct heat use or further electricity generation by steam engines. The start-up time of these systems is longer than that of low-temperature systems, but the advantages of being able to operate the system without external reforming and the higher efficiencies of SOFCs and MCFCs make these systems more suitable for large-scale power plants. The typical needed operating time for the stationary application is approximately 40,000 h.

14.1.2 Transportation Power

Fuel cells have been studied for application in a variety of transportation systems including personal vehicles, buses, trains, ships, and aircraft. There are several variations of how the fuel cell vehicle can be configured. The dominant configurations are enabled by the development of computer-controlled power electronics, commercial introduction of the hybrid electric vehicle technology, and the maturity level of essential components such as hydrogen storage, fuel processing, and electric energy storage. Some of the vehicle configurations that have been introduced by leading automobile developers include (i) direct-hydrogen fuel cell power system and electric motor drive train; (ii) direct-hydrogen fuel cell power system, energy storage, and electric motor drive train; and (iii) onboard fuel processor, energy storage, and electric motor drive train. Fuel cell vehicles can utilize DC as well as AC drive systems, although the latter is more efficient, reliable, and typically is less complicated.

For transport applications, the requirements for fuel cell systems are different from stationary power generation. Compact design for available space and fast response times and start-up times are required for vehicular application of the fuel cells. The PEMFC is regarded as ideally suited for transportation applications because of its compactness, light weight, high power density, high energy conversion efficiency, and low operating temperature (<100°C). Currently, most of the hybrid vehicles or fuel cell vehicles are designed using PEMFC. Various demonstration vehicles such as city bus, vans, and cars are based on a basic conceptual design combining the PEMFC with an electric drive.

However, among all applications for fuel cells, the transportation application involves the most stringent requirements regarding volumetric and gravimetric power density, reliability, and costs. Since a widespread hydrogen retail infrastructure will not be available in the near future, car manufacturers consider a liquid fuel as the best option for a short-term market introduction of a fuel cell propulsion system. Furthermore, the higher energy density of a liquid fuel guarantees a driving range similar to that of internal combustion engine vehicles.

There is a great deal of market opportunity for fuel cell vehicles for passenger use. There are hundreds of fuel cell vehicles in various stages of demonstration and development by major automobile companies worldwide.

Since Ballard unveiled the world's first fuel cell bus in 1993, a number of fuel cell buses have been built and operated worldwide. A typical bus power plant is 250 kW in size and developers have utilized both phosphoric acid and proton exchange membrane fuel cells. Although the technology has made progress over recent decades, there are still a number of issues to be resolved. The development of compact, efficient, cost-effective, high-purity, hydrogen-producing reformer technology is a key requirement. An alternative strategy to relieve the need for onboard reformation of liquid transportation fuels is the storage and direct use of hydrogen. This approach will require significant advances in the storage of hydrogen using metal hydrides, carbon nanotubes, or crash-worthy, high-pressure hydrogen tank technology. A significant development of the hydrogen supply infrastructure would be needed as well.

14.1.3 Portable Applications

For small power applications like laptops, camcorders, individually carried communication equipment, and mobile phones, the requirements of the fuel cell systems are even more specific and stringent than for vehicle applications. The requirements for portable applications are mostly focused on size and weight of the system (as well as the temperature). Currently, Li-based batteries are successful in powering laptop computers, mobile phones, and the like. Li-ion battery and nickel-based batteries especially those based on metal hydrides generally provide higher energy density, higher specific energy, or longer operational time between recharges. However, with recent demands for quick recharge time and higher power density requirements usage, fuel cells with hydrogen or methanol seem attractive.

Low temperatures are necessary and therefore PEMFC and DMFC are suitable for portable applications. Other fuel cells are, therefore, not suitable for this kind of application. The possibilities for fuel cell systems are the combination of fuel cell with hydrogen storage by hydrides or gas cartridges or the direct methanol fuel cell. Portable devices need lower power than other fuel cell applications and, thus, DMFC systems may be well suited for this kind of application. With further technology improvements and better storage systems, PEM fuel cells and DMFC systems will continue to compete in this market. The drivers for the adoption of portable fuel cells depend on the application area. For the military, there is the requirement for a power solution that has a longer run time, higher energy density and lighter weight than the battery packs used today. A similar driver is in place in the consumer electronics industry, with products requiring more power in a smaller volume.

14.1.4 Military Applications

Fuel cells could provide power for most types of military equipment from land and sea transportation to portable handheld devices used in the field; hence, military applications are expected to become a significant market for fuel cell technology. The efficiency, versatility, extended running time, and quiet operation make fuel cells extremely well suited for military applications (Lewis 2003; Defense Update 2004). For military field application such as for field radios and other communication equipment, weight and volume of the power packs are important. For a dismounted soldier, the required mean power is 20 W, and for a 7-day mission, this amounts to 3.4 kWh or a total of 12 MJ of electrical energy. Fuel cells are found to offer significant savings of loads, in weight and volume, compared to conventional power sources, and can provide a higher amount of energy compared to battery packs of the same size and weight. DMFCs with cartridge fuel have been developed for such portable applications (Kukkonen 2006). For example, Direct Methanol Fuel Cell Corporation develops and manufactures disposable methanol fuel cartridges that provide the fuel for fuel cell–powered portable electronics including notebook computers, mobile phones, and military equipment. A power system developed by this company can provide 0.9 Wh of energy per cubic centimeter of 100% methanol fuel, which amounts to more than three times the energy of most current lithium ion batteries. There are also efforts on applications of fuel cell technology for US army vehicles as auxiliary power units, which offer adequate supply of electrical power, to operate digital equipment and extended silent watch requirements.

14.1.5 Landfills and Other Applications

The landfills are the largest source of anthropogenic methane emissions, constituting almost 40% of these emissions each year in the United States. Methane is a potent greenhouse gas (GHG); it is approximately 25 times more effective at trapping radiation in the atmosphere than an equivalent mass of carbon dioxide. Recovery and use of methane from landfills substantially reduces these emissions while capturing their energy value. Typically, an anaerobic digester is used to treat the municipal waste or wastewater to produce a methane-rich gas, which can be used by a fuel cell. Anaerobic digestion is a biochemical process in which several different types of bacteria work together to break down complex organic wastes in stages, resulting in the production of methane-rich gas called biogas. An anaerobic digester is an airtight chamber which maintains a temperature of at least 20°C to promote bacterial activity. The biogas after removal of sulfur and other impurities is then used by the fuel cell. The waste heat from fuel cells can be used to maintain temperature in the digester.

Fuel cells are also considered for application indoors and in deep mines. Besides reducing emissions, fuel cells have the potential to effectively lower total logistics cost since they require minimal refilling and significantly less maintenance than battery-run equipment, whose batteries must be periodically charged, refilled with water, and replaced. In case of forklift, they experience numerous interruptions in current input and output—because of the frequent starting and stopping during use. Fuel cells ensure constant power delivery and performance, eliminating the reduction in voltage output that occurs as batteries discharge.

14.2 Fuel Cell Codes and Standards

Codes and standards are compilations of requirements aimed to protect public health, safety, and general well-being. Codes and standards ensure safety, interchangeability, consistency, and regularity. Almost all industries that rely on construction, engineering components, or equipment are governed by codes and standards. For fuel cell systems, the codes and standards are continually being revised, expanded, and improved such that hydrogen systems, fuel cells, and other equipment are installed correctly and located in an environment in which they can operate safely. In this process, federal, state, and local governments and independent standards development organizations are involved. More than 600 organizations create codes and standards, and sometimes those organizations overlap or conflict. The government's role is to assess all the current and evolving codes to adopt into laws and regulations.

A code is a collection of requirements covering a broad subject matter that is suitable for adoption into law independently of other codes and standards. They specify the minimum acceptable level of safety for a system or component such as buildings and other structures. Codes are passed by nations, states, legislative bodies, and regulatory jurisdictions to control those things under their authority. A standard is a document that contains technical stipulations or other precise criteria designed to be used consistently as a rule or recommendation. In contrast to codes, compliance with standard is not compulsory as they are written. However, compliance can become compulsory when they are referenced by another standard or code that can be adopted into law. For example, standards NFPA 852 "Standard for the Installation of Stationary Fuel Cell Power Systems" and SAE J2600 "Compressed Hydrogen Surface Vehicle Refueling Connection Devices" are referred by other codes and standards and hence are mandatory for installation of fuel cell system power systems and for refueling stations, respectively (US Department of Energy 2009, 2011).

Fuel cell regulatory codes and standards are developing around a number of areas and applications. These include (i) general—product requirements and manufacturing; (ii) stationary and building specific—installation and site requirements, building codes, interconnection requirements, and electric utility requirements; (iii) mobile specific—safety, performance, reliability, recyclability, hydrogen storage, and transport; and (iv) portable specific—hydrogen storage and transport.

There are several agencies and certifying facilities that provide oversight or overviews and support development of fuel cell codes and standards. These organizations are listed in the following:

US Fuel Cell Council: Through its Codes & Standards Working Group, their mission is to become aware of all Codes & Standards activities, gather information from all possible sources, and provide information to the members of the Council in order to encourage participation and commentary. The Codes & Standards Working Group will serve as the communication conduit for member commentary on Codes & Standards at the local, regional, national, and international levels. The Codes & Standards Working Group will also make recommendations to codes and standards bodies with the concurrence of the Council Board.

Code of Federal Regulations: The Code of Federal Regulations (CFR) is a codification of the general and permanent rules published in the Federal Register by the executive departments and agencies of the Federal Government.

National Evaluation Service, Inc.: NES is an independent, not-for-profit organization that conducts a voluntary and advisory program of evaluation for both traditional and innovative building materials, products, and systems. The evaluation reports issued by the NES facilitate the acceptance of new building technology by those in the US building regulatory community as well as by those who design and construct buildings.

National Hydrogen Association: The mission of the NHA is to foster the development of hydrogen technologies and their utilization in industrial and commercial applications and promote the transition role of hydrogen in the energy field. Technology Transition Corporation provides analytical support and demonstration planning consulting services in the areas of vehicles and transportation systems, commercial aircraft, electric utility generation, and energy storage.

A National Resource for Global Standards: NSSN is setting the pace worldwide in promulgating standards information to a broad constituency. Whether it is searching for a specific standard, tracking the status of a new development project, or identifying a contact person who can interpret a standard, NSSN serves as an information repository. In addition, the agencies providing oversight or overviews and support development of fuel cell codes and standards involved include ANSI, ASME, IEEE, NFPA, IEC, ICC, and SAE.

14.2.1 Stationary and Portable Fuel Cell Commercial Systems

Many organizations are working to develop codes and standards needed to prepare for the commercialization of stationary and portable fuel cell systems. Several regulatory bodies control the implementation and compliance of the codes and standards. Table 14.1 lists agencies responsible for leading the support and development of key standards for stationary and portable fuel cell systems.

TABLE 14.1

Standards Development Organizations for Stationary and Portable Fuel Cell Systems

Organization		Responsibility
AGA	American Gas Association	Materials testing standards
API	American Petroleum Institute	Equipment standards for petroleum production, storage, and handling
ASHRAE	American Society of Heating, Refrigerating, and Air-Conditioning Engineers	Promoting sustainability through research, standards writing, publishing, and education
ASME	American Society of Mechanical Engineers	Mechanical and multidisciplinary engineering, design codes, and standards
ASTM	American Society for Testing and Materials	Technical test standards for materials, products, systems
CGA	Compressed Gas Association	Equipment design and performance standards for compressed gas systems and components
CSA	Canadian Standards Association	US and Canadian equipment standards
DOT	Department of Transportation	Federal transportation regulatory agency
FERC	Federal Energy Regulatory Commission	Regulates interstate transmission of electricity, gas, and oil
GTI	Gas Technology Institute	Supplies technical support and training for the energy industry
ICC	International Code Council	Family of model building codes, including the International Fire Code
IEEE	Institute of Electrical and Electronics Engineers	Electrical standards
NERC	North American Electric Reliability Corporation	Produces standards for electric grid operation
NFPA	National Fire Protection Association	Model codes and standards, including the National Electric Code
NIST	National Institute of Standards and Technology	Measurement standards
SAE	Society of Automotive Engineers	Vehicle standards
NARUC	National Association of Regulatory Utility Commissioners	Represents state public service commissions
UL	Underwriters Laboratory	Equipment and performance testing standards

Source: National Renewable Energy Laboratory NREL/FS-560-48614, July 2010, http://www1.eere.energy.gov/hydrogenandfuelcells/codes/standards_organization.html, accessed April 2, 2013.

In the area of stationary and portable fuel cell systems, the regulatory agencies and the organizations involved in the standards developed are listed in Table 14.2 for categories of hydrogen generation, portable fuel cells, stationary and combustion devices, and interface to the fuel cell systems. In Table 14.3, the published list of codes and standards for fuel cell system design, installation, nameplate, testing, performance, subsystems, and electrical interface to grids and panel boards is listed.

14.2.2 Hydrogen Vehicle and Infrastructure Codes and Standards

In Table 14.4 the hydrogen vehicle–related codes and standards are listed. A document from US Department of Energy codes and standards is typically

TABLE 14.2

Regulatory, Codes, and Standards Development Organizations for Stationary and Portable Fuel Cell Systems

Hydrogen Generation	Portable Fuel Cells	Stationary Fuel Cells and Combustion Devices	Interface
Regulators: EPA (emissions) DOT/PMHSA (pipeline) OSHA, state and local government (zoning, building permits) *Electrolyzers:* UL, CSA *Reformers:* UL, CSA, API *Performance Test Procedures:* ASME, CSA *Chemical Hydrides:* UL, CSA, NFPA	***Regulators:*** CPSC, DOT/PHMSA, OSHA, EPA (methanol), state and local government (zoning, building permits) *Handheld Systems:* UL, CSA *Portable Systems:* UL, CSA, CGA *Handheld Fuel Containers:* UL, CSA, CGA *Portable Fuel Containers:* ASME CSA, CGA *Performance Test Procedures:* ASME, CSA, GTI	***Regulators:*** OSHA, state and local government (zoning, building permits) *H2 and Combustion Devices:* UL, CSA *H2 Fueled Turbines:* ASME, UL, CSA, API *FC Systems:* ASME, UL, CSA *FC Installation:* NEPA *FC Performance Test Procedures:* ASME,CSA, GTI	***Installation Piping:*** ASME, CSA, CGI, NFPA, ICC *Storage:* ASME, CSA, CGI, NFPA, API *Compressors Safety Certificate:* UL, CSA *Computer Design, Performance and Safety:* API *Sensors/Detectors:* UL, CSA, NFPA *Fuel Specifications:* API, SAE, CGI, ASTM *Weights/Measures:* ASME, API, NIST *Dispensers:* UL, CSA, NFPA, API, SAE *Non-vehicle Dispensing:* CGI *Codes for Built Environment:* CGI, NFPA, ICC, ASHREA *Interconnection:* IEEE, UL, NFPA

TABLE 14.3

Published Codes for Stationary Fuel Cell Applications

Fuel Cell Power Systems—System Design/Testing

ANSI/CSA America FC1 Fuel cell power systems (United States)
CEN/CENELEC EN 50465 Fuel cell gas heating appliances (European Union)
IEC 62282-3-100 Stationary fuel cell systems—safety (International)
BS EN 62282-3-100:2012 Stationary fuel cell systems—safety (European Union and UK)
JIS C 8801 General rules for phosphoric acid fuel cell power generating system (Japan)
DVGW VP119 Preliminary basic rules for testing fuel cell gas appliances 70 kW (Germany)
JIS C 8821 General rules for small polymer electrolyte fuel cell power systems (Japan)
JIS C 8822 General safety code for small polymer electrolyte fuel cell systems (Japan)
JIS C 8823 Testing methods for small polymer electrolyte fuel cell power systems (Japan)
GB/Z 21743-2008 Stationary proton exchange membrane fuel cell power system (China)
JIS C 8841-1 Small solid oxide fuel cell power systems. Part 1: General rules (Japan)
JIS C 8841-2 Small solid oxide fuel cell power systems. Part 2: General safety codes and safety testing methods (Japan)
GB/T 25447-2010 Proton exchange membrane fuel cell test system and activation system (China)
GB/T 27748.1-2011 Stationary fuel cell power system—Part 1: Safety (China)

Fuel Cell Power Systems—Nameplates

JIS C 8803 Indication of phosphoric fuel cell power facility (Japan)
JIS C 8811 Indication of polymer electrolyte fuel cell power facility (Japan)

Fuel Cell Power Systems—Performance—efficiency, emissions, durability

ASME PTC-50 Performance test code for fuel cell power systems performance (US and other locales)
IEC 62282-3-200 Test method for the performance of stationary fuel cell power plants (International)
BS EN 62282-3-200:2012 Test method for the performance of stationary fuel cell power plants (European Union and UK)
JIS C 8802 Test method for durability of phosphoric acid fuel cell power facility (Japan)
State of California Regulations Emission Regulations
JIS C 8825 Testing methods for EMC of small polymer electrolyte fuel cell power systems (electromagnetic compatibility) (Japan)
JIS C 8824 Testing methods for environment of EMC for polymer electrolyte fuel cell systems (Japan)
JIS C 8841-3 Small solid oxide fuel cell power systems. Part 3: Performance testing methods and environment testing methods (Japan)

Fuel Cell Power Systems—Subsystems—Fuel Cell Modules

CSA International Component Acceptance Service No. 33 Proton exchange membrane fuel cell stacks (United States and Canada)
IEC 62282-2 Ed. 2 (2012-03) Fuel cell modules (International)
BS EN 62282-2:2012 Fuel cell modules (European Union and UK)
CAN/CSA C22.2 No. 62282-2:07 Fuel cell technologies—Part 2: Fuel cell modules (Canada)
JIS C 8831 Safety evaluation test for stationary polymer electrolyte fuel cell stack (Japan)
JIS C 8832 Performance test for stationary polymer electrolyte fuel cell stack (Japan)
GB/T 20042.2-2008 Proton exchange membrane stacks (China)
KS C IEC 62282-2 Fuel cell technologies—Part 2: Fuel cell modules (Korea)

(*continued*)

TABLE 14.3 (Continued)

Published Codes for Stationary Fuel Cell Applications

Fuel Cell Power Systems—Subsystems—Fuel Cell Modules—Subscale Testing

GB/T 20042.3-2009 Proton exchange membrane fuel cell—Part 3: Test method for proton exchange membrane (China)

GB/T 20042.4-2009 Proton exchange membrane fuel cell—Part 4: Test method for electrocatalysis (China)

GB/T 20042.5-2009 Proton exchange membrane fuel cell—Part 5: Test method for membrane electrode assembly (China)

GB/T 20042.6-2011 Proton exchange membrane fuel cell—Test method of bipolar plate properties (China)

GB/T 28817-2012 Single cell test methods for proton exchange membrane fuel cell (China)

GB/Z 27753-2011 Test method for adaptability to operating conditions of membrane electrode assembly used in PEM fuel cells (China)

Working Group #11/IEC 62282-7-1 Single cell test method for polymer electrolyte fuel cells (International)

Fuel Cell Power Systems—Subsystems—Inverters

UL 1741 Standard for inverters, converters, controllers, and interconnection system equipment for use with distributed energy resources (United States and other locales)

JIS C 8826 Testing methods of power conditioner for grid interconnected small fuel cell systems (Japan)

JIS C 8827 Testing procedure of islanding prevention methods for utility interconnected small polymer electrolyte fuel cell power system power conditioners (Japan)

Fuel Cell Power Systems—Installation

ANSI/NFPA 853 Installation of stationary fuel cell power plants (United States)

BS EN 62282-3-3:2008 Stationary fuel cells power systems—Installation (European Union and UK)

CAN/BNQ 1784-000 Canadian hydrogen installation code (Canada)

IEC 62282-3-300 Stationary fuel cell power systems—Installation (International)

Petroleum Association of Japan Fuel Cell System Installation Criteria (Japan)

US Department of Energy Hydrogen and Fuel Cells Permitting Guide (United States)

HYPER Project Installation Permitting Guidance for Hydrogen and Fuel Cells Stationary Applications (EU-Supported Coordinating Activity)

State of South Carolina—Hydrogen (and Fuel Cell) Permitting Act (United States—South Carolina)

GB/T 27748.3-2011 Stationary fuel cell power system—Part 3: Installation (China)

Fuel Cell Power Systems—Electrical Interfaces—With Panel Board

ANSI/NFPA 70: Article 692—Fuel cell systems national electrical code (United States)

Fuel Cell Power Systems—Electrical Interfaces—With Grid

NFPA 110 Standard for standby power systems (United States)

ANSI/IEEE 1547 Interconnecting distributed resources with electric power systems (United States)

IEC/PAS 63547—IEEE standard for interconnecting distributed resources with electric power systems (International)

IEEE 1547.1 Standard for conformance test procedures for equipment interconnecting distributed resources with electric power systems (United States)

IEEE 1547.2 Application guide for IEEE 1547 standard for interconnecting distributed resources with electric power systems (United States)

IEEE 1547.3 Guide for monitoring, information exchange, and control of distributed resources interconnected with electric power systems (United States)

IEEE 1547.4 Guide for design, operation and integration of distributed resources island systems with electric power systems (United States)

IEEE 1547.6 Recommended practices for interconnecting distributed resources with electric power distribution secondary networks (United States)

TABLE 14.4

Hydrogen Vehicle and Infrastructure Codes and Standards

Vehicles	Storage	Dispensing	Infrastructure
Regulators: DOT/NHTS (crashworthiness) EPA (emissions) *General FC Vehicle Safety:* SAE *Fuel Cell Vehicle Systems:* SAE *Fuel System Components:* CSA Standards *Containers:* SAE *Reformers:* SAE *Emissions:* SAE *Recycling:* SAE *Service/Repair:* SAE	***Regulators:*** DOT/ PHMSA (over-road transport, pipeline safety) *Composite Containers:* ASME, CSA International, CGS NFPA *Pipelines:* ASME, API, AGA, CGS *Equipment:* ASME, API, AGA, CGS *Fuel Transfer:* NFPA, API	***Regulators:*** State and local government (zoning, building permits) *Storage Tanks:* ASME, CSA International, CGS NFPA, API *Piping:* ASME,CSA International, CGS NFPA *Dispensers:* UL, CSA International, NFPA *On-site H2 Production:* UL, CSA International, API, CGA *Codes for the Environment:*	**Fuel Specs:** Weights/measures: fueling: Sensors/detectors: Connectors: Communications: Building and fire code requirements:

used for US hydrogen vehicle and infrastructure projects (US Department of Energy 2009). To determine which codes and standards apply to a specific project, one has to identify the codes and standards currently in effect within the jurisdiction where the project will be located. Some jurisdictions also have unique ordinances or regulations that could be applied. Hydrogen vehicle and infrastructure codes and standards can be found in these categories: annual inspections, balance of plant, canopy tops, compressed hydrogen gas storage, compression systems and equipment, design, dispensing, dispensing, operations, maintenance safety, fire safety, liquid hydrogen storage, on-site hydrogen production, operation approvals, setbacks and footprints, outdoor gaseous systems, transportation, and vaporizers.

Additional regulatory guidelines and resources for Hydrogen Vehicle and Infrastructure are listed below:

Additional Regulatory Guidance:

- NFPA 54: National Fuel Gas Code
- NFPA 55: Standard for the Storage, Use, and Handling of Compressed Gases and Cryogenic Fluids in Portable and Stationary Containers, Cylinders, and Tanks, 2005 Edition
- US Department of Labor Occupational Safety & Health Administration (OSHA) 29 CFR 1910 Subpart H – Hazardous Materials 1910.103 – Hydrogen
- NFPA 853: Standard for the Installation of Stationary Fuel Cell Power Plants 2003 Edition

NFPA 70: National Electrical Code Article 692 Fuel Cell Systems

ANSI/IEEE 1547-2003: IEEE Standard for Interconnecting Distributed Resources with Electric Power Systems

ANSI/CSA America FC1-2004 (formerly ANSI Z21.83) American National Standard for Fuel Cell Power Systems

IEC 62282-3-1: Fuel Cell Power Systems – Safety

American Society of Mechanical Engineers PTC-50: Performance Test Code, for Fuel Cell Power System Performance

IEC 62282-3-2 (2006-03): Fuel Cell Power System – Performance

Additional Regulatory Resources:

International Code Council – Model Building and Construction Codes and Standards

National Fire Protection Association – Model Building and Construction Codes and Standards

International Association of Plumbing and Mechanical Officials – Model Building and Construction Codes and Standards

14.2.3 Scope of Key Codes and Standards

Here, the scope of key standards and codes related to fuel cell installations, performance, testing, hydrogen vehicles, and storage is given.

ANSI/CSA AMERICA FC 1-2004 Stationary Fuel Cell Power Systems, Scope:

1.1.1 This standard applies to stationary fuel cell power systems that are packaged, self-contained or factory matched packages of integrated systems (see 1.2.1), which through electrochemical reactions and other processes, generate alternating-current or direct-current electricity. Such systems are herein referred to as fuel cell power system(s).

1.1.2 This standard applies to stationary fuel cell power systems when intended for operation with fuels listed in 1.1.3: (a) at an output voltage not exceeding nominal 600 VAC or 600 VDC; and (b) at a power output not exceeding 10 MW.

1.1.3 Stationary fuel cell power systems examined for compliance with this standard shall be for use with (a) natural gas, methane gas mixtures; and/or (b) liquefied petroleum gases; and/or (c) methanol, ethanol (or other alcohol fuel), kerosene, diesel, gasoline (or other liquid hydrocarbons); and/or (d) hydrogen gas; and/or (e) Zinc particulate conveyed in a non-flammable liquid medium.

1.1.4 Exhibit A contains coverage for electrical equipment and wiring.

1.1.5 Publications of nationally recognized standards and codes referenced herein are listed in Exhibit B, List of Reference Standards.

When applicable, other nationally recognized publications may be utilized to supplement those referenced herein.

1.1.6 If a value for measurement as given in this standard is followed by an equivalent value in other units, the first stated value is to be regarded as the specification. Conversion factors for quantities stated herein are listed in Appendix F, Table of Conversion Factors. SI units stated herein are in accordance with the Standard for Use of the International System of Units: The Modern Metric System, IEEE/ASTM SI 10.

1.1.7 All references to psi throughout this standard are to be considered gauge pressure unless otherwise specified.

ANSI/CSA AMERICA FC 3-2004 Portable Fuel Cell Power Systems, Scope:

1.1.1 This Standard applies to ac- and dc-type portable fuel cell power systems, with a rated output voltage not exceeding 600 V, for commercial, industrial, and residential indoor and outdoor use in nonhazardous locations, in accordance with the Rules of the National Electric Code, ANSI/NFPA 70.

1.1.2 This Standard does not apply to portable fuel cell power systems that are (a) permanently connected (stationary) to either fuel or electric supply; (b) designed to export power to the grid; (c) replacement fuel cell power units for appliances; or (d) fuel cell systems for propulsion.

1.1.3 Fuels considered within the scope of this standard are (a) natural gas; (b) liquefied petroleum/propane gases; (c) liquid alcohols, for example, methanol or ethanol; (d) gasoline; (e) diesel; (f) kerosene; (g) hydrogen; (h) chemical hydrides; and (i) metallic.

1.1.4 The overall design of a portable fuel cell power system anticipated by this standard shall form an assembly of integrated systems, as necessary, intended to perform designated functions, as follows: (a) Fuel Processing System—meters and processes incoming fuel supply for use within the portable fuel cell power system. (b) Air Processing System—meters, conditions, processes, and may pressurize (blowers or turbochargers) incoming air supply for use within the portable fuel cell power system. (c) Thermal Management System—provides cooling and heat rejection to maintain thermal equilibrium within the portable fuel cell power system and may provide for recovery of excess heat and assist in heating the portable fuel cell power system during start-up. (d) Electrical System—conditions output power of the fuel cell to meet the specifications of the output system. (e) Automatic Control System—the assembly of sensors, actuators valves, switches, and control logic components that maintains the portable fuel cell power system control parameters

within manufacturer-specified limits without manual intervention.
(f) Fuel Cell Module—an assembly of cells, separators and cooling plates, manifolds, and a supporting structure that electrochemically converts fuel and oxidant to direct electrical current, heat, water, and other by-products.

1.1.5 These requirements are not intended to prevent the design and construction of a portable fuel cell power system not specifically prescribed in this standard, provided that such alternative has been considered in testing and listing the portable fuel cell power system. In considering alternative designs or construction, the materials or methods used shall be evaluated as to their ability to yield equivalent performance to that prescribed by this standard.

ASME PTC-50 Fuel Cell Power Systems Performance, (ASME 2000); Scope: This Code provides test procedures, methods, and definitions for the performance characterization of fuel cell power systems. Fuel cell power systems include all components required in the conversion of input fuel and oxidizer into output electrical and thermal energy. Performance characterization of fuel systems includes evaluating system energy inputs and electrical and thermal outputs to determine fuel-to-electrical energy conversion efficiency and, where applicable, the overall thermal effectiveness. These efficiencies will be determined to an absolute uncertainty of less than ±2% at a 95% confidence level. (For example, for a calculated efficiency of 40%, the true value lies between 38% and 42%.) This Code applies to all fuel cell power systems regardless of the electrical power output, thermal output, fuel cell type, fuel type, or system application.

Fuel cell power systems contain an assembly of electrochemical cells, which oxidize a fuel to generate direct current electricity. Balance-of-plant subsystems may include controls, thermal management, a fuel processor, and a power conditioner. Some fuel cell power systems may contain additional power generating equipment such as steam generators, gas turbine generators, or micro-turbine generators. The net power output and all the fuel input to the system shall be taken into account in the performance test calculations.

This Code applies to the performance of overall fuel cell power systems. The Code addresses combined heat and power systems, that is, the generation of electricity and usable heat at specific thermal conditions. It does not address the performance of specific subsystems nor does it apply to energy storage systems, such as regenerative fuel cells or batteries. It also does not address emissions, reliability, safety issues, or endurance. This Code contains methods and procedures for conducting and reporting fuel cell system testing, including instrumentation to be used, testing techniques, and methods for calculating and reporting results. The Code defines the test boundary for fuel and oxidant input, secondary energy input, and net electrical and thermal energy output. At these boundaries, this Code provides procedures for measuring temperature, pressure, input fuel flow and composition, electrical power, and thermal output. The Code provides procedures for determination of electrical efficiency or heat rate and overall thermal effectiveness at rated or any other steady-state condition.

ANSI/CSA America FC1-2004 (formerly ANSI Z21.83) American National Standard for Fuel Cell Power Systems, Scope: This standard provides design, construction, operating and quality requirements for fuel cell power systems operating on

- Natural gas, methane gas mixtures
- Liquid petroleum gases
- Methane, ethane, kerosene, diesel, gasoline
- Hydrogen gas

NFPA 853 Standard for the Installation of Stationary Fuel Cell Power Systems, Scope:

1.1.1 This standard shall apply to the design, construction, and installation of stationary fuel cell power systems.

1.1.2 The scope of this document shall include the following: (1) a singular prepackaged, self-contained power system unit; (2) any combination of prepackaged, self-contained power system units; (3) power system units comprising two or more factory-matched modular components intended to be assembled in the field; and (4) engineered and field-constructed power systems that employ fuel cells.

NFPA 70 National Electrical Code (Current Edition: 2011, Next Edition: 2014) 90.2, Scope: This Code covers the installation of electrical conductors, equipment, and raceways; signaling and communications conductors, equipment, and raceways; and optical fiber cables and raceways for the following: (1) public and private premises, including buildings, structures, mobile homes, recreational vehicles, and floating buildings; (2) yards, lots, parking lots, carnivals, and industrial substations; (3) installations of conductors and equipment that connect to the supply of electricity; and (4) installations used by the electric utility, such as office buildings, warehouses, garages, machine shops, and recreational buildings, that are not an integral part of a generating plant, substation, or control center.

ANSI/CSA AMERICA FC 1-2004 PDF, Scope:

1.1.1 This standard applies to stationary fuel cell power systems that are packaged, self-contained, or factory matched packages of integrated systems (see 1.2.1), which, through electrochemical reactions and other processes, generate alternating-current or direct-current electricity. Such systems are herein referred to as fuel cell power system(s).

1.1.2 This standard applies to stationary fuel cell power systems when intended for operation with fuels listed in 1.1.3: (a) at an output voltage not exceeding nominal 600 VAC or 600 VDC; and (b) at a power output not exceeding 10 MW. Exception: Output voltage may exceed nominal 600 VAC up to nominal 15,000 VAC in accordance with A.1.31 (see Exhibit A, Electrical Equipment and Wiring).

1.1.3 Stationary fuel cell power systems examined for compliance with this standard shall be for use with (a) natural gas, methane gas mixtures; and/or (b) liquefied petroleum gases; and/or (c) methanol, ethanol (or other alcohol fuel), kerosene, diesel, gasoline (or other liquid hydrocarbons); and/or (d) hydrogen gas; and/or (e) zinc particulate conveyed in a non-flammable liquid medium.

1.1.4 Exhibit A contains coverage for electrical equipment and wiring.

ASTM WK 4548 Standard Test Method for Determination of Trace Contaminants in Hydrogen and Related Fuel Cell Feed Gases Under Development, Scope: Gas chromatography/mass spectrometry is used to separate and determine contaminants in hydrogen, methanol, and other fuel cell feed gases at part per billion or lower concentrations.

ASTM WK 5847 Practice for Sampling of High-Pressure Hydrogen and Related Fuel Cell Feed Gases, Scope: This practice describes the apparatus and a procedure for the sampling of high-pressure hydrogen and related fuel cell feed gases from fueling nozzles and similar interfaces used in fueling automotive vehicles and stationary appliances.

ASTM WK 6527 Standard Test Method for Ion-Selective Electrode-Based Determination of Ammonia in Hydrogen and Other Fuel Cell Feed Gases, Scope: An ion-selective electrode detector system is used to measure the total ammonium content. Sensitivity from low part per billion up to part per million concentrations is achievable.

ASTM WK 6624 Standard Test Method for Determination of Formaldehyde and Other Carbonyl Compounds in Hydrogen and Other Fuel Cell Feed Gases, Scope: Test method describes a procedure for determining formaldehyde and other carbonyl compounds in hydrogen and other fuel cell feed gases. Sensitivity from low part per billion up to part per million is achievable.

ASTM WK 8150 Test Method for Determination of Ammonia in Hydrogen and Other Fuel Cell Feed Gases by Gas Chromatography and Nitrogen Chemiluminescence Detection, Scope: This test method provides for the determination of ammonia in hydrogen and other gaseous fuels in the ppmv range. Ammonia is a poison for hydrogen fuel cells and its presence needs to be carefully monitored for efficient operation.

ASTM WK24073 Standard Test Method for Determination of Trace Hydrogen Sulfide, Methyl Mercaptan, and Carbonyl Sulfide in Hydrogen Fuel, Scope: This test method describes a procedure for the determination of hydrogen sulfide, methyl mercaptan, and carbonyl sulfide in hydrogen fuel used in fuel cells. Gas chromatography/sulfur chemiluminescence detection is used to separate and determine contaminants in hydrogen at 1 part per billion (ppb) or better concentration.

NFPA 70 National Electrical Code Article 692 Fuel Cell Systems, Scope: The National Electrical Code is incorporated into most state and local political jurisdiction regulations and therefore carries the weight of law. Article 692 is a new section dealing with the electrical installation of fuel cell power plants to a building.

NFPA 853 Standard for the Installation of Stationary Fuel Cell Power Plants, Scope: The 2003 Edition has been expanded to include stationary fuel cells below 50 kW. The new chapter, "Fuel Cell Power Systems 50 kW or Less," gives requirements for both outdoor and indoor installations as well as ventilation and fire protection for these smaller systems.

SAE J 2600 Compressed Hydrogen Vehicle Fueling Connection Devices, Scope: SAE J2600 applies to design, safety, and operation verification of Compressed Hydrogen Surface Vehicle (CHSV) refueling connection devices hereinafter referred to as nozzle and receptacle. CHSV Refueling nozzles and receptacles shall consist of the following components, as applicable. Alternatives may be used as long as the alternative geometries shall not be interchangeable with the standard geometry specified in this SAE Standard and the standard geometry in Section 6 provides insufficient flow rates: (a) receptacle and protective cap (mounted on vehicle) (see Sections 6 and 7); (b) nozzle (see Section 5).

SAE TIR J2601 Fueling Protocols for Light Duty Gaseous Hydrogen Surface Vehicles, Scope: SAE TIR J2601 establishes safety limits and performance requirements for gaseous hydrogen fuel dispensers. The criteria include maximum fuel temperature at the dispenser nozzle, the maximum fuel flow rate, the maximum rate of pressure increase and other performance criteria based on the cooling capability of the station's dispenser. This document establishes fueling guidelines for "non-communication fueling" in the absence of vehicle communication and guidelines for "communication fueling" when specified information is transmitted from the vehicle and verified at the dispenser. The process by which fueling is optimized using vehicle-transmitted information is specified. This document provides details of the communication data transmission protocol. The mechanical connector geometry is not covered in this document.

SAE J2719 Hydrogen Quality Guideline for Fuel Cell Vehicles, Scope: Develop an evolving Hydrogen Fuel Quality guideline for the vehicle refueling interface, which will mature as technology advances toward commercial feasibility. The latest guideline would form the basis of a proposed standard for hydrogen fuel purity for H Powered vehicles, possibly including ICEs.

SAE J2722 Recommended Practice for the Durability Testing of PEM Fuel Cell Stacks, Scope: This recommended practice is intended to serve as a procedure to verify the durability specifications or vender claims of any PEM fuel cell stack subsystem for automotive applications. Definitions, specifications, and methods for the durability characterization of the fuel cell stack subsystem are provided. The durability characterization includes evaluating electrical outputs and other stack characterization over time while controlling fluid inputs and outputs based on the test boundary defined in this document.

SAE J1766 Recommended Practice for Electric and Hybrid Electric Vehicle Battery Systems Crash Integrity Testing, Scope: Electric and Hybrid Electric Vehicles contain many types of battery systems. Adequate barriers between occupants and battery systems are necessary to provide protection from potentially harmful factors and materials within the battery system that can cause injury to occupants of the vehicles during a crash. This SAE Recommended Practice is applicable to all Electric Vehicles and Hybrid Electric Vehicle battery designs, including those described in SAE J1797. The potentially harmful factors and materials addressed by this document include electrical isolation integrity, electrolytc spillage, and retention of the battery system. The purpose of this document is to define test methods and performance criteria that evaluate battery system spillage, battery retention, and electrical system isolation in Electric and Hybrid Electric Vehicles during specified crash tests.

SAE J2579 Technical Information Report for Fuel Systems in Fuel Cell and Other Hydrogen Vehicles, Scope: The purpose of this document is to define design, construction, operational, and maintenance requirements for hydrogen storage and handling systems in on-road vehicles. Performance-based requirements for verification of design prototype and production hydrogen storage and handling systems are also defined in this document. Complementary test protocols (for use in type approval or self-certification) to qualify designs (and/or production) as meeting the specified performance requirements are described.

UL Subject 2265 A, Hand Held or Transportable Fuel Cell Power Units with Fuel Containers—Methanol Fuel Cartridges, Scope: Covers handheld or hand-transportable fuel cells that provide dc outputs rated <60 Vdc and <240 VA and use disposable methanol fuel cartridges. The disposable cartridges are also covered and are limited with a capacity <950 mL, are un-refillable, and limited in pressure to the vapor pressure of the methanol fuel.

UL Subject 2265 C, Handheld or Transportable Fuel Cell Power Units with Fuel Containers—Borohydride Fuel Cartridges, Scope: Covers handheld or hand-transportable fuel cells that provide dc outputs rated <60 Vdc and <240 VA and use un-refillable borohydride alkaline fuel. The fuel is either sodium or potassium borohydride compound or a mixture of these compounds.

UL Subject 2262, Portable PEM Type Fuel Cell Power Plants with or without UPS Features and Portable PEM Type Fuel Cell Modules for Factory Installation in OEM Type Equipment for Indoor Use, Scope: These requirements cover portable PEM fuel cell power systems used indoors. They also cover PEM fuel cell modules intended to be factory installed in complete products. Products covered by these requirements are not intended to be permanently connected to fuel supplies, electrical power sources, electrical appliances, or utilization equipment.

UL Subject 2267, for Installation in Industrial Electric Trucks, Scope: The requirements cover fuel cell power systems intended to be installed in type E, EE, ES and EX industrial trucks used in locations as defined in the Fire Safety Standard for Powered Industrial Trucks including Type Designations, Areas of Use, Conversions, Maintenance and Operations NFPA 505, and the National Electrical Code NFPA 70. The fuel cell power systems are intended to be connected to or integrated with industrial trucks to supply electrical power. These requirements anticipate the use of designs that are fueled by various fuels including hydrogen gas.

14.3 Environmental Effects

14.3.1 Fuel Cell Emissions

Fuel cell power generation systems in operation today achieve 40% fuel-to-electricity efficiency utilizing hydrocarbon fuels. Since fuel cells operate silently, they reduce noise pollution as well as air pollution and when the fuel cell is sited near the point of use, its waste heat can be captured for beneficial purposes (cogeneration). In large-scale building systems, these fuel cell cogeneration systems can reduce facility energy service costs by 20% to 40% over conventional energy service and increase efficiency to 85%. In Figure 14.1, the comparison between emissions for combustion engines and fuel cell is shown. Looking at comparisons with existing fuel systems, the hydrogen fuel cell certainly comes out ahead environmentally.

The environmental impacts of fuel cell use depend upon the source of the hydrogen-rich fuel used. By using pure hydrogen, fuel cells have virtually no emissions except water. However, currently hydrogen is rarely used because of problems with storage and transportation. It is predominantly during the fuel processing stage that atmospheric emissions are released by a fuel cell power plant. When methanol from biomass is used as a fuel, fuel cells have no net emissions of carbon dioxide (CO_2, a GHG) because any carbon released was recently taken from the atmosphere by photosynthetic plants. Any high-temperature combustion, such as that which would take place in a spark ignition engine fueled by methanol, produces nitrous oxides (NO_x), gases that contribute to acid rain. Fuel cells virtually eliminate NO_x emissions because of the lower temperatures of their chemical reactions. Fuel cells, using processed fossil fuels, have emissions of CO_2 and sulfur dioxide (SO_2), but these emissions are lower than those from traditional thermal power plants or spark ignition engines owing to the higher efficiency of fuel cell power plants. Higher efficiencies result in less fuel being consumed to produce a given amount of electricity or to travel a given distance.

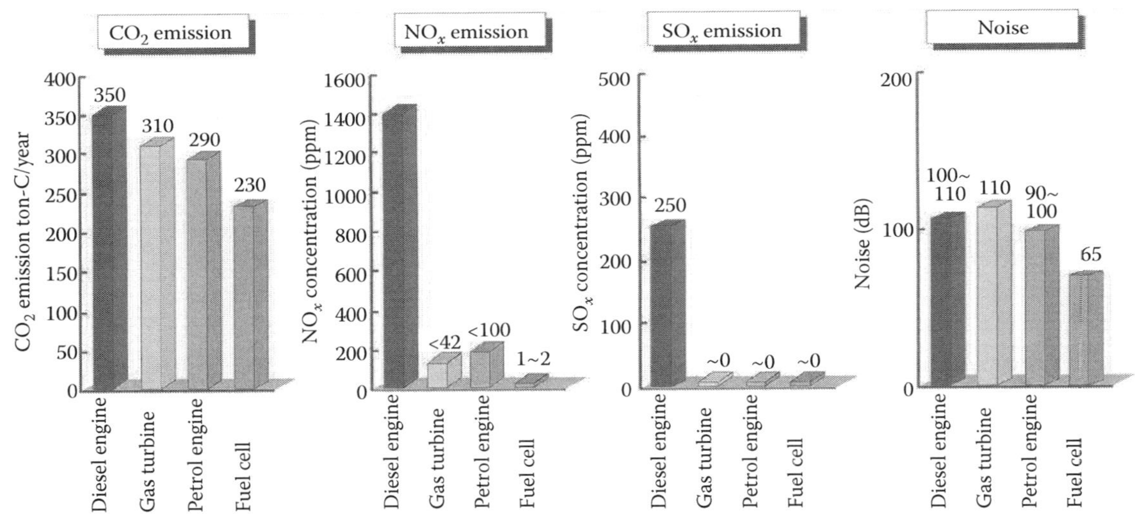

FIGURE 14.1
Comparison of emissions from diesel, gas turbine and petrol engines with fuel cell. (Adapted from Nolan, G. Applications for Fuel cells, By Gerry Nolan, Silicon Chips, Issue 166, Published: July 6, 2002. Available at http://www.siliconchip.com.au/cms/A_30527/article.html, accessed March 28, 2013.)

This corresponds to lower CO_2 and CO_2 emissions. Fuel cell power plants also have longer life expectancies and lower maintenance costs than their alternatives.

When all costs are considered—new technology costs, ongoing upstream costs (e.g., fuel production and distribution), and emission costs, particularly for vehicular applications—fuel cell technology is ahead but it is not a clear-cut conclusion. Fuel cells produce dramatically lower emissions than today's traditional coal, oil, or combustion technologies. While they do still produce some CO_2 when using a hydrocarbon base fuel, the production of NO_x, SO_x, and particulates is negligible. The internal combustion engine is a notoriously inefficient engine, converting only approximately 15%–18% of the heat content of gasoline, diesel, and so on, into useful energy. Fuel cells' 35%–55% efficiency can directly translate into reduced fuel consumption. The US Environmental Protection Agency (EPA) estimates that today, the fuel efficiency of US vehicles is 23.8 mpg, the lowest since 1980—due in large part to the prevalence of light trucks and SUVs. It is expected to reach 31.4 by 2010. According to methanex, a methanol-fueled FCV is approximately 1.74 times more efficient and is therefore expected to attain approximately 55 mpg.

A fuel cell power generation system consists of several components besides the fuel cell such as a fuel processor and a power conditioner/inverter. The fuel processor is the first step of the conversion of fuel into an electrical current. Typically, a fuel processor utilizes a combination of steam reforming (SR) and partial oxidation (POX) methods to convert hydrocarbons (methane, natural gas) into the pure hydrogen necessary as input to the fuel processor. During this process, the fuel processor also should strip the input gas of its pollutants such as carbon and carbon monoxide. The fuel processor is one of the areas in which the greatest environmental threat can occur because of this. There are a number of other considerations to be taken into account when examining the environmental impact and life cycle assessment of fuel cell power generation system such as axillary equipment and their economic and environmental impact (Kordesch and Simader 1995; van Rooijen 2006; Tromp 2002).

In the assessment of environmental risks with options using non-renewable sources, diesel vehicles, diesel hybrids, and fuel cell vehicles using compressed hydrogen centrally produced by fossil sources are more convenient than petrol engine cars. Fuel cell vehicles using methanol and petrol are worse, and the same is valid for engines using hydrogen produced by nonrenewable sources or methanol. In case of the use of natural gas in an internal combustion engine, the results are not significant. Within the options based on fossil energy carriers, the best internal combustion engine option (diesel hybrid) has a clearly better value as the best fuel cell vehicle.

Though fuel cell cars have more efficient energy use in the vehicles than conventional internal combustion engines, the use of alternative fuels require additional conversion stages onboard the vehicles. These fuels, except bio-energy carriers, can only be extracted and produced with a higher use of energy than for conventional fuels, so the additional conversion processes

TABLE 14.5

Used Emission Factors for Fuel Cell Compact Heat and Power Production Installations and Conventional Installations

Fuel Cell CCHPI				Conventional Installation		
Emission	**Unit**	**With NG Reformer**	**With Fuel Oil POX Reformer**	**NG Caloric Power Heating Installation**	**Fuel Oil Low-Temperature Heating Installation**	**District Heat**
CO_2	g/kWh	199	268	199	268	290
CH_4	mg/kWh	30	3	18.1	0.9	810
SO_2	mg/kWh	0	0	1.6	270	50
NO_x	mg/kWh	4.4	20	10	80	550
N_2O	mg/kWh	—	0	3.6	1.3	330
Dust	mg/kWh	0	0	0.5	3.6	30
CO	mg/kWh	9.6	240	30	6.4	150
NMVOC	mg/kWh	1	17	18.1	3.7	89
CO_2 Equiv.	g/kWh		268.1	200.5	268.4	324

Source: Adapted from Oertel, D., and Fleischer, T. Fuel cells impact and consequences of fuel cells technology on sustainable development, Technical Report Series, EUR 20681 EN, March 2003.

Note: CCHPI, compact combined heat and power production installation; NG, natural gas; NMVOC, NO_x, non-methane volatile organic compounds; POX, partial oxidation.

lead to further losses. Vehicles with fuel cell drives emit comparatively low emissions, but the production process of alternative fuels, especially those produced by fossil sources and partially also biogenic fuels, is often connected with higher emissions compared with the production of the conventional fuels such as petrol and diesel.

In the case of a stationary fuel cell system, such as a CHP system, the specific harmful and greenhouse emissions of the different heating systems are compared in Table 14.5. In the case of conventional heating installations, the maximum limits can be the valid emission limit values as well as measured values. As seen from the table, in all cases, the quantity of CO_2 emissions is about the same; CO emissions are relatively high for fuel oil–powered fuel cell compact combined heat and power production installations (mostly determined by the reformer); SO_2 emissions (sulfur content of the fuel oil) for fuel oil–powered low-temperature installations are relatively high, and for natural gas installations, all considered harmful emissions are lower.

14.3.2 Fuel Cell Life Cycle Assessment

The environmental evaluation of the different fuel cell system is performed through the life cycle assessment (LCA) method where an investigation of the complete life cycle is performed to ensure that no environmental aspect is neglected. Higher environmental operation standards for modern energy conversion systems require careful analysis of the upstream and downstream processes, for example, fuel supply or system production. The absolute impact of production is more important than the relative contribution of the production. Often technologies exhibiting good characteristics in the use phase lead to higher absolute environmental impacts in the production phase owing to the use of more sophisticated materials and components. Thus, for fuel cells, this implies that the LCA of producing the systems will be of higher importance.

The two key elements of an LCA of the fuel cell system are the assessment of the entire life cycle of the fuel cell system and the assessment of a variety of environmental impacts because of it. The first step is the goal and scope definition, in which the system, the intended application, the data sources, and system boundaries are described and the functional unit, that is, the reference of all related inputs and outputs, is defined. The criteria for selecting input and output flows or processes have to be specified. In this step, the data quality requirements; time-related and geographical coverage; the consistency, representativity, and uncertainty of the data; and the critical review procedure have to be described.

The life cycle of fuel cells is shown in Figure 14.2. The Life Cycle Inventory (LCI) analysis involves data collection and calculation procedures to quantify relevant inputs and outputs. These input and output flows involve

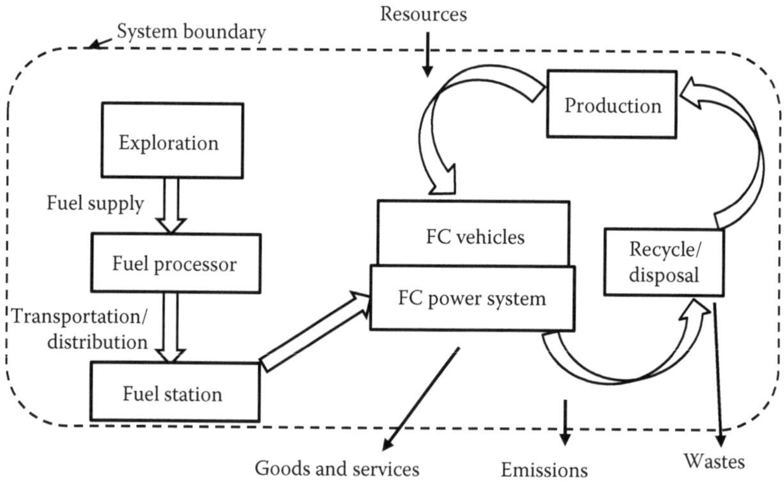

FIGURE 14.2
Inputs and outputs of fuel cell production in terms of its life cycle. (Pehnt, M.: Life cycle analysis of fuel cell system components. In *Handbook of Fuel Cells—Fundamentals, Technology, and Applications* (ISBN 0-471-49926-9), W. Vielstich, A. Lamm, and H.A. Gasteiger (eds.), Vol. 4, pp. 1293–1317, 2003. Copyright Wiley-VCH Verlag GmbH & Co. KGaA. Reproduced with permission.)

consumed or produced goods as well as emissions, waste streams, and so on. It is essential to consider all life cycle stages, that is, system production, operation, and disposal/recycling. Principally, there will be iterative steps leading to additional data requirements. The data collection usually follows the process chain, that is, extraction, conversion, transport, production, use, and disposal or recycling, respectively. The phases may be divided into smaller phases (unit processes). Every unit process of the chain has several incoming and outgoing material and energy flows that are carefully recorded. The main product or the by-products, energy carriers, wastes, and emissions into air, water, or soil, are outputs leaving the system boundaries as shown in Figure 14.2. The potential impacts of the inputs and outputs of the LCI are then determined by the impact assessment, which categorizes and aggregates the input and output flows to the biosphere to so-called impact categories, such as the global warming potential (GWP), by multiplication with characterization factors.

The impact categories for fuel cell systems are as follows: (i) depletion of abiotic resources, such as fossil energy carriers and uranium, metals, or other materials; (ii) depletion of biotic resources as a measure of overexploitation; (iii) GWP, because the emissions of GHGs influence the stability of solar irradiation and adsorption/reflection at the surface; (iv) depletion of stratospheric ozone particularly by chlorinated and brominated compounds, nitrous oxides, and indirectly by the greenhouse effect; (v) acidification; several substances, particularly sulfur dioxide, nitrogen oxides, and, indirectly, ammonia, act as proton sources and acidify soil and water; (vi) eutrophication involves the addition of mineral nutrients to soil and water, which results in shifts in increased algal growth, a reduction in ecological diversity, and, in some instances, a lack of oxygen; mainly nitrogen and phosphorus components contribute to nitrification; (vii) emission of ecotoxic and human toxic substances, such as pesticides, heavy metals, and carcinogenic substances; (viii) emission of radioactive substances; and (ix) other impact categories, such as land use, noise, waste, and odor.

For the overall assessment of mobile fuel cells, fuel production is of high importance. Not only do the questions of storage systems and costs for fuel production or infrastructure considerations have to be answered, but the environmental impacts for the different fuels are of importance as well.

Generally, the following factors are of relevance for the LCA of fuels:

i. The primary energy carrier has an especially high impact on the impact categories of global warming and use of abiotic resources. The change from crude oil to natural gas, for instance, is associated with a decrease in CO_2 intensity owing to the higher hydrogen-to-carbon ratio of natural gas. Switching to renewable primary energy carriers clearly reduces these impacts to low inputs of fossil energy along the production chain.

ii. The efficiencies and impacts of processing are also of importance. Today's crude oil–based fuels exhibit an extremely high energetic efficiency of more than 90%. In contrast, steam or combined reforming of natural gas for hydrogen and methanol production, respectively, has comparatively lower efficiencies. Thus, it is important to distinguish between the production of gasoline in average refineries and marginal plants.

iii. The upstream and downstream processes involving different requirements for transportation or distribution are the third important factor for the assessment of the fuel supply. The possible use of joint products can reduce environmental impacts if there is a market for the by-product.

In the case of hydrogen, roughly 48% of the worldwide production is accomplished by steam reforming of natural gas, 30% by processing crude oil products, 18% by processing coal, and 3% as a by-product of the chloralkali process. The various hydrogen supply paths differ in terms of the distribution paths. Natural gas steam reforming is one of the most common processes. The efficiency of that conversion depends on the use of the steam produced as a by-product. Gasification of biomass and water electrolysis using renewable electricity are attractive options for producing hydrogen with renewable primary energy carriers. However, the potentials of renewable energies must be taken into account because they can be used alternatively in stationary heat and power generation. Therefore, each option of using renewable energy should be assessed in terms of cost, ecoefficiency, and storage requirements.

Fuel cells can be applied in various stationary applications, ranging from 1-kWel systems for domestic heating and for combined heat and power production (CHP) for district heating or large buildings, up to megawatt applications for industrial cogeneration and electricity production without cogeneration. The environmental assessment needs to distinguish between the applications and compare fuel cells to different competitors. The operation of fuel cell power plants leads to minimal direct emissions owing to relatively low (compared to combustion engines or turbines) operating temperatures (leading to almost zero thermal NO_x emissions) and gas clean-up requirements (e.g., for required SO_2 removal). The emissions are typically dependent on the load only. As a first-order approximation, these emissions can be applied to all natural gas reforming stationary plants as long as the fuel, the reformer type, temperature, and fuel utilization are comparable. Generally, these emissions are very low in comparison with emissions from other life cycle stages so that the uncertainty is not very relevant for the total results.

Essential for systems' LCA are the assumed electrical and thermal efficiencies, which differ according to the system and the fuel cell type. The potentially high electrical efficiency of fuel cell power plants is a major advantage of

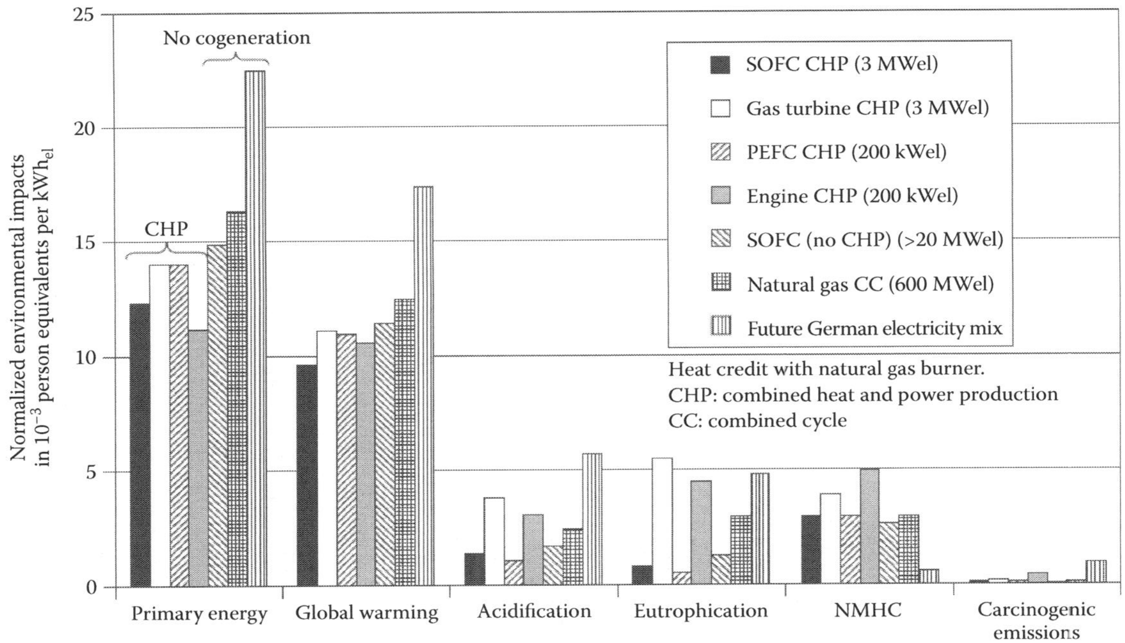

FIGURE 14.3
Environmental impacts associated with primary energy, global warming potential, acidification, eutrophication, non-methane hydrocarbons, and carcinogenic emissions for various techniques employed to generate hydrogen. (Pehnt, M.: Life cycle analysis of fuel cell system components. In *Handbook of Fuel Cells—Fundamentals, Technology, and Applications* (ISBN 0-471-49926-9), W. Vielstich, A. Lamm, and H.A. Gasteiger (eds.), Vol. 4, pp. 1293–1317, 2003. Copyright Wiley-VCH Verlag GmbH & Co. KGaA. Reproduced with permission.)

this system. For each power range, fuel cells offer higher efficiencies than the conventional competitors. With natural gas as a fuel in the low power range, polymer electrolyte fuel cells (PEFCs) have electrical efficiencies on the order of 32.35% for house heating systems and 40% in the 100 kWel range. In some fuel cell systems including large PAFCs, however, especially of the early generations, degradation effects lowered the lifetime efficiency. High-temperature fuel cells offer efficiencies near 50% when used in lower power regimes.

For CHP, the thermal efficiency is also of importance. The thermal efficiencies of conventional systems have been a key parameter for past optimization of the systems. Gas engines, for instance, can reach total efficiencies approaching 100% (LHV) owing to the use of the condensing heat. In practice, more than 90% total efficiency is achievable. Combined cycle CHP plants can also reach thermal efficiencies of 50%, resulting in total efficiencies of nearly 90%. Thermal efficiency is a function of the temperature of the heat medium and the load. Generally, current target values for most fuel cell systems are approximately 80% total efficiency. To successfully compete with conventional systems, future work should also focus on increasing thermal efficiencies by using the reformer exhaust heat and other heat sources.

In Figure 14.3, various environmental impacts of fuel cell energy production including all life cycle stages are compared for competing technologies. The heat produced in cogeneration systems is credited with a modern natural gas burner; if the system produces x kWh of electricity and y kWh of heat simultaneously, the impacts of producing y kWh of heat with a modern natural gas are subtracted from the total impacts because this heat production is substituted by the cogeneration system. It is obvious that high-temperature fuel cells in this application offer significant advantages compared to the competing technologies. Considering the GWP, a SOFC in cogeneration is 12% more efficient than a gas turbine and even 47% more efficient than a German electricity mix. At the same time, a gas turbine in the 3 MWel power range produces less GHGs than a SOFC without cogeneration. Combined heat and power production generally should be promoted. In addition, both the electrical and total efficiency need to be optimized. This is even more important for PEFCs in the 100 kWel range where engine CHPs show total efficiencies exceeding 90% (LHV) because the heat of condensation is used. However, the development of high-efficiency centralized electricity production based on fuel cells decreases the gap between cogeneration and non-cogeneration plants.

14.4 Summary

Fuel cells are a future energy system with a high potential for environmentally friendly energy conversion that can be used in stationary and mobile applications. Depending on the type of fuel cells, stationary applications include small residential, medium-sized cogeneration, or large power plant applications. In the mobile sector, fuel cells, particularly low-temperature fuel cells, can be used for heavy-duty and passenger vehicles, trains, boats, or auxiliary power units for airplanes. Mobile applications also include portable low power systems for various uses. The high efficiency can lead to a significant reduction of fossil fuel use and lower GHG emissions. In addition, the electrochemical nature of the reaction, the low temperature in the reforming steps, and the necessity to remove impurities in the fuel, such as sulfur, result in extremely low local emissions, an important feature especially in highly populated areas. In vehicle applications, particularly at low speed, reductions in noise emissions are expected. Other advantages include the elimination of gear shifts, the higher potential reliability, the compatibility with other electric or electronic devices, and new options with respect to the safety design of vehicles. Thus, clear environmental advantages are expected in the applications of fuel cells.

Fuel cells are promising energy converters for mobile, portable, and stationary applications. For an environmental evaluation of new technologies, however, an investigation of the complete life cycle is necessary to ensure that no environmental aspect is neglected. In stationary applications, the potentially high electrical efficiency of fuel cell power plants, especially high-temperature fuel cells, leads to clear resource and GHG emission advantages compared to the competing technologies. Further advantages could be achieved if not only the electrical but also the total efficiency were simultaneously optimized. This is particularly important for low-temperature fuel cells in CHP applications where some engine CHP plants show total efficiencies of more than 90%. Right now, fuel cells are deployed in various sectors and several industries are developing fuel cell applications. All of the major automakers are working to commercialize a fuel cell car. Fuel cells are powering buses, boats, trains, planes, scooters, forklifts, and even bicycles. There are fuel cell–powered vending machines, vacuum cleaners, and highway road signs. Miniature fuel cells for cellular phones, laptop computers, and portable electronics are on their way to the market. Hospitals, credit card centers, police stations, and banks are all using fuel cells to provide power to their facilities. Wastewater treatment plants and landfills are using fuel cells to convert the methane gas they produce into electricity. Telecommunications companies are installing fuel cells at cell phone, radio, and 911 towers.

References

ASME PTC-50: Fuel Cell Power Systems Performance, 2000. ISBN 0791827801.

Defense Update. Military Applications of Fuel Cell Technologies, Defense Update Magazine Year 2004 Issue 4. Available at http://defense-update.com/products/f/fuel-cells.htm (accessed March 28, 2013).

Kordesch, K. V. and G. R. Simader. Environmental impact of fuel cell technology. *Chemical Reviews* 95: 191–207, 1995.

Kukkonen, C. DMFCC fuel cartridges, 2006. Available at http://www.viaspace.com/ae_cartridges.php (accessed March 28, 2013).

Lewis, N. S., H. Abarbanel, L. Bildsten, P. Dimotakis and F. Dyson. Portable energy for dismounted soldier. JASON, The MITRE Corp. Report JSR-02-135, McLean, VA, June 2003.

Nolan, G. Applications for Fuel cells, By Gerry Nolan, Silicon Chips, Issue 166, Published: July 6, 2002 Available at http://www.siliconchip.com.au/cms/A_30527/article.html (accessed March 28, 2013).

Oertel, D. and Fleischer T. FUEL CELLS Impact and consequences of fuel cells technology on sustainable development. Technical Report Series, EUR 20681 EN, March 2003.

Pehnt, M. Life cycle analysis of fuel cell system components, Vol. 4. In: *Handbook of Fuel Cells–Fundamentals, Technology, and Applications*, Editors W. Vielstich, A. Lamm and H. A. Gasteiger. John Wiley & Sons, Ltd., Chichester, NH, 2003, pp. 1293–1317. ISBN 0-471-49926-9.

Tromp, T. K. Environmental impact of H2 from hydrogen fuel cell on the stratosphere, American Geophysical Union, Fall Meeting, abstract #A72C-0192, 2002.

US Department of Energy. Hydrogen Fuel Station Codes and Standards Citations, 2011. Available at http://www.hydrogen.energy.gov/fueling_stations/index.cfm (accessed March 28, 2013).

US Department of Energy. Hydrogen Vehicle and Infrastructure Codes and Standards Citations, 2009. Available at www.afdc.energy.gov/pdfs/48608.pdf (accessed on June 7, 2013).

van Rooijen, J. A Life Cycle Assessment of the PureCell™ Stationary Fuel Cell System: Providing a Guide for Environmental Improvement, A report of the Center for Sustainable Systems, Report No. CSS06-09, University of Michigan, Ann Arbor, MI, June 30, 2006.

Nomenclature

A: Area or surface area
A_c: Catalyst surface area per unit volume of active layer
A_{carbon}: Area of carbon in electrode
A_{ch}: Cross-sectional area of gas flow channel
A_{MEA}: Surface area of MEA
A_o: Catalyst surface area per unit mass of catalyst particle
A_{Pt}: Area of platinum in electrode
a: Water activity defined as $a = \frac{P_{H_2O}}{P_{sat}}$ or thickness or width of a channel or mass of adsorbent
a_{al}: Thickness of anode active layer
a_{cl}: Thickness of cathode active layer
a_{ct}: Thickness of catalyst layer
a_{el}: Thickness of gas diffusion layer
a: Effective pore radius
b: Height of a channel
C: Molar concentration or conductor or capacitance
$C_{agg,i}$: Gas concentration within agglomerate particle
$C_{i,in}$: Concentration of species i at inlet of gas channel
C_f: Skin friction
C_{fx}: Local skin friction
C^*: Bulk concentration
C_G: Gouy–Chapman capacitance
C_H: Capacitance of the double layer
C_I: Capacitance of the space between the electrode and the inner Helmholtz plane
C_p: Constant pressure specific heat
C_{po}: Constant pressure specific heat of ideal gas
$\bar{C}_p$: Molar constant pressure specific heat of ideal gas
C_v: Constant volume specific heat
$\bar{C}_v$: Molar constant volume specific heat of ideal gas
C_{vo}: Constant volume specific heat of ideal gas
$C_{Ox,surf}$: Concentration of oxidant species
$C_{Red,surf}$: Concentration of reductant species
CO_3^{2-}: Carbonate ion
CO: Carbon monoxide
CO_2: Carbon dioxide
C_i: Species concentration or ion concentration
$C_{i,0}$: Ion concentration at zero electrical potential
$C_{i\infty}$: Free stream concentration of species i

C_{iS}: Species concentration at the surface
$C_{i,ch}$: Species concentration in the channel
$C_{i,el}$: Species concentration at electrolyte interface
$C_{i,ct}$: Species concentration at catalyst surface
c: Concentration or mole fraction or speed of light
c_{ad}: Density of adsorbed species
D: Diffusion coefficient or diameter
D_{ab}: Binary diffusion with two species a and b
D_H: Hydraulic diameter
D_i: Diffusion coefficient of ion in electrolyte
D_o: Reference diffusion constant
D_{ij}: Multi-component diffusion coefficient
D_{ij}^{eff}: Effective diffusion coefficient
D_{agg}^{eff}: Effective diffusion coefficient in agglomerate layer
D_f^{eff}: Effective diffusion coefficient in electrolyte film
$D_{H_2}^{eff}$: Effective diffusion coefficient of hydrogen in anode electrode layer
$D_{O_2}^{eff}$: Effective diffusion coefficient of oxygen in cathode electrode layer
D_{kd}: Knudsen diffusion coefficient
D_{sd}: Surface diffusion coefficient
D_{md}: Molecular or ordinary diffusion coefficient
$D_{\lambda m}$: Water diffusion coefficient in the membrane
D_i^T: Thermal diffusion coefficient
d: Separation between charges
d_p: Pore diameter
d_{Pt}: Platinum particle diameter
$\frac{dq_1}{dq_E}$: Rate of change of specifically absorbed charge with charge on the electrode
E: Voltage or electrical potential or total energy
E_o: Voltage amplitude for sinusoidal potential
E_F: Highest available electron energy level
E_{rev}^o: Reversible potential or reversible voltage
E_{Nernst}^o: Nernst potential
E_{cell}: Cell voltage
e: Electric charge per electron
e^-: Electron charge
F: Faraday constant or force
F_b: Hydrodynamic viscous force used in Nernst–Einstein equation
FF: Photovoltaic fill factor
f: Frequency factor for the reaction constant or friction factor or vector valued function
G: Total Gibbs energy or Gibbs free energy or solar flux
Gr: Grashof number
g: Specific Gibbs function or acceleration due to gravity
$\vec{g}$: Gravity vector

$\bar{g}$: Molar specific Gibbs function
$\bar{g}_f^0$: Gibbs function of formation
Δg: Change of Gibbs function
ΔG_f: Gibbs free energy of formation or free energy of activation for forward reaction
ΔG_b: Free energy of activation for backward reaction
H: Total enthalpy
H_2: Hydrogen molecule
H^+: Hydrogen ion or proton
H_2O: Water molecule
hhv: Higher heating value
h: Specific enthalpy or enthalpy per unit mass or Planck's constant
h_c, h_{conv}: Convection heat transfer coefficient
h_m: Convection mass transfer coefficient
$\bar{h}$: Molar specific enthalpy
h_{RP}: Enthalpy of combustion
$\bar{h}_f^0$: Enthalpy of formation at the reference state
Δh: Change of enthalpy
Δh_f: Enthalpy of formation
I: Total current
i: Total current
i_o: Current amplitude for sinusoidal current
J: Mass charge flux
j: Current density or molar flux or part of imaginary quantity
j_a: Anode current density
j_c: Cathode current density
$j_{agg,a}$: Anode active layer current density
$j_{agg,c}$: Cathode active layer current density
j_i'': Molar diffusion flux
$\vec{j}''_{conv}$: Convection mass flux
j_l, j_L: Limiting current density
j_0: Exchange current density
J_{sc}: Photocurrent density
K: Boltzmann constant or permeability
K_{REF}: Reforming reaction rate equilibrium constant
K_p: Hydraulic permeability
K_{wgs}: Water-gas shift reaction's equilibrium constant
k: Thermal conductivity or reforming rate constant
k_B: Boltzmann's constant
k_b: Reduction reaction rate constant or backward reaction rate constant
k_f: Forward reaction rate constant or fluid thermal conductivity
k_{flow}: Flow loss coefficient
k_1'': Reaction rate constant
L: Length or thickness or inductor or inductance

L_c: Characteristics length
L_e: Lewis number
$L_{e,th}$: Thermal entrance length
$L_{e,h}$: Hydronamic entrance length
$L_{e,c}$: Concentration entrance length
L: Catalyst or platinum loading in the electrode
l: Liquid
lhv: Lower heating value
M: Molecular weight
m: Mass, or hydrogen number, or mass flow rate
$\dot{m}$: Mass consumption rate or mass flow rate
$m''_{i,d}$: Mass diffusion flux
m''_i: Total mass flux
m_{pt}: Platinum catalyst
N: Number of electrical charge or number of ions or number of components in a mixture or number of moles
N_a: Avogadro number
N_e: Total charge
N_c: Number of cells in a stack
Nu: Nusselt number
N''_i: Total mass transport flux
n: Number of moles, number of electrons, or carbon number
$\dot{n}$: Gas consumption rate in moles
$\vec{n}$: Normal unit vector
n_e: Number of electrons transfer per molecule in the electrochemical reaction
n_{drag}: Electro-osmotic drag coefficient
n_{drag}^{sat}: Electro-osmotic drag coefficient for fully hydrated membrane
O^{2-}: Oxide ion
OH^-: Hydroxyl ion
P: Pressure or power or perimeter
Pr: Prandtl number
Pt: Platinum
Pt/C: Ratio of platinum to carbon by weight (Pt/C ratio) in platinum supported on carbon
ΔP: Pressure drop
P_i: Partial pressure of the component *i* in a mixture
$P_{i,in}$: Partial pressure of the component *i* at the inlet of gas channel
P_C: Critical pressure
P_c: Power per cell
P: Equilibrium pressure of adsorbate
P_o: Saturation pressure of adsorbate
P_r: Reduced pressure
P_{sat}: Saturation pressure
P_t: Total power for a stack

p: Pressure or oxygen number
Q: Total heat transfer or heat generation or volume flow rate or mass action expression
Q_L: Heat transfer from low temperature
Q_H: Heat transfer from high temperature
$\dot{Q}$: Volumetric heat generation
q: Heat transfer rate or heat transfer per unit mass, electrical charge
q'': Heat flux or heat transfer rate per unit area
$\underline{R}$: Universal gas constant
R: Gas constant or charge electrical resistance or thermal resistance or production rate
Ra: Rayleigh number
R_a: Radius of diffusing spherical particle used in Stokes–Einstein equation
R_{agg}: Radius of agglomerate particle
Re: Reynolds number
Re_x: Local Reynolds number
Re_{crit}: Critical Reynolds number
Re_D: Reynolds number based on diameter
R_m: Mass transfer resistance or ohmic resistance of membrane
$R_{m,conv}$: Convection mass transfer resistance
$R_{m,diff}$: Diffusion mass transfer resistance
$R_{m,r}$: Reaction mass transfer resistance
R_{ohm}: Ohmic resistance
r: Radial direction or ohmic resistance
S: Total entropy or mass source or mass sink
Sh: Sherwood number
Sh_j: Sherwood number for uniform surface flux
Sh_c: Sherwood number with constant surface concentration
Sc: Schmidt number
S_i: Source or sink term or consumption rate of gas species
s: Specific entropy or entropy per unit mass or adsorption cross section of the adsorbing species
$\bar{s}$: Molar specific entropy
$\bar{s}_f^0$: Absolute entropy
Δs: Change of entropy
Δs_f: Change in absolute entropy
T: Temperature
T_H: High temperature
T_L: Low temperature
T_{dp}: Dew point temperature
T_p: Temperature of product of combustion
T_{sat}: Saturation temperature
T_C: Critical temperature
T_r: Reduced temperature
T_{in}: Temperature at inlet of gas channel

T_∞: Upstream temperature
t: Time
U: Total internal energy
U_a: Particle velocity used in the Nernst–Einstein equation
u: Specific internal energy or x-component velocity
U_c: Characteristics velocity
u_∞: Upstream velocity
u_{RP}: Internal energy of combustion
u_j: Electrical mobility or velocity of migration
V: Volume or velocity
V_c: Cell voltage or applied voltage
$V_{cp,m}$: Compressor motor voltage
$\vec{V}$: Velocity vector
v: y-component velocity or solution velocity or specific volume
$\bar{v}$: Molar specific volume
v: Vapor or adsorbed gas quantity in volume units or reaction rate
v_m: Monolayer adsorbed gas quantity
v_b: Backward reaction rate
v_f: Forward reaction rate
v_{net}: Net reaction
W: Work
w: Work done per unit mass or z-component velocity
W_e: Electrical work
x: Mass fraction, or mole fraction or molar ratio or x coordinate, or input variable
y: Mole fraction or y coordinate or output
Z: Compressibility factor or elevation or impedance
Z_o: Magnitude of sinusoidal impedance
Z_i: Charge number of the charge carrier
z: z coordinate or number of units of electronic charge

Subscript

0: At maximum or reference
a: Anode
ae: Anode electrolyte interface
act: Activation overpotential
acta: Anode activation overpotential
actc: Cathode activation overpotential
ah: Anode channel
amb: Ambient
an: Anode

b: Bulk or backward
CV: Control volume
c: Cathode or convection or cell
ca: Cathode
ce: Cathode electrolyte interface
ch: Cathode channel
ch: Channel
cm: Compressor motor
conc: Concentration
ct: Catalyst
d: Dynamic
des: Desired
dl: Double layer
e: Exit or electrolyte
el: Electrolyte
elec: Electrochemical
FC: Fuel cell
f: Forward
gen: Generation
i: Inlet or index for component of a mixture
in: Inlet
irrev: Irreversible
lim: Mass transport limiting
M,m: Membrane
Max: Maximum
mix: Mixture
mt,a: Anodic mass transfer
mt,c: Cathodic mass transfer
net: Net
OC: Open circuit
OCV: Open circuit voltage
Ohm: Ohmic
Other: Other losses
Ox: Oxidant
P: Product
p: Porous
pm: Porous media
Pt: Platinum density
R: Reactant or resistance
Red: Reductant
r: Reaction
ref: Reference
rev: Reversible
reva: Anode electrode reversible
revc: Cathode electrode reversible

rm: Return manifold
Sys: System
s: Surface
sp: Supply manifold
surf: Surface
T: Temperature or Total
th: Thermal
w: Water

Superscript

c: Cell
in: Inlet
o: Reference state or Reversible
s: Stack

Greek Symbol

α: Charge transfer coefficient
β: Transfer coefficient or beta coefficient
β_o: Fraction of water generation that transports through the anode electrode toward the anode channel
κ: Permeability of porous media
κ_ϕ: Electrokinetic permeability
κ_p: Membrane permeability due to pressure gradient
κ_{H_2O}: Hydraulic permeability coefficient of water
μ: Dynamic viscosity, fuel cell efficiency
μ_b: Dynamic viscosity of liquid used in the Stokes–Einstein equation
μ_e: Effective viscosity
μ_f: Fuel utilization coefficient
η: Efficiency or voltage loss or overpotential
η_i: Current efficiency
η_{fc}: Fuel cell efficiency
$\eta_{fc,v}$: Fuel cell voltage efficiency
$\eta_{act,a}$: Anode activation loss
$\eta_{act,c}$: Cathode activation loss
η_{ohm}: Ohmic loss
$\eta_{ohm,a}$: Anode ohmic loss or overpotential
$\eta_{ohm,c}$: Cathode ohmic loss or overpotential
$\eta_{ohm,e}$: Electrolyte ohmic loss or overpotential

η_{mass}: Mass transfer loss or overpotential
τ: Tortuosity factor or shear stress or time constant
τ_{co}: Charge density required to oxidize a monolayer of CO on Pt
τ_w: Wall shear stress
ϵ: Porosity or the dielectric constant
ε_o: Permittivity of free space
ε_{ab}: Lennard-Jones intermolecular force parameter or constant for intermolecular interactions or energy of interaction
ε_{agg}: Porosity of active layer
λ: Water content or wave length
λ_i^o: Ionic equivalent conductance
λ_m: Water content of the membrane
ρ: Density or resistivity, specific resistivity
ρ_b: Bulk density
ρ_p: Density of porous sample
ρ_e: Electrical resistivity
θ: Contact angle for surface tension
ϕ: Electrical potential or flow variables or porosity
ϕ_a: Real potential for anode
ϕ_c: Real potential for cathode
$\phi_{rev,a}$: Reversible potential for anode
$\phi_{rev,c}$: Reversible potential for cathode
ϕ_f: Fraction of fuel
ϕ: Fraction of non-electrochemical reaction or Phase shift
Ω_D: Dimensionless collision parameter
σ: Proton conductivity or electrical or electron conductivity or Warburg coefficient or surface tension
σ_{ab}: Lennard-Jones intermolecular force parameter or constant for intermolecular interactions or collision diameter
σ_c: Charge conductivity
σ_e: Electronic conductivity
σ_i: Ionic conductivity or electrolyte conductivity
σ_0: Reference conductivity
σ_{H^+}: Proton conductivity
ξ: Stoichiometric factor or fuel utilization factor
δ: Thickness of the boundary layer
ΔP: Pressure drop or differential pressure
ΔG_a: Activation energy for diffusion or barrier energy
Δg: Change of Gibbs function
ΔG_f: Gibbs free energy of formation or free energy of activation for forward reaction
ΔG_b: Free energy of activation for backward reaction
ΔH: Heat of reaction
ω: Radial frequency
Γ: Charge required to reduce monolayer of proton on Pt

Ω: Ohmic
ν: Frequency
γ: Surface tension

Abbreviations

ABS: Acrylonitrile–butadiene–styrene
AC: Alternating current
AES: Auger electron spectroscopy
AFC: Alkaline fuel cell
AFM: Atomic force microscope
AGA: American Gas Association
ANSI: American National Standards Institute
APE: Alkaline polymer electrolyte
API: American Petroleum Institute
APU: Auxiliary power unit
ASHRAE: American Society of Heating, Refrigerating, and Air-Conditioning Engineers
ASME: American Society of Mechanical Engineers
ASR: Area-specific resistance
ASTM: American Society for Testing and Materials
ATR: Autothermal reforming
BET: Brunauer–Emmett–Teller
BV: Butler–Volmer
CB: Conduction band
CC: Combined cycle
CCHPI: Compact combined heat and power production installation
CFD: Computational fluid dynamics
CFR: Code of Federal Regulations
CGA: Compressed Gas Association
CGC: Coal gasifier and cleaner
CHP: Combined heat and power
CL: Catalyst layer
CoG: Center of gravity
CoM: Center of maximum
CPE: Constant phase element
CPOX: Catalytic partial oxidation
CSA: Canadian Standards Association
CSTR: Continuously stirred tank reactor
CV: Cyclic voltammetry
DAFC: Direct alcohol fuel cell
DAS: Data acquisition system

DC: Direct current
DIR: Direct internal reforming
DMC: Dimethyl carbonate
DMFC: Direct methanol fuel cell
DOE: Department of Energy
DOT: Department of Transportation
DTA: Differential thermal analysis
EC: Ethylene carbonate
ECSA: Electrochemical surface area
EDX: Energy dispersive x-ray
EIA: Energy Information Agency
EIS: Electrochemical impedance spectroscopy
EMF: Electromotive force
EPA: Environmental protection agency
FCV: Fuel cell vehicle
FERC: Federal Energy Regulatory Commission
FFT: Fast Fourier transform
FT-IR: Fourier transform infrared
GDL: Gas diffusion layer
GES: Gyner electrochemical systems
GHG: Greenhouse gas
GT: Gas turbine
GTI: Gas Technology Institute
GWP: Global warming potential
HAD: Hydrogen adsorption/desorption
HDS: Hydrodesulfurization
HEV: Hybrid electric vehicle
HFP: Hexafluoropropylene
HFR: High-frequency resistance
HHV: Higher heating value
HTE: High-temperature electrolysis
IC: Internal combustion
ICC: International Code Council
IEC: International Electrotechnical Commission
IEEE: Institute of Electrical and Electronics Engineers
IFD: Information flow diagram
IGBT: Insulated gate bipolar transistor
IHP: Inner Helmholtz plane
IIR: Indirect internal reforming
IPE: Ideal polarized electrode
IR: Infrared
IT: Intermediate temperature
KOH: Potassium hydroxide
LANL: Los Alamos National Laboratory
LCA: Life cycle assessment

LCI: Life cycle inventory
LCM: Calcium-doped lanthanum manganite
LCR: Inductance–capacitance–resistance
LHS: Left-hand side
LHV: Lower heating value
LQE: Linear quadratic estimator
LQR: Linear quadratic regulator
LSCF: Lanthanum strontium cobalt ferrite
LSM: Strontium-doped lanthanum manganite
M: Metal
MCFC: Molten carbonate (electrolyte) fuel cell
MEA: Membrane electrode assembly
MEMS: Micro-electro-mechanical systems
M-H: Metal–hydrogen
MOF: Metal organic framework
MOM: Mean of maximum
MWCNT: Multi-walled carbon nanotube
NARUC: National Association of Regulatory Utility Commissioners
NASA: National Aeronautics and Space Administration
NEMS: Nano-electro-mechanical systems
NERC: North American Electric Reliability Corporation
NES: National Evaluation Service
NFPA: National Fire Protection Association
NHA: National Hydrogen Association
NHE: Normal hydrogen electrode
NIST: National Institute of Standards and Technology
NMR: Nuclear magnetic resonance
NMVOC: NO_x, non-methane volatile organic compounds
NLLS: Nonlinear least squares
NREL: National Renewable Energy Laboratory
NSSN: National Resource for Global Standards
NTP: Normal temperature and pressure (20°C and 1 atm or 1.01325 bar)
OCV: Open circuit voltage
OHP: Outer Helmholtz plane
OM: Optical microscope
OSHA: Occupational safety and health administration
PAFC: Phosphoric acid fuel cell
PBI: Polybenzimidazole
PEM: Proton exchange membrane or polymer electrolyte membrane
PEMFC: Proton exchange membrane fuel cell or polymer electrolyte membrane fuel cell
PEO: Polyethylene oxide
PFD: Process flow diagram
PFSA: Perfluorosulfonic acid
PGM: Platinum group metal

PID: Proportional–integral–derivative
PNGV: Partnership for a new generation of vehicle
POX: Partial oxidation
PROX: Preferential oxidation
PTFE: Polytetrafluoroethylene
PV: Photovoltaic
PVD: Physical vapor deposition
QAPS: Quaternary ammonium polysulfone
RC: Resistance–capacitance
RE: Reference electrode
RFC: Regenerative fuel cell
RHS: Right-hand side
SAE: Society of Automotive Engineers
ScSZ: Scandia-stabilized zirconia
SEM: Scanning electron microscopy
SHE: Standard hydrogen electrode
SIMS: Secondary ion mass spectroscopy
SNG: Syngas (synthetic gas)
SOEC: Solid oxide electrolysis cell
SOFC: Solid oxide fuel cell
SR: Steam reforming
SRFC: Solar regenerative fuel cell
SSSF: Steady-state steady flow
ST: Steam turbine
STC: Standard test conditions
SUV: Sport utility vehicle
SWCNT: Single-walled carbon nanotube
SXAS: Small-angle x-ray scattering spectroscopy
TBP: Triple phase boundaries
TEM: Transmission electron microscopy
TGA: Thermogravimetric analysis
THMS: Thermal heat management system
UFUS: Uniform flow uniform state
UL: Underwriters Laboratory
UV: Ultraviolet
VB: Valance band
VdF: Vinylidene fluoride
WE: Working electrode
WTT: Well to tank
WTW: Well to wheel
XPS: X-ray photoelectron spectroscopy
XRD: X-ray diffraction
YDC: Yttria-doped ceria
YSZ: Yttria-stabilized zirconia

Appendix A: Constants and Conversion Units

TABLE A.1

Fundamental Physical Constants

Avogadro number: $N_a = 6.022136 \times 10^{23}$
Boltzmann constant: $K = 1.380658 \times 10^{-23}$ J/K
Faraday's constant: $F = 96{,}485.34$ C/mol
Gravitational acceleration: $g = 9.80665$ m/s^2
Planck constant: $h = 6.26076 \times 10^{-34}$ J · s
Electron charge: $e^- = 1.6021777 \times 10^{-19}$ C
Electron mass: $m_e = 9.109389 \times 10^{-31}$ kg
Stefan Boltzmann constant: $\sigma = 5.67051 \times 10^{-8}$ W/m^{2}K^4
Universal gas constant: $\bar{R} = 8.31451$ J/mol · K

TABLE A.2

Conversion Factors

Area (A)	
1 mm^2 = 1.0×10^{-6} m^2	1 ft^2 = 144 in.^{2}
1 cm^2 = 1.0×10^{-4} m^2 = 0.1550 in.^{2}	1 in^2 = 6.4516 cm^2 = 6.4516×10^{-4} m^2
1 m^2 = 10.7639 ft^2	1 ft^2 = 0.092903 m^2
Conductivity (k)	
1 W/m · K = 0.577789 Btu/h · R	1 W/m · K = 0.577789 Btu/h · R
Distance	
1 cm = 0.3970 in.	1 in. = 2.54 cm
1 cm = 0.0328 ft	1 ft = 0.3048 m
1 m = 3.28084 ft	
Density (ρ)	
1 kg/m^3 = 0.0624 lbm/ft^3	1 lbm/ft^3 = 16.018 kg/m^3
1 g/cm^3 = 1 kg/L	
Energy (E)	
1 J = 1 N · m = 1 kg · m^2/s^2	1lbf · ft = 1.3558 J = 1.28507×10^{-3} Btu
1 J = 0.7375 lbf · ft	1 Btu = 1.055 kJ = 778.1693 lbf · ft
1 cal = 4.1868 J	
1 J = 2.777778×10^{-7} kWh	
1 J = 9.478134×10^{-4} Btu	
1 J = 6.241506×10^{18} eV	

(continued)

TABLE A.2 (Continued)

Conversion Factors

Specific Heat (C)	
1 kJ/kg · K = 0.2388 Btu/lbm · R	1 Btu/lbm · R = 4.1868 lkJ/kg · K
Heat Transfer Coefficient	
1 W/m^2 · K = 0.17611 Btu/h · ft^2 · R	1 Btu/h · ft^2 · *R* = 5.67828 W/m^2 · K
Mass	
1 kg = 2.204623 lbm	1 lbm = 0.453592 kg
Power	
1 W = 1 J/s = 1 N m/s	1 lbf · ft/s = 1.355818 W
1 kW = 3412.14 Btu/h	1 Btu/s = 1.055056 kW
1 hp (metric) = 0.735499 kW	1 hp (UK) = 0.7457 kW = 550 lbf · ft/s
1 kW = 1.34 hp	
Pressure	
1 Pa = 1 N/m^2 = 1 kg/m · s^2	1 lbf/in.^{2} = 6.894757 kPa
1 bar = 100 kPa	1 atm = 14.69594 lbf/in.^{2} (psi)
1 atm = 1.01325 × 10^5 Pa = 101.325 kPa	
Volume	
1000 cm^3 = 1 L	
1 L = 0.264 gal	
1 L = 3.53 × 10^{-2} ft^3	
Weight	
1 kg = 2.2 lb	

Appendix B: Useful Equations for Fuel Cell Calculations

Thermodynamics

Ideal gas equation of state:

$$P\bar{v} = \bar{R}T$$

where

$$\bar{R} = \text{Universal gas constant} = 8.3145\ \frac{\text{kN}\cdot\text{m}}{\text{kmol}\cdot\text{K}} = 8.3145\ \frac{\text{kJ}}{\text{kmol}\cdot\text{K}}$$

The first law of thermodynamics for a system undergoing a process:

$${}_1Q_2 = (U_2 - U_1) + \frac{1}{2}m\left(V_2^2 - V_1^2\right) + mg(Z_2 - Z_1) + {}_1W_2$$

The first law of thermodynamics for a control volume

$$\dot{Q}_{\text{CV}} + \sum \dot{m}_{\text{i}}\left(h_{\text{i}} + \frac{V_{\text{i}}^2}{2} + gZ_{\text{i}}\right) = \frac{\text{d}E_{\text{CV}}}{\text{d}t} + \sum \dot{m}_{\text{e}}\left(h_{\text{e}} + \frac{V_{\text{e}}^2}{2} + gZ_{\text{e}}\right) + \dot{W}_{\text{CV}}$$

Steady-state steady flow (SSSF) process

$$\sum \dot{m}_{\text{e}} = \sum \dot{m}_{\text{i}}$$

$$\dot{Q}_{\text{CV}} + \sum \dot{m}_{\text{i}}\left(h_{\text{i}} + \frac{V_{\text{i}}^2}{2} + gZ_{\text{i}}\right) = \sum \dot{m}_{\text{e}}\left(h_{\text{e}} + \frac{V_{\text{e}}^2}{2} + gZ_{\text{e}}\right) + \dot{W}_{\text{CV}}$$

The first law of thermodynamics for a reacting system under steady-state steady flow:

$$Q_{CV} + \sum_{R=1} n_R \left(\bar{h}_f^0 + \Delta\bar{h}\right)_R = \sum_{P=1} n_P \left(\bar{h}_f^0 + \Delta\bar{h}\right)_P + W_{CV}$$

Enthalpy of formation and change of enthalpy of a reacting system:

$$H_{RP} = \sum_{P=1}^{N_P} n_P \left(\bar{h}_f^0 + \Delta\bar{h}\right)_P - \sum_{R=1}^{N_R} n_R \left(\bar{h}_f^0 + \Delta\bar{h}\right)_R$$

Change in entropy of a reacting system:

$$\Delta S = \sum_{P=1}^{N_P} n_P \bar{s}_P - \sum_{R=1}^{N_R} n_R \bar{s}_R$$

Change in Gibbs function of reacting system:

$$\Delta G = \sum_{P=1}^{N_P} n_P \bar{h}_f^0 - \sum_{R=1}^{N_R} n_R \bar{h}_f^0 - T\left(\sum_{P=1}^{N_P} n_P \bar{s}_f^0 - \sum_{R=1}^{N_R} n_R \bar{s}_f^0\right)$$

Thermodynamics of a Fuel Cell

Cell voltage:

$$E = \frac{-\Delta\bar{G}}{n_e F}$$

Reversible open circuit cell voltage:

$$E_{rev}^0 = \frac{-\Delta\bar{g}_f^0}{n_e F}$$

Gibbs function of formation:

$$\Delta G_f = \sum_{P} n_P \left(\bar{g}_f^0\right)_P - \sum_{R} n_R \left(\bar{g}_f^0\right)_R$$

For hydrogen–oxygen fuel cell:

$$\Delta\bar{g}_{\mathrm{f}} = \left(\bar{g}_{\mathrm{f}}^{0} + \Delta\bar{g}\right)_{\mathrm{H_2O}} - \left(\bar{g}_{\mathrm{f}}^{0} + \Delta\bar{g}\right)_{\mathrm{H_2}} - \frac{1}{2}\left(\bar{g}_{\mathrm{f}}^{0} + \Delta\bar{g}\right)_{\mathrm{O_2}}$$

The Nernst Equation

$$E = E_{\mathrm{rev}}^{0} - \frac{\bar{R}T}{n_{\mathrm{e}}F}\ln\frac{P_{\mathrm{M}}^{n_{\mathrm{M}}}P_{\mathrm{N}}^{n_{\mathrm{N}}}}{P_{\mathrm{A}}^{n_{\mathrm{A}}}P_{\mathrm{B}}^{n_{\mathrm{B}}}}$$

Generalized Nernst equation with arbitrary number of reactants and products:

$$E = E_{\mathrm{rev}}^{0} - \frac{\bar{R}T}{n_{\mathrm{e}}F}\ln\frac{\prod_{i=1}^{N}P_{Pi}^{n_i}}{\prod_{i=1}^{N}P_{Ri}^{n_i}}$$

The Nernst equation for hydrogen-oxygen fuel cell:

$$E = E_{\mathrm{rev}}^{0} - \frac{\bar{R}T}{2F}\ln\frac{P_{\mathrm{H_2O}}}{P_{\mathrm{H_2}}P_{\mathrm{O_2}}^{1/2}}$$

Fuel Cell Efficiency

Reversible thermodynamic efficiency of a fuel cell:

$$\eta_{\mathrm{fc,rev}} = \frac{\Delta\bar{g}_{\mathrm{f}}}{\Delta\bar{h}_{\mathrm{f}}} = \frac{n_{\mathrm{e}}FE_{\mathrm{rev}}^{0}}{\Delta\bar{h}_{\mathrm{f}}}$$

Voltage efficiency

$$\eta_{\mathrm{fc,v}} = \frac{E}{E_{\mathrm{rev}}^{0}}$$

Current or fuel utilization efficiency

$$\eta_i = \frac{I/n_e F}{\dot{n}_{f,in}} = \frac{1}{\xi_f}$$

Fuel Consumption and Supply Rates

Oxygen Consumption and Supply Rates

Direct oxygen consumption:

For a single fuel cell:

$$n_{O_2} = \frac{I}{n_e F}\left(\frac{\text{mol}}{\text{s}}\right)$$

For a stack:

$$n_{O_2} = \frac{IN_c}{n_e F}\left(\frac{\text{mol}}{\text{s}}\right)$$

In terms of power

$$n_{O_2} = \frac{P_t}{n_e F V_c}\left(\frac{\text{mol}}{\text{s}}\right) \text{for the stack}$$

$$n_{O_2} = \frac{P_t}{n_e F V_c N_c}\left(\frac{\text{mol}}{\text{s}}\right) \text{for a single cell}$$

Mass oxygen consumption rate:

$$\dot{m}_{O_2} = \frac{M_{O_2} P_t}{n_e F V_c}\left(\frac{\text{kg}}{\text{s}}\right) \text{ for the stack}$$

$$\dot{m}_{O_2} = \frac{M_{O_2} P_t}{n_e F V_c N_c}\left(\frac{\text{kg}}{\text{s}}\right) \text{ for a single cell}$$

Oxygen consumption as air

$$\dot{m}_{\text{air}} = \frac{\xi_{\text{air}} M_{\text{air}} P_{\text{t}}}{c_{\text{O}_2} n_{\text{e}} F V_{\text{c}}} \left(\frac{\text{kg}}{\text{s}} \right) \text{ for a stack}$$

$$\dot{m}_{\text{air}} = \frac{\xi_{\text{air}} M_{\text{air}} P_{\text{t}}}{c_{\text{O}_2} n_{\text{e}} F V_{\text{c}} N_{\text{c}}} \left(\frac{\text{kg}}{\text{s}} \right) \text{ for a single cell}$$

In terms of total current:

$$\dot{m}_{\text{air}} = \frac{\xi_{\text{air}} M_{\text{air}} I}{c_{\text{O}_2} n_{\text{e}} F} \left(\frac{\text{kg}}{\text{s}} \right) \text{ for a stack}$$

$$\dot{m}_{\text{air}} = \frac{\xi_{\text{air}} M_{\text{air}} I}{c_{\text{O}_2} n_{\text{e}} F N_{\text{c}}} \left(\frac{\text{kg}}{\text{s}} \right) \text{ for a cell}$$

Hydrogen Consumption and Supply Rates

Mole consumption of hydrogen in terms of total current:

$$n_{\text{H}_2} = \frac{I}{n_{\text{e}} F} \left(\frac{\text{mol}}{\text{s}} \right) \text{ for a single cell}$$

$$n_{\text{H}_2} = \frac{I\, N_{\text{c}}}{n_{\text{e}} F} \left(\frac{\text{mol}}{\text{s}} \right) \text{ for a stack}$$

Hydrogen mole consumption rate:

$$n_{\text{H}_2} = \frac{P_{\text{t}}}{n_{\text{e}} F V_{\text{c}}} \left(\frac{\text{mol}}{\text{s}} \right) \text{ for a stack}$$

$$n_{\text{H}_2} = \frac{P_{\text{t}}}{n_{\text{e}} F V_{\text{c}} N_{\text{c}}} \left(\frac{\text{mol}}{\text{s}} \right) \text{ for a cell}$$

Hydrogen mass consumption rate:

$$\dot{m}_{H_2} = \frac{M_{H_2}P_t}{n_e F V_c}\left(\frac{kg}{s}\right) \text{ for a stack}$$

$$\dot{m}_{H_2} = \frac{M_{H_2}P_t}{n_e F V_c N_c}\left(\frac{kg}{s}\right) \text{ for a cell}$$

Water Production Rate

In terms of total current:

$$n_{H_2O} = \frac{I}{n_e F}\left(\frac{mol}{s}\right) \text{ for a cell}$$

$$n_{H_2O} = \frac{IN_c}{n_e F}\left(\frac{mol}{s}\right) \text{ for a stack}$$

In terms of power:

$$n_{H_2O} = \frac{P_t}{n_e F V_c}\left(\frac{mol}{s}\right) \text{ for a stack}$$

Water mass production rate:

$$\dot{m}_{H_2O} = \frac{M_{H_2O}P_t}{n_e F V_c} \text{ (kg/s) for a stack}$$

$$\dot{m}_{H_2O} = \frac{M_{H_2O}P_t}{n_e F V_c N_c} \text{ (kg/s) for a cell}$$

Heat Generation in a Fuel Cell

Reversible heat generation rate:

$$Q_{gen,rev} = \frac{I}{n_e F}(-T\Delta S) \text{ for a cell}$$

$$Q_{gen,rev} = \frac{I}{n_e F}(-T\Delta S)N_c \text{ for a stack}$$

Irreversibilities' heat generation rate:

$$Q_{\text{gen,irrev}} = I(E_{\text{rev}} - \dot{E}_{\text{c}})\phi_{\text{f}} \text{ for a cell}$$

$$Q_{\text{gen,irrev}} = I(E_{\text{rev}} - \dot{V}_{\text{c}})\phi_{\text{f}} N_{\text{c}} \text{ for a stack}$$

Irreversible heat generation due to ohmic overpotential:

$$Q_{\text{gen,ohm}} = I\eta_{\text{ohm}} = I^2\dot{R}_{\text{c}} \text{ for a cell}$$

$$Q_{\text{gen,ohm}} = I^2 R_{\text{c}} N_{\text{c}} \text{ for a stack}$$

Irreversible heat generation due to activation overpotential (η_{act}):

$$Q_{\text{gen,act}} = I\eta_{\text{act}} \text{ for a cell}$$

$$Q_{\text{gen,act}} = I\eta_{\text{act}} N_{\text{c}} \text{ for a stack}$$

Total heat generation due to electrochemical reaction:

$$Q_{\text{elec,gen}} = \phi_{\text{f}}\,[\dot{n}_{\text{f}}(-T\Delta S) + I(E_{\text{rev}} - V_{\text{c}})] \text{ for a cell}$$

Considering the ohmic heating as the primary contributor to irreversible heat generation:

$$Q_{\text{elec,gen}} = \phi_{\text{f}}\,[\dot{n}_{\text{f}}(-T\Delta S) + I^2 R_{\text{c}}] \text{ for a cell}$$

Considering the activation overpotential as the primary contributor to irreversible heat generation:

$$Q_{\text{elec,gen}} = \left[\frac{I}{n_{\text{e}}F}(-T\Delta S) + I\eta_{\text{act}}\right] \text{ for a cell}$$

Gouy–Chapman Capacitance—The Capacitance for the Diffuse Layer

$$C_{\text{G}} = \left(\frac{2N\varepsilon\varepsilon_0 z^2 e^2}{k_{\text{B}}T}\right)^{1/2} \cosh\frac{zeV}{2k_{\text{B}}T}$$

Butler–Volmer Equation

$$j = j_0 \left\{ \frac{C_{\text{Ox,surf}}}{C_{\text{Ox,bulk}}} e^{-\alpha nF(E-E_0)/RT} - \frac{C_{\text{Red,surf}}}{C_{\text{Red,bulk}}} e^{\beta F(E-E^{\circ})/RT} \right\}$$

Simplified Butler–Volmer equation: very small η_{act}

$$\eta_{\text{act},i} = \frac{RT}{n_{\text{e},i}F}\left(\frac{j}{j_{\text{o},i}}\right)$$

Simplified Butler–Volmer equation: very large η_{act}

$$\eta_{\text{act},i} = \frac{RT}{\alpha_i n_{\text{e},i}F} \ln \frac{j}{j_{\text{o},i}}$$

Simplified Butler–Volmer equation with identical charge transfer coefficient:

$$\eta_{\text{act}} = \frac{RT}{\alpha n_{\text{e}}F} \sinh^{-1}\left(\frac{j}{2j_{\text{o}}}\right)$$

Tafel Equation

$$\eta = \frac{RT}{\alpha F} \ln j_0 - \frac{RT}{\alpha F} \ln j$$

Conservation of Mass for Fluid Flow

$$\frac{\partial \rho}{\partial t} + \nabla \cdot (\rho \vec{V}) = 0$$

Conservation of Momentum for Fluid Flow

Navier–Stokes equation

$$\rho \frac{D\vec{V}}{Dt} = \rho\vec{g} - \nabla P + \frac{\partial}{\partial x_j}\left[\mu\left(\frac{\partial u_i}{\partial x_j} + \frac{\partial u_j}{\partial x_i}\right) - \frac{2}{3}\mu\left(\nabla \cdot \vec{V}\right)\right]$$

Mass Continuity in Porous Media with Consumption

$$\frac{\partial \rho}{\partial t} + \nabla \cdot \rho\vec{V} = S_i$$

Momentum Equation in Porous Media

Darcy's law

$$\vec{V} = \frac{\kappa}{\mu}\nabla P$$

Brinkman's equation

$$\nabla P = -\frac{\mu}{\kappa}\vec{V} + \mu_e \nabla^2 \vec{V}$$

Modified Navier–Stokes equation

$$\rho\left[\frac{\partial \vec{V}}{\partial t} + \left(\vec{V} \cdot \nabla\right)\vec{V}\right] = \rho\vec{g} - \nabla P + \mu\nabla^2\vec{V} - \frac{\mu}{k}\varepsilon\vec{V}$$

Energy and Heat Equation

$$\rho c_p\left[\frac{\partial T}{\partial t} + \left(\vec{V} \cdot \nabla\right)T\right] = \nabla \cdot (k\nabla T) + \dot{Q}$$

Stefan–Maxwell Equation

$$m_i'' = -\rho x_i \sum_{j=1'}^{n} D_{ij} \left\{ \frac{M}{M_j} \left(\nabla x_j + x_j \frac{\nabla M}{M} \right) + (y_j - x_j) \frac{\nabla P}{P} \right\} + D_i^T \frac{\nabla T}{T}$$

Mass Transport Due to Combined Multi-Component Diffusion and Convection

$$m_i'' = -\rho x_i \sum_{j=1'}^{n} D_{ij} \left\{ \frac{M}{M_j} \left(\nabla x_j + x_j \frac{\nabla M}{M} \right) + (y_j - x_j) \frac{\nabla P}{P} \right\} + D_i^T \frac{\nabla T}{T} + x_i \rho \vec{V}$$

For negligible diffusion thermal effect:

$$m_i'' = -\rho x_i \sum_{j=1'}^{n} D_{ij} \left(\nabla y_j + (y_j - x_j) \frac{\nabla P}{P} \right) + x_i \rho \vec{V}$$

In terms of mole fraction:

$$j_i'' = C y_i \sum_{j=1'}^{n} Dij \left(\nabla y_j + (y_j - x_j) \frac{\nabla P}{P} \right) + C \vec{V} y_i$$

Mass Species Transport in a Fuel Cell

$$\left[\frac{\partial C_i}{\partial t} + (\vec{V} \cdot \nabla) C_i \right] = \nabla \cdot (J_i'') + S_i$$

Limiting Current Density

$$i_L = n_c F \frac{C_{i,\mathrm{ch}}}{\left(\frac{1}{h_m} + \frac{a_{\mathrm{el}}}{D_{ij}^{\mathrm{eff}}}\right)}$$

and for diffusion resistance controlled:

$$i_L = n_e F D_{ij}^{\mathrm{eff}} \left(\frac{C_{i,\mathrm{ch}}}{a_{\mathrm{el}}}\right)$$

Mass Transfer or Concentration Loss

$$\eta_{\mathrm{mass}} = \frac{RT}{n_e F}\left(1 + \frac{1}{\alpha}\right) \ln\left(\frac{i_l}{i_l - i}\right)$$

Nernst–Planck's Equation

$$J_i = -\sigma_c \nabla\Phi - D_i \nabla C_i + C_i \vec{V}_i$$

or

$$J_i = -\sigma_c \frac{\mathrm{d}\Phi}{\mathrm{d}x} - D_i \frac{\mathrm{d}C_i}{\mathrm{d}x} + C_i \frac{K_p}{\mu}\frac{\mathrm{d}P}{\mathrm{d}x}$$

Schlogl's Equation

$$\vec{V} = \frac{K_\phi}{\mu} z_f C_f \nabla\phi - \frac{K_p}{\mu}\nabla P$$

Water Transport in an Electrolyte Membrane

Nernst–Planck's equation for water transport in polymer membrane:

$$j_{H_2O} = 2n_d \frac{i}{n_e F} - D_{H_2O} \frac{dc_{H_2O}}{dz} \nabla c_{H_2O} - c_{H_2O} \frac{\kappa_{H_2O}}{\mu_{H_2O}} \frac{dP}{dz}$$

Schlogl equation for water transport in polymer membrane:

$$j_{H_2O} = n_d \frac{i}{n_e F} - c_{H_2O} \frac{k_{H_2O}}{\mu_{H_2O}} \nabla p$$

Infinite Warburg Impedance

$$Z = \frac{\sigma}{\sqrt{\omega}}(1 - j)$$

Warburg coefficient:

$$\sigma = \frac{RT}{n^2 F^2 A\sqrt{2}} \left(\frac{1}{C_{\text{Ox,bulk}}\sqrt{D_{\text{Ox}}}} \frac{1}{C_{\text{Red,bulk}}\sqrt{D_{\text{Red}}}} \right)$$

Finite Warburg equation:

$$Z_0 = \frac{\sigma}{\sqrt{\omega}}(1 - j)\tanh\left(\delta(j\omega / D)^{1/2}\right)$$

Appendix C: Chemical and Thermodynamic Data

TABLE C.1

Salinity (Density of Salt in Saltwater) and Conductivity at 25°C

NaCl Density (W/V)%	Conductivity (mS/cm)	NaCl Density (W/V)%	Conductivity (mS/cm)
0.1	2.0	1.1	19.2
0.2	3.9	1.2	20.8
0.3	5.7	1.3	22.4
0.4	7.5	1.4	24.0
0.5	9.2	1.5	25.6
0.6	10.9	1.6	27.1
0.7	12.6	1.7	28.6
0.8	14.3	1.8	30.1
0.9	16.0	1.9	31.6
1.0	17.6	2.0	33.0

TABLE C.2

Standard Potentials of Electrode Reactions at 25°C

$Li^+ + e^- \rightarrow Li$	–3.045	$Ni^{2+} + 2e^- \rightarrow Ni$	–0.25
$Rb^+ + e^- \rightarrow Rb$	–2.98	$Sn^{2+} + 2e^- \rightarrow Sn$	–0.136
$Cs^+ + e^- \rightarrow Cs$	–2.92	$Pb^{2+} + 2e^- \rightarrow Pb$	–0.126
$K^+ + e^- \rightarrow K$	–2.935	$Sn^{2+} + 2e^- \rightarrow Sn$	–0.14
$Ba^{2+} + 2e^- \rightarrow Ba$	–2.92	$D^+ + e^- \rightarrow 1/2\ D_2$	–0.003
$Sr^{2+} + 2e^- \rightarrow Sr$	–2.89	$H^+ + e^- \rightarrow 1/2\ H_2$	0.000
$Ca^{2+} + 2e^- \rightarrow Ca$	–2.866	$Sn^{4+} + 2e^- \rightarrow Sn^{2+}$	0.15
$Na^+ + e^- \rightarrow Na$	–2.714	$Cu^{2+} + 2e^- \rightarrow Cu$	0.337
$Mg^{2+} + 2e^- \rightarrow Mg$	–2.375	$1/2\ O_2 + H_2O + 2e^- \rightarrow 2OH^-$	0.401
$Be^{2+} + 2e^- \rightarrow Be$	–1.70	$Cu^+ + e^- \rightarrow Cu$	0.52
$Al^{3+} + 3e^- \rightarrow Al$	–1.662	$Hg^{2+} + 2e^- \rightarrow 2Hg$	0.789
$Ti^{2+} + 2e^- \rightarrow Ti$	–1.628	$Ag^+ + e^- \rightarrow Ag$	0.799
$Mn^{2+} + 2e^- \rightarrow Mn$	–1.180	$Pd^{2+} + 2e^- \rightarrow Pd$	0.987

(continued)

TABLE C.2 (Continued)

Standard Potentials of Electrode Reactions at 25°C

$Zn^{2+} + 2e^- \rightarrow Zn$	−0.764	$AuCl_4^- + 3e^- \rightarrow Au + 4Cl$	1.00
$Ga^{3+} + 3e^- \rightarrow Ga$	−0.52	$2Br + 2e^- \rightarrow 2Br^-$	1.065
$Fe^{2+} + 2e^- \rightarrow Fe$	−0.441	$O_2 + 4H^+ + 4e^- \rightarrow 2H_2O$	1.229
$Cd^{2+} + 2e^- \rightarrow Cd$	−0.403	$Cl_2 + 2e^- \rightarrow 2Cl^-$	1.36
$In^{3+} + 3e^- \rightarrow In$	−0.34	$Ce^{4+} + e^- \rightarrow Ce^{3+}$	1.61
$Tl^+ + e^- \rightarrow Tl$	−0.34	$Au^+ + e^- \rightarrow Au$	1.69
$Co^{2+} + 2e^- \rightarrow Co$	−0.277	$F_2 + 2e^- \rightarrow 2F^-$	2.65

TABLE C.3

Properties of Various Ideal Gases at 25°C, 100 kPa

Substance	Formula	Mol.Wt.	R (kJ/kg·K)	ρ (kg/m³)	C_{p0} (kJ/kg·K)	C_{v0} (kJ/kg·K)
Ammonia	NH_3	17.031	0.4882	0.694	2.130	1.642
Benzene	C_6H_6	78.114				
Butane	C_4H_{10}	58.124	0.1430	2.407	1.716	1.573
Carbon dioxide	CO_2	44.010	0.1889	1.775	0.842	0.653
Carbon monoxide	CO	28.011	0.2968	1.13	1.041	0.744
Ethene		28.054				
Ethane	C_2H_6	30.070	0.2765	1.222	1.766	1.490
Ethanol	C_2H_5OH	46.069	0.1805	1.883	1.427	1.246
Ethylene	C_2H_4	29.054	0.2964	1.138	1.548	1.252
Hydrogen	H_2	2.016	4.1243	0.0813	14.209	10.085
Methane	CH_4	16.043	0.5183	0.648	2.254	1.736
Methanol	C_2H_3OH	32.042	0.2595	1.31	1.405	1.146
n-Octane	C_8H_{18}	114.232	0.0727	0.092	1.711	1.38
Oxygen	O_2	31.999	0.2598	1.292	0.922	0.662
Sulfur dioxide	SO_2	64.059	0.1298	2.618	0.624	0.494
Sulfur trioxide	SO_3	80.058	0.1038	3.272	0.635	0.531
Nitrogen	N_2	28.013	0.2968	1.13	1.042	0.745
Nitrous oxide	N_2O	44.013	0.1889	1.775	0.879	0.690
Nitrogen dioxide	NO_2	46.005				
Steam	H_2O	18.050	0.4615	0.0231	1.872	1.410

Source: Sonntag et al.: *Fundamentals of Thermodynamics*. 2003. Copyright Wiley-VCH Verlag GmbH & Co. KGaA. Reproduced with permission.

TABLE C.4

Functional Relationships of Specific Heat with Temperature

Gas	Chemical Formula	Constant Pressure Specific Heat, C_p (kJ/kg·K)
Air		$C_p = 1.05 - 0.365\theta + 0.85\theta^2 - 0.39\theta^3$
Carbon dioxide	CO_2	$C_p = 0.45 + 1.67\theta - 1.27\theta^2 + 0.39\theta^3$
Carbon monoxide	CO	$C_p = 1.10 - 0.46\theta + 1.0\theta^2 - 0.454\theta^3$
Ethane	C_2H_6	$C_p = 0.18 + 5.92\theta - 2.31\theta^2 + 0.29\theta^3$
Ethanol	C_2H_5OH	$C_p = 0.2 - 4.65\theta - 1.82\theta^2 - 0.03\theta^3$
Hydrogen	H_2	$C_p = 13.46 + 4.6\theta - 6.85\theta^2 + 3.79\theta^3$
Methane	CH_4	$C_p = 1.2 + 3.25\theta + 0.75\theta^2 - 0.71\theta^3$
Methanol	CH_3OH	$C_p = 0.66 + 2.21\theta + 0.81\theta^2 - 0.89\theta^3$
Nitrogen	N_2	$C_p = 1.10 - 0.46\theta + 1.0\theta^2 - 0.454\theta^3$
Nitric oxide	NO	$C_p = 0.98 - 0.031\theta + 0.325\theta^2 - 0.14\theta^3$
Nitrous oxide	N_2O	$C_p = 0.49 + 1.65\theta - 1.31\theta^2 + 0.42\theta^3$
Oxygen	O_2	$C_p = 0.88 - 0.0001\theta + 5.4\theta^2 - 0.33\theta^3$
Sulfur dioxide	SO_2	$C_p = 0.37 + 1.05\theta - 0.77\theta^2 + 0.21\theta^3$
Steam	H_2O	$C_p = 1.79 + 0.107\theta + 0.586\theta^2 - 0.20\theta^3$
		$\theta = \frac{T(K)}{100}$

Source: Sonntag et al.: *Fundamentals of Thermodynamics*. 2003. Copyright Wiley-VCH Verlag GmbH & Co. KGaA. Reproduced with permission.

TABLE C.5

Saturation Properties of Water

		Specific Volume (m^3/kg)		Enthalpy (kJ/kg)		Entropy (kJ/kg·K)	
Temp (°C)	Press (kPa)	Sat. Liquid	Sat. Vapor	Sat. Liquid	Sat. Vapor	Sat. Liquid	Sat. Vapor
10	1.2276	0.001000	106.377	41.99	2519.74	0.1510	8.9007
20	2.339	0.001002	57.7897	83.94	2538.06	0.2966	8.6671
30	4.246	0.001004	32.8932	125.77	2556.25	0.4369	8.4533
40	7.384	0.001008	19.5229	167.54	2574.26	0.5724	8.2569
50	12.350	0.001012	12.0318	209.31	2592.06	0.7037	8.0762
60	19.941	0.001017	7.67071	251.11	2609.59	0.8311	7.9095
70	31.19	0.001023	5.04217	292.96	2626.80	0.9548	7.7552
80	47.39	0.001029	3.40715	334.88	2643.66	1.0752	7.6121
90	70.14	0.001036	2.36036	376.90	2660.09	1.1924	7.4790
100	101.3	0.001044	1.67290	419.02	2676.05	1.3068	7.3548
110	143.3	0.001052	1.21014	461.27	2691.47	1.4184	7.2386

(continued)

TABLE C.5 (Continued)

Saturation Properties of Water

		Specific Volume (m³/kg)		Enthalpy (kJ/kg)		Entropy (kJ/kg·K)	
Temp (°C)	Press (kPa)	Sat. Liquid	Sat. Vapor	Sat. Liquid	Sat. Vapor	Sat. Liquid	Sat. Vapor
120	198.5	0.001060	0.89186	503.69	2706.30	1.5275	7.1295
130	270.1	0.001070	0.66850	546.29	2720.46	1.6343	7.0269
140	361.3	0.001075	0.50885	589.11	2733.87	1.7390	6.9298
150	475.9	0.001090	0.39278	632.18	2746.44	1.8417	6.8378
160	617.8	0.001102	0.30706	675.53	2758.09	1.9426	6.7501
170	791.7	0.001114	0.24283	719.20	2768.70	2.0418	6.6663
180	1002.2	0.001127	0.19405	763.21	2778.16	2.1395	6.5857
190	1254.4	0.001141	0.15654	807.61	2786.37	2.2358	6.5464
200	1553.8	0.001156	0.12736	852.43	2793.18	2.3308	6.5078
210	1906.3	0.001173	0.10441	897.75	2798.48	2.4247	6.3584
220	2317.8	0.001190	0.08619	943.61	2802.12	2.5177	6.2860
230	2794.9	0.001209	0.07158	990.10	2803.95	2.6099	6.2146
240	3344.2	0.001229	0.05976	1037.31	2803.81	2.7015	6.1436
250	3973.0	0.001251	0.05013	1085.34	2801.52	2.7927	6.0729
260	4688.6	0.001276	0.04220	1134.35	2796.89	2.8837	6.0018
270	5498.7	0.001302	0.03564	1184.49	2789.65	2.9750	5.9301
280	6411.7	0.001332	0.03017	1235.97	2779.53	3.0667	5.8570
290	7436.0	0.001366	0.02557	1289.04	2766.13	3.1593	5.7821
300	8581.0	0.001404	0.02167	1344.01	2748.94	3.2533	5.7044
310	9856.6	0.001447	0.01835	1401.29	2727.27	3.3492	5.6229
320	11,274	0.001499	0.01549	1461.45	2700.08	3.4479	5.5361
330	12,845	0.001561	0.01300	1525.29	2665.85	3.5506	5.4416
340	14,586	0.001638	0.01080	1594.15	2622.01	3.6593	5.3356
350	16,514	0.001740	0.00881	1670.54	2563.92	3.7776	5.2111
360	18,651	0.001892	0.00694	1760.48	2481.00	3.9146	5.0525
370	21,028	0.002213	0.00493	1890.37	2332.12	4.1104	4.7972

Source: Sonntag et al.: *Fundamentals of Thermodynamics*. 2003. Copyright Wiley-VCH Verlag GmbH & Co. KGaA. Reproduced with permission.

TABLE C.6

Enthalpy of Formation and Absolute Entropy of Various Substances at 25°C, 0.1 MPa

Substance	Formula	Mol.Wt.	Enthalpy of Formation, $\bar{h}_f^0$ (kJ/kmol)	Absolute Entropy, $\bar{s}_f^0$ (kJ/kmol·K)	Gibbs Function of Formation, $\bar{g}_f^0$ (kJ/kmol·K)
Ammonia	NH_3	17.031	46,190	192.572	–16,128
Benzene	C_6H_6	78.114	+82,050	269.20	+129,765
Butane	C_4H_{10}	58.124	–126,200	306.647	–15,970
Carbon–graphite (solid)	C	12.011	0	5.740	0
Carbon dioxide	CO_2	44.010	–393,522	213.795	–394,389
Carbon monoxide	CO	28.011	–110,527	197.653	–137,163
Ethene	C_2H_4	28.054	+52,467	219.330	+68,421
Ethane	C_2H_6	30.070	–84,740	229.597	–32,885
Ethanol	C_2H_5OH	46.069	–235,000	282.444	–168,319
Ethanol (liquid)	C_2H_5OH	46.069	–277,380	160.554	–174,890
Hydrogen	H_2	2.016	0	130.678	0
Methane	CH_4	16.043	–74,873	186.251	–50,768
Methanol	CH_3OH	32.042	–201,300	239.709	–162,551
Methanol (liquid)	CH_3OH	32.042	–239,220	126.809	–166,360
n-Octane	C_8H_{18}	114.232	–208,600	466.514	+16,660
n-Octane (liquid)	C_8H_{18}	114.232	–250,105	360.575	+6741
Oxygen	O_2	31.999	0	205.148	0
Sulfur (solid)	S	32.06	0	32.056	0
Sulfur dioxide	SO_2	64.059	–296,842	248.212	–300,125
Sulfur trioxide	SO_3	80.058	–395,765	256.769	–371,016
Nitrogen	N_2	28.013	0	191.609	0
Nitric oxide	NO	30.006	+90,291	210.759	+86,607
Nitrogen oxide	N_2O	44.013	+82,050	219.957	+104,179
Nitrogen dioxide	NO_2	46.005	+33,100	240.034	+51,291
Water (vapor)	H_2O	18.015	–241,826	188.834	–228,582
Water (liquid)	H_2O	18.015	–285,830	69.950	–237,141

Source: Sonntag et al.: *Fundamentals of Thermodynamics*. 2003. Copyright Wiley-VCH Verlag GmbH & Co. KGaA. Reproduced with permission.

Note: Selected values of Gibbs function, *Fundamentals of Classical Thermodynamics*, van Wylen et al. (1994).

TABLE C.7

Ideal Gas Enthalpy and Entropy as a Function of Temperature of Various Substances

Temperature, T (K)	$\bar{h}_T = \left(\bar{h} - \bar{h}^0_{298}\right)$ (kJ/kmol)	$\bar{s}^0_T$ (kJ/kmol·K)
Carbon Dioxide (CO_2)		
0	–9364	0
100	–6457	179.010
200	–3413	199.976
298	0	213.794
300	69	214.024
400	4003	225.314
500	8305	234.902
600	12,906	243.284
700	17,754	250.752
800	22,806	257.496
900	28,030	263.646
1000	33,397	269.299
1100	38,885	274.528
1200	44,473	279.390
1300	50,148	283.931
1400	55,895	288.190
1500	61,705	292.199
1600	67,569	295.984
1700	73,480	299.567
1800	79,432	302.969
1900	85,420	306.207
2000	91,439	309.294
2200	103,562	315.070
2400	115,779	320.384
2600	128,074	325.307
2800	140,435	329.887
3000	152,853	334.170
Carbon Monoxide (CO)		
0	–8671	0
100	–5772	165.852
200	–2860	186.024
298	0	197.651
300	54	197.831
400	2977	206.240
500	5932	212.833
600	8942	218.321
700	12,021	223.067
800	15,174	227.277

(continued)

TABLE C.7 (Continued)

Ideal Gas Enthalpy and Entropy as a Function of Temperature of Various Substances

Temperature, T (K)	$\bar{h}_T = \left(\bar{h} - \bar{h}^0_{298}\right)$ (kJ/kmol)	$\bar{s}^0_T$ (kJ/kmol·K)
Carbon Monoxide (CO)		
900	18,397	231.074
1000	21,686	234.538
1100	25,031	237.726
1200	28,427	240.679
1300	31,867	243.431
1400	35,343	246.006
1500	38,852	248.426
1600	42,388	250.707
1700	45,948	252.866
1800	49,529	254.913
1900	53,128	256.860
2000	56,743	258.716
2200	64,012	262.182
2400	71,326	265.361
2600	78,679	268.302
2800	86,070	271.044
3000	93,504	273.607
Hydrogen (H_2)		
0	–8467	0
100	–5467	100.727
200	–2774	119.410
298	0	130.678
300	53	130.856
400	2961	139.219
500	5883	145.738
600	8799	151.078
700	11,730	155.609
800	14,681	159.554
900	17,657	163.060
1000	20,663	166.225
1100	23,704	169.121
1200	26,785	171.798
1300	29,907	174.294
1400	33,073	176.637
1500	36,281	178.849
1600	39,533	180.949
1800	42,826	182.941

(continued)

TABLE C.7 (Continued)

Ideal Gas Enthalpy and Entropy as a Function of Temperature of Various Substances

Temperature, T (K)	$\bar{h}_T = \left(\bar{h} - \bar{h}^0_{298}\right)$ (kJ/kmol)	$\bar{s}^0_T$ (kJ/kmol·K)
Hydrogen (H_2)		
1700	46,160	184.846
1900	49,532	186.670
2000	52,942	188.419
2200	59,865	191.719
2400	66,915	194.789
2600	74,082	197.659
2800	81,355	200.355
3000	88,725	202.898
Oxygen (O_2)		
0	–8683	0
100	–5777	173.308
200	–2868	193.483
298	0	205.148
300	54	205.329
400	3027	213.873
500	6086	220.693
600	9245	226.450
700	12,499	231.465
800	15,836	235.920
900	19,241	239.931
1000	22,703	243.579
1100	26,212	246.923
1200	29,761	250.011
1300	33,345	252.878
1400	36,958	255.556
1500	40,600	258.068
1600	44,267	260.434
1700	47,959	262.673
1800	51,674	264.797
1900	55,414	266.819
2000	59,176	268.748
2200	66,770	272.366
2400	74,453	275.708
2600	82,225	278.818
2800	90,080	281.729
3000	98,013	284.466

(*continued*)

TABLE C.7 (Continued)

Ideal Gas Enthalpy and Entropy as a Function of Temperature of Various Substances

Temperature, T (K)	$\bar{h}_T = \left(\bar{h} - \bar{h}^0_{298}\right)$ (kJ/kmol)	$\bar{s}^0_T$ (kJ/kmol·K)
Nitrogen (N_2)		
0	−8670	0
100	−5768	159.812
200	−2857	179.985
298	0	191.609
300	54	191.789
400	2971	200.181
500	5911	206.740
600	8894	212.177
700	11,937	216.865
800	15,046	221.016
900	18,223	224.757
1000	21,463	228.171
1100	24,760	231.314
1200	28,109	234.227
1300	31,503	236.943
1400	34,936	239.487
1500	38,405	241.881
1600	41,904	244.139
1700	45,430	246.276
1800	48,979	248.304
1900	52,549	250.234
2000	56,137	252.075
2200	63,362	255.518
2400	70,640	258.684
2600	77,963	261.615
2800	85,323	264.342
3000	92,715	266.892
Nitric Oxide (NO)		
0	−9192	0
100	−6073	177.031
200	−2951	198.747
298	0	210.759
300	55	210.943
400	3040	219.529
500	6059	226.263
600	9144	231.886
700	12,308	236.762

(continued)

TABLE C.7 (Continued)

Ideal Gas Enthalpy and Entropy as a Function of Temperature of Various Substances

Temperature, T (K)	$\bar{h}_T = \left(\bar{h} - \bar{h}^0_{298}\right)$ (kJ/kmol)	$\bar{s}^0_T$ (kJ/kmol·K)
Nitric Oxide (NO)		
800	15,548	241.088
900	18,858	244.985
1000	22,229	248.536
1100	25,653	251.799
1200	29,120	254.816
1300	32,626	257.621
1400	36,164	260.243
1500	39,729	262.703
1600	43,319	265.019
1700	46,929	267.208
1800	50,557	269.282
1900	54,201	271.252
2000	57,859	273.128
2200	65,212	276.632
2400	72,606	279.849
2600	80,034	282.822
2800	87,491	285.585
3000	94,973	288.165
Nitrogen Dioxide (NO_2)		
0	−10,186	0
100	−6861	202.563
200	−3495	225.852
298	0	240.034
300	68	240.263
400	3927	251.342
500	8099	260.638
600	12,555	268.755
700	17,250	275.988
800	22,138	282.513
900	27,180	288.450
1000	32,344	293.889
1100	37,606	298.904
1200	42,946	303.551
1300	48,351	307.876
1400	53,808	311.920
1500	59,309	315.715
1600	43,319	265.019

(continued)

TABLE C.7 (Continued)

Ideal Gas Enthalpy and Entropy as a Function of Temperature of Various Substances

Temperature, T (K)	$\bar{h}_T = \left(\bar{h} - \bar{h}^0_{298}\right)$ (kJ/kmol)	$\bar{s}^0_T$ (kJ/kmol·K)
Nitrogen Dioxide (NO_2)		
1700	46,929	267.208
1800	50,557	269.282
1900	54,201	271.252
2000	57,859	273.128
2200	65,212	276.632
2400	72,606	279.849
2600	80,034	282.822
2800	87,491	285.585
3000	94,973	288.165
Water (H_2O)		
0	−9904	0
100	−6617	152.386
200	−3282	175.488
298	0	188.835
300	62	189.043
400	3450	198.787
500	6922	206.532
600	10,499	213.051
700	14,190	218.739
800	18,002	223.826
900	21,937	228.460
1000	26,000	232.739
1100	30,190	236.732
1200	34,506	240.485
1300	38,941	244.035
1400	43,491	247.406
1500	48,149	250.620
1600	52,907	253.690
1700	57,757	256.631
1800	62,693	259.452
1900	67,706	262.162
2000	72,788	264.769
2200	83,153	269.706
2400	93,741	274.312
2600	104,520	278.625
2800	115,463	282.680
3000	126,548	286.504

Source: Sonntag et al.: *Fundamentals of Thermodynamics*. 2003. Copyright Wiley-VCH Verlag GmbH & Co. KGaA. Reproduced with permission.

TABLE C.8

Enthalpy of Combustion of Hydrocarbon Fuels at 25°C, 0.1 MPa

Substance	Formula	Liquid Water in the Products (kJ/kmol)	Gas Water in the Products (kJ/kmol)
Ammonia	NH_3	–45,720	192.572
Benzene	C_6H_6	+82,050	219.957
Butane	C_4H_{10}	–126,200	306.647
Carbon–graphite (solid)	C	0	5.740
Carbon dioxide	CO_2	–393,522	213.795
Carbon monoxide	CO	–110,527	197.653
Ethene	C_2H_4	+52,467	219.330
Ethane	C_2H_6	–84,740	229.597
Ethanol	C_2H_5OH	–235,000	282.444
Ethanol (liquid)	C_2H_5OH	–277,380	160.554
Hydrogen	H_2	0	130.678
Methane	CH_4	–74,873	186.251
Methanol	CH_3OH	–201,300	239.709
Methanol (liquid)	CH_3OH	–239,220	126.809
n-Octane	C_8H_{18}	–208,600	466.514
n-Octane (liquid)	C_8H_{18}	–250,105	360.575
Oxygen	O_2	0	205.148
Sulfur (solid)	S	0	32.056
Sulfur dioxide	SO_2	–296,842	248.212
Sulfur trioxide	SO_3	–395,765	256.769
Nitrogen	N_2	0	191.609
Nitric oxide	NO	+90,291	210.759
Nitrogen oxide	N_2O	+82,050	219.957
Nitrogen dioxide	NO_2	+33,100	240.034
Water (vapor)	H_2O	–241,826	188.834
Water (liquid)	H_2O	–285,830	69.950

Source: Sonntag et al.: *Fundamentals of Thermodynamics*. 2003. Copyright Wiley-VCH Verlag GmbH & Co. KGaA. Reproduced with permission.

TABLE C.9

Collision integral parameters for gas properties

kT/ε	Ω_μ	kT/ε	Ω_μ	kT/ε	Ω_D
1.0	1.593	4.0	0.9699	9	0.8380
1.5	1.315	4.5	0.9462	10	0.8244
2.0	1.176	5.0	0.9268	20	0.7436
2.5	1.093	6.0	0.8962	30	0.7010
3.0	1.039	7.0	0.8727	40	0.6723
3.5	0.9996	8.0	0.8538	50	0.6510

Source: Bird et al.: *Transport Phenomena*, 2nd Edition. 2002. Copyright Wiley-VCH Verlag GmbH & Co. KGaA. Reproduced with permission.

Index

Page numbers followed by f and t indicate figures and tables, respectively.

E

H